Arthropod Borne Diseases

Carlos Brisola Marcondes
Editor

Arthropod Borne Diseases

Springer

Editor
Carlos Brisola Marcondes
Department of Microbiology
Immunology and Parasitology
Center of Biological Sciences
Federal University of Santa Catarina
Florianópolis, SC, Brazil

ISBN 978-3-319-79190-6   ISBN 978-3-319-13884-8 (eBook)
DOI 10.1007/978-3-319-13884-8

Softcover reprint of the hardcover 1st edition 2016

Printed on acid-free paper

This Springer imprint is published by Springer Nature
The registered company is Springer International Publishing AG
The registered company address is: Gewerbestrasse 11, 6330 Cham, Switzerland

This book is dedicated to
Maria Helena, Mariana, and Melina, for a life of happiness
Insects, for a life of work

Foreword

The purpose of this book is to provide easily accessible and readily understandable information related to arthropod-borne diseases. It contains 38 chapters written by more than 50 experts from several countries and provides information on vector and pathogen biology, diagnosis, treatment and prevention as well as ecology and epidemiology, topics essential to detect their occurrence and initiate timely intervention. The repeated evolution of blood feeding by wide variety of arthropods has resulted in their conflicting relationship with animals, including humans, by allowing them to serve as vectors of many pathogens, cause allergic reactions or provide a nuisance through their bites. Diseases caused by pathogens transmitted by arthropods such as malaria and dengue fever have strongly influenced human history and continue to be important in tropical and temperate countries. Climate change and the increased mobility of humans and domestic animals have resulted in the emergence of disease in unexpected situations, such as chikungunya and zika viruses in the New World, visceral leishmaniasis in Finland and sleeping sickness in the USA.

The overarching objective of this book is to prevent suffering and save lives by providing information accessible to health professionals and laypersons as well as students, scientists and researchers. Even though directed to solve problems, an effort was made to "guard against stripping the science in lieu of its poetry," to enhance the interest and stimulate research.

Preface

Arthropod parasites have always had a "conflicting" relationship with animals including humans. Due to their need to feed on blood or other host tissues, they serve as vectors of pathogenic viruses, bacteria, protozoa, and helminths, cause severe allergic reactions, or simply provide nuisance due to their bites.

Some of the most important diseases affecting human history, such as malaria, nagana, yellow fever, and plague, are arthropod borne and remain public health problems today. Others such as Chikungunya and Zika viruses recently have emerged as problems of global proportions. The growing knowledge of pathogenic mechanisms, diagnosis, treatment, and prophylaxis has not been well divulged, and therefore many public health providers remain scarcely informed and believe these diseases do not occur in their day-to-day practice. As a consequence, many preventable deaths have occurred in both endemic and non-endemic countries and regions. As an example, in Brazil, the probability of death by malaria in an infected patient is 1000 times greater in non-endemic areas than in the endemic Amazonian region. Recent reports of deaths by malaria caused by *Plasmodium falciparum* in Rio de Janeiro and of delay in the correct diagnosis of malaria caused by *P. vivax* in Florianopolis emphasize the need for adequate health provider awareness. The more exotic findings of a fatal congenital case of Chagas disease, recently diagnosed in Spain, and 244 cases of African trypanosomiasis in Europe and the USA from 1902 to 2012 (see Chap. 18) emphasize the need of a good anamnesis, always including the travel history of the patient, and of a comprehensive knowledge of arthropod-borne diseases. It should be remembered that "the death of one man is a tragedy; the death of millions is a statistic."[1]

Infectious patients, agents, and vectors are mobile and may be undiagnosed in new areas; therefore it is dangerous to affirm that "this does not occur here." For example, several dogs and sand flies have recently been found infected by *Leishmania infantum* in the Santa Catarina Island in Santa Catarina, one of the few Brazilian states where no visceral leishmaniasis previously had been reported. The apparent nonoccurrence of diseases may sometimes reflect "the happiness that results from ignorance."

Because human beings and other animals frequently share arthropod parasites and the pathogens they transmit, it would not be sensible to analyze separately parasites of medical and veterinary importance. The finding of

[1] Attributed to Josef Stalin

West Nile virus in alligators in Florida and of rickettsiae in ticks from snakes in Thailand emphasizes the complexity of studies on arthropod-borne disease. The old Jacob's (1957) expression "sea of *Toxoplasma* around us" should also be applied to these diseases.

Arthropod-borne diseases have been associated by lay persons and even by health professionals with tropical climates and forests, but they are widely distributed in the world; at least 17 arboviruses have been found in Canada, and malaria was previously a health problem in the UK and Finland.

Predominance of occurrence of several arthropod-borne diseases in poor people living in underdeveloped countries has made them neglected diseases, but for humanitarian and practical reasons, they should not be considered "someone else's problem." In 2012, 2200 cases of dengue were reported in Madeira Island, and some autochthonous cases occurred in France and Croatia. Dengue and Chikungunya fever (also transmitted by *Aedes aegypti*) have occurred in other countries of continental Europe, and international travels of humans and pets have introduced several pathogens into unexpected places, as shown, for example, by the recent finding of autochthonous visceral leishmaniasis in dogs in Finland.

Environmental changes as a consequence of human activities (e.g., deforestation and pollution) may influence the transmission and the distribution of diseases. Previous exposure to pathogens and nutritional status can influence the risks for disease. Knowledge of the clinical presentation resulting from arthropod-borne pathogen infections and of the associated conditions is essential for adequate diagnosis, treatment, and prevention. This knowledge must include information on the epidemiology of these diseases, current and potential distribution, and observation of conditions of life and previous travels.

The objective of the present book is to furnish to health workers and the general public information useful to solve problems related to most arthropod-borne diseases in the world. The case of Lyme disease, in which two worried mothers of children from the American city of Old Lyme, Connecticut, which were affected by a mysterious disease, found similar cases in the public health records[2] and informed the health authorities of the occurrence of 12 cases in the small community, unleashed the discovery of the causative agent, the natural hosts, and the tick vectors and illustrates the value of a well-informed general public.

I expect this book will contribute to saving lives and to improving health in many regions, most of them already affected by other serious problems, such as famine and wars, and frequently related diseases. Most diseases transmitted or caused by arthropods are included in this book; however, dracunculiasis, a helminthiasis caused by *Dracunculus medinensis* and transmitted by ingestion of crustaceans, is briefly referred in Chaps. 3 and 24 but not described in detail because of its trend toward eradication. Venomous arthropods (spiders, scorpions, bees, etc.) were not included, due to the availability of other recently published books. This book is focused on

[2]Dr. M. P. Deane (pers commun 1986), watching a lecture on Lyme disease, praised the availability of good libraries and statistics in small US cities.

diseases and must be accompanied and supplemented by other works on medical and veterinary entomology, if identification of vectors and more details of their biology are needed.

Even being a book directed to solve health problems, an effort was made to "guard against stripping our science of its share of poetry," as suggested by Marc Bloch for history, trying to enhance interest and stimulate research.[3]

Florianópolis, SC, Brazil Carlos Brisola Marcondes

[3] See: http://buscatextual.cnpq.br/buscatextual/visualizacv.do?id=K4783901J2.

About the only genuine sporting proposition
that remains unimpaired… is the war against
these ferocious little fellow creatures, which lurk in the
dark corners and stalk us in the bodies of rats, mice, and
all kinds of domestic animals; which fly and crawl with the
insects, and waylay us in our food and drink and even in our love

Hans Zinsser (1935)

Acknowledgments

To Springer, for acceptance of the publication of this work, and to all authors, for believing in the usefulness of such work and for collaborating in this arduous task.

To Ronald Sherman, for the review and suggestions for maggot therapy section. To Amer Al-Jawabreh, Caryn Bern, Henk D. F. H. Schallig, Jane da Silva, José Diomedes Barbosa Neto, Kwan Poo Chang, Martin J. R. Hall, Naomi Dunning-Foreman, Paola Marchesini, Pedro Luiz Tauil, R. A. Sherman, Salman Singh, Shikha Arora, and Wes Watson, for sending publications, figures, and/or information useful for the book. To Alan Kahan (Entomological Society of America), Amy Ellis-Thompson (Maney Publishing), José Venzal, Luiz Henrique Guimarães (Federal University of Bahia), Natalino Yoshinari, Societé Française de Ophtalmologie, PLOS, US Centers for Disease Control and Prevention (CDC), US National Oceanic and Atmospheric Administration, and Wikimedia Commons for authorization to utilize photos or tables. To Leonardo Vincensi for the drawings of the examples of Chilopoda and Diplopoda.

To all authors, for their effort, with so many tasks, especially to Pedro Marcos Linardi, my professor and friend for 40 years, for delivering his chapter in so difficult familiar conditions.

Contents

Contributors

Leucio Camara Alves Department of Veterinary Medicine, Federal Rural University of Pernambuco, Recife, Brazil

Maria Paloma Silva de Barros Department of Microbiology, Aggeu Magalhães Center of Research (CPqAM/Fiocruz), Oswaldo Cruz Foundation, Recife, PE, Brazil

Carla de Lima Bicho Department of Biology, Center of Biological Sciences and Health, State University of Paraiba, Campina Grande, PB, Brazil

Emanuele Brianti Department of Veterinary Sciences, University of Messina, Messina, Italy

Eric P. Caragata René Rachou Center of Research, Belo Horizonte, MG, Brazil

Keith H. Carter Department of Communicable Disease and Health Analysis, Pan American Health Organization/World Health Organization, Washington, DC, USA

Cédric Chesnais Joint Research Unit 233 "TransVIHMI", Institut de Recherche pour le Développement, Montpellier University, Montpellier, France

Marta S. Contigiani Arbovirus Laboratory, Institute of Virology "Dr. J. M. Vanella", National University of Córdoba, Córdoba, Argentina

Iliano V. Coutinho-Abreu Vector Molecular Biology Section, Laboratory of Malaria and Vector Research, National Institute of Allergy and Infectious Diseases, National Institutes of Health, Rockville, MD, USA

James Lee Crainey Leônidas e Maria Deane Institute (Fiocruz Amazônia), Manaus, Amazonas, Brazil

Filipe Dantas-Torres Department of Immunology, Aggeu Magalhães Research Center, Recife, PE, Brazil

Department of Veterinary Medicine, University of Bari, Valenzano, Italy

João Carlos Pinto Dias René Rachou Center of Research, Oswaldo Cruz Foundation, Belo Horizonte, MG, Brazil

Luis A. Diaz Arbovirus Laboratory, Institute of Virology "Dr. J. M. Vanella", National University of Córdoba, Córdoba, Argentina

Institute for Biological and Technical Researches, CONICET-UNC, Córdoba, Argentina

John Patrick Ehrenberg Western Pacific Regional Office of the World Health Organization, Merida, Yucatan, Mexico

Rainier P. Escalada Department of Communicable Disease and Health Analysis, Pan American Health Organization/World Health Organization, Washington, DC, USA

Alexsandra R.M. Favacho Laboratory of Hantaviroses and Rickettsioses, Oswaldo Cruz Institute, Oswaldo Cruz Foundation, Rio de Janeiro, RJ, Brazil

Gilberto Fontes Federal University of São João del Rei, Campus Centro Oeste, Divinópolis, Minas Gerais, Brazil

Annunziata Giangaspero Department of Science of Agriculture, Food and Environment, University of Foggia, Foggia, Italy

Raquel M. Gleiser Centro de Relevamiento y Evaluación de Recursos Agrícolas y Naturales, Instituto Multidisciplinario de Biología Vegetal (IMBIV), Universidad Nacional de Córdoba (UNC), Consejo Nacional de Investigaciones Científicas y Técnicas (CONICET), Córdoba, Argentina

Edmundo C. Grisard Department of Microbiology, Immunology and Parasitology, Center of Biological Sciences, Federal University of Santa Catarina, Florianópolis, SC, Brazil

Hilary Hurd Centre for Applied Entomology and Parasitology, School of Life Sciences, Keele University, Keele, Staffordshire, UK

Tereza Cristina Leal-Balbino Department of Microbiology, Aggeu Magalhães Center of Research (CPqAM/Fiocruz), Oswaldo Cruz Foundation, Recife, PE, Brazil

Hermenio C. Lima Division of Dermatology, Department of Medicine, Michael G. DeGroote School of Medicine, McMaster University, Hamilton, ON, Canada

Pedro Marcos Linardi Department of Parasitology, Institute of Biological Sciences, Federal University of Minas Gervais, Belo Horizonte, Minas Gerais, Brazil

Susan E. Little Department of Veterinary Pathobiology, Center for Veterinary Health Sciences, Oklahoma State University, Stillwater, OK, USA

Sérgio Luis Bessa Luz Leônidas e Maria Deane Institute (Fiocruz Amazônia), Manaus, Amazonas, Brazil

Cláudio Mafra Laboratory of Parasitology and Molecular Epidemiology, Department of Biochemistry and Molecular Biology, Federal University of Viçosa, Viçosa, MG, Brazil

Carlos Brisola Marcondes Department of Microbiology, Immunology and Parasitology, Center of Biological Sciences, Federal University of Santa Catarina, Florianópolis, SC, Brazil

Maria Angélica Monteiro de Mello Mares-Guia Laboratory of Hantaviroses and Rickettsioses, Oswaldo Cruz Foundation, Oswaldo Cruz Institute, Rio de Janeiro, RJ, Brazil

Jansen Fernandes Medeiros Fundação Oswaldo Cruz - Fiocruz Rondônia, Porto Velho, Rondônia, Brazil

Luiz C. Miletti Department of Animal and Foods Production, Center of Agroveterinary Sciences, State University of Santa Catarina, Lages, SC, Brazil

Hélio Amante Miot Department of Dermatology, UNESP Medical School, Botucatu, SP, Brazil

Carlos Emmanuel Montandon Laboratory of Parasitology and Molecular Epidemiology, Department of Biochemistry and Molecular Biology, Federal University of Viçosa, Viçosa, MG, Brazil

Luciano A. Moreira René Rachou Center of Research, Belo Horizonte, MG, Brazil

Gerson Azulim Müller Federal Institute of Education, Science and Technology Farroupilha, Panambi, RS, Brazil

Maria Ogrzewalska Laboratory of Hantaviroses and Rickettsioses, Oswaldo Cruz Institute, Oswaldo Cruz Foundation, Rio de Janeiro, RJ, Brazil

Maria Betânia Melo de Oliveira Department of Biochemistry, Federal University of Pernambuco, Recife, PE, Brazil

Domenico Otranto Department of Veterinary Medicine, University of Bari, Valenzano, Italy

Felipe Arley Costa Pessoa Leônidas e Maria Deane Institute (Fiocruz Amazônia), Manaus, Amazonas, Brazil

Kim Picozzi School of Biomedical Sciences, Centre for Infectious Diseases, College of Medicine and Veterinary Medicine, University of Edinburgh, Edinburgh, UK

Sébastien Pion Joint Research Unit 233 "TransVIHMI", Institut de Recherche pour le Développement, Montpellier University, Montpellier, France

William K. Reisen Department of Pathology, Microbiology and Immunology, School of Veterinary Medicine, Center for Vectorborne Diseases, University of California, Davis, CA, USA

Eliana Maria Mauricio da Rocha Federal University of São João del Rei, Campus Centro Oeste, Divinópolis, Minas Gerais, Brazil

Tatiana Rozental Laboratory of Hantaviroses and Rickettsioses, Oswaldo Cruz Institute, Oswaldo Cruz Foundation, Rio de Janeiro, RJ, Brazil

Vladimir da Mota Silveira-Filho Department of Biology, University of Pernambuco, Garanhuns, PE, Brazil

Prabhjot Singh Department of Communicable Disease and Health Analysis, Pan American Health Organization/World Health Organization, Washington, DC, USA

Marise Sobreira Department of Microbiology, Aggeu Magalhães Center of Research (CPqAM/Fiocruz), Oswaldo Cruz Foundation, Recife, PE, Brazil

Lorena I. Spinsanti Arbovirus Laboratory, Institute of Virology "Dr. J. M. Vanella", National University of Córdoba, Córdoba, Argentina

Mário Steindel Department of Microbiology, Immunology and Parasitology, Center of Biological Sciences, Federal University of Santa Catarina, Florianópolis, SC, Brazil

Dietmar Steverding Norwich Medical School, University of East Anglia, Norwich, UK

Patricia H. Stoco Department of Microbiology, Immunology and Parasitology, Center of Biological Sciences, Federal University of Santa Catarina, Florianópolis, SC, Brazil

Laura B. Tauro Laboratory of Arbovirus, Institute of Virology "Dr. J. M. Vanella", National University of Córdoba, Córdoba, Argentina

Patricia Jacqueline Thyssen Department of Animal Biology, Institute of Biology, State University of Campinas, UNICAMP, Campinas, SP, Brazil

Donato Traversa Faculty of Veterinary Medicine, Teaching Veterinary Hospital, University of Teramo, Teramo, Italy

Gerrit Uilenberg "A Surgente", Cargèse, France

Jesus Valenzuela Vector Molecular Biology Section, Laboratory of Malaria and Vector Research, National Institute of Allergy and Infectious Diseases, National Institutes of Health, Rockville, MD, USA

Pedro F.C. Vasconcelos National Reference Laboratory of Arboviruses, Evandro Chagas Institute, Ananindeua, Brazil

National Institute of Science and Technology for Viral Hemorrhagic Fevers, Ananindeua, Brazil

PAHO/WHO Collaborating Center for Arbovirus Research and Diagnostic Reference, Evandro Chagas Institute, Ministry of Health, Ananindeua, Brazil

Department of Pathology, Pará State University, Belém, Pará, Brazil

Part I

Introduction

Generalities and Importance of Arthropod-Borne Diseases

1

Carlos Brisola Marcondes

Abstract

Arthropod-borne diseases are very important for human and animal health. A brief introduction is developed on the importance and needs for the control of such diseases.

Keywords

Arthropod-borne diseases • Malaria • Sleeping sickness • Nagana

Arthropods are a very large and important group of animals, present in every place in the planet. They seem to exist since at least the Cambrian "explosion" (ca. 550 million years ago), and fossil insects were present in the Devonian, at 350 million years ago. Comparing their antiquity and ubiquity to ours, it is too anthropocentric to refer to them as "intruders in our world." It is necessary to learn how to live with (or among) them to minimize conflicts and protect the health of humans and domesticated animals and not cause excessive damage to the environment.

Several insects, all ticks, and some mites developed the capacity of sucking blood, causing problems to hosts, not only by the direct effect of bites and theft of blood but also due to the transmission of virus, bacteria, protozoa, and helminths. By the adaptation to transmission through vectors, these living beings got access to the hosts, without direct exposition to desiccation, rain, direct sunlight, etc.

There are currently almost 15,000 species of bloodsucking insects (Lehane 2005), ca. 880 of ticks (Jongejan and Uilenberg 2004), and many mites adapted to bloodsucking, and several lice and mites can feed on the skin and annexes. Most bloodsucking insects belong to Diptera, and several of the most important diseases are transmitted[1] by arthropods. They have caused great mortality and misery to mankind, sometimes influencing economy and history, and their control is very important for public health. For example, the mortality of malaria in the world was estimated at 584,000 deaths (367,000–755,000) with 184 million of cases in 2013 and was the tenth cause of mortality in the world, in 2012. The estimated

C.B. Marcondes (✉)
Department of Microbiology, Immunology and Parasitology, Center of Biological Sciences, Federal University of Santa Catarina,
88040-900 Florianópolis, SC, Brazil
e-mail: cbrisolamarcondes@gmail.com

[1] Although arthropods actually transmit agents able to cause diseases and not diseases, transmission of diseases is utilized in this book for practicability.

C.B. Marcondes (ed.), *Arthropod Borne Diseases*, DOI 10.1007/978-3-319-13884-8_1

annual incidence of cutaneous leishmaniasis in the world is 690,900, while that of visceral form (VL) is 202,200, in 98 countries and three territories on five continents (Alvar et al. 2012). Mortalities from VL, probably underreported, varied from 1.5 to 7.2 %, but more careful surveys indicated case fatality rates, respectively, of 10 and 20 % in South Sudan and India.

Several arthropod-borne diseases are very important for domestic animals and trypanosomiasis, for example, prejudice cattle breeding in 10 million km^2 of arable land in Africa (Itty 1996); nagana and sleeping sickness, respectively, cause a damage >US$4.5 billion/year and 50,000 deaths/year (Kabayo 2002). Babesiosis, several arboviruses, mange, bloodsucking flies, and ticks strongly influence the production of food. Eighty percent of cattle in the world are exposed to tick bites, with an impact of US$7.3/head year (FAO 1984). Long contact of human and domestic animal populations with diseases related or not to arthropods and availability of animals capable of domestication have contributed for the advantage of Eurasian populations (Diamond 1999).

Arthropod-borne diseases are frequently associated with poorness and precariousness of living conditions, contributing to aggravate such conditions, and this has contributed to make them neglected diseases. However, for humanitarian reasons and because they can cause disease to all people, they deserve great attention. For example, the association of dengue to precarious neighborhood is controversial, at least in Rio de Janeiro and other Brazilian cities (Teixeira et al. 2003; Machado et al. 2009).

Epidemiology of diseases is much diversified, and its study must be developed before they occur, to prevent suffering and deaths, and all control measures must be based in solid knowledge. As an example, an outbreak of yellow fever in Botucatu, in the Brazilian state of São Paulo in 2009, with 28 cases and 11 deaths, induced the vaccination of one million people that caused five deaths; a careful calculation of proportion of people to be vaccinated (Ribeiro et al. 2015) and/or a more complete knowledge of the mosquito fauna in the area would probably have minimized the mortality.

The sudden occurrence of an outbreak of Chagas disease linked to ingestion of sugarcane juice, in 2005, in the non-endemic Brazilian state of Santa Catarina (Steindel et al. 2008) emphasized the need of thorough studies of sylvatic cycle of *Trypanosoma cruzi* in the area. However, local researchers, after developing preliminary studies on this cycle in the 1980s, were induced to work in other fields of research, due to the nonavailability of resources. The recent cases of dermal leishmaniasis cases in the Santa Catarina Island justified the usual comment of the author in classes, pointing to the forest and saying that there is probably some *Leishmania* there, which someday would infect a human being and be perceived.

Many problems caused by arthropods are related to inadequate management of natural environment, promoting proliferation of vectors and their contact with parasites and human populations. The dream of "a world without flies" just after the discovery of organochlorine insecticides was not fulfilled, and most arthropod-borne diseases, after a time of successful control, continued to be very frequent and widely distributed. Control of such diseases must be based on solid knowledge of their many aspects and on the ecology of vectors and reservoirs.

References

Alvar J, Vélez ID, Bern C et al (2012) Leishmaniasis worldwide and global estimates of its incidence. Plos One 7, e35671

Diamond J (1999) Guns, germs and steel: the fate of human societies. W. W. Norton and Co., New York/London, p 457

FAO (1984) Ticks and tick-borne disease control: a practical field manual, vol I, Tick control. Food and Agriculture Organization of the United Nations, Rome, p 299

Itty P (1996) Profitability, efficiency and comparative advantage of African cattle meat and milk production: the case of trypanotolerant village cattle production. Agric Econ 14:33–44

Jongejan F, Uilenberg G (2004) The global importance of ticks. Parasitology 129:S3–S14

Kabayo JP. 2002. Aiming to eliminate tsetse of Africa. Trends Parasitol 18:473–475

Lehane MJ (2005) The biology of blood-sucking in insects. Cambridge University Press, Cambridge

Machado JP, Oliveira RM, Souza-Santos R (2009) Spatial analysis of dengue occurrence and living conditions in Nova Iguaçu, Rio de Janeiro state, Brazil. Cad Saúde Pública 25:1025–1034

Ribeiro AF, Tengan C, Sato HK et al (2015) A public health risk assessment for yellow fever vaccination: a model exemplified by an outbreak in the state of São Paulo, Brazil. Mem Inst Oswaldo Cruz 1:1–5

Steindel M, Pacheco LK, Scholl D et al (2008) Characterization of *Trypanosoma cruzi* isolated from humans, vectors, and animal reservoirs following an outbreak of acute human Chagas disease in Santa Catarina state, Brazil. Diagn Microbiol Infect Dis 60:25–32

Teixeira MG, Barreto ML, Costa MCN et al (2003) Dynamics of dengue virus circulation in a metropolitan area of Brazil. Epid Serv Saúde 12:87–97

Epidemiology

2

William K. Reisen

Abstract

Concepts required to understand the epidemiology and ecology of vector-borne diseases are presented in sections that cover components of the transmission cycle, modes of transmission by the vector, different types of transmission cycles based on vertebrate hosts, vector incrimination, interseasonal maintenance mechanisms, and surveillance. Minimal components required for transmission of a vector-borne pathogen include a competent vertebrate host and an arthropod vector, a virulent pathogen, and a suitable environment. The efficiency of transmission depends on the frequency of contact between host and vector and is delineated by blood meal acquisition behavior by the vector and environmental conditions that drive the system. Transmission cycles mostly have evolved from sylvan zoonoses comprised of a diverse variety of hosts and vectors to urbanized anthroponoses comprised of human hosts and a limited number of vectors that frequently rest and blood feed in houses. Vector incrimination is dependent upon the diagnosis of frequent field infection, degree of competent host contact, and vector competence determined experimentally. Many vector-borne pathogens appear to have evolved in the tropics, but have become a serious public, veterinary, or wildlife health problem after invading temperate latitudes. Here, interseasonal maintenance becomes a key element for pathogen persistence, delineates endemicity, and delimits distribution in time and space. Because outbreaks of vector-borne disease occur intermittently even in endemic areas, surveillance programs are required to track cases and the pathogen within transmission cycles in time and space to inform public health policy and provide operational decision support to direct intervention.

W.K. Reisen (✉)
Department of Pathology, Microbiology and Immunology, School of Veterinary Medicine, Center for Vectorborne Diseases, University of California, Davis, CA 95616, USA
e-mail: wkreisen@ucdavis.edu

C.B. Marcondes (ed.), *Arthropod Borne Diseases*, DOI 10.1007/978-3-319-13884-8_2

Keywords
Epidemiology • Ecology • Transmission cycles • Landscape • Vector incrimination • Vectorial capacity • Surveillance

2.1 Introduction

Epidemiology (etymology: epi = upon, demos = people, logos = study) developed as a science through the investigation of outbreaks of infectious diseases. As a modern discipline, it deals primarily with the natural history and spread of diseases within human and animal populations. Vector-borne diseases comprise a subset of infectious diseases and are caused by parasites that are transmitted by arthropods. The transmission cycle minimally consists of a competent arthropod vector, a susceptible vertebrate host, and a virulent parasite interacting within a favorable environment (Fig. 2.1). Where these factors come together in time and space is known as the *nidus* (literally nest or center) of infection (Pavloskiy 1966). The spread of parasites by arthropods is complex, because in addition to the myriad of interactions between the vertebrate host and the parasite to cause disease, an arthropod is required to distribute the parasite among uninfected hosts. The ecology, behavior, and physiology of the arthropod as well as environmental factors such as temperature and rainfall impact the transmission processes by affecting the rate of parasite acquisition and maturation within the arthropod host and distribution of the parasite to new hosts. Failure of any one of the basic cycle components may result in local parasite extinction.

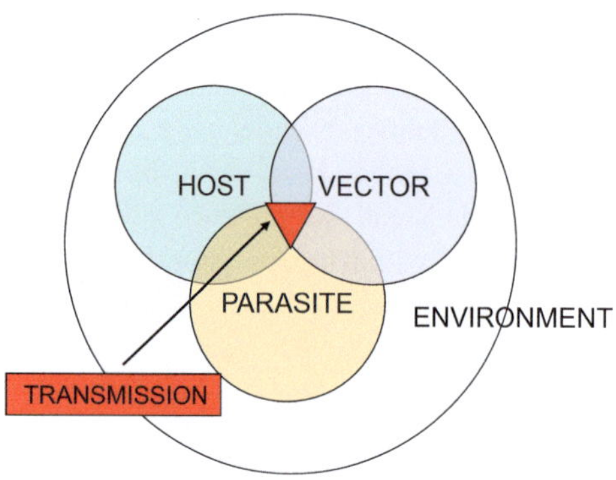

Fig. 2.1 Four required components of a vector-borne transmission cycle: (*1*) susceptible vertebrate host, (*2*) competent vector, (*3*) infectious parasite, and (*4*) suitable environment. Transmission occurs when all four components come together in time and space (Adapted from Reisen 2010)

With an epidemiological context, a *vector* is an arthropod responsible for distributing a parasite (not the disease) among vertebrate hosts. *Disease* is the response of the host to infection with the parasite. A *parasite* in this chapter is any organism, including viruses, bacteria, protozoa, helminths, and arthropods, which is dependent upon a host for its survival. Parasites may or may not cause disease. When a parasite injures its host and causes disease, it is referred to as a *pathogen*. A *vector-borne disease*, therefore, is an illness caused by a pathogen that is transmitted by an arthropod vector. *Facultative parasites* may have both free-living and parasitic forms, whereas *obligate parasites* are totally dependent upon their host(s) to provide their requisites for life. *Ectoparasites* live on or outside the host, whereas *endoparasites* live inside the host. When interacting with their hosts, ectoparasites produce an *infestation* that typically remains topical or peripheral, whereas endoparasites produce an *infection* when they invade host tissues or cells. The occurrence and severity of disease depend upon host–parasite interaction after infection and are often related to the immune response of the host. A host carrying a parasite is *infected*, whereas an infected host that is capable of transmitting a parasite is *infective* or *infectious*. A host capable of parasite maintenance without severe clinical symptoms is a *carrier*.

Understanding the epidemiology of arthropod-borne disease requires knowledge of the ecology, physiology, immunology, and genetics of parasite, arthropod vector, and vertebrate host populations and how they interact within their environment. Although specific methods of investigation vary considerably among the vast array of vectors and vector-borne parasites, overarching concepts unify the information necessary to understand the epidemiology of vector-borne diseases. Information on the disease typically evolves chronologically from the discovery of the parasite as the causative agent to identifying its mode of transmission by the arthropod vector(s) among vertebrate hosts and to monitoring, forecasting, and intervention. During the discovery period, clinical case definition and diagnosis are established enabling the tracking of human and/or veterinary cases in time and space, and the causative agent is identified, leading to the development of specific laboratory diagnosis and perhaps indicating that an arthropod may be responsible for transmission. The incrimination of the vector(s) requires a combination of field and laboratory investigation that measures abundance in time and space, blood-feeding patterns, field infection rates, and vector competence. Although short-term studies rapidly may determine the mode(s) of transmission, delineating the components of transmission cycles and interseasonal maintenance mechanisms typically requires years of careful ecological investigation and laboratory experimentation. Effective surveillance and control programs are best implemented after maintenance, amplification and epidemic transmission patterns have been described. Practically, discovery rarely progresses in the orderly fashion outlined above, and frequently, monitoring and management of cases progress more rapidly than the discovery of the vector or the mode(s) of transmission.

The current chapter details, in a general sense, the mechanisms by which arthropods serve as vectors of parasites that cause human or veterinary diseases. Although some examples are presented, details of specific parasites and their vectors will be described by others in subsequent chapters. This chapter revises and updates a similar presentation (Reisen 2009) that was published in Medical and Veterinary Entomology (Mullen and Durden 2009). Some overlap and redundancy between chapters were unavoidable as many of the definitions and concepts have not changed.

2.2 Components of Transmission Cycles

The components of a *transmission cycle* of an arthropod-borne disease include (1) a parasite that can multiply within both vertebrate and invertebrate host tissues, (2) a vertebrate host (or hosts) which develops a level of infection with the parasite that is infectious to the vector, (3) a competent arthropod vector that is able to acquire the parasite from the infected host and is capable of transmission, and (4) a suitable environment (Fig. 2.1).

2.2.1 The Parasite

A wide variety of human and animal parasites exploit arthropods as a means of transmission, including multiple families of viruses, bacteria, protozoa, and helminths. Vector-borne parasites have by necessity evolved mechanisms for tolerating high constant body temperatures and for evading the complex responsive immune systems of vertebrate hosts, as well as for tolerating variable body temperatures and avoiding the very different innate immune mechanisms of arthropod vectors. In addition, the parasites must locate and replicate in very different target organs and then develop mechanisms for transmission either by producing stages in vertebrates that aggregate where vectors blood feed or in vectors that can be deposited or injected during blood feeding. Asexual parasites such as viruses and bacteria employ the same life form to infect both vertebrate and arthropod hosts, whereas more highly evolved parasites such as protozoa and helminths have very different life stages within their vertebrate and arthropod hosts. In addition, some asexual parasites such as the plague bacillus at times may bypass the arthropod host and be transmitted directly from one vertebrate host to another by aerosol droplets or contact.

Among sexually reproducing parasites, the host in which gametocyte union occurs is the *definitive host*, whereas the host in which asexual reproduction occurs is the *intermediate host*. Vertebrates or arthropods can serve as either definitive or intermediate hosts, depending upon the life cycle of the parasite. For example, humans are the definitive host for the filarial worm, *Wuchereria bancrofti*, because adult male and female worms mate within the human lymphatic system, whereas the mosquito vector, *Culex quinquefasciatus*, is the intermediate host where larval worms develop and transform without reproduction. In contrast, humans are the intermediate host of the *Plasmodium* protozoan that causes malaria, because only asexual reproduction occurs in the human host; haploid gametocytes produced in the human host unite in the gut of the definitive *Anopheles* mosquito host, after which transformation and asexual reproduction occur in a series of events that eventually produces infectious sporozoites in the salivary gland.

Disease is the response of the host to infection with the parasite and can occur in either vertebrate or arthropod hosts. *Immunity* includes all properties of the host that confer resistance to infection and plays an important role in determining host suitability and the extent of disease or illness. Some species or individuals within species populations have *natural (or innate) immunity* and are refractory to infection. Innate immunity does not require that the host has previous contact with the parasite. For example, humans do not become infected with avian malaria parasites such as *Plasmodium relictum*, even though infective *Culex* mosquito vectors feed frequently on humans and probably transfer sporozoites. Conversely, mosquitoes do not become infected with measles or the AIDS viruses that infect humans, even though these viruses undoubtedly are ingested by mosquitoes blood feeding on viremic human hosts.

2.2.2 The Vertebrate Host

One or more *primary vertebrate hosts* are essential for the maintenance of parasite transmission, whereas *secondary* or *incidental hosts* are not essential and may or may not contribute to parasite amplification. *Amplification* refers to the general increase in the number of parasites present in a given area. An *amplifying host* increases the number of parasites and theoretically then the number of infected vectors. Amplifying hosts often do not remain infected for long periods of time and may develop severe disease. A *reservoir or maintenance host* supports parasite development, remains infected for long periods, and serves as a source of vector infection, but usually does not develop acute disease (similar to the *carrier* defined above). Humans can be the primary, if not, the only, hosts in diseases called *anthroponoses* such as malaria or be incidental hosts in *zoonoses* such as West Nile virus (WNV) with the basic cycle involving birds or mammals (Fig. 2.2).

Attributes of a suitable vertebrate host include vector accessibility, parasite susceptibility, and parasite transmissibility.

2.2.2.1 Accessibility

The vertebrate host must be abundant and fed upon frequently by vectors. Seasonality, daily activity rhythms, and habitat selection determine availability in time and space to host-seeking or questing vectors. For example, in New Jersey, USA, the avian hosts of eastern equine encephalomyelitis virus (EEEV) generally begin nesting in wetlands coincidentally with the emergence of the first spring generation of the mosquito vector, *Culiseta melanura*, thereby bringing EEEV, susceptible avian hosts, and mosquito vectors together in time and space (Crans et al. 1994). Conversely, although many of winter resident sparrows are highly susceptible to WNV, they arrive and are present during winter when the vectors enter diapause/quiescence and therefore are not important in transmission (Reisen et al. 2010).

Diel activity patterns also may be critical. Certain vectors host-seek at night along flight paths delineated by landscape features and blood feed upon those host species that are found along those flight paths. Similarly, ticks quest on vegetation along certain trails, at a certain height, and at certain times of the day when they are most likely to contact a suitable host. Epidemics of vector-borne diseases frequently have been associated with increases in the accessibility of human hosts to vectors during wars, natural disasters, environmental changes, or human migrations.

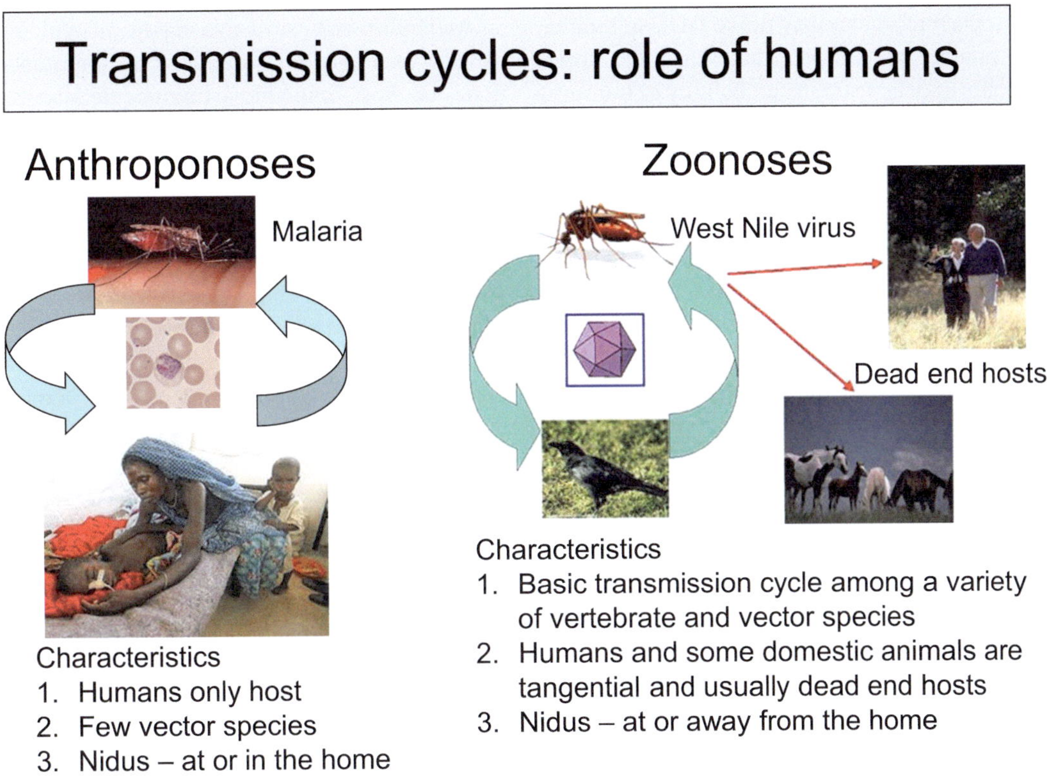

Fig. 2.2 Characteristics of anthroponosis and zoonosis transmission cycles using malaria and West Nile virus as respective examples (From Wikimedia)

For example, in 1952 a wet spring and flooding by the Kern River followed by a strong earthquake that destroyed most of the homes in Bakersfield, California, forced people to sleep out of doors and led to a large western equine encephalomyelitis virus epidemic (Reeves and Hammon 1962).

2.2.2.2 Susceptibility

Once exposed, a primary host must be susceptible to infection and permit the development and reproduction of the parasite. *Dead-end hosts* do not support a level of infection sufficient to infect vectors, although these hosts may become extremely ill and succumb. Ideal reservoir hosts permit parasites to survive in the peripheral circulatory system (or other suitable tissues) in adequate numbers for a sufficiently long time period to be an effective source for vector infection. Asexual parasites such as viruses and bacteria typically produce intensive infections that yield large numbers of infectious organisms for relatively short periods during which the host either succumbs to infection or develops protective immunity. In the case of WNV, for example, 1 ml of blood from an infected American crow may contain as many as 10^{10} virus particles during both day and night for a 2–5-day period (Komar et al. 2003); birds that survive such infections typically develop long-lasting protective immunity. In contrast, highly evolved parasites produce comparatively few individuals during a longer period. *Wuchereria bancrofti*, for example, maintains comparatively few microfilariae in the blood stream (usually <10 microfilariae per ml of blood), which circulate most abundantly in the peripheral blood during periods of the day when the mosquito vectors blood feed (Weerasooriya et al. 1998). However, because both the worms and the human host are long lived, transmission is enhanced by repeated exposure rather than by an intense parasite presentation over a short period of time. In fact, infection with >25 microfilariae per female mosquito may prove fatal, so limiting the number of parasites that infect the vector actually may increase the probability of transmission (Subramanian et al. 1998).

2.2.2.3 Transmissibility

Suitable numbers of susceptible vertebrate hosts must be available to become infected and thereby maintain the parasite population. Transmission rates typically decrease concurrently with a reduction in the number of susceptible (i.e., non-immune) individuals remaining in the host population. The *epidemic threshold* refers to the number of susceptible individuals required for epidemic transmission to occur, whereas the *endemic threshold* refers to the number of susceptibles required for parasite persistence. These numerical thresholds vary depending on the immunology and dynamics of infection in the host population and relate to the basic reproductive rate of the parasite, R_0 (May 1983). Therefore, suitable hosts must be abundant and either not develop lasting immunity or have a relatively high reproductive rate to ensure the rapid recruitment of susceptibles into the population. In the case of malaria, for example, the parasite elicits an immune response that rarely is completely protective, and the host remains susceptible to reinfection. In contrast, vector-borne virus infections of birds typically produce life-long protection, but bird life expectancy is short and the population replacement rate rapid, thereby ensuring the constant and rapid renewal of susceptible hosts.

2.2.2.4 Vertebrate Immunity

Some individuals infected with parasites recover and in the process acquire immunity. This acquired immunity ranges from transient to life long and may provide partial to complete protection. A partial immune response may permit continued infection, but may reduce the severity of disease, whereas a complete response results in parasite elimination and usually prevents reinfection. Acquired immunity may be humoral and result in the rapid formation of antibodies or may be cellular and result in the activation of T cells and macrophages.

Antibodies consist of five classes of proteins called *immunoglobulins* that have specific functions in host immunity. Immunoglobulin G (abbreviated IgG) is most common, comprising over 85 % of the immunoglobulins present in the sera of normal individuals. The IgGs are relatively small proteins, typically develop to high concentration several weeks after infection, and may persist at detectable and protective levels for years. Therefore, parasites such as yellow fever virus that induce long-lasting immunity are good candidates for vaccine development. In contrast, immunoglobulin M (IgM) is a large macroglobulin that appears shortly after infection, is very specific for the infecting parasite, but decays rapidly relative to IgG. For the laboratory diagnosis of many diseases, serum samples are tested during periods of acute illness and convalescence, 2–4 weeks later. The presence of elevated concentrations of IgM presumptively implies a current or very recent infection, whereas a fourfold or greater increase in parasite-specific IgG concentration in paired acute and convalescent sera provides confirmatory diagnosis. T cells and macrophages are several classes of cells that are responsible for the recognition and elimination of parasites. In long-lived vertebrate hosts, acquired immunity may decline over time, eventually allowing reinfection. Time here ranges from months in the case of human malaria to decades in the case of yellow fever virus.

Clinically, the host response to infection ranges from inapparent or asymptomatic to mildly symptomatic to acute. Generally it is beneficial for the parasite if the host tolerates infection and permits parasite reproduction and/or development without becoming severely ill and dying before vectors can be infected. However, the fitness of many parasites is contingent upon elevated virulence that causes high parasitemia and frequent mortality in the vertebrate host, but facilitates transmission by increased infection probability in the vector. Generally, the susceptibility of the vector to infection dictates the concentration levels or *virulence* of parasites required in the vertebrate host to infect vectors and complete the transmission cycle.

2.2.2.5 Risk Factors

A risk factor for human exposure is a variable statistically associated with an increased or decreased risk of disease and may depend on a variety of factors, including place of residence, socioeconomic status, occupation, age, and sex. Typically risk factors are correlational and frequently pathogen specific, closely interconnected, and studied as model covariates paired with serological or pathogen survey data. For vector-borne disease, risk may be related to frequency of exposure to the vector as well as innate susceptibility to pathogen infection and disease. Risk related to the place of residence varies with scale, pathogen endemicity, and vector exposure over coarse geographical areas delimited by landscape or political boundaries or over fine scale at microhabitat levels. For example, the USA is currently endemic for West Nile virus, whereas Brazil is not, whereas residents of houses within villages living close to mosquito larval habitat typically are at higher risk of pathogen transmission than houses furthest from these habitats. Socioeconomic status frequently is interrelated with residence, with poorer urban areas having higher risk of pathogen transmission due to inconsistent or no potable water delivery, open sewage drains, and poor solid waste management creating vector and rodent problems, and poor housing construction allowing vector access and harborage. Occupation often delimits the extent and type of exposure, related to time spent outdoors in wood gathering, agriculture or tending domestic animals, or spent indoors at the place of work or residence. The impact of age may be biased for exposure by occupation, but disease may be related to the response of the host to infection as well as exposure history. Even closely related pathogens may have markedly different age impacts which are poorly understood. For example, infection with Japanese encephalitis virus typically leads to neuroinvasive disease in children, whereas genetically related West Nile virus (WNV) most frequently expresses neuroinvasive disease in the elderly. Similarly sex-related disease risk is confounded by exposure as well as the response of the host to infection. In many areas, men typically engage in outdoor activities such as entering forests for hunting or logging thereby exposing them to pathogens such as sylvan yellow fever virus cycling among forest mosquitoes and primates, whereas women may spend more time in and around the home thereby exposing them more frequently to diurnal endophagic mosquitoes such as *Aedes aegypti*[1] and dengue virus. In addition, certain hormones at high levels such as testosterone may serve as an immunosuppressant and facilitate more frequent disease in males.

2.2.3 The Arthropod Vector

Literally, a *vector* is a "carrier" of a parasite from one host to another. The degree of contact between the vertebrate host and vector ranges from intermittent (e.g., mosquitoes) to continuous (e.g., sucking lice). Frequently the host provides the vector not only food in the form of blood or other tissues but also a habitat or place in which to live such as a nest or burrow. Blood feeding by the vector is important, because it brings parasite, vector, and vertebrate host together in time and space and ultimately is responsible for the transmission of parasites from infectious to susceptible vertebrate hosts. The vector usually must take at least two blood meals during its lifetime to transmit a parasite, the first to acquire the infection and the second to transmit it. Blood meals are taken to provide the arthropod with nutrients necessary for reproduction as well as metabolism and metamorphosis. The *gonotrophic cycle*, or literally "reproductive feeding cycle" of the arthropod, includes the sequence of questing or searching for a vertebrate host, blood feeding, blood meal digestion, egg maturation, and oviposition. *Parous* females have completed one or more gonotrophic cycles and

[1]Although several taxonomic modifications have been proposed on Culicidae genera, mostly splitting *Aedes* by Reinert et al. (2009) (and accepted by CBM), this taxonomy has been used in the chapters according to authors' preference. To facilitate utilization by health personnel, all new Aedine genera can be considered *Aedes* (CBM).

therefore have a greater probability of being infected with and transmitting parasites than *nulliparous* females that have not reproduced and are host seeking or blood feeding for the first time. Unlike parasites that are transmitted directly from host to host, parasites transmitted by arthropods generally have replaced the free-living or environmentally resistant stages with those that can multiply and develop within the arthropod and be transmitted during the blood-feeding process.

An effective arthropod vector generally exhibits characteristics that complement those listed above for the vertebrate host and includes host selection, infection, and transmission.

2.2.3.1 Host Selection

Blood feeding by arthropods has arisen evolutionarily at least 21 times in different branches of the phylum Arthropoda and is the most important behavior for bringing vector, host, and parasite together in time and space. Most blood meals are used for extra protein and lipids for egg development, but they also can be used for water balance and nutrition (Clements 1992). An effective vector must be abundant and feed frequently upon infective vertebrate hosts when stages of the parasite are circulating in the peripheral blood stream or other tissues accessible to the vector. Difficulty in locating the vertebrate host depends upon the degree of intimacy and duration of contact and ranges from constant contact by ectoparasites such as sucking lice to intermittent contact by vectors such as mosquitoes that live away from the host. In the latter situation, the vector can either wait for the arrival of the host such as ticks questing by along animal trails or flea awaiting the arrival of the host at burrows or nests or alternatively search actively for the host as do female mosquitoes or flies. Patterns of host selection or encounter determine the types of parasites to which vectors are exposed. *Anthropophagic*[2] (literally "human-eating") vectors such as some *Anopheles* mosquitoes feed selectively on humans and therefore are important in the transmission of human parasites such as malaria (Fig. 2.2). Anthropophagic vectors which readily enter houses to feed on humans or to rest on the interior surfaces of the house are termed *endophilic* (literally, "inside loving"), whereas vectors which rarely enter houses are termed *exophilic* (i.e., "outside loving"). *Zoophagic* ("animal-eating") vectors feed primarily on vertebrates other than humans, with *mammalophagic* vectors blood feeding primarily on mammals and important in the maintenance of mammalian parasites and *ornithophagic* vectors feeding primarily on avian hosts and important in the maintenance of avian parasites. Frequently host selection is dependent upon how diurnal rhythms and flight paths of questing vectors lead to encounters with available potential vertebrate hosts. Juxtaposition of landscape features, vector questing patterns, and vertebrate behavior enable these encounters (Reisen 2010).

2.2.3.2 Infection

To transmit the parasite, the vector must be able to penetrate the host skin, be susceptible to infection, and survive long enough for the parasite to complete multiplication and/or development to the infective stage. Not all arthropods that ingest parasites become infected or support parasite maturation, dissemination, and transmission. Usually infection of the vector is dose dependent and requires that the number of parasites ingested exceeds a minimal threshold. In addition, some arthropod species that are susceptible to infection under laboratory conditions seldom feed on vertebrate hosts infective with the parasite and/or survive long enough to allow parasite development in nature. *Aedes albopictus*, for example, readily becomes infected with WNV in the laboratory, but this mosquito is not considered an important vector in nature because it does feed consistently on birds, the primary host for WNV. The *transmission rate* is the number of new infections produced per unit of time and is dependent upon the frequency of blood feeding by the vector and the duration of the *extrinsic incubation period* or the time required for parasite development to the infective stage. Because arthropod vectors are poikilothermic, transmission rates frequently are

[2]The suffix "phagic" here has also been replaced by "philic," meaning "loving" (CBM).

dependent upon ambient temperature and therefore proceed more rapidly at tropical than temperate latitudes and progress most rapidly during summer (Walton and Reisen 2014). The frequency of host contact and therefore the transmission rate also depend upon the life history of the vector. For example, epidemics of malaria in the tropics transmitted by an *Anopheles* mosquito that feeds at 2 day intervals progress faster than epidemics of Lyme disease at temperate latitudes where the spirochetes are transmitted to humans principally by the nymphal stage of the *Ixodes* tick vector that may have one generation and one blood meal per life stage per year.

2.2.3.3 Transmission

Once infected, the vector must exhibit a high probability of refeeding on one or more susceptible hosts to ensure the transmission of the parasite. Diversion of vectors to non-susceptible or "dead-end" hosts dampens transmission effectiveness. The term *zooprophylaxis* (literally, "animal protection") originally described the diversion of host-seeking *Anopheles* infected with human malaria parasites from humans to cattle, a "dead-end" host for the malaria parasite. The dead-end host typically exhibits innate immunity where host tissues do not permit parasite growth to levels suitable to infect additional vectors. In some instances infection of the dead-end host results in serious illness, because the host–parasite relationship has not coevolved to the point of tolerance by the vertebrate host. WNV, for example, can cause serious illness in humans which are considered to be a dead-end host, because the virus rarely produces a viremia sufficient to infect mosquitoes. With zoonoses such as WNV in species diverse ecosystems, the *Culex* mosquito vector may blood feed on a wide assortment of avian hosts with a wide variety of competence thereby dampening amplification transmission, whereas in simple suburban/urban ecosystems with comparatively low avian diversity, *Culex* may blood feed on a limited number of competent hosts thereby enabling efficient transmission. Here, the loss in infective bites to dead-end hosts is called the *dilution effect* and is similar to zooprophylaxis.

2.2.3.4 Vector Immunity

After locating a suitable host, the skin, including hair and feathers, forms the first obstacle to blood meal acquisition by arthropods. This problem has been overcome by competent vectors with the coevolution of a varied assortment mouthparts designed to penetrate the skin and complex saliva that contains an assortment of analgesics to limit detection and anticoagulants, vasodilators, and antiplatelet-aggregating compounds to enhance tissue fluid and blood flow. As mentioned, not all vectors that ingest parasites with the blood meal become infected. Because arthropods typically are short-lived, they rely mostly upon genetically determined *innate immunity* to resist infection (Schmidt et al. 2008). Large parasites such as helminths may be destroyed by the cibarial armature during ingestion of the blood meal. Once in the midgut or *mesenteron*, the natural microbial flora may negatively impact parasites, limiting infection. However, the primary defense against most parasites seems to involve the midgut epithelium, where infections must be initiated prior to the formation of the *peritrophic* (literally, "around food") *matrix* during blood digestion. Penetration of the epithelium involves receptor–ligand interactions, physical burrowing, or digestion of host cells. Response by the vector to these processes includes lytic and melanization events which reduce the number of parasites successfully entering the body cavity or *hemocoel*. Unless the epithelium has been compromised, viruses multiply within focal infections of mesenteronal cells before escaping into the hemocoel (Hardy et al. 1983). Bacteria and some parasites either remain within the digestive tract (plague bacillus/*Leishmania*) or enter the hemocoel (*Borrelia* spirochete/malaria). In the hemocoel, parasites enter the open circulatory system, but then must evade *hemocytes* (several types of blood cells, including mostly granulocytes) that upon recognition of parasites respond by phagocytosis, encapsulation/melanization, and/or the secretion of immune factors. Transmission of most parasites depends on infection of the salivary glands and then injection with the saliva into the host during blood feeding. Some mosquitoes have salivary gland infection and escape barriers for some

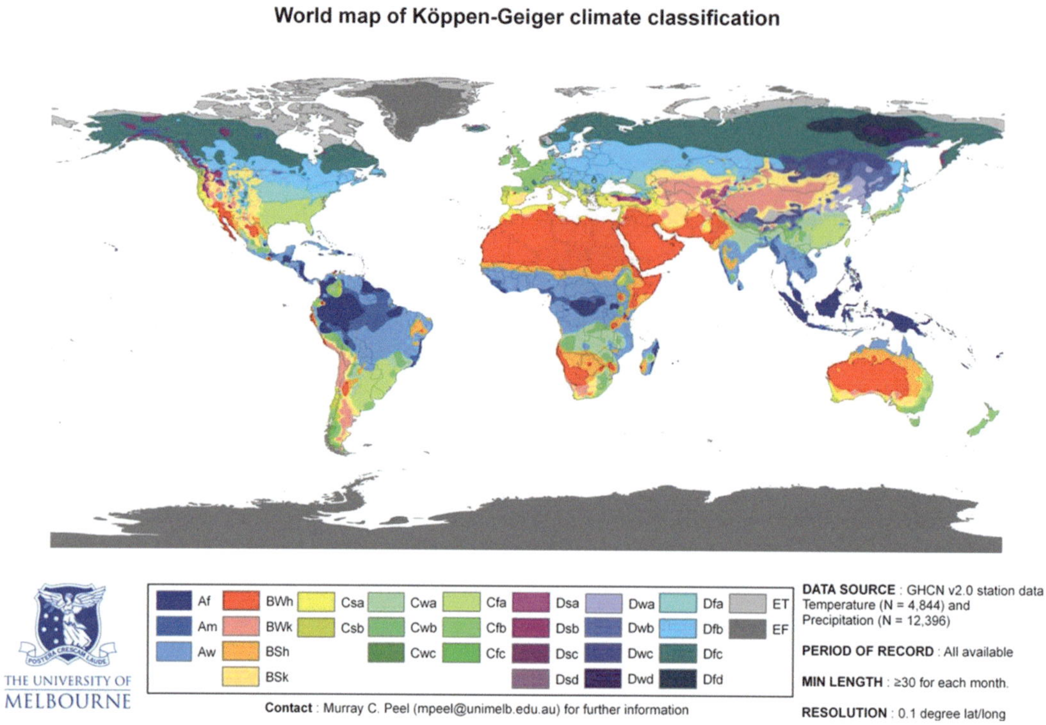

Fig. 2.3 Köppen–Geiger global climate classification (Downloaded from http://people.eng.unimelb.edu.au/mpeel/koppen.html; reference: Peel MC, Finlayson BS, and McMahon. 2007. Updated world map of the Köppen–Geiger climate classification. Hydrol. Earth Syst, Sci. 11: 1633–1644)

arboviruses thereby blocking transmission, but these mechanisms remain poorly understood.

2.2.4 The Environment

In an epidemiological context, the environment includes all things surrounding the transmission cycle, including climate, habitat, and plant and animal communities (Reisen 2010). Climate[3] is based on long-term patterns of temperature and rainfall and can be classified globally using schemes such as the Köppen–Geiger system based partly on the response of plant communities to temperature, moisture, soil, and slope (Fig. 2.3). Plant communities, in turn, delineate spatial suitability and often the distribution of vertebrate hosts, vectors, and their parasites.

Climate and the epidemiological environment are dynamic, with short-, mid-, and long-term variations, cycles, and trends imposed upon annual seasonality. In general, seasonal extremes between hot and cold increase as a function of latitude and/or elevation. Short-term climate variations in terms of days or weeks essentially define *weather* and include changes due to the interaction of global forces such as the jet stream modified by local landscape features such as oceans, mountains, and lakes. In recent years, scientists have recognized the importance of mid-term cycles in sea surface temperature on weather patterns. The Walker oscillation in the Pacific Ocean has been most intensively studied. Its cyclic variations are driven by interactions between the sea surface temperatures and the depth of the thermocline in the eastern Pacific off the western coast of North America (Fig. 2.4). The resulting *El Niño/Southern Oscillation* has multiyear cycles that strongly influence the distribution of the upwelling of warm moist air to

[3] More information on the influence of climate on diseases may be found in Chap. 37 (Geoprocessing) (CBM).

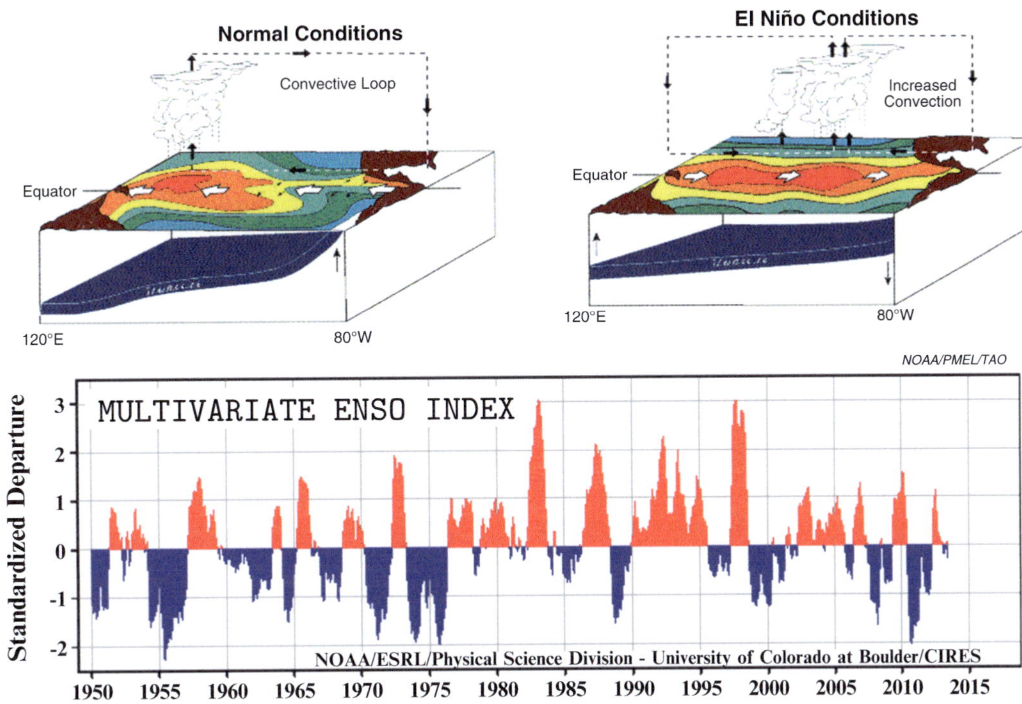

Fig. 2.4 Walker oscillation showing the upwelling of rain clouds during normal and El Niño conditions and time series changes in the ENSO index anomalies for the Pacific. (Graphics downloaded from the US Climate Prediction Center, http://www.cpc.ncep.noaa.gov/products/analysis_monitoring/ensocycle/enso_cycle.shtml)

form rain clouds and subsequently temperature and rainfall patterns in the Pacific and the New World as well as the Indian Ocean and parts of eastern Africa. These oscillations and the resulting temperature and rainfall anomalies have a strong, but spatially variable, influence on vector and host populations and their parasites. In the central prairies of North America, for example, the incidence of human infection with WNV is linked closely with hot and dry summers (Reisen et al. 2006), whereas outbreaks of Rift Valley fever virus in Kenya are linked to above-normal rainfall, both associated with El Niño events (Linthicum et al. 1999).

Superimposed on short-, annual, and mid-term variations are long-term trends such as *global warming*. Linked closely to the dramatic increase in the earth's human population has been the expanding need for resources for fuel, housing, and food (O'Neill et al. 2010). Carbon-based fuel consumption, whether a dung cooking fire in a developing country or a jet engine airplane, releases carbon dioxide and other chemicals into the air cumulatively creating a "greenhouse" effect that is gradually warming the earth. On a global scale, warming has varied spatially, being most dramatic at northern latitudes, where it has led to melting of the polar and Greenland ice caps and montane glaciers. Warming temperatures and changing rainfall patterns have altered the receptivity of these areas to vector-borne diseases due to milder and shorter winters and longer and warmer transmission seasons, leading to range extensions by vector species and outbreaks in susceptible populations. Examples include the invasion of Canada and the northern USA by WNV (Brault and Reisen 2014) and the African highlands by falciparum malaria (Chaves and Koenraadt 2010), both pathogens of tropical origin now creating outbreaks at new latitudes and elevations, respectively. In combination with anthropogenic changes such as urbanization, deforestation, and expansion of agroecosystems, vectors and parasites have extended their distributions, leading to the emergence of new public health problems (Brault and Reisen 2014).

By definition, an *emerging disease* has shown a significant increase in incidence, severity, and/or distribution within recent history and threatens to remain or increase as a health problem in the future. Historically, human movement frequently has led to changes in pathogen distributions and public health. Expanding trade with Asia along the Silk Road during the 1300s, for example, was followed by plague epidemics which killed 30–60 % of the European population (Kupferschmidt 1993). Recently, an expanding human population, an increased rapid travel, the globalization of commerce, and a variety of anthropogenic factors, including global warming and urbanization, have produced conditions conducive for the emergence and/or resurgence of new as well as old infectious diseases, including those transmitted by vectors (Institute of Medicine 2003). For many emerging diseases, a cascade of historical events altering human demography, vector or pathogen distributions, anthropogenic environmental change, and/or the genetics of the parasites have enabled the emergence of parasites, leading to major human, domestic animal, or wildlife health problems.

2.2.4.1 Human Demography

Although the global per capita population growth or birth rate has slowed in recent years, the earth's human population continues to increase at a constant rate and has tripled during the last generation. Demographers project final stabilization of the earth's human population at 10–12 billion or almost a 30–40 % increase above the 7.5 billion realized in 2012. Per capita growth rates have been variable, with countries in Europe showing declines, whereas others in Africa showing marked increases. Growth rates frequently are linked to socioeconomic factors and childhood mortality. The transition from high birth–high death rate to low birth–low death rate changes population age structure and therefore the population receptivity to some vector-borne diseases that exert a greater health burden on young or old age groups. In addition, since 2012 more than 50 % of the human population now resides in rapidly expanding cities, where municipal services such as piped water and sewage have not kept abreast with growth. These conditions are especially conducive for the establishment of large *peridomestic* (living in or around human habitations) vertebrate and vector populations making them receptive for the introduction of new vectors and parasites. More humans living in close proximity linked by rapid travel has enabled the rapid transit of parasites, especially anthroponoses utilizing humans as reservoir or amplifying host. Historically, the large-scale displacement of ethnic groups or trade has created patterns of frequent and repeated travel that may be exploited by vectors and pathogens (Tatem 2009).

2.2.4.2 Globalization of Commerce

The globalization of commerce, originally by the sailing ships of the European colonial era and recently by rapid international exchange of goods by both sea and air, has established conditions suitable for the inadvertent transport of both vectors and the parasites they transmit. *Aedes aegypti* and *Ae. albopictus*, for example, both lay drought-resistant eggs in dark areas such as water barrels or tires that collect rainwater, enabling the transport of the immature stages, the circumglobal establishment of these effective arbovirus vectors, and an ongoing dengue virus pandemic of global proportions. Of additional concern is the wide-scale, and often illegal, trade in exotic pets that may bring with them parasites that may escape into new geographical areas.

2.2.4.3 Anthropogenic Change

A rapidly expanding human population and its growing need for food, fuel, and shelter have markedly altered the environment in ways conducive to the increase of vector-borne pathogens. The large-scale movement of the rural poor to urban centers to find jobs in developing countries and an unchecked human population growth have caused the large-scale unplanned *urbanization* in previously agrarian societies, the overwhelming of municipal infrastructure, the creation of inadequate housing, a lack of potable water that frequently must be stored for domestic use, and an absence of adequate waste management systems.

This has created conditions suitable for the increased abundance and distribution of peridomestic mosquitoes such as *Ae. aegypti* and the *Culex pipiens* complex that preferentially rest in houses and feed on humans (Gubler and Meltzer 1999). The reduction of species diversity and the increase of successful peridomestic commensals such as crows and house sparrows have enabled the highly efficient urban transmission of zoonoses such as WNV by *Cx. pipiens* complex mosquitoes (Brault and Reisen 2014). Ecosystem simplification by agriculture and urbanization has made parasite amplification more rapid and transmission more efficient than in complex sylvan ecosystems, because there is less diversion of pathogens to noncompetent vertebrate or invertebrate hosts (Bonds et al. 2012). Some pathogens such as dengue and yellow fever viruses have transitioned from tropical zoonoses to anthroponoses and become *urbanized* (Weaver and Reisen 2010). Yellow fever virus, for example, persists naturally in the gallery forests of Africa and South America in cycles involving canopy primates and mosquitoes. When humans enter the forest or when primates raid invading agriculture, the virus transitions from canopy to village ecosystems where it is transmitted among humans by peridomestic *Ae. aegypti*. Human travel brings the virus to an urban *Ae. aegypti*—human transmission cycle—and the virus becomes urbanized, i.e., persists in the urban environment. Urbanization of viruses requires the presence of large human populations, because infection produces lasting immunity and therefore the continued need for nonimmunes for transmission to be maintained.

An *ecotone* is the transition area between biomes or landscapes dominated by different ecosystems and may appear as a gradual blending or a sharply delineated boundary. The alteration of landscapes and the spatial expansion of human and associated domestic animal populations into or adjacent to natural areas have concurrently expanded ecotone habitat and enabled the expansion of some zoonoses. Movement of housing into wooded areas of the NE USA, for example, has increased the forest ecotone; expanded browse vegetation thereby increasing deer, mice, and associated tick vector populations; and produced the ongoing and expanding Lyme disease epidemic (Brownstein et al. 2005).

2.3 Modes of Transmission

The transmission of parasites by vectors may be either vertical or horizontal. *Vertical transmission* is the passage of parasites directly to subsequent life stages or generations within the vector population. Some vertical transmission systems may require a blood meal to trigger molting by the arthropod, but the host is not required for the infection process of the next stage. *Horizontal transmission* describes the passage of parasites between vector and vertebrate hosts and can be accomplished by vertically infected arthropods.

2.3.1 Vertical Transmission

Three types of vertical transmission are possible within vector populations: transstadial, transgenerational, and venereal.

2.3.1.1 Transstadial Transmission

Transstadial transmission (literally, across "stadia or stages") is the sequential passage of parasites acquired during one life stage through the molt to the next stage(s) or stadium. Transstadial transmission is essential for the survival of parasites transmitted by mites and hard ticks that blood feed once during each life stage and die after oviposition. Lyme disease spirochetes, for example, that are acquired by larval ticks must be passed transstadially to the nymphal stage before they can be transmitted to other vertebrates.

2.3.1.2 Transgenerational Transmission

Transgenerational transmission (across generations) is the passage of parasites by an infected parent to its offspring in the next generation. Some parasites may be maintained transgenerationally for multiple generations, whereas others require horizontal transmission during each generation for amplification. Transgenerational transmission normally occurs *transovarially*

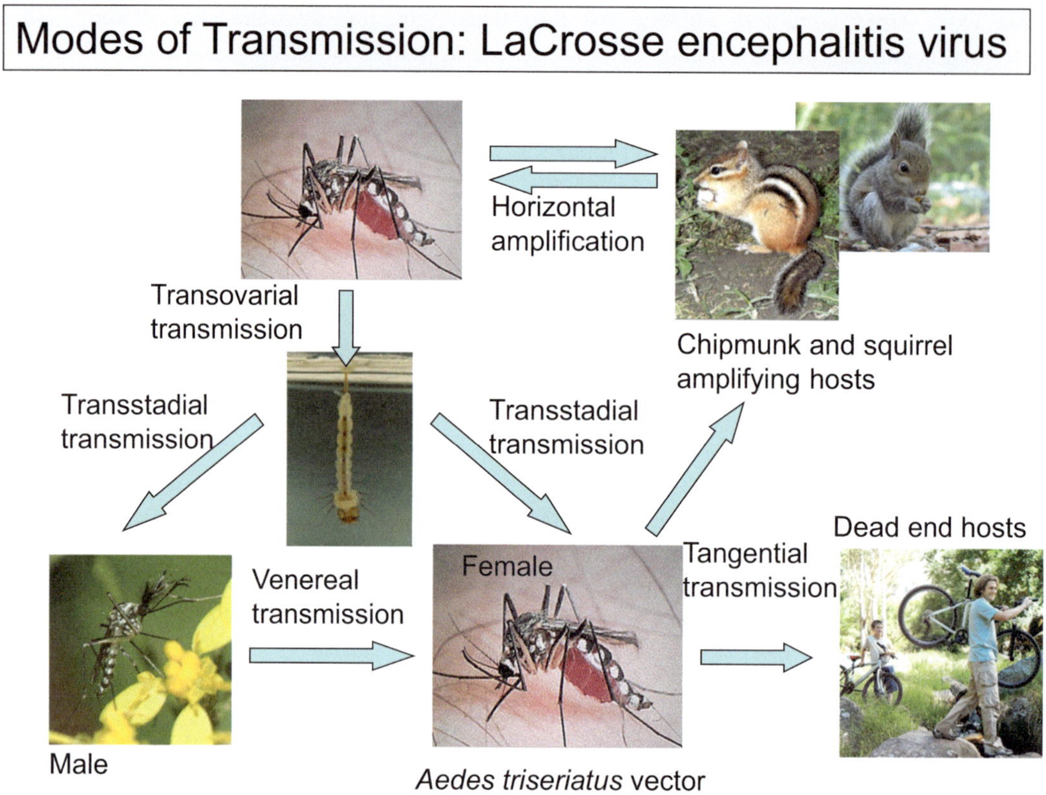

Fig. 2.5 La Crosse encephalitis virus transmission, showing multiple modes of transmission

(across the ovary) after the parasites infect the ovarian germinal tissue and then transstadially to the next reproductive or blood-feeding stage. In true transovarial transmission, most of the progenies are infected. Other parasites do not actually infect the ovary, and, although they are passed on to progeny, transmission is vertical but not truly transovarial. This latter situation is less efficient and usually only a small percentage of the progenies are infected. Transgenerational transmission in vectors such as mosquitoes also must include transstadial transmission, because the immature life stages do not blood feed.

2.3.1.3 Venereal Transmission

Venereal transmission is the passage of parasites between males and females of vector species during mating and is relatively rare. Venereal transmission usually is limited to transovarially infected males who infect females during insemination which, in turn, infect their progeny during fertilization.

La Crosse virus (Fig. 2.5) is an example of a vertically maintained parasite where the arthropod host serves as the reservoir and the vector (DeFoliart 1983). This arbovirus is maintained transovarially by transgenerational transmission within clones of infected *Aedes triseriatus* mosquitoes in the eastern USA and is amplified by horizontal transmission among squirrels and chipmunks. Because this temperate mosquito rarely has more than two generations per year, the virus spends long periods in infected vectors and relatively short periods in infected vertebrate hosts. Female mosquitoes infected vertically or horizontally transmit their infection transovarially to first-instar larvae. These larvae pass virus transstadially through the four larval stadia and the pupal stage to emerging adults. These transgenerationally infected females then take a blood meal and oviposit infected eggs, often in the same tree hole from which they emerged. Some blood meal hosts such as chipmunks become highly viremic and amplify the number of infected *Ae. triseriatus* when uninfected females feeding on these rodents become horizontally infected. Venereal transmission of virus from transgenerationally infected males to uninfected females has been demonstrated in the laboratory and may serve to establish new clones of infected females in nature. In contrast, scrub typhus rickettsia (*Orientia tsutsugamushi*) is vertically maintained transgenerationally within infected clones of the chigger mite *Leptotrombidium* and is transmitted to vertebrate hosts by bite of the larval stage; however, it has not been possible to experimentally establish new infectious colonies of mites in the laboratory by blood feeding (Frances 2005).

2.3.2 Horizontal Transmission

Horizontal transmission is essential for the maintenance of almost all vector-borne parasites and is accomplished by either anterior (biting) or posterior (defecation) routes. *Anterior-station transmission* occurs when parasites are liberated from the mouthparts or salivary glands during blood feeding (e.g., malaria parasites, encephalitis viruses, filarial worms). *Posterior-station* (or *stercorarian*) *transmission* occurs when parasites remain within the gut and are transmitted via contaminated feces. The trypanosome protozoan that causes Chagas' disease, for example, develops to the infective stage within the hindgut and is discharged onto the host skin when the triatomine bug vector *Rhodnius* defecates during blood feeding. Irritation resulting from salivary proteins introduced into the host during feeding causes the host to scratch the bite and rub the parasite into the wound. Louse-borne relapsing fever and typhus fever rickettsia also employ similar posterior-station modes of transmission.

There are four types of horizontal transmission, depending upon the role of the arthropod in the life cycle of the parasite: mechanical, multiplicative, developmental, and cyclodevelopmental.

Mechanical transmission occurs when the parasite is transmitted to vertebrate hosts without amplification or development within the vector, usually by blood feeding with contaminated mouthparts. Arthropods that are associated intimately with their vertebrate hosts and feed at frequent intervals have a greater probability of transmitting parasites mechanically. The role of the arthropod, therefore, may be little more than an extension of contact transmission between vertebrate hosts (Lindsay and Scudder 1956). Eye gnats (genus *Hippelates*), for example, have rasping, sponging mouthparts and feed at the mucous membranes of a variety of vertebrate hosts making them an effective mechanical vector of the bacteria and viruses which cause conjunctivitis or "pink eye." Pink eye also may be transmitted from infected to susceptible hosts by direct contact. Mechanical transmission also may be accomplished by contaminated mouthparts if the vector is interrupted while blood feeding and then immediately refeeds on a second host in an attempt to complete the blood meal. *Nonviremic transmission* is a special form of nonpropagative transmission where infectious vectors are able to transmit viruses through the host directly to concurrently feeding uninfected vectors without host infection or parasite replication. With *Ixodes ricinus* ticks, this occurs through the skin with viruses such as tick-borne encephalitis virus when multiple ticks feed adjacent to one another (Labuda et al. 1993). With mosquitoes, this has been demonstrated experimentally for WNV when infectious *Culex* vectors inject large quantities of virus directly into the circulatory system of small vertebrate hosts such as laboratory mice or house finches, and uninfected vectors feed concurrently or shortly after this initial feed (Higgs et al. 2005).

Multiplicative (or *propagative*) *transmission* occurs after the parasite multiplies asexually within the vector and is transmitted after a suitable incubation period is completed, allowing the parasite to disseminate to the salivary glands. In this case, the parasite does not undergo *metamorphosis* (transformation or development), and the form transmitted is indistinguishable from the form ingested with the blood meal. Arboviruses such as dengue virus, for example, are not transmitted until the virus replicates within and passes through the mosquito vector midgut, is disseminated throughout the hemocoel, and then enters and replicates within the salivary glands. The number, but not the form, of the viruses changes during these processes, and the number of virus particles transmitted may be less than the number ingested with the blood meal. Likewise the titer of virus required to infect the vertebrate host typically is far less than required to infect the vector.

Developmental transmission occurs after parasite metamorphosis, but not multiplication, within the vector. Microfilariae of *Wuchereria bancrofti*, for example, are ingested with the blood meal, penetrate the mosquito gut, move to the flight muscles where they molt twice, and then move to the mouthparts where they remain until deposited on the skin during blood feeding. These worms do not reproduce asexually within the mosquito vector; i.e., the number of worms available for transmission is always equal to or less than the number ingested.

Cyclodevelopmental transmission occurs when the parasite metamorphoses and reproduces within the arthropod vector. In the life cycle of the malaria parasite, for example, haploid gametocytes that are ingested by the *Anopheles* mosquito vector with the blood meal unite within the mosquito midgut and then change to an invasive form (ookinete) that penetrates the gut and forms an asexually reproducing stage (oocyst) on the outside of the gut wall. Following replication, the oocyst ruptures and liberates the infective forms (sporozoites) that move through the hemocoel to the salivary glands from where they are transmitted when the mosquito vector expectorates during a subsequent blood meal.

The *extrinsic incubation period* is the time interval between vector infection and parasite transmission and originally was recognized by the time period between clinical cases when the parasite was "away" from the vertebrate host. The *intrinsic incubation period* is the time from infection to the onset of symptoms (or infectiousness) in the vertebrate host. Repeated lag periods of consistent duration between clusters of new cases at the onset of epidemics were noticed by early epidemiologists who coined the term "extrinsic incubation." These intervals actually represent the combined duration of extrinsic and intrinsic incubation periods.

The duration of the extrinsic incubation period in the arthropod is temperature dependent (Walton and Reisen 2014). The rate (or inverse of the time from infection to transmission) increases as a linear function of ambient temperature above a minimal threshold. After being ingested by the mosquito vector, arboviruses such as WNV, for example, must enter and multiply in cells of the midgut, escape the gut, be disseminated throughout the hemocoel, and then infect the salivary glands, after which the virus may be transmitted by bite (Hardy et al. 1983). Under hot summer conditions, this process may be completed within 6–7 days, and the *Culex* vector mosquito is capable of transmitting virus during the next blood meal. In contrast, under cool conditions transmission may be delayed for more than 2 weeks or until the third or fourth blood meal. Therefore, the number of bites by the vector until transmission occurs is a convenient way to express the thermodynamics between the duration of the gonotrophic cycle and the extrinsic incubation periods. Some parasites may alter vector behavior and thereby increase the frequency of vector blood feeding and transmission. The plague bacillus, for example, remains within and eventually blocks the foregut of the flea vector, *Xenopsylla cheopis* (Kupferschmidt 1993). Regurgitation occurs during blood feeding and causes vector starvation, stimulating frequent blood-feeding attempts and more transmission at progressively more closely spaced intervals before the vector succumbs to starvation.

2.3.3 Transmission Cycles

Transmission cycles essentially describe the ecology or epidemiology of the parasites and vary considerably depending upon their complexity and the role of humans as hosts (Fig. 2.2). A vector-borne *anthroponosis* is a disease resulting from a parasite that can be transmitted among humans by one or more anthropophagic vectors.

Malaria, dengue, some forms of filariasis, and louse-borne typhus are examples of anthroponoses. Humans serve as reservoir hosts for these parasites, which may persist for years as chronic infections. Vectors of anthroponoses selectively blood feed upon humans and are intimately associated within domestic or peridomestic environments. Widespread transmission of an anthroponosis with an increase in the number of human disease cases during a specified period is an *epidemic*. When human cases reappear consistently over time and space, transmission is said to be *endemic*.

Zoonoses are diseases of animals caused by parasites that also may infect humans (Fig. 2.2). In most vector-borne zoonoses, humans and some domestic animals are not an essential component of the transmission cycle, but rather become infected when bitten by an infective vector that fed previously on an infectious animal host. Although humans frequently become ill, they rarely circulate sufficient numbers of parasites to infect vectors and therefore are *dead-end hosts*. The *enzootic transmission cycle* (literally "in animals") is the basic or primary animal cycle. When levels of enzootic transmission escalate, transmission may become *epizootic* (an outbreak of disease among animals). Transmission from the enzootic cycle to dead-end hosts such as humans is called *tangential transmission* (i.e., at a tangent from the basic cycle). Different or the same vector species may be responsible for enzootic, epizootic, and tangential transmission. *Bridge vectors* transmit parasites tangentially between enzootic and dead-end host species. Human involvement in zoonoses may depend on the establishment of a secondary *amplification cycle* among vertebrate hosts inhabiting the peridomestic environment.

Lyme disease, caused by infection with the spirochete *Borrelia burgdorferi*, is an example of a tick-borne zoonosis (Fig. 2.6) that is now

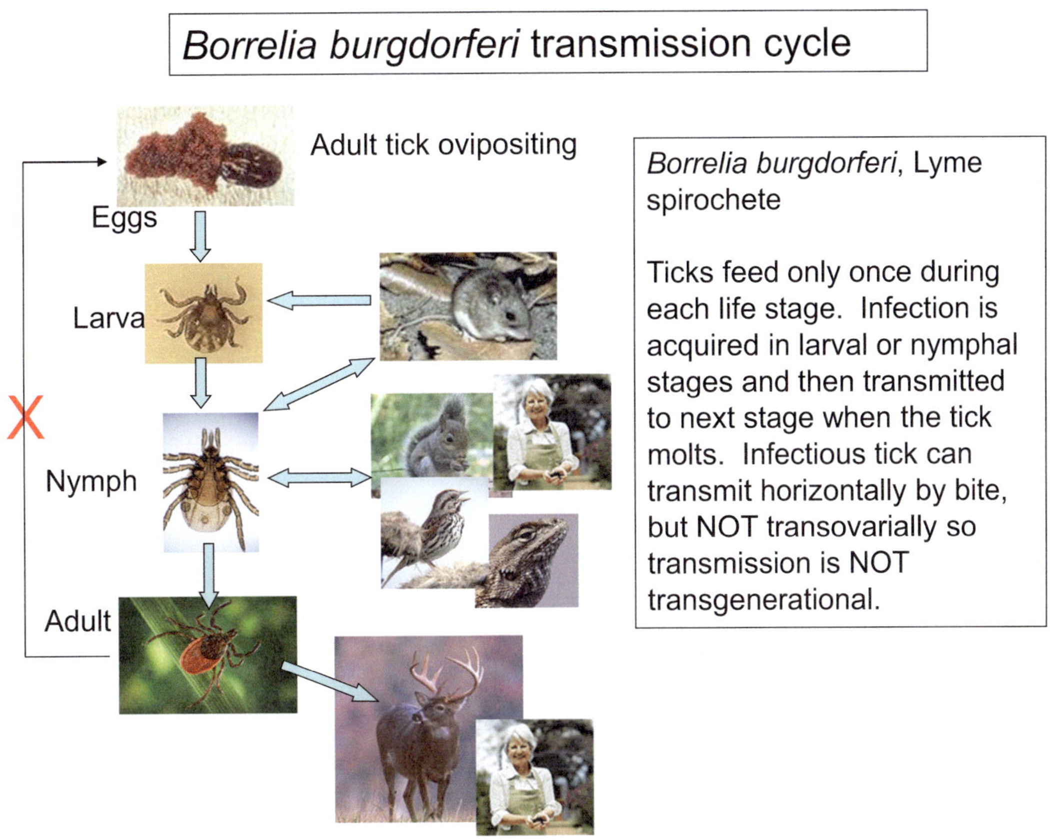

Fig. 2.6 Lyme disease spirochete, *Borrelia burgdorferi*, transmission cycle showing transstadial transmission

epidemic in eastern North America (Lane et al. 1991). If left untreated, the spirochete causes serious chronic disease in humans, presenting a variety of symptoms that may include arthritis and fatigue. The vectors are principally ticks in the *Ixodes ricinus* complex including *scapularis* in eastern and *pacificus* in western USA. Hard ticks require blood meals for both molting and reproduction. Larval ticks acquire *Borrelia* blood feeding on mice during summer that have infectious spirochetemias (number of spirochetes in the blood), maintain infections during winter, and then pass this infection transstadially to the nymphal stage the following spring. Nymphal ticks subsequently transmit their infection to a variety of hosts including rodents, squirrels, lizards, birds, and humans, but, if uninfected, also may acquire *Borrelia* during blood feeding. Lizards, some birds, and humans are refractory or "dead-end" hosts, and their infection may actually reduce the rate of *Borrelia* amplification. Infected nymphs also pass their infection transstadially to the adult stage, and adults may transmit to large mammals such as deer and perhaps humans during blood feeding, although deer seems to be refractory to infection. There is minimal evidence to support vertical transmission of *Borrelia* to the eggs, and therefore infected larval ticks and perhaps mice seem to be the reservoirs of infection. The changing landscape and reforestation of eastern USA, accompanied by large increases in whitetail deer and *Peromyscus* mouse populations, and the construction of housing adjacent to or within wooded areas have combined to create epidemiological situations conducive to large-scale outbreaks of Lyme disease. Infected immature tick populations residing in suburban gardens and lawns greatly increase the risk of transmission to humans.

2.4 Vector Incrimination

To understand the epidemiology and focus control on vector-borne disease, it is essential to establish which arthropod(s) is the primary vector(s) responsible for parasite transmission. Partial or incomplete vector incrimination has resulted in the misdirection of control efforts at arthropod species that do not play a substantial role in either enzootic maintenance or epidemic transmission. Vector incrimination combines field and laboratory investigations that measure field infection rates, vector competence, and vectorial capacity.

2.4.1 Infection Rates

The collection of infected arthropods in nature is an important first step in identifying potential vectors, because it indicates that the candidate species feeds on vertebrate hosts carrying the infective parasite. Infection data may be expressed as *infection prevalence* (percentage of population infected at one point in time or number of vectors infected/number examined × 100). The more commonly employed term, *infection rate*, usually refers to *infection incidence* and includes the number of new infections per unit of population over a specified time period. When the infection prevalence is low and arthropods are tested in groups or pools, data are expressed as a *minimum infection rate* (number of pools of vectors positive/total specimens tested/unit of time × 100 or 1000). Minimum infection rates are relative values, with ranges delineated by pool size. For example, minimum infection rates of vectors tested in pools consisting of 50 individuals each must range from 0 to 20 per 1000 females tested. When pool sizes vary, a maximum likelihood estimate (MLE) of the infection rate should be used (Biggerstaff 2003).

It is important to distinguish between infected hosts harboring a parasite and infective hosts capable of transmission. In developmental and cyclodevelopmental vectors, the infective stages may be distinguished by location in the vector, morphology, or biochemical properties. Distinguishing infective from noninfective vectors is difficult, if not impossible, with viral or bacterial infections, because the parasite does not change form, although the location of recovery within the vector can be informative. The ability to transmit may be implied by testing selective body parts such as the legs, cephalothorax, salivary glands, or head. With some tick pathogens, however, parasite movement to the mouthparts does not occur until after attachment. As mentioned previously, the transmission rate is the number of new infected hosts per time period. The annual parasite incidence frequently used in malaria control programs is the number of new human cases detected per 1000 population (or other units of population size) per year. The *entomological inoculation rate* is the number of potentially infective bites per host per unit of time. This frequently is determined from the human- or host-biting rate and the proportion of vectors that are infective and is calculated as bites per human per time period × infectivity prevalence. In malaria, for example, this is the number of *Anopheles* positive for sporozoites biting humans per time period.

2.4.2 Vector Competence

The ability of an arthropod species to become infected with and transmit a parasite is *vector competence*. It typically is determined experimentally by feeding the candidate vector on a vertebrate host circulating the infective stage of the parasite, incubating the blood-fed arthropod under suitable temperature and at times humidity conditions, refeeding the arthropod on a noninfected susceptible vertebrate host, and then examining this host to determine if it became infected. This in some ways is similar to *Koch's postulates* used in microbiology. Because it often is difficult to maintain natural vertebrate hosts in the laboratory and control the concentration of parasites in their peripheral circulatory system, laboratory model hosts or more frequently artificial feeding systems are used to expose the vector to a known quantity of parasites and evaluate infection and transmission. *Susceptibility to infection* may be expressed as the percentage of arthropods that became infected among those blood feeding. When the arthropod is fed on a range of parasite concentrations, susceptibility may be expressed as the *median infectious dose* (ID_{50}) or the concentration of parasites required to infect 50 % of the blood-fed arthropods. The ability to transmit may be expressed either as the percentage of blood feeding or infected blood-feeding females that transmitted parasites or the percentage of recipient hosts that became infected when multiple arthropods were used per feeding.

Failure of a blood-fed arthropod to become infected with or transmit a parasite may be attributed to the presence of one or more barriers to infection (also see earlier vector immunity Sect. 2.2.3.4). For most parasites, the arthropod midgut provides the most important barrier (Hardy et al. 1983). Many viruses will grow in refractory or nonvector species, after they are inoculated into the hemocoel, thereby bypassing this *gut barrier*. *Arboviruses* (viruses transmitted by arthropods) must infect and replicate within the midgut epithelium and then escape through the basal lamina into the hemocoel, so it is possible to have *midgut infection* and/or *escape barriers*. After penetrating and escaping from the midgut, the parasite then must multiply and/or mature and be disseminated to the salivary glands or mouthparts. Arthropod cellular or humoral immunity may clear the infection at this point creating a *dissemination barrier*. Even after dissemination to the salivary glands, the parasite may not be able to infect or be transmitted from the salivary glands due to the presence of *salivary gland infection* or *escape barriers*, respectively. For parasites transmitted via the posterior station, vector competence may be expressed as the percentage of infected vectors passing infective stages of the parasite in their feces.

2.4.3 Vectorial Capacity

The formula for *vectorial capacity* summarizes quantitatively the epidemiological attributes of the vector relative to parasite transmission (Garrett-Jones 1970) within the Ross–MacDonald model (Macdonald 1957). Although originally developed for the *Anopheles* vectors of human malaria parasites and therefore most appropriate for anthroponoses, this model provides a framework to conceptualize how the components of transmission interact.

Vectorial capacity is expressed by the formula:

$$C = ma^2\left(P^n\right)V / -\ln P$$

where C = vectorial capacity or new infections per infection per day, ma = bites per human per day, a = human-biting habit, P = probability of daily survival, n = extrinsic incubation period in days, and V = vector competence or innate transmission efficiency. This function comprises the entomological portion of the Ross–MacDonald model used to estimate the basic malaria reproductive rate, R_o.

The *biting rate*, *ma*, frequently is estimated by collecting vectors as they attempt to blood feed and is expressed as bites per human per day or night (e.g., ten mosquitoes per human per night). The human-biting habit, *a*, combines vector-feeding frequency and host selection. *Feeding frequency* is the inverse of the duration of the gonotrophic cycle. Host selection patterns are determined by testing blood-fed vectors to determine what proportion feeds on humans or the primary reservoir. Therefore, if the blood-feeding frequency is 2 days and 50 % of host-seeking vectors feed on humans, $a = (1/2 \text{ days}) \times (0.5) = 0.25$. In this example, $ma^2 = 10$ bites/human/night $\times 0.25 = 2.5$. a is repeated, because infected vectors must refeed to transmit.

The probability of the vector surviving through the extrinsic incubation period of the parasite, P^n, requires information on the probability of daily survival, P, and the duration of the extrinsic incubation period, n. P is estimated either vertically by determining the age structure of the vector population or horizontally by marking cohorts and monitoring their death rate over time. In mosquitoes, P may be estimated vertically from the parity rate (proportion of parous females/number examined). In practice, $P = (\text{parity rate})^{1/g}$, where g is the length of the gonotrophic cycle. The duration of the extrinsic incubation period may be estimated from ambient temperature from data gathered during vector competence experiments by testing the time from infection to transmission for infected vectors incubated at different temperatures. Continuing our example, if $P = 0.8$ (or 80 % of the population survives per day) and $n = 10$ days, then the duration of infective life $P^n/{-\ln P} = 0.8^{10}/{-\ln 0.8} = 0.48$ days. In addition it is useful to also account for vector competence, V. For this example, we will assume that 90 % of vectors become infected and 90 % of infected females are capable of transmission, so $V = 0.9 \times 0.9 = 0.81$. Therefore, vectorial capacity $C = 2.5 \times 0.48 \times 0.81$ or 0.97 parasite transmissions per infective host per day. When C approaches 1, the parasite can be maintained at a steady state. When $C > 1$, the parasite will increase and outbreaks may ensue, whereas when $C < 1$, the parasite may become focally extinct unless there are mechanisms in place for persistence without transmission or repeated introduction (see earlier section).

2.5 Interseasonal Maintenance

An important aspect of the ecology of vector-borne parasites is the mechanism(s) by which they persist between transmission seasons or outbreaks. Parasite transmission typically is most efficient when weather conditions are suitable for vector activity, reproduction, and population growth, and warm temperatures expedite parasite replication and dissemination within the vector and shorten the gonotrophic cycle increasing the frequency of host–vector contact. At temperate latitudes, winter temperatures frequently decrease to levels below which the parasite cannot replicate within the vector, vectors are driven into reproductive arrest, and vertebrate hosts may migrate or hibernate. Similar problems may face parasites at tropical latitudes, when transmission is interrupted by prolonged dry seasons. This seasonality which is characteristic of most vector-borne parasites may be due to either the periodic amplification of a constantly present parasite or to the consistent reintroduction of parasites following annual focal extinction.

There are several mechanisms that may allow parasite maintenance during unfavorable periods.

2.5.1 Continued Transmission

During periods of unfavorable weather, vectors may become active intermittently but continue to transmit parasites, although transmission rates may be slowed by cool temperature, low vector abundance, or infrequent blood feeding. At temperate latitudes with cool winters, the frequency of blood feeding and rate of parasite maturation in the vector are diminished. At tropical latitudes, widespread transmission may be terminated during extended dry seasons that reduce vector abundance and survival, even though blood-feeding activity may continue at refugia near permanent water sources. In both instances, transmission may be restricted spatially and involve only a small portion of the vertebrate host population. Therefore, human infections during adverse periods may be highly clumped, because vector and parasite dispersal is limited at this time.

2.5.2 Dormant Infected Vectors

Many vectors enter a state of dormancy as non-blood-feeding immatures or adults. Vertically infected vectors typically remain infected for life and therefore may maintain parasites during periods when horizontal transmission is interrupted. La Crosse encephalitis virus (Fig. 2.5), for example, is maintained during winter and drought periods within transovarially infected eggs of its vector, *Aedes triseriatus*. Infected eggs of this tree-hole mosquito may remain dormant and infected for several years and are able to withstand winter cold, summer heat, and extended dry periods. Here, the mosquito is the reservoir host. Inundation of eggs during the spring rains produces adult mosquitoes that are infected at emergence. In contrast, WNV may overwinter in vertically infected adult female *Culex* mosquitoes that have entered *diapause* (from the Greek *diapauein* to "pause," here insect hibernation). Termination of diapause, warming temperature, and the renewal of blood feeding initiate the transmission cycle among birds and host-seeking female *Culex*. Tick-borne parasites, such as Lyme *Borrelia*, persist through winter within infected immature stages of the tick vector. Similarly, vectors that inhabit the nests of migratory hosts such as cliff swallows often remain alive and infected for extended periods until their hosts return.

2.5.3 Infected Vertebrate Hosts

Parasite maintenance may be accomplished by infected reservoir hosts that either continue to produce stages infective for vectors or harbor inactive stages of the parasite that relapse or recrudesce during the season when vectors are blood feeding. Adult filarial worms, for example, continue to produce microfilariae throughout their lifetime, regardless of the population dynamics or seasonality of the mosquito vector. In contrast, some Korean strains of *Plasmodium vivax* malaria overwinter as dormant stages in the liver of the human host and then relapse in spring concurrent with the termination of diapause by the *Anopheles* mosquito vector(s) (Brunetti et al. 1954).

2.5.4 Focal Extinction

Alternatively, parasites may become regionally extinct during unfavorable weather periods and then be reintroduced from refugia. Two possible mechanisms may allow the reintroduction of parasites:

Migratory vertebrate hosts. Many bird species overwinter in the tropics and return to temperate or subarctic breeding sites each spring, potentially bringing with them infections acquired at southern latitudes. However, finding evidence of infection to support this mechanism has been difficult (Reisen et al. 2010). It also is possible that the stress of long flights and ensuing reproduction triggers relapses of chronic infections. In addition, many large herbivores migrate annually between summer (or wet) and winter (or dry) pastures bringing with them an array of parasites as well as infestations of vectors such as ticks. Rapid long-range human or commercial transportation is another possible mode for both vector and parasite introduction along frequently used routes (Tatem 2009).

Weather fronts. Infected vectors may be carried to long distances as “aerial plankton” by prevailing weather fronts (Sellers 1980). Consistent weather patterns such as the sweep of the southeastern monsoon from the Indian Ocean across the Indian subcontinent may passively transport infected vectors such as some *Anopheles* species hundreds of kilometers. Historically, the onset of western equine encephalitis virus activity in the north central USA and Canada has been attributed to the passive dispersal of infected mosquitoes by storm fronts as has the repeated introduction of Japanese encephalitis virus into northern Australia (Kay and Farrow 2000).

2.6 Surveillance

Tracking endemic and introduced vector-borne pathogens provides a challenge for health and surveillance (from French surveiller, “to watch over”) programs. Information on the number of cases over time and space in the human population can be gathered from morbidity and mortality records maintained by state or national governmental agencies. *Morbidity data* are records of illness that frequently are recognized clinically, whereas *mortality data* are records of the cause of death. These data vary greatly in their quality and timeliness, depending upon the disease, the accuracy of determining the cause of illness or death, and the rapidity of reporting. In the USA, the occurrence of confirmed cases of many vector-borne diseases including yellow fever, plague, malaria, and encephalitis by law must be reported to state health authorities. The frequency of case detection and accuracy of reporting systems are dependent on the type of surveillance employed and the ability of the medical or veterinary community to recognize suggestive symptoms and request appropriate confirmatory laboratory tests. In addition, some laboratory tests vary in their specificity and sensitivity complicating the interpretation of laboratory results. Cases may be classified as suspect or *presumptive* based on the physician’s clinical diagnosis or results from nonspecific laboratory tests or *confirmed* based on a diagnostic rise in specific antibodies, the direct observation (or isolation) of the parasite from the case, or the results from an accepted assay such as demonstration of IgM in sera from presumptive WNV case. However, infections with many arthropod-borne parasites such as Lyme disease and WNV frequently are asymptomatic, present variable clinical symptoms, and therefore remain largely undiagnosed and underreported.

Surveillance for clinical cases may be active or passive. *Active surveillance* involves *case detection* by health workers who seek out and test suspect cases that fit a predetermined case definition. In malaria control programs, for example, a field worker may visit every household systematically and collect blood films from all persons with a current or recent fever. Fever patients are treated presumptively with antimalarial drugs, and these suspect cases then are confirmed by detection of malaria parasites in a blood smear or other samples. Confirmed cases are revisited; additional medication administered, if necessary; and a follow-up blood sample is taken. This active surveillance provides true population base infection rates. Frequently active surveillance is done by notification of hospitals to be aware of certain parasites thought to be actively present in the population.

Most surveillance programs rely on *passive case surveillance* and utilize clinical human or veterinary cases detected at medical or veterinary clinics. In this system individuals seeking medical attention at primary healthcare organizations are diagnosed by an attending healthcare professional who requests appropriate confirmatory laboratory tests. However, because many arthropod-borne diseases present a variety of nonspecific symptoms (e.g., headache, fever, general malaise, arthralgia), cases frequently may be missed or not specifically diagnosed. This *iceberg effect* (Fig. 2.7) is common among arthropod-borne viruses such as WNV, where ca. 80 % of the infections are asymptomatic and ca.

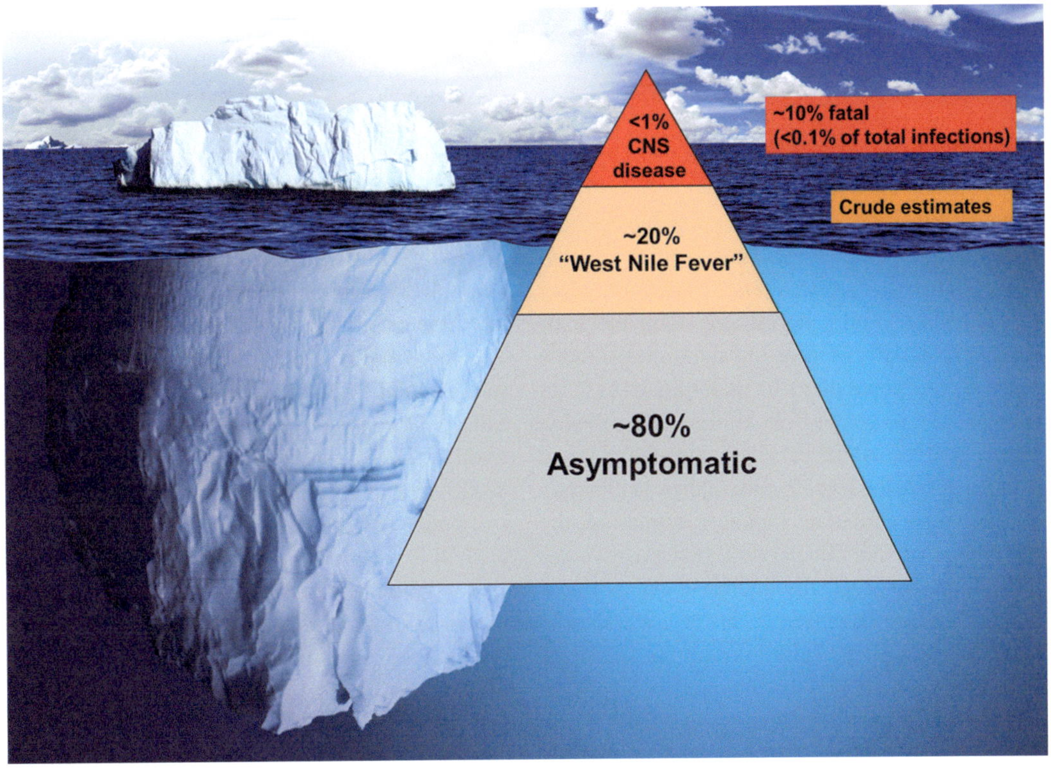

Fig. 2.7 Pyramid of West Nile virus disease, showing that only the "tip of the iceberg" of infections becomes clinical disease reported to surveillance programs (Iceberg image from https://littlemissholley.files.wordpress.com/2011/08/iceberg-logo.jpg)

20 % produce febrile disease of varying severity, only some of which are diagnosed; <1 % of infections develop neuroinvasive disease with severe symptoms that are frequently diagnosed. As illustrated, most of the infections remain "below the surface" and go undetected. This problem is critical because often human data is the only surveillance information available for intervention decision support. Because many health decisions are driven by economics, underserved populations may appear to be devoid of infection/disease and therefore not receive preventive public health programs such as mosquito control.

The reporting system for clinical cases of vector-borne diseases must be evaluated carefully when interpreting surveillance data. This evaluation should account for the disease, its frequency of producing clinically recognizable symptoms, the official case definition, the sensitivity and specificity of confirmatory laboratory tests, and the type and extent of the reporting system. Usually programs that focus on the surveillance of a specific disease and employ active case detection provide the most reliable epidemiological information. In contrast, broad-based community healthcare systems that rely on passive case detection typically produce the least reliable information, especially for relatively rare vector-borne diseases with nonspecific symptoms. Unfortunately, in the modern era, the extent of diagnosis and therefore the sensitivity of surveillance frequently are dependent upon the extent of medical insurance coverage and the physician's decision to provide a definitive diagnosis.

In epidemiology, the *population* is the number of individuals at risk for infection in a given geographical area at a given time. *Incidence* is the number of new cases per unit of population per unit of time and is derived from two or more successive samples spaced over time *(longitudinal survey)*. *Prevalence* is the frequency of both old and new infections among members of a population. Prevalence typically is determined by a single point-in-time estimate *(cross-sectional survey)* and frequently is expressed as the percentage of the population tested that was infected or presented antibodies indicative of recent or previous infections. The level of parasite endemicity in a population may be graded as *hypoendemic* (low), *mesoendemic* (medium), or *hyperendemic* (high), depending upon the incidence and stability of infection and/or the immune status of the population. In malaria surveys, for example, the percentage of children with palpable spleens and the annual parasite incidence are used to characterize the level of endemicity. In endemic disease with continuous annual transmission, the percentage of individuals with sera positive for IgG antibodies typically increases as a linear function of age or residence history, whereas in hypoendemic disease with intermittent transmission, this function may be disjunct with specific age groups expressing elevated positivity rates. The occurrence of an above-normal number of human infections or cases is termed an *epidemic*. Health agencies, such as the World Health Organization (WHO), typically monitor incidence data to establish baseline criteria necessary to classify the level of endemicity and to decide when an epidemic is underway. A geographically widespread epidemic on a continental or global scale is called a *pandemic*.

Serological surveys (or *serosurveys*) are a useful epidemiological tool for determining the cumulative infection experience of a population with one or more parasites, host-related factors affecting the efficiency or risk of transmission, and reinfection rates. Antibodies are especially useful when the duration of infection with a parasite is brief, thereby limiting the chance of detecting infection. When coupled with morbidity data, serosurveys provide information on the ratio of apparent to inapparent infections. *Random sampling* representatively collects data from the entire population and may provide ecological information retrospectively by analysis of data collected concurrently with each serum sample. This information may assign risk factors for infection such as sex, occupation, and residence history or may help in ascertaining age-related differences in susceptibility to disease. *Stratified sampling* targets a specific cohort, subpopulation, or geographical area. Although stratified samples may have greater sensitivity in detecting rare or contiguously distributed parasites, the data is not always readily extrapolative to infection or disease trends in the entire population. Repeated serological testing of the same individuals within a population can determine the time and place of infection by determining when individuals first become *seropositive*, i.e., serologically positive with circulating antibodies against a specific parasite. This change from seronegative to seropositive is called a *seroconversion* and frequently is used to detect infection in sentinel animals.

Forecasting human infection usually is accomplished by monitoring environmental factors, vector abundance, the level of transmission within enzootic and/or amplification cycles, and the numbers of human or domestic animal cases (Kramer 2012). As a general rule, the accuracy of forecasting is related inversely to the time and distance of the predictive parameter from the onset of human cases. Surveillance activities to validate or correct forecasts typically include the time series monitoring of environmental conditions, vector abundance, enzootic transmission rates, and clinical cases.

2.6.1 Environmental Conditions

Unusually wet or warm weather may indicate favorable conditions for vector activity or population increases and concurrently parasite transmission. Parameters frequently monitored include temperature, rainfall, snow pack (predictive of vernal flooding), and agricultural irrigation schedules. These are usually antecedent factors useful in the long-range prediction of risk. Remotely sensed data from satellite imagery is useful in showing changes in landscape vegetation.

2.6.2 Vector Abundance

Standardized vector monitoring at fixed sites and time intervals can be used to compare relative changes in temporal and spatial vector abundance that validate environmental predictions and is useful in detecting an increased risk of parasite transmission. Trap type and placement are critical in effectively sampling vectors such as mosquitoes for surveillance purposes (Meyer and Reisen 2003). Devices are deployed in or adjacent to habitats that maximize their effectiveness in collecting the target mosquito species. Sampling in a systematic fashion over time can produce historical data useful in determining anomalous increases or decreases in abundance. Extraordinary increases in vector abundance and survival may accurately forecast increased enzootic transmission and, in turn, epidemics.

2.6.3 Enzootic Transmission Rates

For zoonoses, systematically monitoring the level of parasite infection in vector or vertebrate populations provides direct evidence that the parasite is present and being actively transmitted. The level of enzootic transmission usually is directly predictive of the risk to humans or domestic animals. Enzootic transmission activity may be monitored by measuring vector infection rates, vertebrate host infection rates, sentinel seroconversion rates, and the number of clinical cases.

2.6.3.1 Vector Infection Rates

Sampling vectors and testing them for parasites determine the level of infection in the vector population. When vectors are tested individually, prevalence data are expressed as percentages; e.g., ten females infected per 50 tested are a 20% infection rate. When combined with abundance estimates, infection rates also may be expressed as the *entomological inoculation rate* or the number of infectious vectors per sampling unit per time interval; e.g., 100 bites per human per night × 0.2 infectious rate = 20 infective bites per human per night. When infection rates are low, vector populations large, and sampling independent of vector age, vectors usually are tested in lots or *pools*. It is statistically advantageous to keep pool size constant and therefore the chance of detecting infection the same. Because there may be >1 infected vector per pool, infection rates are expressed as a minimum infection rate = positive pools/total individuals tested × 100 or 1000. There are several new mathematical approaches to calculating these estimates that rely on maximum likelihood estimation methods.

2.6.3.2 Vertebrate Host Infection Rates

Introduced zoonoses, such as sylvatic plague in North American rodents, WNV in American crows, or yellow fever virus in South American monkeys, frequently produce mortality that may be used to track epizootics of these parasites over time and space. Large numbers of dead American crows have been a hallmark of the ongoing WNV epidemic in North America, and counts of reports by the public have been used for surveillance purposes to indicate recent transmission as well as to forecast human risk using predictive spatial models (Carney et al. 2011). In contrast, endemic zoonoses rarely result in vertebrate host mortality. Collecting and testing reservoir or amplifying hosts for infection is necessary to monitor the level of enzootic parasite transmission. Stratified sampling for these parasites (directly by parasite isolation or indirectly by seroprevalence) usually focuses on the young of the year to determine ongoing or recent transmission.

Monitoring the incidence of newly infected individuals in a population over time is necessary to detect increased transmission activity. Because many parasites are difficult to detect or are only present for a limited time period, sampling frequently emphasizes the monitoring of seropositivity or seroprevalence. Testing for IgM antibody, which rises rapidly after infection, is parasite specific, and decays relatively quickly can indicate the level of recent infection, whereas monitoring IgG antibody documents the population's historical experience with the parasite. Sampling, marking, releasing, recapturing, and resampling wild animals are most useful in providing information on the time and place of infection in free-roaming animal populations.

2.6.3.3 Sentinel Seroconversion Rates

Sentinels typically are animals that can be monitored over time to quantify the prevalence of enzootic parasite transmission. Trapping wild animals or birds is labor intensive, and determining seroprevalence may provide little information on the time and place of infection, especially if the host is a bird species that has a large home range. To circumvent this problem, caged or tethered natural hosts or suitable domestic animals of known infection history are placed in sensitive habitat and repeatedly bled to detect infection. A suitable sentinel should be fed upon frequently by the primary vector species, be easy to diagnose when infected, be unable to infect additional vectors (i.e., not serve as an amplifying host), not succumb to infection, and be inexpensive to maintain and easy to bleed or otherwise sample for infection. In the USA, flocks of seronegative chickens are placed at farmhouses or other suitable localities, housed in standard coops, and then bled weekly or biweekly to determine seroconversions to viruses such as WNV that have a natural mosquito–bird transmission cycle. Small blood samples taken on filter paper can be tested by a semiautomated enzyme-linked immunosorbent assay (ELISA) to detect seroconversions. Because the chickens are confined and the date of seroconversion known, the time and place of infection are determined, while the number seroconverting estimates the intensity of transmission.

2.6.4 Clinical Cases

Detecting infection among domestic animals may be an important indication that an epizootic is underway and that the risk of human infection has become elevated. Domestic animals such as horses and pigs often are more exposed to vectors than humans and therefore provide a more sensitive indication of parasite transmission, unless they are protected by widespread vaccination campaigns. Clinical human cases in rural areas in close association with primary transmission cycles may be predictive of future epidemic transmission in urban settings.

2.6.4.1 Intervention

Vector-borne diseases frequently affect only a small percentage of the human population or do not impart long-lasting protective immunity (e.g., malaria), thereby making vector control the intervention of choice. Integrated vector control programs attempt to maintain vector abundance below thresholds necessary for the transmission of parasites to humans or domestic animals. When these programs fail, personal protection by repellents or insecticide-impregnated clothing, bed nets, or curtains is often the only recourse. Vaccination may be a viable alternative method of control for specific vector-borne diseases, if the vaccine imparts lasting immunity as in the case of yellow fever or Japanese encephalitis viruses. However, many parasites such as malaria have evolved to the point where infection elicits a weak immune response that provides only short-term and marginal protection without continued reinfection. The need for continued revaccination at short intervals severely limits their global usefulness, especially in developing countries where delivery systems are rudimentary. Although breakthroughs in chemotherapy have been useful in case management, it still remains the mandate of the medical/veterinary entomologist to devise strategies which combine epidemiological and ecological information to devise ways to effectively reduce or eliminate the risk of disease.

References

Biggerstaff BJ (2003) Pooled infection rate. http://www.cdc.gov/ncidod/dvbid/westnile/software.htm 1–5

Bonds MH, Dobson AP, Keenan DC (2012) Disease ecology, biodiversity, and the latitudinal gradient in income. PLoS Biol 10, e1001456

Brault AC, Reisen WK (2014) Environmental purtubations that influence arboviral host range: insights into emergence mechanisms. In: Singh SK (ed) Viral infections and global change. Wiley, Hoboken, pp 57–75

Brownstein JS, Holford TR, Fish D (2005) Effect of climate change on Lyme disease risk in North America. Ecohealth 2:38–46

Brunetti R, Fritz RF, Hollister AC, Jr (1954) An outbreak of malaria in California. Am J Trop Med Hyg 3:779–788

Carney RM, Ahearn SC, McConchie A, et al (2011) Early warning system for West Nile virus risk areas, California, USA. Emerg Infect Dis 17:1445–1454

Chaves LF, Koenraadt CJ (2010) Climate change and highland malaria: fresh air for a hot debate. Q Rev Biol 85:27–55

Clements AN (1992) The biology of mosquitoes, vol. 1. Development, nutrition and reproduction. Chapman and Hall, New York

Crans WJ, Cassamise DF, McNelly JR (1994) Eastern equine encephalomyelitis in relation to the avian community of a coastal cedar swamp. J Med Entomol 31:711–728

DeFoliart GR (1983) *Aedes triseriatus*: vector biology in relationship to the persistence of La Crosse virus in endemic foci. In: Calisher CH, Thompson WH (eds) California serogroup viruses. Alan R. Liss, New York, pp 89–104

Frances SP (2005) Potential for horizontal transmission of *Orientia tsutsugamushi* by chigger mites (Acari: Trombiculidae). Int J Acarol 31:75–82

Garrett-Jones C (1970) Problems in epidemiological entomology as applied to malariology. Misc Publ Entomol Soc Am 7:168–180

Gubler DJ, Meltzer M (1999) Impact of dengue/dengue hemorrhagic fever on the developing world. Adv Virus Res 53:35–70

Hardy JL, Houk EJ, Kramer LD et al (1983) Intrinsic factors affecting vector competence of mosquitoes for arboviruses. Ann Rev Entomol 28:229–262

Higgs S, Schneider BS, Vanlandingham DL, et al (2005) Nonviremic transmission of West Nile virus. Proc Natl Acad Sci USA .102:8871–8874

Institute of Medicine (2003) Emerging infections, microbial threats to health in the United States. The National Academies Press, Washington

Kay BH, Farrow RA (2000) Mosquito (Diptera: Culicidae) dispersal: implications for the epidemiology of Japanese and Murray Valley encephalitis viruses in Australia. J Med Entomol 37:797–801

Komar N, Langevin S, Hinten S, et al (2003) Experimental infection of North American birds with the New York 1999 strain of West Nile virus. Emerg Infect Dis 9:311–322

Kramer VL (2012) California State mosquito-borne virus surveillance and response plan. http://westnile.ca.gov/resources.php

Kupferschmidt H (1993) Epidemiology of the plague. Changes in the concept in research of infection chains since the discovery of the plague pathogen in 1894. Gesnerus Suppl 43:1–222

Labuda M, Jones LD, Williams T et al (1993) Efficient transmission of tick-borne encephalitis virus between cofeeding ticks. J Med Entomol 30:295–299

Lane RS, Piesman J, Burgdorfer W et al (1991) Lyme borreliosis: relation of its causative agent to its vectors and hosts in North America and Europe. Ann Rev Entomol 36:587–609

Lindsay DR, Scudder HI (1956) Nonbiting flies and disease. Annu Rev Entomol 1:323–346

Linthicum KJ, Anyamba A, Tucker CJ, et al (1999) Climate and satellite indicators to forecast Rift Valley fever epidemics in Kenya. Science 285:397–400

Macdonald G (1957) The epidemiology and control of malaria. Oxford University Press, London

May RM (1983) Parasitic infections as regulators of animal populations. Am Sci 71:36–46

Meyer RP, Reisen WK (2003) Vector and Vector-borne Disease Committee. Integrated mosquito surveillance guidelines. Sacramento, California: Mosq. Vector Control Assoc. Calif.

Mullen GR, Durden LA (2009) Medical and veterinary entomology, 2nd edn. Academic/Elsevier, Burlington

O'Neill BC, Dalton M, Fuchs R et al (2010) Global demographic trends and future carbon emissions. Proc Natl Acad Sci 107:17521–17526

Pavloskiy EN (1966) Natural nidality of transmissible diseases with special reference to the landscape ecology of zooanthroponses. University of Illinois Press, Urbana

Reeves WC, Hammon WM (1962) Epidemiology of the arthropod-borne viral encephalitides in Kern County, California, 1943–1952. Univ. Calif. Berkeley Publ. Publ. Hlth. 4:1–257

Reinert JF, Harbach RE, Kitching IJ (2009) Phylogeny and classification of tribe Aedini (Diptera: Culicidae). Zool J Linn Soc 157:700–794

Reisen WK, Fang Y, Martinez VM (2006) Effects of temperature on the transmission of West Nile virus by Culex tarsalis (Diptera: Culicidae). J Med. Entomol. 43:309–317.

Reisen WK (2009) Epidemiology of vector-borne diseases. In: Mullen GR, Durden LA (eds) Medical and veterinary entomology. Academic/Elsevier, Burlington

Reisen WK (2010) Landscape epidemiology of vector-borne diseases. Ann Rev Entomol 55:461–483

Reisen WK, Wheeler SS, Garcia S et al (2010) Migratory birds and the dispersal of arboviruses in California. Am J Trop Med Hyg 83:808–815

Schmidt O, Theopold U, Beckage NE (2008) Insect and vertebrate immunity: key similarities versus differences, 1st edn. Academic, San Diego

Sellers RF (1980) Weather, host and vector- their interplay in the spread of insect-borne animal virus diseases. J Hyg 85:65–102

Subramanian S, Krishnamoorthy K, Ramaiah KD et al (1998) The relationship between microfilarial load in the human host and uptake and development of *Wuchereria bancrofti* microfilariae by *Culex quinquefasciatus*: a study under natural conditions. Parasitology 116:243–255

Tatem AJ (2009) The worldwide airline network and the dispersal of exotic species: 2007–2010. Ecography 32:94–102

Walton WE, Reisen WK (2014) Influence of climate change on mosquito development and blood-feeding patterns. In: Singh SK (ed) Viral infections and global change. Wiley, Hoboken

Weaver SC, Reisen WK (2010) Present and future arboviral threats. Antiviral Res 85:328–345

Weerasooriya MV, Mudalige MP, Gunawardena NK, et al (1998) Microfilarial periodicity of Wuchereria bancrofti and man landing periodicity of the vector Culex quinquefasciatus Say in Matara, Sri Lanka. Ceylon Med J 43:78–83.

Further Reading

Beaglehole R, Bonita R, Kjellstrom T (1993) Basic epidemiology. World Health Organization, Geneva

Bruce-Chwatt LJ (1980) Essential malariology. William Heinemann Medical Books Ltd, London

Davis JR, Lederberg J (2001) Emerging infectious diseases from the global to the local perspective, Workshop summary. National Academies Press, Washington, DC

Garrett-Jones C (1970) Problems of epidemiological entomology as applied to malariology. Misc Publ Entomol Soc Am 7:168–178

Gregg MB (1988) Epidemiological principles applied to arbovirus diseases. In: Monath TP (ed) The arboviruses: epidemiology and ecology, vol 1. CRC, Boca Raton, pp 292–309

Jawetz E, Melnick JL, Adleberg EA (1972) Host-parasite relationships. In: Review of medical microbiology. Lange Medical Publications, Los Altos, pp 128–135

Macdonald G (1957) The epidemiology and control of malaria. Oxford University Press, Oxford

Reisen WK (2009) Epidemiology of vector-borne diseases. In: Mullen GR, Durden LA (eds) Medical entomology, 2nd edn. Academic/Elsevier, Burlington, pp 19–34

Reisen WK (2010) Landscape epidemiology of vector-borne diseases. Ann Rev Entomol 55:461–483

Rice PL, Pratt HD (1992) Epidemiology and control of vector borne diseases. U.S. Dept. Hlth, Educ., & Welfare, Publ. Hlth Serv. Pub. No. 72–8245

3 Arthropod-Borne Diseases and History

Carlos Brisola Marcondes

Abstract

Bloodsucking insects and other arthropods have had great influence on the history of mankind by causing epidemics that have been reported in many books, e.g., the Bible, and anecdotal narratives. They have contributed to defeating armies, like those of Napoleon during the invasions of Palestine, Haiti, and Russia, usually causing more morbidity and mortality than by battles. Insects and other arthropods have also been utilized as weapons, since the Paleolithic until now, and could possibly be dangerous as terrorist devices.

The knowledge of the influence of arthropod-borne diseases in history and the history of their discoveries and epidemiology is very useful for the understanding of disease transmission and prevention. Considering social and climatic changes, the distribution in the past can provide information on their possible occurrence nowadays.

Keywords

History • Discoveries • War • Torture • Terrorism

3.1 Insects in History

In a time when sanitation has become a standard in many countries and societies, it is difficult for most people to imagine life in crowded and dirty houses, a situation aggravated by religious taboos, when personal hygiene was even regarded as a sin.

Disease control may be influenced by observation of conditions associated with transmission, even before knowing the etiologic agents and vectors. For example, the observation of ancient Romans that malaria (*mala aria* means bad air in Medieval Italian) was associated with swamps and marshes stimulated draining activity, turning previously dangerous areas into productive ones. Abandonment of such measures in the Italian peninsula caused a recrudescence of malaria. For similar reasons, Cistercian monks

C.B. Marcondes (✉)
Department of Microbiology, Immunology and Parasitology, Center of Biological Sciences, Federal University of Santa Catarina,
88040-900 Florianópolis, SC, Brazil
e-mail: cbrisolamarcondes@gmail.com

C.B. Marcondes (ed.), *Arthropod Borne Diseases*, DOI 10.1007/978-3-319-13884-8_3

made a serious effort to improve health conditions by canalization of water (Sabbatani 2005).[1] Control of flies was long ago suggested by Muhammad ("If a fly falls into one of your containers [of food or drink], immerse it completely (falyaghmis-hukullahu) before removing it, for under one of its wings there is venom and under another there is its antidote[2]"), indicating the knowledge of transmission of something noxious by them.[3]

In Medieval Europe ("one thousand years without a bath"[4]), it is not surprising that "the war declared in the Middle Ages against the flesh and all cleanliness bore its fruits" (Michelet 1863), and lice and several pests caused great suffering in Europe. Lice and other vermin were banal, even for great men. As an example, Archbishop Thomas à Becket was murdered, in 29 December 1170, and, in the following day, among eight layers of clothes, "as the body grew cold, the vermin that were living in this multiple covering started to crawl out, and, according to the chronicler, 'The vermin boiled over like water in a simmering cauldron, and the onlookers burst into alternate weeping and laughter'" (Zinsser 1935).

There are also many examples of the impact of insect-borne diseases on battles and wars influencing the course of history.

Typhus is supposed to have evolved from a transmission by fleas among rodents, in Asia, to lice, whose ill-adaptation is indicated by their mortality. Vectored by lice, from the campaign of Maximilian against the Turks and through the Thirty Years' War, it caused high mortality "among the wretched populations under conditions ideal for typhus—of famine, abject poverty, homeless wandering, and constant warfare" (Zinsser 1935).

In 1489, louse-borne typhus fever ravaged the army of Ferdinand and Isabella in their trial to oust the Moors from Spain, indicating a double entry in Europe, through Spain and Balkans (Zinsser 1935). In 1527, Charles V of Spain occupied Naples, and, after plague reduced its army to 11,000 malnourished and diseased men, he dreamed of a miracle. An epidemic of typhus, which decimated the 35,000 strong French army laying siege to the city constituted such miracle, and as a result, he was crowned Holy Roman Emperor in 1530. Typhus also influenced Napoleon's advance to Russia, when his armies were decimated by this disease and by cold weather in Poland and Russia, and caused the death of two thirds of British and French soldiers during the Crimean War. In the beginning of the First World War, typhus caused a great epidemic among Austrian armies, jeopardizing the invasion of Serbia and Russia. Lice and typhus, besides a then unknown disease, the trench fever, were very important in the Western Front, and great measures were taken for their control. The difference between epidemiological conditions and efficiency of lice control in Eastern and Western fronts permitted a grand unintentional experiment (Lockwood 2009).

---

[1] In classes of genetics in my undergraduate course in 1972, cervical cancer was associated to promiscuity, and just some years later HPV was incriminated as a cause of such cancer. In 1842, analysis of data from death certificates of women in Verona from 1760 to 1839 already had related such cancer to sexual contacts.

[2] http://www.quran-m.com/firas/en1/index.php/life-sciences/156-new-studies-reveal-scientific-miracles-in-the-fly-hadith.html

[3] There are three chapters in Quran in the honor of insects which are called The Ants (Al Naml), Chapter 27; The Bees (Al Nahl), Chapter 16; and The Spider, Chapter 29. Eight arthropods were mentioned in the Quran: moths (Chapter 101, verse 4), ants (Chapter 27, verse 18), honeybees (Chapter 16, verse 68), flies (Chapter 22, verse 73), mosquitoes or gnats (Chapter 2, verse 26), locusts (Chapter 54 verse 7 and Chapter 7 verse 133), lice (Chapter 7, verse 133), and spiders (Chapter 29, verse 41). God tries to show his ability in creation of living beings and emphasizes that just he can do it as in the case of flies and mosquitoes or gnats (Mycetophilidae). Also God punished sinful people with lice and locusts. God mentions that ants speak with each other, honey which is made by honeybees is used for healing, moths to describe the situation of people on judgment day, and the "house web" which is made by spider is very weak (from Amer Al-Jawabreh, Reza Yaghoobi, and Alon Warburg).

[4] "Nul bain pendant mille ans" and "la guerre que le moyen âge déclara et à la chair, et à la propreté, devait porter son fruit" in the original 1862 version of Michelet's book ("La Sorcière"). The translator of the book to English (F. J. Trotter) recognized to have done some modifications, due to different readers (men in France and "drawing-room ladies, who tolerate grossness only in the theaters and the columns of the newspapers").

Although famine was an important cause of mortality in the 1840s potato crisis in Ireland, typhus, associated to overcrowding, made a great contribution, and, at least 63,500 died of the disease in 1849–1850, with successful potato harvests (Bray 1996).

Several great epidemics of plague occurred in Europe, and the most serious one was in the 1340s, killing one-third of the population living in an area ranging from India to Iceland (15–23.2 million in Europe), and it took 150 years for Europe's population to recover. Possibly started by catapulting infected corpses in the city of Kaffa (or Caffa, today Feodosiya, Ukraine) by a Mongol army, the disease was introduced in Europe by a Genoese boat arriving at the coast of Messina (Sicily) in October 1347 (Zietz and Dunkelberg 2004;[5] Lockwood 2009). This epidemic enforced the end of feudal system, the Inquisition's fury in Spain and the rise of the protestant Reform. From 1348 to 1665, only a few decades were free of plague epidemics in Europe. When plague affected the Russian army besieging the troops of Karl XII of Sweden in Reval (modern Tallinn, in Estonia) in 1710 and corpses were hurled over the walls, they caused an outbreak of the disease in the city. However, as there was no possibility of the people to escape, no repeat of the fourteenth-century pandemics occurred.[6]

Even accompanied by several transmissible diseases ("dysentery, typhoid fever, diphtheria, smallpox, scarlet fever, and a variety of less deadly confederates"), Thirty Years' War was divided in two phases: from 1618 to 1630, when typhus was the chief scourge, and from 1630 to 1648, when plague became the more ascendant (Zinsser 1935).

Based on immunity to HIV-1 on 10% of European population, linked to a mutation on allele *CCR5-Δ32*, before 1000 BC, the success of *quarantine* (isolation for 40 days), and several other factors, it was supposed "Black Death" of the fourteenth century, and other epidemics in Europe before 1600 AD were caused by a hemorrhagic virus, and not by *Yersinia pestis* (Duncan and Scott 2005). An alternative explanation for the great mortality in Black Death, frequently involving entire families, not usual in plague, would be the involvement of lice (McLean and Fall 2010; Drancourt and Raoult 2010) or the transmission by droplets (pneumonic plague) (Bray 1996).

Toussaint l'Ouverture was the leader of a successful revolt of slaves in Haiti, which started in 1791. The army sent by Napoleon Bonaparte to fight the revolt was decimated by yellow fever, a disease caused by a mosquito-borne virus. About 40,000 French soldiers, officers, doctors, and sailors died, and only 3000 returned to France (Lockwood 2009). This disaster stimulated the selling of a great part of the dreamed "French Empire in the New World," the Mississippi Valley ("Louisiana") to the USA for $15 million, which was the price the Americans were willing to pay only for New Orleans. Although yellow fever caused great suffering in North American cities, from Philadelphia to New Orleans (Spielman and D'Antonio 2001), the disease was indirectly very important for the nation building of the USA, because the "Louisiana Purchase" contributed with 2,140,000 km^2 to its territory, where 15 present states are situated,[7] and facilitated the march to West.

In the US-Mexican war in 1846–1847, General Scott had to attack Veracruz as soon as possible, to prevent greater losses due to yellow fever ("La Vómito"), and this and other diseases caused seven times more deaths in the American army than Mexican soldiers.

The failure of the building of a canal between the Pacific and Atlantic Ocean, promoted by

[5]A very complete and interesting review of the history of discoveries on plague etiology and history

[6]The city capitulated to Russian army on 30 September 1710, and from ~20,000 inhabitants in August, about 15,000 had died of the plague by mid-December. Surviving Swedish troops and some citizens were authorized to leave by ship and carried the plague to Finland (http://en.wikipedia.org/wiki/Great_Northern_War_plague_outbreak#Lithuania.2C_Livonia.2C_Estonia). Simultaneous outbreaks occurred in several nearby countries.

[7]*Sans moustiques, les Etats-Unis pourraient être francophones*—that is, "without mosquitoes, the United States might be French-speaking" (Lockwood 2009).

Ferdinand de Lesseps, in the 1880s, was in large part attributable to yellow fever and malaria, with 52,000 deaths among 85,000 workers (Brèss 1986). This failure caused the bankruptcy of the French company financing the project and had a great influence on anti-Semitism in France and on the Dreyfus affair (Tuchman 1982). However, it was only after the discovery of the pathogens responsible for these diseases that the control was improved, and the Americans finally could conclude the Panama Canal. Even Portugal, Spain, France, the UK, and the USA had outbreaks of yellow fever in the beginning of the nineteenth century (Sabbatani and Fiorino 2007). In Brazil, special institutes needed to be developed in the beginning of the twentieth century in order to control yellow fever and other infectious diseases, which seriously jeopardized the progress of the country.

Yellow fever, malaria, and dengue fever hampered the development of Florida. From the time of the First World War, control of mosquitoes and of mosquito-borne diseases has been conducted in the state with great success (Hribar 2013). An additional $2 million was spent in mosquito control during an epidemic of Saint Louis encephalitis in the late 1990s in Florida, and tourism declined by 15%. However, the cost of health care for the 213 people taken ill and of the 11 lives lost is difficult to calculate. Lifetime care of a child who survived a severe eastern equine encephalitis infection exceeded $3 million.

Heracles legendary killing of hydra snake was suggested to be an attempt to reclaim the malaria-infested marshes of Lema country in the Argus region. Empedocles (504 BC) is considered the first malaria engineer, as it is claimed that he drained swamps about Agrigentum (Bray 1996). The disease has been related to "porotic hyperostosis" in old Helenic text. Although malaria was prevalent in Italy from 700 to 350 BC and has not been considered very dangerous in the Republican period, deforestation in the times of Julius and Augustus Caesar caused the dispersion of more dangerous vectors (*Anopheles sacharovi* and *An. labranchiae*), causing a great burden of the disease (Bray 1996). However, it would be an exaggeration to attribute to malaria the decadence of classical Greece and Rome (Bray 1996).

In eras before mechanized armies, huge numbers of horses and cattle were necessary, producing an enormous quantity of dung. Thus, flies and associated diseases like typhoid fever, cholera, and diarrhea proliferated, ravaging armies. This was obvious during the American Civil War, when dysentery and diarrhea took 60,000 lives out of 1.7 million cases in the Union side, and ca. one-fourth of prisoners from both sides died of these diseases. Dead bodies and wounded and amputated soldiers were also attacked by flies. Fortunately, the Confederate surgeons Joseph Jones and J. F. Zacharias noticed that maggots were beneficial in wound healing and the last utilized them for treatment (see Chap. 31). The abundance of horses in New York City (120,000 in the beginning of the twentieth century), producing an immense amount of dung (and many carcasses on the streets), was supposed to be associated to typhoid, cholera, and even to smallpox and yellow fever.[8]

Scrub typhus, the tsutsugamushi (small and dangerous creature, in Japanese) fever, transmitted by a bloodsucking mite, affected American troops during their invasion of New Guinea in the Second World War. An outbreak of louse-borne typhus in Naples in 1943 was only averted by the utilization of DDT, and still a secret of the Allies.

African trypanosomiasis, causing disease in humans and cattle, have greatly affected life on the continent. *Trypanosoma brucei* infects antelopes and other ruminants and, like other subspecies of *T. brucei*, is transmitted by tsetse flies. The "tsetse belt," now including eight million square kilometers from 14° N to 20° S, once stretched as far as into the Egyptian Old Kingdom (3000 BC–2000 BC) up to the Nile delta (>30° N) (Steverding 2008). In the Middle Kingdom (2000 BC–1300 BC) when the course of the river Nile was adjusted, breeding sites of the tsetse flies were probably destroyed and the more efficient zebu cattle (*Bos taurus indicus*) was interbred with trypanotolerant aurochs (*B. primigenius*) gradually replacing the original breeds.

[8]These horses would produce approximately 1277 tons of dung per day and 128 billion of flies per year (http://www.banhdc.org/archives/ch-hist-19711000.html).

Arthropod-borne diseases, like other infectious diseases, are more dangerous to nonimmune people, and this was becoming evident on imperial and exploratory expeditions of Europeans to Africa and Asia. Diseases like malaria and yellow fever prevented and delayed European colonization of the African continent for at least one century, and, for a long period, settlements by European immigrants were mainly at higher altitudes and only in a few locations where strictly enforced measures could improve conditions of living (Spielman and D'Antonio 2001). West Africa, due to malaria and yellow fever, was considered to be the "White Man's Grave,"[9] and the mean mortality in six expeditions to rivers was 49 %, reaching 91 % in the Mungo Park expedition of 1805 to explore the river Niger (Gelfand 1964).

Panic caused by diseases provoked irrational reactions and behavior, including prejudice and massacres. Plague epidemics in the fourteenth century stimulated aggressions against Jews and Muslims, both considered to be heretics and sinners. Because of this persecution, Jews were invited to migrate to East Europe, and soon half of the Jews lived in Poland and Ukraine. Epidemics of yellow fever in New Orleans and other cities in the south of the USA stimulated prejudice of northerners against southerners and of southerners against Afro-Americans and foreigners, causing violence and even lynching (Spielman and D'Antonio 2001). The greater risk of Europeans to suffer from tropical diseases stimulated mutual racial prejudice; some Europeans considered their brains were superior but more delicate while many Africans concluded that the whites were inferior creatures.[10]

King Herod, in his sixties, suffered a terrible degeneration, including suppuration, gangrene, and "worms" in the lesions. Among several hypotheses for his ailment, high blood pressure, associated with kidney failure, may have caused gangrene, and maggots could have developed in his wounds. At the time, maggot infestation was thought to be only found in wicked men and thus Herod's affliction may have been considered as a curse by the likes of Antiochus IV Epiphanes, Agrippa I (grandson of Herod and friend of Caligula), and Judas Iscariot (Montefiore 2011).

3.2 Insects as Weapons

Besides the impact of arthropod-borne diseases on history and wars, insects have also frequently being used as weapons (Lockwood 2009). Already in late Stone Age times, beehives have been employed to dislodge enemies from caves. Other examples are discussed below.

Six of the ten plagues cited in the liberation of Jews from captivity in Egypt may be attributed to insects (Lockwood 2009). The third plague could have been caused by biting midges (Ceratopogonidae) or lice (*Kirin* in Hebrew) (Mumcuoglu 2008, King James Bible) (Exodus 8:16). The fourth ("Yahweh warned:…I will send swarms of flies upon thee,… and the houses of Egyptians shall be full of swarms of flies," Exodus 8:21) could be attributed to biting flies, probably *Stomoxys calcitrans* and horse flies (Tabanidae) arising from rotting vegetation laced with decomposing fish. The fifth plague affected Egyptian's livestock (Exodus 9:2–3) and was possibly caused by blue tongue disease and African horse sickness. One possible explanation for the sixth plague ("boil breaking forth with blains upon man, and upon beast", Exodus 9:8–9) could be that is explained by glanders and anthrax, which can be transmitted by stable flies. The eighth plague was caused by locusts (Exodus 10:3–5), and, the ninth, the darkness, was probably caused by the destruction of cropland by a sand storm. Finally, the tenth plague, killing of the first born, could have been either caused by bubonic plague transmitted by rats and fleas or by a black mold that grew following water damage on cereals in pits contaminated with plagues. Other explanations for these plagues have also been published (Sabbatani and Fiorino 2010).

[9]The term was originally coined for Sierra Leone, Britain's first West African colony, conceived as a home for slaves freed after slavery was declared illegal in Britain (http://myweb.tiscali.co.uk/kenanderson/histemp/whitemans-grave.html).

[10]According to an African song , *The poor white man, faint and weary/Came to sit under our tree/He has no mother to bring him milk/No wife to grind his corn/Let us pity the white man* (Spielman and D'Antonio 2001).

Other biblical books refer to entomological warfare. Joshua 24:12 and Exodus 23:28 referred to the use of hornets as weapons. This use was sufficiently common to be attributed to Almighty (Deuteronomy 7:20) (Lockwood 2009).

The pest of Philistines, after robbing the Ark of the Covenant, has been attributed to plague. Some authors thought that the black rat had only arrived to Egypt in the first or second century A.D. and suggested that the Philistines had hemorrhoids, caused by dysentery, but recent archaeological findings in Egypt indicated the presence of plague about 1350 BC (Sabbatani and Fiorino 2010).

The San Bushmen, from Kalahari Desert, utilize the toxin produced by Chrysomelidae beetles (*Ngwa* or *Kaa*) on their arrows, which can even kill antelopes and giraffes (Lockwood 2009).

In the American Civil War, as part of a tactical maneuver, Confederate General Joseph Johnston withdrew its small force to lure the enemy into a malarigenous swampy area, and, in July 1862, half of the Union force was hospitalized or too sick to fight and they had to withdraw in August (Steiner 1968).

In the First World War, malaria caused 617,150 cases among entente soldiers and 562,096 among central powers soldiers, respectively, with fatality estimates of 0.63 and 4.15 %. The difference was attributed to better management and quinine regimes, besides prewar prevention measures developed by Ronald Ross for entente armies (Brabin 2014). The fatality index varied according to species involved (*P. vivax* or *P. falciparum*), and the ignorance on the permanence of hepatic stages (hypnozoites, see below) of *P. vivax* and adequate prevention of relapses.

Pontine Marshes, near Rome, had been a malarigenous area, but reclamation in the 1930s controlled the disease. In 1943, dikes were destroyed, allegedly by biological warfare by the German retreating army, and an epidemic broke out affecting one-third of the population. However, the evidence that the German army tried to cause an epidemic in this case and of any other biological warfare during the Second World War was considered to be weak (Gueissler and Guillemin 2010; Sabbatani et al. 2013). This does not, of course, mean that the Nazis did not experiment and develop biological weapons. They reality is that they did extensive work on biological weapons testing the preservation of plague bacilli and other pathogens and inoculation in prisoners, as exemplified by the career of Dr. Kurt Blome (1894–1969), with frequent contact with Unit 731 (see below).[11] He, one of the 23 defendants of the Doctor's Trial,[12] and several other Nazi experts on biological warfare furnished information for Fort Detrick's experts[13] (see below).

During the same time, the Japanese undertook intensive research on biological warfare, establishing the infamous Unit 731 to test pathogens in the Chinese (and Westerners, called "white rats"), to utilize insect-borne pathogens like plague-infected fleas and to contaminate flies with the cholera bacterium. Starting by the injection of plague bacilli in three Communist prisoners and their vivisection in 1935, many ghastly experiments[14] and trials to initiate epidemics were developed. Special bombs with cholera bacteria and house flies were launched in two northern Chinese provinces, causing a huge mortality in the population.[15] However, at the end of the war, the North Americans and the Soviets pardoned unit 731 researchers and General Ishii Shiro, the chief of the Japanese biological weapon program, in exchange for their research data (Lockwood 2009).

Insects have been, and probably will continue to be, tested as biological weapons by several countries, like the USA, Canada, France, and Russia. Accusations of utilization of insects and insect-borne diseases as biological arms, always difficult to prove or disprove, were exchanged between the USA and communist countries, and research on defensive and offensive use have been carried out in Fort Detrick, Maryland (Lockwood 2009). Release experiments with

[11] http://en.wikipedia.org/wiki/Kurt_Blome

[12] http://en.wikipedia.org/wiki/Doctors%27_trial

[13] http://en.wikipedia.org/wiki/U.S._intelligence_involvement_with_German_and_Japanese_war_criminals_after_World_War_II

[14] Many Chinese prisoners ("maruta") were maintained in cages, infected by plague bacillus, and bled to infect millions of fleas (Lockwood 2009).

[15] The Chinese provinces of Yunnan and Shandong were called "the Hiroshima and Nagasaki of China" (Lockwood 2009).

uninfected mosquitoes were conducted in Florida, and a huge production of mosquitoes was created. However, several Cuban accusations of intentional introduction of insects and pests in Cuba remained unproven.

Serious projects for the production of entomological arms were developed in WW2, including the Colorado potato beetle by Germans. French scientists were developing studies on the utilization of this beetle but were interrupted by the German invasion. The deliverance of a cargo of ticks and beetles by an American plane to the UK stimulated the production of these beetles by Germans (Lockwood 2009). The liberation of these beetles by Germans in southern Germany (near Speier, in October 1943), during the war, and American plans to disperse rubber pellets with scent attractive to flies and bacteria ("contaminated pseudo-poops") by planes in Morocco, to cause disease to German Army, are examples of somewhat ridiculous trials of biological war in WW2 (Lockwood 2009). The sudden transference of German soldiers to Stalingrad, much more dangerous than Spanish Morocco, lent to cancelation of the above American plans.

In Vietnam War, utilization of herbicides by the US force indirectly caused scrub typhus, through replacement of forest by grassland, with subsequent proliferation of rats and mites. Vietcong used scorpions in tunnels and the ferocious tropical honeybee (*Apis dorsata*) in forests to attack enemy soldiers, and intense research was carried out to produce pheromones to redirect the attacks of the bees against the Vietcong (Lockwood 2009).

3.3 Insects for Torture and Terrorism

Arthropods have frequently been used for torture. During the "Great Game,"[16] two Englishmen were tortured in Bokhara by staying for several months in a pit with assassin bugs (Reduviidae) and sheep ticks (probably *Dermacentor marginatus*), before being beheaded. Siberian and Apache tribes also utilized this practice, exposing nude victims, respectively, to thousands of bites of mosquitoes and black flies and to ants. The CIA, who used insects to get information from a suspected member of al-Qaeda (Lockwood 2009), has also recently employed this.

Insects are potentially also useful for terrorist actions, in the "arboterrorism" (from "arthropod-borne terrorism") (Tabachnick et al. 2011). For example, eggs of *Aedes* mosquitoes contaminated with the Rift Valley fever virus, glued to a small piece of paper, can be easily dropped into the water supply systems and thus start an epidemic. In addition, insects are a good way to protect and disperse pathogenic agents, like the plague bacterium and arboviruses. Lockwood (2009) analyzes several possibilities and occurrences of bioterrorism and the unpreparedness of the USA (and other countries) for such danger. The discovery of the agent of mysterious encephalitis in New York City ("the best prepared city in the nation"), initially attributed to Saint Louis virus, spent 35 days, with precious collaboration of an ornithologist of the Bronx zoo, examining dead birds, and finally it was shown to be West Nile virus (Lockwood 2009). It subsequently expanded to almost all the country (and the American continent[17]).

3.4 History of Medical Entomology

Although the involvement of arthropods in the life cycles of *Protospirura muris*, *Dracunculus medinensis*, and *Dipylidium caninum* was discovered in the 1860s, medical entomology was born in 1878,[18] when Patrick Manson demon-

[16] This name (Большая игра, Bolshaya Igra, in Russian) has been utilized to designate the dispute for the Central Asia, the so-called Stans, developed from 1837 to 1907, between Great Britain and Russia. The expression was created by Captain Arthur Conolly (one of the victims in Bokhara) and popularized by Rudyard Kipling, in "Kim."

[17] A human case, the first in Brazil, was observed in the state of Piauí in the end of 2014.

[18] The year 1878 has usually been considered as the beginning of medical entomology, but P. Manson published a report on the role of mosquitoes in the transmission of *W. bancrofti* in *China Customs Medical Reports* in September 1877 (Russell 1955), or alternatively in January 1878 (Service 1978).

strated the role of *Culex* mosquitoes in the transmission of *Wuchereria bancrofti*. Aware of the observations of A. P. Fedchenko on the life cycle of *D. medinensis*, he supposed that after the development of filarial worms in mosquitoes, they were released into water and were ingested by humans, and only in 1899–1900 the direct transmission from mosquitoes to humans was demonstrated.

Working in China, P. Manson also noticed differences between quantities of microfilariae in the blood of two Chinese assistants, one who was sleeping at night while the other due to his sick mother was working in the evenings. When Dr. T. S. Cobbold presented his observations on periodicity of microfilariae to Linnean Society of London, they caused heckling and ridiculing, but P. Manson finally confirmed this periodicity for *W. bancrofti* and *Dirofilaria immitis* (Service 1978).

In 1890, Theobald Smith and Frederick Kilbourne had demonstrated the transmission of *Babesia bigemina*, an intraerythrocytic parasite like *Plasmodium*, by ticks, but malaria researchers were not aware of this study, probably because it was published in an American government agricultural document.

Hippocrates proposed the miasmatic origin of malaria thoroughly accepted until the nineteenth century, and, Giovanni Maria Lancisi (1654–1720), physician to the pope and professor at the Sapienza University, in Rome, published *De noxiis paludum effluviis, eorumque remediis* ("On the noxious effluvia of paludi and on their remedies") in 1717, suggesting the possible role of mosquitoes in the transmission of parasites.

Although some knowledge about mosquitoes and other insects came from Aristotle, their relationship with certain diseases was discovered very late. Louis-Daniel Beauperthuy,[19] in 1854, suspected of mosquitoes as vectors of malaria by, but it was first thought that they obtained some infective material from marshes.

In 1807, the surgeon John Crawford suggested that malaria was caused "by eggs introduced without our knowledge into our bodies" by biting "animalcules and insects like mosquitoes and fleas." Other similar proposals followed, ridiculed by experts' opinion of the association of malaria to "bad air." Laveran (1884) described the discovery of a parasite of malaria in humans and reviewed previous studies involving injection of material from marshes in rabbits and other experimental animals.

In 1881 and 1898, respectively, Albert F. A. King[20] and Robert Koch suggested a role for mosquitoes in the transmission of malaria parasites. Ronald Ross, in 1897–1898, stimulated by P. Manson and alternating with studies on *Plasmodium* of birds, which developed in "gray" mosquitoes (*Culex*), elucidated the transmission of human *Plasmodium* by "dappled-wing mosquitoes" (*Anopheles*). However, he had no access to areas with many human cases of malaria, and, G. B. Grassi and A. Bignami, after suggesting in 1896 that mosquitoes might transmit malaria by inoculation, produced in 1898 the final proof, transmitting the parasite between infected patients and uninfected individuals by *Anopheles claviger*. The incrimination of *Anopheles* as vectors of *Plasmodium* involved a dispute, sometimes acrimonious, between R. Ross and A. Grassi, both having strong personalities, in "one of the least attractive episodes in the whole history of malariology"; the dispute, finally decided by R. Koch, was fundamental to the nomination of the first to Nobel prize in 1902[21] (Capanna 2006).

[19]This brilliant scientist, born in Guadeloupe in 1808, also associated, in 1853, the "striped-legged mosquito" (*Aedes aegypti*, "zancudo bobo") to yellow fever (Agramonte 1908).

[20]Albert F. A. King (1841–1914) presented and elaborated his hypothesis on the role of mosquitoes in the transmission of malaria to the Philosophical Society of Washington in 1882 (also proposing to protect Washington from malaria encircling the city with a wire screen as high as the Washington Monument) and published it in 1883. He served in both sides in the American Civil War and helped the president Lincoln just after he was shot.

[21]Nobel Prize in 1902 should possibly have been shared by R. Ross and these two Italian workers, and the 1928 one on exanthematic typhus, given to C. Nicolle, forgot Henrique Rocha Lima, who described *Rickettsia prowazekii* (http://www.invivo.fiocruz.br/cgi/cgilua.exe/sys/start.htm?infoid=752&sid=7).

Variations on the efficiency of anopheline mosquitoes to transmit the parasites originated the concept of "anophelism without malaria" and stimulated the study of taxonomy and biology of mosquitoes; F. V. Theobald soon published his huge five-volume "A Monograph of the Culicidae of the World."

Although Grassi and his colleagues had suggested in 1893 and 1894 an exoerythrocytic cycle, after the influential German scientist Schaudinn allegedly saw the penetration of red blood cell by sporozoites (Desowitz 2000), 40 years were spent to confirm such cycle. Finally, only in 1982 the existence of hypnozoites in *P. vivax* was proven, which explained the long prepatent period of this malaria species (Cox 2004). The history of the discovery of the malaria cycle illustrates the great effort necessary and the damage a small mistake can make for the progress of knowledge.

The physician and cosmograph Cosme Bueno reported in 1765 the association by Indians from Peruvian Amazonia of cutaneous leishmaniasis and sand flies, both called *uta*.[22] André Pressat and the Sergent brothers (Edmond[23] and Etienne) proposed, respectively, in 1904 and 1905, the transmission of *Leishmania* by phlebotomine sand flies. Among several discoveries on the etiology and transmission of arthropod-borne diseases, like the transmission of *Trypanosoma berberum* (=*T. evansi*) among dromedary camels by *Atylotus* horse flies (in Arabic *debab* for disease and vector[24]) and *Oestrus ovis* infestation (in Berber *thimni*) (De la Tribonnière 2000), the Sergent brothers proved in 1921 the role of *Phlebotomus papatasi* in the transmission of *Leishmania major*. In the following year, Henrique B. Aragão successfully infected a dog inoculating material of *L. braziliensis* from *Phlebotomus intermedius* (=*Nyssomyia intermedia*) in the animal's nose, in a focus at Rio de Janeiro.

Elucidation of transmission of *Leishmania* causing visceral leishmaniasis by sand flies was a saga in the first half of the twentieth century. After following false trails by experimenting almost any bloodsucking animals (except leeches and vampire bats) and strongly suspecting of bedbugs, three teams (in Calcutta and Assam in India, and Shantung Province in China) in the 1920s and 1930s finally concentrated on sand flies. After many frustrated trials, involving hundreds of laboratory rodents and human voluntaries, effective transmission was obtained after infected sand flies received sugar from raisins, promoting an adequate multiplication of the protozoa (Killick-Kendrick 2013).

The discovery of other arthropod-borne diseases was also driven by competition between researchers. The race for the discovery of the causative agent of plague between Shibasaburo Kitasato and Alexandre Yersin, who both arrived in Hong Kong in June 1894 to find the plague pathogen, is an example of such fruitful competition. Masanori Ogata proposed the transmission of the plague bacterium by fleas in 1897, based on his research in Taiwan, where he infected mice by the injection of an emulsion of internal organs of dead fleas collected from plague-infected rats. In addition, Paul-Luis Simond, after epidemiological studies of an Indian epidemic, proposed in 1898 the central role of fleas in the transmission of plague (Zietz and Dunkelberg 2004).

Discoveries on the transmission and etiological agent of exanthematic typhus (ET) before and during WW1 also involved competition between Charles Nicolle, who had discovered the role of lice on the transmission of the "filterable virus" and several American and Mexican researchers, running for a $25,000 reward (=2 years of salary in an American university) to clarify "tabarillo"

[22] "The narrow valleys are very unhealthy where two kinds of maladies are noted; these diseases are also present in other cool provinces. One is verruga [Carrion's disease], which happens to be very troublesome and dangerous if not accompanied by cutaneous eruptions. The other results in corrosive ulcers, located on the face, is very difficult to cure, and causes the death of some people. It is said that both diseases originate from the bite of a small insect called uta [sand fly]" (translated from Spanish by Herrer and Christensen 1975).

[23] Edmond Sergent (1876–1969) was the director of Pasteur Institute of Algeria for over 60 years (1900–1963), with 592 publications (Dedet 2006).

[24] This observation of laypeople, like that on *uta*, stimulates the adaptation of the nineteenth-century Russian *Narodnik* proposal: "Go to the people."

(Mexican typhus). Howard T. Ricketts, after explaining the transmission of Rocky Mountain spotted fever, participated in the competition, being very careful to not be spied upon by his colleagues; he published four papers on ET in JAMA in 1910 and died of this disease in the same year. Stanislaus von Prowazek and Henrique Rocha Lima made important studies during the Balkan wars and on Russian prisoners in 1914–1915. The later finally described the bacteria *Rickettsia prowazekii* and definitely established its role of lice (Silva 2011). Both were infected by the bacteria, and von Prowazek died in January 1915. After the corroboration of the role of lice in the transmission of typhus, delousing (=*entlausung*[25] and *depouillage*) was utilized by all armies in WW1, and good conditions of hygiene in trenches were reason for pride among soldiers.

The transmission among humans of *Borrelia recurrentis* in an outbreak of relapsing fever in an oasis of North Africa occurred only in the winter. Mosquitoes and other flying insects were not biting at this season, fleas were absent in the region, and bedbugs were present only in houses of Europeans. After examining soft ticks (Argasidae) and lice, Edmond Sergent and Henry Fol proved the vectorial role of the later insects in 1907–1908 (Dedet 2006).

The Scottish missionary and explorer David Livingstone suggested that nagana is caused by the bites of tsetse flies. The transmission of *T. brucei* by tsetse flies was demonstrated by David Bruce in 1895 and was followed by the observation of trypanosomes (*T. brucei gambiense*), which were classified as worms by R. M. Forde, but some months later correctly identified as flagellates by J. E. Dutton. In 1902, D. Bruce proved the transmission of trypanosomes causing sleeping sickness in humans by tsetse flies but thought that the transmission was mechanical. After F. K. Kleine had described the cyclical transmissionof trypanosomes in 1909, D. Bruce described the full life cycle of the parasite in the insect host. In the meantime, other animal pathogenic trypanosome species (*T. congolense* and *T. vivax*) were described, including the second human pathogenic trypanosome species (now *T. brucei rhodesiense*).

The elucidation of the life cycle of the parasite causing American trypanosomiasis (Chagas disease) is exceptional, because Carlos Chagas, sent to control malaria in the north of the Brazilian state of Minas Gerais, noticed the strange symptoms of Chagas disease, described the parasite, the vectors and some reservoir hosts, and obtained preliminary clinical data.[26] Chagas' discovery, starting with the examination of insects (the "wrong end"), where he found epimastigotes ("chritidias"), instead of trying to discover the causative organism of a disease, was probably due to the fact that he carried out the research virtually by himself (Lewinsohn 2003). This is in contrast to the discovery of other diseases, like malaria and leishmaniasis, whose cycles were elucidated during several decades and involved many researchers (Coutinho and Dias 1999).

The Cuban Carlos Finlay in 1881 proposed a role of *Aedes aegypti* in the transmission of the etiological agent of yellow fever (YF), in the so-called Havana theory. After invading Cuba in February 1898, Americans had difficulty to control YF, and after contacting Dr. Finlay such a role was studied by American and British scientists. Tests developed by Brazilian workers Adolpho Lutz and Emilio Ribas, in which Émile Marchoux, Paul-Louis Simond, and Alexandre Salimbeni, visiting Rio de Janeiro, were very interested, definitely established the role of *Ae. aegypti*, against other ways for contamination.

Discoveries on the role of *Ae. aegypti* on the transmission of YF virus and of *Anopheles* on malaria permitted the control of such diseases in the building of Panama Canal (Parker 2008) and the control of YF in Belém (Frahia Neto 2012) and Rio de Janeiro.

In the following decades, the role of insects and other arthropods in the transmission of many

[25] Lockwood (2009) included a detailed (and vivid) description of delousing in a German prisoner camp in WW1.

[26] A campaign developed by Brazilian colleagues probably jeopardized his nomination, twice pondered, for Nobel Prize (Lewinsohn 2003).

protozoan, helminths, bacteria, and viruses was discovered. After several reports of "acrodermatitis chronica atrophicans (ACA)" and "erythema chronicum migrans" in Europe in the years from 1890 to 1970, Lyme disease was characterized by a careful study conducted in the northeast of the USA, in the 1970s (Steere et al. 1977), leading to the discovery of the vectors, *Ixodes* ticks (Sood et al. 2011).[27]

Even in 2012, the discovery of vectors continued, as exemplified in a very exciting narrative of the finding of the black fly vector of *Onchocerca lupi* in Greece (Otranto et al. 2012).

Henry Shortt (1887–1987), in the day after recommending the Kala-azar Commission to be disbanded, because 3358 bites of *Phlebotomus argentipes* on 35 hamsters did not produce any infection, found an infected one, 511 days after the first bite (Killick-Kendrick 2013). Leonidas M. Deane (1914–1993) discovered the first wild reservoir of visceral leishmaniasis in the Americas examining a dying wild dog in the 1954 Christmas Eve, postponing the travel from Sobral to Fortaleza, the capital of state of Ceará (Deane, pers commun 1974). These facts emphasize the need of persistence for discoveries.

Acknowledgments To Amer Al-Jawabreh, Dietmar Steverding, Domenico Otranto, Jeffrey Lockwood, Kosta Mumcuoglu, and Reza Yaghoob for information useful for this chapter.

References

Agramonte A (1908) An account of Dr. Louis-Daniel Beauperthuy, a pioneer in yellow fever research. Boston Med Surg J 158:927–930

Brabin BJ (2014) Malaria's contribution to World War One—the unexpected adversary. Malaria J 13:497

Bray RS (1996) Armies of pestilence: the effects of pandemics on history. The Lutterworth Press, Cambridge

Brès PLJ (1986) A century of progress in combating yellow fever. Bull WHO 64:775–786

Capanna E (2006) Grassi *versus* Ross: who solved the riddle of malaria? Int Microbiol 9:69–74

Coutinho M, Dias JCP (1999) A descoberta da doença de Chagas/The discovery of Chagas disease. Cad Cien Tecnol 16:11–51

[27]More details on vector incrimination are in Chap. 2.

Cox EG (2004) History of human parasitic diseases. Infect Dis Clin N Am 18:171–188

Dedet JP (2006) Les découvertes d'Edmond Sergent sur la transmission vectorielle des agents de certaines maladies infectieuses humaines et animales. Bull Soc Pathol Exot 100:147–150

De la Tribonnière X (2000) Edmond Sergent (1876–1969) et l'Institut Pasteurd'Algérie. Bull Soc Pathol Exot 93:365–371

Desowitz RS (2000) The fate of sporozoites. Bull WHO 78:1445–1446

Drancourt M, Raoult D (2010) Body lice, *Yersinia pestis* Orientalis, and Black Death. Emerg Inf Dis 16:1650

Duncan CJ, Scott S (2005) What caused the Black Death? Postgrad Med J 81:315–320

Frahia Neto H (2012) Oswaldo Cruz e a febre amarela no Pará. Instituto Evandro Chagas, Belém

Gueissler E, Guillemin J (2010) German flooding of the Pontine marshes in World War II. Polit Life Sci 29:2–23

Gelfand M (1964) Rivers of death. Oxford University Press, London

Herrer A, Christensen HA (1975) Implication of *Phlebotomus* sand flies as vectors of bartonellosis and leishmaniasis as early as 1764. Science 190:154–155

Hribar LJ (2013) Influence and impact of mosquito-borne diseases on the history of Florida, USA. Life Excit Biol 1:1–16

Killick-Kendrick R (2013) The race to discover the insect vector of kala-azar: a great saga of tropical medicine 1903–1942. Bull Soc Pathol Exot 106:131–137

Laveran A (1884) Traité des fièvres palustres. Octave Doin Editions, Paris

Lewinsohn R (2003) Prophet in his own country: Carlos Chagas and the Nobel Prize. Persp Biol Med 46:532–549

Lockwood JA (2009) Six-legged soldiers: using insects as weapons of war. Oxford University Press, New York, Recently published in Portuguese by Editora da UFSC (trad.: CBM)

McLean RG, Fall MW (2010) Body lice, *yersinia pestis* orientalis, and black death. Emerg Inf Dis 16:1649–1650

Michelet J (1863) La sorcière: the witch of the Middle Ages. Simpkin, Marshall, and Co., London. 403 p (translated from the French 1862 edition)

Montefiore SS (2011) Jerusalem: the biography. Phoenix, London

Mumcuoglu KI (2008) Human lice: *Pediculus* and *Pthirus*. In: Raoult D, Drancourt M (eds) Paleomicrobiology: past human infections. Springer, Berlin Heidelberg, pp 215–222

Otranto D, Dantas-Torres F, Papadopoulos E et al (2012) Tracking the vector of *Onchocerca lupi* in a rural area of Greece. Emerg Inf Dis 18:1196–1200

Parker M (2008) Panama Fever: the epic story of one of the greatest human achievements of all time—the building of the Panama Canal. Double Day, New York

Russell PF (1955) Man's mastery of malaria. Oxford University Press, London, New York, Toronto

Sabbatani S (2005) Attempts to fight paludism and malaria in the Middle ages. Role of Benedictine and Cistercian monks in the rise of monastic medicine and in land reclamation during the Middle Ages. Infez Med 13:196–207

Sabbatani S, Fiorino S (2007) La febbre amarilla. Inf Med 2:129–141

Sabbatani S, Fiorino S (2010) The plague of the Philistines and other pestilences in the Ancient World: exploring relations between the religious-literary tradition, artistic evidences and scientific proofs. Inf Med 3:199–207

Sabbatani S, Fiorino S, Manfredi R (2013) The re-introduction of malaria in the Pontine Marshes and the Cassino district during the end of World War II. Biological warfare or global war tactics? Inf Hist Med 21:320–342

Service MW (1978) Patrick Manson and the story of Bancroftian filariasis. In: Willmott S (ed) Medical entomology centenary symposium proceedings. Royal Society of Tropical Medicine and Hygiene, London, pp 11–14

Silva AFC (2011) A trajetória científica de Henrique da Rocha Lima e as relações Brasil-Alemanha (1901–1956). Dissertation, Federal University of Rio de Janeiro, Rio de Janeiro

Sood SK, O'Connell S, Weber K (2011) The emergence and epidemiology of Lyme borreliosis in Europe and North America. In: Sood SK (ed) Lyme borreliosis in Europe and North America: epidemiology and clinical practice, 1st edn. Wiley, New Jersey, pp 1–35

Spielman A, D'Antonio M (2001) Mosquito: the story of man's deadliest foe. Hyperion, New York

Steere AC, Malawista SE, Snydman DR et al (1977) Lyme arthritis: an epidemic of oligoarticular arthritis in children and adults in three Connecticut communities. Arthritis Rheum 20:7–17

Steiner PE (1968) Disease in the Civil War: natural biological warfare in 1861–1865. Charles C Thomas, Springfield

Steverding D (2008) The history of African trypanosomiasis. Paras Vectors 1:3. doi:10.1186/1756-3305-1-3

Tabachnick WJ, Harvey WR, Becnel JJ et al (2011) Countering a bioterrorist introduction of pathogen-infected mosquitoes through mosquito control. J Am Mosq Contr Assoc 27:165–167

Tuchman B (1982) The proud tower: a portrait of the world before the war. Bantam Books, New York

Zietz BP, Dunkelberg H (2004) The history of the plague and the research on the causative agent *Yersinia pestis*. Int J Hyg Environ Health 207:165–178

Zinsser H (1935) Rats, lice and history. Little Brown & Co, Boston, A fascinating and very informative book, available for download

Blood Sucking, Vector-Parasite Relationship, and Transmission of Diseases

4

Carlos Brisola Marcondes, Iliano V. Coutinho-Abreu, Jesus Valenzuela, and Hilary Hurd

Abstract

Blood feeding is very important for disease causing and transmission. It evolved by several ways, and its characteristics are analyzed, besides the vector-parasite relationship and its influence on transmission. The importance of saliva and digestive process to transmission, immunity, and the development of vaccines is analyzed. The influence of several parasites on the behavior of vectors, facilitating disease transmission, is described.

Keywords

Blood feeding • Hematophagy • Vector-parasite relationship

Several arthropods have evolved adaptations for blood sucking. The comprehension of the mechanisms for such a complex method of feeding is important to understand the mechanisms of causing and transmitting diseases.

4.1 Evolution and Occurrence of Blood Sucking in Arthropods

Many arthropods have independently evolved the capacity to suck blood, obtaining a valuable source of energy, protein, and several other useful substances. This polyphyletic origin of blood feeding created several mechanisms for the finding, biting, and digesting of blood, besides the contra-measures to defenses of hosts.

Blood feeding evolved at least six times in approximately 15,000 species of 400 genera of arthropods during Cretaceous and Jurassic eras (145–65 MYA) (Balashov 1984; Ribeiro 1995).

C.B. Marcondes (✉)
Department of Microbiology, Immunology and Parasitology, Center of Biological Sciences, Federal University of Santa Catarina, 88040-900 Florianópolis, SC, Brazil
e-mail: cbrisolamarcondes@gmail.com

I.V. Coutinho-Abreu • J. Valenzuela
Vector Molecular Biology Section, Laboratory of Malaria and Vector Research, National Institute of Allergy and Infectious Diseases, National Institutes of Health, Rockville, MD 20852, USA
e-mail: iliano.vieiracoutinhoabreugomes2@nih.gov; jvalenzuela@niaid.nih.gov

H. Hurd
Centre for Applied Entomology and Parasitology, School of Life Sciences, Keele University, Keele, Staffordshire ST5 5BG, UK
e-mail: h.hurd@keele.ac.uk

C.B. Marcondes (ed.), *Arthropod Borne Diseases*, DOI 10.1007/978-3-319-13884-8_4

Some insects became associated with vertebrate nests, feeding on detritus with their chewing mouthparts, and evolved to feed on discarded skin tissues, hairs, and feathers. Probably by accident, they fed on living and dead tissues of the host, including blood, evolved, and, due to advantages gained from this material, were selected for blood feeding. This evolution probably occurred in lice and fleas, and a future adaptation to blood feeding by some chewing lice ("Mallophaga"), some of them (e.g., *Menacanthus stramineus*) already ingesting blood and even causing damage to the skin of hosts, can be speculated.

Other insects, with mouthparts already adapted for sucking, previously feeding on plants or other insects, developed the capacity to pierce the skin, getting a much more nutritive meal. The evolution from plant feeding occurred in mosquitoes, sand flies, biting midges, and black flies, and blood feeding in bugs evolved from predation on other insects.

The blood feeding of ticks was probably developed in the late Cretaceous, with a major role of the evolution of this group taking place in Australia. A great radiation of ticks was probably related to that of mammals and birds at that period, but a multiple origin of blood feeding in ticks, starting by feeding on amphibians in Devonian times, has also been suggested (Mans and Neitz 2004).

Four insect orders include blood feeders: Phthiraptera (lice), Siphonaptera (fleas), Hemiptera (Heteroptera: Triatominae, Cimicidae, etc.) and Diptera (Ceratopogonidae, Culicidae, Phlebotominae, Simuliidae, Tabanidae and some Rhagionidae and Muscoid flies). Some moths (order Lepidoptera) feed on blood (Noctuidae: *Calyptra*), potentially transmitting pathogens (Plotkin and Goddard 2013).

Blood feeding is restricted to females in almost all hematophagous Diptera, to adults in blood-feeding flies and in all fleas and to all mobile stages in Heteroptera and lice. Larvae of a fly, *Auchmeromyia luteola*, suck blood from humans sleeping on the floor in some parts of Africa. All ticks (ca. 870 species in Acari: Ixodida) suck blood exclusively. Some myiasis-causing maggots probably evolved from bloodsucking larvae.

Besides damage directly and indirectly caused by blood feeding, some insects can cause annoyance and transmit diseases by feeding on tears, like some eye gnats (Chloropidae), *Musca autumnalis* (Muscidae) (see Chap. 31) and Lepidoptera (ca. 100 species in several animals and 23 on humans).

4.2 Mechanisms for Finding, Feeding, and Digesting Blood

Several bloodsucking insects have a delay period after emergence before the first blood meal, probably because time is necessary to mature the female reproductive system, and it is more secure to be off the host. When the insect is hungry, it develops an appetitive behavior, influenced by an internal programming, usually with a circadian rhythm. Advantages and disadvantages of feeding on diurnal or nocturnal period have been listed (Gibson and Torr 1999). If the insect is not near the potential host, it can wait for its arrival or actively look for the host.

Insects localize hosts by olfaction and vision, and resistance of hosts, by movement, aggression, and other forms of defensive behavior is very important for the choice of host. These defensive behaviors include head shaking, bill rubbing, and foot stamping seen in birds such as egrets and heron and the swishing of tails, flicking of ears, stamping, and shuddering of the skin observed among mammals. It has also been suggested that gathering in herds offers some protection from biting flies, at least to those animals in the center of the herd. The preference of *Anopheles gambiae* for human adults, compared to children, could be attributed to the size, but calves are preferred to adult cows by this mosquito. The low attraction of European *Anopheles* for humans has been considered an explanation for the low level of transmission of malaria in this continent (Lehane 2005).[1]

The knowledge of the body region preferred by hematophagous insects may be useful for

[1]A very informative and interesting book (CBM).

protection, and colors of clothes or skin can influence the attraction for some insects. For example, some biting flies prefer darker parts of cattle's bodies (see Chap. 31), and mosquitoes usually prefer to bite through black clothes.

The association of fleas to host nests is probably important for their preference for mammals, and 74 % of flea species are associated to rodents, unrivaled as home builders.

Mosquitoes, fleas, lice, and bloodsucking bugs, whose mouthparts are long and well adapted to localize blood vessels and suck directly from them or from pools formed by their rupture, are called *solenophagous*.

Other insects (*telmophagous*) are small[2] (black flies, biting midges, and sand flies) or ill-adapted, causing painful bites (Tabanidae), sucking blood from a small pool produced under the skin by the mouthparts. Several Muscid flies are opportunistic, sucking blood from wounds, sometimes disturbing and interrupting the feeding by other flies. The tip of the proboscis of *Musca crassirostris* is expanded, like that of *M. domestica*, and its small teeth are used for rasping the skin, sucking the liberated blood, whereas *M. autumnalis* rasps wounds, to get blood.

The distinction between feeding methods of insects is important for protection against bites. Clothes are useful for protection against telmophagous insects, but not necessarily against solenophagous ones, which bite through several kinds of cloth, or walk under clothes, in the case of fleas and lice.

Ticks are pool feeders, and to get the blood they must penetrate the skin and damage blood vessels. Slow feeders (Ixodidae) cause a lesion developing gradually to a hematoma, mostly produced by degranulation of neutrophils and inflammation (Ribeiro 1987), while fast feeders (nymphae and adults of Argasidae) penetrate deeply into the host skin, and in consequence the blood continues to flow long after the tick has stopped feeding (Binnington and Kemp 1980). These arthropods have a peculiar mode of feeding, absorbing, and digesting intracellularly parts of the blood and regurgitating other parts to the host, including previously obtained material from the gut. This has great importance for the transmission of some pathogens, like those of spotted fever and other diseases, and ticks must be removed as soon as possible from the skin.

4.3 Blood Feeding and Transmission of Disease

Transmission of parasites by insects can be *mechanical*, when there is no modification of the parasites in vectors, or *biological*, when multiplication and/or modification occur.

Due to the worldwide importance of biologically transmitted diseases, the mechanical transmission of parasites by arthropods has been overlooked. However, several important diseases are mechanically transmitted. Botulism, enteric diseases, neural lymphomatosis of the chickens (Marek's disease), and nosocomial diseases are examples of diseases transmitted mechanically from feces, host secretions, and other detritus and yaws, myxomatosis, and hepatitis B may be transmitted directly from a host to another by arthropods (Foil and Gorham 2004).

Efficiency of mechanical transmission depends on the quantity of parasites acquired by the vector, its behavior, and persistence of parasites on the vector and the quantity discharged in the following host. The mechanical transmission of blood parasites is much more efficient for horse flies, which can retain 10^{-7} ml of blood after biting, and frequently try to bite a host just after being interrupted, than by mosquitoes, which retain only 10^{-9} ml and usually give up the biting when interrupted. The possibility of mechanical transmission is strongly influenced by the quantity of parasites in the blood. For example, the low concentration of HIV in the blood precludes its transmission by all insects, but the higher concentration of several viruses (Myxoma virus, VEE, etc.) and bacteria (e.g., *Francisella tularensis*) permits their transmission even by horse flies and sometimes by mosquitoes (Foil and Gorham 2004).

[2]Since lice and fleas, not bigger than black flies, are solenophagous, body size is not decisive for the mode of imbibing blood from hosts, more determined by evolution.

In *propagative* transmission, virus and bacteria only reproduce (i.e., propagate) in the arthropod vector. The transmission of filarial worms, whose microfilariae develop but not reproduce in vectors, is *cyclo-developmental*. Protozoan parasites, having reproduction and modification in vectors, have a *cyclopropagative* transmission. More details on the types of transmission were included in Chap. 2.

4.3.1 Biological Transmission of Parasite via Blood Feeding

Diverse parasitic organisms have taken advantage of the habit of blood feeding to hitch a ride to their next host, passing into a vector in the blood meal and back to a new host during another feeding episode.

However, this transmission from vertebrate host to vector and back again is not a passive affair. The majority of these parasites engage in complex molecular and behavioral interactions with their vector that have been hypothesized to increase their chance of survival and completion of their life cycle (this concept and examples that may support it have been discussed in several reviews, e.g., Hurd 2003). In order for such a parasite to complete its life cycle, its vector must engage in a minimum of two blood feeding episodes: one to inadvertently acquire the parasite and a second one to pass it on to the next host. Between these two encounters, the parasite may pass through developmental stages and/or undergo sexual or asexual reproduction. This may occur within the gut lumen, as in the case of *Trypanosoma cruzi*, infecting triatomine bugs or *Leishmania* spp. infecting sand flies; in tissues such as the Malpighian tubules, for example, the filarial nematodes *Wuchereria bancrofti* and *Dirofilaria immitis*, or the thoracic muscles, e.g., or *Onchocerca volvulus* or, in the case of the malaria parasites, *Plasmodium* spp. beneath the basal lamina of the gut wall. In most cases, once infective, vector-transmitted parasites migrate to the salivary glands or mouth parts of their vector, ready for transmission when their vector next takes a meal. Transmission will only succeed if the vector survives this intrinsic period, but biting insects are vulnerable to attack from predators such as birds and spiders and those that oviposit in water have a high mortality at this stage. Most importantly, due to host defensive behavior, blood feeding is a risky strategy and is the most likely cause of adult mortality, for example, for mosquitoes (Edman and Scott 1987). Thus insects that take fewer blood meals and are more cautious during the period before the parasite is mature will make better vectors.

However, once a parasite is infective to the next host, the chance of transmission occurring will be increased if the vector takes more risks attempting to feed and feeds more often. Several studies have shown that the presence of a parasite changes the blood-feeding behavior of its vector in ways that should indeed make it a better vector. A look at the results of investigations of several aspects of the blood-feeding behavior of infected vectors will serve to illustrate these changes.

4.3.2 Obtaining Blood from an Infected Host

As previously outlined in this chapter, blood feeding consists of several discrete stereotyped behavioral steps that can be aborted at any stage (Hamilton and Hurd 2002). In addition to exogenous factors like light intensity and temperature, host factors including odor, visual clues, and temperature stimulate host-seeking behavior (Takken et al. 2001). If hosts infected with vector-transmitted parasites send out exaggerated signals, they are likely to be more attractive than uninfected neighbors, thus enhancing transmission prospects for the parasite, providing that the infective stage of the parasite is present in the blood. There is some evidence that this may be the case for malaria-infected hosts.

In the 1980s, it was shown that rats infected with a rodent malaria at the time of peak production of gametocytes (the stage that is infective to mosquitoes) were more attractive to the vector than uninfected animals (Day and Edman 1983) and similar findings were published two decades later, using a different rodent malaria parasite (Ferguson and Read 2004). Some evidence was also found to support the view that dogs infected with *Leishmania* and oxen infected with *Trypanosoma* were also more attractive than uninfected individuals (these and more examples have been reviewed (Clements 1999; Hamilton and Hurd 2002; Hurd 2010). Recent studies also showed that birds are more attractive to mosquitoes during the chronic stage of infection and this was so for both uninfected and malaria-infected mosquitoes and that people infected with the malaria parasite *Plasmodium vivax* are more attractive when gametocytes are present (Batista et al. 2014). However, none of these experiments conclusively demonstrate changes in host attractiveness that operates over long distances, nor do they demonstrate what the proximal cues for this attraction were. In humans, malaria infections cause a variety of symptoms that might attract mosquitoes including increased temperature during fever episodes, sweating that produces skin volatiles, anemia with reduced blood viscosity and makes blood feeding easier, and hyperventilation which would increase CO_2 output (Nacher 2005). The most conclusive study to date was conducted in the field in Kenya. A comparison of the number of mosquitoes attracted toward groups of children that were noninfected and infected with asexual (noninfective stages) or gametocyte stages demonstrated that those with gametocytes that were infective to mosquitoes were twice as attractive than the other groups and that increased attractiveness disappeared after antimalarial treatment (Lacroix et al. 2005). The design of these experiments leads to the conclusion that odor profiles may change in humans when malaria gametocytes are present in the blood. Interestingly, pregnant women are twice as attractive to mosquitoes possibly, due to increased heat and volatiles production, and are, probably consequentially, more prone to malaria infection (Ansell et al. 2002). It is possible that this behavior could be useful to the parasite as congenital transmission of malaria can occur.

There is also some evidence that the odor from hamsters infected with *Leishmania infantum* is more attractive to the sand fly vectors than uninfected ones when presented in an olfactometer (e.g., O'Shea et al. 2002), and preliminary work showed that this is also the case with mice infected with a rodent malaria, with attractiveness increasing as parasitemia increased (see Hamilton and Hurd 2002).

Once a mosquito has arrived in the vicinity of a prospective host, factors such as temperature and defensive behavior may influence host choice. We know of no evidence that infection-induced elevation in host temperature makes them more attractive to vectors; indeed, there is evidence to the contrary in the case of arbovirus-infected chickens, Rift Valley fever-infected lambs and malaria-infected mice (Clements 1999).[3] Lethargy caused by infection decreases host defensive behavior and thus reduces the risky aspect of blood feeding. There is some evidence that this does actually influence vector choice in favor of infected hosts, but studies have produced conflicting conclusions (Moore 1993).

Finally, the quality of the blood of infected host may be altered, and this, in turn, may affect the vector's decision to continue feeding, or the amount of blood taken up. Many chronic parasitic infections cause a reduction in host hematocrit which will affect the viscosity of the blood and hence how easily it will flow upfeeding stylets. For instance, mosquitoes feeding on mice infected with malaria have been shown to spend less time probing and therefore feed more quickly. During the initial period of mouse infection, mosquitoes have also been shown to imbibe more blood, although as parasitemia increases and the hematocrit falls further, more time was spent trying to obtain enough red blood cells and when mice were severely anemic poorer quality meals were obtained (Taylor and Hurd 2001). Interestingly it has been proposed that tsetse fly feeding is enhanced on oxen infected with *Trypanosoma congolense* because the parasites attach to the wall of the microvasculature and cause vasodilatation. The greater the quantity of blood imbibed and the shorter the time it takes to obtain a blood meal will benefit the vector and

[3] See especially chapter 7 (Blood-feeding and its epidemiological significance, pp. 677—769).

the parasite as there is less likelihood of the vector being swatted by the host if it feeds quickly. It will also increase the chances that the vector becomes infected as more parasites may be imbibed in a larger blood meal and thus imbibing a large blood meal also benefits the parasite.

4.3.3 Pathogen Development in the Vector

The insect midgut is the only part of the alimentary tract not protected by a chitinous cuticle, and it is the organ where food digestion and absorption take place. This portion of the digestive tract of most insects, as well as other invertebrates, secretes a peritrophic matrix (PM), which is a noncellular semipermeable layer separating the contents of the gut lumen from the digestive epithelial cells. By compartmentalizing the midgut lumen, the PM regulates the contact of molecules with the microvilli and also allows the recycling of digestive enzymes. Thereby, the PM protects the midgut epithelium from abrasion, toxic compounds, and pathogens.

Although the insect PM is an important line of defense against pathogen infection, some pathogens can bypass this barrier by taking advantage of distinct mechanisms. Some pathogens escape from entrapment within the PM by reaching the midgut epithelium before PM synthesis or after PM breakdown (Pimenta et al. 1997). Most viruses and microfilaria are capable of attaching to and/or crossing the midgut epithelium before secretion of the PM whereas *Leishmania* survives in the endoperitrophic space until the sand fly PM breaks down. On the other hand, other pathogens can escape from PM entrapment by secreting enzymes capable of digesting structural components of the PM. *Plasmodium*, the causative agent of malaria, disrupts the PM scaffolding by secreting a chitinase: an enzyme that cleave PM chitin. This allows the parasites to penetrate and cross the mosquito PM.

During the digestive process, pathogens are also exposed to the action of harmful proteolytic enzymes. It is noteworthy that parasites can manipulate the expression of digestive enzymes, as seen for *Leishmania* sand fly pairs. The expression profile of sand fly midgut proteases was lower in *Leishmania*-infected midguts than in uninfected midguts. The expression profile of other molecules such as microvilli like proteins was also affected by the presence of the parasite in the gut of the insect; nevertheless, its significance for vector competence has yet to be demonstrated. The assays, on the other hand, showing that knocking down the expression of a trypsin gene in sand flies led to increased *Leishmania* load conclusively demonstrates that digestive enzymes are harmful to parasites (Sant'anna et al. 2009). Despite this barrier, some parasites have ways to evade or even take advantage to the proteolytic system so as to develop in their insect vectors.

Leishmania is taken up in the blood as amastigotes and develop into the promastigote forms in the sand fly midgut. Although both of these forms are resistant to the midgut proteolytic activity, the transitional stage between such forms is susceptible to these enzymes. Intriguingly, the *Leishmania* transitional stage survives proteolysis due to the presence of the PM (Pimenta et al. 1997). This structure limits the flow of proteolytic enzymes within the PM, reducing the *Leishmania* transitional stage contact with the harmful proteases, and allowing *Leishmania* survival (Coutinho-Abreu et al. 2013). Thereby, the sand fly PM protects *Leishmania* development early on, but it acts as a barrier later on in digestion as the parasite needs to escape from the PM so as not to be eliminated along with the feces (Pimenta et al. 1997; Coutinho-Abreu et al. 2010, 2013). As aforementioned, *Plasmodium* secretes a chitinase that allows its penetration and escape from the PM. In fact, this chitinase is secreted as an inactive enzyme that is activated by the mosquito midgut proteases. Hence, *Plasmodium* takes advantage of the midgut proteolytic system to escape from the PM and develop in the mosquito vector.

Upon escaping from the endoperitrophic space, parasites take different routes to complete their development. Most viruses and *Plasmodium* parasites cross the midgut epithelium to reach and invade the salivary glands so as to be inoculated into the host skin along with saliva. For *Plasmodium* parasites, mosquito midgut invasion triggers production of reactive oxygen (ROS) and nitrogen oxygen (NOS) species, such as nitric oxide (NO) by inducible nitric oxide synthase (iNOS), which are important effector molecules against pathogens. NO can be transformed in other free radicals via spontaneous or enzymatic reaction, and these radicals also can react with endogenous structures, causing damage and apoptosis. Mosquito midgut cells take advantage of this dual role of ROS/NOS as a strategy to eliminate *Plasmodium* parasites. Before invading midgut cells, *Plasmodium* is not detected by iNOS due to the cross-linking of microvilli proteins by the dual oxidase enzyme (Kumar et al. 2010). This limits the influence of immune elicitors in the midgut epithelium (Kumar et al. 2010). Parasite invasion of mosquito midgut cells triggers NO production by iNOS and cell death, killing and eliminating *Plasmodium* along with the dead cell. Protein nitration seems to be the main mechanism used by midgut epithelium cells to kill *Plasmodium*, as the expression of enzymes that share hydrogen peroxide (H_2O_2) as substrate (peroxidase and catalase) in the midgut are fine-tuned in favor of the peroxidase-dependent protein nitration pathway. At the midgut basal lamina, *Plasmodium* oocysts are exposed to the hemolymph complement-like system, which is the most potent immune defense mechanism of anopheles mosquitoes against plasmodium parasites. The three main components of the complement system (TEP1, APL1C, and LRM1) opsonize and kill the oocysts, leading to either parasite lysis or melanization (Blandin et al. 2009). Nonetheless, some strains of *Plasmodium* bear evasion mechanisms that prompt parasite escaping to adjacent cells or to the basal lamina so as to resume development and migration toward the salivary glands.

Leishmania, on the other hand, uses a surface-expressed lipopolysaccharide (LPG) to anchor itself onto protein receptors present on the sand fly epithelium microvilli, for instance, the *P. papatasi* galectin (PpGalec), so as not to be flushed away along with the fecal pellets (Kamhawi et al. 2004). Once the digested blood is defecated, the parasites migrate forward to the anterior midgut and differentiate in the metacyclic infective forms. In the anterior midgut, the chitinous lining of the sand fly stomodeal valve seems to be degraded probably by a *Leishmania*-secreted chitinase. Due to the damages, the stomodeal valve-mediated bloodsucking mechanism may be disrupted, leading to parasite regurgitation at the biting site.

4.3.4 Importance of Saliva on the Transmission of Parasites, Immunity and Vaccines

Once a blood-feeding insect has selected and landed in the host and starts probing to get a blood meal it faces a major challenge, the host hemostatic system, which gets immediately activated to prevent blood loss. As mammals have evolved this mechanism of protection, insects have been able to outsmart the host by developing mechanisms to counteract this hemostatic response. Insects have in their saliva very potent pharmacologically active components that can disarm the host's blood coagulation cascade, vasoconstriction and platelet aggregation, and inherent redundancy of the hemostatic system. The presence of anticoagulants, vasodilators, and inhibitors of platelet aggregation in various blood-feeding insects including relevant vectors of disease has been well documented (Ribeiro and Francischetti 2003). In terms of anticoagulants, the presence of a Factor Xa inhibitor was first described in the saliva of the mosquito *Aedes aegypti*, this salivary anticoagulant protein belongs to the serpin family of serine protease inhibitors. Interestingly, in *Anopheles* mosquitoes, a different type of anticoagulant exist, this salivary anticoagulant named *anophelin,* first identified in the saliva of *Anopheles albimanus,* is a small and novel inhibitor of thrombin, a key enzyme of the blood coagulation cascade. The salivary anticoagulant from sand flies was recently identified, and it is a novel salivary protein named lufaxin that inhibits specifically Factor Xa, and it has no homology to other inhibitors found in blood feeders (Collin et al. 2012).

Platelet aggregation represents one of the first responses of the hemostatic system and therefore a major roadblock for blood feeders. To counteract platelets, the saliva of blood-feeding insects has a salivary apyrase, an enzyme that destroys ADP, an important agonist of platelet aggregation. Having an apyrase in their saliva, blood-feeding arthropods can prevent platelet aggregation and all the downstream effects induced by platelets and its derived factors. All blood-feeding insects studied to date have apyrase in their saliva; however, the protein responsible for this activity is different in the different insects.

Besides classical anti-hemostatic components (anticoagulant, antiplatelet, and vasodilator), blood-feeding insects were recently reported to have a variety of molecules acting in other key host molecules and processes (Calvo et al. 2009). The D7 protein from the mosquito *Anopheles gambiae* was shown to bind biogenic amines a function never described in insect's salivary proteins, furthermore, the D7 protein from the mosquito *Aedes aegypti* was shown to bind cysteinyl leukotrienes with high affinity. Another set of new molecules recently described are the yellow related proteins from the saliva in sand flies. These abundantly expressed salivary proteins were recently shown to bind biogenic amines including norepinephrine and serotonin.

In addition to the roles in hemostasis and inflammation, salivary proteins from different insects have been shown to act in the host immune system by altering the response of key cells of the immune system including macrophages and T cells (Abdeladhim et al. 2014). The cumulative effect of salivary anti-hemostatic and immunomodulatory activities has been shown in some systems to have a profound effect on the pathogen that is transmitted by the vector. The first evidence came from work with salivary glands of sand flies, in this work the saliva of sand flies was shown to enhance *Leishmania* infection and to exacerbate disease outcome. Similar effects have been shown with the saliva of mosquitoes and the virus they transmit.

One of the most interesting observations regarding insect saliva in parasite transmission, specifically from sand fly saliva, is that if an animal produces an immune response to sand fly salivary proteins, these animals become protected against leishmaniasis. These finding prompt the concept that salivary proteins can work as a vaccine or a component as a vaccine against the parasite the insect is transmitting. Important to note, these salivary proteins do not cross react with *Leishmania* proteins, and it has been proposed that the protective mechanism is related to a cellular immune response induced by the sand fly salivary proteins that produces an inhospitable site for the parasite and also creates an immunological environment that favors the induction of a protective immunity against the parasite. Various salivary vaccine candidates from the saliva of sand flies have been identified including the salivary protein PpSP15 from *Phlebotomus papatasi* (Valenzuela et al. 2001) that protects rodents against cutaneous leishmaniasis induced by *L. major*, LJM19 from *Lutzomyia longipalpis* that protected hamsters against visceral leishmaniasis induced by *L. infantum*, LJM17 and LJL143 from *Lutzomyia longipalpis* that produced a protective immune response against *L. infantum* in dogs, and Linb-11 from *Nyssomyia intermedia* that protected rodents against cutaneous leishmaniasis induced by *L. braziliensis*. Importantly, it was recently shown that sand fly salivary proteins can induce an immune response in humans living in endemic areas, and that this immune response is similar to the protective immune response observed in animal models (Oliveira et al. 2013).

4.3.5 Passing a Parasite Back to Another Host

Once a parasite has been imbibed by the correct vector, it must complete the next developmental stage before its life cycle is completed by transmission to a new host. Investigations of both infected mosquitoes and sand flies have shown that various aspects of the feeding behavior of infected vectors changes as the infection progresses (Hamilton and Hurd 2002; Hurd 2010).

Malaria parasites must survive the intrinsic period during which fertilization of the female gametocytes takes place in the gut lumen, and the resulting ookinete traverses the midgut epithelium to form an oocyst on the inner midgut wall. The oocyst grows and divides asexually to produce thousands of sporozoites that migrate to the salivary glands. During this period (usually 10–14 days), mosquitoes are likely to complete at least three gonotrophic cycles, each one involving at least one blood-feeding attempt and resulting in the production and oviposition of a batch of eggs. Transmission success for the malaria parasite is thus dependent upon the infected mosquito surviving host contact associated with blood feeding, the hazards of oviposition in water bodies and other dangers such as predation, at least until sporozoites are in the salivary glands. Of all of these hazards, blood feeding is likely to be the most dangerous. As a result of modeling different scenarios for mosquito survivorship, it was predicted that a 20% decrease in feeding mortality could result in a 60% increase in force of infection. Any decrease in feeding behavior of infected mosquitoes during the developmental phase of the infection is thus very likely to increase the chances of transmission; as would an increase in blood feeding once sporozoites are in the salivary glands ready to be injected into the next host with the saliva.

A recent study of mosquitoes infected with sporozoites of a human malaria species did produce data that showed that human odor was significantly more attractive to these mosquitoes than to uninfected ones (Smallegange et al. 2013). In a more detailed study (Cator et al. 2013), attention was focused upon the foraging behavior of mosquitoes infected with rodent malaria and attraction to host odors and to host location was measured. Odor-mediated responses are controlled by the olfactory system with odor receptors present in the maxillary palps. Electropalpograms were recorded during oocyst infection and when sporozoites were present in the hemolymph or the salivary glands. To do this these investigators measured the electrophysiological responses to a known stimulant, 1-octen-3-ol, that was blown over the palps. Compared with the uninfected control group, the responsiveness of maxillary palps was significantly decreased when oocysts were developing but increased significantly at the sporozoite stage. This suggested a decrease in host-seeking behavior would take place during the parasite's intrinsic period, and an increase in foraging would occur once the parasite was infective to the next host.

Results from long-range host-seeking tests, where the attraction to a human positioned at the end of a large enclosure was monitored, paralleled results from the electropalpograms. Thus, the prediction that changes in olfactory responses would alter host seeking was confirmed. Unexpectedly, these changes occurred whether mosquitoes that had been exposed to infected blood actually became infected or not. When mosquitoes were then immune challenged by injecting them with heat-killed *Escherichia coli*, a similar change in the responsiveness of maxillary palps to the odorant occurred, although the magnitude of the changes was less than those caused by malaria infection. These intriguing results suggest that the observed changes in mosquito-feeding behavior are not specific to malaria infection and may thus represent part of the mosquitoes' reaction to infection, rather than a manipulation by the parasite. This is discussed in more detail below.

Once an infected mosquito has located a potential host, it must evade the host's defenses. It has been shown that the persistence in attempting to feed in the face of a defensive response is also downregulated when oocysts are present, and the reverse happens when sporozoites have reached the salivary glands. These experiments were performed using mosquitoes infected with rodent malaria and a human arm as the potential food source. Avian oocyst infections have also been shown to lower the threshold value for blood meal size, making mosquitoes less likely to refeed if interrupted, whereas the threshold value was increased in sporozoite-infected mosquitoes giving rise to increased foraging if a meal was interrupted. Increased feeding persistence was also observed when blood-feeding sand flies containing the infective stage of *Leishmania* were interrupted.

Once a host has been selected, the vector will engage in probing or biting activity, usually spitting saliva into the wound. In the case of many vector-transmitted parasites, this is the feeding phase when transmission back to the host begins. If a vector makes several attempts to probe or probes more than one host before completing its blood meal, there is more likelihood that transmission will occur. There are several examples of vector insects that exhibit changes in probing and/or biting behavior once they are infected with transmissible stages of a parasite.

4.4 The Manipulation Hypothesis

The alterations in aspects of vector-feeding behavior discussed above have been interpreted as examples of adaptive manipulation on behalf of the parasite. All of these changes would appear to facilitate the transmission prospects of the parasite, either to its vector or back again to its vertebrate host. Indeed, multiple feeding episodes that should give rise to multiple infections from one vector have, on occasion, been demonstrated. However, great care must be taken if we are to interpret these examples as true manipulation.

The criteria required to fulfill such an explanation are rigorous (e.g., Poulin 2010) and require the discovery of a mechanism for manipulation that involves a product of the parasite genome (known as a manipulator molecule) that is involved in changing the mosquito phenotype. Facilitated transmission could, in fact, occur as a result of the response of the host or vector to the presence of the parasite, for example, as a by-product of an immune response to the parasite or mechanical damage caused by it. Much more research is required, particularly at the vector stage of parasite life cycles, to determine what causes changes in responses to odors, feeding persistence, biting, and probing before we can understand the evolutionary pressures that are driving enhancement of vector transmission (Lefèvre et al. 2006). The recent analysis of the changes in the head proteome of infected mosquitoes and tsetse flies is a good start in this direction (Lefèvre et al. 2007a, b), as is an analysis of the salivary gland proteins in a malaria-infected mosquito (Shandilya et al. 1999). Whatever the cause, a better understanding of when and how infection influences the blood-feeding behavior of vector insects will increase our understanding of transmission ecology and better inform us for the use of control strategies such as insecticide treated bed nets or the management of infected people carrying gametocytes (Cator et al. 2012).

References

Abdeladhim M, Kamhawi S, Valenzuela JG (2014) What's behind a sand fly bite? The profound effect of sand fly saliva on host hemostasis, inflammation and immunity. Infect Genet Evol 28:691–703

Ansell J, Hamilton KA, Pinder M et al (2002) Short-range attractiveness of pregnant women to *Anopheles gambiae* mosquitoes. Trans R Soc Trop Med Hyg 96:113–116

Balashov YI (1984) Arthropods and their hosts, and its influence on vector potential. Annu Rev Entomol 29:137–156

Batista PA, Costa EFM, Silva AA (2014) *Anopheles darlingi* (Diptera: Culicidae) displays increased attractiveness to infected individuals with *Plasmodium vivax* gametocytes. Parasit Vectors 7:251

Binnington KC, Kemp DH (1980) Role of tick salivary glands in feeding and disease transmission. Adv Parasitol 18:315–339

Blandin SA, Wang-Sattler R, Lamacchia M et al (2009) Dissecting the genetic basis of resistance to malaria parasites in *Anopheles gambiae*. Science 326: 147–150

Calvo E, Mans BJ, Ribeiro JM et al (2009) Multifunctionality and mechanism of ligand binding in a mosquito antiinflammatory protein. Proc Natl Acad Sci 106:3728–3733

Cator LJ, George J, Blanford S et al (2013) 'Manipulation' without the parasite: altered feeding behaviour of mosquitoes is not dependent on infection with malaria parasites. Proc R Soc B 280(1763):20130711

Cator LJ, Lynch PA, Read AF et al (2012) Do malaria parasites manipulate mosquitoes? Trends Parasitol 28:466–470

Clements AN (1999) The biology of mosquitoes, vol 2. CABI Publishing, Oxford

Collin N, Assumpção TC, Mizurini DM et al (2012) Lufaxin, a novel factor Xa inhibitor from the salivary gland of the sand fly *Lutzomyia longipalpis* blocks protease-activated receptor 2 activation and inhibits inflammation and thrombosis *in vivo*. Arterioscler Thromb Vasc Biol 32:2185–2198

Coutinho-Abreu IV, Sharma NK, Robles-Murguia M et al (2013) Characterization of *Phlebotomus papatasi* peritrophins, and the role of PpPer1 in *Leishmania major* survival in its natural vector. PLoS Negl Trop Dis 7, e2132

Coutinho-Abreu IV, Sharma NK, Robles-Murguia M et al (2010) Targeting the midgut secreted PpChit1 reduces *Leishmania major* development in its natural vector, the sand fly *Phlebotomus papatasi*. PLoS Negl Trop Dis 4, e901

Day JF, Edman JD (1983) Malaria renders mice susceptible to mosquito feeding when gametocytes are most infective. J Parasitol 69:163–170

Edman JD, Scott TW (1987) Host defensive behaviour and the feeding success of mosquitoes. Insect Sci Appl 8:617–622

Ferguson HM, Read AF (2002) Why is the effect of malaria parasites on mosquito survival still unresolved? Trends Parasitol 18:256–261

Ferguson HM, Read AF (2004) Mosquito appetite for blood is stimulated by *Plasmodium chabaudi* infections in themselves and their vertebrate hosts. Malar J 3:12

Foil LD, Gorham JR (2004) Mechanical transmission of disease agents by arthropods. In: Eldridge BF, Edman JD (eds) Medical entomology: a textbook on public health and veterinary problems caused by arthropods. Kluwer Academic Publ, Dordrecht/Boston/London, pp 461–514

Gibson G, Torr SJ (1999) Visual and olfactory responses of haematophagous Diptera to host stimulus. Med Vet Entomol 13:2–23

Hamilton JCG, Hurd H (2002) Parasite manipulation of vector behaviour. In: Lewis EE, Cambell JF, Sukhdeo MVK (eds) The behavioural ecology of parasites. CAB International, London, pp 259–281

Hurd H (2003) Manipulation of medically important insect vectors by their parasites. Annu Rev Entomol 48:141–161

Hurd H (2010) Parasite-mediated enhancement of transmission by haematophagous insects. In: Takken W, Knols BGJ (eds) Ecology and control of vector-borne diseases: olfaction in vector-host interactions. Academic Publishers, Wageningen, pp 349–364

Kamhawi S, Ramalho-Ortigao M, Pham VM et al (2004) A role for insect galectins in parasite survival. Cell 119:329–341

Kumar S, Molina-Cruz A, Gupta L et al (2010) A peroxidase/dual oxidase system modulates midgut epithelial immunity in *Anopheles gambiae*. Science 327: 1644–1648

Lacroix R, Mukabana WR, Gouagna LC et al (2005) Malaria infection increases attractiveness of humans to mosquitoes. PLOS Biol 3:e298

LefèvreT KJC, Renaud F et al (2006) New prospects for research on manipulation of insect vectors by pathogens. PLoS Pathog 2:633–635

Lefèvre T, Thomas F, Ravel S et al (2007a) *Trypanosoma brucei brucei* induces alteration in the head proteome of the tsetse fly vector *Glossina palpalis gambiensis*. Insect Mol Biol 16:651–660

Lefèvre T, Thomas F, Schwartz A et al (2007b) Malaria *Plasmodium* agent induces alteration in the head proteome of their *Anopheles* mosquito host. Proteomics 7:1908–1915

Lehane MJ (2005) The biology of blood-sucking in insects. Cambridge University Press, Cambridge

Mans BJ, Neitz AWH (2004) Adaptation of ticks to a blood-feeding environment: evolution from a functional perspective. Insect Biochem Mol Biol 34:1–7

Moore J (1993) Parasites and the behavior of biting flies. J Parasitol 79:1–16

Nacher M (2005) Charming the mosquito: do malaria symptoms increase the attractiveness of the host for the vector? Med Hypotheses 64:788–791

Oliveira F, Traoré B, Gomes R et al (2013) Delayed-type hypersensitivity to sand fly saliva in humans from a leishmaniasis-endemic area of Mali is Th1-mediated and persists to midlife. J Invest Dermatol 133: 452–459

O'Shea B, Rebollar-Tellez E, Ward RD et al (2002) Enhanced sandfly attraction to *Leishmania*-infected hosts. Trans R Soc Trop Med Hyg 96:117–118

Pimenta PF, Modi GB, Pereira ST et al (1997) A novel role for the peritrophic matrix in protecting *Leishmania* from the hydrolytic activities of the sand fly midgut. Parasitology 115:359–369

Plotkin D, Goddard J (2013) Blood, sweat, and tears: a review of the hematophagous, sudophagous, and lachryphagous Lepidoptera. J Vector Ecol 38:289–294

Poulin R (2010) Parasite manipulation of host behavior: an update and frequently asked questions. Adv Study Behav 41:151–186

Ribeiro JMC (1987) Role of saliva in blood-feeding by arthropods. Annu Rev Entomol 32:463–478

Ribeiro JMC (1995) Blood-feeding arthropods: live syringes or invertebrate pharmacologists? Infect Agents Dis 4:143–152

Ribeiro JMC, Francischetti IMB (2003) Role of arthropod saliva in blood feeding: sialome and post-sialome perspectives. Annu Rev Entomol 48:73–88

Sant'anna MR, Diaz-Albiter H, Mubaraki M et al (2009) Inhibition of trypsin expression in *Lutzomyia longipalpis* using RNAi enhances the survival of *Leishmania*. Parasit Vectors 2:62

Shandilya S, Gakhar SK, Adak T (1999) *Plasmodium* infection-induced changes in salivary gland proteins of the malaria vector *Anopheles stephensi* (Diptera: Culicidae). Jap J Inf Dis 52:214–216

Smallegange RC, van Gemert GJ, van de Vegte-Bolmer M et al (2013) Malaria infected mosquitoes express enhanced attraction to human odor. Plos One 8, e63602

Takken W, van Loon JJA, Adam W (2001) Inhibition of host-seeking response and olfactory responsiveness in *Anopheles gambiae* following blood feeding. J Insect Physiol 47:303–310

Taylor PJ, Hurd H (2001) The influence of host haematocrit on the blood feeding success of *Anopheles stephensi*: implications for enhanced malaria transmission. Parasitology 122:491–496

Valenzuela JG, Belkaid Y, Garfield MK et al (2001) Toward a defined anti-*Leishmania* vaccine targeting vector antigens: characterization of a protective salivary protein. J Exp Med 194:331–342

Part II

Diseases Transmitted by Arthropods

General Aspects on Arboviruses

5

Marta S. Contigiani, Luis A. Diaz,
and Lorena I. Spinsanti

Abstract

Arboviruses do not represent a based-related phylogenetic group, but they are all transmitted by arthropods. Fifty arboviruses pathogenic for animals (including humans) have been reported, belonging to the families *Asfarviridae*, *Bunyaviridae*, *Flaviviridae*, *Orthomyxoviridae*, *Rhabdoviridae*, *Reoviridae*, and *Togaviridae*. A wide variety of hematophagous arthropods transmits the arboviruses: biting midges, cimicid bugs, mosquitoes, sand flies, and ticks. The vector acquires the infection by blood feeding from a viremic host and transmits the virus to a new host by the oral route (inoculating infected saliva). However, vertical transmission routes already exist (transovarial, venereal). Arbovirus can be amplified by a diverse range of vertebrate hosts including birds, rodents, equines, humans, and monkeys. Through biological evolution and cultural development, human beings were able to modify the environments according to their needs. Thus, deforestation has produced new areas for agriculture, livestock, farming activities, and urbanization. These anthropogenic activities have produced great changes to host and vector communities and population abundance, sometimes driving emergence and reemergence of arboviruses. In this chapter, we give a general view for most important aspects of arboviruses, their classification, transmission and maintenance mechanisms, ecology, and emergence process.

Keywords

Arbovirus • Vector • Hosts • Transmission networks • Emergence • Environmental changes

M.S. Contigiani (✉) • L.I. Spinsanti
Arbovirus Laboratory, Institute of Virology
"Dr. J. M. Vanella", National University of Córdoba,
Enfermera Gordillo Gomez S/N, Ciudad
Universitaria, Córdoba, Argentina
e-mail: martascontigia@hotmail.com;
l_spinsanti@yahoo.com.ar

L.A. Diaz
Arbovirus Laboratory, Institute of Virology
"Dr. J. M. Vanella", National University of Córdoba,
Enfermera Gordillo Gomez S/N, Ciudad
Universitaria, Córdoba, Argentina

Institute for Biological and Technical Researches,
CONICET-UNC, Córdoba, Argentina
e-mail: adrian.diaz@conicet.gov.ar

C.B. Marcondes (ed.), *Arthropod Borne Diseases*, DOI 10.1007/978-3-319-13884-8_5

Arboviruses[1] do not represent a based-related phylogenetic group, but they are all transmitted by arthropods (arboviruses = arthropod-borne viruses). A total of 50 arboviruses pathogenic for animals (including humans) have been reported, belonging to the families *Asfarviridae* (genus *Asfivirus*: tick-borne African swine fever virus), *Bunyaviridae* (tick-borne Nairobi sheep disease virus, Crimean-Congo hemorrhagic fever virus (CCHFV); mosquito-borne Rift Valley fever virus (RVFV), La Crosse virus (LACV), snowshoe hare virus, and Cache Valley virus (CVV); biting midges-borne Aino, Akabane, Main Drain, Shuni, and Schmallenberg viruses), *Flaviviridae* (genus *Flavivirus*: mosquito-borne Dengue virus (DENV), Japanese encephalitis virus (JEV), Murray Valley encephalitis virus (MVEV), St. Louis encephalitis virus (SLEV), Tembusu virus, Usutu virus (USUV), Yellow fever virus(YFV), West Nile virus(WNV), Zika virus (ZIKV); tick-borne encephalitis, Omsk hemorrhagic fever, Kyasanur Forest disease, and Tyuleniy viruses), *Orthomyxoviridae* (tick-borne Thogoto virus), *Rhabdoviridae* (sandfly/mosquito-borne bovine ephemeral fever, vesicular stomatitis-Indiana, vesicular stomatitis-New Jersey, vesicular stomatitis-Alagoas, and coccal viruses), *Reoviridae* (biting midges-borne African horse sickness, Kasba, bluetongue, epizootic hemorrhagic disease of deer, Ibaraki, equine encephalitis, Peruvian horse sickness, and Yunnan viruses), and *Togaviridae* (genus *Alphavirus*: Chikungunya virus (CHKV); Eastern, Western, and Venezuelan equine encephalitis viruses (EEEV, WEEV, VEEV); Getah, Mayaro, Middelburg, Sindbis, and Semliki Forest viruses). Excepting *Asfivirus* (African swine fever virus), all arboviruses are enveloped RNA viruses (Hubálek et al. 2014).

[1]The word "arbovirus" (originally "arborvirus") was created by illustrious pioneer William C. Reeves, as described in his nice review (Reeves 2001) (CBM).

5.1 Biology of Transmission

Transmission of arboviruses from a viremic to a susceptible vertebrate host is based on biological transmission through the bite of an infected arthropod. Biological transmission requires an active viral replication in the arthropod vector (Kuno and Chang 2005). Although this is the main transmission pathway, there are other alternative ways such as host-to-host transmission (Rift Valley encephalitis virus) (Daubney et al. 1931), vector to vector transmission (transovarial and venereal transmission—La Crosse virus) (Hughes et al. 2006), and mechanic transmission (*Bluetongue virus*) (EFSA 2011).

5.1.1 Arbovirus Infection in the Vector

A wide variety of hematophagous arthropods transmits the arboviruses: biting midges (Ceratopogonidae), cimicid bugs (Cimicidae), mosquitoes (Culicidae), sand flies (Psychodidae, Phlebotominae), and ticks (Ixodidae and Argasidae) (Hubálek et al. 2014). The arthropod vector acquires the virus from a viremic host during its feeding process (Weaver and Barrett 2004). After that, the virus infects the posterior midgut cells and replicate. Then the virus disseminates elsewhere through hemolymph and neurons to other organs. Finally, the virus reaches the salivary glands where it produces infection and replicates (Fig. 5.1). Along with the saliva, virus can get into the new susceptible vertebrate host and start the transmission cycle again. Throughout this process, virus must pass over several infections and escape barriers (midgut cell infection, mesenteronal escape, salivary gland) (Kenney and Brault 2014).

The time spanned from viral infection and viral transmission is known as *extrinsic incubation period* and is strongly influenced by temperature: the higher the temperatures, the shorter extrinsic incubation period. Not all the ingested viruses can reach the salivary glands. Therefore, a minimal viral load is needed to produce a successful viral infection in the vector. This is called

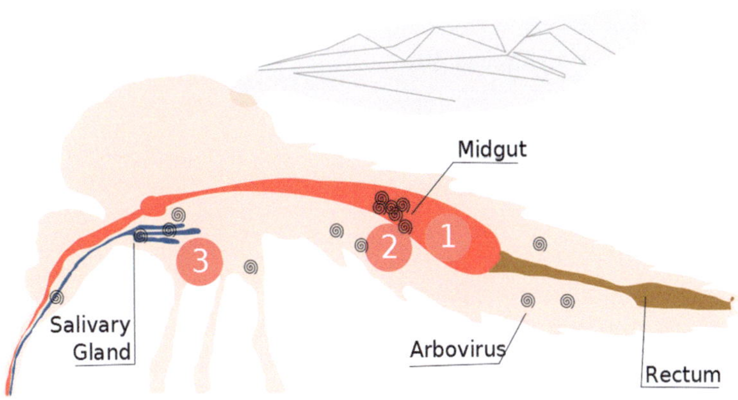

Fig. 5.1 Arbovirus infection process in the arthropod vector. Arboviruses get into the arthropod through digestive tube. Viral particles attached to the midgut membrane infecting those cells (*1*: midgut infection barrier). After replication, viral particles must escape through basal lamina of midgut epithelium (*2*: midgut escape barrier). Outside the midgut, viral particles spread through the entire vector organism. Arbovirus must infect salivary gland (*3*) in order to get out the vector and infect an amplification host

minimal infection threshold (MIT). It is assumed that high viremias in host are related to high MIT in vectors and vice versa. The intrinsic ability to transmit an arbovirus is genetically determined and it is known as *vector competence*. On the other hand, when we consider environmental and ecological factors that influence vector population dynamics, we talk about *vectorial capacity* (Higgs and Beatty 2005). The vectorial capacity (VC) can be defined and quantified as the average number of potentially infective bites that will ultimately be delivered by all the vectors feeding on a single host in 1 day. Basically, VC is the product of three components: feeding, survival rate, and length of the extrinsic incubation period. The vector capacity of populations is therefore a dynamic, fluctuating process that should be estimated in the context of a particular population at a particular point in time.[2]

5.1.2 Arbovirus Infection in Host

The hosts are in charge of amplifying the arbovirus developing viremias, making them high enough to infect vector arthropods. This viremia must be higher than the MIT for the particular vector. Mammals and birds are the main vertebrate hosts for most of the arboviruses (Scott 1988). For example, birds are main amplifiers for Japanese encephalitis, Saint Louis encephalitis, and West Nile, Eastern, and Western equine encephalitis viruses (McLean and Bowen 1980; McLean et al. 2001; Komar et al. 2003; Hassan et al. 2003; Komar et al. 1999; Reisen et al. 2003). However, rodents are suitable hosts for Venezuelan Equine encephalitis virus and monkeys for yellow fever in wild areas (Weaver et al. 2004b; Barrett and Monath 2003).

It is hypothesized that small mammals (e.g., rodents) and birds are more effective hosts for viral maintenance since their life cycles are short, ensuring that a new generation of susceptible individuals will be available at least once a year. Due to the high fertility rate of these animals, a great number of potential reservoirs are always available. Therefore, transmission and amplification are not going to be disrupted by a high prevalence of antibodies in the host population. By contrast, large mammals are used as hosts for viruses whose vectors have long life cycles, such as ticks (Potosi virus serogroup) (Karabatsos 1985).

For most arboviruses (SLEV, USUV, WNV, JEV, EEEV, VEEV, WEEV), human beings are dead-end hosts, which means that viremias are not high enough to infect the arthropod vector. Therefore, humans are not necessary for virus maintenance and represent just an accident during

[2] These concepts are more thoroughly analyzed in Chap. 2.

the biological transmission among vectors and hosts. However, CHIKV, DENV, YFV, and ZIKV are exceptions, given that these viruses can replicate and generate viremia titers in the human host high enough to infect vector mosquitoes (Morris 1988; Scott 1988; Reisen and Monath 1989).

The arboviruses cause an acute self-limited infection after a few days of viremia, and the host immune system clears the viral infection producing neutralizing antibodies. These antibodies will protect the vertebrate from future infections (memory cells). Most of arbovirus infections in hosts are unapparent but sometimes can manifest clinical signs, like encephalitis (WNV, SLEV, JEV, ZIKV), abortions (BUN, *Bluetongue virus* (BTV), and Schmallenberg (SBV) viruses), and hemorrhages (DENV, YFV, TBEV, CCHFV) (Solomon and Mallewa 2001; Labuda and Nuttall 2004; Maclachlan 2011; Turtle et al. 2012; Oncü 2013; Wernike et al. 2014).

The old paradigm that good viral reservoirs are not affected by infection is not completely true for the arbovirus system. For example, corvids amplify WNV but are highly susceptible to its infection. Before dying, they produce high viremia able to infect a bunch of mosquito vectors. Moreover, ill birds lose their antimosquito bite behavior, increasing mosquito bites chances (Reisen et al. 2006). Another example, Rift Valley fever virus, produces high mortality rate in calves, and VEEV epizootic subtypes produce diseases and death in equines; in both viruses, the last are also amplifier hosts (Weaver et al. 2004b; Antonis et al. 2013).

5.2 Biology of Maintenance (Network Figures)

Several intrinsic and extrinsic requirements (physiological/behavioral/ecological/environmental) must be fulfilled for species to be considered a vector or a host (Table 5.1). Ecosystems are inherently variable across time and space. The intrinsic characteristics are not modified by time; however, there might be exceptions, such as certain selective processes that affect the population, determining susceptibilities to the differential infection among hosts and vectors, e.g., environmental stress and detrimental nutrition affect both vector and host competence (Kramer and Ebel 2003; Reisen et al. 2003). In contrast, the extrinsic factors, such as species availability, density, and abundance, are modified in time and space. If we analyze certain ecosystems where the arbovirus is maintained by the network of interactions between hosts and vectors, we are likely to see that the cycles, part of that network, change over time.

In ecological terms, infectious agents can be classified as generalists or specialists according to the number of host/vector they can infect. Specialist arboviruses are those transmitted by specific species of host/vector. Thus, as a result of centuries of coadaptation, CHIKV, DENV, and YFV viruses are particularly efficient in being transmitted by *Aedes aegypti/Ae.*[3] *albopictus* mosquitoes and amplified by humans in urban environments (Weaver and Reisen 2010). In certain cases, some viruses make a switch over to other species and have the ability of being transmitted by another species of host/vector. For example, due to a special mutation, CHIKV is transmitted by an alternative mosquito species: *Ae. albopictus* (Tsetsarkin et al. 2007). On the other hand, those viruses maintained in nature by more than one host/vector species are considered

Table 5.1 Intrinsic and extrinsic characteristics fulfilled by a vector-host of an arbovirus

Characteristics	Vectors	Hosts
Intrinsic	Viral replication	
	Susceptibility to viral infection	
	Host feeding preference	
	Behavior	
Extrinsic	Abundance and dispersal	
	Seasonal breeding patterns	
		Attractiveness to mosquito vector
	Host feeding selection	
	Distribution	

[3]Special abbreviations for mosquito genera proposed by Reinert (2009) were utilized.

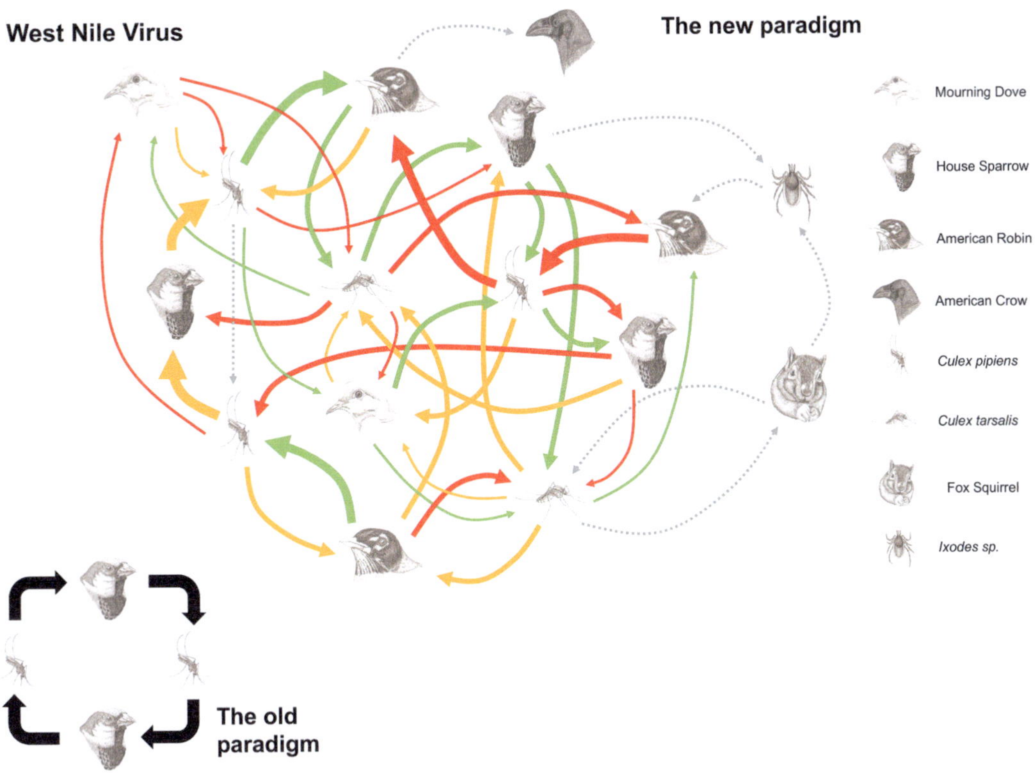

Fig. 5.2 Hypothetical transmission networks for West Nile virus in the USA. Foot figure: the *arrows* represent the viral flow between the vectors and hosts involved in the arbovirus maintenance network. The thickness of the arrow represents the amount of existing virus between the particular connection of host and vector (which is determined by the vector-host preference, vector-host population density, vector and host competence). The *spotted line arrows* represent alternative transmission way (venereal and/or transovarial transmission), hosts (mammals), and vectors (ticks). The *colored arrows* represent the season in which the vector-host relation takes place (*green*, Spring; *red*, Summer; *orange*, Fall) (From Diaz et al. 2013)

as generalists, such as SLEV, WNV, JEV, EEV, WEEV, and VEEV. The analysis of the maintenance of these viruses is more related to a transmission network than to a transmission cycle (Fig. 5.2) (Diaz et al. 2013).

In systems like the arboviruses, characterized by multi-host-vector interactions, the ecological dynamics may alter the epidemiological patterns and scenarios. Many studies have recently focused on the effects of biodiversity over arbovirus activity (Ezenwa et al. 2006, 2007; Swaddle and Calos 2008; Allan et al. 2009; Loss et al. 2009; McKenzie and Goulet 2010). It has been hypothesized that a high diversity of host would result in a reduced viral activity. This could be caused by two different mechanisms: firstly, due to a decrease in the probability that the vector comes into contact with the host of higher competition, resulting from a decrease in the relative abundance of the host with the higher competition or an increase in the relative abundance of hosts of low competition. Secondly, higher host diversity could increase interspecific interactions, such as predation and competition, thus potentially regulating the abundance of the most competent host. Empirical studies aimed to test the hypothesis that higher host diversity reduces transmission rate and viral concentration in the ecosystem had produce inconclusive results. While some studies have found support for this hypothesis, others have not found a relation between host diversity and virus transmission (local spatial scale vs. regional/national scale) (Ezenwa et al. 2006; Allan et al. 2009).

Moreover, diversity effects on transmission dynamics vary over time. These evidences emphasize the need to consider the diversity of hosts and vectors within an ecosystem when analyzing virus dynamics (Kilpatrick 2011). Therefore, when assessing the suitability of an ecosystem to an arbovirus, it is important to know and consider not only the diversity (abundance, richness) but also the species composition of vectors and hosts (Diaz et al. 2013).

Hosts are not always necessary for arbovirus maintenance, since there are alternative ways for arbovirus perpetuation. During dispersion through the hemolymph, arboviruses are able to infect the reproductive cells and be transmitted to the next generation through eggs (vertical transmission). La Crosse virus is a well-studied case. Almost 80 % of the new generation of mosquitoes *Aedes triseriatus* are infected at birth. This transovarial transmission allows the virus to survive during the period of low vertical transmission. Some of the newborn infected mosquitoes are males and can infect females by venereal transmission, another alternative way for arbovirus maintenance in nature (Kuno and Chang 2005; Lequime and Lambrechts 2014).

5.2.1 Overwintering

Arboviruses are mainly maintained by vectorial transmission between hosts and vectors. In temperate areas, mosquito vector abundance decreases during winter, endangering arbovirus maintenance. However, it was observed that arboviruses remain endemic to those areas, indicating the existence of overwintering mechanisms. Several strategies have been postulated to explain this phenomenon: transovarial transmission and perpetuation in the mosquito vector population, vectorial transmission by alternative vector such as ticks, annual reintroduction by displacement of birds or bats, and viral permanence in nondiapausing infected females (White et al. 2005; Brown et al. 2010).

5.3 Pathogenesis

Animals infected with arboviruses regularly develop a febrile disease accompanied by various nonspecific symptoms; however, additional severe syndromes may occur: neurological diseases (meningitis, encephalitis, encephalomyelitis), hemorrhagic symptoms, abortions and congenital disorders, or vesicular stomatitis. Certain arboviral diseases cause significant economic losses in domestic animals, such as Eastern, Western, and Venezuelan equine encephalitides, West Nile encephalitis, Nairobi sheep disease, Rift Valley fever, Akabane fever, Schmallenberg disease, African horse sickness, bluetongue, vesicular stomatitis, and African swine fever (Hubálek et al. 2014).

Arboviruses penetrate the tissues of vertebrates by inoculation of contaminated saliva of hematophagous arthropod vectors (mosquitoes, ticks, sandflies, and biting midges). Infecting viruses can either enter directly to the bloodstream through capillaries or infect susceptible skin cells (macrophages, epithelial cells, fibroblasts) (Kenney and Brault 2014). Viral replication occurs near the entry site (vascular endothelium or regional lymph nodes). The primary viral replication is local, low, and time-limited. After that, viruses spread to the lymph nodes nearby where they replicate and disperse through the lymphatic and vascular system, producing a primary viremia. A second viral replication into the vascular system produce a higher and long-lasting viremia, allowing viruses to reach target organs (brain, neurothropic arboviruses [WNV, SLEV, JEV, EEEV, WEEV, VEEV, LACV, BUNV]; vascular and blood cells, hemorrhagic arbovirus [YFV, CCHFV, RVFV]; joints, arthritogenic arboviruses [DENV, CHKV, MAYV]) (Iranaia et al. 2013). Arboviruses associated with the production of central nervous system (CNS) disease can go through the brain barrier. These viruses have a high tropism for endothelial cells forming the CNS barrier, thereby facilitating the entry of the virus into the brain (Suen et al. 2014). Moreover, other orthobunyaviruses (Akabane virus (AKAV), Aino virus (AINV), Bunyamwera virus (BUNV)) have a high tropism for endothelial cells associated with placental trophoblast infect cells and affect fetal development, resulting in neural and skeletal muscle malformations and abortion.

During the secondary viremia period, arthropod vectors can get infected during the feeding process. The period spanned between virus entry and the onset of symptoms is known as *intrinsic incubation period* (Kenney and Brault 2014).

Viral replication (duration, viral loads) at the entry sites and dissemination to target organs are regulated by the innate immune host system, which modulates the acquired specific immune response. The initial response to infection by an arbovirus is the production of type I interferon (IFN). Its main function is associated with a short-term restriction of viral replication (Lazear and Diamond 2014). Moreover, antigen-presenting cells infiltrate to the lymphatic ganglion initiating the acquired immune response. As soon as IFN production declines, a strong cell-mediated immune response initiates the clearing of virus from the bloodstream and replication sites. In order to resolve the viral infection in target organs, the activation of the cell-mediated immune response is needed. During this stage, the production of cytokines stimulates the activation and proliferation of T cells (Schmid et al. 2014). The B lymphocytes develop a key role during the viral clearing by the production of specific antibodies. The IgM-mediated immune response eliminates viruses from tissues, recovering the organism from the infection. These antibodies are used in the viral diagnosis as recent infection markers. The IgG antibodies appear in sera between 7 and 14 days after onset of symptoms and maintain stable levels for years. They neutralize infecting viruses and stimulate antibody-mediated phagocytosis by monocytic-phagocytic cells along with the complement system. Moreover, the IgG antibodies confer protective defense for future specific arbovirus infections.

In the case of newborn animals, because the interferon system has not fully matured, the main line of defense against viral infection is given by the generation of neutralizing antibodies (NTAb). Bunyaviruses are successfully adapted to counteract cellular immunity; this is carried out mainly by the viral NSs protein, which acts by inhibiting both the production of IFN and the ability of infected cells to respond to IFN.

5.4 Diagnosis

In the diagnosis of human and/or animal cases with clinical symptoms (febrile syndrome and/or neurological, hemorrhagic fever, congenital malformations, and abortion) compatible with arbovirus infection, it is essential to know the proper epidemiological situation in the region and to be very strict with diagnostic criteria in order to specifically identify the causative agent of the disease.

The differential diagnosis must be made including other arboviruses and other pathogens (viruses, bacteria, and parasites) for both humans and domestic and wild animals that are circulating in the region and produce the same syndrome.

Arbovirus infection diagnosis can be performed by isolation, viral genome detection, antigen detection, or serology. First three methodologies are usually employed during the first days of onset of symptoms, when there is some possibility of recovering virus genome or antigen detection. On the other hand, serological techniques are applied in samples obtained in a later period (Fig. 5.3).

For viral isolation in humans, the most suitable samples are cerebrospinal fluid (CSF) and blood/serum/plasma of the early days of onset of symptoms or autopsy materials. The best type of sample in animals is obtained from the blood and, in the case of abortions and/or death, brain, and if possible, spleen, kidney, and heart samples are most appropriate. It is essential to succeed in isolating samples that are taken under aseptic conditions, completely refrigerated, and sent to the laboratory as soon as possible. Isolation can be performed using in vivo systems (suckling mice) or in vitro systems (cell cultures). Subsequent viral identification is carried out using techniques such as neutralization (NT), hemagglutination

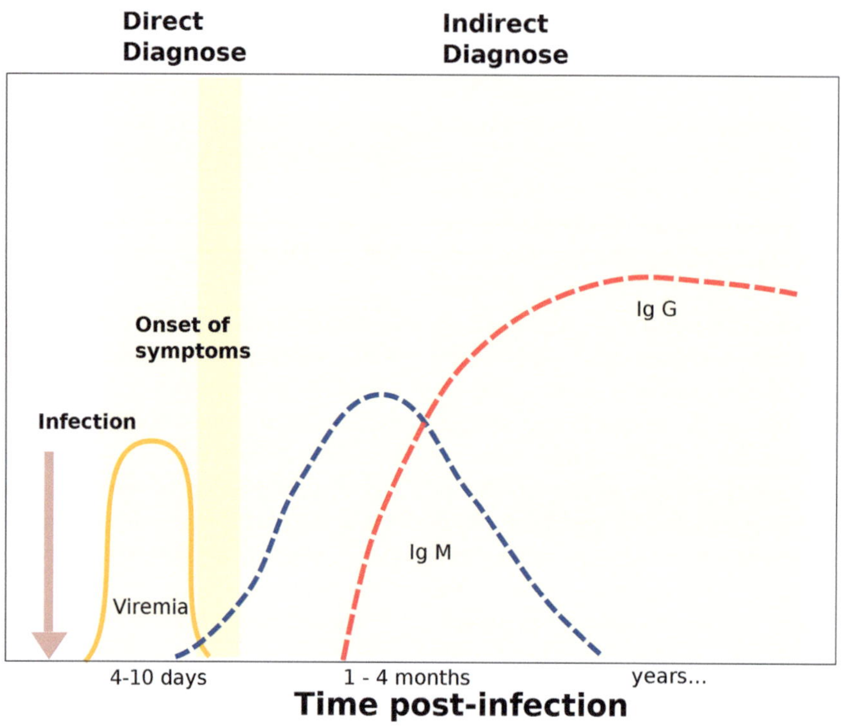

Fig. 5.3 Kinetics of antigen (viremia) and antibodies (IgM and IgG) to arbovirus infection in humans

inhibition (HI), immunofluorescence (IF), and enzyme immunoassay (ELISA), which are also used for serological diagnosis (Fig. 5.3).

Genome amplification is performed by RT-PCR (qualitative) and real-time PCR (qualitative and quantitative). Recently, metagenomics analysis has been increasingly used for the study of new viruses through next-generation sequencing techniques (454 sequencing, Solexa, Ion Torrent, and SOLiD), allowing access to the entire viral genome.

The most specific serological technique used in arbovirus diagnosis (the "gold standard") is the NT test, which detects neutralizing antibodies. To obtain a confirmatory diagnosis by serology, it is necessary to have paired serum samples: the first sample obtained during the first days of illness and another one obtained at least 15 days later. Seroconversion or a fourfold increment in antibody titer between samples must be demonstrated. If the first sample was obtained later, a detection of a fourfold fall antibody titer can be useful. The presence of high titers of IgG antibodies when a single sample is available can be significant if the rate of infection is not common in the region and particularly if serological surveys show low or null prevalence of antibodies to the virus in the population.

5.5 Arbovirus Emergence

Arboviruses are emerging/reemerging infectious agents worldwide; CHIKV, DENV, YFV, SLEV, WNV, and ZIKV are some examples of this phenomenon. Although not fully understood, several factors are thought to promote reemergence, for example, environmental disturbances from anthropogenic activities, climatic changes affecting vector and host population fluctuations, human movements through airplanes, animal trade and migration, and genetic mutations that cause spillovers. Globalization of arbovirus-induced diseases is a concern because it is also accompanied by the resistance of vectors to insecticides, resistance of pathogens to drugs, unprecedented population growth in developing countries, and uncontrolled urbanization in tropical areas where vector-borne diseases occur most frequently. Conditions necessary for arbovirus outbreaks include adequate populations of amplifying hosts (potentially humans in some settings), vector mosquitoes, and appropriate climatic conditions for transmission (Vasconcelos et al. 2001; Weaver and Barrett 2004; Weaver and Reisen 2010; Coffey et al. 2013).

Historically, most of the arbovirus transmission cycles frequently remained silent in rural and suburban areas. Currently, they have shifted significantly to urban areas, infecting humans and domestic animals and causing diseases and outbreaks. This emergence phenomenon is due to a combination of factors, such as human population explosion, invasion to non-urbanized areas, changes in land use, and climate changes that produce disorders, affecting the biology and population dynamics of both hosts and their vectors (Weaver and Barrett 2004).

Human civilization is under global expansion, building settlements everywhere. In most developing countries, this expansion is taking place without environmental management and urban planning. For example, new settlements around urban periphery have no water service, so people create temporary water reservoirs that represent breeding sites for vector mosquitoes like *Ae. aegypti* and *Ae. albopictus*, increasing the risk for YFV, DENV, and CHKV outbreaks (Kuno 1995).

The air traffic increased 136 times worldwide between 1950 and 2007. This huge mass of travelers and associated animal trade promotes the movements of potential vectors, vertebrate host, and humans. This human activity increases the chances for virus invasion to novel areas (Tatem et al. 2006). In 1999, West Nile virus reached the Americas, spreading to most countries. This exotic virus was able to invade and establish in a new ecosystem. Since the American biota was not immunized to this new pathogen, thousands of encephalitis and deaths in humans, equines, and birds were reported. This viral invasion represented a real conservation threat for several wild bird species whose populations decreased after WNV introduction (Chancey et al. 2015). Endemic to Asian and African countries until 2004, CHKV reached the European and American continents due to human movements basically for tourism. Human beings are amplification hosts for this *Alphavirus*, allowing its establishment in new areas where *Ae. aegypti* and *Ae. albopictus* mosquito vectors are abundant (Weaver and Forrester 2014).

Through biological evolution and cultural development, human beings were able to modify the environments according to their needs. Thus, deforestation has produced new areas for agriculture, livestock, farming activities, and urbanization. These anthropogenic activities have produced great changes to host and vector communities and population abundance, sometimes driving emergence and reemergence of arboviruses. JEV was originated in Malaysia and dispersed further to other Asian countries where it is endemic. This rapid and widespread expansion of JEV was apparently closely associated with increases in human populations, rice crops, and pig farming (Le Flohic et al. 2013). The frequency of polyarthritis by MAYV infection in humans is closely associated with the increment of human activities in the forested Amazon areas (Muñoz and Navarro 2012). Likewise, fever cases by Oropouche virus (*Orthobunyavirus*) are linked to urbanization of forested areas in Brazil.

Except for *Asfivirus*, all known arboviruses are RNA viruses. Thus, they have an intrinsic capacity to mutate and adapt rapidly to new biological contexts or to acquire new biological features. Accordingly, they can spillover a new vector and host or become more virulent for humans and other animals (Weaver and Barrett 2004). For example, CHKV acquired mutations at envelope protein that allowed it to be transmitted by a new mosquito vector *Ae. albopictus*, facilitating its establishment in new areas where the original vector *Ae. aegypti* does not occur (Weaver and Forrester 2015). It was hypothesized that epizootic VEEV strains emerged from enzootic subtypes that mutated naturally. Thus, they became able to produce neuroinvasion in equines (Weaver et al. 2004a).

Emergence of arboviruses is a multicausal event rather than a single-cause event. For example, SLEV reemerged in South America, affecting central Argentina and southern Brazil. Ecoepidemiological research carried out in Argentina demonstrated the main factor for its reemergence was the introduction of a new, more virulent strain coupled with environmental changes that led to the increment in *Culex quinquefasciatus* mosquito vector and *Zenaida auriculata* amplifying host (Díaz et al. 2011, 2012).

References

Allan B, Langerhans R, Ryberg W et al (2009) Ecological correlates of risk and incidence of West Nile virus in the United States. Oecologia 158:699–708

Antonis AF, Kortekaas J, Kant J et al (2013) Vertical transmission of Rift Valley fever virus without detectable maternal viremia. Vector Borne Zoonotic Dis 13: 601–606

Barrett AD, Monath TP (2003) Epidemiology and ecology of yellow fever virus. Adv Virus Res 61:291–315

Brown CR, Strickler SA, Moore AT et al (2010) Winter ecology of Buggy Creek virus (Togaviridae, Alphavirus) in the Central Great Plains. Vector Borne Zoonotic Dis 10:355–363

Chancey C, Grinev A, Volkova E et al (2015) The global ecology and epidemiology of West Nile virus. Biomed Res Int 2015:376230

Coffey LL, Forrester N, Tsetsarkin K et al (2013) Factors shaping the adaptive landscape for arboviruses: implications for the emergence of disease. Future Microbiol 8:155–176

Daubney R, Hudson J, Garnham P (1931) Enzootic hepatitis or Rift Valley fever: an undescribed virus disease of sheep, cattle and man from East Africa. J Pathol Bacteriol 34:545–579

Díaz LA, Albrieu Llinás G, Vázquez A et al (2012) Silent circulation of St. Louis encephalitis virus prior to an encephalitis outbreak in Cordoba, Argentina (2005). PLoS Negl Trop Dis 6, e1489

Diaz LA, Flores FS, Quaglia A et al (2013) Intertwined arbovirus transmission activity: reassessing the transmission cycle paradigm. Front Physiol 3:493

Diaz LA, Nemeth NM, Bowen RA et al (2011) Comparison of Argentinean Saint Louis encephalitis virus non-epidemic and epidemic strain infections in an avian model. PLoS Negl Trop Dis 5, e1177

EFSA (2011) Panel on Animal Health and Welfare (AHAW); Scientific Opinion on bluetongue serotype 8. EFSA J 9(5):2189 [51 pp]. doi:10.2903/j.efsa.2011.2189, Available online: www.efsa.europa.eu/efsajournal.htm

Ezenwa V, Godsey M, King R et al (2006) Avian diversity and West Nile virus: testing associations between biodiversity and infectious disease risk. Proc Biol Sci 273:109–117

Ezenwa V, Milheim L, Coffey M et al (2007) Land cover variation and West Nile virus prevalence: patterns, processes, and implications for disease control. Vector Borne Zoonotic Dis 7:173–180

Hassan HK, Cupp EW, Hill GE et al (2003) Avian host preference by vectors of eastern equine encephalomyelitis virus. Am J Trop Med Hyg 69:641–647

Higgs S and Beaty BJ (2005) Natural cycles of Vector-borne pathogens. In "Biology of Diseases Vectors". Marquardt WC (ed). Elsevier, San Diego, CA, p. 167-185

Hubálek Z, Rudolf I, Nowotny N (2014) Arboviruses pathogenic for domestic and wild animals. Adv Virus Res 89:201–275

Hughes MT, Gonzalez JA, Reagan KL et al (2006) Comparative potential of *Aedes triseriatus*, *Aedes albopictus*, and *Aedes aegypti* (Diptera: Culicidae) to transovarially transmit La Crosse virus. J Med Entomol 43:757–761

Iranaia AM, Cruz-Oliveira C, DaPoian AT (2013) Molecular mechanisms involved in the pathogenesis of alphavirus-induced arthritis. Biomed Res Int 2013:973516

Karabatsos N (1985) International catalogue of arboviruses including certain other viruses of vertebrates. American Society Tropical Medicine and Hygiene, San Antonio, TX

Kenney JL, Brault AC (2014) The role of environmental, virological and vector interactions in dictating biological transmission of arthropod-borne viruses by mosquitoes. Adv Virus 89:39–83

Kilpatrick A (2011) Globalization, land use and invasion of West Nile virus. Science 334:323–327

Komar N, Dohm DJ, Turell MJ et al (1999) Eastern equine encephalitis virus in birds: relative competence of European starlings (*Sturnus vulgaris*). Am J Trop Med Hyg 60:387–391

Komar N, Langevin S, Hinten S et al (2003) Experimental infection of North American birds with the New York 1999 strain of West Nile virus. Emerg Infect Dis 9:311–322

Kramer LD, Ebel GD (2003) Dynamics of flavivirus infection in mosquitoes. Adv Virus Res 60:187–232

Kuno G (1995) Review of the factors modulating dengue transmission. Epidemiol Rev 17:321–335

Kuno G, Chang G (2005) Biological transmission of arboviruses: reexamination of and new insights into components, mechanisms, and unique traits as well as their evolutionary trends. Clin Microbiol Rev 18: 608–637

Labuda M, Nuttall PA (2004) Tick-borne viruses. Parasitology 129(Suppl):S221–S245

Lazear HM, Diamond MS (2014) New insights into innate immune restriction of West Nile virus infection. Curr Opin Virol 11:1–6

Le Flohic G, Porphyre V, Barbazan P et al (2013) Review of climate, landscape, and viral genetics as drivers of the Japanese encephalitis virus ecology. PLoS Negl Trop Dis 7, e2208

Lequime S, Lambrechts L (2014) Vertical transmission of arboviruses in mosquitoes: a historical perspective. Infect Genet Evol 28:681–690

Loss S, Hamer G, Walker E et al (2009) Avian host community structure and prevalence of West Nile virus in Chicago, Illinois. Oecologia 159:415–424

Maclachlan NJ (2011) Bluetongue: history, global epidemiology, and pathogenesis. Prev Vet Med 102:107–111

McKenzie V, Goulet N (2010) Bird community composition linked to human West Nile virus cases along the Colorado front range. Ecohealth 7:439–447

McLean RG, Bowen GS (1980) Vertebrate hosts. In: Monath TP (ed) St Louis encephalitis. American Publich Health Association, Washington, DC, pp 381–450

McLean RG, Ubico SR, Docherty DE et al (2001) West Nile virus transmission and ecology in birds. Ann N Y Acad Sci 951:54–57

Morris CD (1988) Eastern equine encephalomyelitis. In: Monath TP (ed) The arboviruses: epidemiology and ecology, vol 3. CRC Press, Boca Raton, FL, pp 1–20

Muñoz M, Navarro JC (2012) Mayaro: a re-emerging Arbovirus in Venezuela and Latin America. Biomedica 32:286–302

Oncü S (2013) Crimean-Congo hemorrhagic fever: an overview. Virol Sin 28:193–201

Reeves WC (2001) Partners: serendipity in arbovirus research. J Vector Ecol 26:1–6

Reinert JF (2009) List of abbreviations for currently valid generic-level taxa in family Culicidae (Diptera). Eur Mosq Bull 27:68–76

Reisen WK, Barker CM, Carney R et al (2006) Role of corvids in epidemiology of west Nile virus in southern California. J Med Entomol 43:356–367

Reisen WK, Chiles RE, Martinez VM et al (2003) Experimental infection of California birds with western equine encephalomyelitis and St. Louis encephalitis viruses. J Med Entomol 40:968–982

Reisen WK, Monath TP (1989) Western equine encephalomyelitis. In: Monath TP (ed) The arboviruses: epidemiology and ecology, vol 3. CRC Press, Boca Raton, FL, pp 89–137

Schmid MA, Diamond MS, Harris E (2014) Dendritic cells in dengue virus infection: targets of virus replication and mediators of immunity. Front Immunol 5:647

Scott TW (1988) Vertebrate host ecology. In: Monath TP (ed) The arboviruses: epidemiology and ecology, vol 3. CRC Press, Boca Raton FL, pp 257–280

Solomon T, Mallewa M (2001) Dengue and other emerging flaviviruses. J Infect 42:104–115

Suen WW, Prow NA, Hall RA et al (2014) Mechanism of West Nile virus neuroinvasion: a critical appraisal. Viruses 6:2796–2825

Swaddle JP, Calos SE (2008) Increased avian diversity is associated with lower incidence of human West Nile infection: observation of the dilution effect. PLoS One 3, e2488. doi:10.1371/journal.pone.0002488

Tatem AJ, Hay SI, Rogers DJ (2006) Global traffic and disease vector dispersal. Proc Natl Acad Sci U S A 103:6242–6247

Tsetsarkin KA, Vanlandingham DL, McGee CE, Higgs S (2007) A single mutation in chikungunya virus affects vector specificity and epidemic potential. PLoS Pathog 3(12), e201

Turtle L, Griffiths MJ, Solomon T (2012) Encephalitis caused by flaviviruses. QJM 105:219–223

Vasconcelos P, Travassos da Rosa A, Rodrigues S et al (2001) Inadequate management of natural ecosystem in the Brazilian Amazon region results in the emergence and reemergence of arboviruses. Cad Saude Publica 17:155–164

Weaver SC, Anishchenko M, Bowen R et al (2004a) Genetic determinants of Venezuelan equine encephalitis emergence. Arch Virol Suppl 18:43–64

Weaver SC, Barrett AD (2004) Transmission cycles, host range, evolution and emergence of arboviral disease. Nat Rev Microbiol 2:789–801

Weaver SC, Ferro C, Barrera R et al (2004b) Venezuelan equine encephalitis. Ann Rev Entomol 49:141–174

Weaver SC, Forrester NL (2015) Chikungunya: evolutionary history and recent epidemic spread. Antiviral Res 120:32–39

Weaver SC, Reisen WK (2010) Present and future arboviral threats. Antiviral Res 85:328–345

Wernike K, Conraths F, Zanella G et al (2014) Schmallenberg virus-two years of experiences. Prev Vet Med 116:423–434

White DM, Wilson WC, Blair CD et al (2005) Studies on overwintering of bluetongue viruses in insects. J Gen Virol 86(Pt 2):453–462

Flavivirus

6

Marta S. Contigiani, Luis A. Diaz, and Lorena Spinsanti

Abstract

The *Flavivirus* genus currently includes 53 viral species, most of them transmitted by mosquitoes and tick bites and others with not known vector and specific insect flavivirus. Flaviviruses comprise important arthropod-transmitted human pathogens, including yellow fever virus (YFV), Dengue virus (DENV), Japanese encephalitis virus (JEV), West Nile virus (WNV), and tick-borne encephalitis virus (TBEV). DEN viruses are globally expanded and are the most important mosquito-borne disease after malaria. Also with a worldwide distribution, YFV still cause mortality events around the world. Besides human vaccination campaign, enzootic foci remain active triggering emerging events for this pathology. West Nile virus, endemic in Old World countries, was introduced to American continent by 1999. It became an important human and veterinary pathogen due to its virulence in human beings, horses, and wild birds (especially in corvids). Antigenically closely related, St. Louis encephalitis virus co-circulates with WNV in American continent. Since 2002, it reemerged in the south cone of South America as a human encephalitis etiological agent. Tick-borne encephalitis virus still causes human encephalitis outbreaks in Eastern Europe and Russia. Zika virus with endemic activity in Africa and Polynesia became an emerging flavivirus with outbreaks reported in South America (Brazil, Colombia, Easter Island, and several other areas). In this chapter, authors will focus in general aspects of flaviviruses and their diseases, their spatial and temporal distributions patterns, ecology, epidemiology, pathogeny, and phylodynamics. Moreover, factors affecting their epidemiology and emergence are emphasized. Due to the

M.S. Contigiani (✉) • L. Spinsanti
Laboratory of Arbovirus,
Institute of Virology "Dr. J. M. Vanella",
National University of Córdoba,
Enfermera Gordillo Gomez S/N,
Ciudad Universitaria, Córdoba, Argentina
e-mail: martascontigia@hotmail.com;
l_spinsanti@yahoo.com.ar

L.A. Diaz
Laboratory of Arbovirus, Institute of Virology "Dr. J. M. Vanella", National University of Córdoba, Enfermera Gordillo Gomez S/N, Ciudad Universitaria, Córdoba, Argentina

Institute for Biological and Technical Researches, CONICET-UNC, Córdoba, Argentina
e-mail: adrian.diaz@conicet.gov.ar

C.B. Marcondes (ed.), *Arthropod Borne Diseases*, DOI 10.1007/978-3-319-13884-8_6

great importance and amount of knowledge on dengue and yellow fever, these diseases deserved separated chapters (Chaps. 7 and 8).

Keywords
Flavivirus • Mosquito-borne encephalitis virus • Tick-borne encephalitis virus

6.1 Virus Structure

The flavivirus virion is a small icosahedral enveloped particle of 40–60 nm in diameter, which includes a 30 nm core. Structural analysis of mature flavivirus virions revealed that the virus possesses an icosahedral envelope organization and a spherical nucleocapsid core (Fig. 6.1). Mature virions contain three structural proteins: the capsid protein C, the membrane protein M, and the envelope protein E. Multiple copies of the C protein (11 kDa) encapsulate the RNA genome to form the viral nucleocapsid (Jones et al. 2003). The nucleocapsid is surrounded by a host-cell-derived lipid bilayer, in which copies of M and E are anchored (Fig. 6.1a). The M protein is a small (approximately 8 kDa) proteolytic fragment of its precursor form prM. The E protein is approximately 50 kDa and is the main protein present in the virion envelope. The role of the E protein is to recognize the host-cell receptors and is the main target of neutralizing antibodies. Certain mutations in the E protein can cause the lack of virulence of the virion.

6.1.1 Genome Organization and Protein Functions

The *Flavivirus* genome consists of a single-stranded positive-sense RNA fragment of approximately 11,000 nt (nucleotides). The viral genome works as a single RNA messenger molecule that possesses a single-open reading frame (ORF) (Fig. 6.1b). The ORF is flanked by noncoding regions (NCR). These NCRs form an RNA secondary structure, which could be involved in the processes of genome amplification, translation, and packaging.

NS1 is a homodimer that participates in the replication process of viral RNA; it is one of the first viral proteins to be secreted and induces a strong humoral immune response. NS2A is a small hydrophobic protein (22 kDa) thought to intervene in the recovery of the RNA templates associated with viral polymerase, to module the host's antiviral response and cleavage of the NS1–NS2A junction after translation, and to play a functional role in the viral replicase complex. NS2B is a small membrane-associated protein and acts as a cofactor necessary for the NS2B–NS3 complex serine-protease activity, which cleaves the viral polyprotein at the NS2A/NS2B, NS2B/NS3, NS3/NS4A, and NS4B/NS5 junctions. NS3 is a big cytoplasmic protein (70 kDa) that intervenes in several enzymatic activities (protease, helicase) involved in the polyprotein processing and viral RNA replication. Interestingly, NS3 also appears to be involved in the virus assembly through mechanisms that are independent of the enzymatic functions outlined above. NS4A and NS4B are small hydrophobic membrane-associated proteins of 16 and 27 kDa, respectively. Based on its subcellular localization, both proteins may intervene in the viral RNA replication. NS5 is the biggest (103 kDa) and most conservative protein throughout the flavivirus genera. It has a methyltransferase activity on its N-terminal region and an RNA-dependent RNA polymerase (RdRp) activity on the C-terminal motifs.

6.1.2 Replication Cycle

The virus enters the cell via a mechanism known as receptor-mediated endocytosis, with protein E being the viral receptor for recognition and cell adsorption. Once the endocytic vesicle is formed,

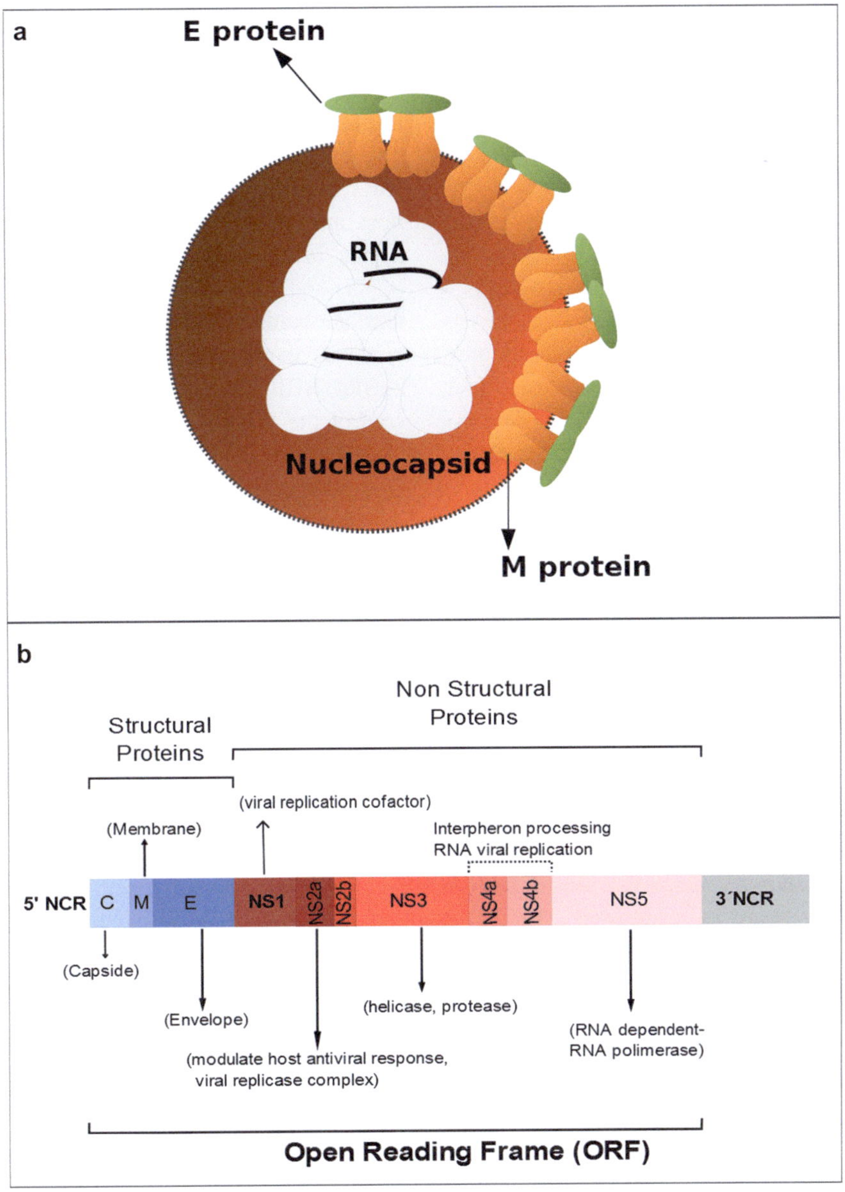

Fig. 6.1 Graphic representation of flavivirus virion (**a**) and their genome (**b**)

the endosome acid pH generates structural alterations in protein E, producing endosome–virion membrane's fusion and releasing the nucleocapsid to the cytoplasm, although direct fusion of viral envelope with the cellular membrane has been also observed. After the viral uncoating process, replication starts. Viral RNA replication occurs in perinuclear spots and implies the synthesis of a single complementary negative strand that works as a template for the positive strand molecule synthesis. The genomic viral RNA is used directly as a messenger and is completely translated from its 5′ end to produce a big precursor polyprotein that is later cleaved to generate each individual viral protein. The new viral particle assembly occurs in the endoplasmic reticulum (ER), where immature particles are generated (containing prM) and then transported to the cell exterior through the exocytic pathway. Evidence was found that the acid pH in the trans-Golgi network (TGN) produces conformational changes in the prM-E complex that are necessary for particle maturation. Once the cleavage of the complex is produced by the action of cellular furins, the mature particles are released through exocytosis.

6.2 Pathogenesis and Pathology

The genus *Flavivirus* contains some of the most important arboviral pathogens of man. Hemorrhagic disease, encephalitis, biphasic fever, flaccid paralysis, and jaundice are typical manifestations of diseases in human beings after infections by mosquito-borne or tick-borne flaviviruses. The genus includes several etiological agents of encephalitis, the most significant being Japanese encephalitis virus (JEV), West Nile virus (WNV), and tick-borne encephalitis virus (TBEV) (Turtle et al. 2012).

Most exposed individuals will not develop disease, but a minority will develop a severe illness with a significant chance of permanent neurological damage or death. In recent years, incidences of patients with neurological complications from non-encephalitic viruses, such as Dengue virus, have been reported. The factors involved are numerous, including complex interactions between virus and host. In many cases, it appears that the immune response, while crucial to contain the virus and limit the spread to the brain, is also responsible for causing neurological damage. Innate responses can limit viral replication but may also be responsible for generating pathological levels of inflammation. Neutralizing antibody responses are protective but take time to develop. The role of T cells is less clear and may be either protective or pathogenic (Turtle et al. 2012).

Many infections with these viruses are asymptomatic or cause a nonspecific febrile illness. Neurological presentations include aseptic meningitis, febrile convulsions in children, encephalitis, or myelitis—which may cause a polio-like flaccid paralysis. JEV tends to infect children because it is more abundant; hence, children are more likely to acquire disease by this virus, whereas adults are more likely to be affected by WNV, SLEV, and MVEV.

The pathogenesis of flaviviruses encephalitis remains poorly understood. The neurotropic viruses may employ different neuroinvasive pathways, depending on viral strain, immune status of host, and route of infection. Many aspects, such as the invasion of the central nervous system (CNS) and the intervention of the immune system, are being studied since they play an important role in the breakdown of the blood–brain barrier (BBB). Inflammation may be necessary to control the virus but also a mechanism that is used to enter the CNS, compromising the BBB (Turtle et al. 2012).

After the bite of an infected mosquito, the primary site of infection is the epidermis. The initial viral replication occurs inside the dendritic cells in the dermis (DCs) and keratinocyte residents (target cell of WNV). From here, the virus spreads to local lymph nodes where continuous replication produces a primary viremia, reaching visceral organs and eventually CNS. How the neurotropic flaviviruses gain access to the CNS remains unknown, but interactions at the BBB are probably critical. Proposed mechanisms include active replication within endothelial cells and passive transfer across the BBB or within leukocytes that migrate across the barrier (Turtle et al. 2012; Ong et al. 2014).

Flaviviruses interact with pathogen-recognition receptors that detect RNA viruses, such as Toll-like receptors (TLR-3, TLR-7) and the retinoic acid-inducible gene (RIG-I). TNF-α production mediated by TLR-3 and the induction of ICAM-1 (vascular endothelial intracellular adhesion molecule) with cellular infiltration are WNV-infection-induced processes with the aim of breaking the BBB. Besides entering the CNS through a hematogenous pathway, WNV can infect the motor neurons of the anterior horn of the spinal cord (axonal transport). This entry pathway can be independent or not of the hematogenous one. Once established in the brain, the virus has many mechanisms to induce neuronal damage and cell death, leading to clinical disease. There is a strong inflammatory response, with perivascular cuffing of lymphocytes and macrophages and increased glial cells. It is still not clear whether brain injury is due to virus-induced cell death, apoptosis, or pathological inflammation. Other forms of access, where the BBB is absent, are the choroid plexus, circumventricular organs, and posterior pituitary. Access by olfactory pathway (may be by aerosol exposure) has been associated with JEV as a result of the increase of the BBB permeability (Turtle et al. 2012).

6.3 Mosquito-Borne Viruses

6.3.1 Neurotropic Flaviviruses

6.3.1.1 Ilheus Virus

The Ilheus virus (ILHV) was isolated from mosquitoes of the genera *Aedes*[1] and *Psorophora* near Ilheus (Bahia, Brazil) in 1944 (Laemmert and Hughes 1947). It was later found in other mosquito species and in a variety of birds in different Latin American countries. The transmission cycle in nature involves mosquitoes and birds. The infection may be asymptomatic or cause a DENV-like febrile syndrome. The reports of encephalitis have been infrequent. There are records of human infections in Brazil, Trinidad, Colombia, Panama, Bolivia, Argentina, and Ecuador. No encephalitis outbreaks attributed to ILHV have been reported (Pauvolid-Corrêa et al. 2013; Sabattini et al. 1998).

Another South American endemic flavivirus that causes encephalitis is Rocio virus (ROCV). Currently, it is classified as a genotype of ILHV (ICTV 2014). This virus was the cause of outbreaks of encephalitis occurred in Sao Paulo, Brazil, between 1975 and 1977 (Lopes et al. 1978). During that outbreak, hundreds of cases were reported. ROCV was isolated from *Psorophora ferox* mosquitoes and the sparrow is supposed to be its main vertebrate host.

[1]Although several taxonomic modifications have been proposed on *Culicidae* genera, mostly splitting *Aedes* by Reinert (2009) (and accepted by CBM), this taxonomy has been used in the chapters according to authors' preference. To facilitate utilization by health personnel, all new aedine genera can be considered *Aedes* (CBM).

6.3.1.2 Bussuquara Virus

The Bussuquara virus (BSQV) was originally isolated from a sentinel monkey in Brazil (Gomes and Causey 1959). It is widely distributed in Central America, the Caribbean, and South America. Its transmission cycle includes *Culex* mosquitoes and rodents. In Panama, it was recovered from a patient with undifferentiated febrile syndrome. In Argentina, it was isolated from a suspected case of Argentinean hemorrhagic fever in 1963. There are serological records in birds, horses, cattle, and humans with very low prevalence (Gubler et al. 2007; Sabattini et al. 1998).

6.3.2 Japanese Encephalitis Virus

Japanese encephalitis virus (JEV) is the main cause of arbovirus encephalitis worldwide. It is an endemic virus widely distributed in Asia, including Japan, China, Taiwan, Korea, the Philippines, Far Eastern Russia, all of Southeast Asia, India, Papua New Guinea, and the Torres Strait of Australia (Mackenzie 2005). Its ancestor would have originated in the region of Indonesia–Malaysia, where all genotypes (I, II, III, IV, and V) circulate. There is no evidence of differences in virulence among genotypes, and all form a single serotype (Solomon et al. 2003).

The virus is maintained in an enzootic cycle between *Culex* mosquitoes, with *Cx. tritaeniorhynchus*[2] being the most important vector and with vertebrate host being pigs and poultry (mainly Ardeidae family (herons and egrets) and possibly ducks). Pigs develop high viremias and are considered the main amplifying host. These animals play a significant role during epidemic activity due to their close proximity to humans. Horses are susceptible to develop encephalitis with high mortality; therefore, in endemic areas, they are vaccinated. Both horses and humans are dead-end hosts because they do not develop a viremia high enough to infect vector mosquitoes (Fig. 6.2).

[2]Special abbreviations for mosquito genera proposed by Reinert (2009) were utilized.

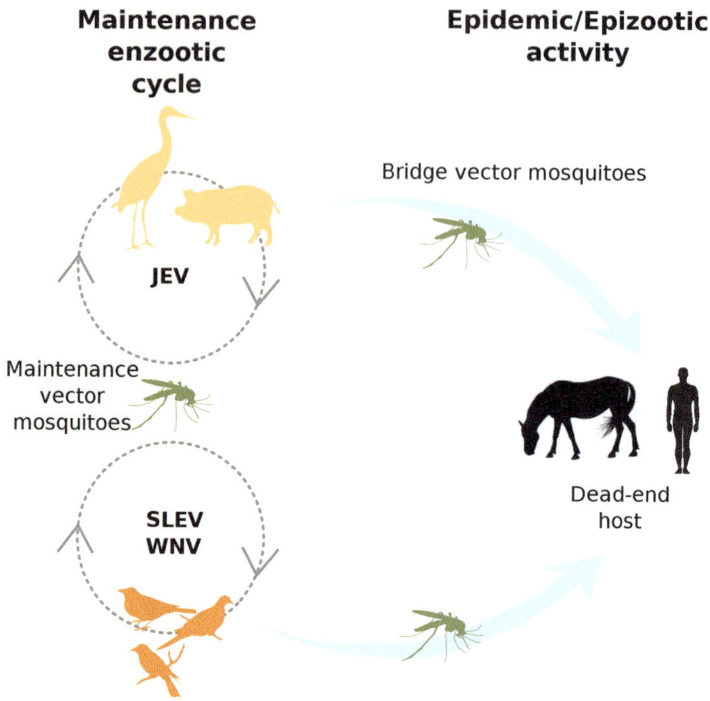

Fig. 6.2 Maintenance and transmission cycles for mosquito-borne *Flavivirus* that cause neurologic symptoms and use birds as amplifiers hosts. *SLEV* St. Louis encephalitis virus, *JEV* Japanese encephalitis virus, *WNV* West Nile virus

The transmission and the high incidence of human cases occur mainly in rural areas, often associated with breeding sites of the vector and hosts (rice crops and flooding areas). In some regions of Asia, these ecological conditions may occur near or occasionally in urban centers. In temperate and subtropical areas of Asia, viral transmission is seasonal; outbreaks usually occur in the rainy season and human disease usually peaks in the summer and fall. In tropical areas, the transmission periods vary with the monsoon rains and irrigation practices and may be prolonged even throughout the year.

In endemic countries, where adults have acquired immunity by natural infection, JEV causes disease primarily in children. About 50,000 cases and 10,000 deaths occur annually throughout Asia, although the disease records are underestimated. Since 1980, the incidence in Japan, South Korea, and Taiwan has decreased due to the use of the vaccine in children along with other preventive measures. For most travelers to Asia, the risk of infection is extremely low but varies according to destination, length of stay, season, and activities. Prolonged stay in rural areas makes travelers as susceptible as a resident individual. Outbreaks in endemic areas affect all age groups, but young children and elderly are in greater risk of disease, as observed for SLEV and WNV.

The inapparent: apparent infection ratio is 200:1–300:1. Among the factors that influence this ratio are age, virulence of viral strain, and cross-protective immunity to other flaviviruses especially dengue (Dutta et al. 2013; Gubler et al. 2007).

6.3.2.1 Clinical Symptoms

Most human infections with JEV are asymptomatic; <1 % of people infected with JEV develop clinical disease. Acute encephalitis is the most commonly recognized clinical manifestation of JEV infection. Milder forms of disease, such as aseptic meningitis or undifferentiated febrile illness, can also occur. The incubation period is 5–15 days. Illness usually begins with sudden onset of fever, headache, and vomiting. Mental status changes, focal neurologic deficits, generalized weakness, and movement disorders may develop over the next few days. Common clinical laboratory findings include moderate leukocytosis, mild anemia, and hyponatremia. Cerebrospinal fluid (CSF) typically has a mild to moderate pleocytosis with a lymphocytic predominance, slightly elevated protein, and normal ratio of CSF to plasma glucose. The case fatality ratio is approximately 20–30 %. Among survivors, 30–50 % have serious neurological, cognitive, or psychiatric sequelae (Gubler et al. 2007; Gould and Solomon 2008).

6.3.2.2 Treatment and Prevention

There is no specific antiviral treatment for JEV infection; therapy consists of supportive care and management of complications. Minocycline has proved to be effective in preventing animal mortality after viral infection and is scheduled for a randomized double-blind phase II clinical trial (Mishra et al. 2009). There are several vaccines, but they have limitations. The only WHO licensed vaccine is inactivated, is costly, and is limited by concerns over side effects; it also requires multiple doses to maintain efficiency and immunity (Shlim and Salomon 2002). Another inactivated vaccine has recently been developed in VERO cells and appears to be the most promising one (Tauber et al. 2007). Many efforts are underway to develop recombinant vaccines, some of which are already at various stages of clinical trials.

6.3.3 Murray Valley Encephalitis

Murray Valley encephalitis virus (MVEV) is the leading cause of arboviral encephalitis in Australia. The MVEV (JEV serocomplex member) is enzootic in the northern territory and northern parts of Western Australia and Queensland, as well as in Papua New Guinea. The main mosquito vector is *Cx. annulirostris*, which feeds mainly on birds (waterfowl) but also on mammals, including man. Other vectors are *Cx. bitaeniorhynchus*, *Cx. sitiens*, and *Ae. normanensis*.

Most infections are asymptomatic and children are the most affected age group. The prodrome includes fever, frontal headache, nausea, vomiting, dizziness, confusion, and convulsions, which can progress to a neurological syndrome, such as encephalitis, in approximately 5 days. The case fatality rate is high (15–32 %), with neurological sequelae in approximately 50 % of survivors.

MVEV has been responsible for six encephalitis outbreaks in southeastern Australia in the 1900s, the last being in 1974 with 58 cases. Since then, the clinical cases have been limited largely to the western and central parts of the country north. However, in 2011, for the first time since 1974, there was an outbreak in that region with 17 confirmed cases. This outbreak showed the change in the epidemiology of MVEV in the last 15 years, with most human cases affecting nonaboriginal people, a few children, and nonresident individuals. During the summer–autumn 2011, there was a record rainfall in many areas of Australia, which was associated with flooding and an increase in the number of mosquitoes and subsequent activity of the virus. This outbreak showed that there is a permanent risk of encephalitis by MVEV in the densely populated southeast Australia. There are no vaccines or antiviral agents available against MVEV (Selvey et al. 2014).

6.3.4 St. Louis Encephalitis Virus

St. Louis encephalitis virus (SLEV) was isolated for the first time during a human encephalitis outbreak in St. Louis, Missouri (USA), in 1933. SLEV is distributed exclusively in the Americas, where it is widely extended. In the USA, SLEV has been one of the main causes of arbovirus encephalitis epidemics until the introduction of WNV in 1999. Most of the clinical cases have been reported by the states of Texas, Florida, southeastern states, and Ohio River basin (Diaz et al. 2013). From 2004 to 2013, an average of seven cases was reported annually in North America (range 1–12) (CDC 2015).

Urban epidemics of encephalitis due to SLEV in other American countries are rare, focal, or of small magnitude or remain undetected. This low incidence in Central and South America may be due to an inadequate case notification system, laboratory diagnosis deficiencies, attenuated viral strains spreading, and/or enzootic cycles involving mosquitoes that do not feed on humans frequently. However, since 2002, SLEV has reemerged as human encephalitis etiological agent, producing outbreaks and isolated cases in Argentina and Brazil. In Cordoba province (central Argentina), a human encephalitis outbreak caused by SLEV was reported in 2005, with 47 cases (predominantly among people 60 years and older) and nine deaths. It was the first outbreak in South America (Spinsanti et al. 2008; Mondini et al. 2007).

SLEV strains generally cluster according to their geographic origin and can be divided into eight genotypes (I to VIII). Genotypes I and II circulate mainly in North America, and genotypes III to VIII circulate mainly in Central and South America. Genotype V is also prevalent in Florida.

SLEV is maintained in nature by transmission between different mosquito *Culex* spp. and passeriform and columbiform birds. The members on this transmission network vary according to the geographic localization and time of the year. SLEV ecology is well characterized for the USA, whereas for the rest of the American continent, it remains practically unknown, except for Argentina, where research has provided advances in SLEV ecological characterization. For this region, *Culex quinquefasciatus* is assumed to be the principal vector, and the main avian hosts are Picui ground doves (*Columbina picui*) and eared doves (*Zenaida auriculata*) (Fig. 6.2).

In the east and west of the USA, SLEV transmission cycles are differentiated epidemiologically based on ecological determinants of virus transmission. In the eastern states, main vectors belong to the *Culex pipiens* complex (*Cx. pipiens* and *Cx. quinquefasciatus*), and their main hosts are house sparrows (*Passer domesticus*). These peridomestic habit species develop frequently in rich organic material water, such as sewers and peridomestic water reservoirs. These mosquitoes are spread in densely populated urban and suburban environments, especially under deficient sanitary conditions. In western states of the USA, SLEV has an endemic transmission pattern; the main vector mosquito is *Cx. tarsalis*. This species reproduces in flooded and irrigated soils and in industrial or urban residual water.

Humans and domestic mammals are excluded from the basic transmission cycle because the viremia titers are insufficient to infect vector mosquitoes. The SLEV epidemiology is driven by climatic, entomological, virus, and host factors, which form a still not fully understood complex network of interactions (Diaz et al. 2013).

6.3.4.1 Clinical Manifestations

The clinical spectrum of a SLEV infection can comprise from an unspecific febrile syndrome, as in influenza, to a disease affecting the CNS. The incubation period can last from 4 to 21 days. The onset of symptoms is characterized by general discomfort, fever, chills, cephalea, anorexia, nausea, myalgia, sore throat, or cough, followed by the neurological or meningeal signs after 1–4 days. Less than 1 % of SLEV infections are clinically apparent. Disease severity increases with age, with people over 60 years old being the most frequently reported cases of encephalitis. Mortality rate increases with age (over 75 years). Underlying medical conditions, such as diabetes, hypertension, chronic alcoholism, and arteriosclerosis, predispose to severe infection and subsequent death. There are no vaccines or antiviral agent available against SLEV (Diaz et al. 2013).

6.3.5 Usutu Virus

Usutu virus (USUV), a member of JEV complex, is phylogenetically closely related to WNV. It was isolated from *Culex* mosquitoes in Natal, South Africa, in 1959. Later, it was found in other mosquito genera, such as *Mansonia* and *Aedes*, and in rodents and birds in other countries in Africa. This virus has recently been introduced in Europe, first detected in birds in Italy, in 1996, and later in Vienna, Austria, in 2001, where it caused hundreds of deaths of birds. Other countries, such as the Czech Republic, Germany, Poland, Spain, England, Hungary, and Switzerland, have reported serological and/or viral activity in birds and humans. Cases of human disease are rare. Recently, a case with rash and fever in Central African Republic was reported, and the virus was detected in patients with neuroinvasive disease in Italy (Ashraf et al. 2015).

6.3.6 West Nile Virus

West Nile virus (WNV) was first recognized as a human pathogen in 1937, when it was isolated from a febrile woman in Uganda (West Nile district). Later on, its activity was reported in Africa, Asia, Australia, and Europe. It was not considered a human concern until the end of 1990. Outbreaks by WNV were infrequent, sporadic, and associated with low pathogenicity and mild neurological disease. The first outbreak in humans was reported in Israel in 1950. Other outbreaks were detected in southern France and South Africa. After that, larger outbreaks were reported in Algeria (50 cases and 2 deaths in 1994), Romania (393 cases and 17 deaths in 1996), Tunisia (173 cases in 1997), Russia (318 cases in 1999), and Israel (417 cases in 2000) (Chancey et al. 2015).

Considered as a flavivirus from the Old World, WNV was first detected in the New World in 1999, causing a human outbreak in New York City. After its introduction, the virus successfully established and dispersed throughout the USA. Since its introduction, the virus has caused human and equine encephalitis outbreaks every year and has become a causal factor for the decline of wild bird populations (Roehrig 2013).

By 2002, WNV reached Mexico and the Caribbean islands. Today, it is active in most countries of the Americas. Contrarily to the USA, no human encephalitis outbreaks by WNV were reported in Latin American countries. Few epizootic events in equines have been documented in Argentina and Mexico (Diaz et al. 2011).

Several Mediterranean countries reported WNV cases in animals and humans from 2009 to 2013. During this period, WNV cases were reported in Algeria, Bulgaria, Croatia, Former Yugoslav Republic of Macedonia, Greece, Hungary, Israel, Italy, Kosovo, Montenegro, Morocco, Occupied Palestinian Territories, Portugal, Romania, Russia, Serbia, Spain, Tunisia, Turkey, and Ukraine (Chancey et al. 2015).

WNV is maintained through biological transmissions in which *Culex* spp. mosquitoes are involved as vectors and Passeriformes birds as hosts. The most comprehensive knowledge regarding the ecology of WNV is based on studies carried out in the USA, where most aspects of its transmission dynamics have been analyzed. The main vector species include *Cx. pipiens*, *Cx. restuans*, *Cx. quinquefasciatus*, *Cx. nigripalpus*, and *Cx. tarsalis*, whereas only a few non-*Culex* species have been considered potential vectors, such as *Aedes albopictus* and *Ae. vexans* (Turell et al. 2005). Regarding hosts, the main species involved are the American robin (*Turdus migratorius*), the Northern Cardinal (*Cardinalis cardinalis*), the house sparrow (*Passer domesticus*), the blue jay (*Cyanocitta cristata*), the northern mockingbird (*Mimus polyglottos*), the western scrub jay (*Aphelocoma californica*), the American crow (*Corvus brachyrhynchos*), and the black-billed magpie (*Pica hudsonia*) (Kilpatrick et al. 2007) (Fig. 6.2). Geographical and seasonal variations in hosts and vectors were observed across the USA as well as in other countries (Argentina, France, and Spain). Alternative transmission mechanisms contribute to the maintenance of WNV in nature. For instance, in vertical vector transmission, the virus is transmitted to its descendants through an infected female mosquito. This mode tends to have low transmission rates but gains importance in mild temperate areas as an overwinter mechanism (Anderson and Main 2006; Unlu et al. 2010). Direct transmission among hosts is another alternative mechanism observed in several bird species maintained under laboratory conditions, where virus transmissions occur both due to the intake of WNV-infected food, as in the great horned owl (*Bubo virginianus*), the American crow, the common grackle, the house finch, and the house sparrow, and to direct contact with partners (as in the ring-billed gull, the blue jay, the black-billed magpie, and the American crow) (Komar et al. 2003). Besides mosquitoes and birds being the key participants in the maintenance of the virus in nature, other animals, such as fox squirrels (*Sciurus niger*) and ticks (*Dermacentor andersoni*, *De. variabilis*, *Ixodes scapularis*), might play a fundamental role (Anderson et al. 2003; Root et al. 2006).

6.3.6.1 Clinical Manifestations

As a rule for flavivirus infections, most of WNV infections in humans are asymptomatic; however, during recent outbreaks in Europe, Israel, and the USA, approximately 20 % of infections have resulted in a mild flu-like illness. Symptoms include an abrupt onset of fever, headache, myalgia, nausea, fatigue, weakness, vomiting, and diarrhea. A generalized, maculopapular rash may also develop 5–12 days after onset of illness. Symptoms usually develop 2–14 days after virus infection. The illness typically lasts for 2–5 days. The neuroinvasive infection by WNV is characterized by altered mental status (disorientation, obtundation, stupor, and coma) encephalitis, meningitis, and/or poliomyelitis-like flaccid paralysis with long-term neurological sequelae. Less frequent complications include myocarditis, acute inflammatory demyelinating polyradiculoneuropathy (Guillain–Barré syndrome), and respiratory failure and death resulting from infection of the brainstem and cranial portion of the cervical spinal cord. The fatality rate in patients with neurological symptoms is approximately 10 %. Neurological disease is more common in elderly and immunocompromised patients (Gray and Webb 2014).

Treatment of humans and other animals with WNV disease is symptomatic. No clear benefits were found with the administration of interferon, ribavirin, human immunoglobulin, corticosteroids, or antisense oligomers.

6.3.6.2 Prevention and Control

The important focus of prevention in most arbovirus diseases is environmental control of mosquito vectors. These measures include removal of any potential mosquito breeding sites, such as containers or old tires that collect standing water. Mosquito control can be based on biological control (mosquito fish, crustaceans, entomopathogenic bacteria, and fungi), environmental management, and chemical insecticides (larvicides and pesticides).

Mosquito avoidance and repellents are another line of defense against infection (*N,N*-diethyl-*m*-toluamide (DEET) or *p*-menthane-3,8diol-oil of lemon eucalyptus). Remaining indoors during dawn and dusk helps humans and domestic animals avoid exposure to *Culex* mosquitoes during their peak feeding periods. The use of fans and mosquito traps can also reduce mosquito exposure (Bellini et al. 2014). Due to the potential severity of WNV infection in horses, vaccination is recommended in North America. There are several commercially available vaccines for equine use that contain either killed virus or modified-live recombinant virus. No licensed human vaccine exists to date.

6.3.7 Zika Virus

The Zika virus (ZIKV) was isolated from a *Rhesus* monkey in the Zika forest (Kampala, Uganda) during 1947. It was first isolated from humans in 1952, in Uganda and in Tanzania. The activity of ZIKV was reported in Africa (Burkina Faso, Cameroon, Central African Republic, Egypt, Ethiopia, Gabon, Ivory Coast, Kenya, Nigeria, Senegal, Sierra Leone, Somalia, Tanzania, Uganda), Asia (India, Indonesia, Malaysia, the Philippines, Thailand, Vietnam), in the Pacific (Micronesia/Yap, French Polynesia (FP), New Caledonia, Cook, and Easter Island), and Oceania (Ioos et al. 2014). By 2014, ZIKV reached South American territory. At the moment, 17 human cases by ZIKV were diagnosed in Brazil (eight in Bahia, eight in Rio Grande do Norte, and one in São Paulo states) (ProMED mail 2015).[3] This virus received little attention until its sudden emergence in Yap Island (Micronesia) in 2007, which affected about 5000 people (Duffy et al. 2009), and its largest epidemic in FP by 2013,

[3]Dozens of cases of Guillain-Barré syndrome have been reported from the Brazilian state of Bahia, probably associated to Zika virus (http://www.saude.ba.gov.br/novoportal/index.php?option=com_content&view=article&id=9453:sindrome-de-guillain-barre-perguntas-e-respostas&catid=13:noticias&Itemid=25, accessed in Sept 14, 2015) (CBM).

where an estimated of 29,000 people consulted for ZIKV infection, which revealed its epidemic potential (Cao-Lormeau et al. 2014).

ZIKV ecology is poorly known. The transmission is mostly vectorial by sylvatic and urban mosquitoes of the genus *Aedes* (*Ae. aegypti*, *Ae. africanus*, *Ae. albopictus*, *Ae. apicoargenteus*, *Ae. furcifer*, *Ae. luteocephalus*, *Ae. polynesiensis*, and *Ae. vitattus*) (Hayes 2009). Ecological, epidemiological, and laboratory evidences support the role of *Ae. aegypti* as the main urban vector of ZIKV activity in Asian countries. *Aedes hensilii* was the predominant mosquito species present during the outbreak in Yap Island. However, ZIKV was not detected in any mosquitoes on the island during the outbreak (Duffy et al. 2009). The virus sylvatic host has not been completely identified, but some authors suggest a primate host. Anti-Zika antibodies have been reported in various animals such as big mammals (orangutans, zebras, elephants, etc.) and rodents (Darwish et al. 1983). Although viral loads in humans were not quantified at this moment, the epidemic potential of ZIKV in urban ecosystems clearly indicate human beings are amplifying host.

ZIKV infection can cause a mild chikungunya and dengue-like syndrome, including mild fever; headache; edema of extremities; retro-orbital pain; rash in face, thorax, and extremities; arthralgia; myalgia; vertigo; conjunctivitis; and digestive disorders. Symptoms appear after an incubation period of a few days after the bite of an infected mosquito and usually last 3–12 days (Duffy et al. 2009; Ioos et al. 2014; Oehler et al. 2014). Asymptomatic presentations are frequent in ZIKV infection. This clinical similarity with other, more commonly diagnosed arboviral infections, such as chikungunya (CHIKV) and dengue (DENV), might delay the diagnosis and/or lead to underestimation of ZIKV infections. From 2007 to 2015, the outbreaks of ZIKV infection have been associated with different consequences in each region. During the ZIKV outbreak in FP (2013), there was an increase in the number of cases of Guillain–Barré syndrome (Willison et al. 2016). In addition, by the end of 2015 and the beginning of 2016, the number of cases of confirmed microcephaly in fetus and newborns from Brazil remained at least five times higher than the number of annual cases reported before 2015 (Victora et al. 2016). ZIKV was specifically detected in the brain tissue of a fetus with microcephaly and also in the amniotic fluid the other patients whose babies were diagnosed with microcephaly (Calvet et al. 2016).

The viremic period in humans could be short, from the third to the fifth day after onset of symptoms. No commercial kit is currently available for the detection of antibodies specifically related to ZIKV. The Centers for Disease Control and Prevention developed an ELISA to detect IgM and IgG anti ZIKV. Cross-reactive results in sera from patients in the convalescent phase occurred more frequently among patients with evidence of previous flavivirus infections than among those with apparent primary ZIKV infections. Cross-reactivity was more frequently noted with DENV than with YFV, JEV, MVEV, or WNV. Neutralizing antibody developed as early as 5 days after illness onset. The plaque reduction neutralization assay has generally improved specificity over immunoassays. PCR tests can be conducted on samples obtained less than 10 days after illness onset (Hayes 2009).

There is no specific treatment or vaccine. The treatment is symptomatic, combining acetaminophen and antihistaminic drugs. Prevention against the infection relies on individual protection against bites and eradication of mosquitoes (antivectorial prevention). Deltamethrin could currently be the insecticide warranting a satisfactory result in aerosol application. Individual protection includes wearing long and light-colored clothes and using skin repellents and mosquito bed nets.

6.4 Tick-Borne Viruses

The family *Flaviviridae* includes several tick-borne viruses affecting humans. There are two phylogenetically distinguishable groups of tick-borne viruses, the TBE complex viruses (including Karshi virus, KFDV, Langat virus, Looping-ill virus, OHFV, POWV, Royal Farm virus, and TBEV) and the Tyuleniy group (including Meaban virus, Saumarez Reef virus, and Tyuleniy virus). The viruses in the TBE group are mammalian viruses, whereas the

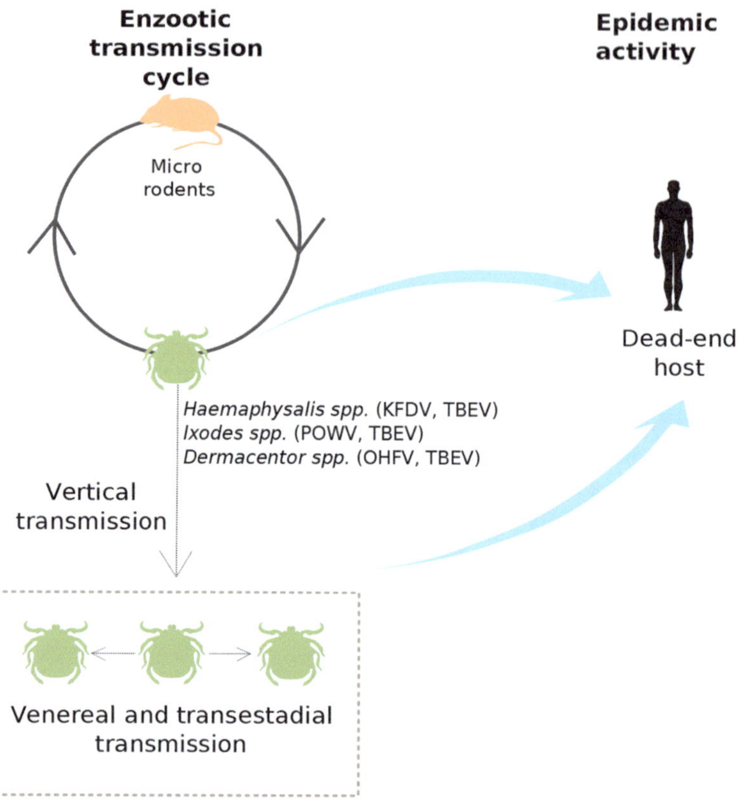

Fig. 6.3 Transmission cycle for tick-borne *Flavivirus*. *KFDV* Kyasanur Forest diseases virus, *TBEV* tick-borne encephalitis virus, *POWV* Powassan virus, *OHFV* Omsk hemorrhagic fever virus

members of the Tyuleniy group are seabird viruses. Gadgets Gully and Kadam viruses do not fit in either group. Tick-borne viruses are transmitted by a variety of hard and/or soft ticks (Gubler et al. 2007).

6.4.1 Tick-Borne Encephalitis Virus

The tick-borne encephalitis virus (TBEV), isolated in 1937, is the etiological agent of tick-borne encephalitis. Three different subtypes were detected according to its geographic distribution: European, Siberian, and Far Eastern (formerly known as Russian Spring Summer encephalitis virus, RSSEV). The European subtypes are found from the Ural Mountains (Russia) to the Alsace region (France), Scandinavia to the north, and parts of the Mediterranean basin along the Adriatic coast to the south. The Far Eastern subtype has a distribution from the Baltic countries to northeastern China and northern Japan, and the Siberian subtype is found in Siberia and the Baltic states. Thus, in parts of the Baltic countries, the three subtypes overlap (Gubler et al. 2007).

Hard ticks of the Ixodidae family, which thrive in the rich and moist undergrowth of forest floors, are the main vector and reservoir of TBEV. In Western Europe, the main vector is *Ix. ricinus*,[4] and *Ix. persulcatus* is the vector in eastern Eurasia. The distribution of the two species overlaps in eastern parts of the Baltic. In northeastern China and northern Japan, the vector is *Ix. ovatus* and, on the coast of the Adriatic Sea, *Ix. gibosus* would be involved in the transmission cycle. *Haemaphysalis* and *Dermacentor* spp. ticks are involved in the transmission cycle. The main hosts are small rodents, in particular *Apodemus* species, with humans being accidental hosts (dead-end host) (Fig. 6.3). Large animals serve as feeding hosts for ticks but have no role in viral maintenance. Human infection occurs frequently in rural areas and during peri-

[4]Special abbreviations for tick genera proposed by Dantas-Torres (2008) were utilized.

ods of increased activity of the vector (spring–summer). Person to person transmission has not been reported, but vertical transmission from infected mother to fetus has been proven. Infection can also occur by drinking raw milk from infected goats, sheep, or cows.

The modes of transmission in vertebrates include three mechanisms, such as biological transmission, direct transmission, and non-viremic transmission, and in ticks transstadial and vertical transmissions. Biological transmission takes place when infected ticks feed on a vertebrate host. In non-viremic transmission, the virus is transferred from infected ticks to uninfected when fed simultaneously and in close proximity on the same non-viremic host. Transstadial (between stages larva–nymph–adult) and vertical transmissions are important for the maintenance of virus (ticks have a long life) (Gubler et al. 2007).

In Europe and Asia, between 10,000 and 15,000 TBE cases are reported annually. TBE is endemic in 27 European countries and is a reportable disease in only 16 countries (Bogovic and Strle 2015). Russia has the largest number of reported cases, and western Siberia the highest incidence in the world. Other European countries with known endemic activity are Austria, Croatia, Czech Republic, Denmark, Estonia, Finland, France, Germany, Hungary, Italy, Latvia, Lithuania, Norway, Poland, Romania, Slovakia, Slovenia, Sweden, and Switzerland. Asian countries with cases or viral activity include China, Japan, Kazakhstan, Kyrgyzstan, Mongolia, and South Korea.

The population density of the vectors and infection rates in endemic areas are highly variable, and rates higher than 40 % have been reported for the tick vector in Siberia. In Central Europe, the peaks of human infection occur 3–4 weeks after the maximum activities of adult ticks (May–June, September–October), whereas in the region of Scandinavia, there is an infection peak in late summer.

The incidence and severity of infection is higher in adults ≥50 years. Most cases occur in low areas (<760 m above sea level). In the last 30 years, the geographical range has extended to new areas, with the virus being found at altitudes of 1500 m or higher. This is probably due to a complex combination of changes in the diagnosis and monitoring, human activities (increasing outdoor activities such as hiking, camping, fishing, cycling, training, etc.), and socioeconomic and ecological factors. The risk is negligible for those who remain in urban or non-forested areas and consume pasteurized products (Gubler et al. 2007; Bogovic and Strle 2015).

6.4.1.1 Clinical Manifestations

All age groups can be infected, but mild symptoms are generally seen in children and the most severe ones in elderly people. The incubation period ranges from 7 to 14 days. The clinical syndrome varies among three subtypes. Seventy percent of the infected individuals with European subtype develop a biphasic febrile illness similar to the beginning of an Influenza, which can last up to 7 days without neurological signs. However, in some patients with fever, an involvement of CNS such as meningitis or meningoencephalitis with a case fatality rate of 1–2 % is observed. The severe forms of the disease are more frequently caused by TBEV Far Eastern subtype. The onset is more gradual and includes fever, headache, anorexia, nausea, vomiting, and photophobia. Neurological syndromes are observed in 5–50 % of cases and include aseptic meningitis, meningoencephalitis, encephalomyelitis, and postencephalitic syndrome. Although rare, hemorrhagic manifestations have been observed in fatal cases in the region of Novosibirsk (Russia). Chronic infections with this subtype have been described and the virus has been isolated from the brain of patients for recurrent episodes. Cognitive or neuropsychiatric problems, headaches, balance disorders, dysphasia, and hearing impairment may persist for a long period. The case fatality rate is 5–35 %. Infection with the Siberian subtype is less severe than with Far Eastern subtype but is more severe than the European subtype, with a mortality rate of 1–3 % (Růžek et al. 2013).

6.4.1.2 Treatment and Prevention

Tick-borne encephalitis should be considered in a traveler who developed generalized febrile illness that progresses to a neuroinvasive form within 4 weeks of arrival from a TBEV endemic area. Support treatment is supplied to avoid complications and relieve symptoms.

Prevention measures include vaccination and avoidance of vector contact. Inactivated vaccines are available in Europe and Russia. Immunogenicity studies suggest that they would get cross-protection against the three subtypes. The travelers should note that the full immunization scheme lasts more than 6 months (Růžek et al. 2013; Bogovic and Strle 2015).

6.4.2 Powassan

Powassan virus (POWV) is a member of TBEV antigen complex. It is widely distributed in North America and Russia. The first case was identified in Powassan, Ontario, Canada, in 1958. It is maintained through biological transmission between *Ixodes* ticks and small- and medium-sized rodents. Humans are considered dead-end hosts. There are two POWV lineages with different transmission cycles: lineage I (POWV) and lineage II (deer tick virus; DTV). POWV is associated with *Ix. cookei* and *Ix. marxi*, which rarely bites humans, and DTV, associated with *Ix. scapularis*, which often feeds on humans and transmits Lyme disease. Both lineages are associated with human disease. In North America, there are three enzootic transmission cycles: *Ix. cookei* and marmots, *Ix. marxi* and squirrels, and *Ix. scapularis* and white-footed mice. In Russia, the virus has been isolated from *Ix. persulcatus* and *Ha. longicornis* (Fig. 6.3).

POWV is frequently found in the central and northeastern US states, Great Lakes region, and southern Canada, but it has also been isolated in Colorado and California Lakes region and southern Canada, but it has also been isolated in Colorado and California. The initial symptoms include headache, fever, and nuchal rigidity. Symptoms also include vomiting, respiratory distress, seizures, lethargy, memory impairment, somnolence, hemiplegia, paralysis, and wasting. It can also cause meningitis and encephalitis. Approximately 10 % of encephalitis cases are fatal (Gubler et al. 2007).

From 2004 to 2013, Powassan virus neuroinvasive disease cases have been reported in Maine (2), Massachusetts (1), Minnesota (20), New Hampshire (1), New Jersey (1), New York (17), Pennsylvania (1), Virginia (1), and Wisconsin (13) (CDC 2015).

6.4.3 Other Tick-Borne Viruses

6.4.3.1 Kyasanur Forest Disease Virus

Kyasanur Forest disease virus (KFDV) was isolated in 1957 from a sick primate in Kyasanur Forest, Shimoga District, Karnataka, India. In nature, the virus is transmitted between vertebrate and *Ha. spinginera* ticks (Fig. 6.3). The human disease has an incubation period of 3–8 days followed by chills, headache, malaise, and high fever for 5–12 days. Hemorrhagic manifestations and neurological syndromes may also occur; mortality rate is 3–5 %. Between January and March 2014, an outbreak by KFDV in Kannangi and Konandur (India) was recorded (Yadav et al. 2014).

Another virus closely related to KFDV that produces hemorrhagic disease in humans is the Alkhurma virus. It was isolated in 1994 from a patient with a diagnosis of DHF in Saudi Arabia. Individuals who are most exposed include animal handlers in sheep trading, butchers, and meat consumers. One person was infected by drinking raw camel milk (Gubler et al. 2007).

6.4.3.2 Omsk Hemorrhagic Fever Virus

Omsk hemorrhagic fever is a prevalent disease in parts of western Siberia (Russia). Symptoms include fever, headache, nausea, myalgia, and hemorrhagic manifestations. Some patients may develop pneumonia, nephrosis, and meningitis. The virus is mainly transmitted by the *De. reticulatus* tick, and humans (especially hunters) are infected by contact with infected muskrats (Fig. 6.3). The introduction of this species to Siberia in the 1930s probably led to the emergence of this virus in the region, coupled with ecological disturbance caused by humans (Růžek et al. 2010).

References

Anderson JF, Main AJ, Andreadis TG et al (2003) Transstadial transfer of West Nile virus by three species of ixodid ticks (Acari: Ixodidae). J Med Entomol 40:528–533

Anderson JF, Main AJ (2006) Importance of vertical and horizontal transmission of West Nile virus by *Culex pipiens* in the Northeastern United States. J Infect Dis 194:1577–1579

Ashraf U, Ye J, Ruan X et al (2015) Usutu virus: an emerging flavivirus in Europe. Viruses 7:219–238

Bellini R, Zeller H, Van Bortel W (2014) A review of the vector management method o prevent and control outbreaks of West Nile virus infection and the challenge for Europe. Parasit Vectors 7:323

Bogovic P, Strle F (2015) Tick-borne encephalitis: a review of epidemiology, clinical characteristics, and management. World J Clin Cases 3:430–441

Calvet G, Aguiar RS, Melo AS et al (2016) Detection and sequencing of Zika virus from amniotic fluid of fetuses with microcephaly in Brazil: A case study. Lancet Infect Dis 16:653–660

Cao-Lormeau VM, Roche C, Teissier A et al (2014) Zika virus, French Polynesia, South Pacific, 2013. Emerg Infect Dis 20:1085–1086

Center for Diseases Control and Prevention (2015). Saint Louis Encephalitis. Epidemiology and Geographic distribution. http://www.cdc.gov/sle/technical/epi.html

Chancey C, Grinev A, Volkova E et al (2015) The global ecology and epidemiology of West Nile Virus. Biomed Res Int 2015:376230

Dantas-Torres F (2008) Towards the standardization of the abbreviations of genus names of ticks (Acari: Parasitiformes: Ixodida). Vet Parasitol 154:94–97

Darwish MA, Hoogstraal H, Roberts TJ et al (1983) A sero-epidemiological survey for certain arboviruses (Togaviridae) in Pakistan. Trans R Soc Trop Med Hyg 77:442–445

Diaz LA, Quaglia A, Flores FS et al (2011) Virus West Nile en Argentina: un agente infeccioso emergente que plantea nuevos desafíos. Hornero [online] 26:5–28, http://www.scielo.org.ar/scielo.php?script=sci_arttext&pid=S0073-34072011000100002&lng=es&nrm=iso

Diaz LA, Spinsanti L, Contigiani M (2013) St. Louis encephalitis. In: Singh S, Růžek D (eds) Neuroviral infections. Taylor and Francis Group, LLC, CRC Press, Boca Raton, London, New York, pp 239–260

Duffy MR, Chen TH, Hancock WT et al (2009) Zika virus outbreak on Yap Island, Federated States of Micronesia. N Engl J Med 360:2536–2543

Dutta K, Nazmi A, Basu A (2013) Japanese encephalitis virus and human CNS infection. In: Singh S, Růžek D (eds) Neuroviral infections. Taylor and Francis Group, LLC, CRC Press, Boca Raton, London, New York, pp 193–210

Gomes G, Causey OR (1959) Bussuquara, a new arthropod-borne virus. Proc Soc Exp Biol Med 101:275–279

Gould EA, Solomon T (2008) Pathogenic flaviviruses. Lancet 371:500–509

Gray TJ, Webb CE (2014) A review of the epidemiological and clinical aspects of West Nile virus. Int J Gen Med 7:193–203

Gubler DJ, Kuno G, Markoff L (2007) Flaviviruses. In: Knipe DM, Howley PM, Griffin DE et al (eds) Fields virology, 5th edn. Lippincott Williams & Wilkins Publishers, Philadelphia, PA, pp 1153–1252

Hayes EB (2009) Zika virus outside Africa. Emerg Infect Dis 15:1347–1350

ICTV (2014) Virus Taxonomy: 2014 Release. http://www.ictvonline.org/virusTaxonomy.asp

Jones CT, Ma L, Burgner JW et al (2003) Flavivirus capsid is a dimeric alpha-helical protein. J Virol 77: 7143–7149

Ioos S, Mallet HP, Leparc Goffart I et al (2014) Current Zika virus epidemiology and recent epidemics. Med Mal Infect 44:302–307

Kilpatrick AM, Ladeau SL, Marra PP (2007) The ecology and impact of West Nile virus in the Western Hemisphere. Auk 124:1121–1136

Komar N, Langevin S, Hinten S et al (2003) Experimental infection of North American birds with the New York 1999 strain of West Nile virus. Emerg Infect Dis 9:311–322

Laemmert HW, Hughes TP (1947) The virus of Ilheus encephalitis; isolation, serological specificity, and transmission. J Immunol 55:61–67

Lopes OS, Sacchetta LA, Coimbra TLM et al (1978) Emergence of a new arbovirus disease in Brazil. II. Epidemiological studies on 1975 epidemic. Am J Epidemiol 108:394–401

Mackenzie JS (2005) Emerging zoonotic encephalitis viruses: lessons from Southeast Asia and Oceania. J Neurovirol 11:434–440

Mishra MK, Ghosh D, Duseja R et al (2009) Antioxidant potential of Minocycline in Japanese Encephalitis Virus infection in murine neuroblastoma cells: correlation with membrane fluidity and cell death. Neurochem Int 54:464–470

Mondini A, Cardeal IL, Lázaro E et al (2007) Saint Louis encephalitis virus, Brazil. Emerg Infect Dis 13:176–178

Oehler E, Watrin L, Larre P et al (2014) Zika virus infection complicated by Guillain-Barré syndrome—case report, French Polynesia, December 2013. Eurosurveillance 19:20720

Ong RY, Lum FK, Ng LFP (2014) The fine line between protection and pathology in neurotropic flavivirus and alphavirus infections. Future Virol 9(3): 313–330

Pauvolid-Corrêa A, Kenney JL, Couto-Lima D et al (2013) Ilheus virus isolation in the Pantanal, west-central Brazil. PLoS Negl Trop Dis 7, e2318. doi:10.1371/journal.pntd.0002318

ProMED mail (2015) ZIKA VIRUS - BRAZIL (02): (São Paulo). International Society of Infectious Diseases. Archive Number: 20150524.3382529. http://www.promedmail.org/

Reinert JF (2009) List of abbreviations for currently valid generic-level taxa in family Culicidae (Diptera). Eur Mosq Bull 27:68–76

Roehrig JT (2013) West Nile virus in the United States—a historical perspective. Viruses 5:3088–3108

Root JJ, Oesterle PT, Nemeth NM et al (2006) Experimental infection of fox squirrels (*Sciurus niger*) with West Nile virus. Am J Trop Med Hyg 75:697–701

Růžek D, Yakimenko VV, Karan LS et al (2010) Omsk haemorrhagic fever. Lancet 376:2104–2113

Růžek D, Bartosz B, Günther G (2013) Tick-borne encephalitis. In: Singh S, Růžek D (eds) Neuroviral infections. Taylor and Francis Group, LLC, CRC Press, Boca Raton, London, New York, pp 211–238

Sabattini MS, Avilés G, Monath TP (1998) Historical, epidemiological and ecological aspects of arbovirus in Argentina: Flaviviridae, Bunyaviridae and Rhabdoviridae. In: Travassos da Rosa APA, Vasconcelos PFC, Travassos da Rosa JFS (eds) An overview of arbovirology in Brazil and neighboring countries. Instituto Evandro Chagas, Belém, pp 113–134

Selvey LA, Dailey L, Lindsay M et al (2014) The changing epidemiology of Murray Valley encephalitis in Australia: the 2011 outbreak and a review of the literature. PLoS Negl Trop Dis 8, e2656

Shlim DR and Solomon T (2002) Japanese encephalitis vaccine for travelers: exploring the limits of risk. Clin Infect Dis 35:183–188

Solomon T, Ni H, Beasley DW et al (2003) Origin and evolution of Japanese encephalitis virus in southeast Asia. J Virol 77:3091–3098

Spinsanti LI, Díaz LA, Glatstein N et al (2008) Human outbreak of St. Louis encephalitis detected in Argentina, 2005. J ClinVirol 42:27–33

Tauber E, Kollaritsch H, Korinek M et al (2007) Safety and immunogenicity of a Vero-cell-derived, inactivated Japanese encephalitis vaccine: a non-inferiority, phase III, randomised controlled trial. Lancet 370:1847–1853

Turell MJ, Dohm DJ, Sardelis MR et al (2005) An update on the potential of North American mosquitoes (Diptera: Culicidae) to transmit West Nile Virus. J Med Entomol 42:57–62

Turtle L, Griffiths MJ, Solomon T (2012) Encephalitis caused by flaviviruses. Q J Med 105:219–223

Unlu I, Mackay AJ, Roy A et al (2010) Evidence of vertical transmission of West Nile virus in field-collected mosquitoes. J Vector Ecol 35:95–99

Victora CG, Schuler-Faccini L, Matijasevich A et al (2016) Microcephaly in Brazil: How to interpret reported numbers? Lancet 387:621–624

Willison HJ, Jacobs BC, van Doorn PA (2016) Guillain–Barre syndrome. Lancet 388(10045):717–727

Yadav PD, Shete AM, Patil DY et al (2014) Outbreak of Kyasanur Forest disease in Thirthahalli, Karnataka, India, 2014. Int J Infect Dis 26:132–134

Dengue

7

Pedro F.C. Vasconcelos

Abstract

Dengue is an infectious disease and represents a major public health problem in tropical and subtropical countries with crescent concern due the severe illness and deaths worldwide. Indeed, the WHO estimates an annual occurrence of approximately 392 million infections in more than 100 countries of four of the five continents, where disease is endemic. Among these, around 90 million are symptomatic infections that result in almost half million hospitalization and 50,000 deaths. Dengue is, presently, the leading widespread mosquito-borne disease in the world affecting or placing under risk almost half of the world population (~3 billion people) living in the endemic areas, especially Latin America and Asia-Pacific regions, but the absence of information in Africa is of concern. Epidemics of dengue disease in general are accompanied by patient's suffering, massive economic losses due the combat of the vectors and hospitalization, and strained health services, as consequence of the increase by at least 30-fold in the incidence of disease in the last five decades. It is remarkable that the increase in air travel in the world has facilitated the disease spread globally. Therefore, a further global strategy for controlling and preventing dengue is necessary and should be jointly driven by WHO and its agencies, and the development of vaccines is expected to help on this objective. Only a global effort including the community will allow diminishing the burden of dengue in the world.

Keywords

Dengue • Hemorrhage • *Aedes aegypti* • Culicidae

P.F.C. Vasconcelos (✉)
National Reference Laboratory of Arboviruses, Evandro Chagas Institute, Rodovia BR 316, Km 07, S/N, 67030-000 Ananindeua, Brazil

National Institute of Science and Technology for Viral Hemorrhagic Fevers, Rodovia BR 316, Km 07, S/N, 67030-000 Ananindeua, Brazil

PAHO/WHO Collaborating Center for Arbovirus Research and Diagnostic Reference, Evandro Chagas Institute, Ministry of Health, Rodovia BR 316, Km 07, S/N, 67030-000 Ananindeua, Brazil

Department of Pathology, Pará State University, Belém, Pará, Brazil
e-mail: pedrovasconcelos@iec.pa.gov.br

C.B. Marcondes (ed.), *Arthropod Borne Diseases*, DOI 10.1007/978-3-319-13884-8_7

7.1 Introduction

Dengue virus (DENV) is endemic in more than 100 countries around the world mainly in the tropical and subtropical regions. There are four DENV serotypes (DENV1, DENV2, DENV3, and DENV4), and all have been associated with dengue fever (DF) and severe dengue, and, in general, these serotypes are hyperendemic in almost all affected countries. It is estimated that approximately 92 million dengue cases and 390 million infections by DENV occur annually affecting people living in more than 100 countries especially in the tropical and subtropical regions affecting the continents of Asia, Africa, Oceania, and America; it has been estimated that approximately three billion people live in areas under risk of dengue (Bhatt et al. 2013). The World Health Organization (WHO) estimates the occurrence of around half million severe cases and 50,000 deaths each year, mostly in Southeast Asia and Latin America, but these numbers are underestimated, since very few information on cases is available on African endemic countries and also due the asymptomatic infections that occur in numbers almost unknown in many large urban areas. These numbers incriminate dengue as the most widespread arbovirus and arthropod-borne disease in the world (Teixeira et al. 2002; Bhatt et al. 2013; PAHO 2014).

DENV is transmitted in urban sets by the primatophilic peri-domestic mosquito *Aedes aegypti*; other *Aedes*[1] mosquito species such as *Ae. albopictus* and *Ae. polynesiensis* have been incriminated as potential secondary vectors especially in Southeast Asia. These urban mosquito species are living near or inside the human dwellings and use any recipients containing clean water to breed. Infection becomes possible when a female *Ae. aegypti* mosquito bites a febrile viremic human, and further 10–14 days later, it becomes infective and ready to transmit the DENV to other people during another bite. Indeed, it is too common the infection of several people living in the same house and also their neighbors. In general, these clusters are responsible for a quick spread of dengue disease transmission due the increased number of human infection, in which they represent source of DENV to uninfected mosquitoes (Vasconcelos et al. 2013; PAHO 2014).

[1]Although several taxonomic modifications have been proposed on Culicidae genera, mostly splitting *Aedes* by Reinert et al. (2009) (and accepted by CBM), this taxonomy has been used in the chapters according to authors' preference. To facilitate utilization by health personnel, all new aedine genera can be considered *Aedes* (CBM).

7.2 The Agent and Its Genome

DENV is a member of the Flaviviridae, genus *Flavivirus*. Among flaviviruses, the DENV complex is constituted by four DENV serotypes (DENV1, DENV2, DENV3, and DENV4), and all of them can cause dengue fever and severe dengue, the clinical presentation of dengue disease. As other flaviviruses, DENV serotypes are single strand plus sense non-segmented RNAs. The entire genome consists of a single open reading frame (ORF) of approximately 10.5 kilobases (kb). The ORF is flanked by two noncoding regions (NCR) at the terminal ends 5′ (5′NCR) and 3′ (3′NCR). These NCRs are intensely involved in genome regulation of translation, RNA replication, interaction with viral and host cell proteins, and packaging during the virus replication in the infected cells. The genomic RNA is infectious to cells and is directly responsible for RNA synthesis during its replication cycle. After replication, the virus particles are released by exocytosis (Chambers et al. 1990; Lindebach and Rice 2003).

It is well established that the ORF is cotranslationally and translationally cleaved during replication period, which originates ten viral proteins, three structural (C, capsid; M, membrane and its precursor prM; and E, envelope) and seven non-structural [NS] (NS1, NS2A, NS2B, NS3, NS4A, NS4B, and NS5). C is the major structural protein and responsible for RNA encapsidation, prM is mainly associated to fusion process, and E is the hemagglutinin, which is responsible for receptor binding, the acid pH-dependent fusion property, and endocytosis of viral particle, and finally is also the most immunogenic viral protein, which induces production of neutralizing antibodies (Chambers et al. 1990).

NS1 is associated with RNA replication and fixing the complement; it also plays a role in the viral neuroinvasiveness; during infection it is quickly produced, and thus, it is early detected in the blood of sick people. NS2A helps the virus particle production and inhibit interferon signaling. NS2B is a cofactor to the NS3 serine protease activity. NS3 is a multifunctional protein and is closely associated with polyprotein processing, RNA replication, virus assembly, mature capsid, and several enzymatic actions such as helicase, serine protease, and RNA triphosphatase activities. NS4A and NS4B are small hydrophobic proteins implicated in the RNA replication and inhibition of interferon signaling pathway. NS5 is the largest *Flavivirus* NS protein; it is multifunctional phosphoprotein with RNA capping, methyltransferase, and RNA-dependent RNA polymerase activities; it blocks the Jak-STAT pathway of Interferon signaling and finally is associated with induction of nuclear activity of interleukin-8 (IL-8) (Chambers et al. 1990; Lindebach and Rice 2003; Vasilakis et al. 2011).

7.3 Pathogenesis and Pathology

The pathogenesis of dengue disease is a controversial matter, and no consensus has been achieved during the Pan American Health Organization (PAHO)/WHO dengue expert meetings. It is clear that in the last two decades, strong advances were obtained on several aspects of the disease pathogenesis and physiopathology. Moreover, many different animal models were developed, which furnished lots of information helping to the understanding of many cellular and humoral response during the course of disease that probably play a key role in its pathogenesis. Presently it is clear and consensual that the pathogenesis of dengue is complex and perhaps will never be completely understood. Nonetheless, many different theories have been proposed to explain the pathogenesis of dengue; it is also clear that all of them failed to (Rothman et al. 2014).

The most commented and discussed theories include the antibody-dependent enhancement (ADE), the origin of virus strains as primordial to explain the disease severity, the antigenic sin phenomena, and a plural causality origin, among others (Rosen 1977; Halstead 1981; Bravo et al. 1987; Kouri et al. 1987). None of them is accepted by the scientific community as definitive to explain the pathogenesis of dengue. The consequence of this is an open large field to study dengue pathogenesis and its consequences for patients.

The major problem in dengue has been the absence of a good animal model reproducing the human dengue disease. Indeed, although new studies have revealed the marmosets as excellent model to study antigenic and antibodies responses, as well as the viral load, the pathology changes of dengue infections in these New World monkey tissues did not completely reproduce the disease presented by human beings and, therefore, fail to mimetize the dengue disease, as has been since long time defined, for example, to yellow fever virus (Omatsu et al. 2011; Ferreira et al. 2014). Several mouse models were also developed, but again failed to reproduce human changes, and in general the animals are interferon immunodeficient, humanized, or transgenic, with a few exceptions (Balsitis et al. 2010; Mota and Rico-Hesse 2011; Diniz et al. 2013).

Pathology changes have been described elsewhere, both in Asia and the Americas, and describe the final events of severe fatal dengue disease, failing to demonstrate in details the intrinsic events associated with the disease outcome (Bhamarapravati et al. 1967; Pagliari et al. 2014). Therefore the look for an ideal animal model is a priority to a best understanding of disease pathogenesis, clinical studies (vaccine testing and therapeutic approaches), etc. Very interesting reviews were published elsewhere in the last decade, and for more details their reading is recommendable.

7.4 Clinical Syndromes

In the past dengue infection was classified as dengue fever (DF) and dengue hemorrhagic fever (DHF), and this latter is without or with dengue shock syndrome (DSS) (Nimmannitya et al. 1969; Halstead 1989). Nonetheless, this classification was used and useful for several decades, the WHO convened during a meeting in Geneva to review this classification of dengue, and a group of clinicians evaluated and considered to be more practical the adoption of a new classification for dengue disease, as dengue fever, and severe dengue with or without complication. This new classification put together all severe dengue pictures including atypical presentation such as myocarditis, hepatitis, encephalopathy, and/or encephalitis, which are severe pictures but in many cases do not meet the DHF/DSS diagnostic criteria, and although they are severe and frequently fatal, they were not classified as DHF/DSS using the old classification. Figure 7.1 shows a schematic flowchart of dengue pictures using the new WHO dengue clinical picture classification (Miagostovich et al. 1997; WHO 2009).

7.4.1 Dengue Without (Dengue Fever) and with Warning Signs

In the typical dengue fever, a probable dengue case is typically characterized by an illness of acute onset with fever (>38.5 °C), headache, nausea, vomiting, rash, muscle and joint pains, leukopenia with increased lymphocytes, and a tourniquet test positive. In general, this picture remains for at least 5–7 days, and patient's recovery is uneventful. Convalescence is prolonged for 2 or more weeks.

In some patients, however, after the febrile period, a sudden worsening of the picture is observed characterized for the presence of warning signs (a group of clinical findings in general absent in almost all patients with uncomplicated pictures) including abdominal pain or tenderness, lethargy or restlessness, accumulation of fluid (ascites, pericarditis, and/or pleural effusion), persistent vomiting, enlargement of the liver, and mucosal bleeding, and at laboratory level two major findings are frequently observed, the thrombocytopenia and increase of hematocrit (indicating hemoconcentration). The platelets are systematically below 100,000 per milliliter and in many case below 20,000 per mL or even less than 10,000 per mL. In small children several of these warning signs cannot be easily achieved, but irritability and frequent crying should be considered, especially if these findings were not previously observed (WHO 2009).

7.4.2 Severe Dengue

In general the severe pictures of dengue are characterized as complication of classic dengue fever cases; these severe cases are diagnosed during the defervescence period, i.e., just after the febrile period.

Indeed, severe dengue pictures are characterized by three clinical aspects: (1) severe plasma leakage with fluid accumulation in natural cavities (pericarditis, ascites, and pleural effusion) and in many important organs especially the lungs with consequent respiratory distress and hypotension that if an early supportive treatment with plasma expanders is not immediately used, the consequence is the shock; (2) severe bleeding in different sites; and (3) severe organ impairments, including the liver (hepatitis characterized by Aspartate aminotransferase (AST) and Alanine aminotransferase (ALT) increased with values that reach 1000 IU or even more), heart (myocarditis), CNS (encephalopathy or encephalitis), pancreas (pancreatitis), and other stricken organs (WHO 2009).

7.5 Diagnosis

The diagnosis of dengue virus infection is possible by different methods depending on time of disease since the onset.

Indeed, during the *acute phase*, in the first 5 days, the virologic methods are more effective to diagnosis dengue infection. Thus the attempts of virus isolation into cell cultures and the molecular approaches as RT-PCR or real-time RT-PCR are more indicated.

During the *convalescence period* or after the fifth day of disease, the serologic tests are more indicated. In fact, the ELISA tests for detection

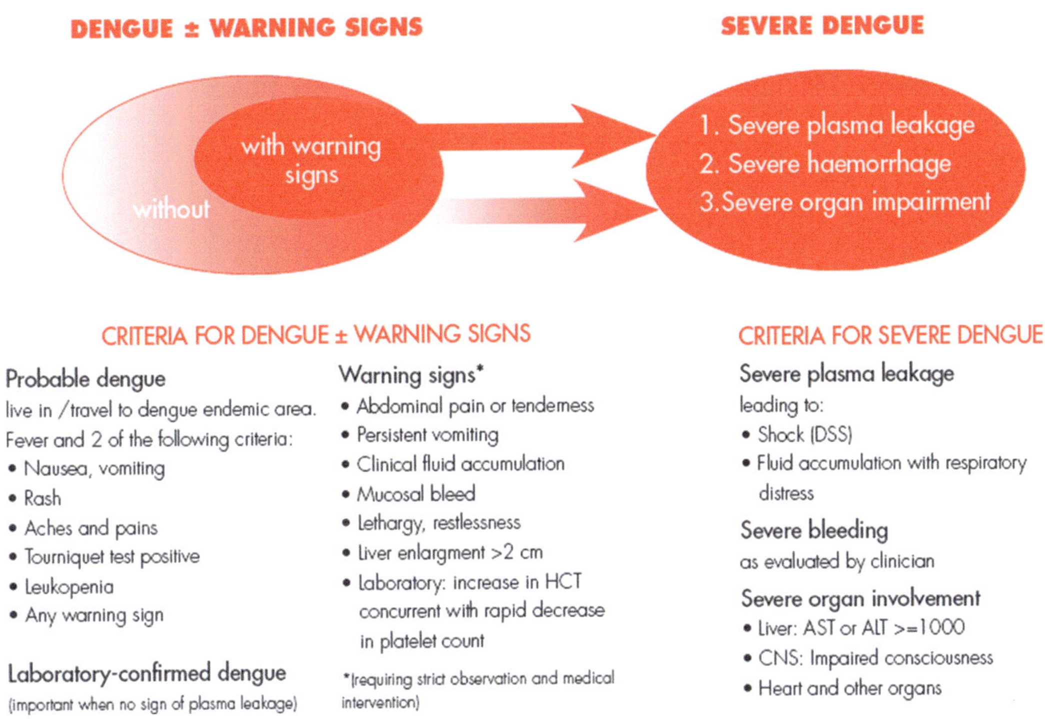

Fig. 7.1 Suggested dengue case classification and levels of severity. Reprinted from WHO. Dengue: guidelines for diagnosis, treatment, prevention and control. Chapter 1, page 11, ©2009

of IgM are the best choice to have a presumptive diagnostic of dengue infection, especially in patients with a compatible clinical picture. The detection of IgM is the most useful and practical serologic method since a single blood sample obtained after the fifth day since disease onset is enough to give a presumptive diagnostic of dengue. Other serologic methods such as hemagglutination inhibition (HI), complement fixation (CF), and plaque reduction neutralization test (PRNT) only are available on specialized reference laboratories and to be conclusive need to have two samples collected in the acute and convalescent phase with an interval of at least 2 weeks. An alternative method is the IgG-ELISA, which like HI, CF, and PRNT needs to have two serologic samples. A conclusive diagnostic using these techniques is only possible when serologic conversion occurs , i.e., the increase in the titers of at least four times between the acute and convalescent blood samples (Vasconcelos et al. 2013).

7.6 Treatment

There are no specific drugs for treatment of dengue disease, and therefore, treatment is exclusively supportive. Despite the absence of specific therapeutics, the support treatment is effective and should be used as recommended by WHO/PAHO in the guidelines (WHO 2009; Fig. 7.2). When well used, the supportive treatment is effective and will avoid unnecessary deaths due to cavity effusion in consequence of accumulation of fluid (ascites, pericarditis, and/or pleural effusion), hemorrhages, hypotension, and shock. Almost all endemic countries have adopted the WHO recommendations for treatment of severe dengue with very good results when initiated in the due time. The decision to initiate the treatment is exclusively a physician decision, and to facilitate the WHO established a flowchart (reproduced below) that helps in the decision and the more appropriated moment to do for.

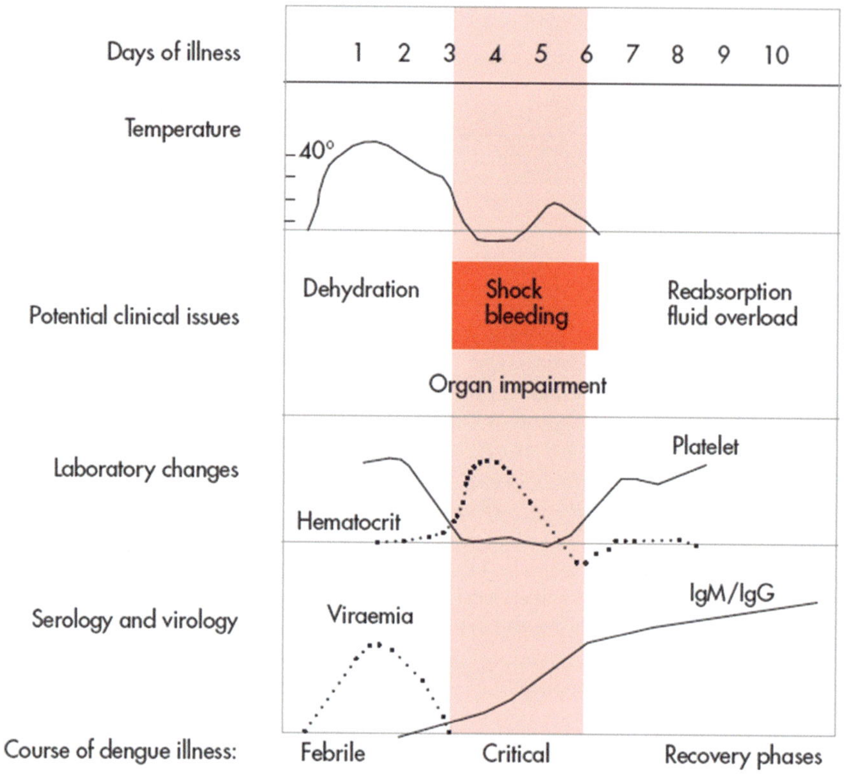

Fig. 7.2 The course of dengue illness. Reprinted from WHO. Dengue: guidelines for diagnosis, treatment, prevention and control. Chapter 2, page 25, ©2009

7.7 Epidemiology

DENV serotypes are widespread throughout four of the five continents, excepting Europe. Indeed, according to the WHO, approximately three billion people live in areas with transmission or under risk of DENV due the infestation of DENV vectors (Bhatt et al. 2013; PAHO 2014). To occur DENV transmission, a combination of several factors including the virus, the vector, the host, and the environment is necessary. After introduction of a DENV serotype in a place it is necessary a time to virus replication, when occurs a silent spread before becoming evident the epidemic situation, the so-called lag phase (Travassos da Rosa et al. 2000). The house-to-house transmission is the most common mechanism of virus spread due to human movement in an area or city. The long-distance transmission and spread of DENV serotypes are made principally by viremic people presenting symptoms of disease or unapparent infection through airplane traveling (Nunes et al. 2012, 2014).

The DENV mosquito vector activities are diurnal. In fact, *Ae. aegypti* prefers to lay its eggs in many different artificial apparatus that can act as (small or big) containers of water. Only the female mosquito is hematophagous and, therefore, is the only responsible for disease spread in a community. The containers used for water storage are the most common breeding sites into or near the human dwellings, while in outdoor environments, the vehicle tires and trash in general are the preferable sites used by the DENV vectors.

As transmission is commonly acquired in or around human dwellings, it is not uncommon that two or more people living in the same home become ill; this efficient mechanism of virus transmission is strongly associated with epidemic spread of DENV in a place, and this behavior is important for the success for epidemic transmission of DENV by the *Ae. aegypti* mosquito. This is apparently also observed for chikungunya virus transmission but not for yellow fever virus (Dussart et al. 2012; Monath and Vasconcelos 2015; Teixeira et al. 2015).

People become viremic at the last 2 days of incubation period, which in general ranges from 3 to 6 days, through 3 to 5 days of the febrile period. During this time, the DENV circulate in the peripheral blood, and if a noninfected mosquito bites a viremic person, it can become infected, and subsequently, the spread of virus will occur easily when in next blood meal it reaches a susceptible person (Dussart et al. 2012; Vasconcelos et al. 2013).

Presently dengue is the most important and widespread mosquito-borne disease, and year by year the disease is becoming more pandemic. This characteristic contrasts with dengue situation in the first decades of the last century. Indeed, following the end of the World War II, especially in Asia, the disease transmission exploded due to the huge quantity of war trashes and the ideal conditions to disease widespread enhanced by the modern lifestyle implemented with plastic disposable and containers used respectively to food and water storage, particularly in the tropical and subtropical metropolis, and the rising in regional and global travel. Interestingly, this pattern was followed initially in Asia and then in the Americas, with more severe dengue clinic pictures, increase of hospitalization, and increased case fatality rates.

The endemicity of dengue in Southeast Asia and Latin America plays a pivotal role in the disease spread and, of course, of the four DENV serotypes (Vasilakis et al. 2011). The hyperendemicity of the four DENV serotypes is also associated with the emergence of new and most virulent genetic lineages (genotypes) of all serotypes. Presently six DENV1 genotypes, four for DENV2, four for DENV3, and three for DENV4 are recognized. Based on clinic-epidemiologic information and nucleotide sequencing findings, it is consensual that the most virulent genotypes are involved with more severe epidemics and clinical presentations, including the emergence of escape mutant associated with atypical dengue cases. It is also consensus among the scientific community that the emergence/introduction of new genetic lineages in a place/region/city/country has provoked changes in the clinical pattern of disease presentation (Rico-Hesse 1990; Rico-Hesse et al. 1997; Whitehead et al. 2007).

Dengue has emerged as a potential threat in Latin America in the late 1970s and followed the spread of *Ae. aegypti* infestation in the region. In fact, as a single example of dengue emergence in the region, it is the DENV1 epidemic registered in Cuba in 1979. It was followed by the 1981 DENV2 epidemic in the country with the occurrence of much more severe cases than reported in previous decades for the Americas and with the report of 159 deaths due to severe dengue (previous DHF/DSS) (Guzmán et al. 1990; WHO 2009; Pinheiro and Corber 1997). After these epidemics, the spread of DENV serotypes was continuous and affected from the southeast USA (Florida and Texas states) through northern Argentina in annual report hundreds of thousands of cases in the 1980s to millions of cases in the present century. The increase of the numbers of reported cases was accompanied by a sharp increase of disease severity and, also, of human deaths due to dengue, although the WHO Global Strategy for the prevention and control of dengue 2012–2020 has a specific goal to reduce dengue mortality in 50% and morbidity in 25% by 2020 (PAHO 2014).

7.7.1 Dengue in the World

Dengue is the most widespread vector-borne disease in the world, which has an estimated occurrence of almost 400 million infections and 100 million cases of disease annually (Bhatt et al. 2013). Indeed, it is endemic in all continents excepting Europe, where sporadic outbreaks have been diagnosed especially in the Mediterranean area and the Portuguese island of Madeira, and a continuous and systematic surveillance is active to control these episodes and avoid the advance of disease to an endemic status in the continental territory.

Traditionally, Asia is the continent with more occurrence of dengue cases and highest disease severity, and this apparently occurs due Asia be hyperendemic for decades. In fact, the hyperendemicity of dengue in the continent, especially in the Southeast Asian countries, has resulted in explosion of cases, sharp increase of disease severity and hospitalization, and obviously deaths due to dengue. Since the 1950s the WHO has reported an important involvement of dengue disease as the more frequent cause of death especially in children.

Regarding Africa, dengue is a neglected disease and only recently has been considered as important etiologic agent of disease and deaths in the African continent, where statistics are poorly obtained and have not been used in the scientific context. Nonetheless, Bhatt et al. (2013), using mathematic models, reported cases, and deaths by viral hemorrhagic fevers, among other variables, have estimated the annual occurrence of almost 100 million infections in Africa. Certainly, efforts are necessary to improve the disease surveillance for increasing the report of dengue in Africa in order to have feasible numbers of dengue in the continent (Bhatt et al. 2013).

Similar to other Latin American countries, Brazil experienced its first dengue fever outbreak in 1982 in Boa Vista, Roraima state in the Amazonian region, which was caused by DENV1 and DENV4 (Osanai et al. 1983). This focal outbreak was followed 4 years later by another epidemic in 1986 in Rio de Janeiro state, this time caused only by DENV1 (Schatzmayr et al. 1986). This large epidemic probably has caused around one million cases in Rio de Janeiro and neighbor municipalities based on an epidemiologic survey (Figueiredo et al. 1990). In this same year and the following ones, the DENV1 spread to other states causing several epidemics, and millions of people became sick with dengue. In 1990 the DENV2 was isolated in Rio de Janeiro state, and in the same year, the virus caused an outbreak in Araguaina, Tocantins state (Miagostovich et al. 2003; Vasconcelos et al. 1993); in Ceará, Alagoas; and in other Brazilian states; the first human severe cases (DHF/DSS) in higher scale were recognized in Brazil (Vasconcelos et al. 1995). A randomized serologic survey carried in Fortaleza and São Luís showed the occurrence of more than 600,000 and 400,000 infections in the Ceará and Maranhão state capitals, respectively (Vasconcelos et al. 1998, 1999). Similar situation was observed in a longitudinal study conducted in Salvador, where people of some districts of it were accompanied for several years, and it was demonstrated that the number of silent DENV infection was more frequent and with high prevalence, especially the secondary infections than previously expected (Teixeira et al. 2002). Later the DENV3 was associated with epidemics in Rio de Janeiro state and then quickly spread into the country, and DENV4 was reintroduced in Roraima state in 2010 (Temporão et al. 2011; Nunes et al. 2012).

7.8 Control and Prevention

Presently, there is no available and/or licensed vaccine to prevent dengue disease. Several vaccine candidates in different development stages are being proposed by different producers and using many different approaches including chimeric live-attenuated vaccines (using as backbone the yellow fever vaccine virus 17D or dengue 2 and dengue 4), inactivated vaccines, DNA vaccines, and others.

The WHO-has registered several initiatives to the development of dengue vaccines and presently they are in different stages of development (preclinic; phase I, phase II, or phase III) Recently a chimeric tetravalent dengue vaccine has been licensed in a few countries including Brazil, Mexico, Philippines and others. Therefore, on a perspective view, it is estimated that soon, a WHO-licensed dengue vaccine is expected to be available for public health use. Even if we have a WHO-licensed vaccine, for prompt use, a challenge to surpass is the capacity to produce the Dengue vaccine to attend the world demand on large scale and in quantity enough to prevent dengue disease in great part of the endemic countries. Indeed, the more recent

WHO estimative of people living at risk of dengue reaches around three billion inhabitants. Thus, presently it is not possible to any manufacturers to produce in the next 10 years at least one billion doses of a vaccine that will need two or three doses (Lam et al. 2011).

Since a vaccine is still a distant dream, the vector control measures are the only feasible action that can prevent dengue disease. Nonetheless, the crescent insecticide resistance especially to the organophosphorates is a great challenge to be considered ahead and may be debated for all governments of endemic countries and the WHO. Individual measures are not practical and are expensive.

The dengue vectors present large ability to reproduce, and all initiatives to control transmission should target *Ae. aegypti*, the main vector, in the habits of its immature and adult stages in the household and immediate vicinity, as well as other places where human activity can facilitate the vector-human contact (WHO 2009). It is noteworthy that *Ae. aegypti* replicates in many different recipients, where clear water can be storage either for human usage or for decorative pants, and also in discard recipients. Indeed, in many places due to the deficiency on solid waste collection mainly of discard food and beverage containers, used tires, blocked gutters, and buildings under construction, these materials and places are receptive to mosquito that lays eggs. The consequence is the index levels of mosquito to increase sharply, and transmission occurs locally, since *Ae. aegypti* in general does not fly far (around 100 m) and the feed occurs almost entirely on human indoors and outdoors mainly during daylight hours (WHO 2009).

The PAHO/WHO has proposed an integrated strategy to promote vector control, where five key actions are important to the management process, namely (WHO 2009; PAHO 2014):

1. Advocacy, social mobilization, and legislation: all three elements are essential to proceed the vector control, but the establishing of regulative and legislative is mandatory.
2. Collaboration within the health sector and with other sectors: the strengthening of collaboration between public health and private sectors is fundamental for the action process; planning and decision-makers play an important role, as well as the communication of managers of programs for vector control.
3. Integrated approach to disease control: this will allow a rational use of financial resources through the application of a multi-disease control approach, including the use of chemical and nonchemical control methods.
4. Evidence-based decision-making: the strategies for local vector control, epidemiology, and resources should be guided by operational research and subject to routine monitoring and evaluation to avoid losses of financial resources.
5. Capacity building: the development of infrastructure to have enough financial and training of human resources both at local and national levels is fundamental, based on a situation analysis.

It is important to emphasize that control of *Ae. aegypti* population is mainly achieved by elimination of container habits that enhance oviposition and, of course, facilitate the development of *Ae. aegypti* aquatic stages. Without this principle on the mind, it is impossible to control this mosquito. In other words, the human lifestyle brings the mosquito vectors to their dwellings and nearest of them, and the consequence is the explosive dengue epidemics and more recently chikungunya virus and Zika virus epidemics in the dengue endemic areas. There are currently two different promising entomological approaches that are being used in the field to combat *Ae. aegypti*.

The first utilizes transgenic mosquitoes, which have been modified to produce larvae that die during early development, resulting to smaller mosquito populations and less frequent disease transmission.

The other initiative, which has shown to be more feasible, is the use of *Wolbachia*-infected *Ae. aegypti* mosquitoes. *Wolbachia* is a bacterium that can greatly inhibit DENV infection and replication when present in *Ae. aegypti* mosquitoes (see Chap. 38). *Wolbachia*-based control programs have been successfully deployed in the field in Queensland, Australia (Walker et al. 2011; Muller et al. 2012). Presently, a pilot experiment is ongoing in the municipality of Niteroi and in the Ilha do Governador district of Rio de Janeiro city in Rio de Janeiro state, Brazil. Unfortunately, we do not have the preliminary results yet.

References

Balsitis SJ, Williams KL, Lachica R et al (2010) Lethal antibody enhancement of dengue disease in mice is prevented by Fc modification. PLoS Pathog 6, e1000790

Bhamarapravati N, Tuchinda P, Boonyapaknavik V (1967) Pathology of Thailand haemorrhagic fever: a study of 100 autopsy cases. Ann Trop Med Parasitol 61:500–510

Bhatt S, Gething PW, Brady OJ et al (2013) The global distribution and burden of dengue. Nature 496(7446):504–507. doi:10.1038/nature12060

Bravo JR, Guzmán MG, Kouri GP (1987) Why dengue haemorrhagic fever in Cuba? 1. Individual risk factors for dengue haemorrhagic fever/dengue shock syndrome (DHF/DSS). Trans R Soc Trop Med Hyg 81:816–820

Chambers TJ, Hahn CS, Galler R et al (1990) Flavivirus genome organization, expression and replication. Ann Rev Microbiol 44:649–688

Travassos da Rosa APA, Vasconcelos PFC, Travassos da Rosa ES et al (2000) Dengue epidemic in Belém, Pará, Brazil, 1996–97.Emerg Infect Dis 6:298–301

Diniz DG, Foro CAR, Sosthenes MCK et al (2013) Aging and environmental enrichment exacerbate inflammatory response on antibody-enhanced dengue disease in immunocompetent murine model. Eur J Inflammation 11:719–731

Dussart P, Baril L, Petit L et al (2012) Clinical and virological study of dengue cases and the members of their households: the multinational DENFRAME project. PLoS Negl Trop Dis 6(1):e1482, 1–13.

Ferreira MS, Castro PH, Silva GA et al (2014) *Callithrix penicillata*: a feasible experimental model for dengue virus infection. Immunol Lett 158(1-2):126–133. doi:10.1016/j.imlet.2013.12.008

Figueiredo LT, Cavalcante M, Simões MC (1990) Dengue serologic survey of schoolchildren in Rio de Janeiro, Brazil, in 1986 and 1987. Bull Pan Am Health Organ 24:217–225

Guzmán MG, Kouri GP, Bravo J et al (1990) Dengue hemorrhagic fever in Cuba, 1981: a retrospective seroepidemiologic study. Am J Trop Med Hyg 42:179–184

Halstead SB (1981) The pathogenesis of dengue: molecular epidemiology in infectious disease. Am J Epidemiol 114:632–648

Halstead SB (1989) Antibody, macrophages, dengue virus infection, shock, and hemorrhage: a pathogenetic cascade. Rev Infect Dis 11:S830–S839

Kouri GP, Guzmán MG, Bravo JR (1987) Why dengue haemorrhagic fever in Cuba? 2. An integral analysis. Trans R Soc Trop Med Hyg 81:821–823

Lam SK, Burke D, Capeding MR et al (2011) Preparing for introduction of a dengue vaccine: recommendations from the 1st Dengue v2V Asia-Pacific Meeting. Vaccine 29:9417–9422

Lindebach BD, Rice CM (2003) Molecular biology of flaviviruses. Adv Virus Res 59:23–61

Miagostovich MP, Ramos RG, Nicol AF et al (1997) Retrospective study on dengue fatal cases. Clin Neuropathol 16:204–208

Miagostovich MP, Sequeira PC, Dos Santos FB et al (2003) Molecular typing of dengue virus type 2 in Brazil. Rev Inst Med Trop Sao Paulo 45:17–21

Monath TP, Vasconcelos PFC (2015) Yellow fever. J Clin Virol 64:160–173

Mota J, Rico-Hesse R (2011) Dengue virus tropism in humanized mice recapitulates human dengue fever. PLoS One 6, e20762

Muller DA, Frentiu FD, Rojas A et al (2012) A portable approach for the surveillance of dengue virus-infected mosquitoes. J Virol Methods 183:90–93. doi:10.1016/j.jviromet.2012.03.033

Nimmannitya S, Halstead SB, Cohen SN et al (1969) Dengue and Chikungunya virus infection in man in Thailand, 1962–1964. I. Observations on hospitalized patients with hemorrhagic fever. Am J Trop Med Hyg 18:954–971

Nunes MRT, Faria NRF, Medeiros DBA et al (2012) Phylogeography of dengue virus serotype 4, Brazil, 2010–2011. Emerg Infect Dis 18:1858–1864

Nunes MRT, Palacios G, Faria NR et al (2014) Air travel is associated with intracontinental spread of dengue virus serotypes 1–3 lineages in Brazil. PLoS Negl Trop Dis 8, e2769

Omatsu T, Moi ML, Hirayama T et al (2011) Common marmoset (*Callithrix jacchus*) as a primate model of dengue virus infection: development of high levels of viraemia and demonstration of protective immunity. J Gen Virol 92:2272–2280. doi:10.1099/vir.0.031229-0

Osanai CH, Travassos da Rosa AP, Tang AT et al (1983) Dengue outbreak in Boa Vista, Roraima. Preliminary report. Rev Inst Med Trop Sao Paulo 25:53–54

Pagliari C, Quaresma JAS, Fernandes ER et al (2014) Immunopathogenesis of dengue hemorrhagic fever: contribution to the study of human liver lesions. J Med Virol 86:1193–1197. doi:10.1002/jmv.23758

PAHO (2014) Meeting report: state of the art in the prevention and control of dengue in the Americas. Pan American Health Organization, Washington, 54 p

Pinheiro FP, Corber SJ (1997) Global situation of dengue and dengue haemorrhagic fever, and its emergence in the Americas. World Health Stat Quart 50:161–169

Rico-Hesse R (1990) Molecular evolution and distribution of dengue viruses type 1 and 2 in nature. Virology 174:479–493

Rico-Hesse R, Harrison LM, Salas RA et al (1997) Origins of dengue type 2 viruses associated with increased pathogenicity in the Americas. Virology 230:244–251

Rosen L (1977) The Emperor's New Clothes revisited, or reflections on the pathogenesis of dengue hemorrhagic fever. Am J Trop Med Hyg 26:337–343

Rothman AL, Medin CL, Friberg H et al (2014) Immunopathogenesis versus protection in dengue virus infections. Curr Trop Med Rep 1:13–20

Schatzmayr HG, Nogueira RM, Travassos da Rosa AP (1986) An outbreak of dengue virus at Rio de Janeiro, 1986. Mem Inst Oswaldo Cruz 81:245–246

Teixeira MG, Barreto ML, Costa MCN et al (2002) Dynamics of dengue virus circulation: a silent epidemic in a complex urban área. Trop Med Int Health 7(9):757–762

Teixeira MG, Andrade A, Costa MC et al (2015) Chikungunya outbreak in Brazil by African genotype. Emerg Infect Dis 21:906–908

Temporão JG, Penna GO, Carmo EH et al (2011) Dengue vírus serotype 4, Roraima state, Brazil. Emerg Inf Dis 17:938–940. doi:10.3201/eid1705.101681

Vasconcelos PFC, Travassos da Rosa ES, Travassos da Rosa JFS et al (1993) Epidemia de febre clássica de dengue causada pelo sorotipo 2 em Araguaína, Tocantins, Brasil. Rev Inst Med Trop Sao Paulo 35:141–148

Vasconcelos PFC, Menezes DB, Melo LP et al (1995) A large epidemic of dengue fever with hemorrhagic cases in Ceará State, Brazil. Rev Inst Med Trop S Paulo 37:253–255

Vasconcelos PFC, Lima JWO, Travassos da Rosa APA et al (1998) Epidemia de dengue em Fortaleza, Ceará, 1994: Inquérito soro-epidemiológico aleatório. Rev Saúde Públ 32:447–454

Vasconcelos PFC, Lima JWO, Raposo ML et al (1999) Inquérito soro-epidemiológico aleatório na Ilha de São Luís, durante epidemia de dengue no Maranhão. Rev Soc Bras Med Trop 32:171–179

Vasconcelos PFC, Rodrigues SG, Oliveira CS et al (2013) Dengue. In: Leão RNQ, Bichara CNC, Fraiha Neto H et al (eds) Medicina tropical e infectologia na Amazônia. Samauma Editora, Belém, pp 511–530

Vasilakis N, Cardosa J, Hanley KA et al (2011) Fever from the forest: prospects for the continued emergence of sylvatic dengue virus and its impact on public health.Nat Rev Microbiol 9:532–541. doi:10.1038/nrmicro2595

Walker T, Johnson PH, Moreira LA et al (2011) The wMel *Wolbachia* strain blocks dengue and invades caged *Aedes aegypti* populations. Nature 476:450–453. doi:10.1038/nature10355

Whitehead SS, Blaney JE, Durbin AP et al (2007) Prospects for a dengue virus vaccine. Nat Rev Microbiol 5:518–528

WHO (2009) Dengue: guidelines for diagnosis, treatment, prevention and control. World Health Organization, Geneva

Yellow Fever

8

Pedro F.C. Vasconcelos

Abstract

Yellow fever remains as an important threat in sub-Saharan Africa and South America, the regions where it is endemic; the yellow fever virus has periodically stricken human and nonhuman primates' population, its urban and jungle hosts, when epidemics and/or epizootics are registered, respectively. These periodic expansions and retractions in both endemic regions are not completely understood, but several factors play a role including a heavy rainy season, the number of vectors, and the level of immunity coverage conferred by 17D vaccine use. The mosquito vectors are different for urban and jungle cycles for Africa and South America, and several sylvatic *Aedes* and *Haemagogus* species are its vectors, respectively. In urban settings the only vector is the peri-domestic *Aedes aegypti*. Nonetheless, only in Africa a third intermediate (rural) cycle is vectored by *Aedes simpsoni*. The diagnostic of yellow fever is made by molecular approaches (RT-PCR and real-time RT-PCR), virus isolation (cell culture), serology (IgM-ELISA), and immunohistochemical assay on tissue of fatal cases. Treatment is only supportive, and the prevention should be focused on vaccination of people living in or moving to endemic areas. These aspects are discussed in this chapter. Although several taxonomic modifications have been proposed on Culicidae genera, mostly splitting *Aedes* mosquitoes into different genera, this taxonomy putting all *Aedes* in a single genus has been used in the yellow fever chapter. To facilitate utilization by health personnel, all new aedine genera can be considered *Aedes* (CBM).

P.F.C. Vasconcelos (✉)
National Reference Laboratory of Arboviruses,
Evandro Chagas Institute,
Rodovia BR 316, Km 07, S/N, 67030-000
Ananindeua, Brazil

National Institute of Science and Technology for
Viral Hemorrhagic Fevers,
Rodovia BR 316, Km 07, S/N, 67030-000
Ananindeua, Brazil

PAHO/WHO Collaborating Center for Arbovirus
Research and Diagnostic Reference, Evandro Chagas
Institute, Ministry of Health,
Rodovia BR 316, Km 07, S/N, 67030-000
Ananindeua, Brazil

Department of Pathology, Pará State University,
Belém, Pará, Brazil
e-mail: pedrovasconcelos@iec.pa.gov.br

C.B. Marcondes (ed.), *Arthropod Borne Diseases*, DOI 10.1007/978-3-319-13884-8_8

Keywords
Yellow fever • *Aedes aegypti* • Hemorrhage • Black vomit • Arbovirus

8.1 Introduction

Yellow fever (YF) is considered an African disease that is caused by the prototype member of the genus *Flavivirus* (family Flaviviridae), the yellow fever virus (YFV). Flaviviruses are single-strand positive-sense RNA viruses, with approximately 70 serotypes (Gubler et al. 2007); the majority of them are transmitted by arthropods (mosquitoes and ticks). Yellow fever is currently endemic in tropical regions of Africa and South America, where sporadic cases, limited outbreaks, or large epidemics periodically are detected in both continents (Monath et al. 2012; Vasconcelos 2010; Barrett and Higgs 2007). These periodic expansions and retractions are multifactorial and are not completely understood; however, heavy rainy season, numbers of vectors, and unvaccinated people play an important role in these periodic episodes of reemergence (Jentes et al. 2011).

The YFV has limited host range to cause infection, and the nonhuman primates (NHP) constitute the primary host, and during outbreaks/epidemics in the forest, they also play as amplifier source of virus to the mosquito vectors into the jungle (*Haemagogus* and *Sabethes* in South America and *Aedes (Stegomyia)* in Africa), the so-called sylvatic yellow fever. It also has been supposed to occur transovarial transmission in these vectors, and susceptible humans are only infected when intrude in these ecosystems. In the urban set, humans are infected by being bitten by the urban mosquito vectors, *Aedes aegypti*; when infected people in the jungle cycle during incubation period or viremia phase go to urban areas and serve as viremic host for interhuman transmission, this urban cycle is called *urban yellow fever* (Fig. 8.1). In Africa, an additional cycle that is common to occur in the Savannah connects the jungle and urban cycles and is vectored by several *Aedes* spp. mosquitoes. Thus the epidemiology of YF in Africa is complex and often involves both sylvatic and domestic vector species during interhuman transmission, but the role of each one during transmission in suburban areas is still yet to be determined. Some authors consider this responsible for the higher numbers of infection in Africa at least ten times higher when compared with South America; consequently, the occurrence of epidemics in Africa is much more common than in the endemic area in the Americas. In the past, yellow fever was a serious threat, and the disease was responsible for large numbers of deaths from the eighteenth century to the early twentieth century, when epidemics occurred repeatedly in the major urban centers in the North and South America, as well as in the Caribbean and Central America and in Europe. In general, introductions of YFV occurred in the coastal towns and then cities distant from endemic areas in North America, the Caribbean, and Europe.

The identification in 1900 by Walter Reed that *Aedes aegypti* mosquitoes are responsible for urban transmission of disease was soon followed by efforts to control the vector using an efficient vector control plan that resulted in a decline in yellow fever outside the tropical, endemic zone in the first two decades of the twentieth century. These efforts were complemented with the development of a live attenuated YF vaccine by Max Theiler in the 1930s. The demonstration that the 17D vaccine was efficient to prevent further yellow fever in the 1940s resulted in the decline of the number of cases of disease, and the periodic upsurges of YF activity in endemic regions were only reported in countries without an efficient 17D routine immunization programs (Monath et al. 2012).

8.2 The Virus

YFV is the prototype species of the genus *Flavivirus* (family Flaviviridae). Flaviviruses are single-strand plus-sense RNA viruses with approximately 70 members. Flaviviruses include

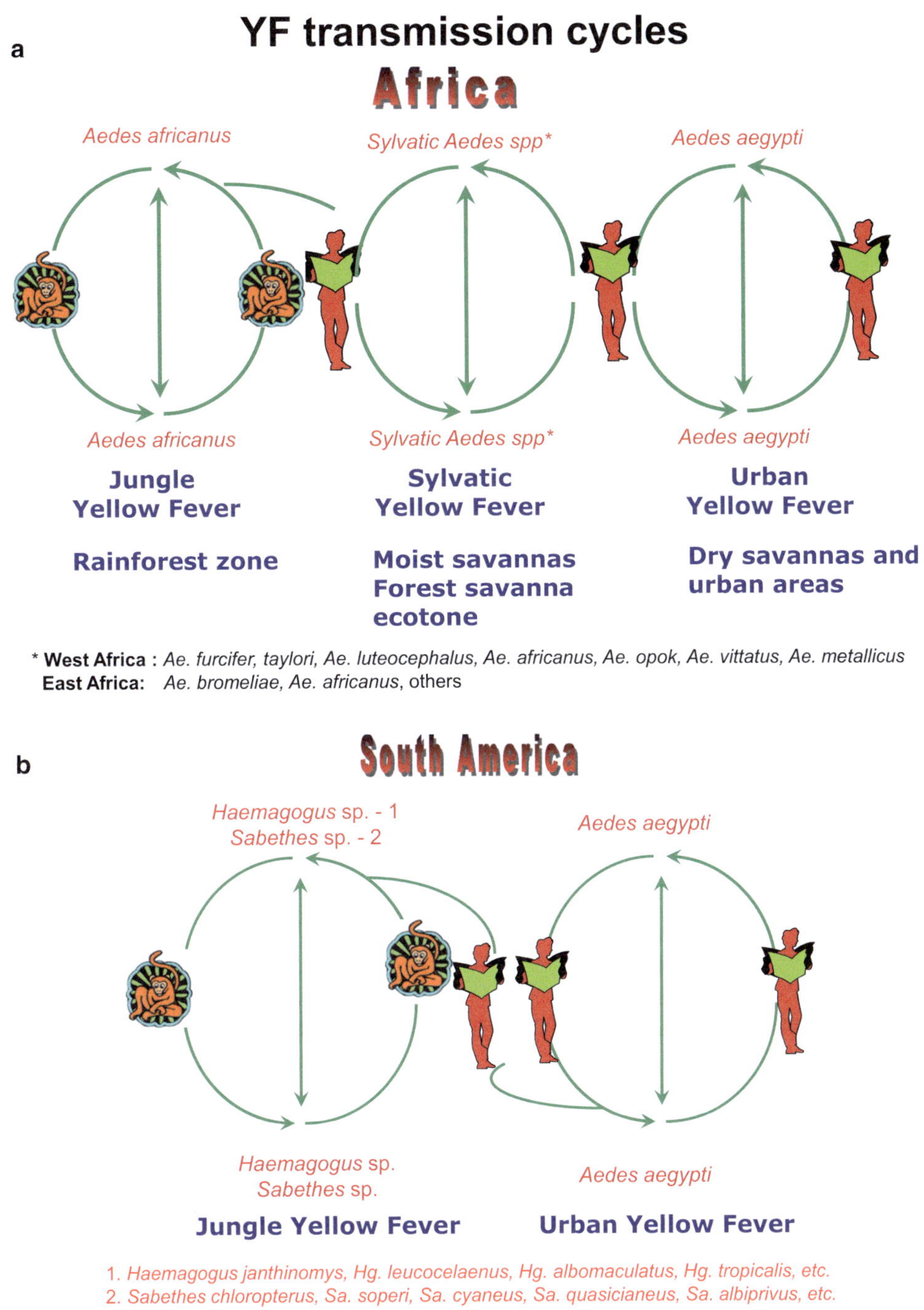

Fig. 8.1 Potential yellow fever virus cycles in Africa (**a**) and in the Americas (**b**). Observe the major complexity of the YFV cycle in Africa represented by the large biodiversity of potential vectors and the presence of an intermediate cycle, which is not observed in the Americas

several important viruses causing human diseases such dengue, Saint Louis encephalitis, tick-borne encephalitis, West Nile, Zika, and other arthropod-borne viruses, as well as Kyasanur Forest disease, a nonvectored *Flavivirus* (see Chaps. 6 and 7).

YFV was the first flavivirus to have its complete genome sequenced and showed a pattern of three structural (C, PrM, and E) and seven nonstructural (NS1, NS2A, NS2B, NS3, NS4A, NS4B, and NS5) genes, which is found in all other members of the Flavivirus genus (Rice et al. 1985; Chambers et al. 1990).

YFV has a single serotype and several genotypes (genetic lineages), but the diversity and dispersal of its lineages are tremendous. The partial and complete genome sequencing of YFV strains in both endemic regions are consistent. In fact, two genotypes were recognized in South America (genotypes SA1 and SA2). It is interesting to observe that SA1 genotype is widely distributed, while the genotype SA2 is restricted to Bolivia, Peru, and nearest regions of Brazil and Ecuador, and they were derived from West Africa, dating from the slave trade period (Vasconcelos et al. 2004; Bryant et al. 2007). Five genotypes are found in Africa: two West Africa WA1 (West/Central) and WA2 (West), two East Africa EA1 (East) and EA2 (East/Central), and genotype Angola. The higher occurrence of genotypes in Africa, especially in East African region, reflect higher heterogeneity of virus and suggest that YFV evolved separately in the African continent and partially explain the epidemiology of disease there. The information available supports the concept that YF virus circulates in discrete foci, with periodic emergences limited by accessible corridors containing vectors and susceptible hosts, which are responsible for maintenance of YFV in nature (Robertson et al. 1996; Monath 2001; Mutebi et al. 2001).

8.3 Clinical Picture

The classic clinical picture caused by the YFV is of hemorrhagic fever that is clinically characterized by an abrupt onset of fever, headache, chills, myalgia, and arthralgia followed few days later by jaundice, hemorrhagic signs, torpor, shock, and death 7–10 days after onset. Nowadays, it is well known that this picture is responsible for around 10 % of all YFV infections. So the clinic spectrum of yellow fever infection ranges from asymptomatic infection and mild and moderate illness to severe disease, which can result in death due to multi-organ failure (Monath 2001; Vasconcelos et al. 2004; Monath et al. 2012).

In general, the mild pictures and asymptomatic infections are common among children that received specific antibodies from the mothers during pregnancy and adults with previous *Flavivirus* infections especially viruses related to the YFV, while severe pictures are more frequently observed among people naïve of contact with other flaviviruses. The picture of severe yellow fever is dramatic with patient presenting aside from above-described symptoms, hematemesis, melena, genital hemorrhages, and other bleeding sites. Patient's death occurs in around 50 % of patients due to liver failure, acute renal insufficiency, severe hemorrhagic diathesis in average 7–10 days of disease, or cardiac arrest in general 2 or 3 weeks of illness during the convalescent period (Vasconcelos et al. 2004; Monath et al. 2012; Monath and Vasconcelos 2015).

8.4 Diagnosis

The diagnosis of yellow fever can be done by different methods depending of the time of disease and collection of sample. Indeed, virologic approaches are used in samples collected up to 7 days of disease, while serologic procedures in general are preferred when blood samples are obtained after a week of disease. Immunohistochemical assays are used to detect specific viral antigens in liver paraffin-embedded fragments (Monath 2001; Monath and Vasconcelos 2015).

8.4.1 Virologic Tests

Attempts of virus isolation should be made on clinical specimens (blood, serum, liver, and other viscera fragments) and also in lots of mosquito vectors using basically two systems: cell culture (VERO, C6/36, BHK21, etc.) and suckling mice. Both systems present very good susceptibility to YFV and isolation of suspected cases around 4–7 days postinfection and are confirmed by specific tests, usually immunofluorescence assay (IFA) using monoclonal antibodies and RT-qPCR.

Other tests such as plaque reduction neutralization test (PRNT), complement fixation (CF), and hemagglutination inhibition (HI) assays can also be used to characterize YFV isolates. The direct detection of YFV can be done using conventional RT-PCR or RT-qPCR. The later one is more specific, sensible, and automatized, minimizing risk of cross contamination and should be considered the golden standard virologic procedure, as demonstrated in several studies where this assay showed to be superior to virus isolation techniques for detecting of YFV (Weidmann et al. 2010; Nunes et al. 2011; Monath and Vasconcelos 2015).

8.4.2 Serologic Tests

In general, serology is used to detect specific antibodies against YFV during surveillance or following detection of a yellow fever suspect case. Several techniques can be used to confirm infection, but the most common assay is the IgM-Enzyme linked immunosorbent assay (IgM-ELISA) that can detect specific IgM against YFV on serum samples of suspect cases with five or more days of illness, and IgM can be detected by at least 3 months after natural infection, but there are reports of detection of specific YFV IgM antibodies more than 1 year after vaccination using 17D vaccines. Therefore, a well-documented clinical history should be obtained to conclude any suspect case. Other tests such as HI, CF, IFA, and PRNT can be also used, but it is necessary to have paired samples (an acute phase up to 7 days postinfection and another convalescent phase 2–3 weeks after collection of the first sample). This is a limiting factor to confirm yellow fever, especially on those severe cases with fatal outcome, since no convalescent sample will be available (Monath 2001; Vasconcelos et al. 2001a; Monath and Vasconcelos 2015).

8.4.3 Histopathology and Immunohistochemical Assay

The study of pathogenesis of yellow fever in human beings are limited due to ethic concerns, and, therefore, much of our knowledge on pathogenecity and physiopathology is consequence of experimental model studies in monkeys and hamsters (Monath and Barrett 2003; Quaresma et al. 2005, 2013). Wild-type YFV is primarily viscerotropic, with the liver being the main target organ. Nonetheless, the lymph nodes, heart, kidney, and spleen and probably other organs are also directly or indirectly injured by YFV especially during the course of severe clinical pictures. Characteristically, yellow fever causes an intense damage in all infected tissues and particularly in the liver, and description of pathology changes was recognized since the last years of the nineteenth century. Midzonal necrosis is pathognomonic finding and together with steatosis (micro- and macro-goticular) and Councilman bodies (now recognized as apoptotic hepatic cells) constitutes the major findings in the injured liver in yellow fever.

Immunohistochemical assay is used to detect YFV-specific antigens. The large amounts of YFV antigens help to explain the severity of cases, since the hepatic acini are plenty of viral antigens. The finding of antigens in the liver is confirmatory of yellow fever naturally acquired or following severe adverse event postvaccination with 17D vaccines (Monath and Barrett 2003; Quaresma et al. 2005, 2013).

In the last decade, JAS Quaresma et al. focused several studies on the in situ immunopathogenesis of yellow fever in liver tissues during fatal disease. These superb studies resulted in the improvement of our knowledge, and their main findings include: (1) Apoptosis is the main mechanism of cellular death in fatal yellow fever and probably also in severe nonfatal pictures. (2) The transforming growth factor beta (TGF-β) is probably responsible for the induction of apoptosis and for downregulation of inflammatory infiltrates, a feature observed in the livers of fatal cases despite the extensive area of the liver tissue infected and showing necroapoptosis. TGF-β is a well-known potent apoptotic inductor and also acts as prominent anti-inflammatory cytokine during the course of disease. (3) Several other cytokines were present in the liver in yellow fever; the more important were tumor necrosis factor alpha (TNF-α) and interferon gamma (IFN-γ), but their role in livers of fatal cases should be lower than TGF-β, since the immunostaining of cells was less evident and more circumscribed than TGF-β. (4) The immune response is prominently Th1 oriented. (5) The pro-inflammatory cytokine IL-6 is markedly expressed in the injured area. (6) The inflammatory infiltrate is mainly constituted by CD4+ lymphocytes (helper cells) and CD8+ lymphocytes (cytotoxic cells) and by NK cells with rare neutrophils and plasmocytes in the affected areas. (7) The cellular inflammatory response is more intense in the portal tract followed by the midzone and is scarce in the central vein area. (8) The viral antigens are found mainly in the liver cells, a strong signal of a direct cytopathic (apoptotic) effect of YFV on hepatic cells, ending with damage on hepatocytes and Kupffer cells. (9) The apoptosis in the liver tissue is much more prominent than necrosis and is induced by TGF-β and expressed by Fas/FasL. The apoptotic cells correspond to the Councilman bodies and represent hepatocytes showing eosinophilic degeneration (Quaresma et al. 2005, 2006a, b, c, 2007, 2013).

8.5 Treatment

No specific drugs are available to treat yellow fever, and treatment is exclusively supportive to combat symptoms/signs during the course of disease. This is a problem, since in general patient with yellow fever is recognized in rural sometimes forested remote areas with no hospital support, and when they are removed to tertiary hospitals, they are on critical phase with severe bleeding and liver and renal failure, and obviously, this explains the high case-fatality rates (CFR) registered virtually in all endemic countries.

The interferon-α produced by DEF201 (an adenovirus containing a gene for the production of human IFN) showed good results in a hamster model when administered 24 h after infection and has been proposed to be used in cases of accidental exposition to YFV. However, this hypothetic alternative treatment would be only tested in such cases, and obviously, the 17D vaccination of all medical and laboratory personnel working on yellow fever patients or YFV is mandatory. Similarly, the intravenous use of globulin containing high levels of specific antibodies against YFV in the same model presented good results and can be an alternative in accidental exposition (Julander et al. 2011).

Many drugs or small molecules have been described as potentially alternative drugs for specific treatment of yellow fever, and as this is beyond the scope of this chapter, we will only cite them, and people interested should consult an excellent review on this (Monath 2008), they are:

1. Ribavirin (1-β-d-ribofuranosyl-1H-1,2,4-triazole-3- carboxamide), a purine nucleoside, is active against YFV in vitro but at relatively high concentrations, but no clinic experience is available with it; (2) RNAi, similar to ribavirin, has demonstrated good action on YFV both in vitro and in the hamster model, but none was tested on human patient. Some authors have proposed the use of suppressor doses of corticosteroid on severe yellow fever to treat shock, and that showed good results, but the number of cases is too short to determine if high doses (200–300 mg/day) (Monath 2008) are enough to improve survival of severe yellow fever patients. In conclusion, there are no licensed drugs for treatment of yellow fever.

8.6 Epidemiology

The annual reporting rate of yellow fever cases is underestimated due to the surveillance of YF is predominantly passive, and, therefore, the true incidence is unknown in South America and in Africa, the YFV endemic regions, and consequently, the disease has a typical pattern of iceberg (Fig. 8.2). Only sporadically when major epidemics are registered that are deeply investigated; we have numbers nearest to the real prevalence of disease in the endemic areas. Indeed, during non-epidemic periods, when the official numbers decrease dramatically, the estimative of cases of disease/infection is unreal, and the incidence in these periods became anecdotal. In fact, some papers have reported the annual occurrence of 200,000 cases without scientific/epidemiologic support (Monath and Vasconcelos 2015).

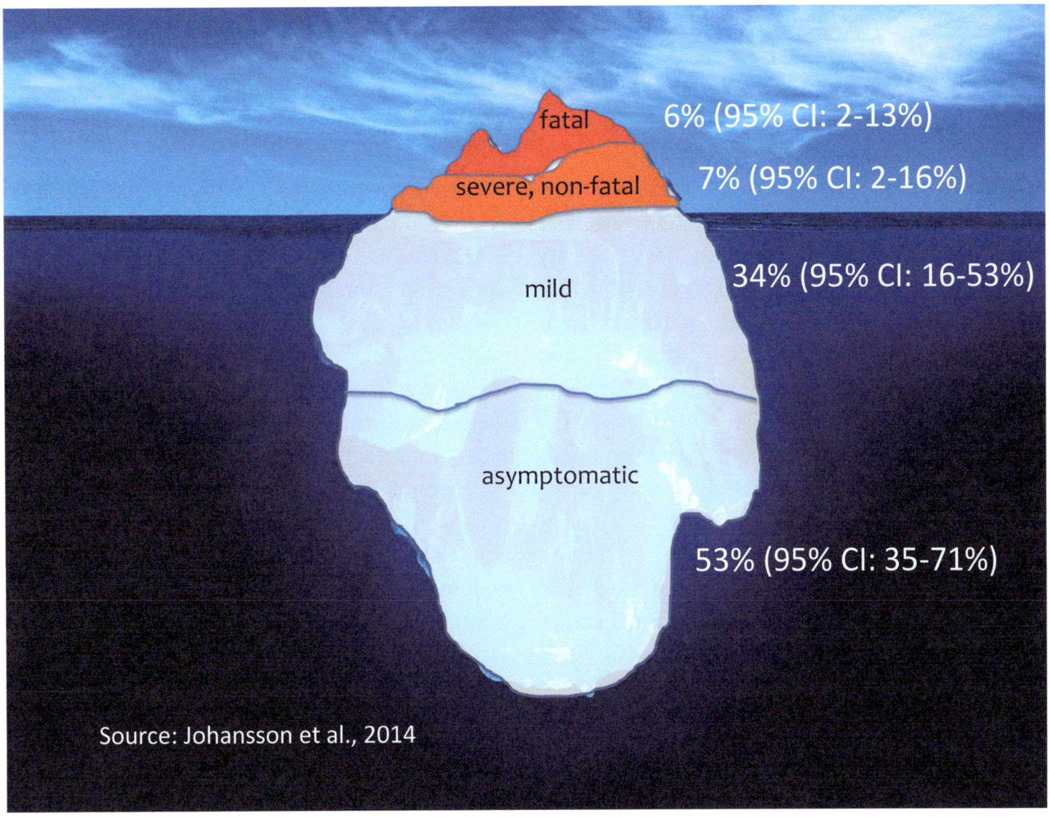

Fig. 8.2 The yellow fever iceberg, showing the typical clinic presentation of disease and the average of each clinical picture based on reported cases and published articles. Adapted from Johansson et al. (2014), authorized by Michael A. Johansson

The number of cases annually reported has dramatically reduced when active laboratory-based surveillance activities were used to investigate suspect clinical cases in Africa. Indeed, only 1–2 % of suspect yellow ever cases in Africa were confirmed by laboratory, demonstrating that 98 % or more of all reported cases were not yellow fever in fact; this is due in Africa that just one laboratory is responsible to confirm YFV infection in entire continent. In South America, where a well-structured laboratory network is established since the 1990s, more than 95 % of reported cases were confirmed by laboratory procedures or were epidemiologically linked to laboratory-confirmed cases. Nonetheless, it is estimated that a great number of unrecognized cases of disease are not reported, which increases underreport of disease, especially the mild cases and unapparent infections. The Pan American Health Organization (PAHO) and World Health Organization (WHO) consider the report of a single yellow fever case as an emergency, due to the risk of a further spread, since the occurrence indicates current transmission of the virus and poses a true concern for spread especially in areas with low vaccination coverage. Thus, both the PAHO and WHO recommend a mass vaccination campaign in response to prevent further cases and spread of disease to other areas (WHO 2010a).

Between 2007 and 2010, 57 million people were vaccinated against YF in ten countries at risk in Africa, and during the same period, 17 million people were protected through emergency vaccination (WHO 2010b). In the 1990s, despite 50 years of use and over 500 million doses distributed, new safety concerns about the live attenuated YF 17D vaccine have come to light, revealing that in rare circumstances the vaccine can cause a disease similar to parental wild-type virus. This fact has modified vaccine policy and regulations in some circumstances, as will be discussed later on.

Despite the availability of vaccines since the 1940s, large epidemics occurred in areas without a background of naturally acquired or artificial immunity. Dramatic upsurges in YFV activity occurred in Africa in the 1960s and the late 1980s each involving >100,000 cases, and recent outbreaks affected southern Brazil, Paraguay and Argentina (2007–2009), Uganda (2010), and Sudan and Ethiopia (2012–2013). Although the absence of an immune barrier in the human population is a key factor, the underlying reasons for virus amplification remain unclear and are multifactorial, involving deterministic (vector density and competence, viral virulence) and stochastic factors. Expansions of YFV activity have sometimes been associated with the emergence of a new virus lineage (Souza et al. 2010), but the lack of information about biological correlates of genetic change make it difficult to assign causality. Perturbations of weather, particularly prolonged increases in rainfall and high temperatures have been associated with outbreaks of YF in South America (Vasconcelos et al. 2001b) and Africa.

In humans, YF is a severe acute illness with fever, nausea, vomiting, epigastric pain, hepatitis with jaundice, renal failure, hemorrhage, shock, and death in 20–60 % of cases. Yellow fever is the prototypical viral hemorrhagic fever and shares many pathophysiological features with unrelated diseases associated with a similar syndrome, except that the severity of hepatic dysfunction is generally greater in YF patients. The lower case fatality in Africa (~20 %) than in South America (40–60 %) (Tuboi et al. 2007) suggests that genetic factors determine lethality of the infection, a subject that deserves further study. Interestingly, the neutralizing antibody response to YF 17D vaccine is statistically higher in Caucasians than in African–Americans (Monath et al. 2002) possibly indicating genetic resistance to YF in the latter. Some New World monkeys, notably *Alouatta* (howling monkeys), are also susceptible to lethal infections, and epizootics associated with monkey deaths may precede the occurrence of human cases, a useful surveillance tool. In contrast, almost all African nonhuman primates have inapparent, viremic infections. This reflects the origin of yellow fever virus (YFV) in Africa several 1,000 years ago and a balanced coevolution of virus and hosts. Based on these factors and genetic analyses, YFV was introduced into the Americas from West Africa during the slave trade about 400 years ago (Bryant et al. 2007) and rapidly invaded a new ecological niche involving local hosts and vectors, much like another flavivirus, West Nile, and the *Alphavirus* chikungunya did after their more recent introductions into the Americas.

A serious concern for the future is whether YFV could be introduced by a viremic air traveler to *Ae. aegypti*-infected areas outside the endemic zone and particularly India and Southeast Asia. The recent spread of another virus transmitted in a human–*Aedes* cycle, chikungunya virus in islands of the Indian Ocean, India, southern Europe, and the Caribbean and South American countries, shows the risk. Although the WHO maintains an emergency YF 17D vaccine stockpile, an extensive outbreak outside the endemic regions could create a significant shortfall in vaccine supply at global level (WHO 2010a).

8.6.1 Geographic Distribution of Yellow Fever Cases/Outbreaks in South America

In the last 8 years, an intense YFV circulation was observed in several countries. Indeed, a large epizootic circulation of YFV has stricken Brazil especially Pará and Goiás states which is widespread to areas in contiguous states of Goiás and Mato Grosso do Sul (Central Brazil), and also outside the endemic region including countries such as Paraguay and Argentina, that reported yellow fever cases after a silence of YFV circulation for 34 and 41 years, respectively (Cardoso et al. 2010; Souza et al. 2010; Vasconcelos 2010; Goenaga et al. 2012; Moreno et al. 2013; Almeida et al. 2014; Romano et al. 2014). Several outbreaks were also registered in Colombia and

Peru, both endemic countries, the latter being responsible for almost 50 % of all YF cases reported in the Americas (Monath 2001).

In 2008, cases of jungle yellow fever were diagnosed in San Izidro and San Pedro Departments, and almost simultaneously also, an outbreak with 14 cases was laboratory confirmed in the district of Laurelty in the metropolitan area of Asunción, the Paraguayan capital. The episode represented a dangerous occurrence to a large urban area. The effective vector control measures and powerful vaccine coverage interrupted transmission in Asunción. For over 40 years, urban yellow fever had not been recognized in urban sets. It is important to mention that before the occurrence of cases, *Aedes aegypti* Breteau and house indices were approximately 30 % in the affected area. In the same year, five yellow fever cases in the Argentinean Misiones Province were reported, and an investigation showed that the cases were associated with a new vector, *Sabethes albiprivus*, in disease transmission (Goenaga et al. 2012).

In Brazil, the spread of YF in 2008–2009 was impressive. First cases were recognized in the municipalities of Afua and Breves in State of Pará, in the first months of 2008 and almost simultaneously, as reported in the states of Tocantins and Goiás. Another explosive epizootic with human cases was reported in 2008–2009, when cases were registered almost at the same time in Central (states of Goiás and Mato Grosso do Sul), Southeast (São Paulo State), and Southern (Rio Grande do Sul State) Brazil. The occurrence of cases after this epidemic/epizootic has been low. The vector responsible for transmission in Brazil changes according to the region; for example, in the endemic area, the main vector is *Haemagogus janthinomys*, while in the Southern region and in the state of São Paulo, the most important vector is *Hg. leucocelaenus* (Souza et al. 2010; Almeida et al. 2014). Different species of *Sabethes* mosquitoes have been considered secondary vectors based on YFV isolation from their tissues. Thus, *Sa. chloropterus*, *Sa. cyaneus*, *Sa. quasicyaneus*, *Sa. belisarioi*, *Sa. glaucodaemon*, and other more play a role in the YFV maintenance cycle even in the absence of *Haemagogus* mosquitoes as observed in 2008 in Argentina (Vasconcelos et al. 2003; Goenaga et al. 2012). In South America, the annual average of cases reported is limited and in general does not reach more than 100; however, the CFR frequently is around or even more 50 % in almost endemic countries.

8.6.2 Geographic Distribution of Yellow Fever Cases/Outbreaks in Africa

Yellow fever is endemic in tropical and subtropical regions of Africa, and it is believed YFV has an African origin. In these areas, cases have been reported almost yearly. However, it is interesting to highlight that in some years, up to 5,000 cases were reported to the WHO, while in others no case has been officially registered. This means that an incomplete and weak surveillance is in general the rule in the continent. Large epidemics have been recognized since the 1960s in both West and East African countries. Indeed, in West Africa the largest epidemics were registered in Nigeria (1986–1994) and before in Ghana (1977–1983) and later in Guinea (2000–2005). The impact of epidemic of Nigeria was tremendous with a conservative estimation occurrence of 20,000 cases and 5,000 deaths; both urban and jungle epidemiologic forms were reported, but cases were predominantly of urban yellow fever with transmission predominantly throughout the period by *Ae. aegypti* (Monath et al. 1972). In East Africa, the most recent important epidemics were reported in Sudan during the years of 2003, 2005, and 2012–2013. The last epidemic had a great impact in terms of public health in a country divided due to civil wars. Before, we have reports of bigger epidemics in Ethiopia (1961, 1962, 1966), when several thousands of cases were notified. It is important to emphasize that in Central African countries, the epidemics are limited in terms of number of cases and deaths (Robertson et al. 1996; Garske et al. 2013; Markoff 2013).

In Nigeria, during the reemergence of yellow fever in the 1990s, the incidence (based on official reports) was 3–4 per 100,000. While, during occurrence of large outbreaks in Liberia

(1995) and Guinea (2000), the incidence rates were 10–14 per 100,000. Nonetheless, it is important to mention that where investigations were conducted, such as in Nigeria (1986–1987) and the Gambia (1978–1979), the attack rates were of 3,000 per 100,000 inhabitants (Germain et al. 1980; Robertson et al. 1996; Monath et al. 1972, 2012; Sall et al. 2010; Garske et al. 2013).

In Africa the number of annual yellow fever cases reported to the WHO is variable, but in the years with YFV great activity, it can reach 5,000 or even more, while the case-fatality rate (CFR) is of around 20 %, much lower than in South America, where the CFR accomplish 50 % or more in some countries. In the present century, yellow fever outbreaks/epidemics were reported in Guinea (2000–2001, 2005), Liberia (2000–2001, 2004), Côte d'Ivoire (2001–2003), Ghana (2003), Senegal (2002–2003, 2006), Burkina Faso (2004), Mali (2004), Togo (2006), Ethiopia (2013), Uganda (2010), and southern Sudan (2003, 2005, 2012–2013). An increase of incidence was noticed in the Central African Republic, Congo, and Chad by 2008, all of them located in Central African region. In other countries the number of reported cases has decreased dramatically, but this is unclear if occurred due to improvement in the surveillance or because disease has not correctly diagnosed/recognized. In fact, the yellow fever reemergence in Sudan and Ethiopia in 2013 can signalize the second option (Germain et al. 1980; Sall et al. 2010; Garske et al. 2013; Markoff 2013).

In Africa the main vector of jungle yellow fever is *Ae. africanus* and a related species, *Ae. opok*. Nonhuman primates are the host, but contrarily to South America where animals die of disease, African monkeys are resistant to YFV and only rarely they succumb due to yellow fever. On the other hand, humans exposed to vectors may become infected, but the mosquito–human continued transmission is rare. Different vectors are recognized in different regions of Africa. Indeed, in West Africa the tree-hole breeding anthropophilic mosquitoes *Ae. furcifer-taylori*, *Ae. luteocephalus*, and *Ae. metallicus* together with the main vector *Ae. africanus* have been associated with transmission of YFV. Urban yellow fever is only transmitted by the domestic vector *Ae. aegypti*. In East Africa, *Ae. aegypti* is not anthropophilic and does not appear to play a role in transmission. The principal vector in both delimited and very large outbreaks is *Ae. bromeliae*, a species of the *Ae. simpsoni* complex (Barrett and Higgs 2007).

Although outside of the scope of this chapter, a small text on the absence of YFV in Asia will be discussed below. Remains as a big mystery why yellow fever has never been established in the Asian continent, since the occurrence of other *Aedes*-vectored flaviviruses supports the idea that YFV maintenance should be feasible in Asia. Indeed, Asia has the most susceptible NHP to YFV and large amounts of *Aedes aegypti* documented by annual description of dengue and chikungunya epidemics. Therefore, all conditions to maintain a YFV cycle, but many authors have hypothesized that the endemicity of dengue and other flaviviruses protects against yellow fever, since heterologous antibodies to dengue, Japanese encephalitis and other flaviviruses limit the viremia below the threshold of mosquito infection preventing autocthounous circulation of YFV Asia (Mutebi et al. 2001; Monath et al. 2012).

8.7 Prevention and Control

The vaccination of all people living in or traveling to endemic areas is the only available measure to prevent yellow fever. Actually two 17D substrains are used to produce yellow fever vaccines in six countries, with no differences in the immunogenicity or safety of the vaccines: (1) 17DD used in Brazil (producer—Oswaldo Cruz Foundation), a derived lineage with approximately 85 more passages than original 17D strain, and (2) 17D-204 (five different producers in the USA, France, Belgium, Russia, and China). Both vaccines are manufactured as extracts of infected chicken embryos[1] to the same WHO

[1] Allergy to egg products precludes the utilization of this vaccine.

required standards, with the use of seed lots to avoid (or minimize) risk of mutation to revert the viscerotropism of its sylvatic YFV, the Asibi YFV strain. Annually approximately 20 million doses of vaccines are produced and distributed, but in certain years (when virus circulation has increased), more than 50 million doses can be delivered. The WHO maintains a stockpile of around eight million doses to be used during emergencies, as occurred in 2008 in Paraguay and Brazil and are common to occur in Africa. In general, 17D vaccines are recommended in routine immunization of children in endemic regions with 12 months of age, during occurrence of epidemics, and in catch-up campaigns in Africa to improve vaccine coverage.

In general, 17D vaccine is well tolerated, and rare adverse events have been described. Since the early use of 17D vaccine up to nowadays, only two distinct groups, neurologic or viscerotropic adverse events, have been described. In experimental animals, as mice, the 17D viruses are neurotropic, and infected animals via intracerebral develop fatal encephalitis; in rhesus monkeys similar pattern has been observed, but changes have been described in both the spinal and brain consistent with panneurotropism. These monkeys are used as a model to test neurovirulence of 17D vaccines. In humans, neurotropic averse events (yellow fever vaccine-associated neurotropic disease or YEL-AND) caused by 17D vaccines are rare (incidence of 0.8 per 100,000 vaccinees in the USA and 0.2 per 100,000 in Europe). The mechanism of neuroinvasion is unknown, but it is hypothesized that the entrance of 17D virus occurs during the viremia through the blood–brain barrier. For this cause, and due to the immaturity of the blood–brain barrier, vaccination of infants below 6 months of age is contraindicated.

Viscerotropic adverse events were more recently recognized and resulted in a new paradigm that the safety of 17D vaccines is not 100 %. Indeed, since the report of seven original cases (six of them are fatal) of acute multi-organ failure resembling wild-type yellow fever caused by 17D vaccines (Chan et al. 2001; Martin et al. 2001; Vasconcelos et al. 2001a), more than 60 cases of yellow fever vaccine-associated viscerotropic disease or YEL-AVD were reported with an extremely high CFR of 63 %, but these cases are also rare. A good review on YEL-AVD and YEL-AND was recently published (Gershman et al. 2012).

The overall incidence of YEL-AVD in the USA has been estimated at 0.4 per 100,000 but is much higher in elderly (1.0–2.3 per 100,000) (Monath et al. 2012). Similar rates were also estimated in areas of Brazil in populations without previous vaccination during mass campaigns to abort outbreaks. Although the higher incidence of YEL-AVD is in older people, fatal outcome is more frequent in young women persons (Seligman 2011). It is important to mention that YEL-AVD has showed higher CFR than for wild-type YF disease. This apparent discrepancy probably is due to underlying conditions associated with immune downregulation that many affected patients present. It has been hypothesized YEL-AVD is due to a host susceptibility rather than 17D mutations causing an increase in virus virulence. The risk factors recognized for YEL-AVD include thymus disorders/thymectomy, autoimmune diseases (for details, see Monath et al. 2012), genetic factors, and advanced age. Older people vaccinees present delayed antibody response and prolonged viremia following 17D vaccination, possibly due to some acquired impairment of innate immune responses.

In conclusion, although safe, the risk–benefit of the use of 17D vaccines should be carefully evaluated due to the occurrence of YEL-AND and YEL-AVD that when diagnosed represent a serious threat to the people vaccinated with 17D viruses.

The adoption of individual protection such as the use of repellent is important, but vaccination is mandatory to prevent yellow fever. Revaccination is recommended after 10 years of previous 17D shot. Due to the shortage of 17D vaccines, actually the WHO does not recommend revaccination for people living in endemic areas, just for travelers (Jentes et al. 2011), but some endemic countries like Brazil disagree with this recommendation and as manufacturer of 17DD YFV vaccine have maintained the policy of revaccination after each 10 years living in Brazil.

The USA apparently will also adopt revaccination for all citizens living in the country that need to travel or move to the endemic region.

References

Almeida MAB, Cardoso JA, Santos E et al (2014) Surveillance for yellow fever virus in non-human primates in southern Brazil, 2001–2011: a tool for prioritizing human populations for vaccination. PLoS Negl Trop Dis 8, e2741. doi:10.1371/journal.pntd.0002741

Barrett AD, Higgs S (2007) Yellow fever: a disease that has yet to be conquered. Annu Rev Entomol 52:209–229

Bryant JE, Holmes EC, Barrett AD (2007) Out of Africa: a molecular perspective on the introduction of yellow fever virus into the Americas. PLoS Pathog 3, e75

Cardoso JC, Almeida MAB, Santos E et al (2010) Yellow fever virus isolation from *Haemagogus leucocelaenus* and *Aedes serratus* mosquitoes, southern Brazil, 2008. Emerg Infect Dis 16:1918–1924

Chambers TJ, Hahn CS, Galler R et al (1990) Flavivirus genome organization, expression, and replication. Annu Rev Microbiol 44:649–688

Chan RC, Penney DJ, Little D et al (2001) Hepatitis and death following vaccination with 17D-204 yellow fever vaccine. Lancet 358:121–122

Garske T, Van Kerkhove MD, Yactayo S et al (2013) Yellow fever in Africa: estimating the burden of disease and impact of mass vaccination from outbreak and serological data. PLoS Med 11(5), e1001638. doi:10.1371/journal.pmed.1001638

Germain M, Francy DB, Monath TP et al (1980) Yellow fever in the Gambia, 1978–1979: entomological aspects and epidemiological correlations. Am J Trop Med Hyg 29:929–940

Gershman MD, Staples JE, Bentsi-Enchill AD et al (2012) Viscerotropic disease: case definition and guidelines for collection, analysis, and presentation of immunization safety data. Vaccine 30:5038–5058

Goenaga S, Fabbri C, Dueñas JC et al (2012) Isolation of yellow fever virus from mosquitoes in Misiones province, Argentina. Vector Borne Zoonotic Dis 12:986–993

Gubler DJ, Kuno G, Markoff L (2007) Flaviviruses. In: Knipe DM, Howley PM, Griffin DE et al (eds) Fields virology, 5th edn. Lippincott Williams & Wilkins Publishers, Philadelphia, PA, pp 1153–1252

Jentes EJ, Poumerol G, Gershman MD et al (2011) The revised global yellow fever risk map and recommendations for vaccination, 2010: consensus of the Informal WHO Working Group on Geographic Risk for Yellow Fever. Lancet Infect Dis 11:622–632

Johansson MA, Vasconcelos PF, Staples JE (2014) The whole iceberg: estimating the incidence of yellow fever virus infection from the number of severe cases. Trans R Soc Trop Med Hyg 108:482–487

Julander JG, Ennis J, Turner J et al (2011) Treatment of yellow fever virus with an adenovirus-vectored interferon, DEF201, in a hamster model. Antimicrob Agents Chemother 55:2067–2073

Markoff L (2013) Yellow fever outbreak in Sudan. N Engl J Med 368:689–691

Martin M, Tsai TF, Cropp B et al (2001) Fever and multisystem organ failure associated with 17D-204 yellow fever vaccination: a report of four cases. Lancet 358:98–104

Monath TP (2001) Yellow fever: an update. Lancet Infect Dis 1:11–20

Monath TP (2008) Treatment of yellow fever. Antiviral Res 78:116–124

Monath TP, Barrett ADT (2003) Pathogenesis and pathophysiology of yellow fever. Adv Virus Dis 60: 343–395

Monath TP, Vasconcelos PF (2015) Yellow fever. J Clin Virol 64:160–173. doi:10.1016/j.jcv.2014.08.030

Monath TP, Onejeme S, Okweke G et al (1972) Surveillance of yellow fever in Nigeria, 1970–1971. Niger Med J21:178–186

Monath TP, Nichols R, Archamboult W et al (2002) Comparative safety of two yellow fever 17D vaccines (Arilvax and YF-Vax) in a phase III multicenter, double-blind clinical trial. Am J Trop Med Hyg 66:533–541

Monath TP, Gershman M, Staples EJ et al (2012) Yellow fever vaccine. In: Plotkin SA, Orenstein WA et al (eds) Vaccines, chapter 36, 6th edn. Saunders Elsevier, Philadelphia, pp 870–896

Moreno ES, Spinola R, Tengan CH et al (2013) Yellow fever epizootics in non-human primates, São Paulo state, Brazil, 2008–2009. Rev Inst Med Trop Sao Paulo 55:45–50

Mutebi J-P, Wang H, Li L et al (2001) Phylogenetic and evolutionary relationships among yellow fever virus isolates in Africa. J Virol 75:6999–7008

Nunes MRT, Palacios G, Nunes KNB et al (2011) Evaluation of two molecular methods for the detection of Yellow fever virus genome. J Virol Meth 174:29–34

Quaresma JAS, Barros VLRS, Fernandes ER et al (2005) Reconsideration of histopathology and ultrastructural aspects of the human liver in yellow fever. Acta Trop 94:116–127

Quaresma JAS, Barros VLRS, Pagliari C et al (2006a) Revisiting the liver in human yellow fever: virus-induced apoptosis in hepatocytes associated with TGF-β, TNFα, and NK cells activity. Virology 345:22–30

Quaresma JAS, Barros VLRS, Pagliari C et al (2006b) Immunohistochemical examination of the role of FAS Ligand and lymphocytes in the pathogenesis of human liver Yellow fever. Virus Res 116:91–97

Quaresma JAS, Duarte MIS, Vasconcelos PFC (2006c) Midzonal lesion in the yellow fever: a specific pattern of liver injury caused by direct virus action and by immune anti-inflammatory response. Med Hypotheses 47:618–621

Quaresma JAS, Barros VLRS, Fernandes ER et al (2007) Hepatocyte lesions and cellular immune response in yellow fever infection. Trans R Soc Trop Med Hyg 101:161–168

Quaresma JAS, Pagliari C, Medeiros DBA et al (2013) Immunity and immune response, pathology and pathologic changes. Progresses and challenges in the immunopathology of yellow fever. Rev Med Virol 23:305–331

Rice C, Lenches EM, Eddy SR et al (1985) Nucleotide sequence of yellow fever virus: implications for flavivirus gene expression and evolution. Science 229(4715):726–733

Robertson SE, Hull BP, Tomori O et al (1996) Yellow fever: a decade of reemergence. JAMA 276:1157–1162

Romano AP, Costa ZA, Ramos DG (2014) Yellow fever outbreaks in unvaccinated populations, Brazil, 2008–2009. PLoS Negl Trop Dis 8(3), e2740. doi:10.1371/journal.pntd.0002740

Sall AA, Faye O, Diallo M et al (2010) Yellow fever exhibits slower evolutionary dynamics than dengue virus. J Virol 84:765–772

Seligman SJ (2011) Yellow fever virus vaccine—associated deaths in young women. Emerg Infect Dis 17:1891–1893

Souza RP, Foster PG, Sallum MA et al (2010) Detection of a new yellow fever virus lineage within the South American genotype I in Brazil. J Med Virol 82:175–185

Tuboi SH, Costa ZGA, Vasconcelos PFC et al (2007) Clinical and epidemiologic characteristics of yellow fever in Brazil: analysis of reported cases, 1998–2002. Trans R Soc Trop Med Hyg 101:169–175

Vasconcelos PFC (2010) Yellow fever in Brazil: thoughts and hypotheses on the emergence in previously free areas. Rev Saude Publica 44:1144–1149

Vasconcelos PF, Luna EJ, Galler R et al (2001a) Serious adverse events associated with yellow fever 17DD vaccine in Brazil: a report of two cases. Lancet 358:91–97

Vasconcelos PFC, Costa ZG, Travassos da Rosa ES et al (2001b) An epidemic of jungle yellow fever in Brazil, 2000. Implications of climatic alterations in disease spread. J Med Virol 65:598–604

Vasconcelos PFC, Sperb AF, Monteiro HAO et al (2003) Isolations of yellow fever virus from *Haemagogus leucocelaenus* in Rio Grande do Sul State, Brazil, in the Southern Cone. Trans R Soc Trop Med Hyg 97:60–62

Vasconcelos PFC, Bryant JE, Travassos da Rosa APA et al (2004) Genetic divergence and dispersal of yellow fever virus, Brazil. Emerg Infect Dis 10:1578–1584

Weidmann M, Faye OUS, Faye OUM et al (2010) Improved DNA probe based assay for the detection of African and South-American Yellow Fever Virus strains. J Clin Virol 48:187–192

WHO (2010a) WHO working group on technical specifications for manufacture and evaluation of yellow fever vaccines, Geneva, Switzerland, 13–14 May 2009. Vaccine 28:8236–8245

WHO (2010b) Yellow fever surveillance and outbreak response: revision of case definitions, October 2010. Wkly Epidemiol Rec 85:465–472

9 Togaviridae

Marta S. Contigiani and Luis A. Diaz

Abstract

Alphavirus constitutes one of the two genera included in the family Togaviridae. This genus contains 31 viral species (with different variants and strains), grouped into seven antigenic complexes. Depending on the geographical location where they were isolated, this genus is divided into alphavirus in the New World (including Eastern equine encephalitis virus, Venezuelan equine encephalitis, and Western equine encephalitis, which cause encephalitis in humans and other mammals) and the Old World (chikungunya virus, o'nyong-nyong virus, Ross River virus, Semliki Forest virus, and Sindbis virus causing syndrome characterized by fever, rash, and arthralgias, which rarely cause mortality). However, Sindbis virus (the prototype alphavirus) causes encephalomyelitis in mice, and Ross River virus and chikungunya virus (CHIKV) are also neuroinvasive and cause neurological disease in humans. Alphaviruses are responsible for several medically important emerging diseases and are also significant veterinary pathogens. Due to the aerosol infectivity of some alphaviruses and their ability to cause severe, sometimes fatal neurological diseases, they are also of biodefense importance. Likewise, they are of interest for their potential use in gene therapy. This chapter will describe general aspects of alphavirus, with emphasis on their pathology, ecology, epidemiology, clinical, diagnosis, treatment, prevention, and control measures.

Keywords

Alphavirus • Equine encephalitis • Chikungunya

M.S. Contigiani (✉)
Laboratory of Arbovirus, Institute of Virology "Dr. J. M. Vanella", Enfermera Gordillo Gomez s/n, Ciudad Universitaria, National University of Cordoba, Córdoba, Argentina
e-mail: martascontigia@hotmail.com

L.A. Diaz
Laboratory of Arbovirus, Institute of Virology "Dr. J. M. Vanella", Enfermera Gordillo Gomez s/n, Ciudad Universitaria, National University of Cordoba, Córdoba, Argentina

Institute for Biological and Technical Researches, CONICET-UNC, Córdoba, Argentina
e-mail: adrian.diaz@conicet.gov.ar

C.B. Marcondes (ed.), *Arthropod Borne Diseases*, DOI 10.1007/978-3-319-13884-8_9

9.1 Introduction

The family Togaviridae includes two genera: *Rubivirus* (rubella virus, which only infects humans and is maintained by human-to-human transmission) and *Alphavirus* (arboviruses) (ICTV 2014). Most of the last are mosquito-associated arboviruses but some are tick borne.

Many alphaviruses are of human and veterinary concern because they cause diseases of public health and economic importance (Hubálek et al. 2014). Clinical symptoms caused by alphavirus infection vary according to the viral species (Griffin 2007).

Alphavirus genera account for a total of 31 viral species (ICTV 2014). Their members have been detected in all continents, except for Antarctica, mainly due to the absence of an arthropod vector. Many alphaviruses were firstly isolated in tropical regions from Africa and South America and in some Asian countries (Calisher and Karabatsos 1988). Several factors, such as changes in viral genetics, host and/or vector population, and climate changes, facilitated the expansion and transmission of alphavirus in the past several decades.

Based on their original geographic distribution, alphaviruses are clustered into two main groups: New World and Old World. The former includes mainly neurotropic alphaviruses pathogenic for humans and equines (i.e., Eastern equine encephalitis virus, Western equine encephalitis virus, Venezuelan equine encephalitis complex viruses). The exception is the Mayaro virus, which is associated with febrile illness and arthralgia (a dengue- and chikungunya-like disease).

The Old World alphaviruses are mainly associated with febrile illness, eruption, and arthralgia in humans. They rarely cause death (i.e., chikungunya virus, CHIKV; o'nyong-nyong virus, ONNV; Ross River virus, RRV; Getah virus, GETV; Middelburg virus, MIDV; Semliki Forest virus, SFV; Sindbis virus, SINV). GETV, MIDV, SFV, and SINV can cause neurological affections in equines (Table 9.1) (Griffin 2007).

9.2 Viral Agent

The prototype virus in the genera is SINV. All members have a similar virion structure, replication strategy, and molecular characteristics (Griffin 2007).

Table 9.1 Biological features and geographic distribution of virus belonging to *Alphavirus* genus

Antigenic complex	Viral specie	Antigenic subtype	Antigenic variety	Associated clinical syndrome	Distribution
Barmah Forest	Barmah Forest virus (BFV)			EFA	Australia
Eastern equine encephalitis (EEE)	EEEV	I, II, IV		EFE	North America and Caribbean
	Madariaga virus	III			South and Central America
Middelburg	Middelburg virus (MIDV)			ENR	Africa
Ndumu	Ndumu virus (NDUV)			ENR	Africa
Semliki Forest	Semliki Forest virus (SFV)			EF	Africa
	Chikungunya virus (CHIKV)			EFA	Africa, Asia, Europe
	O'nyong-nyong virus (ONNV)			EFA	Africa
	Getah virus (GETV)			ENR	Asia
	Bebaru virus (BEBV)			ENR	Malaysia
	Ross River virus (RRV)	Sagiyama		EFA	Australia, Oceania
	Mayaro virus (MAYV)			EFA	Central and South America, Trinidad
	UNA virus (UNAV)			ENR	South America

(continued)

Table 9.1 (continued)

Antigenic complex	Viral specie	Antigenic subtype	Antigenic variety	Associated clinical syndrome	Distribution
Venezuelan equine encephalitis (VEE)	VEEV	I	AB	EFE	American continent
			C		Northern South America
			D		South and Central America
			E		Mexico and Central America
	Mosso das Pedras virus (MEDV)		F	EFE	South America (Brazil)
	Everglades virus (EVEV)	II		EFE	Florida (USA)
	Mucambo virus (MUCV)	III	A, C, and D	EFE	South America
	Tonate virus (TONV)		B		Brazil, USA
	Pixuna virus (PIXV)	IV		ENR	Brazil, Argentina
	Cabassou virus (CABV)	V		ENR	French Guyana
	Rio Negro virus (RNV)	VI		ENR	Argentina
Western equine Encephalitis (WEE)	Sindbis virus (SINV)	Sindbis		EFA	Africa, Europe, Asia, Australia
		Babanki		EFA	Africa
		Ockelbo		EFA	Europe
		Kyzylagach		ENR	Azerbaijan, China
	Whataroa virus (WHAV)			ENR	New Zealand
	AURA virus (AURAV)			ENR	South America
	WEEV	Several		EFE	West coast of North America, South America
	Highlands J virus (HJV)			ENR	East coast of North America
	Fort Morgan virus (FMV)	Buggy Creek		ENR	West coast of North America
Trocara	Trocara virus (TROV)			ENR	South America
Eilat virus		I–V		ENR	Israel
	Salmon pancreas disease virus (SPDV)			Pancreatic Disease (Salmon)	Atlantic Ocean
	Sleeping disease virus			Sleeping disease (trout)	Europe
	Southern elephant seal virus (SESV)			ENR	Australia

The alphaviruses are enveloped, single positive-stranded RNA viruses with icosahedral symmetry. Virions are 60–70 nm in diameter and are sensitive to ether and chloroform (Fig. 9.1a). Viral genome is organized in two modules: two thirds of the methylated 5′ end encodes for non-structural proteins (nsPs). The polyadenylated 3′ end third encodes the structural proteins (E1, E2, E3, and 6K) (Fig. 9.1b). Nonstructural proteins have a key role during transcription and replication of viral RNA. Viral genome (10–12 kb length) is included inside the viral capsid made by C protein (capsid protein). Both elements compose the nucleocapsid surrounded by a lipoprotein envelope. The viral envelope is a host cell-derived structure on which glycoproteins E1

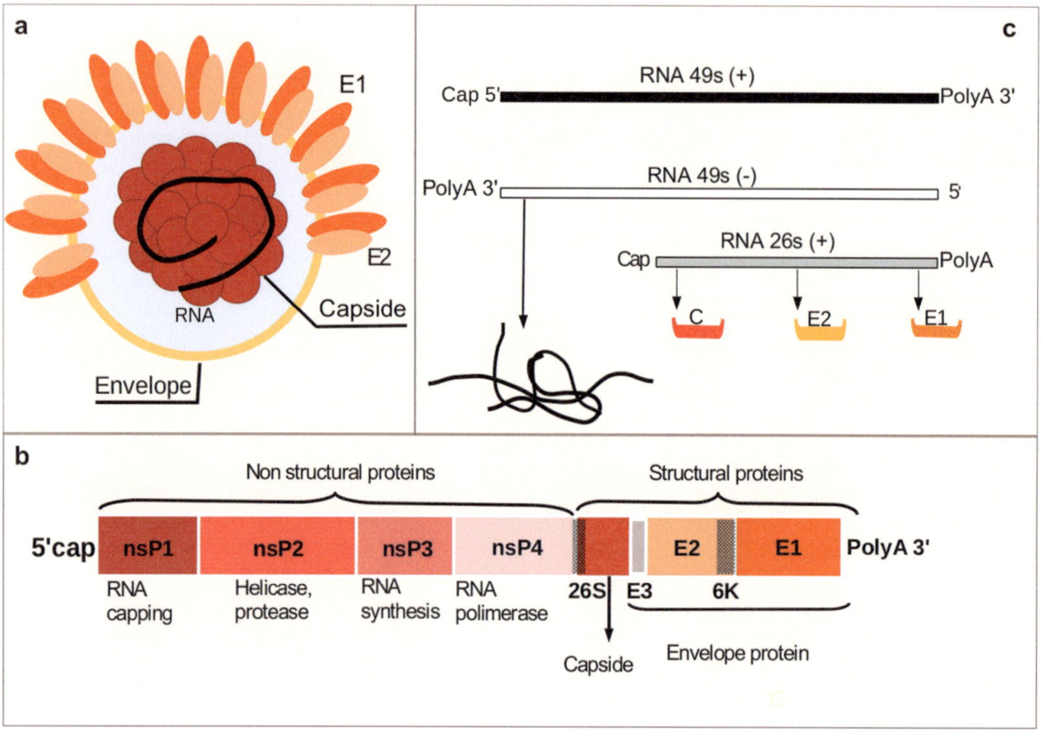

Fig. 9.1 A schematic representation of an *Alphavirus* virion (**a**) and its RNA genome (**b**) alphavirus replication cycle (**c**)

and E2 are embedded. The E2 glycoprotein has main epitopes for neutralizing antibodies (Griffin 2007, Jose et al. 2009).

Alphavirus virions have the ability to agglutinate avian red cells. This trait is based on E1 and E2 proteins and has been used for viral quantification. It also allows the use of the hemagglutination inhibition assay (HIA) for the serological screening in alphavirus infections. Neutralizing antibodies react to epitopes localized mainly in E2 protein and are viral type specific. The HIA has been an extremely useful technique for analyzing phylogenetic relationships and clustering alphavirus in serogroups. Up to date, eight serocomplexes have been detected in the *Alphavirus* genera, including the latest Trocara virus complex. The BF, EEE, MID, NDU, and TROC serogroups have only one viral member, whereas other serogroups include several viral species (Table 9.1). Moreover, subtypes can be identified for particular viral species (Calisher and Karabatsos 1988, Powers et al. 2001).

At the molecular level, the C and E1 amino acid sequences are more conservative than the E2 protein, and, in general, antigenic clusters overlap with the molecular classification.

Viral species are defined by a combination of genetic, ecological, and antigenic information. In general, they have a different transmission cycle and 23 % and 10 % of nucleotide and amino acid divergence, respectively (Griffin 2007).

9.3 Replication Cycle

Viral infection in a vertebrate host cell starts with the inhibition of the macromolecular synthesis of the cell. The cell biosynthetic mechanism is driven by virus proteins and converted into a viral particle factory.

The viral RNA acts as a messenger RNA in the production of viral nsPs. It is also transcripted by a viral polymerase into a negative-sense complementary strain RNA, which is used as a template during the synthesis of subgenomic RNA 26S and 49S (Fig. 9.1c). The smaller subgenomic RNA encodes for structural proteins. Nucleocapsids are produced in the cytoplasm and released by rotation through the plasmatic membrane of the cell host.

Alphaviruses can replicate in a wide range of vertebrate species and types of cells. Suckling albino mice are highly susceptible to most alphaviruses. These viruses are also well isolated and replicate in a wide variety of cellular cultures, such as primary chick embryo fibroblast, Vero (green monkey kidney), BHK (suckling hamster kidney), and C6/C36 (mosquitoes) (Griffin 2007).

9.4 Neurotropic Alphaviruses

9.4.1 Eastern Equine Encephalitis Virus (EEEV)

9.4.1.1 Viral Pathogenesis in Humans and Equines

Most EEEV infections in *humans* are inapparent. In those patients who get sick, signs and symptoms start between 3 and 10 days postinfection and the disease lasts from 1 to 2 weeks. Symptoms range from febrile illness, with cephalea and myalgia, followed by recovery to encephalitis, coma, and death. Death generally occurs 2–10 days after the onset of encephalitis (Go et al. 2014).

Cerebrospinal fluid (CSF) can appear clear, with white blood cells (2–2000 μl), with predominance of polymorphonuclear leukocytes, and the disease starts with replacement of mononuclear cells. CSF protein level is high, whereas glucose is normal or low. Blood can be observed in CSF. Electroencephalogram analyses can provide additional indirect evidence of EEEV infection. Computed tomography examination can be normal or show only edema, whereas magnetic resonance images are frequently abnormal and show focal lesions commonly in the basal ganglia, thalamus, and brainstem (Steele and Twenhafel 2010).

EEEV represents the most virulent and pathogenic alphavirus in the USA, being a serious concern for the public health. Mortality rate ranged from 30 to 40 %. People under 10 years old are the most susceptible to develop encephalitis and death. The highest mortality rates are observed in elderly population as well. Sequelae like paralyses and mental retardation are observed in 35–80 % of survivors, particularly in children.

EEEV causes encephalitis in *equines*. Disease starts with fever, anorexia, and colic followed by virus invasion of CNS, causing encephalitis and myelitis associated with abnormal behavior (stepping, disequilibrium, tendency to walk in circles), somnolence, paralysis, and seizures before death. Mortality rate ranges from 80 to 90 %. In the USA, the disease has been detected in pheasants, turkeys, dogs, and pigs. Birds generally do not get sick but develop high viremias. Those birds that get sick and die develop systemic illness with viscerotropic compromise with no encephalitis except for pheasants, which develop encephalitis. Recently, wild deers (*Odocoileus virginianus*) with neurological disorders have been detected to be infected with EEEV. This finding has placed this virus as a concern for wildlife conservation (Zacks and Paessler 2010).

9.4.1.2 Epidemiology

EEEV is widely distributed in the Americas. It has been detected all along the eastern coast, from Canada and the USA, through the Gulf of Mexico and Central American and Caribbean islands, to South America (Go et al. 2014).

In the USA, this virus is a main cause of neurological disease in domestic animals and humans. Severe disease is observed in humans, pigs, dogs, equines, and pheasants. Diseases by EEEV in equines were first recognized in 1931 in the USA. However, some reports of encephalomyelitis outbreaks from 1845 to 1912 are available (Scott and Weaver 1989). The first EEEV strain was isolated from the brain of an ill horse in New Jersey and Virginia during a wide extended epizooty in 1933. Moreover, the first fatal human cases were reported at the same places where equine encephalitis cases were documented in 1938. Although human cases by EEEV are not common in the USA, the average annual cases amount to 5 (ranging between 0 and

15). The states of Florida, Georgia, Massachusetts, and New Jersey have a high number of cases. Recently, human outbreaks of encephalitis were reported in northern states along the east coast of the USA (New Hampshire and Maine) and Canada. Likewise, simultaneous EEEV and VEEV outbreaks were reported in 2010 in Panama (Lubelczyk et al. 2013, Molaei et al. 2013, Vander Kelen et al. 2012, Yu et al. 2015).

In South America, EEEV was first isolated during an equine encephalitis outbreak in Argentina in 1936. Numerous epizootics by EEEV, some of them with thousands of cases, have been reported in Argentina, Brazil, Colombia, and Venezuela. The last EEEV epizootic reported in Argentina was in 1988. In Brazil, new equine encephalitis cases from southeastern areas were recorded during 2003, 2006, and 2009. During the equine epizootics in Argentina and Brazil, the high seroprevalence was detected in human population without human clinical cases. Up to date, only human cases were reported in Brazil (1956) and Trinidad (1970) (Sabattini et al. 1998, Scott and Weaver 1989, de Novaes Oliveira et al. 2014).

The comparison of EEEV strains from North and South America shows molecular, epidemiological, and ecological differences. Based on molecular evidences, four lineages can be distinguished: lineage I detected in North America and three lineages detected in South America (lineage II, Brazil, Guatemala, and Peru; lineage III, Argentina, Brazil, Colombia, Equador, Guayana, Panama, Peru, Trinidad, and Venezuela; lineage IV, Brazil). These clusters show that South American strains are more diversified and heterogeneous and that their antigenicity and geographic distribution are correlated. North American strains are associated with high virulence in humans and equines, whereas South American strains are pathogenic for equines, being attenuated for humans (Bhavez et al. 1999, Young et al. 2008, Weaver et al. 2012).

In North America, EEEV is endemically maintained by birds and ornithophilic mosquitoes (*Culiseta melanura*) in freshwater swamps. Under favorable ecological and environmental conditions, the virus can be transmitted to humans and domestic animals. Several mosquito species (e.g., *Aedes*[1] *sollicitans*, *Culex salinarius*, *Ae. canadensis*) with a wide host preference range have been incriminated as bridge vectors (Fig. 9.2). In the USA, human and equine cases are seasonal (late summer-early fall); however, in southeastern areas cases are reported year-round. The overwinter mechanism in the temperate area is unknown. Annual reintroduction through migrating birds and/or infecting mosquitoes from subtropical areas has been postulated. On the other hand, no transovarian transmission has been documented (Weaver et al. 2012, Go et al. 2014).

In the Caribbean and Central and South America, viral transmission occurs endemically, and several mosquito species are involved as enzootic vectors. Epidemiological data is scarce because of the lack of human cases by EEEV infection. Multiple isolations were made from *Culex (Melanoconion)*[2] spp. mosquitoes collected in Central and South America. Serological studies indicate that small mammals and birds would serve as maintenance host. Moreover, bats and marsupials infected by EEEV were detected (Go et al. 2014).

In Venezuela, EEEV enzootic strains were isolated from sentinel hamsters in 1975 and from equines in 1976 during an epizootic period. Circulating strains belong to lineage III.

In Brazil, equine epizootic events were reported in north, northeastern, and southeastern areas. Viral strains have been isolated from monkeys, chickens, sentinel hamsters, birds, mosquitoes (*Culex* spp., *Ae. taeniorhynchus*), and ill equines. The three South American lineages circulate in Brazil and are maintained by birds and rodents (hosts) and the mosquitoes *Culex pedroi* (enzootic vector) and *Ae. taeniorhynchus* (epizootic vector). In 2003, two equine encephalitis

[1]Although several taxonomic modifications have been proposed on Culicidae genera, mostly splitting *Aedes* by Reinert et al. (2009) (and accepted by CBM), this taxonomy has been used in the chapters according to authors' preference. To facilitate utilization by health personnel, all new aedine genera can be considered *Aedes* (CBM).

[2]This is a group of mosquitoes (160 spp.) particularly difficult to identify (CBM).

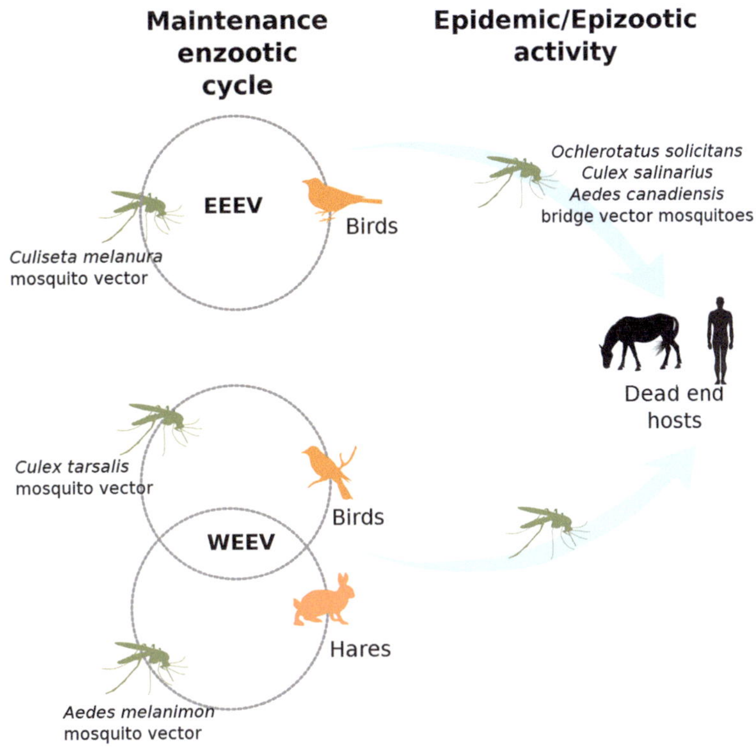

Fig. 9.2 Maintenance and transmission cycles for Eastern and Western equine encephalitis viruses

cases were reported in the southeastern region. The two last encephalitis cases in equines have been detected in 2006. Only one human case was reported 50 years ago.

In Argentina, EEEV was first isolated in a sick horse in Buenos Aires province in 1930; however, the virus was identified in 1953. EEEV has produced important periodic epizooties. Although mixed epizooties (EEEV/WEEV) have occurred, most of them have been produced by one viral agent. This recognized situation in Argentina is exceptional and has only been observed in Guayana. So far, EEEV has never been isolated from mosquitoes and its possible vector in the maintenance cycle is unknown. Serological studies incriminate wild and domestic birds in its viral cycle. Paradoxically, no infected bird was detected during equine epizootics in Santiago del Estero (1981) and Chaco (1988) provinces, suggesting that maintenance cycle occurred elsewhere or that migratory birds could have been involved (Weaver et al. 2012).

9.4.2 Western Equine Encephalitis Virus (WEEV)

The WEEV along with six viral species makes up the Western equine encephalitis antigenic complex (Table 9.1). WEEV, Highlands J virus (HJV), Fort Morgan virus (FMV), and AURA virus (AURAV) circulate in the Americas, each of them showing a particular ecological niche. HJV, WEEV, and FMV belong to a lineage that originated from recombination of a Sindbis-like virus and EEEV (Zacks and Paessler 2010, Go et al. 2014).

9.4.2.1 Pathogenicity in Humans and Animals

Neurological diseases caused by WEEV have signs and symptoms similar to those of EEEV, but less severe. The spectrum ranges from asymptomatic infection to lethal forms, with intermediate stages characterized by fever, headache, or aseptic meningitis. In the USA, an asymptomatic/symptomatic infection ratio is strongly age dependent, being more severe in young children. The initial symptoms are fever, chills, headache, nausea, and vomiting. Respiratory symptoms are occasional and may last from 3 to 5 days; then, the neurological signs and symptoms appear (tremors, lethargy, irritability, stiff neck, photophobia, dizziness, and altered mental status). CSF shows pleocytosis with 100–1500 cells/μl predominantly neutrophils at first and subsequently with mononuclear cells. Most cases have complete recovery. In adult individuals, remission of signs and symptoms occurs between 5 and 10 days, recovery is generally rapid, and sequelae are rare. In children under 1 year of age, 60% of survivors have brain damage, and in some of them, the disease is progressive with persistent infection.

In *equines* (horse, mules, and ponies), the disease is characterized by fever and encephalitis, which is often manifested as depression, anorexia, and paralysis of the lips and legs. The incubation period is 3–12 days. This is the form of the disease known as "lethargic"; the sick animal may or may not progress to the so-called "furious" form, which is where the vulgar designation "crazy horse" is derived, and it is how this disease is known in the field.

Signs of CNS involvement include falling head or ears, chewing movements, excessive salivation, incoordination, locomotion in circles, inability to stand, flabby lips, apparent blindness, ataxia, involuntary movements, irritability, hyperexcitability, and seizures. In the terminal state, side-lying prostrate animals show pedaling movements, nystagmus, paralysis, difficult breathing, and coma (Griffin 2007, Steele and Twenhafel 2010).

9.4.2.2 WEEV Distribution and Ecoepidemiology

WEEV is distributed from Canada to Argentina and is the only member of the complex pathogenic for humans. Epizootic and enzootic viral subtypes have been recognized for this virus. Viral strains isolated from ill horses and humans are considered pathogenic strains (epizootic strains), whereas those isolated from mosquitoes during interepizootic periods are attenuated for humans and equines (enzootic strains). Four viral lineages have been identified, two lineages endemic to South America and the other two widely distributed in the Caribbean and North and South America.

WEEV was isolated for the first time during equine epizootics in California (USA) in the summer of 1930. Previous evidences suggest that thousands of equines would have died by this virus in 1912. After that, in 1938 WEEV was isolated from the brain of a fatal human case confirming its virulence for humans. In Argentina, similar facts were reported: in 1908 an equine encephalitis outbreak took place and in 1933 public health personnel isolated the virus from an ill horse.

In the west coast of the USA, the transmission cycle includes *Cx. tarsalis* mosquitoes as enzootic and epizootic/epidemic vector and Passeriformes birds (mainly house sparrows, *Passer domesticus*) as primary amplifying hosts. Secondary amplifying hosts which *Cx. tarsalis* frequently feed on include Passeriformes birds, chickens, and pheasants. An alternative *Cx. tarsalis* bird transmission cycle can occur between *Ae. melanimon* mosquitoes and black-tailed jackrabbit (*Lepus californicus*). The latter develops a nonfatal infection with long viremia. Humans, equines, and other mammals are dead-end hosts (Fig. 9.2). Although *Cx. tarsalis* is mainly considered an ornithophilic vector, its host feeding preference changes from summer to fall, with mammals becoming the most frequent host. This evidence suggests the role of *Cx. tarsalis* as a bridge host, driving viral activity from birds to humans during early fall.

Viral isolations and serological studies have detected natural infections in chickens and other domestic birds, rodents, rabbits, reptiles, and amphibians; these two last animals are suspected to act as overwinter hosts for WEEV in temperate regions. In some regions of South America, most mosquito species of which WEEV have been isolated are mammalophilic. Small mammals (rodents and rabbits) are frequently found infected by WEEV. The opposite is observed in other areas where birds are frequently seropositive.

In the temperate area of Argentina, important equine epizooties were recorded from the beginning of the twentieth century, simultaneously with that caused by WEEV. Equine epizootics were recorded regularly every 5–10 years, with a mortality rate of 1–40 %. The last outbreak occurred in the summer of 1988–1989. During interepizootic periods, sporadic equine cases were reported, suggesting small focal outbreaks or iatrogenic cases due to inadequately inactivated vaccines. During the 1982–1983 epizootic, epizootic viral strains were isolated from *Ae. albifasciatus*, *Anopheles albitarsis*, *Mansonia* spp., and *Psorophora pallescens* mosquitoes. In the years preceding this epizooty, enzootic strains were isolated from *Cx. (Melanoconion) ocossa*. This prototype enzootic strain (AG80-646) is attenuated for equines, which shows it is different from WEEV. The natural transmission cycle is supposed to be integrated by *Cx. (Mel.) ocossa* mosquitoes and cricetid rodents, in which antibodies have been found (Fig. 9.3). For epizootic viral strains isolated in Argentina, the existence of two transmission cycles is assumed. The primary cycle might be called maintenance cycle, with unknown vectors and hosts, which would be analog to that observed in the USA between *Cx. tarsalis* mosquitoes and house sparrows. The other one, called amplifying cycle, has *Ae. albifasciatus* mosquitoes and mammals of Caviidae and Leporidae families as potential hosts. In this cycle, birds have had no apparent role in the maintenance of WEEV, since none or low seroprevalence was detected. Experimentally inoculated horses developed viremias high enough to infect vector mosquitoes, indicating their potential role as hosts.

Besides equine encephalitis epizootics reported in Argentina, outbreaks have been documented in Brazil and Paraguay (Go et al. 2014, Hubálek et al. 2014).

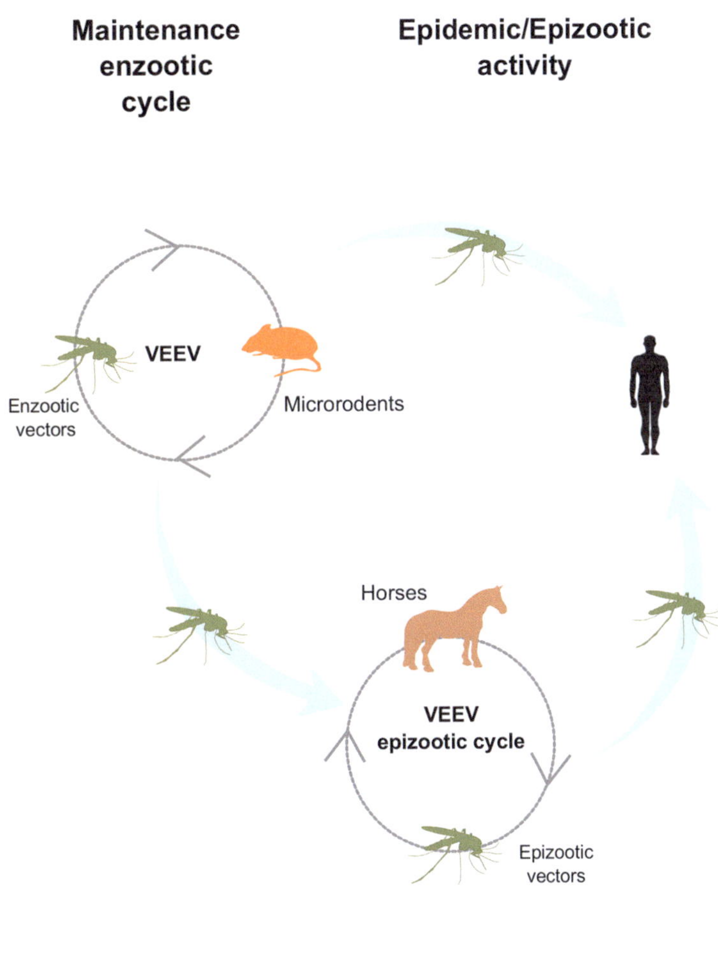

Fig. 9.3 Maintenance and transmission cycle of enzootic and epizootic subtypes of Western equine encephalitis virus

9.5 Other WEEV Virus Complex in the Americas

9.5.1 Highlands J Virus (HJV)

The Highlands J virus (HJV), identified in the eastern USA, was initially considered a variant of WEEV, but now is considered a separate entity. It was isolated from a corvid (*Cyanocitta cristata*) in Florida in 1960. This virus is enzootic in the eastern USA and is maintained by *Cs. melanura* as the primary vector and migratory birds as primary hosts. HJV may occasionally cause encephalitis in horses but is recognized as pathogenic for turkeys, pheasants, partridges, ducks, and other birds (Weaver et al. 1997, Griffin 2007, Steele and Twenhafel 2010).

9.5.2 Fort Morgan Virus (FMV)

The FMV and its closely related Buggy Creek virus (BCV) are not recognized as pathogenic for humans and pets. Both were isolated from swallows and sparrows in eastern Colorado and Oklahoma. It is interesting to note that BCV strains were isolated from bedbugs (*Oeciacus vicarius*) collected from bird nests during the summer in Colorado. These bugs could keep the virus during the winter and transmit it to the newborn birds in the early spring to restart the bird-mosquito cycle (Griffin 2007, Allison et al. 2015).

9.5.3 AURA Virus

The AURA virus was isolated from *Culex (Melanoconion)* sp. and *Ae. serratus*[3] later in northern Brazil in 1959. In Argentina, it was isolated from *Ae. serratus* captured in the province of Misiones in 1966. There is no evidence of a pathogenic role for humans. This virus is serologically related to Sindbis virus (found in the Old World) and WEEV (present in the Americas). AURAV shares substantial sequence identity and genome organization with Sindbis virus and important antigenic epitopes are preserved in both viruses. However, despite their close relationship, there is a significant divergence, sharing 73 % of the amino acid sequence of the nonstructural proteins and 62 % identity in structural proteins. By contrast, WEEV glycoproteins are more closely related to Sindbis virus than to AURA virus.

AURA virus has been demonstrated to be the least neurovirulent virus for mice among the enzootic viruses in WEEV complex. A characteristic demonstrated for the virus isolated in Argentina is the ability to replicate and produce experimental myocarditis in mice.

The geographic distribution of AURA virus obtained from epidemiological surveys in Argentina is greater in subtropical provinces (Chaco and Corrientes) than in temperate ones (Cordoba, Santa Fe, and Santiago del Estero) (Sabattini et al. 1998, Griffin 2007).

9.5.4 Venezuelan Equine Encephalitis Virus Complex

Venezuelan equine encephalitis virus (VEEV) is an important human and animal zoonotic pathogen, which causes periodic outbreaks of highly debilitating disease in the Americas. VEEV was first isolated in the Guajira region, Venezuela, from the brain of a sick horse during an outbreak of equine encephalitis in 1938. In the following years, VEE-related viruses were isolated and identified in many locations in South America, Central America, the Caribbean islands, and the southern regions of the USA. Isolates of these viruses related to VEEV were originally classified into subtypes I through IV, forming the VEEV complex by using a short incubation hemagglutination inhibition (HI) test. Subsequently, Cabassou virus (CABV) and Rio Negro virus (RNV) were isolated and were found to be within VEEV antigenic complex, becoming subtypes V and VI, respectively. Hence, VEEV complex consists of six serological subtypes, which are currently considered as viral species, differing in

[3]This is a species complex, whose females are very difficult to differentiate (CBM).

their epidemiological characteristics and pathogenicity for humans and Equidae (mainly horses) (Table 9.1) (Steele and Twenhafel 2010, Weaver and Reisen 2010, Hubálek et al. 2014).

The analysis of the genetic relationships of different viruses included in VEEV complex has been consistent with the classification scheme in serocomplex. However, recent analyses of the nucleotide sequence of the 3′ terminal region that encodes the nsP4 and the complete 26S mRNA sequence of the genome region have suggested some rearrangements. Thus, the virus subtype II (EVE) would be more related to the IAB, IC, ID, and IE subtypes, whereas subtype IF would be more related to Rio Negro virus (subtype VI) isolated only in Argentina to date.

VEEV subtypes are divided into two epidemiological groups: epidemic/epizootic viruses and enzootic viruses (Table 9.2). Subtypes IAB and IC belong to the group of epidemic/epizootic viruses, which emerge periodically causing outbreaks that affect humans and equines, with high morbidity and mortality rates.

Since the isolation and characterization of VEEV in Venezuela in 1938 during an epidemic in horses, sporadic epidemics have been reported. Among them we can highlight the one that occurred between 1969 and 1972, affecting South and Central America, Mexico, and Texas, and the epizootic/epidemic in 1995 in Colombia and Venezuela after more than 20 years of silence, with about 50,000 equine cases and 75,000 human cases (Aguilar et al. 2011).

In Argentina, there are records of epizootic VEEV strains isolated from sick horses. However, the origin of these cases is not clear, and they may have been introduced by inactivated vaccines produced by private laboratories. Handling of these strains has caused laboratory infections, some of them serious.

Table 9.2 Members of the Venezuelan equine encephalitis virus complex

Serologic classification (subtype)	Virus/prototype strain	Origen	Source	Current distribution
IAB	VEEV – Trinidad donkey	Trinidad, 1943	Donkey	North, Central, and the north of South America
IC	VEEV – P676	Venezuela, 1963	Mosquito	North of South America
ID	VEEV – 3880	Panama, 1961	Human	South and Central America
IE	VEEV – Mena II, 68U201	Guatemala, 1968	Hamster	Central America and Mexico
IF	Mosso das Pedras virus – 78 V3531	Brazil, 1978	Mosquito	Brazil
II	Everglades virus (EVEV) – Fe3-7c	Florida, 1963	Mosquito	Florida (USA)
IIIA	Mucambo virus (MUCV) – BeAn8	Brazil, 1954	Monkey	North of South America and Trinidad
IIIB	Tonate virus – CaAn410d	French Guiana, 1973	Birds	French Guiana
IIIC	Bijou Bridge virus	Colorado, 1974	Mosquito	West of the USA (Colorado)
IIID	MUCV – 71D1252	Peru, 1971	Mosquito	Peru
	MUCV – V407660	Peru, 1998	Rodent, mosquito, and human	Peru
IV	Pixuna virus (PIXV) – BeAr35645	Brazil, 1961	Mosquito	Brazil
V	Cabassou virus (CABV) – CaAr508	French Guiana, 1968	Mosquito	French Guiana
VI	Rio Negro virus (RNV) – AG80-663	Argentina, 1980	Mosquito	Argentina

Transmission cycle for these subtypes involves equines, which act as highly efficient amplification hosts, developing high viremia titers, and mosquito vectors (Fig. 9.3). The equine virulence and the viremia induction are the most important labels of the epizootic phenotype (Weaver et al. 2004, Greene et al. 2005, Griffin 2007).

Although infection by epidemic/epizootic strains has been observed in humans, sheep, dogs, bats, rodents, and some birds, no major epidemics have been recorded in the absence of equine outbreak. Humans develop substantial viremia titers, but they probably do not act as amplifying hosts due to a lower exposure to mosquito bites. Because of this, transmission by humans should not be discarded. Infected people develop high titers of virus with both epizootic and enzootic strains (Weaver et al. 2004).

The epidemic/epizootic strains are opportunistic in the use of mosquito vectors during outbreaks. Field studies have indicated that more than one mosquito species can be involved in viral transmission during an outbreak. The vectors involved for this virus are mosquitoes of the genera *Aedes* and *Psorophora* including mosquitoes with peridomestic urban habits, such as *Ae. aegypti* and *Ae. albopictus*. Therefore, the requirements for the occurrence of human-mosquito-human transmission are met. *Aedes sollicitans* and *Ae. taeniorhynchus* also exhibit high infection rates, depending on the region of the Americas. *Aedes taeniorhynchus* is probably the most important epizootic vector in South America (Weaver et al. 2004, Weaver and Reisen 2010, Aguilar et al. 2011).

9.5.4.1 Pathogeny in Humans and Other Animals

Enzootic strains produce mild clinical disease, whereas the clinical forms of epizootic virus infection can be serious. The infection entry is generally through the skin, mosquito bites, but may occur through the airway (human cases of infection by aerosols have been recorded in laboratory accidents). The disease is almost undistinguishable clinically from other viral diseases such as influenza and dengue. Moreover, several Venezuelan equine encephalitis infections at early stage have been diagnosed as dengue. The disease usually begins abruptly 2–5 days after exposure (sometimes the incubation period is only 1 day), with chills, severe headache, fever, myalgia, retro-orbital pain, nausea, vomiting, and sore throat. Eighty percent of the infections are mild and last only 3–5 days. In many cases the febrile course is biphasic.

Neurological symptoms appear with headache and vomiting 4–10 days after the onset of clinical symptoms. Central nervous system disorders ranged from drowsiness to frank encephalitis with disorientation, convulsions, paralysis, coma, and death. This condition of the central nervous system is most common in children. Five percent of VEEV-infected children under 15 years old may develop neurological conditions, whereas in children of less than 5 years of age, this estimation may reach 35% of infected individuals. In many cases there are sequelae, such as mental retardation, epilepsy, learning difficulties, hydrocephalus, personality changes, and paralysis. In young adults, the disease is relatively benign. Adults over 50 years of age are more likely to develop encephalitis.

Laboratory analyses often show leukopenia. The pathological damages observed in fatal cases are myocarditis, focal hepatic necrosis, inflammation, and generalized lymphoid depletion. In pregnant women infected during pregnancy, an increase in abortions and children born with congenital malformations (especially at the central nervous system level) was found.

When enzootic strains infect *horses*, they produce asymptomatic infections or cause a short fever, low level of viremia, and mild clinical symptoms. Enzootic virus infection can immunize horses against epizootic strains. The clinical symptoms produced by epizootic strains is characterized by fever, depression, and diarrhea, leading to death 6–8 days after infection. The animals have viremia titers well above the vector infection threshold. The virus can be recovered from ocular and nasal lavage and urine. Progressive leukopenia is observed until death. Fatalities in pancreatic necrosis and cell depletion in the bone marrow, lymph nodes, and spleen were observed. The brain of horses with encephalitis has inflammation in cerebrovascular endothelial cells, edema, and extravasation of blood and leukocyte infiltration (Zacks and Paessler 2010, Taylor and Paessler 2013).

9.5.4.2 Treatment

Treatment of these diseases is only symptomatic, since no specific treatment is available.

9.5.4.3 Origin of Epidemics

Venezuelan equine encephalitis epidemics or epizootics have occurred in intervals of approximately 10–20 years in areas with livestock in many places of South America, when epizootic mosquito populations increase. For several years, the main enigma regarding VEEV epidemiology was to detect the source of epidemic/epizootic virus and the mechanisms of persistence between outbreaks. To explain this, several hypotheses have been postulated: (a) the emergence of epizootic subpopulations within enzootic VEEV populations, (b) the initiation of outbreaks by the administration of improperly inactivated vaccines, (c) the emergence of subtypes IAB and IC from cryptic transmission cycles, (d) the maintenance of epizootic strains in latent infections in horses or other animals, and (e) the periodic emergence of epizootic IAB, IC, and IE strains from the evolution (mutations) of enzootic VEEV progenitors. The last hypothesis has been supported by extensive phylogenetic analyses and is the most accepted so far. Remarkable similarities between epizootic IC and enzootic ID viruses have been documented. A total of 15 amino acid differences were identified between strains of both viruses isolated in Venezuela, and two of them were located within the E2 glycoprotein region, which is suspected to be the major determinant of equine virulence and amplification potential. Some studies have reported that a single amino acid substitution in the position E2-213 in ID Venezuelan strains (enzootic) resulted in a change in the equine viremia phenotype, generating high-titered viremia in horses and increasing the ability to infect the epidemic mosquito vector *Ae. taeniorhynchus* (Weaver et al. 2004, Aguilar et al. 2011, Weaver et al. 2012, Medina et al. 2015).

9.5.4.4 Enzootic Subtypes

Subtypes ID, IE, and IF and from II to VI have not been originally associated with epidemics or epizootics. They are called enzootic subtypes because they complete their cycle in wild habitats, involving mosquito vectors and rodent hosts (Fig. 9.3). These viruses are attenuated and unable to amplify in equines, but most of them can cause illness in humans. Only subtype IE has been related to an encephalitis outbreak in horses in Mexico, but equines were unable to amplify the virus. Because of this, equines may be dead-end hosts for these strains.

Enzootic strains are involved in active transmission cycles in tropical and subtropical areas of the Americas (Table 9.2). In endemic areas, mosquito isolates are mainly made from the mosquitoes *Culex* (*Melanoconion*) spp., which live in swampy areas and breed near aquatic plants. These mosquitoes feed on a variety of rodents, birds, and other vertebrates. However, isolations have also been performed from the genera *Aedes*, *Mansonia*, *Psorophora*, *Haemagogus*, *Sabethes*, *Deinocerites*, and *Anopheles*. Wild birds are susceptible to infection, but mammals (mainly rodents) are the most probable hosts, as shown by viral isolations, levels of viremia, serology, and disease resistance. Enzootic transmission cycles for VEEV ID, VEEV IE, EVEV, MUCV, and Tonate virus (Bijou Bridge virus) have been described; all of them – with the exception of Bijou Bridge – are maintained in cycles involving rodents and mosquitoes of the subgenus *Culex* (*Melanoconion*). In the western USA, Bijou Bridge virus is transmitted to birds by the bedbug *O. vicarious* (Aguilar et al. 2011, Weaver and Reisen 2010).

In Argentina, activity of Rio Negro (subtype VI) virus, which was isolated from *Cx.* (*Mel.*) *delpontei* and rodents of the genus *Akodon* in the subtropical area of Chaco and Formosa provinces, has been known for more than 3 decades. It is associated with the production of acute febrile disease and has been only recognized in this country (Contigiani et al. 1993, Cámara et al. 2003, Pisano et al. 2013, Pisano et al. 2012)

Pixuna (subtype IV) virus, first isolated from *Anopheles nimbus* mosquitoes in 1961 in northern

Brazil, has been also detected in Argentina (Pisano et al. 2010a, b). Enzootic circulation of multiple subtypes within a region (i.e., Argentina, Peru, and Venezuela) could result in the emergence of epizootic/epidemic strains. The interaction among strains and genetic mechanisms of adaptation to new hosts are postulated as mechanism for emergence of epizootic strains.

In Peru, epizootic/epidemic subtype IAB was recorded in the 1940s. Studies to determine the origin of this subtype led to the isolation of enzootic subtypes (ID and IIIC) from sentinel hamsters and mosquitoes (Aguilar et al. 2011).

In Brazil, the VEEV subtype IF circulates in the southeast and causes febrile illness and diarrhea in humans. It has also been isolated in the Amazon region (Calisher et al. 1982).

9.5.5 Control and Prevention

The neurotropic alphavirus should be continuously monitored through an active surveillance system that includes serological and virological surveillance (detection of antibodies and/or viral agent hosts and sentinel animals), clinical-epidemiological surveillance (by recording every disease suspected to be caused by an alphavirus), and entomological surveillance (mosquito collection, taxonomic identification, and identification of viral agent by RT-PCR and/or viral isolation).

When available, preventive measures should include the use of vaccines. Available vaccines are generally of two types: attenuated and inactivated virus vaccines.

For EEEV and WEEV, inactivated virus vaccines for veterinary use containing both viruses are prepared in Argentina. In the USA, a similar vaccine for veterinary and laboratory personnel use is prepared.

An attenuated virus vaccine for VEEV (TC83 strain) has been developed; it has been successful in immunizing horses and controlling epizootics in Venezuela. It is also used in the USA to protect laboratory personnel and military people. In humans, it can produce systemic reactions (fever, myalgias, leukopenia) so it is not used to immunize the general population. There is an inactivated virus vaccine at experimental level which is effective in humans and produces minimal side effects. There are attempts to develop a new vaccine against VEEV using recombinant techniques. Argentina is free from VEEV epizootic subtypes, so the use of this virus vaccine is not allowed, and vaccination is considered to be the cause of possible introduction (Go et al. 2014).

9.6 Polyarthritis Alphaviruses

The alphaviruses that cause joint disorders are chikungunya virus (Africa, Asia, Europe, America), o'nyong-nyong virus (central Africa), Ross River virus and Barmah Forest viruses (Australia and the Pacific Islands), Sindbis virus (cosmopolitan), and Mayaro virus (South America).

9.6.1 Pathogeny of Arthritogenic Alphaviruses

After subcutaneous inoculation through the bite of an infected mosquito in the skin, these alphaviruses disseminate in the host organism through the bloodstream. The liver, spleen, muscle, and lymph nodes are sites of primary replication, allowing an efficient virus spread. Langerhans cells facilitate virus delivery to the lymph nodes. Leukopenia in acute phase of the disease is a very common hematologic alteration in alphavirus infection, suggesting primary replication of the virus in the leukocytes. Interferon (IFN) program is early activated, but the alphaviruses developed several mechanisms to inhibit this antiviral response.

The acute phase of the disease involves virus replication followed by an inflammatory response in the target tissues, which is characterized by an extensive infiltration of lymphocytes (CD4+ and CD8+ T lymphocytes), NK cells, neutrophils, and macrophages (the main component). The increase in the levels of several proinflammatory cytokines and chemokines in the site of infection and in the plasma is associated with myositis and arthralgia/arthritis. The secretion of metalloproteinases (MMPs) in the joint tissue may also contribute to articular damage. Persistence and severity of the symptoms may be related to the persistence of the virus or its products in the target cells with the subsequent accumulation of inflammatory mediators such as IL-6. A question that remains open is whether an autoimmune process is associated with the persistence of the inflammatory response, as observed for rheumatoid arthritis (Assunção-Miranda et al. 2013).

9.6.2 Treatment

There is no antiviral drug therapy available for these kinds of pathologies (polyarthritis of viral origin), with symptomatic treatment during acute phase of illness being only indicated. Patients are treated with nonsteroidal anti-inflammatory drugs, fluids, and medicines to relieve symptoms of fever and aching, such as ibuprofen, naproxen, acetaminophen, or paracetamol.

9.6.3 Sindbis Virus (WEEV Complex)

The Sindbis virus (SINV), with four subtypes, has been detected in Africa, Asia, Australia, and Europe. It was originally isolated from *Cx. univittatus* mosquitoes collected in Sindbis village, Nile Delta, Egypt, in 1952. SINV is maintained mainly between ornithophilic *Culex* spp. mosquitoes, but *Cs. morsitans*, *Coquillettidia richiardii*, *Mansonia africana*, *Aedes* spp., and *An. hyrcanus* are also involved. Birds act as hosts; rodents, amphibians, and bats have also been incriminated but less frequently. There are records of encephalitis cases in horses in South Africa. In humans it produces febrile illness with arthralgia and rash. Its geographic distribution includes Africa, Israel, Asian Turkey, India, Indonesia, Australia, New Zealand (Whataroa strain), China, Central Asia, Azerbaijan, Sweden, Finland, Russia, infrequently Italy (Sicily), Slovakia, and Germany (Assunção-Miranda et al. 2013, Hubálek et al. 2014).

9.6.4 Semliki Forest Complex

9.6.4.1 Mayaro Virus

The Mayaro virus (MAYV), with activity reported in the Americas, belongs to the Semliki Forest complex along with CHIKV, SFV, ONNV, and RRV (Old World viruses that produce the same clinical syndrome in humans) (Table 9.1).

This virus was first isolated from a sick man in Trinidad in 1954. It was subsequently isolated from humans with undifferentiated febrile illness in northern Brazil. It was also recovered from humans, primates, and wild mosquitoes in Bolivia, Brazil, Colombia, Venezuela, and Peru. Serological studies show that human infection is common in forested areas of northern South America. Activity has also been detected in Central America and human infections have recently been reported in Mexico. MAYV clinical cases in South America are sporadic and have been reported in individuals with history of recent activity in jungle areas. This is related to its *Haemagogus* mosquito vectors, which inhabit forested areas, preferably in the canopy. In recent years, travel-related infections by MAYV imported from South America to Europe and the USA have been increasingly reported.

At the molecular level, two MAYV genotypes can be identified: genotype D (of wide distribution) and L (of restricted distribution), which differ by more than 15 % in terms of their nucleic acid. However, that nucleotide divergence within each genotype did not exceed 5.9 % for genotype D and 3 % for genotype L. These data suggest that different patterns of transmission may be involved in the maintenance and evolution of MAYV.

The transmission cycle of MAYV could be similar to that of yellow fever virus, including monkeys as hosts and *Haemagogus* mosquitoes as vectors. Human outbreaks in Brazil in 1978 and 1991 contributed to the epidemiological knowledge of this virus. Antibodies were detected in a high percentage in *Callithrix* monkeys and viral isolations were made from *Haemagogus janthinomys* mosquitoes. It was further found that MAYV can infect and be transmitted by *Ae. aegypti*, raising the possibility of urban outbreaks.

Demographic changes and the result of human activities (urbanization, deforestation, intensive farming) that produce dramatic changes in the environment may alter the frequency of clinical cases of this virus.

In humans, MAYV produces an acute febrile disease that can be confused with dengue. The clinical manifestation lasts between 3 and 5 days, and it is characterized by fever, headache, myalgia, rash, joint pain, and, less often, arthritis. No specific treatment or vaccine is available for MAYV infection. Studies aimed at having a live-attenuated vaccine are being developed (Assunção-Miranda et al. 2013, Muñoz and Navarro 2012, Weise et al. 2014).

9.6.4.2 UNA Virus

The epidemiology of UNAV is little known. It was first isolated from *Psorophora ferox* mosquitoes collected in the state of Pará, Brazil. It is widely distributed with low prevalence of infection in tropical and subtropical regions of Central and South America.

It is the only member of the Semliki Forest complex with documented activity in Argentina, where it was isolated in 1964 from a dead horse and a febrile one. In addition, antibodies against this virus were detected in birds and horses. Human infections were detected, suggesting they are contemporaneous to viral isolations. Another serological study in howler monkeys detected activity in Argentina and Paraguay, suggesting that this virus would be endemic to those regions and that these primates could be involved in the maintenance cycle. Similar epidemiological characteristics were described for MAYV.

UNAV strains, grouped into a single genotype, exhibit marked genetic diversity between them (about 28 %). UNAV and MAYV share 45 % of nucleotide identity. The genetic diversity pattern of UNAV suggests that this virus moved to new ecological niches, where it established maintenance enzootic foci following different evolutionary paths (Díaz et al. 2007, Powers et al. 2006).

9.6.4.3 Chikungunya Virus

The chikungunya virus (CHIKV) is an emerging virus that causes an acute febrile illness with arthralgia and rash of sudden onset. It belongs to the SFV complex, as UNAV and MAYV, in American continent (Table 9.1). The virus was first isolated from a feverish human in Tanzania in 1953. The disease by CHIKV is known as "chikungunya fever," "epidemic arthritis chikungunya," or just "chikungunya." This name from the language Makonde means "sickness of the folded walker" or "that which bends up." In India, it is known as "Aakyda," meaning "hard man." These names refer to the stooped posture as a result of the symptom of arthritis (Figueiredo and Figueiredo 2014, Caglioti et al. 2013).

- *Pathology in humans*. The incubation period in CHIKV infection is on average 2–4 days (range 1–12 days). The onset of illness is abrupt, with high fever, headache, back pain, myalgia, and arthralgia. Joint pains that occur in almost 80 % of patients can be severe and affect the extremities (ankles, wrists, phalanges), knees, elbows, etc., but this condition is rarely seen in children. Although this clinical manifestation is overcome in weeks or months, cases of persistence for several years and even chronic arthritis were observed. Joint pain and inflammation mainly affect symmetrically the small joints (such as those from fingers, wrists, and tarsus), but eventually occur in the large joints (such as those from knees and shoulders) and may also involve several joints simultaneously (polyarthralgia/polyarthritis). Besides rash and arthritis, myalgia is a very common symptom during alphavirus infection, also demonstrating the virus tropism for the muscular tissue.

Between 2 and 5 days after onset of symptoms, a maculopapular and erythematous rash mainly on the face, trunk, and limbs is displayed. These symptoms may remain until the tenth day after onset. Furthermore, it can cause conjunctivitis, pharyngitis, and lymphadenopathy. In children, localized petechiae and gingivorragia can be observed. The peak incidence of serious infections was estimated in less than 0.02 %. The typical clinical presentations are myocarditis, central nervous system involvement, hemorrhagic manifestations, and possible teratogenic effects in neonatal CHIK.

On the other hand, there is no evidence that the lethality is produced by a direct action of CHIKV. The causes of severe clinical manifestations and deaths observed in recent virus epidemics in India and Reunion are unknown, but may be due to circulating strains of different virulence (Assunção-Miranda et al. 2013, Caglioti et al. 2013).

- *Epidemiology*. CHIKV is endemic to Africa, India, Southeast Asia, and the Philippines. During the period between 1960 and 1980, the virus was isolated repeatedly from different regions of Africa and Asia. Likewise, major epidemics occurred in India and the Philippines. In 2004, the virus emerged in the Indian Ocean islands (Lamu, Commodores, Seychelles, and others), where the vector *Ae. aegypti* is the most prevalent. Later, it spread to other areas where *Ae. albopictus* is the dominant vector (Reunion Island, Mauritius) and caused explosive epidemics with millions of cases in India after a silent period of more than 30 years. In 2006, imported cases by CHIKV were reported in different countries in Europe, with France and Italy having higher risk, due to the number of visitors coming from endemic regions, mainly India. In 2007, a native localized outbreak was reported for the first time in Europe, specifically in northeastern Italy, where *Ae. albopictus* was the vector. Between 2006 and 2011, imported cases were reported in the USA, Canada, Caribbean islands, Brazil, Oceania, and Australia associated with travelers from India and Indian Ocean islands (Coffey et al. 2014, Morrison 2014).

In late 2013 the Pan American Health Organization (PAHO)/World Health Organization (WHO) reported the first indigenous cases of CHIKV infection in the Caribbean islands (Guadeloupe, British Virgin Islands, Martinique, St. Barthélemy, and San Martin). In 2014, CHIKV local transmission has been identified in 41 countries and territories in the Caribbean, Central America, South America, and North America (Weaver 2014, Powers 2015).

The mosquitoes (*Ae. aegypti* and *Ae. albopictus*) that transmit the virus are found in a large part of the Americas, including areas of the USA. Since chikungunya virus is new to the Americas, most people in the region are not immune, meaning that they can be infected and the virus spread to other mosquitoes.

Genetic studies have identified three different lineages for CHIKV: Eastern, Central, and Southern African (ECSA), West African, and Asian genotypes. Recent isolates made in the Indian Ocean area form a distinct cluster within the large ECSA genotype. The genotypes should indicate independent evolutionary processes associated with isolated geographic regions. As expected in an RNA virus, there are differences among strains. In Central and West Africa, CHIKV is maintained in a sylvatic cycle between nonhuman primates and forest mosquitoes *Aedes* spp., from which it was repeatedly isolated. Wild mosquito species primarily involved in the sylvatic cycle (depending on ecological and geographical conditions) are *Ae. furcifer*, *Ae. taylori*, *Ae. luteocephalus*, *Ae. africanus*, and *Ae. neoafricanus*; epidemic vectors would involve *Ae. furcifer* and *Ae. taylori. Aedes aegypti* was associated as vector in urban epidemics from East Africa and the Indian Ocean.

The mosquito *Ae. albopictus* is strongly associated with humans but also with peridomestic habitats and therefore responsible for the large urban epidemics in Reunion, India, and Southeast Asia. In Asia, *Ae. aegypti* was also incriminated as the main vector. Besides these mosquito vector species, other peridomestic mosquitoes which have been detected in abundance include *Ae. albopictus*, *Ae. vittatus*, and *An. stephensi*.

Little is known about vertebrates that may be involved in maintenance cycles. In urban populations, epidemic periods, the virus is maintained by the mosquito-human-mosquito cycle. In non-epidemic periods, reservoirs may be nonhuman primates or other unidentified vertebrates. The epizootic in monkeys occurs when immunity at populational level is very low. These animals develop viremia but have no serious clinical manifestations (Coffey et al. 2014, Morrison 2014, Caglioti et al. 2013).

- *Control and prevention*. There is no vaccine available to prevent chikungunya fever. An experimental live-attenuated virus vaccine (TSI-GSD-218) obtained from a strain isolated from a patient with CHIKV from Thailand and attenuated by serial passages in MRC-5 cells was tested in human volunteers. This vaccine demonstrated to be a good inductor of neutralizing antibody response that endures overtime. Effective preventive measures consist of individual protection against mosquito bites and vector control of *Ae. aegypti* and *Ae. albopictus*. Sick people should be kept protected from potential mosquito bites, either by staying in the home or using insecticide or repellent during the first 4–6 days after infection, to avoid infections by new vectors and consequent amplification of the disease.

Other arthritogenic alphaviruses are ONNV, RRV (Semliki Forest virus complex), SINV (WEEV complex), and BFV (Barmah Forest virus).

9.6.4.4 O'Nyong-Nyong Virus (ONNV)

ONNV was detected for the first time in Uganda (Africa) in 1959, during a major central African epidemic of ONN fever that began in northern Uganda and spread to Kenya and Tanzania (1959–1962). It is genetically and serologically related to CHIKV and is restricted to the African continent. Typical clinical features include low-grade fever, symmetrical polyarthralgia, lymphadenopathy (particularly of the posterior cervical region), and a generalized papular or maculopapular exanthem. Although some patients suffered from prolonged joint pain during the recovery phase, no fatal cases or permanent sequelae were observed. During 1996 and 1997, south-central Uganda experienced the second ONN fever epidemic ever recognized.

Anopheline mosquitoes (*An. funestus*) were involved as the primary epidemic vectors. Vertebrate studies conducted during the epidemic failed to incriminate rodents or mongooses as amplifying hosts of the virus.

A laboratory-confirmed case of an ONNV infection imported into Europe was recently reported. This patient most likely was infected in the eastern part of Kenya (Kisumu region). Because of the serological and clinical similarities of ONNV and CHIKV infections, infections in travelers may have been wrongly diagnosed as CHIK. Similarly to other arboviruses, especially CHIKV and dengue viruses, ONNV might have the potential to spread to areas outside Africa. There are no known invasive anopheline vectors for ONNV in Europe, but it was demonstrated that the species *Ae. aegypti*, found in some parts of Europe, might be a competent vector for ONNV (Assunção-Miranda et al. 2013).

9.6.4.5 Ross River Virus (RRV)

Ross River virus (RRV) is endemic and enzootic in Australia and Oceania (Papua New Guinea, the Solomon Islands, and the South Pacific islands) and Southeast Asia, where it caused major epidemics. It was first isolated from *Ae. vigilax* mosquitoes collected near Townsville in northern Queensland in 1959. In Australia, RRV disease is the most widely spread mosquito-borne disease. In 2011, RRV infection accounted for 63 % (5149 cases) of all mosquito-borne disease records. This virus causes a nonfatal, but prolonged and debilitating disease known as epidemic polyarthritis or RRV disease. The incubation period may be as long as 21 days or as short as 3 days (usually 7–9 days).

The epidemiology of RRV disease is complex because transmission cycles are driven by various mosquito species and vertebrate hosts within a variety of disparate geoclimatic regions. The natural vertebrate hosts include marsupial animals (e.g., kangaroos, wallabies) and possibly other animals (e.g., dogs, cats, horses, possums). More than 30 mosquito species have been implicated as vectors of RRV, with *Ae. vigilax* and *Ae. camptorhynchus* being significant in the coastal regions and *Cx. annulirostris* being common in tropical areas and temperate regions that are subject to flooding or irrigation during summer. Species such as *Ae. notoscriptus* may be important in semirural and urban areas (Tong et al. 2008).

9.6.4.6 Barmah Forest Virus (BFV) (BFV Complex)

Barmah Forest virus (BFV) was first isolated from a pool of *Cx. annulirostris* mosquitoes collected in southeastern Australia in 1974. Although the animal reservoir remains unknown, native animals, such as kangaroos, brushtails, and wallabies, could be involved in the cycle of infection. The mosquitoes that can spread the virus are *Ae. vigilax*, *Cx. annulirostris*, *Ae. normanensis*, and *Ae. notoscriptus*. Vector competence studies showed that *Ae. notoscriptus* could act as an efficient vector of BF in urban environments. BFV infection cannot be spread from person to person. Increases in travel and trade will undoubtedly increase the risk of BFV introduction to other continents. At present, BFV has been the cause of outbreaks of human disease in Australia (Jacups et al. 2008).

9.7 Other Alphaviruses Pathogenic for Animals

9.7.1 Getah Virus (GETV) (SFV Complex)

GET virus was first isolated from *Cx. gelidus* mosquitoes near Kuala Lumpur (Malaysia) in 1955. Disease in animals (horses) was first recognized in Japan, 1978. GETV is widely distributed in the countries of Southeast Asia and in northern Australia along the Pacific Ocean. The mosquito species implicated as vectors are *Cx. gelidus*, *Cx. tritaeniorhynchus*, *Cx. fuscocephala*, *Ae. vexans nipponensis*, *Ae. nigripes*, *Ae. communis*, and *Ae. excrucians*. Horses, pigs, and wild boars would act as vertebrate hosts. The virus is known to be pathogenic for horses (often racehorses) and pigs. In horses, the disease is characterized by depression, anorexia, fever, nasal discharge, urticarial rash, edema of the hind limbs, swelling of the submandibular lymph nodes, and lymphocytopenia (experimentally confirmed). It can also cause abortions in pigs. The virus (very closely related to Ross River virus) has not been linked to illness in humans; however, neutralizing antibodies to GETV have been identified in human sera and in birds in Malaysia, in northern Australia, and in China. Outbreaks of GETV infection were first recorded in racehorses at two training centers in Japan in 1978. Since then, several outbreaks of the disease have been reported in Japan especially, in horse racing tracks (1991–1997), and one was reported from India in 1990 (Hubálek et al. 2014).

9.8 Concluding Remarks

Pathogenic alphaviruses for humans can produce very different clinical manifestations. The same virus can cause symptoms not previously recorded as a result of changes at the genomic level that cause changes in its virulence and disease with different clinical characteristics.

The co-circulation of multiple viral species belonging to the same serological complex in a particular geographic region could favor the emergence of new pathogenic variants for humans and animals of veterinary importance.

The biological characteristics of transmission and maintenance of these viruses make their infections independent of socioeconomic development of a country or region. In addition, its spread is favored by increasing circulation by tourism or commerce.

The current scenarios of emergence and reemergence observed worldwide indicate the need to intensify surveillance systems through monitoring of vectors and hosts, as well as deeper understanding of the biology, epidemiology, and pathogenesis of these zoonoses. Finally, the knowledge of natural genetic variability of circulating strains in each region will lead to improved diagnosis, treatment, and prevention of these diseases.

References

Aguilar PV, Estrada Franco JG, Navarro López R et al (2011) Endemic Venezuelan equine encephalitis in the Americas: hidden under the dengue umbrella. Future Virol 6:721–740

Allison AB, Stallknecht DE, Holmes EC (2015) Evolutionary genetics and vector adaptation of recombinant viruses of the western equine encephalitis antigenic complex provides new insights into alphavirus diversity and host switching. Virology 474:154–162

Assunção-Miranda I, Cruz-Oliveira C, Da Poian AT (2013) Molecular mechanisms involved in the pathogenesis of alphavirus-induced arthritis. Biomed Res Int 2013:973516

Bhavez CL, Powers AM, Chavez CL et al (1999) Genetic and antigenic diversity among eastern equine encephalitis viruses from North, Central, and South America. Am J Trop Med Hyg 61:579–586

Caglioti C, Lalle E, Castilletti C et al (2013) Chikungunya virus infection: an overview. New Microbiol 36:211–227

Calisher CH, Karabatsos N (1988) Arbovirus serogroups: definition and geographic distribution. In: Monath TP (ed) The arboviruses: epidemiology and ecology, vol I. CRC, Boca Raton, pp 19–57

Calisher CH, Kinney RM, de Souza LO et al (1982) Identification of a new Venezuelan equine encephalitis virus from Brazil. Am J Trop Med Hyg 31:1260–1272

Cámara A, Díaz G, Vega V et al (2003) Seroprevalence of antibodies to Venezuelan equine encephalitis complex (subtypes IAB and VI) in humans from General Belgrano Island, Formosa, Argentina. Rev Inst Med Trop Sao Paulo 45:201–204

Coffey LL, Failloux AB, Weaver SC (2014) Chikungunya virus-vector interactions. Viruses 11:4628–4663

Contigiani MS, de Basualdo M, Cámara A et al (1993) Presence of antibodies against Venezuelan equine encephalitis virus subtype VI in patients with acute febrile illness. Rev Argent Microbiol 14:212–220

de Novaes Oliveira R, Iamamoto K, Silva ML et al (2014) Eastern equine encephalitis cases among horses in Brazil between 2005 and 2009. Arch Virol 10:2615–2620

Díaz LA, Díaz MP, Almirón WR et al (2007) Infection by UNA virus (Alphavirus; Togaviridae) and risk factor analysis in black howler monkeys (*Alouatta caraya*) from Paraguay and Argentina. Trans R Soc Trop Med Hyg 10:1039–1041

Figueiredo MLG, Figueiredo LTM (2014) Emerging alphaviruses in the Americas: Chikungunya and Mayaro. Rev Soc Bras Med Trop 6:677–683

Go YY, Balasuriya UB, Lee CK (2014) Zoonotic encephalitides caused by arboviruses: transmission and epidemiology of alphaviruses and flaviviruses. Clin Exp Vaccine Res 1:58–77

Greene I, Paessler S, Austgen L et al (2005) Envelope glycoprotein mutations mediate equine amplification and virulence of epizootic Venezuelan equine encephalitis virus. J Virol 14:9128–9133

Griffin DE (2007) Alphaviruses, Chapter 31. In: Knipe DM, Howley PM (eds) Fields virology, 5th edn. Lippincott Williams and Wilkins, Philadelphia, PA, pp 1024–1067

Hubálek Z, Rudolf I, Nowotny N (2014) Arbovirus pathogenic for domestic and wild animals. Adv Virus Res 89:201–275

ICTV (2014) Virus Taxonomy: 2014 Release. http://www.ictvonline.org/virusTaxonomy.asp

Jacups SP, Whelan PI, Currie BJ (2008) Ross River virus and Barmah Forest virus infections: a review of history, ecology, and predictive models, with implications for tropical northern Australia. Vector Borne Zoon Dis 2:283–297

Jose J, Snyder JE, Kuhn RJ (2009) A structural and functional perspective of alphavirus replication and assembly. Future Microbiol l7:837–856

Lubelczyk C, Mutebi JP, Robinson S et al (2013) An epizootic of eastern equine encephalitis virus, Maine, USA in 2009: outbreak description and entomological studies. Am J Trop Med Hyg 88:95–102

Medina G, Garzaro DJ, Barrios M et al (2015) Genetic diversity of venezuelan alphaviruses and circulation of a Venezuelan Equine Encephalitis virus subtype IAB strain during an interepizootic period. Am J Trop Med Hyg 93:7–10

Molaei G, Andreadis TG, Armstrong PM et al (2013) Vector-host interactions and epizootiology of eastern equine encephalitis virus in Massachusetts. Vector Borne Zoonotic Dis 5:312–323

Morrison TE (2014) Reemergence of chikungunya virus. J Virol 20:11644–11647

Muñoz M, Navarro JC (2012) Mayaro: a re-emerging arbovirus in Venezuela and Latin America. Biomedica 2:286–302

Pisano MB, Dantur MJ, Ré VE et al (2010a) Cocirculation of Rio Negro Virus (RNV) and Pixuna Virus (PIXV) in Tucumán province, Argentina. Trop Med Int Health 7:865–868

Pisano MB, Ré VE, Díaz LA et al (2010b) Enzootic activity of pixuna and Rio Negro viruses (Venezuelan Equine Encephalitis complex) in a neotropical region of Argentina. Vector Borne Zoonotic Dis 2:199–201

Pisano MB, Spinsanti LI, Díaz LA et al (2012) First detection of Rio Negro virus (Venezuelan equine encephalitis complex subtype VI) in Córdoba, Argentina. Mem Inst Oswaldo Cruz 107:125–158

Pisano MB, Oria G, Beskow G et al (2013) Venezuelan equine encephalitis viruses (VEEV) in Argentina: serological evidence of human infection. PLoS Negl Trop Dis 7, e2551. doi:10.1371/journal.pntd.0002551

Powers AM (2015) Risks to the Americas associated with the continued expansion of chikungunya virus. J Gen Virol 96(Pt 1):1–5

Powers AM, Brault AC, Shirako Y et al (2001) Evolutionary relationships and systematics of the alphaviruses. J Virol 75:10118–10131

Powers AM, Aguilar PV, Chandler LJ et al (2006) Genetic relationships among Mayaro and Una viruses suggest distinct patterns of transmission. Am J Trop Med Hyg 75:461–469

Sabattini MS, Avilés G, Monath T (1998) Historical, epidemiological and ecological aspects of arboviruses in Argentina: Togaviridae, Alphavirus. In: Travassos da Rosa APA, Vasconcelos PFC, Travassos da Rosa JFS (eds) Overview of arbovirology in Brazil and neighbouring countries. Instituto Evandro Chagas, Belém, pp 135–152

Scott TW, Weaver SC (1989) Eastern equine encephalomyelitis virus: epidemiology and evolution of mosquito transmission. Adv Virus Res 37:277–328

Steele KE, Twenhafel NA (2010) Pathology of animal models of alphavirus encephalitis. Vet Pathol 5:790–805

Taylor KG, Paessler S (2013) Pathogenesis of Venezuelan equine encephalitis. Vet Microbiol 167(1–2):145–150

Tong S, Dale P, Nicholls N et al (2008) Climate variability, social and environmental factors, and Ross river virus transmission: research development and future research needs. Environ Health Perspect 12:1591–1597

Vander Kelen PT, Downs JA, Stark LM et al (2012) Spatial epidemiology of Eastern Equine Encephalitis in Florida. Int J Health Geogr 11:47

Weaver SC (2014) Arrival of chikungunya virus in the New World: prospects for spread and impact on public health. PLoS Negl Trop Dis 8(6), e2921

Weaver SC, Reisen WK (2010) Present and future arboviral threats. Antiviral Res 2:328–345

Weaver SC, Kang W, Shirako Y et al (1997) Recombinational history and molecular evolution of western equine encephalomyelitis complex alphaviruses. J Virol 1:613–623

Weaver SC, Ferro C, Barrera R et al (2004) Venezuelan Equine Encephalitis. Annu Rev Entomol 49:141–174

Weaver SC, Winegar R, Manger ID et al (2012) Alphaviruses: population genetics and determinants of emergence. Antiviral Res 3:242–257

Weise WJ, Hermance ME, Forrester N et al (2014) A novel live-attenuated vaccine candidate for Mayaro fever. PLoS Negl Trop Dis 8, e2969

Young DS, Kramer LD, Maffei JG et al (2008) Molecular epidemiology of eastern equine encephalitis virus, New York. Emerg Infect Dis 14:454–460

Yu GY, Wiley MR, Kugelman JR et al (2015) Complete coding sequences of eastern equine encephalitis virus and Venezuelan equine encephalitis virus strains isolated from human cases. Genome Announc 3:e00243-15. doi:10.1128/genomeA.00243-15

Zacks MA, Paessler S (2010) Encephalitic alphaviruses. Vet Microbiol 140:281–286

Bunyaviruses

10

Marta S. Contigiani, Luis A. Diaz,
and Laura B. Tauro

Abstract

The family Bunyaviridae, the most numerous family of RNA viruses, was established in 1975. In these days, it is integrated by more than 350 worldwide viral species, grouped into five genera: *Orthobunyavirus*; *Hantavirus*; *Nairovirus*; *Phlebovirus*, which infect vertebrates; and *Tospovirus* which are viruses that infect plants. From these four genera that infect vertebrates, only *Hantavirus* is not considered an arthropod-borne virus (arbovirus). *Orthobunyavirus* is the largest one, including 48 viral species, which in turn have different variants and strains. This family includes important viral pathogens, which cause diseases of veterinary and human concern such as Rift Valley fever (*Phlebovirus*), Akabane (*Orthobunyavirus*) and Nairobi sheep disease (*Nairovirus*), Oropouche (*Orthobunyavirus*), and Crimean-Congo hemorrhagic fever (*Nairovirus*). Many bunyaviruses are considered emergent, since they have increased their influence in new populations and geographic areas around the world. Examples of this are the viruses Crimean-Congo hemorrhagic fever and Rift Valley fever that emerged in parts of Europe due to the migration of their vectors possibly under the influence of climate change. The study of these emerging viruses is of great importance, whereas they have no preventive treatments and/or therapy, and it is necessary to know deeply its behavior in order to implement efficient control measures. This chapter deals aspects of molecular and cellular biology, cycle of transmission, ecological and epidemiological aspects, pathogenesis, clinical aspects, and diagnosis of the major bunyavirus species of medical and veterinary concern in different regions of the world.

M.S. Contigiani (✉) • L.B. Tauro
Laboratory of Arbovirus, Institute of Virology "Dr. J. M. Vanella", Enfermera Gordillo Gomez s/n. Ciudad Universitaria, National University of Córdoba, Córdoba, Argentina
e-mail: martascontigia@hotmail.com; lauratauro@gmail.com

L.A. Diaz
Laboratory of Arbovirus, Institute of Virology "Dr. J. M. Vanella", Enfermera Gordillo Gomez s/n. Ciudad Universitaria, National University of Córdoba, Córdoba, Argentina

Institute for Biological and Technical Researches, CONICET-UNC, Córdoba, Argentina
e-mail: adrian.diaz@conicet.gov.ar

C.B. Marcondes (ed.), *Arthropod Borne Diseases*, DOI 10.1007/978-3-319-13884-8_10

Keywords
Bunyaviridae • *Orthobunyavirus* • *Phlebovirus* • *Nairovirus* • Mosquitoes

10.1 Introduction

The family Bunyaviridae was established in 1975 to cover a large number of arthropod-borne viruses that share morphological, morphometric, and antigenic features (Schmaljohn and Nichol 2006). The family name derives from Bwamba (Uganda, South Africa), where it was isolated for the first time from mosquito *Aedes* sp., Bunyamwera virus (BUNV), the prototype virus of the family (Monath and Heinz 1996). The classification of members of this family has historically relied on genetic, structural, and antigenic relationships between different viruses.

Currently, this family comprises 97 viral species distributed worldwide and is grouped into five genera: *Orthobunyavirus*, *Nairovirus*, *Phlebovirus*, *Hantavirus*, and *Tospovirus* (Table 10.1) (Plyusnin et al. 2012). The first four genera infect animals, whereas the latter infects many plant species (Plyusnin et al. 2012). The different members of these families exhibit diverse modes of transmission; however, within each family, they are similar. Members of the genera *Orthobunyavirus*, *Phlebovirus*, and *Nairovirus* are transmitted by arthropods (mosquitoes, ticks, sand flies, and biting midges) and kept in a vector-vertebrate cycle (Plyusnin et al. 2012).

Members of this family are widely distributed in the world and have public health importance because of their pathogenicity for both humans and domestic and wild animals (Table 10.2).

10.2 Virus Structure

Bunyavirus are enveloped viruses with a helical and oval or spherical capsid of 80–120 nm of diameter. The envelope consists of a lipid bilayer of 5–7 nm in which two viral glycoproteins called Gn and Gc are embedded. These glycoproteins are heterodimers that extend across the surface forming projections or spikes of 5–10 nm in length, which can be observed by electron microscopy (Fig. 10.1). The chemical composition is about 2 % RNA, 58 % protein, 33 % lipid, and 7 % carbohydrate, varying among species. Treatment with lipid solvents or nonionic detergents removes lipid envelope resulting in loss of infectivity of these viruses (Schmaljohn and Nichol 2006).

10.2.1 Viral Genome

The genome consists of single-stranded RNA of negative polarity divided into three segments called small (S), medium (M), and large (L), reflecting the relative length of nucleotides (Fig. 10.1) (Schmaljohn and Nichol 2006).

Genetic organization of the segments is similar in all genera, each one possessing non-translated

Table 10.1 Bunyaviridae family taxonomy

Genus	Viral species	Virus prototype	Geographical distribution	Vector/host
Orthobunyavirus	48	Bunyamwera	Africa, America, Asia, Australia, Europe	Mosquitoes, *Culicoides*
Phlebovirus	9	Fiebre del Valle de Rift	Africa, America, Asia, Europe	Sand flies, mosquitoes, ticks
Nairovirus	7	Hemorrhagic fever de Crimean-Congo	Africa, America, Asia, Europe	*Culicoides*, ticks, mosquitoes, flies
Hantavirus	24	Hantaan	Africa, America, Asia, Europe	Rodents
Tospovirus	9	Marchitez manchada del tomate	Africa, America, Asia, Europe	Thrips

Table 10.2 Medical and veterinary importance of viral species of *Orthobunyavirus*, *Phlebovirus*, and *Nairovirus* genera

Genus	Viral species	Geographical distribution	Vector	Disease	Affects to
Orthobunyavirus	Akabane	Africa, Asia, Australia	Mosquitoes	Congenital malformations, abort	Animals
	Aino	Africa, Asia, Australia	*Culicoides*, mosquitoes	Congenital malformations, abort	Animals
	Bunyamwera	Africa, America	Mosquitoes	Febrile syndrome, nervous syndrome, Congenital malformations, abort	Animals, human
	Bwamba	Africa	Mosquitoes	Febrile syndrome	Human
	California encephalitis	America, Asia	Mosquitoes	Nervous syndrome	Human
	Caraparu	America	Mosquitoes	Febrile syndrome	Human
	Guaroa	Central America and South America	Mosquitoes	Febrile syndrome	Human
	Inkoo	Europe, Asia	Mosquitoes	Nervous syndrome	Human
	Ilesha	Africa	Mosquitoes	Febrile syndrome, nervous syndrome	Human
	Kairi	America	Mosquitoes	Febrile syndrome	Animals
	Jamestown Canyon	North America	Mosquitoes	Nervous syndrome	Human
	La Crosse	North America	Mosquitoes	Nervous syndrome	Human
	Madrid	America	Mosquitoes	Febrile syndrome	Human
	Main Drain	America	*Culicoides*, mosquitoes	Congenital malformations, febrile syndrome	Animals
	Marituba	America	Mosquitoes	Febrile syndrome	Human
	Oriboca	America	Mosquitoes	Febrile syndrome	Human
	Oropouche	Central America and South America	*Culicoides*, mosquitoes	Febrile syndrome	Human
	Schmallenberg	Europe	*Culicoides*, mosquitoes	Congenital malformations, abort	Animal
	Tahyna	Africa, Asia, Europe	Mosquitoes	Febrile syndrome	Human
	Wyeomyia	Central America and South America	Mosquitoes	Febrile syndrome	Human
Phlebovirus	Candiru	South America	Sand flies	Febrile syndrome	Human
	Punta Toro	South America	Sand flies	Febrile syndrome	Human
	Rift Valley fever	Africa, Asia	Mosquitoes	Febrile syndrome	Human
	Sandfly fever virus	Asia, Europe	Sand flies	Febrile syndrome	Human
Nairovirus	Crimean-Congo hemorrhagic fever	Africa, Asia, Europe	Ticks	Hemorrhagic fever	Human
	Nairobi sheep disease	Africa, Asia	Ticks	Hemorrhagic disease	Human

regions (NTRs) located at the terminal 5′ and 3′ ends surrounding a single transcription unit. These NTRs are highly conserved within each genus and have specific sequences in their terminal ends, which are involved in RNA synthesis and packaging. The viruses within each genus in this family have segments of similar lengths and a common strategy for the general expression of their products (Schmaljohn and Nichol 2006).

Viruses belonging to *Orthobunyavirus*, *Hantavirus*, and *Nairovirus* genera have shown genetic reassortment, which may occur in cells

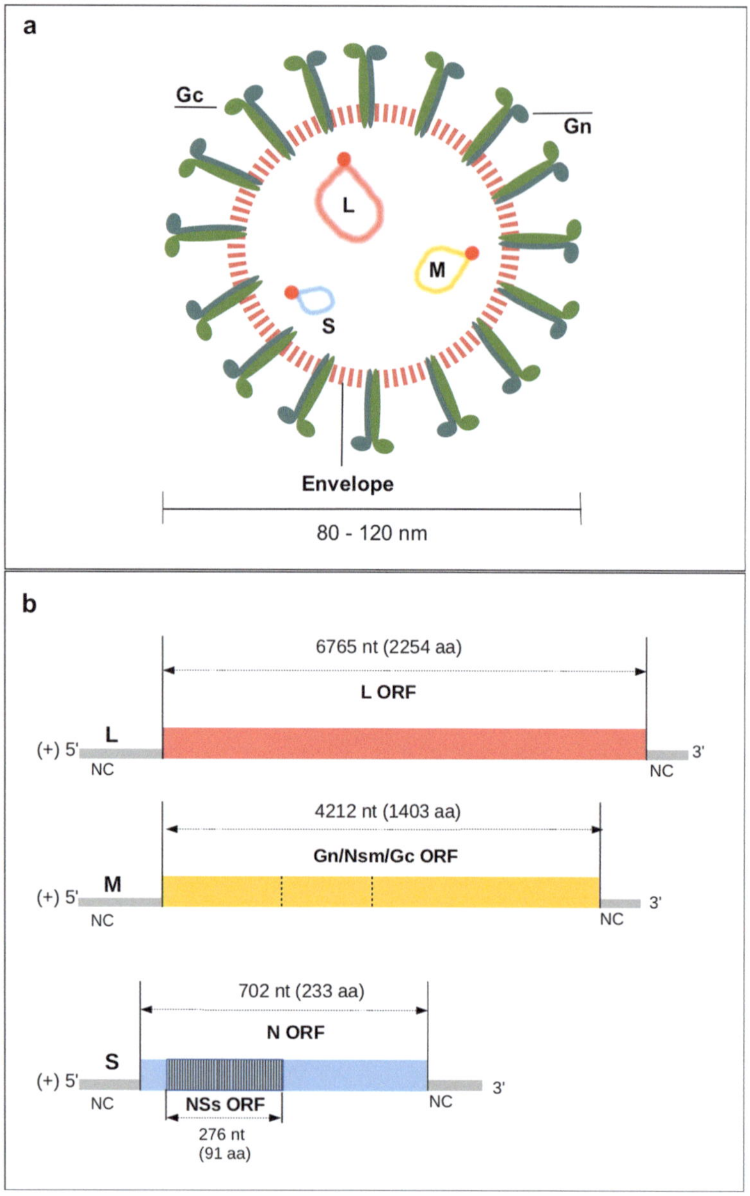

Fig. 10.1 Schematic representation of a virion (**a**) and genomic RNAs (**b**) belonging to prototypic members of family Bunyaviridae

with dual infection by related viruses (same genus) (Reese et al. 2008). Two viruses replicate in the same cell, where the segments of the genome can be rearranged and to be packaged into the virion. For example, in regions where two genotypes of La Crosse virus (LCV) are circulating, a third genotype containing genome segments from the other two genotypes was detected (Reese et al. 2008).

10.2.2 Viral Proteins

The replication strategy varies among genera, being in the negative sense in *Orthobunyavirus*, *Nairovirus*, and *Hantavirus* and ambisense in *Phlebovirus* (Schmaljohn and Nichol 2006).

The S segment encodes the N (nucleocapsid) protein. Its main role is to encapsidate the viral RNA replication products to form ribonucleoprotein complexes. Moreover, this segment encodes a nonstructural protein called NSs in the most of orthobunyaviruses and phleboviruses. Its primary role is to modulate the antiviral response of the host cell. In orthobunyaviruses, the N and NSs proteins are translated from the same messenger RNA (mRNA) encoded by the S segment; however, in phleboviruses, these proteins are translated from different mRNA transcripts (Schmaljohn and Nichol 2006).

The M segment encodes two envelope glycoproteins called Gn and Gc. They are involved in the virus attachment to the cell and the fusion of the viral and cellular membrane. Some viruses from *Orthobunyavirus* and *Phlebovirus* genera encode a third protein, NSm. Its function is unclear and varies with the virus; for example, in orthobunyaviruses, NSm participates in the assembly of viral particles, while in phleboviruses, it would be involved in the regulation of apoptosis (Schmaljohn and Nichol 2006).

Finally, the L segment encodes a single protein called L or RNA-dependent RNA polymerase (RdRp) using a negative-sense transcription-translation strategy. This protein has a weight of 200 KDa and functions as viral transcriptase (Schmaljohn and Nichol 2006).

10.2.3 Viral Replication

As in most of RNA viruses, in bunyaviruses, the replication process takes place in the cytoplasm of the host cell. The first step is the attachment of the virus to the cell surface, a process that is mediated by the Gn and Gc glycoproteins and cellular receptors (Schmaljohn and Hooper 2001). The function of these two glycoproteins varies among viruses of the same genus, and for some of these viruses, the union to mammalian cells is mediated by Gc, whereas in mosquito cells, it is mediated by Gn, although in some cases also by Gc. Furthermore, the Gn glycoprotein has been associated with the ability of the bunyaviruses (excepting *Hantavirus*) to produce high viral loads in peripheral circulation. It has also been associated with their ability to invade nervous tissue. The host cell receptors involved in this step are unknown for most of bunyaviruses. However, evidence suggests that members of the integrin (in hantaviruses) and lectins (in phleboviruses) families are incriminated. In a second step, the virion enters into the target cell by endocytosis, and as a result, it is enclosed into an endosome. It is believed that acidification of the endosome induces a conformational change in Gn and Gc proteins facilitating fusion of viral and cell membranes. This produces the emergence of viral genome into the cytoplasm, where the L protein expressed by the viral genome acts as polymerase (Schmaljohn and Hooper 2001).

The primary transcription of the mRNA from viral RNA (vRNA) of negative polarity (mold) occurs in the cytoplasm. This process is mediated by the viral polymerase and the N protein; the latter protein is involved in the initiation of transcription. The next step is the translation of mRNA into the necessary structural and nonstructural proteins for viral replication; in the case of L and S segments, the translation is performed in free ribosome, whereas in the M segment, it occurs in ribosomes attached to the membrane of the rough endoplasmic reticulum (Schmaljohn and Hooper 2001).

Then the viral replication process takes place, which is mediated by the viral polymerase. Right after that, the process of morphogenesis and maturation of viral particle assembly occurs in the Golgi apparatus. Virions are packed in vesicles from the Golgi apparatus and are released through exocytosis mechanism from the cells as new mature viral particles (Schmaljohn and Hooper 2001).

Differences in the replication process were observed between vertebrate and invertebrate host cells, particularly in mosquitoes. Apparently, some processes associated with transcription regulation of the viral genome into mRNA are tightly controlled by the mosquito ovary metabolism. These processes allow a very effective transovarian transmission, thereby generating an infected progeny (Schmaljohn and Hooper 2001).

10.3 Clinical Manifestations

Most of infections by members of this family in humans and in domestic and wild animals are asymptomatic; however, there are some bunyaviruses that produce various kinds of pathologies

(Beaty and Calisher 1991). Clinical manifestations are diverse and include febrile syndrome, affection of the nervous system (encephalitis, meningitis), hemorrhagic fever, congenital malformations, abortions, and death (Table 10.2) (Beaty and Calisher 1991; Pinheiro et al. 1981). The most common symptoms in the bunyavirus (BUNV, Oropouche virus (OROV), Rift Valley fever virus (RVFV)) are the febrile syndrome. It usually starts suddenly and acutely with high fever, headache, myalgia, retro-orbital pain, nausea, vomiting, and anorexia. Symptoms usually persist for 3–5 days, and in the most severe cases between 7 and 10 days, normally without sequelae. CNS infection is caused by viruses like BUNV, LCV, and California encephalitis virus (CEV) with the most common being encephalitis. The beginning is dominated by symptoms like high fever, holocranial headache, fatigue, stiff neck and followed by myalgia, nausea, vomiting, weakness, and conjunctival congestion. Then, the neurological manifestations characterized by mental confusion, meningeal irritation, motor problems and reflexes, and brain injuries appear. The most severe forms evolve to coma and death. In some cases, neurological sequelae such as paresthesias, motor incoordination, balance problems, visual, olfactory and/or hearing, and memory loss occur. On the other hand, some viruses act as teratogens, producing different birth defects during gestation, such as arthrogryposis, scoliosis, torticollis, hydranencephaly, hydrocephalus, porencephaly, microcephaly, and cerebral and muscular hyperplasia. This is very common in ruminants caused by BUNV, AKAV, and AINV infections but also has been observed in humans caused by BUNV.

10.4 Bunyaviruses of Public Health Concern

10.4.1 *Orthobunyavirus*

This is the largest genus of the Bunyaviridae family and consists of 48 viral species with their strains/isolates distributed worldwide (Table 10.2) (Plyusnin et al. 2012). The delimitation of each virus within each genus is quite difficult because of the little or null virus or strain biochemical and molecular characterization (Schmaljohn and Nichol 2006). Based on serological criteria (FC, NT, and IHA), most of these viruses have been placed in 18 serogroups; however, some viral species are not included in any of them (Schmaljohn and Nichol 2006).

The 18 serogroups are Anopheles A, Anopheles B, Bakau, Bunyamwera, Bwamba, group C, California, Gamboa, Guama, Koongol, Minattian, Nyando, Olifantsvlei, Potosi, Simbu, Tete, Turlock, and Wyeomyia (Schmaljohn and Nichol 2006). Within each serogroup, the antigenic relationships between different members vary greatly, and one reason is the occurrence of natural reassortments among them. Therefore, different viruses within each serogroup may be more or less related to the others and to members of other serogroups, depending on the serological technique used for the study.

The main vectors of orthobunyaviruses are mosquitoes of the genera *Culex* and *Ochlerotatus*; small mammals and birds are regarded as possible vertebrate hosts. Humans and domestic mammals in general are dead-end hosts. Humans serve as host for the maintenance of orthobunyaviruses only in very few cases (i.e., OROV) (Soldano and González-Scarano 2005).

10.4.2 Bunyamwera Serogroup

10.4.2.1 Bunyamwera Virus (BUNV)

BUNV was originally isolated from *Aedes*[1] sp. in Uganda (Africa) in 1943 during a yellow fever virus (YFV) outbreak (Smithbum et al. 1946). Subsequently, it was also recovered from humans with febrile syndrome (Uganda, Nigeria, and South Africa) and hemorrhagic fever (Somalia and Kenya). It is widely distributed across sub-Saharan Africa and is an important etiological

[1] Although several taxonomic modifications have been proposed on Culicidae genera, mostly splitting *Aedes* by Reinert et al. (2009) (and accepted by CBM), this taxonomy has been used in the chapters according to authors' preference. To facilitate utilization by health personnel, all new aedine genera can be considered *Aedes* (CBM).

agent of acute febrile syndrome and CNS pathologies in humans. High prevalence of neutralizing antibodies (NTAb) (>82%) in humans was detected in some regions (Smithbum et al. 1946). This virus was also isolated from various mosquito *Aedes* species, so it is believed to be the primary vector in nature (Smithbum et al. 1946). Moreover, NTAb against BUNV has also been detected in domestic animals, monkeys, rodents, and birds. Although laboratory inoculations showed that rodents, bats, and primates develop enough viremia to infect mosquito, whether any of these would meet the requirements to be considered a host in nature is unknown (Smithbum et al. 1946).

In the Americas, the first *Orthobunyavirus* isolated closely related to BUNV is the Cache Valley virus (CVV), recovered from *Culiseta inornata* (USA, 1956) (Holden and Hess 1959). Currently, according to the classification of the International Committee on Taxonomy of Viruses, CVV is considered a strain or an isolate of BUNV (Plyusnin et al. 2012). This virus is endemic to Canada, the USA, Mexico, and Argentina.

Several strains have been recovered from different species of mosquitoes of the genera *Anopheles*, *Culex*, *Culiseta*, *Ochlerotatus*, and *Psorophora* in Argentina, Brazil, Colombia, Ecuador, Mexico, Panama, and the USA (Plyusnin et al. 2012). Neutralizing antibodies against this virus have been detected in humans, domestic animals (cows, horses, goats, and sheep), and wildlife (moose, deer, caribou, hares, birds, and mice) across all America (McConnell et al. 1987; McLean et al. 1987).

BUNV is considered the causative agent of neural pathologies (encephalitis, meningitis, and febrile syndrome) and CNS defects in humans (Campbell et al. 2006; Mangiafico et al. 1988; Sexton et al. 1997). The first record of infection by BUNV was detected in a human with febrile syndrome in Brazil (1980). Later, in Panama (1985), another strain was recovered from a soldier who had a clinical presentation of malaise, fever, muscle pain, and throat; the patient recovered perfectly after 10 days without apparent sequelae (Mangiafico et al. 1988). In the USA, BUNV was isolated from three severe human cases; two were fatal meningitis and encephalitis (Sexton et al. 1997; Campbell et al. 2006). In Argentina (2009), a fever case was confirmed as a BUNV infection by serology. Interestingly, this case was presumptively diagnosed as dengue fever virus (Tauro et al. 2012). Finally, various seroepidemiological studies suggest that BUNV could be the etiologic agent of congenital malformations in humans, because a correlation was established between the occurrence of malformations in newborns and detection of NTAb against this virus in their mothers (Edwards 1993).

In the USA, BUNV was found associated with the occurrence of febrile and neurologic disease in domestic and wild animals. The subclinical infections result in most of the cases. Clinical manifestations can range from short periods of malaise, fever, loss of appetite, and lack of mobility to encephalitis. Congenital malformations in animals that become infected when being pregnant were also reported (Edwards 1993).

The association between infection with BUNV and the occurrence of congenital malformations were studied for the first time in Texas, USA (1987). Several BUNV strains were isolated during a congenital malformation outbreak in sheep. The observed malformations included CNS and skeletal muscle, such as arthrogryposis, scoliosis, torticollis, hydranencephaly, hydrocephalus, porencephaly, microcephaly, and cerebral and muscular hyperplasia. A high percentage of abortions were also recorded. Neutralizing antibodies were detected in 100% of sheep with malformed offspring (Chung et al. 1991; McConnell et al. 1987; McLean et al. 1987). Experimental inoculations in sheep have revealed that the teratogenic effect of the virus depends on the stage of pregnancy in which maternal infection occurs. Thus, if infection occurs between 28 and 36 days, the CNS and skeletal muscle occur defects; however, if it occurs between days 37 and 42, it is evident only in the skeletal muscle. Infection after 50 days of pregnancy does not cause injury, and after 76 days, the fetus is immunocompetent and produces antibodies (Chung et al. 1991).

During the 1950s and 1960s, two strains were recovered from equine encephalitis cases in Guyana and Colombia; in Argentina, two new

BUNV strains were isolated from equine encephalitis and one from equine abortion in 2013 (Santamartin et al. 1973; Spence and Downs 1968; Tauro et al. 2013).

There are no vaccines or treatments to protect animals from BUNV infection. One of the possible solutions is to make reproductive crosses outside the period of greatest activity of vectors, thus reducing the risk of infection during pregnancy. However, in places where winter is moderate or there are unexpected weather changes, vector activity can be extended, increasing the risk of infection.

10.4.2.2 Kairi Virus (KRIV)

KRIV is a virus exclusively found in the Americas. It was first isolated in 1955 from *Oc. scapularis* mosquitoes collected in Trinidad (Anderson et al. 1961). KRIV was later recovered in the Amazon region from *Oc. scapularis* mosquitoes and monkeys. In Mexico (Yucatan Peninsula), a strain of KRIV was isolated from mosquitoes *Oc. taeniorhynchus* during a study conducted in 2007; this is the most current isolation of this virus and the only record of activity of KRIV in North America.

In Argentina, this virus was isolated from *Ochlerotatus* sp. mosquito (Córdoba province) in 1966 (Sabattini et al. 1998). KRIV NTAb has been detected in humans, domestic animals (cows, horses, goats, and sheep) and wildlife (birds and rodents) in different provinces. Later, in 1974, another strain was recovered from febrile horses from the province of Buenos Aires. NTAb were detected in 67.6% of cohabiting horses. However, since horses infected with Western equine encephalitis virus also occurred in the same area simultaneously, it is difficult to attribute the observed febrile syndrome to KRIV infection (Calisher and Shape 1998).

Serological cross-reactions were observed between KRIV and BUNV. Therefore, it is essential to include both viruses in differential diagnosis and serological surveillance. There are no vaccines currently available against KRIV.

10.4.2.3 Main Drain Virus (MDV)

This virus is limited to North America and was first isolated in California from the biting midge *Culicoides variipennis* (Schmaljohn and Nichol 2006). It was also recovered from the brain of an equine that presented incoordination and ataxia, stiff neck, inability to swallow, fever, and tachycardia. In nature, MDV is maintained through transmission between this *Culicoides* vector and rabbits and rodents as hosts (Schmaljohn and Nichol 2006). Experimental studies in sheep during early gestation showed the teratogenic potential of this virus (Edwards et al. 1997). As for the rest of the viruses belonging to this serogroup currently, there are no vaccines on the market.

10.4.2.4 Ilesha Virus (ILEV)

ILEV was isolated for the first time from a 9-year-old girl with fever and rash in the town of Ilesa, West of Nigeria in 1957 (Okuno 1961). Later, different strains were recovered from patients mostly with fever and erythema in Central and Eastern Africa; there was also a case of meningoencephalitis reported in the Central African Republic and a case of hemorrhagic fever in Madagascar (Morvan et al. 1994). The main vector of ILEV is *Anopheles gambiae*, which was recovered in the Central African Republic (Digoutte et al. 1980). Serological studies in Nigeria detected high prevalence of infection in humans (children and adults) and domestic animals (cattle and goats).

10.4.3 Simbu Serogroup

10.4.3.1 Oropouche Virus (OROV)

OROV was first isolated from a rural worker in Trinidad in 1955 (Anderson et al. 1961). This virus is associated with several major epidemics of febrile illness in the Amazon region of Brazil and Peru and in Panama. The high prevalence of antibodies detected in inhabitants of forest and rural regions of Amazon region suggests an endemic circulation of this virus. In terms of geographic distribution, the activity of OROV has been documented in Argentina, Brazil, Panama, and Peru (Pinheiro et al. 1981). In humans, OROV infection can produce Oropouche fever (OROF), which manifests as an acute clinical infection, along with headache, myalgia, arthralgia among other systemic manifestations (LeDuc and Pinheiro 1988). The presence of rash is rare. Some patients present meningitis. The incubation period is 4–8 days and symptoms last from 2 to 7 days. Viremia in humans may be detected between 2 and 3 days after the onset of symptoms. Recovery from the disease is complete without apparent sequelae even in the most severe cases. No mortality records produced by this virus were documented (Pinheiro et al. 1981). Due to a large number of OROF cases recorded in the last 45 years in humans, OROV is considered the most important emergent viral disease in tropical areas of Central and South America.

OROV has been isolated from various vertebrate sources (humans, monkeys, and edentates) and arthropods (mosquitoes *Cx. quinquefasciatus*, *Coquillettidia venezuelensis*, *Oc. serratus*, and biting midges *Culicoides paraensis*). Studies have shown that OROV is maintained in nature by two cycles (Fig. 10.2): an enzootic cycle characterized by monkeys, sloths, and birds as potential hosts and an unknown vector. In urban settings, an alternative epidemic cycle can take place where human generated high viremia enough to infect *C. paraensis*. This arthropod is the principal urban vector involved in OROV epidemics (Pinheiro et al. 2013).

Recent phylogenetic studies indicate the presence of three different OROV lineages: lineage I, which is the prototype strain recovered in Trinidad and in the western Amazon; lineage II, which contains strains isolated in Peru and eastern Amazon; and lineage III, which includes isolates made in Panama, and more recently, in 2000, isolated strains from *Callithrix* sp. marmosets in the Brazilian state of Minas Gerais (Brazil). These studies suggest movements of strains through Amazon regions of Brazil and Peru (Pinheiro et al. 2013).

In the Brazilian Amazon, OROV is the second most common arbovirus after DENV. Between 1960 and 1980 thousands of people became infected with OROV in the Amazonian areas of Pará state and nearby regions. Later, viral activity was recorded in the states of Acre, Amapá, Maranhão, and Rondônia. In 2003 and 2004, two epidemics were recorded in the state of Pará, and strains belonging to lineages I and II were recovered.

The recent isolation of OROV from *Callithrix* sp. in the southeastern state of Minas Gerais is a fact to consider. Southeastern Brazil is one of the most populated areas of the Americas, with important cities like Belo Horizonte, Rio de Janeiro, and São Paulo. The proximity of OROV to this area represents a serious risk for their inhabitants due to its potential urbanization and development of serious outbreaks (Pinheiro et al. 2013).

In the North of Argentina (Formosa, Jujuy, Salta, and Tucumán), OROV IgM antibodies were detected in febrile humans (North). Molecular phylogenetic analyses indicated the presence of a fourth monophyletic lineage (lineage IV). It was not possible to isolate the virus so far (Pinheiro et al. 2013).

10.4.3.2 Akabane Virus (AKAV)

AKAV is widely distributed in tropical and temperate regions of Oceania, Asia, Middle East, and Africa and is transmitted by different species of biting midges (Kurogi et al. 1987). This virus has great economic importance because it affects

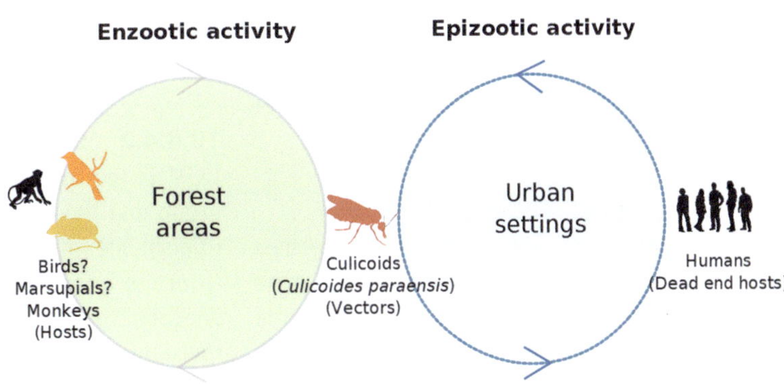

Fig. 10.2 Enzootic and epidemic cycles of Oropouche virus in tropical regions of South America

cattle, sheep, and goats (Kurogi et al. 1987). Sporadic outbreaks with abortions, premature births, and congenital anomalies characterized by arthrogryposis and hydrancephaly were reported, with Australia, Israel, Taiwan, Korea, and Turkey being among the most affected countries by such outbreaks in ruminants (Kurogi et al. 1987). Infection by AKAV in adult animals is usually subclinical, but in cattle, infections are particularly associated with encephalitis production (Kurogi et al. 1987).

Currently, there are two vaccines to immunize domestic animals: a monovalent (AKAV) and a trivalent one (AKAV-AINV-Chuzan virus (CHUV)) (Hechinger et al. 2013). CHUV belongs to the family Reoviridae (genus *Orbivirus*) and is teratogen for ruminants. Given the current epidemiological situation in countries where these viruses circulate, vaccination of young females is an important measure to eliminate the risk of infection of animals during fetal development (Hechinger et al. 2013).

10.4.3.3 Aino Virus (AINV)

This virus has been detected by serological surveys and isolates in Japan, Korea, Taiwan, Israel, Turkey, and Australia. AINV is transmitted by mosquitoes and biting midges and affects ruminants and birds (Tsuda et al. 2004).

Infection in domestic mammals generally causes a subclinical and short viremia. In pregnant animals, the virus can infect the fetus by invading the CNS and/or skeletal muscle tissues, causing hydrancephaly, hydrocephalus, microcephaly, encephalomyelitis, or arthrogryposis (Tsuda et al. 2004). It is very common to observe mixed outbreaks by both AINV and AKAV, so the differential diagnosis is essential to identify the etiologic agent causing disease in animals. The only vaccine available against this virus is trivalent (AKAV-AINV-CHUV) (Hechinger et al. 2013).

10.4.3.4 Schmallenberg Virus (SBV)

SBV has been the *Orthobunyavirus* with greatest veterinary relevance in Europe in the last years (Wu et al. 2014). This virus was originally recovered from clinical samples obtained from cattle with fever, diarrhea, loss of appetite, and decreased milk production in Germany and the Netherlands during the summer and fall of 2011 (Hoffmann et al. 2012). Later, as in infections by other members of the Simbu serogroup, SBV was associated with the occurrence of abortions, stillbirths, and congenital malformations (Wu et al. 2014).

The disease was first observed in cattle, sheep, and goats, but it was also detected in deer, bison, alpaca, moose, and other wild ruminants infected with this virus. Currently, isolations and outbreaks by SBV have been reported in a large area of Europe (Germany, the Netherlands, Belgium, Luxembourg, France, the UK, Italy, and Spain). *Culicoides* species are involved in the transmission of SBV in nature (Rasmussen et al. 2012).

As an effective tool for controlling disease caused by SBV, different inactivated vaccines have been developed and tested; some of which are marketed in some countries, like France and the UK (Wu et al. 2014).

10.4.4 California Serogroup

10.4.4.1 California Encephalitis Virus (CEV)

CEV serogroup prototype was isolated from mosquitoes *Oc. dorsalis* and *Cx. tarsalis* in California in 1943 (Hammon and Reeves 1952). During 1945, also in California, this virus was listed as the etiologic agent causing disease in two children and one adult hospitalized for encephalitis. NTAb were detected in 11 % of patients admitted to the same hospital with nervous syndrome. The last record of isolation for this virus was in 2001. Antibodies against this virus have been detected in rabbits, hares, squirrels, horses, and cows (Schmaljohn and Nichol 2006)

10.4.4.2 La Crosse Virus (LVC)

LCV is the most pathogenic virus in this serogroup and the virus of greatest epidemiological importance in the USA. It was first isolated from the brain of a 4-year-old boy who died of encephalitis in the county of La Crosse (Wisconsin, USA) (Rust et al. 1999).

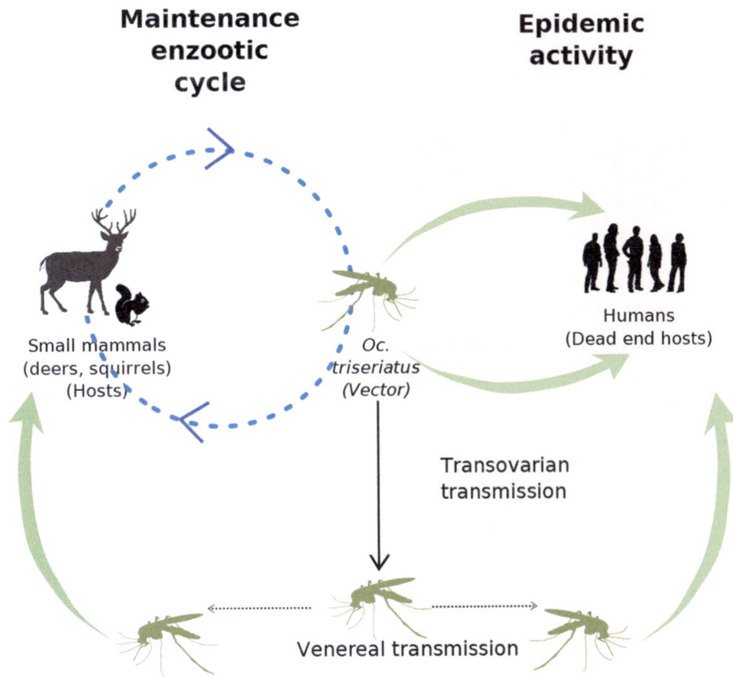

Fig. 10.3 Transmission and maintenance cycle of La Crosse virus in the USA

It is mainly transmitted by the mosquito *Oc. triseriatus*, in which there has been transovarian transmission as a strategy for the winter. Mammal species, such as squirrels (*Sciurus carolinensis*), marmots (*Tamias striatus*), and foxes, act as virus sources in summer, becoming important hosts for LCV amplifiers (Fig. 10.3).

LCV infection is considered a major cause of encephalitis and meningitis in children in the USA; however, in adults, the virus causes only mild fever without neurological involvement (Rust et al. 1999). The incubation period is about 3–7 days, and in most patients, it starts with a rapid rise in temperature followed by stiff neck, lethargy, headache, nausea, and vomiting which generally sharpen the seventh day postinfection. Stroke occurs in 50 % of the patients, and up to 30 % can reach the coma. About 65 % of patients develop meningitis with the presence of mononuclear cells and/or polymorphonuclear cells in the cerebrospinal fluid. Neurological sequelae such as epilepsy occurred in 10–15 % of children who had stroke during the acute phase of the disease. Additionally, about 2 % of patients remain with permanent paresis (Rust et al. 1999).

In 1958, a strain of LCV previously called Snowshoe virus (SSHV) was isolated from the snowshoe hare (*Lepus americanus*) in Montana, USA. This strain has a wide distribution, covering Alaska, Canada, North of USA, and some parts of Asia. The natural cycle develops through mosquitoes and wild animals. In Canada, the snowshoe hare is considered an important host, although other wild animals may be also involved in the cycle. There is evidence of infection in at least 16 species of wild animals (rabbits, rodents, carnivores, and ungulates) and four species of domestic animals (chickens, horses, cows, and dogs). Furthermore, infection was detected in several species of mosquitoes belonging to the genera *Aedes*, *Ochlerotatus*, *Culex*, and *Culiseta*. Human infection with this strain has been associated with the occurrence of encephalitis and meningitis. Domestic and wild animals are often found infected but rarely develop disease (Hubálek et al. 2014).

10.4.4.3 Jamestown Canyon Virus (JCV)

JCV is responsible for encephalitis in humans in the USA, where it is widely distributed. It was isolated from the mosquito *Culiseta inornata* in

California in 1961 (Grimstad 1988). It is mainly transmitted by the bite of *Cs. inornata* and several species of *Ochlerotatus*, in which transovarian transmission was observed. The white deer (*Odocoileus virginianus*) is the amplifier host in nature. This virus recently was isolated from vesicular lesions in horses (Grimstad 1988).

Even though this virus causes neurological disease like LCV in humans, the main difference is related to the severity of the disease, depending on age of the patient. Therefore, LCV infection is always more severe in children, whereas JCV is more severe in adults. The disease can generally range from asymptomatic through a feverish syndrome to fatal cases of encephalitis. Given that there are serological cross-reactions among different virus serogroups and co-circulate in some regions, it is important to make a differential diagnosis to determine the virus causing the disease (Grimstad 1988).

10.4.4.4 Guaroa Virus (GROV)

GROV was first isolated from a human without signs of disease in Colombia in 1959; subsequently, numerous strains of this virus have been recovered from febrile patients in Brazil and mosquitoes in Colombia, Panama, and Brazil. Several studies suggest that GROV is widely distributed in Central and South America, since antibodies have been detected in humans in Brazil, Argentina, Peru, and Guatemala (Groseth et al. 2015).

Since most isolates were made from the mosquito *An. neivai*, it has been incriminated as a potential vector of GROV.

In Brazil, GROV infections were detected in patients with fever and other symptoms, such as headache, myalgia, and prostration. Moreover, a strain was isolated from a liver biopsy performed on a patient with paralysis. The last records of the disease are those obtained in Peru and Bolivia, where 17 cases were confirmed in patients with acute self-limited febrile syndrome; nine were recovered in Peru in 2007 and two from Bolivia in 2007 and 2009. The most common clinical symptoms in patients with GROV were chills, malaise, bone pain, headache, retro-orbital pain, myalgia, and arthralgia (Groseth et al. 2015).

10.4.4.5 Tahyna Virus (TAHV)

TAHV is the first arbovirus isolated in Europe from mosquitoes. Published evidences confirm its circulation in Asia, Africa, and Europe (Hubálek et al. 2014).

In nature, its main vector is *Ae. vexans*, a mosquito species from which most of viral strains were isolated; field work complementary to the laboratory evidence has demonstrated that other mosquito species may also act as vectors: *Oc. cantans*, *Oc. caspius*, *Oc. dorsalis*, *Ae. cinereus*, *Oc. sticticus*, and *Culiseta annulata*. Rabbits, hares, hedgehogs, and pigs are considered the primary hosts (Hubálek et al. 2014).

Infection with this virus is associated mainly with the production of febrile syndrome in children, particularly in European countries (Hubálek et al. 2014).

10.4.4.6 Inkoo Virus (INKV)

INKV was first isolated in Finland in 1964. This virus is mainly distributed in Europe and Russia, and it is transmitted by *Oc. communis* and *Oc. punctor* in Scandinavia. In Russia, it was isolated from *Oc. hexodontus* and *Oc. punctor*. The disease associated with INKV infection has not been well characterized, but a few records suggest this virus as a causative agent of CNS disease in children and young adults in Finland (Putkuri et al. 2007)

10.4.4.7 Group C Serogroup

This serogroup includes Marituba virus (MTBV), Oriboca virus (ORIV), Caraparu virus (CARV), and Madrid virus (MADV); all of them were described for the first time in the Brazilian Amazon during the 1950s. Isolates have been made from humans, wild animals (mainly rodents, marsupials, and bats), and mosquitoes. Geographically, group C is present in American countries, including the USA, Mexico, Panama, Honduras, Guatemala, Trinidad, Brazil, Peru, Ecuador, Venezuela, and French Guiana (Plyusnin et al. 2012).

These viruses have been associated with a human disease that generally occurs as a self-limited dengue-like illness with fever, headache, myalgia, nausea, vomiting, and weakness and lasts for 2–5 days. Despite the public health concern posed by these viruses, they have received little attention, and studies are scarce (Pinheiro and da Rosa 1994)

10.4.5 Wyeomyia Serogroup

10.4.5.1 Wyeomyia Virus (WYOV)

Different strains of WYOV have been recovered exclusively in Central and South America (Colombia, Trinidad, Panama, and Brazil) from different species of mosquitoes collected in rainforest areas of the New World (*Wyeomyia melanocephala*, *Ps. albipes*, *Wy. pilicauda*, *Sabethes soperi*, *Wyeomyia* spp., *Sabethes* spp., and *Sa. glaucadaemon*). The prototype strain of the virus was first isolated from a human and mosquitoes *Wy. melanocephala* and *Ps. albipes* in Colombia in 1940 (Plyusnin et al. 2012).

The range of vertebrate hosts has not been defined. Evidences support involvement of rodents and birds as host. Two viral strains were isolated from *Proechimys guyannensis* and *P. iheringi*. Antibodies were detected in birds (Chowdhary et al. 2012).

The role of WYOV as causal agent of disease in humans is not entirely clear: only two strains have been recovered from ill patients, one with febrile syndrome and the other with a fever accompanied by symptoms of encephalitis (Chowdhary et al. 2012).

10.4.6 Bwamba Serogroup

10.4.6.1 Bwamba Virus (BWAV)

BWAV is transmitted by mosquitoes of the species *An. gambiae*, *An. funestus*, *Ae. furcifer*, *Aedes* spp., *An. coustani*, and *Mansonia uniformis*. The main anthropophilic species of mosquitoes acting as vectors are *An. funestus* and *An. gambiae*. Several strains of BWAV have been recovered from humans in Nigeria, Cameroon, Central African Republic, Kenya, Tanzania, and South Africa (Lutwama et al. 2002).

The disease developed by BWAV was described as a relatively severe generalized infection of short duration (4–5 days) and benign, since no fatalities have been recorded so far. Often, when a meningeal involvement occurs, it is accompanied by rash. Other common symptoms are fever, headache, and arthralgia. In some cases, digestive tract diseases, especially diarrhea, also occur. The convalescent stage is characterized by a marked fatigue that lasts 8–10 days. The viremia in humans is of short duration (2–4 days), which makes virus isolation very difficult (Lutwama et al. 2002).

10.4.6.2 *Phlebovirus*

Viruses belonging to this genus are widely distributed throughout the world, except in Australia, and are the most diverse with respect to the arthropod vectors involved in its natural cycle (Table 10.2) (Plyusnin et al. 2012).

This genus has nine virus species with more than 70 antigenically different strains or isolates divided in complexes according to the transmitted vectors. While most viruses are transmitted by sand flies of the genera *Phlebotomus*, *Lutzomyia*, and *Sergentomyia*, some are transmitted by mosquitoes (Rift Valley fever virus – RVFV) and ticks (Sandfly[2] fever Naples virus – SFNV) (Plyusnin et al. 2012).

10.4.6.3 Rift Valley Fever Virus (RVFV)

RVFV was first identified in Great Rift Valley in Kenya in 1931 during an outbreak that affected thousands of small ruminants causing abortions in sheep and mortality in newborn lambs. This virus is characterized by producing large explosive epidemics across Africa, as in the Arabian Peninsula. RVFV mainly affects ruminants (goats, sheep, cows, and camels) and humans. In ruminants, particularly sheep, RVFV infection is characterized by a high rate of abortion, high mortality rate in young animals (~70 %), and significant mortality in adult animals (20–30 %) (Arum et al. 2015).

[2]Even being "sandfly fever" more usual than "sand fly fever" in the scientific literature, the last form was utilized for the sake of standardization in American English (CBM).

In humans, the virus is transmitted by mosquito bites or exposure to blood and body fluids or spray when handling infected animals themselves. Clinical infection in humans is self-limited and is manifested by fever, headache, weakness, muscle pain, and photophobia. In a very small number of human patients (1–3 %), the virus causes hepatitis, retinitis, blindness, encephalitis, and hemorrhagic fever (Arum et al. 2015).

In nature, RVFV is transmitted by various species of mosquitoes of the genera *Aedes* s.l., *Culex*, and *Anopheles*. Ruminants act as amplifier hosts in epizootic/epidemic cycles, transmitting the virus to other mosquitoes and therefore causing amplification and virus spread as well (Arum et al. 2015).

Due to the ability of RVFV to infect and replicate in different mosquito species, the virus shows great potential to spread to other parts of the world, such as America, Asia, and Europe, which could cause serious economic and health problems (Arum et al. 2015).[3]

Currently, an attenuated virus vaccine has been proven safe for both adults and young animals and very effective in sheep, cows, and monkeys. However, it cannot be used in pregnant individuals because it has teratogenic effect on the fetus (Miller et al. 2015).

10.4.6.4 Sandfly Fever Naples Virus (SFNV)

Numerous strains of SFNV are responsible for most of the cases of febrile syndrome produced by sand flies in Europe and Asia. Previously classified as a separate species, currently Toscana is the strain with greatest impact on health. It was isolated from *Phlebotomus perniciosus* collected in Monte Argentario (Grosseto, Italy) in 1971 (Plyusnin et al. 2012). SFNV is widely distributed, covering Italy, France, Spain, Slovenia, Turkey, Portugal, Sweden, and Greece. It is transmitted by *Ph. perniciosus* and *Ph. perfiliewi*, which also act as reservoirs. Transovarian transmission has been demonstrated in laboratory studies and by virus isolation in *Phlebotomus* males. Although this virus has been isolated from human patients and wild animals, it is not known which vertebrates are involved in its maintenance. In humans, the infection produces mostly a moderate febrile illness, with neurological involvement (aseptic meningitis), sometimes with complete recovery. After an incubation period of up to 2 weeks, the onset of symptoms is abrupt (headache, fever, nausea, vomiting, and myalgia). Other frequent symptoms are leukopenia, stiff neck, and low level of consciousness, tremors, and paresis. Viremia developed in both humans and laboratory animals is low and short-lasting.

10.4.6.5 *Phlebovirus* in the Americas

In the American continent, there is little information regarding epidemiology and ecology of phleboviruses. Punta Toro virus (PTV) and Candiru virus (CANV) are associated with febrile syndrome in humans (Plyusnin et al. 2012). PTV was isolated in Panama and Colombia from sand flies and humans. Seroprevalence detected in Panama ranged from 5 % in children to 27–40 % in adults. PTV produces an acute febrile illness lasting 2–5 days in humans. In the Brazilian Amazon, numerous strains of CANV have been recovered from arthropods and a human febrile case. The disease caused by these Amazonian phleboviruses is acute, self-limited flu-like, and lasts for 2–5 days. It is characterized by a sudden start with high fever, frontal headache, back pain, myalgia, retro-orbital pain, and photophobia (Palacios et al. 2011).

10.4.7 *Nairovirus*

Nairoviruses (Table 10.2) have a worldwide distribution and are transmitted almost exclusively by ticks, but few representatives have been recovered from biting midges, mosquitoes, and flies (mechanical vector). There are seven viral species with numerous strains that exhibit a large geographical distribution, probably due to widely dispersed amplifier host, mostly birds.

[3] This virus (and CCHFV, see below) is cited as a potential for bioterrorism (Lockwood 2009) (see Chap. 3) (CBM).

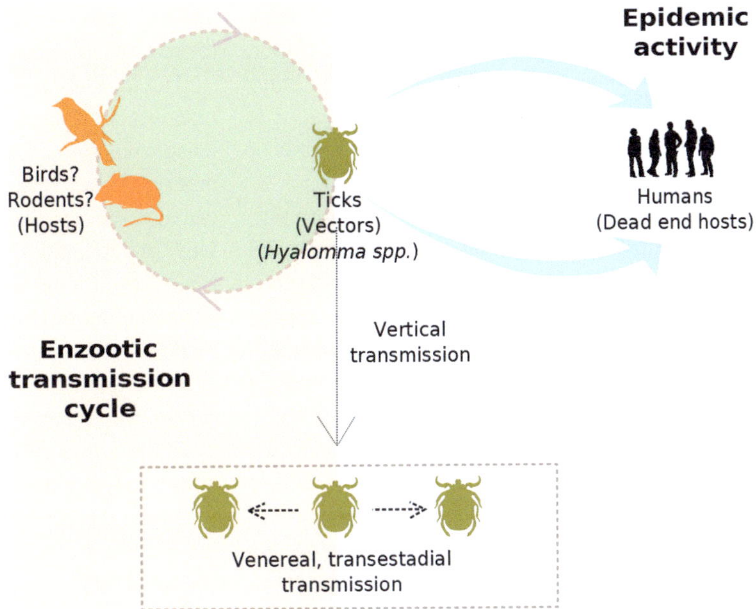

Fig. 10.4 Transmission cycle of Crimean-Congo hemorrhagic fever virus in Congo

10.4.7.1 Crimean-Congo Hemorrhagic Fever Virus (CCHFV)

CCHFV is the *Nairovirus* of health importance. This virus is present across many geographic regions of Asia, Africa, and Europe, overlapping with the distribution of its vector: *Hyalomma* ixodid ticks. Although the virus has been isolated from more than 30 different tick species, *Hyalomma marginatum* is considered the main vector. Transovarian, transestadial, and venereal transmission of the virus has been observed in ticks (Fig. 10.4). Domestic and wild animals that serve as hosts for ticks often develop significant viremia and antibody without appreciable clinical manifestations (Appannanavar and Mishra 2011).

CCHFV is transmitted to humans by the bite of an infected tick or by direct contact with blood or tissues of an infected person. The virus first replicates locally in the entry area, and then it spreads via the lymphatic and hematic systems to the liver, reticuloendothelial system, or other target organs where it replicates massively. CCHFV produces cytopathic effects on endothelial cells, hepatocytes, and macrophages. The disease is characterized by endothelial damage determined by capillary fragility that is evidenced by immune complex formation and complement activation. Moreover, the reduction of thrombocytopenia and thrombosis in bone marrow is observed. Tissue damage in the liver causes the release of procoagulant factors, leading to circulatory collapse by intravenous coagulations. Liver injury also prevents synthesis and replacement of coagulation factors, ultimately leading to hemorrhage (Appannanavar and Mishra 2011).

The incubation period is short (1–13 days), depending on the infection route. The disease onset is sudden, with chills, fever, severe headache, irritability, sore throat, retro-ocular pain, photophobia, myalgia, and back pain. Other usual initial symptoms are nausea, sore throat, vomiting, abdominal pain, and diarrhea. During the first 2 days, the fever appears intermittently, and the patient has mood swings, confusion, and aggression, followed by lassitude, depression, and somnolence between the second and fourth day, accompanied by flushing, tachycardia, and mild hypotension. Between the fourth and fifth day, bleeds appear (epistaxis, hematemesis, hematuria, and gingival and vaginal bleeding). In the most severe cases, patients may develop hepatorenal and pulmonary collapse from the fifth day and can progressively get drowsiness and stupor, leading to coma and death. Case fatality rates ranged from 30 to 50 %. Slow recovery begins at day 9 or 10. Full recovery can take a month or longer (Appannanavar and Mishra 2011).

The infections occur most often among field-workers after being bitten by an infected tick, and to a lesser extent, among workers exposed to contaminated tissues from livestock (e.g., slaughterhouse), and between health workers through contact with the body fluids of infected patients (Appannanavar and Mishra 2011).

Prophylaxis must be based on surveillance programs involving ticks and animals through which endemic or high-risk areas can be defined. The application of chemical products to animals for tick control at a large scale is quite limited because of the development of chemical resistance (Appannanavar and Mishra 2011).

Alternative strategies such as vaccines against ticks, semiochemicals, and the use of biocontrol agents are currently being developed. The main mode of personal protection is to avoid contact with ticks or contaminated tissues, which requires the use of proper clothing (gloves, closed work clothes) and repellents on clothing (Appannanavar and Mishra 2011).

Even though several vaccines have been developed, their use has been rather restricted geographically. The development of a safe and effective vaccine (inactivated virus) is currently limited by the low availability of necessary infrastructure for development and potential demand due to the number of cases that exist today (Appannanavar and Mishra 2011).

10.4.7.2 Nairobi Sheep Disease Virus (NSDV)

NSDV causes a severe hemorrhagic disease in domestic animals with high morbidity and mortality. It was first isolated in 1917 in Nairobi (Africa) and is endemic in Eastern and Central Africa and India. In nature, NSDV is transmitted by ixodid ticks *Rhipicephalus appendiculatus*, the main vector in Africa and *Haemaphysalis intermedia* in India. Goats and sheep are the only host mammals known for NSDV in nature; other animals like cows and horses are refractory to the disease (Marczinke and Nichol 2002).

The incubation period is of 4–15 days. Clinical manifestations begin with high fever (40–41°C), leucopenia, and anorexia; animals are listless and motionless, with gastrointestinal bleeding and hemorrhagic mucopurulent nasal discharge, conjunctivitis, and painful dyspnea. In pregnant animals, NSDV can cause abortions. Animals that survive become immune (Marczinke and Nichol 2002).

NSDV has a limited effect on animals born in enzootic areas because of the acquired immunity across the maternal antibodies. However, the virus has a great impact, causing significant economic losses when infected animals are brought to new free areas of this virus or healthy animals are introduced in enzootic areas (Marczinke and Nichol 2002)

Human infections are rare and generally subclinical; mild flu symptoms, fever, and arthralgias are manifested in rare cases. There are no vaccines available (Marczinke and Nichol 2002).

References

Anderson CR, Downs WG, Aitken TH (1961) Oropouche virus: a new human disease agent from Trinidad, West Indies. Am J Trop Med Hyg 10:574–578

Appannanavar SB, Mishra B (2011) An update on Crimean Congo Hemorrhagic Fever. J Glob Infect Dis 3:285–292

Arum SO, Weldon CW, Orindi B et al (2015) Distribution and diversity of the vectors of Rift Valley fever along the livestock movement routes in the northeastern and coastal regions of Kenya. Parasit Vectors 28:294. doi:10.1186/s13071-015-0907-1

Beaty BJ, Calisher CH (1991) Bunyaviridae–natural history. Curr Top Microbiol Immunol 169:27–77

Calisher CH, Shape RE (1998) Bunyaviridae: the bunyaviruses. In: Balows A, Hauster WJ, Lenette EH (eds) Laboratory diagnosis of infection diseases: principles and practice. Springer Verlag, New York, pp 626–633

Campbell GL, Mataczynski JD, Reisdorf ES et al (2006) Second human case of Cache Valley virus disease. Emerg Inf Dis 12:854–856

Chowdhary R, Street C, Travassos da Rosa A et al (2012) Genetic characterization of the Wyeomyia group of orthobunyaviruses and their phylogenetic relationships. J Gen Virol 93:1023–1034

Chung SI, Livingston CW Jr, Jones CW et al (1991) Cache Valley virus infection in Texas sheep flocks. J Am Vet Med Assoc 199:337–340

Digoutte JP, Salaun JJ, Robin Y et al (1980) Les arboviroses mineures en Afrique Centrale et Occidentale. Med Trop 40:523–533

Edwards JF (1993) Cache Valley Virus. Vet Clin North Am Food Anim Pract 10:515–524

Edwards JF, Karabatsos N, Collisson EW et al (1997) Ovine fetal malformations induced by in utero inoculation with Main Drain, San Angelo, and LaCrosse viruses. Am J Trop Med Hyg 56:171–176

Grimstad PR (1988) California group virus disease. In: Monath TP (ed) The arboviruses: epidemiology and ecology. CRC, Boca Raton, pp 99–136

Groseth A, Wollenberg KR, Mampilli V et al (2015) Spatiotemporal analysis of Guaroa virus diversity, evolution, and spread in South America. Emerg Infect Dis 21:460–463

Hammon WN, Reeves WC (1952) California encephalitis virus, a newly described agent. Calif Med 77:303–309

Hechinger S, Wernike K, Beer M (2013) Evaluating the protective efficacy of a trivalent vaccine containing Akabane virus, Aino virus and Chuzan virus against Schmallenberg virus infection. Vet Res 5:114

Hoffmann B, Scheuch M, Höper D et al (2012) Novel orthobunyavirus in Cattle, Europe, 2011. Emerg Infect Dis 44:469–472

Holden P, Hess AD (1959) Cache Valley virus, a previously undescribed mosquito-borne agent. Cache Valley virus, a previously undescribed mosquito-borne agent. Science 130:1187–1188

Hubálek Z, Rudolf I, Nowotny N (2014) Arboviruses pathogenic for domestic and wild animals. Adv Virus Res 89:201–275

Kurogi H, Akiba K, Inaba Y et al (1987) Isolation of Akabane virus from the biting midge *Culicoides oxystoma* in Japan. Vet Microbiol 15:243–248

LeDuc JW, Pinheiro FP (1988) Oropouche fever. In: Monath TP (ed) The arboviruses: epidemiology and ecology. CRC, Boca Raton, pp 1–14

Lockwood JA (2009) Six-legged soldiers: using insects as weapons of war. Oxford University Press, New York

Lutwama JJ, Rwaguma EB, Nawanga PL et al (2002) Isolations of Bwamba virus from south central Uganda and northeastern Tanzania. Afr Health Sci 2:24–28

Mangiafico JA, Sanchez JL, Figueiredo LT et al (1988) Isolation of a newly recognized Bunyamwera serogroup virus from a febrile human in Panama. Am J Trop Med Hyg 39:593–596

Marczinke BI, Nichol ST (2002) Nairobi sheep disease virus, an important tick-borne pathogen of sheep and goats in Africa, is also present in Asia. Virology 303:146–151

McConnell S, Livingston CW Jr, Calisher CH et al (1987) Isolation of Cache Valley virus in Texas, 1981. Vet Microbiol 13:11–18

McLean RG, Calisher CH, Parham GL (1987) Isolation of Cache Valley virus and detection of antibody for selected arboviruses in Michigan horses in 1980. Am J Vet Res 48:1039–1041

Miller MM, Bennett KE, Drolet BS et al (2015) Evaluation of the efficacy, potential for vector transmission, and duration of immunity of MP-12, an attenuated Rift Valley fever virus vaccine candidate, in sheep. Clin Vaccine Immunol 22:930–937. doi:10.1128/CVI.00114-15

Monath T, Heinz F (1996) Bunyaviruses. In: Fields BN, Knipe DM, Howley PM (eds) Fields virology, 3rd edn. Lippincott-Raven, Philadelphia, pp 1228–1236

Morvan JM, Digoutte JP, Marsan P et al (1994) Ilesha virus: a new aetiological agent of haemorrhagic fever in Madagascar. Trans R Soc Trop Med Hyg 88:205

Okuno T (1961) Immunological studies relating two recently isolated viruses, Germiston virus from South Africa and Ilesha virus from West Africa, to the Bunyamwera group. Am J Trop Med Hyg 10:223–226

Palacios G, Tesh R, Travassos da Rosa A et al (2011) Characterization of the Candiru antigenic complex (Bunyaviridae: Phlebovirus), a highly diverse and reassorting group of viruses affecting humans in tropical Am. J Virol 85:3811–3820

Pinheiro FP, da Rosa APA (1994) Part F. Group C bunyaviral fevers. In: Beran GM (ed) Handbook of zoonoses. CRC, Boca Raton, pp 212–214

Pinheiro FP, Travassos da Rosa AP, Travassos da Rosa JF et al (1981) Oropouche virus. A review of clinical, epidemiological, and ecological findings. Am J Trop Med Hyg 30:149–160

Pinheiro FP, da Rosa Travassos AP, Nunes MRT et al (2013) Febre do Oropouche. In: de Queiroz Leão RN (ed) Medicina Tropical e Infectologia na Amazônia. Samura Editorial, Belém, pp 575–589

Plyusnin A, Beaty BJ, Elliott RM et al (2012) Bunyaviridae. In: King AMQ, Adams MJ, Carstens EB et al (eds) Virus taxonomy: classification and nomenclature of viruses: ninth report of the International Committee on Taxonomy of Viruses. Elsevier/Academic, London, pp 725–741

Putkuri N, Vaheri A, Vapalahti O (2007) Prevalence and protein specificity of human antibodies to Inkoo virus infection. Clin Vaccine Immunol 14:1555–1562

Rasmussen LD, Kristensen B, Kirkeby C et al (2012) Culicoids as vectors of Schmallenberg virus. Emerg Infect Dis 44:1204–1206

Reinert JF, Harbach RE, Kitching IJ (2009) Phylogeny and classification of tribe Aedini (Diptera: Culicidae). Zool J Linn Soc 157:700–794

Reese SM, Blitvich BJ, Blair CD et al (2008) Potential for La Crosse virus segment reassortment in nature. Virol J 5:164–169

Rust RS, Thompson WH, Mathews CG et al (1999) La Crosse and other forms of California Encephalitis. J Child Neurol 14:1–14

Sabattini MS, Avilés G, Monath TP, da Travassos Rosa APA (1998) Historical, epidemiological and ecological aspects of arbovirus in Argentina: Flaviviridae, Bunyaviridae and Rhabdoviridae. In: Travassos da Rosa APA, Vasconcelos PFC, da Travassos Rosa JFS (eds) Overview of arbovirology in Brazil and neighbouring countries. Instituto Evandro Chagas, Belém, pp 127–129

Santamartin C, MacKensie R, Trapido H et al (1973) Encefalitis equina venezolana en Colombia, 1967. Bol Ofic Saint Panam 74:108–144

Schmaljohn CS, Hooper JW (2001) Bunyaviridae: the viruses and their replication. In: Knipe DM, Howley PM (eds) Fields virology, 4th edn. Lippincott Williams & Wilkins, Philadelphia, pp 1581–1502

Schmaljohn CS, Nichol ST (2006) Bunyaviridae. In: Knipe DM, Howley PM (eds) Fields virology, 5th edn. Lippincott Williams & Wilkins, Philadelphia, pp 1741–1789

Sexton DJ, Rolling PE, Breitschwerdt EB et al (1997) Life-threatening Cache Valley virus infection. N Engl J Med 336:547–549

Smithbum KC, Haddow AJ, Mahaffy AF (1946) A neurotropic virus isolated from Aedes mosquitoes caught in the Semliki forest. Am J Trop Med Hyg 26:189–208

Soldano SM, González-Scarano F (2005) Emerging infectious diseases: the Bunyaviridae. J Neurovirol 11:412–423

Spence L, Downs WG (1968) Virological investigations in Guyana, 1956-1966. West IndianMed J 17:83–89

Tauro LB, Venezuela RF, Spinisanti LI et al (2012) First case of human infection with a Bunyamwera serogroup virus in Argentina. J Clin Virol 54:98–99

Tauro LB, Lucca E, Marino B et al (2013) Detección de virus Bunyamwera en equinos con síndrome neurológicos en la provincia de Santa Fe, Argentina. XXXIII Reunión Científica Anual, Sociedad Argentina de Virología, Buenos Aires, 10 December 2013

Tsuda T, Yoshida K, Ohashi S et al (2004) Arthrogryposis, hydranencephaly and cerebellar hypoplasia syndrome in neonatal calves resulting from intrauterine infection with Aino virus. Vet Res 35:531–538

Wu XD, Ge SQ, Zhang YQ et al (2014) A review of research on Schmallenberg virus. Bing Du Xue Bao 30:694–703

Plague and Tularemia

11

Tereza Cristina Leal-Balbino,
Maria Betânia Melo de Oliveira,
Maria Paloma Silva de Barros, Marise Sobreira,
and Vladimir da Mota Silveira-Filho

Abstract

Plague and tularemia are diseases caused by gram-negative bacteria *Yersinia pestis* and *Francisella tularensis*, respectively. Transmission occurs mainly through vectors, affecting humans and other mammals, causing fever and lymphadenopathy. These infectious diseases present very similar clinical and epidemiological characteristics; therefore, it is appropriate to study them jointly. On the other hand, there also are important differences as to their clinical and epidemiological aspects, as well as their etiologic agents. Among the reservoirs common to both microorganisms, rodents are worthy of mention, as well as lagomorphs and wild and domestic carnivores. The primary mode of plague transmission is through the bite of infected fleas, while transmission of tularemia is caused through the bite of several hematophagous arthropods. These diseases can be found throughout the world: the plague persists in natural foci in countries Africa, Asia, Europe, and the Americas, while tularemia is restricted to countries in the northern hemisphere. Treatment for both diseases consists of similar antimicrobials, where streptomycin and gentamicin are the drugs of choice. Without prompt treatment, these diseases can cause serious illness and even death. Bacterial culture and inoculation of animals infected with plague and tularemia should be carried out in a biosafety level 3 (BSL3) laboratory. Both diseases are categorized as international public health emergencies, given their ability to cause significant impact on public health and their potential for rapid international spread.

Keywords

Yersinia pestis • *Francisella tularensis* • Gram-negative bacteria • Infectious diseases

T.C. Leal-Balbino (✉) • M.P.S. de Barros
M. Sobreira
Department of Microbiology, Oswaldo Cruz Foundation, Aggeu Magalhães Research Center (CPqAM/Fiocruz), Recife, PE, Brazil
e-mail: cristina@cpqam.fiocruz.br

M.B.M. de Oliveira
Department of Biochemistry, Federal University of Pernambuco, Recife, PE, Brazil

V.M. Silveira-Filho
Department of Biology, University of Pernambuco, Garanhuns, PE, Brazil

C.B. Marcondes (ed.), *Arthropod Borne Diseases*, DOI 10.1007/978-3-319-13884-8_11

11.1 Introduction

Plague and tularemia are caused by the gram-negative bacteria *Yersinia pestis* and *Francisella tularensis*, respectively. These bacteria were both initially classified as belonging to the genus *Pasteurella*; however, biochemical and genetic differences were sufficient to subdivide them into two genera: *Yersinia* in honor of Alexandre Yersin, the discoverer of the bacillus responsible for plague, and *Francisella* in honor of Edward Francis, a scholar of tularemia.

The clinical and epidemiological similarities between the two diseases justify presenting them together. They are vector-transmitted zoonoses that occasionally affect humans, causing fever, lymphadenopathy, and pulmonary impairment in infected individuals. Plague transmission occurs primarily through the bite of infected fleas, while tularemia is vectored by blood-sucking arthropods (i.e., fleas, ticks, mites, and mosquitoes). Rodents, lagomorphs, and wild and domestic carnivores are reservoirs for both bacterial strains.

The Centers for Disease Control and Prevention (CDC) classifies both bacterial species as biosafety level 3 and within category A of the bioterrorism agent/disease list, considering the severity of disease and the potential use of these pathogens as agents of bioterrorism.[1]

According to the World Health Organization (WHO), 12,503 cases of human plague were documented between 2004 and 2009, causing 843 deaths across 16 countries in Africa, Asia, and the Americas, with a case fatality rate of 6.7 %. Africa was the continent most affected; eight countries reported 12,209 cases in total, with 814 deaths. Asia reported 149 cases in four countries with 23 deaths, while in the Americas, 145 cases were reported in two countries, resulting in six deaths.

Between 1900 and 2010, 999 suspected cases of human plague were reported in the United States, and over 80 % of these cases were the bubonic form. In the last decade, an average of seven cases of human plague was reported each year (range: 1–17 cases per year). Cases are reported in both men and women, although historically infection is more common among men, likely due to outdoor activities that put them at higher risk.

Tularemia affects countries in the northern hemisphere, principally North America, South America (Venezuela), and most countries in Europe and Asia (including China and Japan). In 2000, the WHO indicated 56 cases of tularemia in Pristina, Kosovo. Tularemia was more common in the United States in the early twentieth century than today; the case frequency has declined since 2003, and currently, there are fewer than 200 cases reported per year, despite occurring in every state except Hawaii. Infection occurs sporadically throughout the year with prevalence in adult males, possibly due to greater involvement in outdoor activities such as hunting. Cases are also reported in children, especially in the summer, probably due to high seasonal abundance of ticks and horse flies.

11.2 History

11.2.1 Plague

Plague probably originated in the Central Asian Plateau and has been responsible for high mortality rates in some time periods. Numerous references in art, literature, and historical monuments attest to the horrors and devastation of the plague. During the Christian era, three major pandemics were characterized. The first, called the Plague of Justinian (542–605 AD), originated in Egypt and spread throughout the civilized world, reaching Asia, Africa, and Europe. The second, the dreaded Black Death, began in Asia and extended throughout Europe and North Africa, persisting from the fourteenth to sixteenth centuries and killing a quarter of the European population. The third, the contemporary pandemic, originated in 1891 in Yunnan, China and, in 1894, extended to Hong Kong and quickly spread across the continents via sea transport, leaving natural foci scattered among all continents

[1] See Chap. 3 for more information on utilization in biological warfare.

with the exception of Australia (Perry and Fetherston 1997).

Three relevant events occurred throughout the history of plague: first, the etiologic agent was isolated in 1894 by Alexandre Yersin; second, fleas were discovered as agents of disease transmission in 1898 by Paul-Louis Simond; last, the pandemics created natural foci in various (previously unaffected) regions of the world.

11.2.2 Tularemia

During research on the plague in endemic areas, tularemia was discovered and initially identified as a kind of plague called "pseudo-plague." In 1911, the etiologic agent, initially named *Bacterium tularense*, was isolated from squirrels in the city of Tulare, California (USA). The syndrome was characterized in 1912, with the first human cases observed in 1914. Studies of this zoonotic disease progressed in all aspects thereafter, with much of the work conducted by researcher Edward Francis. Because of his commitment to research on this zoonosis, the name of the bacillus was changed to *Francisella tularensis* in his honor (Eigelsbach and McGann 1984).

Tularemia has been reported to occur in Japan since 1837, described as a disease acquired by ingestion of hare meat. The degree of pathology to humans was discovered in the laboratory after accidental infection. The existence of tularemia was then acknowledged in 1928, after long being mistaken for a type of plague. From 1929 to 1945, it was also acknowledged in Norway, Sweden, Turkey, Austria, Czechoslovakia, and France (Eigelsbach and McGann 1984).

11.3 Etiology

11.3.1 Plague

The causative agent of plague, *Yersinia pestis*, belongs to family Enterobacteriaceae. The genus *Yersinia* consists of 17 species that can be differentiated by their pathogenicity. *Yersinia pseudotuberculosis* and *Y. enterocolitica* are enteropathogenic, transmitted by oral-fecal route and causing a clinical condition called Yersiniosis (Perry and Fetherston 1997). One of these pathogens, *Y. pseudotuberculosis*, is considered the ancestor of *Y. pestis* and is thought to have diverged approximately 1.5–20 thousand years ago (Achtman et al. 1999).

Yersinia pestis is a gram-negative bacillus, short and ovoid (0.5–0.8 mm diam. × 1–3 mm length), and having intense bipolar coloration when impregnated with specific dyes (Fig. 11.1a1). *Yersinia pestis* is aerobic or facultative anaerobic, non-sporulating, and does not ferment lactose, sucrose, or rhamnose; it acidifies glucose, mannitol, and salt without producing gas and does not hydrolyze urea. The results of nitrate reduction and glycerol fermentation tests are variable and important for the identification of different geographic varieties or biovars.

Yersinia pestis grows well on ordinary media, such as peptone agar base and blood agar base (BAB). Growth can occur at temperatures ranging from 4 to 37°C, but 28°C is ideal, producing slow growth. After 48 h, colonies measure from 1 to 2 mm in diameter, whitish with convex shape, shiny, translucent, and non-mucoid. In broth (brain-heart infusion, or BHI), this microorganism exhibits flocculent (i.e., clumped) growth, resulting in nonuniform medium turbidity (Perry and Fetherston 1997).

Yersinia pestis is cold resistant and is well preserved in corpses and in sputum of patients with pneumonic plague and in flea droppings. It can be eliminated by common disinfectants like antiseptic soap and hypochlorite or by sun exposure, boiling water (for a few minutes), or moist heat at 60 °C (40 min). Cultures can be maintained on peptone agar at 4 °C for decades, but the safest way is via lyophilization.

Yersinia pestis strains are considered highly phenotypically homogeneous, having a serotype and phagotype. The biovars and geographical varieties have acquired recent classifications that reflect global diversity based on single nucleotide polymorphism (SNP) type and biochemical characteristics. In accordance with the International Code of Nomenclature of Bacteria (ICNB), *Y. pestis* has been subdivided into two

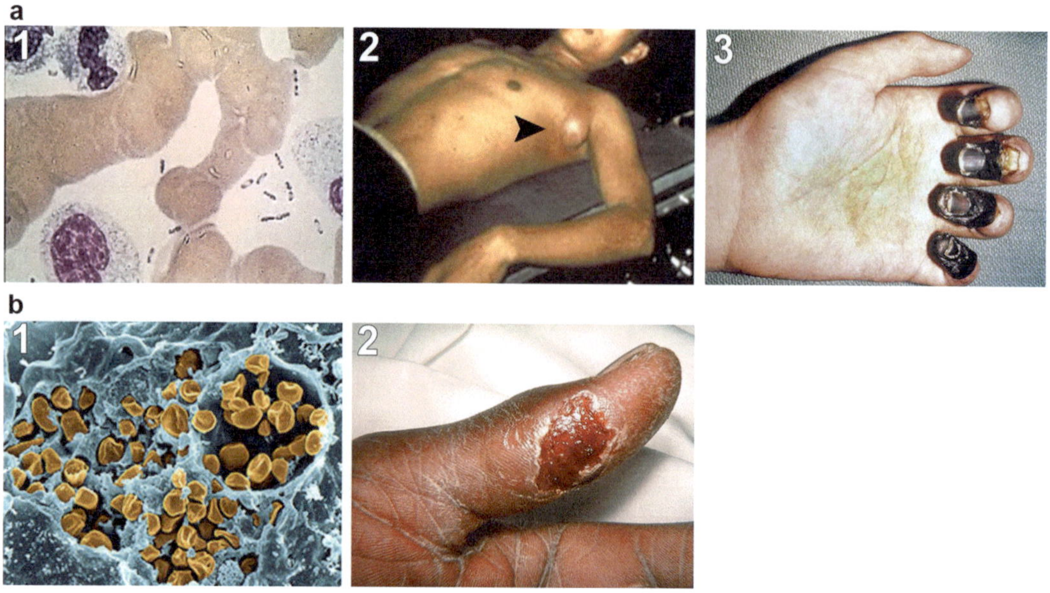

Fig. 11.1 *Yersinia pestis* bacillus and intense bipolar coloration when impregnated with specific dyes (*a1*), bubonic plague (*a2*), septicemic plague (*a3*) image of *Francisella tularensis* bacillus (*b1*), ulceroglandular tularemia (*b2*). *Source*: (**a**) CDC/Plague 2012 and (**b**) CDC/Tularemia 2011

major subspecies: *Y. pestis pestis* and *Y. pestis microtus. Yersinia p. microtus* comprises of a group of biovars that typically do not cause disease in humans (bv. caucasica, bv. angola, bv. talassica, bv. qinghaiensis, bv. xilingolensis, bv. altaica, bv. hissarica, and bv. ulegeica). *Yersinia pestis pestis* includes four biovars (bv. intermedium, bv. antiqua, bv. medievalis, and bv. orientalis) (Platonov et al. 2013).

The *Y. pestis* genome consists of a single circular chromosome of about 4.6 Mb with a GC content of 47 %, with approximately 4200 genes and three prototypical plasmids: pPCP1, PMT1, and pCD1 (9.5, 100, and 70 kb, respectively). Its pathogenicity is attributed to several factors, including the presence of invasins, adhesins, and toxins, as well as intracellular and antibiotic resistance encoded by plasmid and chromosomal genes (Perry and Fetherston 1997; Parkhill et al. 2001; Deng et al. 2002).

In addition to the murine toxin, the pPCP1 plasmid appears to play an essential role in the transmission of *Y. pestis* by fleas. The PMT1 plasmid codes for one of the envelope proteins (F1) of *Y. pestis*, and is highly immunogenic for humans and animals, making it useful in diagnostic tests such as enzyme-linked immunosorbent assay (ELISA). The pCD1 plasmid, also present in other pathogenic *Yersinia*, is responsible for a complex type III secretion system (TTSS), which neutralizes host antibacterial defenses.

The absence of one of these plasmids or the presence of atypical or cryptic plasmids has been observed in *Y. pestis* strains worldwide. These events may be fostered by the presence of insertion sequences (IS1541A, IS100, IS285, IS1661) in the genome of the bacterium, as well as by insertion of plasmid sequences on the chromosome, recombination events, genetic reorganization, and inversion of genome segments (Carniel 2008).

The *Y. pestis* chromosome contains a region called the pgm locus (102 kb), which is composed of two segments. The first segment is a high-pathogenicity island (HPI) associated with iron acquisition, involved in the synthesis of siderophores (Yersiniabactin), and essential for the dissemination of bacteria in mammals; the other segment is the hemin storage (HMS) and is involved in the colonization and blockage of the flea proventriculus by the production of biofilm. The pgm locus is unstable and may delete or block other gene segments resulting in variations in virulence. This deletion is attributed to homologous recombination resulting from the presence of IS in this region (Prentice and Rahalison 2007).

The complete genomes of various *Y. pestis* strains have been sequenced (Parkhill et al. 2001; Deng et al. 2002). Genome comparisons, including between ancestral and current strains, have elucidated mechanisms of pathogen evolution and adaptation for emerging countries and reemerging infections and have contributed to an understanding of the evolution of pathogenicity of the plague (Seifert et al. 2013; Wagner et al. 2014). They have also been useful for reconstruction of the history of the plague, observation of diversity in characteristics, and identification of new virulence factors. However, future experiments involving full characterization of the pathogenicity and lifestyle of *Y. pestis* are still needed.

11.3.2 Tularemia

The causative agent of tularemia, *Francisella tularensis* (*Francisella*: Francisellaceae), is classified into three subspecies, all of them pathogenic to humans. The first is *F. tularensis tularensis*, also called type A, which is highly virulent and occurs only in North America. It is distributed into two subpopulations (A1 and A2), with the A1 population further subdivided into A1a and A1b, each having have different geographic distributions, hosts and vectors. *Francisella tularensis holarctica* is also called type B. It is less virulent and endemic to many countries in the northern hemisphere. *Francisella tularensis mediasiatica* is equivalent to *holarctica* in virulence and can be found only in Central Asia. Although species within genus *Francisella* generally differ in virulence and geographic distribution, they have a high degree of antigenic and genomic similarity (Molins et al. 2014).

Tularemia caused by A1b strains usually exhibit severe clinical progression and are associated with high mortality rates, in contrast to infections with strains A1a, A2, or type B. Recently, subspecies A1b was also discovered in Tasmania and Australia.

Francisella tularensis is a gram-negative coccobacillus (Fig. 11.1b1) that is pleomorphic and non-spore-forming, strictly aerobic, oxidase-negative, and H_2S-positive and that ferments carbohydrates without the presence of gas. This species requires specific cultures rich in cystine and cysteine (e.g., agar blood glucose-cysteine) for growth, incubated at 37 °C for 2–4 days.

Although considered an obligatory intracellular pathogen in vivo, *F. tularensis* can be cultured in vitro, in which case proper environmental maintenance is strongly dependent on temperature. *Francisella tularensis* can survive for months at temperatures below 0 °C in water, straw, grain, and carcasses. However, survival is only possible for a few days at temperatures above 10 °C. The bacteria are killed at 56 °C in 10 min or at 45 °C in 60 min. It can be stored in the laboratory in glycerol at −80 °C or lyophilized.

Some *Francisella* species genomes have been sequenced, including *F. tularensis* (Larsson et al. 2005; Antwerpen et al. 2013). Subspecies *tularensis* has a high degree of genetic similarity with other subspecies. The *F. tularensis* genome consists of a circular chromosome of approximately 1.8 MB, with a GC content of 32.2 % and approximately 1800 genes.

The *F. tularensis* chromosome contains genes that are important for cell maintenance, carbohydrate metabolism, amino acid biosynthesis, metabolic transport, energy metabolism, DNA modifications/restrictions, and maintenance of environmental homeostasis.

Francisella tularensis pathogenicity involves several virulence factors that are important for the development of the disease, including growth in macrophages, degradation of the phagosome membrane and release into the cytosol by the action of the phospholipase family proteins, ammonia production, production of pili, and acquisition of iron. Pathogenic *Francisella* genomes contain a "Francisella pathogenicity island" (FPI) of 33.9 kb. The FPI carries 16–19 genes composing a type VI secretion system. Deletions of many of these genes within the FPI generate mutants that do not grow in macrophages and with severely attenuated virulence in mice.

Additionally, five types of insertion elements (ISFtu1–ISFtu5) have been described in the *F. tularensis* genome. The ISs are responsible for the inactivation of genes in the *Francisella* genome, as well as duplication of the FPI (Kingry and Petersen 2014).

The complete genomes of various *F. tularensis* strains and subspecies have been sequenced (Larsson et al. 2005; Antwerpen et al. 2013). The comparison among different genomes has elucidated aspects of bacterial evolution and biogeographical distribution and increased our knowledge of pathogenic factors. Access to complete genome information has also improved microbial diversity research and our understanding of the epidemiology and expansion of the disease. However, studies are still needed in some areas, such as pathogenicity factors, biodiversity, and taxonomy.

11.4 Epidemiology

11.4.1 Plague

Yersinia pestis has two distinct life cycles. The natural life cycle is known as a zoonotic-sylvatic (Fig. 11.2a), wherein the bacterium is transmitted primarily by the bite of infected fleas. When this vector encounters a similar host in urban setting, the cycle is called zoonotic-urban (Fig. 11.2b). If at some point in these

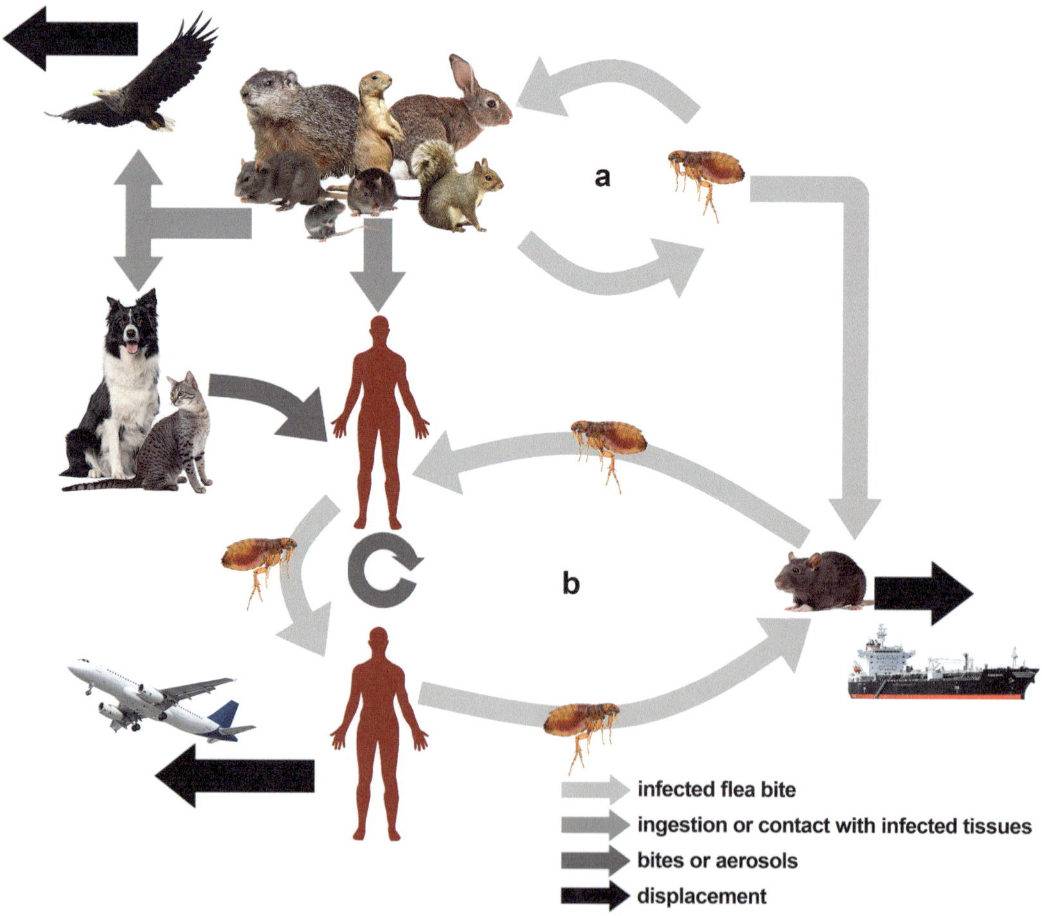

Fig. 11.2 Plague epidemiological cycles. (**a**) Sylvatic zoonotic cycle and (**b**) urban zoonotic cycle

cycles a human or other mammal (dog, cat, rabbit, camel, etc.) comes into contact with any infectious component (fleas, tissues, aerosols), they can become infected with bacteria and develop symptoms of the plague.

The main sources of plague infection in nature are wild rodents, where seemingly every region has a distinct endemic reservoir fauna. It is estimated that roughly 200 species are involved in plague epidemiological cycles. Some rodent species (e.g., *Rattus rattus*) may present a higher resistance, resulting in persistence of infection in regions such as Madagascar. Other species are subject to heavy mortality during epidemics (e.g., *Cynomys* and *Necromys* sp.) or even targets of poaching (e.g., *Marmota* spp.), increasing the risk of contamination and spread of infection among domestic animals and humans.

The primary means of plague transmission is by bite from infected fleas. More than 200 flea species have been found infected with *Y. pestis*. Among these, the "rat flea" (*Xenopsylla cheopis*) has wide geographic distribution, is highly susceptible to infection, and is thus considered the primary plague vector (Eisen and Gage 2009). Other recognized vector species include *X. brasiliensis* in Africa, India, and South America; *X. astia* in Indonesia and Southeast Asia; *X. vexabilis* in the Pacific Islands; *Nosopsyllus fasciatus*, with nearly worldwide distribution in temperate climates; *Malaraeus telchinus*, *Oropsylla montana, Opisocrostis* spp., and *Hoplopsyllus anomalus* in the United States; *Stivalius cognatus* in Indonesia; and *Polygenis* spp. in foci of Brazil and Peru. Fleas in the genus *Ctenocephalides* are also considered possible vectors in Africa and Brazil.

When infected hosts die, fleas abandon the cadaver to seek new sources of food; this new host needs not be the preferred host type, resulting in possible parasitism and infection of humans. There are three different mechanisms of transmission, which may be more or less important according to flea vector species and epizootiological stage (Hinnebusch and Erickson 2008). Biological transmission depends on bacterial proliferation and biofilm production in the intestine and proventriculus of the flea, where growth must be sufficient to block (completely or partially) these organs. The plague bacilli are then regurgitated during attempted feeding on a new host. Early transmission does not depend on the development of the bacteria in the fleas, but it too results from regurgitation of infectious material during the biting of an uninfected host. Mechanical transmission occurs when the flea feeds on a septicemic host and soon thereafter feeds on a new host, inoculating the new host with bacteria residing outside the flea mouthparts.

As described above, the HMS locus is important for colonization of the digestive tract of the flea. Phospholipase D is required for bacterial survival in biofilm and in the flea midgut, apparently because it prevents/blocks bacterial autolysis by subproducts of the digested plasma. Phospholipase D is encoded by the *ymt* gene, which is located on one of the exclusive *Y. pestis* plasmids (PMT1) essential for bacillus transmission by fleas. Other environmental and bacterial factors involved in flea colonization and biofilm formation are poorly understood.

In addition to acting as plague vectors, fleas act as reservoirs of zoonosis, playing an important role in the maintenance of disease due to their ability to live for months within rodent holes/habitats. This microclimate favors *Y. pestis* survival, which is eliminated with flea feces and deposited on the ground. Recent studies indicate that blow flies may act as mechanical vectors of *Y. pestis*, but the epidemiological impact of these organisms is still being discussed.

Pet dogs and pet cats may carry *Y. pestis*-infected fleas and can also develop the infection by acting as predators (e.g., by eating infected rodents). Dogs usually do not develop clinical manifestations, while cats may show the nodal forms, including pneumonic and pharyngeal disease. If they survive the infection, these animals will produce specific antibodies for up to a year, making them important sentinels for epidemiological surveillance in plague outbreaks. Birds in general are refractory to infection, but birds of prey are important in the epidemiological cycle because they may transport infected rodent carcasses and fleas over long distances. Transmission between humans can occur via aerosols due to accidents involving contaminated tissues (e.g., during field work or from lab material) or during utilization of plague bacteria as a weapon of biological warfare. Displacement of humans or animals infected with plague (or carcasses) should also be taken into consideration, because of the potential for transmission and formation of allochthonous epidemics. Transmission between humans by vectors occurs rarely and is characteristic of homes with ectoparasite infestations, such as lice (*Pediculus humanus*) and fleas (*Pulex irritans*). In the Andes, the habit of killing these insects with the teeth causes a syndrome characterized by cervical buboes, peritonsillar abscesses, and pneumonia.

11.4.2 Tularemia

Francisella tularensis infection has been demonstrated in more than 250 vertebrates (mammals, birds, reptiles, amphibians) and invertebrates (arthropods) around the world. Thus, there is no exclusive reservoir/host association for the disease. The fauna involved in epidemiological cycles varies from region to region, where mammalian hosts (e.g., rabbits, hares, muskrats, prairie dogs, skunks, raccoons, mice, rats, squirrels, sheep, cattle, and cats) are more commonly associated with human infection risk (Harik 2013).

Two biological cycles are currently described: terrestrial and aquatic (Carvalho et al. 2014). Tularemia type A (*F. tularensis* subsp. *tularensis*) is more associated with the terrestrial disease cycle (Fig. 11.3a), with wild rabbits and hares

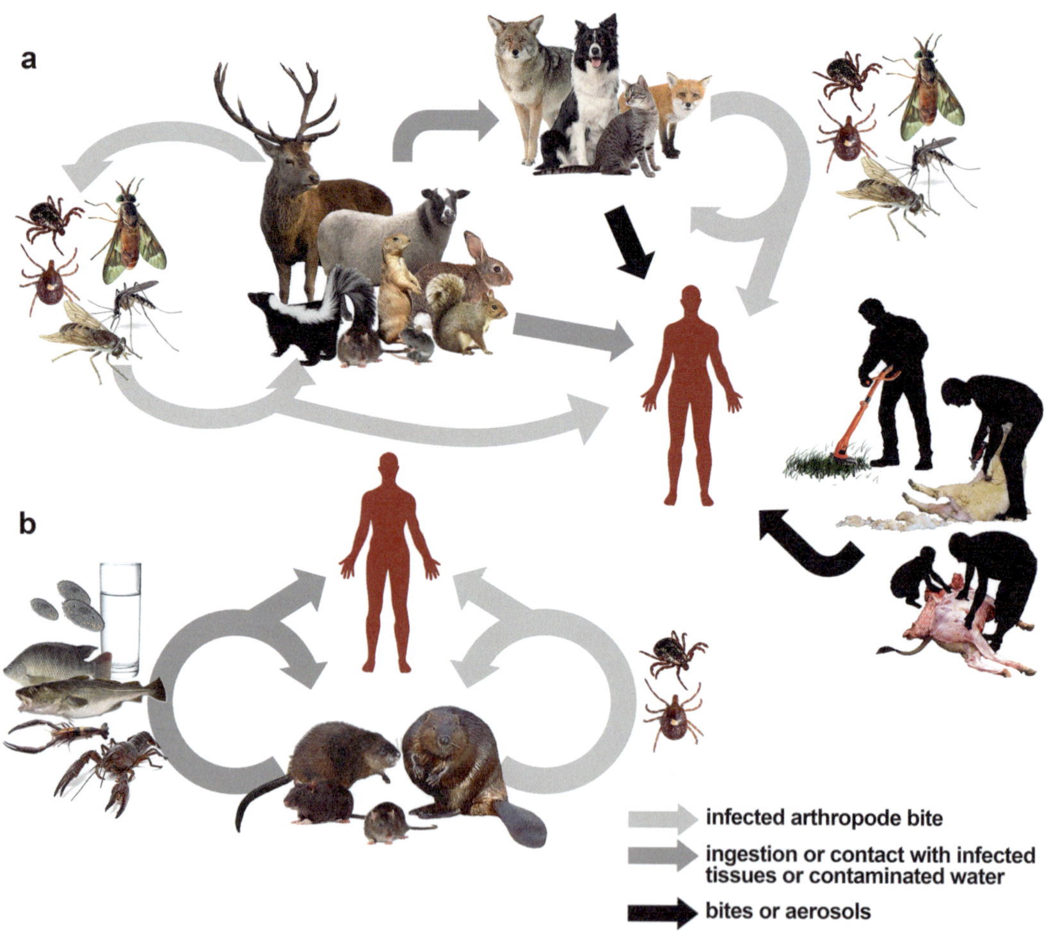

Fig. 11.3 Tularemia epidemiological cycles. (**a**) Terrestrial cycle and (**b**) aquatic cycle

acting as main hosts and arthropod vectors (ticks and horse flies) as the agents spreading the pathogen. Tularemia type B (*F. tularensis holarctica*) is most often associated with aquatic cycles (Fig. 11.3b) but can also be transmitted by ticks. In this cycle, muskrats, beavers, and voles spread the bacteria in their aquatic habitat, and humans become infected directly by ingesting contaminated water or infected animals (e.g., fish and crayfish). There is also evidence that *F. tularensis* may persist in water sources in association with intracellular protozoa, aiding epidemiological maintenance.

In North America, ticks of the genera *Dermacentor* (dog tick and wood tick), *Amblyomma americanum* (lone star tick), and *Chrysops* spp. (deer flies) are the main vectors of tularemia. The animal reservoirs consist of lagomorphs, squirrels, prairie dogs, beavers, groundhogs, deer, sheep, and occasionally humans. Wild predators, such as foxes, coyotes, birds of prey, and snakes, as well as domestic dogs and cats (by nature of carnivorous behavior), typically become infected by coming into contact with infected tissues or prey.

In ticks, bacteria penetrate the midgut and migrate to the salivary glands, where they are inoculated in a new host during a blood meal. Transovarial infection has been confirmed in ticks, and they are known to remain infective during all phases of their life cycle (larva, nymph, adult). It is believed that mites and fleas are also involved in the transmission of tularemia but with lower epidemiological impact.

In Europe and in other regions of the world, the ticks *Ixodes ricinus*, *Haemaphysalis concinna*, and *Dermacentor* sp. are involved in the transmission of tularemia. Mosquitoes of the genera *Aedes*, *Culex*, and *Anopheles* are also implicated; these hematophagous insects act as mechanical vectors for the spread of bacteria, which occur via contaminated mouthparts during blood feeding after previous contact with an infected host.

Another way in which humans acquire *F. tularensis* is through contact with contaminated tissues or carcasses or by inhaling aerosols from contaminated land, grain, or hay during hunting activities, taxidermy, shearing, field work, land clearing, grass cutting, during the trading of wild animals, or via laboratory accidents. Interhuman transmission has not been documented for tularemia, even in cases of the pneumonic form. As with the plague, epizootics occurring in rodents and lagomorphs typically precede human cases, but the detection of infection sources can be extremely difficult due to host diversity.

11.5 Clinical Manifestations

The manifestation of these diseases depends on host inoculation route. The strength of invasion and the ability of the bacterium to infect and cause disease vary among different clinical presentations.

11.5.1 Plague

Clinically, human plague has three main forms: bubonic, septicemic, and pneumonic.

11.5.1.1 Bubonic Plague

Bubonic plague is the most common form of the disease, contracted after the bite of an infected flea. When bacilli are inoculated, buboes form in the lymph nodes, frequently in the inguinal-crural (70 %), axillary (20 %), and cervical (10 %) regions. In the absence of appropriate treatment, mortality rate ranges between 40 and 70 % (Koirala 2006; Prentice and Rahalison 2007). The buboes have a reddish color, glossy surface and are very painful (Fig. 11.1a2). The disease presents with chills, headache, fever, myalgia, anorexia, nausea, vomiting, generalized pain, and tachycardia (Prentice and Rahalison 2007). The incubation period is between 2 and 6 days and faster in some cases (Carniel 2008).

11.5.1.2 Septicemic Plague

Septicemic plague is characterized by severe toxemia and an abundance of bacilli in the blood. Bacteremia can result in necrosis due to accumulation of bacilli and their toxins in the limbs (Fig. 11.1a3), being lethal in 30–50 % of cases with improper treatment. In general, 10–20 % of cases result from progression of improperly treated bubonic plague (Koirala 2006). Clinically, septicemic plague resembles other types of septicemia caused by gram-negative bacteria. It presents a clinical picture with abrupt temperature rise, chills, arrhythmia, hypotension, nausea, vomiting, asthenia, and mental disturbance. This form has a high mortality if left untreated (Koirala 2006; Prentice and Rahalison 2007).

11.5.1.3 Pneumonic Plague

In clinical form, pneumonic plague is the most severe and dangerous, due to the high degree of contagion, high mortality rates (approximately 100 % if untreated and >50 % with antibiotic therapy), and high capacity to cause epidemics by spread of aerosols. Infection may occur by the introduction of bacilli into respiratory mucosa or secondarily by progression of untreated bubonic and septicemic forms (Carniel 2008). This form presents sudden onset with severe and rapid symptom progression, with abrupt temperature rise, chills, arrhythmia, hypotension, nausea, vomiting, and mental disturbance (Koirala 2006, Prentice and Rahalison 2007). Initially, signs of infection and pulmonary symptoms are either mild or absent and later arise with chest pain, shortness of breath and fast, bloody, and sputum rich in plague bacilli. Without early treatment, symptoms include delirium and coma and may cause death.

11.5.2 Tularemia

Depending on the route of entry of the bacteria, the disease can manifest itself in several ways (Ellis et al. 2002). Ulceroglandular tularemia is the most common form of the disease, which typically results from the bite of an infected arthropod vector or through direct contact with the tissues of infected animals. An ulcer can form and persist for several months at the site of infection (Fig. 11.1b2). After an incubation period of 3–6 days, infected persons have sudden onset of flu-like symptoms, mainly chills, fever, headache, and general aches/pains. Bacteria are spread through the lymphatic system to local lymph nodes where swelling occurs, the result often resembling plague buboes (Ellis et al. 2002). *Francisella tularensis* can be transmitted from the site of infection to other organs/tissues, such as the spleen, liver, lungs, kidneys, gut, central nervous system, and skeletal muscles, with less than 3 % mortality (Evans et al. 1985). A rare variation of this clinical presentation is oculoglandular tularemia, wherein the entrance of the bacterium is the conjunctiva, usually as a result of hand-eye contamination. The disease is characterized by the appearance of ulcers or nodules on the conjunctiva, with untreated infection spreading to local lymph nodes (Ellis et al. 2002).

Oropharyngeal or gastrointestinal tularemia occurs through ingestion of contaminated food or water (Berdal et al. 2000). The symptoms are described as a sore throat, enlarged tonsils, and formation of a yellow-white pseudomembrane. This form is usually accompanied by swelling of cervical lymph nodes. This clinical presentation may vary from mild diarrhea to fatal acute disease with intestinal ulceration (Ellis et al. 2002).

Pulmonary and respiratory tularemia is the most severe form, with the inhalation of aerosols as the (rare) route of infection. Any cases of natural inhalation often appear from agricultural activities that involve manipulation of hay, a common habitat of infected rodents. In most cases, this form of tularemia occurs as a clinical progression from other forms (Ellis et al. 2002).

11.6 Diagnosis

11.6.1 Plague

The diagnosis of plague can be accomplished through bacteriological, serological, and molecular techniques. These techniques can be used separately or in combination, which allows greater precision of results. For plague diagnosis, the isolation of the bacteria is still considered the gold standard (BCCDC, BC Centre for Disease Control 2013). However, not all samples from human cases are appropriate for this type of diagnosis. Automated systems for biochemical testing are not considered effective in the identification of *Y. pestis*; however, biochemical tests performed with the API20E kit are accepted.

Laboratory investigation consists of bacterioscopy of smears stained using specific methods, direct immunofluorescence (DIF) with specific antibodies for F1, use of antiplague phage plaques on agar (by determining a lysis area to confirm the culture), and inoculation into laboratory animals (BCCDC, BC Centre for Disease Control 2013).

Depending on observed clinical manifestations, different sample types (e.g., bubo fluid, blood, sputum, cerebrospinal fluid, etc.) are collected for diagnostic purposes. If the patient has the bubonic form of the disease, blood and bubo fluid are collected. In cases of pneumonic plague, blood and sputum fluid are collected. In recently obtained infection, it is possible to obtain material from the mouth or from nasal secretions or through a venous or bubo puncture. Samples are packaged in the Cary-Blair transport medium, and bacilli may be maintained in these preparations for several weeks at room temperature. In rodents, the blood, liver, or spleen samples are collected. If the animal is decomposing, the femur is obtained to sample bone marrow, where plague bacilli are abundant. These techniques can also be used to identify *Y. pestis* in flea vectors (BCCDC, BC Centre for Disease Control 2013).

In cases where bacterial culture presents negative results but disease is still suspected, serological testing is recommended to confirm diagnosis. Serological diagnosis is mainly by hemagglutination (HA) assays, typically via hemagglutination inhibition (HI) tests. F1-sensitive sheep erythrocytes are used for the detection of anti-F1 antibody in blood serum of patients or people at risk of contact with animals. For diagnosis of plague in humans, it is recommended to collect two serum samples: one in the acute phase of the disease and another after 3 weeks. In most patients, seroconversion occurs within 1–2 weeks of the onset of symptoms, occasionally earlier (~5 days). The disease is confirmed when there is a fourfold increase in reaction titer.

The HA/HI assay is also used for diagnosis and serological surveillance of serum samples from live-captured rodents and for domestic and wild carnivore predators of rodents. The enzyme-linked immunosorbent assay (ELISA) is useful for diagnosis of human disease, used in detection of IgM or IgG and for F1 antigen capture. However, these techniques are not universally applicable in serological surveillance activities, as they require the use of different (nonhuman) species-specific antibodies, such as from rodents, dogs, cats, etc., for use in human sera tests.

Molecular techniques have some advantages over bacteriological techniques in plague diagnosis. Molecular assays are rapid techniques that eliminate the need for cultivation of samples and in which detection is possible even when bacteria are not viable. Several protocols based on polymerase chain reaction (PCR) and its variations (nested PCR, multiplex PCR, real-time PCR) have been used for this purpose. These tests are used for diagnosis in human or animal material, by means of identifying virulence genes present in the three plasmids and/or on the chromosome of *Y. pestis* (Dalmasso et al. 2014).

11.6.2 Tularemia

Tularemia diagnosis can also be done by bacteriological, serological, and molecular methods. *Francisella tularensis* is a challenging pathogen in relation to nutrients and has slow growth. Despite these characteristics, bacterial isolation remains the most recommended method, since it allows a conclusive diagnosis of infection and serves as a valuable resource for molecular epidemiology (Hepburn and Simpson 2008).

Appropriate sample type for use in laboratory diagnosis depends on patient clinical manifestations and can include sore throat swabs, blood or urine samples, biopsy, aspiration or scraping of ulcers, lymph node tissue, affected corneal tissue, sputum samples, and bronchial or pleural lavage. Samples of vectors, carcasses, natural host or reservoir feces, and water are all recommended for epidemiological investigations (BCCDC, BC Centre for Disease Control 2013).

Different sample types require different preparations for transportation. Skin lesion samples are stored in the Amies agar with activated charcoal. Sputum, pharyngeal secretions, and lymph node aspirate may be packed, transported, and stored at −80 °C until culture procedures for isolation of the microorganism are performed. Traditional blood transport systems are used for blood samples.

Alternatively, specimens can be plated in appropriate solid media immediately after collection, and sealed plates may then be forwarded to a biosafety level 3 (BSL3) laboratory. Bacteria grow well on chocolate agar medium supplemented with cystine/cysteine (CHAB) II or GC agar with 1 % hemoglobin and 1 % IsoVitaleX plus antibiotic when the specimen is subject to contamination by other microorganisms. Growth can be visualized after 24–48 h at 37 °C, and small, transparent colonies can be identified by slide agglutination (5 min) or by immunofluorescence using specific antibodies (IgG3, IgG2a, IgG1).

Commercially available biochemical test systems for the identification of *F. tularensis* fail, and conventional biochemical tests are limited by the nutritional requirements of the bacteria. Currently, conventional techniques and quantitative PCR are used to identify genes encoding *F. tularensis* outer membrane proteins. PCR can also detect the microorganisms directly on tissues, including skin lesions, lymph nodes, spleen, liver, kidneys, heart, lungs, oropharynx, and meninges. Recently, we developed a "toolbox" which includes all the tools, equipment, and solutions required for conventional PCR identification of bioterrorism agents in the field; the development of this system is a breakthrough for detection of this pathogen in outbreak situations.

In terms of clinical management and treatment, identification of *F. tularensis* at the species level is sufficient. In other situations, it is important to extend the studies to distinguish between the various *F. tularensis* subspecies, particularly subsp. *tularensis* and *holarctica*. This is because when they are endemic, the difference in virulence between these two particular subspecies impacts treatment and control methods, as well as the types of molecular tests that should be used.

Tularemia diagnosis can also be performed by serological methods. Antibody production in response to *F. tularensis* infection is commonly demonstrated by several methods, including agglutination tests, immunofluorescence assays, ELISA, Western blot (WB), and microarray. The most commonly used method is the microscopic agglutination (MA); however, this method has some limitations (Chaignat et al. 2014).

After *F. tularensis* infection, antibodies appear following the end of the 2nd week of onset of symptoms, reaching titers of 1:320–1:1280; antibodies decline 2–3 months after treatment. In some cases, even after clinical recovery, antibody titers may persist for several years. Two serum samples should be collected for proper diagnosis, one in the acute phase of the disease and another in a few weeks.

Another utilized technique corresponds to ELISA but allows differentiation of immunoglobulin class; however, in tularemia, this does not differentiate between an acute episode and prior infection. Alternately, there are commercially available immunochromatographic methods, which detect specific IgM, IgG, and IgA antibodies against *F. tularensis* lipopolysaccharide (LPS). These methods have high sensitivity and specificity and represent an alternative to presumptive diagnoses of tularemia in remote areas (Sharma et al. 2014).

All diagnostic procedures for both plague and tularemia should be performed in laboratories that have specialized facilities, due to the risk that these pathogens present (Fig. 11.4). Cultures and animal inoculations must be conducted in biosafety level 3 (BSL3) laboratories, while serologic tests can be performed in level 2 (NB2).

11.7 Treatment

The treatment of plagueearly and intensive due to the gravity and speed of progression of these diseases and the possibility for *Y. pestis* and *F. tularensis* use as bioterrorism agents. For plague, patient isolation is recommended during the first 48 h of treatment for pneumonic forms to minimize the risk of contamination.

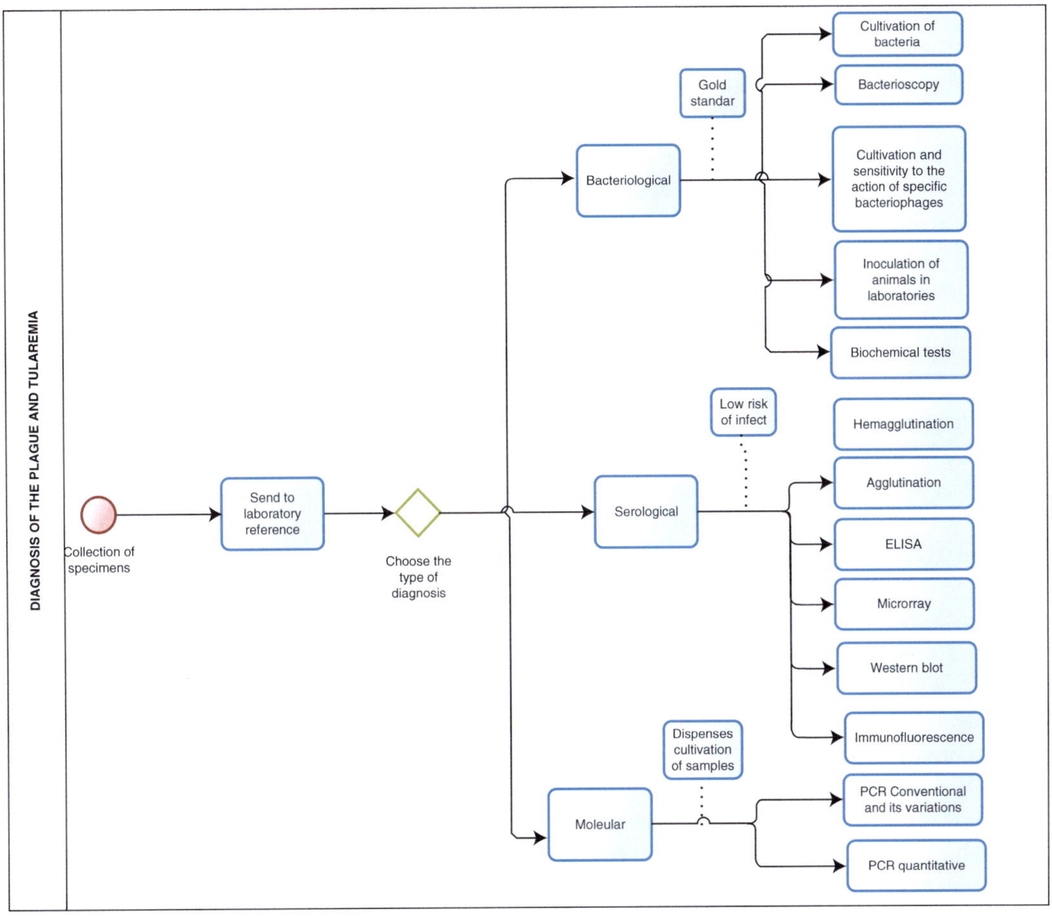

Fig. 11.4 Diagram of the different methods used for the diagnosis of plague and tularemia

Both diseases may be treated with the antimicrobial aminoglycosides streptomycin and gentamicin, which are basically functionally equivalent. Both antibiotics are considered the drugs of first choice. In cases of plague, streptomycin and gentamicin use is recommended for a period of 10 days. Streptomycin is considered the gold standard, especially in cases of pneumonic plague, whereas gentamicin is prescribed for all forms of the disease, especially for pregnant women and children. Currently, gentamycin is the drug of choice for all clinical forms of plague and for tularemia (BCCDC, BC Centre for Disease Control 2013). Tularemia treatment should begin within 24 h of first symptoms and should continue for 14 days to reduce the risk of relapse (Harik 2013).

Tetracycline and chloramphenicol are second choice treatments. Tetracycline is prescribed in uncomplicated cases of plague, while chloramphenicol can be used for all clinical forms. These treatments are effective for tularemia if used for 14–21 days. In the absence of these antibiotics, sulfonamides may be used in uncomplicated cases of plague (BCCDC, BC Centre for Disease Control 2013).

The beta-lactam antibiotics (penicillins, cephalosporins, cephamycins, oxacephems, carbapenems, and monobactams), macrolides (erythromycin, clarithromycin, roxithromycin, and miocamycin), and azalides (azithromycin) should not be employed in the treatment or prophylaxis of plague or tularemia (Harik 2013). Despite the antibiotic susceptibility demonstrated for strains of *Y. pestis*, these antibiotics are ineffective in vivo. Their use may thereby increase the risk of complications (sepsis, pneumonia, and meningitis), including death.

Prophylaxis is recommended for individuals who were exposed to aerosols containing *Y. pestis* or *F. tularensis*. In these situations, doxycycline should be initiated immediately after exposure or up to a maximum of 24 h postexposure. However, other tetracyclines and chloramphenicol are also recommended for exposure to patients with pneumonic plague and after contact with infected fleas or bacteria. The fluoroquinolones (ofloxacin, levofloxacin, and ciprofloxacin) may be used in accordance with sensitivity profile and in the case of terrorist attack with *Y. pestis* (BCCDC, BC Centre for Disease Control 2013).

11.8 Prevention and Control

For plague and tularemia, adequate prevention through immunization is not yet possible. The available plague vaccine is rarely used because the immunity is short-lived, ineffective against bubonic plague, and fails to protect against the primary pneumonic form. For tularemia, a vaccine developed from live-attenuated microorganisms (derived from an avirulent strain) is used in the United States. This vaccine is still being evaluated for effectiveness, and its utilization is limited to professionals with high risk of exposure, such as laboratory workers that handle the microorganism (CDC 2012; WHO 2003).

For both of these zoonosis, prevention and control measures should be based on knowledge of regional characteristics, including ecological and epidemiological aspects; the presence of hosts/reservoirs and vectors; and degree or type of human exposure to infection sources.

For people who live, work, or vacate in areas of zoonosis occurrence, some measures should be taken to avoid contracting the infection and to prevent spread (CDC 2012; WHO 2003). They are:

- Eliminate sources of food and shelter for rodents near homes, work places, and camps/recreational areas. When present in these environments, employ methods to eliminate them. Vectors should be eliminated prior to rodents, since without their usual host vectors tend to invade the home environment.
- Control ectoparasites (fleas, ticks, and horseflies) of pet animals, and prevent pets from coming into contact with wild animals or their carcasses. Animals kept outdoors are more exposed to this contact.
- Avoid areas infested by arthropod vectors (fleas, ticks, and horse flies). In the case of unavoidable exposure, use appropriate repellent, long-sleeved shirts, long pants tucked inside socks, closed shoes, and wide-brimmed hat to protect the face and neck. Check for fleas and ticks on clothing or skin, and if present, remove them.
- Avoid contact with wild rodents and their nests/burrows, droppings, and carcasses. Raising guinea pigs in or near the home for human consumption is particularly risky and should be avoided. Avoid keeping exotic animals (hamsters, gerbils, chinchilla, ferret) as “pets,” as they may be carriers of various infectious agents including but not limited to *Y. pestis* and *F. tularensis* and may carry infected arthropod vectors into homes, thereby contaminating dogs, cats, and humans.
- Be careful when handling lagomorphs, rodents, and other wild animals caught during hunting activities. When removing or abrading the skin of these animals, wear gloves, appropriate mask that covers the nose and mouth and eye protection (such as goggles). Wash hands, utensils, and surfaces thoroughly after contact with hunting-acquired raw meat, and thoroughly cook meats prior to consumption.

A public health surveillance system, when properly designed, can identify risk areas, prevent human infection, promote early detection of human cases or outbreaks, reduce mortality, and prevent the occurrence of epidemics (Table 11.1).

A global outbreak alert was established by the WHO to support the implementation of the international aspects of the International Health Regulations. This alert system depends on systematic monitoring of information concerning suspected outbreaks and on providing support for outbreak responses. This support

Table 11.1 Recommended actions for control of plague and tularemia

	Action	Plague	Tularemia
Ecological surveillance	Systematic survey of natural foci	Investigate the rodent and flea populations and the presence of bacteria among them	Investigate environmental sources of infection: water, soil, vegetation, arthropods, and hosts
	Investigation of epizootics	Collect and identify rodent carcasses for laboratory analysis Actively search for suspicious related human cases	Collect and identify susceptible rodents; collect water and soil samples from the area where dead animals are found for laboratory analysis
Human surveillance	Investigation of human cases	Early diagnosis and treatment	
	Immediate notification	Mandatory for doctors and/or local public health service laboratories Occurrence of pneumonic plague and increasing frequency of the pulmonary form of tularemia may indicate an act of bioterrorism	
	Isolation of patients	For 48 h after beginning treatment and only for the pneumonic form	Not necessary
	Disinfection	Sputum, secretions, and fomites	Secretions of ulcers, conjunctival sac, and lymph nodes
	Chemical prophylaxis	Recommended for individuals exposed to patients with pneumatic plague or after exposure to infected fleas or direct contact with bacteria	Recommended for individuals exposed to aerosols containing *F. tularensis*

involves the provision of technical assistance to contain the national and international public health consequences of outbreaks and is offered by the WHO upon request and acceptance by member states.

References

Achtman M, Zurth K, Morelli G et al (1999) *Yersinia pestis*, the cause of plague, is a recently emerged clone of *Yersinia pseudotuberculosis*. Proc Natl Acad Sci U S A 96(24):14043–14048

Antwerpen MH, Schacht E, Kaysser P et al (2013) Complete genome sequence of a *Francisella tularensis* subsp. *holarctica* strain from Germany causing lethal infection in common marmosets. Genome Announc 1(1):e00135-12. doi:10.1128/genomeA.00135-12

BCCDC, BC Centre for Disease Control (2013) Available in www.bccdc.ca/dis-cond/a-z/_p/Plague/overview/Plague.htm

Berdal BP, Mehl R, Haaheim H et al (2000) Field detection of *Francisella tularensis*. Scand J Infect Dis 32:287–291

Carniel E (2008) Pathogenic *Yersinia*: stepwise gain of virulence due to sequential acquisition of mobile genetic elements. In: Hensel M, Schmidt H (eds) Horizontal gene transfer in the evolution of pathogenesis, vol 8. University Press, Cambridge, pp 193–216

Carvalho CL, Lopes de Carvalho I, Zé-Zé L et al (2014) Tularemia: a challenging zoonosis. Comp Immunol Microbiol Infect Dis 37:85–96

Centers for Disease Control and Prevention (CDC) (2011) Available in www.cdc.gov/tularemia/

Centers for Disease Control and Prevention (CDC) (2012) Available in www.cdc.gov/plague/

Chaignat V, Djordjevic-Spasic M, Ruettger A et al (2014) Performance of seven serological assays for diagnosing tularemia. BMC Infect Dis 14:234. doi:10.1186/1471-2334-14-234

Dalmasso A, Civera T, Filipello V et al (2014) Multiplex primer-extension assay for identification of *Yersinia* species. Mol Biol Rep 41(10):6329–6333

Deng W, Burland V, Plunket G III et al (2002) Genome sequence of *Yersinia pestis* KIM. J Bacteriol 184:4601–4611

Eigelsbach HT, McGann V (1984) Francisella. In: Krieg WR, Holt JG (eds) Bergey's manual of systematic bacteriology, vol 1. Williams and Wilkins, Baltimore, MD, pp 394–399

Eisen RJ, Gage KL (2009) Adaptive strategies of *Yersinia pestis* to persist during inter-epizootic and epizootic periods. Vet Res 40(2):1

Ellis J, Oyston PC, Green M et al (2002) Tularemia. Clin Microbiol Rev 15(4):631–646

Evans ME, Gregory DW, Schaffner W et al (1985) Tularemia: a 30-year experience with 88 cases. Medicine (Baltimore) 64(4):251–269

Harik NS (2013) Tularemia: epidemiology, diagnosis, and treatment. Pediatr Ann 42:288–292

Hepburn MJ, Simpson AJ (2008) Tularemia: current diagnosis and treatment options. Expert Rev Anti Infect Ther 6:231–240

Hinnebusch BJ, Erickson DL (2008) *Yersinia pestis* biofilm in the flea vector and its role in the transmission of plague. Curr Top Microbiol Immunol 322:229–248

Kingry LC, Petersen JM (2014) Comparative review of *Francisella tularensis* and *Francisella novicida*. Front Cell Infect Microbiol 4:1–12

Koirala J (2006) Plague: disease, management, and recognition of act of terrorism. Infect Dis Clin North Am 20:273–287

Larsson P, Oyston PCF, Chain P et al (2005) The complete genome sequence of *Francisella tularensis*, the causative agent of tularemia. Nat Genet 37:153–159

Molins CR, Delorey MJ, Yockey BM et al (2014) Virulence difference between the prototypic Schu S4 strain (A1a) and *Francisella tularensis* A1a, A1b, A2 and type B strains in a murine model of infection. BMC Infect Dis 14:67. doi:10.1186/1471-2334-14-67

Parkhill J, Wren BW, Thomson NR et al (2001) Genome sequence of *Yersinia pestis*, the causative agent of plague. Nature 413(6855):523–527

Perry R, Fetherston JD (1997) *Yersinia pestis* – etiologic agent of plague. Clin Microbiol Rev 10:35–66

Platonov ME, Evseeva VV, Dentovskaya SV et al (2013) Molecular typing of *Yersinia pestis*. Mol Gen Mikrobiol Virusol 2:3–12

Prentice MB, Rahalison L (2007) Plague. Lancet 369(9568):1196–1207

Seifert L, Harbeck M, Thomas A (2013) Strategy for sensitive and specific detection of *Yersinia pestis* in skeletons of the Black Death pandemic. PLoS One 8(9), e75742. doi:10.1371/journal.pone.0075742

Sharma N, Hotta A, Yamamoto Y et al (2014) Serosurveillance for *Francisella tularensis* among wild animals in japan using a newly developed competitive enzyme-linked immunosorbent assay. Vector Borne Zoonotoic Dis 14:234–239

Wagner DM, Klunk J, Harbeck M et al (2014) *Yersinia pestis* and the plague of Justinian 541–543 AD: a genomic analysis. Lancet Infect Dis 14:319–326

World Health Organization (WHO) (2003) Human plague in 2000 and 2001. Wkly Epidemiol Rec 78:130–135

Suggested Reading

Almeida AMP, Alves LC, Amaral RL et al (2003) Transmission of *Yersinia pestis* cultures with different plasmid content from *Xenopsylla cheopis* to *Calomys callosus*. Parasitol Res 89:159–162

Almeida AMP, Tavares C, Leal-Balbino TC (2005) Peste. In: Coura JR (ed) Dinâmica das doenças infecciosas e parasitárias, vol 126, 2nd edn. Guanabara Koogan, Rio de Janeiro, Brasil, pp 1509–1522

Dunkelberg H (2004) The history of the plague and the research on the causative agent *Yersinia pestis*. Int J Hyg Environ Health 207:165–178

Leal-Balbino TC, Leal NC, Lopes CV et al (2004) Differences in the stability of the plasmids of *Yersinia pestis* cultures in vitro: impact on virulence. Mem Inst Oswaldo Cruz 99:727–732

Rickettsial Infections, Bartonella Infections, and Coxiellosis

12

Maria Ogrzewalska, Tatiana Rozental,
Alexsandra R.M. Favacho,
and Maria Angélica Monteiro de Mello Mares-Guia

Abstract

About 20 years ago, various diseases caused by bacteria *Rickettsia*, *Orientia*, *Ehrlichia*, *Anaplasma*, *Coxiella*, and *Bartonella* were all denominated as rickettsial infections. Nowadays, based on genetic analysis, these bacteria are reorganized and considered as distinct entities. The order *Rickettsiales* includes at the moment two families: *Anaplasmataceae* and *Rickettsiaceae*, and the family *Bartonellaceae*, as well as *Coxiella burnetii*, are no more considered rickettsiae. Diseases caused by bacteria from the *Anaplasmataceae* family are presented in Chap. 15, and here diseases caused by *Rickettsia* and *Orientia* are presented. Although *Bartonella* spp. and *Coxiella* spp. are not anymore considered rickettsial disease, they are often still studied within the field of rickettsiology and thus are also presented here.

Keywords

Spotted fever • *Rickettsia* • *Bartonella* • *Coxiella* • Q fever • Typhus • Ectoparasites

12.1 Rickettsial Diseases (Rickettsioses)

12.1.1 Introduction

Rickettsial diseases (rickettsioses) are diseases caused by bacteria from the genera *Rickettsia* and *Orientia*. Currently, there are 26 species within the genus *Rickettsia* with validated and published names. However, some species are divided into subspecies, and more than 70 strains remain unidentified or incompletely described pathogens. Moreover, each year still new species or stains are being discovered. The name of the genus

M. Ogrzewalska (✉) • T. Rozental • A.R.M. Favacho
M.A.M. de Mello Mares-Guia
Laboratory of Hantaviroses and Rickettsioses,
Oswaldo Cruz Foundation, Oswaldo Cruz Institute,
Av. Brasil 4365, Manguinhos, 21040-360
Rio de Janeiro, RJ, Brazil
e-mail: maria.ogrzewalska@ioc.fiocruz.br;
rozental@ioc.fiocruz.br; afavacho@ioc.fiocruz.br;
amguia@ioc.fiocruz.br

C.B. Marcondes (ed.), *Arthropod Borne Diseases*, DOI 10.1007/978-3-319-13884-8_12

Rickettsia was initially given by Brazilian infectologist Henrique da Rocha Lima (1879–1956) to intracellular microorganisms found in lice, *Rickettsia prowazekii*, the agent of epidemic typhus, in honor of American pathologist Howard Taylor Ricketts (1871–1910) and Czech zoologist and parasitologist Stanislaus Josef Mathias von Prowazek (1875–1915), who both died of typhus while studying the disease (see Chap. 3). Ricketts is remembered for discovering the causative organism and the transmission route of Rocky Mountain spotted fever, while von Prowazek and Rocha Lima discovered that typhus was transmitted to humans by the feces of lice, and the last described the bacteria. The genus *Orientia* on the other hand is represented only by one species, *Orientia tsutsugamushi*, the causative agent of scrub typhus. The name comes from the Orient, the area where the organism is widely distributed, and from "tsutsugamushi" that means in Japanese "dangerous mite".

12.1.2 Characterization

They are obligate intracellular bacteria which means that they can reproduce only within animal cells. The members of *Rickettsiaceae* family are maintained in nature through complex cycles involving reservoir in mammals and arthropod vectors such as ticks, mites, and lice. Except for *R. prowazekii*, humans are only accidental hosts for rickettsiae and usually get infected through bites or feces of infected arthropods. Some of these diseases are benign for humans, while others may be fatal. In some cases different isolates of the same species vary in virulence for the same host. Some rickettsioses, such as epidemic typhus caused by *R. prowazekii*, have been described since the sixteenth century, while others have been discovered recently. Many other species exist in arthropods without presently being associated with human disease.

12.1.3 Taxonomy and Life Cycle

The rickettsiae are rod-shaped, small (0.3–0.5 μm by 0.8–2.0 μm), Gram-negative, non-spore-forming bacteria. The genus *Rickettsia* has been divided into four major groups based on their antigenic and genetic characteristics: (1) the spotted fever group (SFG), (2) the typhus group (TG), (3) the *Rickettsia bellii* group, and (4) the *Rickettsia canadensis* group. However, this classification is changing as new data are available. The *Rickettsia* with medical importance belongs to the SFG and the TG. The SFG rickettsiae are mainly associated with mites and ticks while the TG rickettsiae with insects such as fleas and lice.

12.1.3.1 Typhus Group *Rickettsia*

The typhus group includes *Rickettsia prowazekii* and *Rickettsia typhi*.

- *Rickettsia prowazekii* has been found worldwide and causes epidemic typhus, known also as louse-borne typhus fever, typhus exanthematicus, classical typhus fever, European typhus, or jail fever. Due to the difficulty of making a precise clinical diagnosis of epidemic typhus and the similarity of its symptomatology to other fevers prevalent in the past, its origins remain uncertain; however, it is possibly one of the oldest pestilential diseases of mankind and responsible for millions of deaths worldwide. In most parts of the world, humans are the only reservoir host for *R. prowazekii*. The primary vector in person-to-person transmission is the human body louse (*Pediculus humanus*). Lice become infected when they feed on the blood of infected patients and later infected lice inoculate *R. prowazekii* via bite into healthy humans. The pathogen is also present and able to survive in the feces of infected lice for weeks; thus, the transmission may occur by inhalation or contact of infected feces or even by crushing lice when the pathogen enters into a contact with the mucous membranes of the mouth and eyes. Infections can become latent and later recrudesce (referred to as Brill-Zinsser disease). Humans with recrudescent typhus are capable of infecting lice and spreading the disease. Epidemics of typhus usually occur where louse populations are high; therefore, infections are typically seen not just in populations living in unsanitary, crowded conditions and prisons and refugee camps but also in the homeless louse-infested population in developed countries, and outbreaks are often associated with wars, famines, floods, and other disasters. Most epidemics occur during the colder months. In North America *R. prowazekii* has been recently found in flying squirrels (*Glaucomys volans*), and the disease (referred as indigenous epidemic typhus or sylvatic typhus) may be acquired from flying squirrel-parasitizing arthropods.

- *Rickettsia typhi* causes murine typhus, known also as endemic typhus, Mexican typhus, or flea-borne typhus. Murine typhus usually occurs sporadically rather than in epidemics and is widely distributed, especially throughout the tropics and subtropics. Rats (*Rattus norvegicus* and *R. rattus*) are the animal reservoirs. The infection is spread when fleas, mainly the oriental rat flea (*Xenopsylla cheopis*), feed on infected rats. *Rickettsia typhi* infects the gut epithelial cells and is excreted in the feces. Transmission to humans occurs when the skin, respiratory tract, or conjunctiva is contaminated with infected flea feces. Infection via inhalation of dust from rat-infested buildings may occur and also the organism can be transmitted by fleabites alone. Endemic areas are primarily urban settings associated with commensal rats and their fleas. Seaports and coastal areas are favored; however, rural areas may also be affected. People whose occupation or living conditions bring them into close contact with rats are primarily affected, especially in food-storage areas or granaries. Most human cases are acquired indoors where rats are present.

12.1.3.2 Spotted Fever Group *Rickettsia*

The spotted fever group (SFG) includes etiologic agents of various diseases such as Rocky Mountain spotted fever, Mediterranean spotted fever, tick typhus, or rickettsialpox. With the exception of etiologic agent of rickettsialpox (*Rickettsia akari*) which is transmitted by mites and *Rickettsia felis* transmitted by fleas, the SFG rickettsiae are transmitted to humans by the bite of infected hard ticks (Acari, Ixodidae). They share characteristic clinical features including fever, headache, rash, and sometimes eschar formation at the site of the bite. The arthropod vector species and associated human clinical diseases vary depending on the geographical locations (Table 12.1).

The typical life cycle of a SFG rickettsia is complex and involves arthropods and mammalian species and is influenced by environmental conditions. Bacteria infect and multiply in almost all organs of their arthropod hosts, mainly infecting ticks. When the ovaries and oocytes of an adult tick female tick are infected, rickettsiae may be transmitted transovarially (hereditary transmission) to at least some of its offspring (larvae). Thus, once an egg is infected, all subsequent life stages of the tick will be infected, and rickettsiae infecting the ticks' salivary glands can be transmitted to vertebrate hosts during feeding. However, it is known that some bacteria, for example, *R. rickettsii*, are partially pathogenic to ticks; therefore, the infection rate drops in the tick population with each tick generation because mortality rates are higher among infected than among uninfected arthropods. Under these conditions, a so-called amplifier vertebrate host is required to maintain the bacterium active in nature. The amplifier hosts (usually small- and middle-sized mammals), when inoculated by infected ticks with pathogens, maintain the bacteria in their bloodstream for some days or weeks, when new, not infected ticks get infected with bacteria giving continuation for the rickettsial infection among the tick population.

The number of the main mammalian hosts has a direct effect on tick presence. In other words, the higher the host abundance, the higher the chances that ticks will find the hosts to complete their life cycles, increasing the ticks' populations. If additionally this mammalian host is also a good amplifier host for rickettsiae, its presence influences the infection in these ticks and consequently increases the risk for humans being bitten by infected tick. If it also happens that these amplifier hosts are increasing its abundance in disturbed and modified environments, when other potential hosts for ticks disappear, the risk of acquiring disease by humans is increasing dramatically. For example, the higher risk for humans of acquiring *R. rickettsii* in Southern Brazil is strongly associated with the numerous and uncontrolled populations of

Table 12.1 Rickettsial agents of human diseases

Group	Species	Disease	Main vector(s)	Geographic distribution	Epidemiological characteristics and risk factors
Typhus fever	*Rickettsia prowazekii*	Epidemic typhus, sylvatic typhus	Human body louse *Pediculus humanus*, flying squirrel ectoparasites	Worldwide	Risk factors: conditions of overcrowding and poor hygiene
	Rickettsia typhi	Endemic (murine) typhus	Rat fleas *Xenopsylla cheopis*	Worldwide	Ricks factors: contact with rats and their fleas
Spotted fever group	*Rickettsia aeschlimanni*	Unnamed tick-borne rickettsiosis	Ticks *Hyalomma* spp., *Amblyomma* spp., *Rhipicephalus* spp.	Africa, Southern Europe, Asia	First clinical description in 2002, few human cases reported
	Rickettsia africae	African tick bite fever	Ticks *Amblyomma variegatum*, *Amblyomma* spp., *Rhipicephalus* spp.	Sub-Saharan and North Africa	Risk factors: outdoor activities, safari, ecotourism, hunting, bushwalking, camping, etc. Outbreaks and clustered cases common
	Rickettsia akari	Rickettsialpox	House mouse mites *Liponyssoides sanguineus*	Worldwide	Outbreaks occur after contact with infected rodents and their mites
	Rickettsia australis	Queensland tick typhus	Ticks *Ixodes* spp.	Australia	Risk factors: outdoor activities, safari, ecotourism, hunting, bushwalking, camping, etc.
	Rickettsia conorii subsp. *caspia*	Astrakhan fever	Ticks *Rhipicephalus sanguineus*	Europe, sub-Saharan Africa	Risk factors: proximity and contact to infected dogs infested with dog ticks
	Rickettsia conorii subsp. *conorii*	Mediterranean spotted fever, Kenyan tick typhus	Ticks *Rhipicephalus sanguineus*	Sub-Saharan, North Africa, Europe	Disease occurs in urban and rural settings. Cases are generally sporadic. Risk factor: proximity to infected dogs infested with dog ticks
	Rickettsia conorii subsp. *indica*	Indian tick typhus	Ticks *Rhipicephalus sanguineus*	Europe, Asia	Risk factors: proximity and contact to infected dogs infested with dog ticks
	Rickettsia conorii subsp. *israelensis*	Israeli tick typhus	Ticks *Rhipicephalus sanguineus*	Europe, Asia, North Africa	Risk factors: proximity and contact to infected dogs infested with dog ticks
	Rickettsia felis	Flea-borne spotted fever, cat flea typhus	Cat flea *Ctenocephalides felis*	Worldwide	Rick factors: contact with cat fleas
	Rickettsia heilongjiangensis	Far-eastern spotted fever	Ticks *Haemaphysalis* spp., *Dermacentor* spp	Asia (Russia, China, South Korea, Japan)	Distribution is most likely much larger
	Rickettsia helvetica	Aneruptive fever	Ticks *Ixodes persulcatus*, *Ixodes ricinus*	Asia, North Africa, Europe	First clinical description in 2002. Serologically confirmed two cases also in Asia. Risk factors: outdoor activities
	Rickettsia honei	Flinders Island spotted fever, Thai tick typhus	Ticks *Bothriocroton hydrosauri*, *Ixodes* spp.	Asia, Australia, Pacific	Rural, related with outdoor activities
	Rickettsia japonica	Oriental or Japanese spotted fever in Japan and South Korea	*Haemaphysalis* spp.	Asia	Rural. Risk factors: agricultural activities, bamboo cutting

	Rickettsia massiliae	Unnamed tick-borne rickettsiosis	Ticks *Rhipicephalus sanguineus*	North, Central and South America, North Africa, Europe, Asia	First clinical description in 2005. Risk factors: proximity and contact to infected dogs infested with dog ticks
	Rickettsia monacensis	Mediterranean spotted fever-like illness	Ticks *Ixodes persulcatus*, *Ixodes ricinus*	Europe	First clinical description in 2007, few cases known but probably misdiagnosed with *R. conorii*
	Rickettsia parkeri	Unnamed tick-borne rickettsiosis	Ticks *Amblyomma* spp., *Dermacentor* spp.	North, Central and South America	First clinical description in 2004. Milder version of Rocky Mountain spotted fever, misdiagnosed with *R. rickettsii*
	Rickettsia philippi	364D rickettsiosis	Ticks *Dermacentor occidentalis*	North and Central America	First clinical description in 2010, few human cases reported
	Rickettsia raoultii	Tick-borne lymphadenopathy (TIBOLA), *Dermacentor*-borne necrosis erythema and lymphadenopathy (DEBONEL), scalp eschar and neck lymphadenopathy after tick bite (SENLAT)	Ticks *Dermacentor* spp.	Europe, Asia	First clinical description in 2006. Peak incidence occurred during III–V and during VIII–XI linked with the activity of ticks in Europe. Risk factors: outdoor activities
	Rickettsia rickettsii	Rocky Mountain spotted fever, Brazilian spotted fever, *febre maculosa*, *fiebre manchada*, Minas Gerais exanthematic typhus, Tobia fever	Ticks *Amblyomma* spp., *Dermacentor* spp., *Rhipicephalus sanguineus*	North, Central and South America	High fatality rate (80%), small outbreaks in endemic areas. The most common rickettsial infection in Americas: up to 2000 cases are reported yearly in the United States, 200 cases/year in Brazil. Rural and urban disease. Endemic areas are associates with environmental modifications in some regions in Brazil
	Rickettsia sibirica subsp. *mongolitimonae*	Lymphangitis-associated rickettsiosis (LAR)	Ticks *Hyalomma* spp.	Europe, sub-Saharan and North Africa, Asia	
	Rickettsia sibirica subsp. *sibirica*	Siberian tick typhus in Russia, China, and Mongolia	Ticks *Dermacentor* spp.	Asia	Various rickettsial strains phylogenetically related to *R. sibirica* and *R. africae*
	Rickettsia slovaca	Tick-borne lymphadenopathy (TIBOLA), *Dermacentor*-borne necrosis, and lymphadenopathy (DEBONEL)	Ticks *Dermacentor* spp.	Asia, Europe, North Africa	Peak incidence occurred during III–V and during VIII–XI linked with the activity of ticks in Europe. Risk factors: outdoor activities
	Rickettsia tamurae	Unnamed	Ticks *Amblyomma testudinarium*	Asia	Human case was reported from Japan and Laos in 2009
Scrub typhus	*Orientia tsutsugamushi*	Scrub typhus, bush typhus	Larval stage of trombiculid mites called "chiggers"	Northern Japan, Southeast Asia, the western Pacific Islands, eastern Australia, China	More than one million cases occur annually. An occupational disease among rural residents engaged in agricultural or gathering activities, tourist at outdoor activities as camping, hiking, etc.

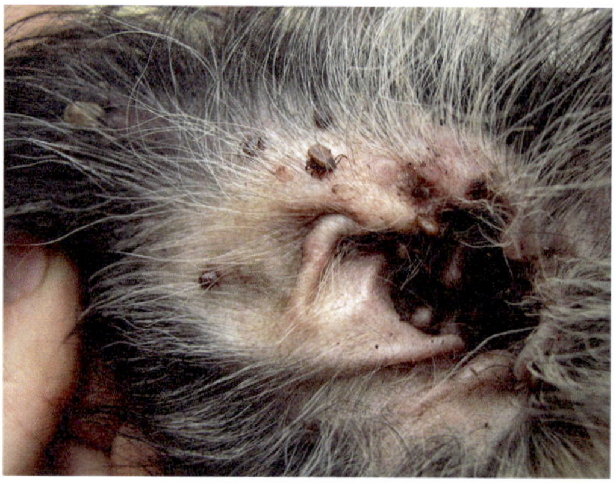

Fig. 12.1 The brown dog tick (*Rhipicephalus sanguineus*) is the main vector of *Rickettsia conorii*, agent of the boutonneuse fever in the Mediterranean Basin, and *Rickettsia rickettsii* agent of spotted fever in the United States and Mexico. Image: M. Ogrzewalska

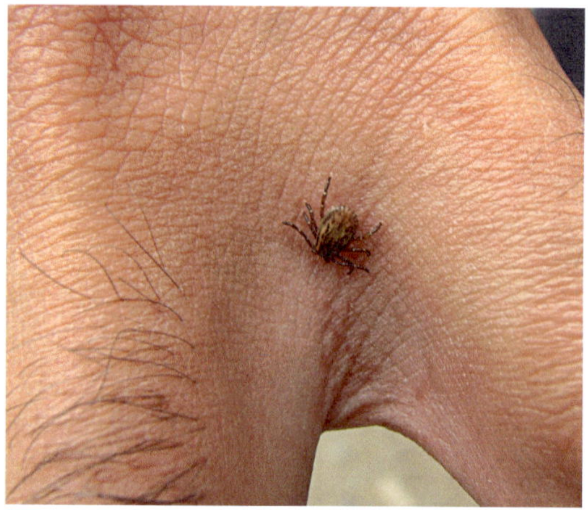

Fig. 12.2 Tick *Amblyomma aureolatum* is a vector of *Rickettsia rickettsii* causing Brazilian spotted fever (BSF) in some regions of Brazil. Human cases associated to these ticks occur when domestic dogs are bitten by adult ticks during incursions into the rain forest and bring them back to human dwellings. Infected ticks drop off from the dog accidentally (scratching, picked by humans) and bite humans inoculating bacteria. Image: M. Ogrzewalska

capybaras (*Hydrochoerus hydrochaeris*) in recreation areas, such as city parks. These rodents, besides being the most important hosts for all parasitic stages of aggressive human biting tick *Amblyomma sculptum* (formerly *A. cajennense*[1]), are efficient amplifier hosts of *R. rickettsii* for this tick. By maintaining and increasing the infected ticks' populations in the proximity of humans, capybaras are playing a crucial role in the epidemiology of spotted fever in Brazil as humans spending time in city parks, where these mammals are present, are frequently attracted by ticks that may be infected with *R. rickettsii*. Consequently, the geographic distribution of *R. rickettsii* is determined by the incidence of its tick amplifier hosts, in this case, capybaras (Figs. 12.1, 12.2, and 12.3).

The number of unnamed and noncultivated *Rickettsia* is growing. New rickettsial species are continuously isolated from ticks around the world, but their pathogenicity for humans remains to be determined; for example, *R. asiatica*, *R. hoogstraalii*, or *R. montanensis* have been only found in ticks. However, some rickettsiae previously thought to be nonpathogenic, later, even sometimes after decades, have been associated with human disease, as in the case of *R. africae*, *R. parkeri*, *R. slovaca*, or *R. helvetica*, which were first isolated from ticks and subsequently from a patient's blood. Thus, it is believed that each *Rickettsia* has a pathogenic potential, provided that its reservoir arthropod is capable of biting humans.

[1]Taxonomic status of *Amblyomma cajennense* was recently reviewed by Nava et al. (2014).

12.1.4 Clinical Manifestations

12.1.4.1 Typhus Group *Rickettsia*

- *Rickettsia prowazekii* (louse-borne epidemic typhus). The disease starts abruptly after 2 weeks of incubation with malaise and nonspecific symptoms as fever, headache, arthralgias, and chills. There is no eschar of inoculation at the site of lice biting. Rash, macular, maculopapular, or petechial, begins at the trunk and spreads to extremities, saving the face, palms, and soles, and is presented in 20–60 % of cases. The central nervous system is affected. At severe cases, multiorgan commitment may occur. Without treatment, the disease is fatal in 10–30 % of patients. Recrudescent typhus or Brill-Zinsser disease occurs in individuals, usually Europeans, who had the primary disease during the World

Fig. 12.3 Capybaras (*Hydrochoerus hydrochaeris*) are amplifier hosts of *Rickettsia rickettsii* for *Amblyomma sculptum* ticks. Capybaras are abundant rodents in anthropized areas in Brazil, close to human settlements in riparian forests and at habitats with water bodies such as urban and peri-urban parks and garden of condominiums. Image: M. Ogrzewalska. (a) capybaras close to humans settelments, (b) capybara (*H. hydrochaeris*)

War II. It can be triggered by factors inherent to the host as lowering immune response, determining a more benign clinical picture without rash and not fatal.

- *Rickettsia typhi* (murine typhus). The disease usually presents with an abrupt onset of symptoms after an incubation period of 1–2 weeks. Fever, headache, chills, and rash are the common symptoms presented. Rash appears at various intervals and is described as macular at most cases, frequently affecting the trunk, rarely seen at the palms, soles, or face. Liver and spleen enlargement and gastrointestinal symptoms can occur. Malaise, anorexia, myalgia, cough, conjunctivitis, nausea, lymphadenopathy, central nervous system abnormalities, gastrointestinal impairment, and enlargement of the liver and spleen are the less common symptoms. Clinical course is usually uncomplicated and fatalities are uncommon (1–5% of nontreated cases). G6PD insufficiency can aggravate or potentiate symptoms, leading to a life-threatening situation, conducting to more severe forms.

12.1.4.2 Spotted Fever Group *Rickettsia*

The best known is the spotted fever caused by *R. rickettsii* (Rocky Mountain spotted fever, Brazilian spotted fever). It is a systemic disease with variable clinical findings. After 1–2 weeks following the bite of an infected tick, an abrupt onset with the classic triad may appear including fever (typically high—39–41 °C), headache (severe frontal), and malaise (sometimes nausea, vomiting). However, acute rickettsial infection develops nonspecific symptoms and is difficult to specifically diagnose because it is similar to many other infectious diseases. The patient may present atypical forms without rash, viscerotropic and/or fulminant. Symptoms may vary with intestinal dysfunction, hepatic impairment, pulmonary involvement, and renal and neurological manifestations. Rash is considered the most important sign, generally absent until the third to fifth day of fever. In approximately 20% of patients, such as children and elderly patients, the absence of rash complicates and delays the diagnosis thus determining a higher number of deaths. Some patients (mainly black patients) can show a fleeting, evanescent, or atypical distribution of rash. Macular, initially pinkish, can develop in 2–5 days for a maculopapular, petechial pattern, reddish, typically reaching at onset of disease, the ends, spreading hereinafter centrally to the trunk. A "classic" rash has a characteristic of petechial lesions and is considered when the distribution includes the palms and soles, although this scenario is not seen in all cases. The presence of petechial rash means that the patient is often severely ill.

Large ecchymotic injury, formed by coalescence of petechiae as a consequence of microvascular damage, can cause necrosis/gangrene.

Clinical manifestations of other SFG Rickettsia are presented here in alphabetic order of species:

- *Rickettsia aeschlimannii*: The patients present typical clinical signs of spotted fever: eschar, high fever, and generalized maculopapular rash.
- *Rickettsia africae* (African tick bite fever): After 5–10 days of the tick bite, the clinical course comprises an abrupt onset of fever, headache, nausea, and myalgia. Eschar is present at most cases (sometimes multiple eschars), and the less common clinical sign is the vesicular or maculopapular cutaneous rash and arthralgia. The infection is symptomatic in less than 50 % of cases. Most of the patients develop a mild to moderate illness and the disease resolves spontaneously within 10 days. Complication is rare and no fatalities are reported.
- *Rickettsia akari* (rickettsialpox): After a mean incubation period of 7 days, a painless papule, reddish at mite bite site, appears. Some days after, it evolves to a vesicular form and ulcerates, forming the eschar. This eschar is formed 3–7 days before onset and persists by 2–3 weeks. Regional lymph nodes are often presented enlarged and tender. The onset of signs and symptoms is sudden with high hectic fever, myalgias (backache), chills, and headache (frontal and severe). Diaphoresis, anorexia, and photophobia may be present also. Within 2–3 days after onset, a generalized papulovesicular rash appears. The severity of rash does not reflect the gravity of the systemic manifestations. Less than 30 % of patients present enanthem.[2] The lesions usually heal by crusting. Generally the disease is described as mild, benign, and self-limited. Untreated illness resolves in 2–3 weeks and death is very rare.
- *Rickettsia australis* (Queensland tick typhus): Sudden onset with fever, headache, and myalgia. Within 10 days, maculopapular or vesicular rash appears. Eschar in identified in approximately 65 % and lymphadenopathy in 71 % of cases, respectively.
- *Rickettsia conorii* (Mediterranean spotted fever, MSF): After a mean incubation time of 6 days, abrupt onset appears with high fever, headache, chills, arthromyalgias, and eschar (rarely multiple) at tick bite site. Exanthem appears 4 days (median) following the onset of fever and is initially macular and scarce becoming maculopapular and disseminated. Rash always presents at the palms and soles (limbs) but seldom at the face and usually is associated to a severe form of MSF. Usually patients will recover within 10 days without sequelae. Severe forms may occur at 5–6 % of cases with neurological manifestations and different organs' involvement.
- *Rickettsia conorii israelensis* (Israeli spotted fever, ISF): After the incubation period of 7–8 days from the tick bite, fever and rash on hands and feet start and then extend to the rest of the skin. Eschar is absent. Up to one third of patients can manifest headache, vomiting, myalgia, and arthralgia. Fatal cases and severe forms may occur, especially in children or in patients with risk factor, as glucose-6-phosphate dehydrogenase (G6PD) deficiency.
- *Rickettsia conorii caspia* (Astrakhan fever, AF): The disease is similar to MSF but eschar is rare (less than 25 % of patients).
- *Rickettsia conorii indica* (Indian tick typhus, ITT): The rash frequently purpuric differs from MSF and eschar is rarely identified.
- *Rickettsia felis* (flea-borne spotted fever): Clinical features are often confused with other febrile illnesses and with those found on patients with murine typhus caused by *R. typhi*. After contact with infected flea feces on wounds or abraded skin or infected fleabites, fever, fatigue, headache, vomiting, anorexia, cough, pharyngitis, and rhinitis are some of the manifestations presented. Patients may develop a cutaneous lesion (eschar) in the site of inoculation, a general maculopapular rash,

[2]Exanthem is a rash in the skin and enanthem is a rash in mucosae (CBM).

pneumonia, and neurological signs. However, clinical findings can be more complex and more severe, although there is limited information at scientific literature.

- *Rickettsia heilongjiangensis* (Far-Eastern tick-borne rickettsiosis): Clinical diagnosis is similar to other spotted fever rickettsioses. Present as mild febrile rash, unique eschar, regional lymphadenopathy, macular or maculopapular rash (faint in most cases) and conjunctivitis.
- *Rickettsia honei* (Flinders Island spotted fever): Sudden onset of fever, headache, arthromyalgias (joint swelling), and cough. A few days later, a maculopapular rash appears with no vesiculation. Eschar presented only in 25 % of patients.
- *Rickettsia japonica* (Japanese spotted fever): Similar to other spotted fever rickettsioses, after tick bite, an eschar is presented and an abrupt onset with high fever, headache, and chills occurs. Macular rash appears after 3–4 days and disappears in 2 weeks. Severe cases can happen including encephalitis, disseminated intravascular coagulopathy, multiorgan failure, and acute respiratory distress syndrome.
- *Rickettsia massiliae*: Patients present fever, necrotic eschar, and maculopapular rash involving the palms and soles.
- *Rickettsia parkeri*: With multiple inoculation eschars, the patient presents fever, headache, myalgias and arthralgias, and maculopapular rash on the trunk and extremities, including the palms and soles. Enlarged lymph nodes are observed near bite sites.
- *Rickettsia sibirica mongolitimonae* (lymphangitis-associated rickettsiosis): Mild disease characterized by a fever, a headache, a discrete rash (maculopapular), an inoculation eschar, a lymphangitis expanding from the inoculation eschar to draining node, and a painful satellite lymphadenopathy.
- *Rickettsia sibirica* (Siberian tick typhus): The incubation period is 4–7 days after the tick bite; eschar appears accompanied by regional lymphadenopathy. Onset is acute with high fever, headache, myalgia and typical maculopapular rash, conjunctivitis, and digestive disturbances. Symptoms are concomitant and can last 6–10 days without treatment. Clinical course is benign.
- *Rickettsia slovaca* (tick-borne lymphadenopathy, TIBOLA, and *Dermacentor*-borne necrosis erythema-lymphadenopathy, DEBONEL): Mean incubation of 7 days, fever (low grade), headache, and rash are uncommon. Chronic fatigue and localized alopecia at the bite site are sequelae presented. Eschar on the scalp can occur, associated with regional painful lymphadenopathy.

12.1.4.3 Scrub Typhus

Orientia tsutsugamushi causes scrub typhus. Eschar is formed by the chigger bite site developing during incubation period (average 10 days). The eschar evolves from a small papule to a blackened crust lesion resembling a cigarette burn. Onset begins abruptly with fever, headache, myalgias, chills, and malaise. Perishable macular rash on the trunk, maculopapular, spreads to extremities and may appear at the end of the first week of disease. Conjunctival suffusion, hearing loss, and cough sometimes with infiltrates are common feature. Severe cases can occur with multiple organ dysfunctions and respiratory failure is the most common cause of death.

12.1.5 Diagnosis

The diagnosis of rickettsial diseases is based primarily on clinical and epidemiological characteristics. Confirmatory techniques provide information that retrospectively validates the accuracy of the clinical diagnosis. The diagnosis of spotted fever is usually serological and retrospective. Serologic assays are commonly available through commercial laboratories. The indirect immunofluorescence antibody assay (IFA) is considered the gold standard and the analysis is conducted with two paired serum samples and obtained within 2–4 weeks. The fourfold increase of the antibody IgM or IgG titer confirms the diagnosis. Serologic diagnosis can only identify genus, and thus the etiology is only possible using the characterization of the agent through molecular technique.

Amplification of specific DNA by polymerase chain reaction (PCR) in blood and/or tissue or even ticks recovered from the patient provides a rapid method for detecting rickettsial disease infection. Currently, molecular diagnostics (PCR sequencing) has been used more frequently and is considered the most suitable and ideal diagnostic tool for identifying severe cases and fulminant ones, where samples for serological pairing are not always available. New techniques (real-time PCR) might offer the advantages of speed, reproducibility, quantitative capability, and low risk of contamination, compared with conventional PCR.

Immunohistochemical staining analysis using immunofluorescence or immunoperoxidase test, even in paraffin-embedded tissue, allows identifying the presence of specific rickettsiae for the spotted fever group. Immunostaining of skin biopsy, preferably a 3 mm fragment derived from a maculopapular rash containing petechial lesions, may aid in diagnosis, before the specific antibiotic therapy is complete within 24 h.

Rickettsial isolation in cell culture from blood, skin biopsy, and/or other organ fragments can provide difficulties in performing the technique, and the risk of infection refrains its use in routine laboratory. The same methodology can be used in tick vectors eventually recovered from patients. This technique is more labor intensive and time consuming than the assays described above. Culture is rarely used for diagnosis and the other methods are used to confirm infection.

12.1.6 Treatment and Prevention

Treatment of patients with potential rickettsioses must be started quickly and should not await confirmatory laboratory testing. Doxycycline is the first-line treatment for adults and children of all ages and should be initiated immediately whenever rickettsioses are suspected. This drug is the first-line treatment for adults and children of all ages (adults, 100 mg every 12 h; children under 45 kg (100 lbs), 2.2 mg/kg body weight given twice a day).

Other drugs such as chloramphenicol, azithromycin, fluoroquinolones, and rifampin may be alternatives. These antimicrobial agents, although they are bacteriostatic, lead to the reduction of toxemia and other clinical signs in 24–48 h and improvement in 2–3 days without complications.

The hospitalization is not necessary in all cases and the treatment should be continued for at least 7–10 days or until no fever for over 24 h. Antibiotics should be administered intravenously in patients with nausea and vomiting. In countries where intravenous doxycycline is not available, in severely ill patients, chloramphenicol should be given, although studies prove that tetracycline is superior. *Rickettsia* spp. are resistant to antibiotics of β-lactam class, aminoglycosides and trimethoprim-sulfamethoxazole, which should not be used.

12.1.7 Prophylaxis and Prevention Measures

There are no vaccines available for rickettsial infections. The only sure way is to prevent infection by avoiding contact with infected arthropods. In the case of lice and flea associated with *Rickettsia*, parasite populations should be reduced, using insecticides to prevent additional exposure. Insecticidal powders are available for body-louse control and treatment of clothing for those at high risk of exposure. Insecticide powders with residual action should be used on rat runs and burrows.

Rickettsia prowazekii is susceptible to 1% sodium hypochlorite, 70% ethanol, glutaraldehyde, and formaldehyde. It can also be inactivated by moist heat (121 °C for a minimum of 15 min) and dry heat (160–170 °C for a minimum of an hour). *Rickettsia typhi* is destroyed by formalin, phenol, and temperatures greater than 56 °C for 30 min.

Reducing exposure to ticks is the most effective way to limit the probability of spotted fever infections. In people exposed to tick-infested environments, prevention measures should be aimed at personal protection:

1. Wear light-colored clothing to allow you to see ticks that are crawling on your clothing.
2. Tuck your pant legs into your socks so that ticks cannot crawl up the inside of your pant legs.

3. Apply repellents to discourage tick attachment. Those repellents containing permethrin can be sprayed on boots and clothing and will last for several days. Repellents containing DEET can be applied to the skin, but will last only a few hours before reapplication is necessary.
4. Prompt careful inspection and removal of crawling or attached ticks upon return from potentially tick-infested areas by searching your entire body for ticks.
5. Remove any tick you find on your body by fine-tipped tweezers by grasping the tick as close to the skin surface as possible and pull upward. Do not twist or jerk the tick; this may cause the mouthparts to break off and remain in the skin.

Ticks may be carried into the household on pets, especially on dogs; thus they should be treated by repellent products recommended by vet and carefully examined daily. Ticks may not only be dangerous by transmitting pathogens that are causing dangerous diseases in pets, but they may leave the animal and bite humans, putting in risk to those who did not suspect being parasitized by ticks during outdoor activities such as small children or elderly people.

12.2 Bartonella Infections

12.2.1 Introduction

Since its discovery, bartonellosis refers to the Carrión's disease (CD), a biphasic disease of the Andes region caused by *Bartonella bacilliformis*. There was a dramatic outbreak of fever and hemolytic anemia among workers based in the upper Rimac valley, during construction of the railway line that would link the capital, Lima, to the city of La Oroya, 4000 m above sea level in the Andes in 1871. During the course of a week, at least 4000 workers from Chile and other regions of Peru died in those episodes, and the disease became known as Oroya fever, although the agent of the syndrome remained elusive.

In 1885, a medical student named Daniel Carrión injected himself with exudate from a verruga lesion, developed Oroya fever, and died. The common etiology of Oroya fever and Peruvian verruga was established, and Carrión is one of the most heroic images in Peruvian medical history. In recognition of his sacrifice, South American bartonellosis is commonly known as Carrión's disease. Finally, in 1905, Alberto Barton solved the riddle of the etiologic agent of CD when he observed intracellular bacteria in blood smears from Oroya fever patients.

Cat scratch disease, caused by *Bartonella henselae*, was firstly discovered in a boy in France, in 1931, associated to superficial wounds caused by cat claws, and trench fever, caused by *B. quintana* and transmitted by body lice, was very frequent in the World War I trenches (see Chap. 3), but became very rare after this, being found in homeless persons in France and the United States.

12.2.2 Taxonomy and Life Cycle

Currently, there are more than 33 recognized species and subspecies and several *Candidatus* spp. within this genus, among which, 14 are associated to human diseases with a worldwide distribution, and that causes an increasingly large number of infectious diseases in humans and animal.

However, bartonellosis is not a reportable disease in human populations in most countries, including those in South America. Therefore, sufficient information to determine the exact incidence or prevalence of *Bartonella* infection is not available. In the United States, it was estimated that 22,000–24,000 humans developed cat scratch disease (CSD) during 1992, of whom 2000 were hospitalized. The estimated annual healthcare cost of the disease was more than $12 million. In Connecticut, which is the only state where the disease is reportable in the United States, the incidence of the disease from 1992 to 1993 was estimated to be 3.7 cases/100,000 persons, whereas in the Netherlands, there was an estimated 2000 cases/years or 12.5 cases/100,000 persons. These observations suggest that several thousand cases of cat scratch disease may occur every year in most part of the world. *Bartonella* human infection seroprevalence varies from 1.4 % (intravenous drugs user, IVDU, from Baltimore, USA) to 77.5 % (children and adults in an area with an outbreak of Carrion disease, Peru) in selected populations in the last decade (1996–2007).

Since the early 1990s, there have been substantial advances in the understanding of the etiology, reservoir potential, vector transmission, and pathogenesis of *Bartonella* infection in a wide range of mammals. In 1993, Brenner and colleagues proposed to unify the *Bartonella* and *Rochalimaea* genera and renamed some species as *B. quintana*, *B. henselae*, *B. vinsonii*, and *B. elizabethae*. As a result of this unification, the transfer of all these organisms from the family Rickettsiaceae to the family Bartonellaceae was required and removed the last family from the order Rickettsiales, based on phylogenetic relationships and the absence of obligate intracellular pathogen between the bartonellas. Nowadays, they belong to the α_2 subgroup of the class Proteobacteria, order Rhizobiales, and are closely related to the genera *Brucella*, *Agrobacterium*, and *Rhizobium*.

Microscopically, all *Bartonella* spp. are Gram-negative bacilli or coccobacilli. Some species, such as *B. bacilliformis* and *B. clarridgeiae*, have flagella, which in the case of *B. bacilliformis* facilitates erythrocyte invasion. *B. henselae* appear to lack flagella. Members of the genus *Bartonella* are small; pleomorphic; facultatively intracellular, fastidious, and aerobic; and catalase-, urease-, nitrate reductase-, and oxidase-negative bacteria.

Bartonella infections have been encountered in all species surveyed, which have extended to members of different orders of mammalian including carnivores, primates, ungulates, rodents, and bats. It is believed that the vector preference for certain hosts can influence the transmission of these organisms and that it is responsible for association of a given *Bartonella* sp. with a specific host, i.e., *B. henselae*, *B. clarridgeiae*, and *B. koehlerae* in cats, *B. alsatica* in wild rabbits, and *B. bacilliformis* and *B. quintana* in humans.

Several hematophagous insects have been implicated in *Bartonella* transmission, including sand flies (*Lutzomyia verrucarum*, *L. peruensis*), human body lice (*Pediculus humanus*), cat fleas (*Ctenocephalides felis*), and, potentially, ticks (*Ixodes pacificus*).

12.2.3 Clinical Manifestations

Bartonella have been recognized as agents causing human disease, and the clinical spectrum of *Bartonella* infection has continually expanded. *Bartonella bacilliformis* (agent of Oroya fever and verruga peruana), *B. henselae* (agent of cat scratch disease, bacillary angiomatosis, bacillary peliosis, endocarditis), and *B. quintana* (agent of trench fever, bacillary angiomatosis, bacteremia, and endocarditis) are the best-known species causing human illness.

The *Bartonella* spp. infection can cause great diversity of clinical manifestations in humans as a fever of unknown origin and recurrent, malaise, fatigue, insomnia, loss of memory, psychiatric disorders, lymphadenopathy, splenomegaly, angiomatosis and bacillary peliosis, endocarditis, hepatitis, osteomyelitis, encephalitis, meningitis, and other neuroretinitides.

Bartonella spp. have pathogenic characteristics, such as to invade cells causing prolonged intraerythrocytic bacteremia in their hosts and to lyse red blood cells. In humans, the infection cycle of *Bartonella* spp. is initiated by colonization of the primary niche. Besides erythrocytes, the endothelial cells represent another target of *Bartonella* in their mammalian hosts. Current opinion is that these cells serve as a primary niche for *Bartonella* prior to its entry in the bloodstream. In this stage, the infection is usually controlled by the immune system, and clinical manifestations are characterized by local lymphadenopathy (i.e., associated with *B. henselae*, *B. quintana*, and *B. alsatica*). *Bartonella* spp. have rapidly cleared from the blood after the initial inoculation but can reappear in the bloodstream. After 4 or 5 days, the bacteria (primarily *B. bacilliformis* and *B. quintana*) are released into the bloodstream and then are entitled to adhere or invade to mature erythrocytes, where intracellular replication occurs. *Bartonella* spp.

can subsequently colonize secondary foci, particularly vascularized tissues such as heart valves, the liver, and the spleen. *Bartonella* spp. can circulate in the blood for the remaining life span of the infected erythrocyte. As a result, bacteremia can last for several weeks to months.

12.2.4 Diagnosis

There is not a standard diagnostic laboratory for infections caused by *Bartonella* spp. Nowadays it is evident that several techniques must be used to avoid simultaneously false-negative results. Diagnostic techniques for infections with *Bartonella* spp. include serology by immunofluorescence (IFA) to detect antibodies in the patient's serum; culture of the pathogen; histopathological examination of lymph nodes or tissue biopsy of the skin, liver, or other affected organs; and molecular biology assays, especially PCR, to amplify *Bartonella* spp. genes from patient's tissue fragments or blood. The blood smear stained with Giemsa method is utilized in the diagnosis of Carrión disease by *B. bacilliformis* to assess the prognosis; even the finding of more coccoid bacteria indicates good evolution.

Serology is particularly important because it allows rapid identification of *Bartonella* infection and should be performed systematically when investigating culture negative or in seroepidemiological studies showing their value in alerting to dispersion of pathogens throughout the world. It is limited by cross-reactions between the different species of *Bartonella* and between genera such as *Coxiella* and *Chlamydia*. Many studies have shown that serological differentiation between *B. henselae* and *B. quintana* through IFA is impossible, since cross-reactivity between these species is very high (95 %). It is also necessary to take into consideration the heterogeneity among strains and genotypes of *Bartonella* spp., differences between the parameters of analysis laboratories, and subjectivity of the readings of the results of IFA, which would result in false positives. Despite these limitations, IFA is the gold standard for the diagnosis of infection, present or previous.

The diagnosis of *Bartonella* infection should be confirmed by culturing the organism from aseptically obtained patient samples, including blood, CSF, lymph nodes, or other tissue aspiration samples, ocular exudates, and from surgical biopsies. The culture liquid *Bartonella* spp. becomes necessary to increase the sensitivity of detection of bacteremia through methods of molecular and is one of the most used methods of diagnosis worldwide. The isolation of most species of *Bartonella* in blood agar plates requires a long incubation period (6–8 weeks) at 35 °C in a water-saturated atmosphere containing 5 % CO_2. The development of a new liquid culture medium called *Bartonella* alpha-Proteobacteria growth medium (BAPGM) that allows the growth of at least seven species of *Bartonella* enabled the improvement of this method as diagnosis. It has been widely used as a pre-enrichment and combined with molecular methods increased the success and sensitivity of culture for diagnosis in both animals and humans. Likewise, for other methods diagnosis, a consensus of choosing primers neither the PCR conditions does not exist. Species-specific PCR has been useful, especially when a particular diagnosis is already suspected.

It is known that ocular manifestations occur in 5–10 % of patients with cat scratch disease (CSD). Several imaging modalities can be used to assist in the diagnosis and management of ocular CSD. They include color fundus photography that allows the clinician to monitor the fundus changes in this disease, fluorescein angiography that demonstrates leakage at the optic nerve in CSD neuroretinitis, and optical coherence tomography that provides confirmation in early stages of neuroretinitis before the formation of a macula star.

12.2.5 Treatment and Prevention

There are few studies on the treatment of bartonellosis. The treatment is more effective for immunodeficient than for immunocompetent patients. Antimicrobials have been used widely in the treatment of bartonellosis. CSD responds poorly to antibiotic treatment. However, a study using azithromycin in CSD showed a benefit in lymph node regression during the first 30 days, as compared with placebo. Available date does not support the use of antibiotics for the treatment of CSD.

In patients with *Bartonella* spp., bacteremia should be treated with 3 mg/kg body weight of gentamicin once daily for 2 weeks in combination with 200 mg of doxycycline, daily for 4 weeks. In immunocompromised patients the antimicrobials clearly indicated are erythromycin, clarithromycin, azithromycin, and doxycycline, either coadministered or not with rifampin, and the period of treatment ranges from 4 to 6 weeks.

In acute febrile phase of Carrión's disease, the preferred treatment has been chloramphenicol. An initial dose of 50 mg/kg/day chloramphenicol for the first 3 days and a subsequent dose of 25 mg/kg/day until the completion of 14 days of treatment have been proposed as the best regimen for the of *B. bacilliformis* bacteremia. Patients with *Bartonella* endocarditis have higher death rate and undergo valvular surgery more frequently than patients with endocarditis caused by other pathogens. Patients with suspected or confirmed *Bartonella* endocarditis should be treated with 3 mg/kg/day gentamicin for 2 weeks in combination with 200 mg of doxycycline daily for 6 weeks. Overall, the treatment of *Bartonella* infections must be adapted based on whether the disease is in the acute or chronic form and based on the infecting *Bartonella* spp.

12.3 Coxiellosis (Q Fever)

12.3.1 Introduction

Historically in 1933, disease of unknown etiology was first identified in slaughterhouse workers in Brisbane, Queensland, Australia. Patients had fever, headache, and malaise and all laboratory tests for the screening of a large number of pathogens were negative. Because it was a disease of unknown etiology, it became known as Q fever (from the word "query") in 1937 by Edward Holbrook Derrick. MacFarlane Burnet and his partner Mavis Freeman, whom Derrick had sent sample of infectious material, isolated a fastidious intracellular bacterium of the animals inoculated. Derrick and his colleagues investigated the epidemiology of the disease, especially the potential role of arthropod vector, and concluded that wild animals were the natural reservoirs of Q fever, while domestic animals are the secondary reservoirs, and an agent in which the disease could be transmitted by ticks or other arthropods.

In 1935, ignoring Derrick's work in Australia, Gordon Davis, in the Rocky Mountain Laboratory in Hamilton, Montana, USA, was investigating the ecology of spotted fever. He noted that ticks collected in Nine Mile, Montana, caused a febrile response in mice that were fed.

In addition, the disease could be transmitted to uninfected guinea pigs by intraperitoneal inoculation of blood collected from infected animals. Analysis of the inflammatory cells of infected mice revealed microorganisms "rickettsial-like," although the disease in the animal would not be compatible to spotted fever. Thus, unlike what occurred in Queensland, although the infectious agent had been proven, the disease was unknown in the United States.

The connection between the groups in Montana and Brisbane arose when a laboratory-acquired Q fever infection occurred in the Rocky Mountain Laboratory in 1938. Rolla Eugene Dyer, Director of the National Institutes of Health, became infected with the organism that the laboratory was working with. A febrile illness was reproduced in guinea pigs inoculated with Dyer's blood, and rickettsiae were identified in spleen samples from the infected animals. Also, cross-immunity was demonstrated between microorganisms isolated from Dyer's blood and the Nine Mile agent. Dyer then established a definitive link between the Nine Mile agent and the Australian Q fever agent. The etiologic agent of Q fever was first named *Rickettsia burnetii*. However, in 1938, Cornelius B. Philip proposed the creation of a new genus called *Coxiella* and the renaming of the etiologic agent as *C. burnetii*, a name that honors both Cox and Burnet, who had identified the Q fever agent as a new rickettsial species.

12.3.2 Taxonomy and Life Cycle

Q fever is a cosmopolitan zoonosis caused by *Coxiella burnetii*, a small Gram-negative pleomorphic obligate intracellular bacterium. Q fever was first described in slaughterhouse workers in Queensland, Australia, and is considered a major public health problem in Europe (especially in the Netherlands and France), Australia, and, more recently, North America, where a growing number of cases have been reported in the last 5 years and around 3 % of the general adult population is serologically reactive.

Q fever is thought to occur in almost all countries of the world, except in New Zealand and in Antarctica. The noninclusion of Q fever in a list of reportable diseases in most countries has led to significant gaps in the knowledge of its epidemiology. As such, Q fever epidemiology has been largely characterized from defined outbreak investigations and serological surveys of human and animal populations conducted in the Northern Hemisphere and Australia. Human infection with *C. burnetii* has been typically considered an occupational disease associated with ruminants. However, there are at least two distinct cycles that enable the perpetuation of *C. burnetii* infection:

1) The *urban/rural cycle* with the participation of domestic animals.
2) A *wild cycle*, with the participation of animals such as marsupials, rodents, and lagomorphs, among others, as well as some ectoparasites, especially ticks. The relations between the two epidemiological cycles are outlined in Fig. 12.4.

The aerosol route is the primary mode of transmission in human and domestic animals. Aerosol contamination can occur directly from the parturition fluids from infected animals, infecting newborn animals, placenta, and wool. *Coxiella burnetii* is highly resistant to destruction in nature, can be dispersed in wind, and survives for several weeks in areas where animals are present. Thus, Q fever can occur in patients whose epidemiological history does not identify any obvious contact with animals.

Intake, especially the intake of raw milk, is a less likely route of *C. burnetii* transmission that remains a point of controversy, as well as the transmission from person to person, which is an extremely rare event; sporadic cases of human Q fever have been identified: (i) after contact with an infected patient (in an obstetrician after performing an abortion), (ii) transmission via the placenta resulting in congenital infections, (iii) during the

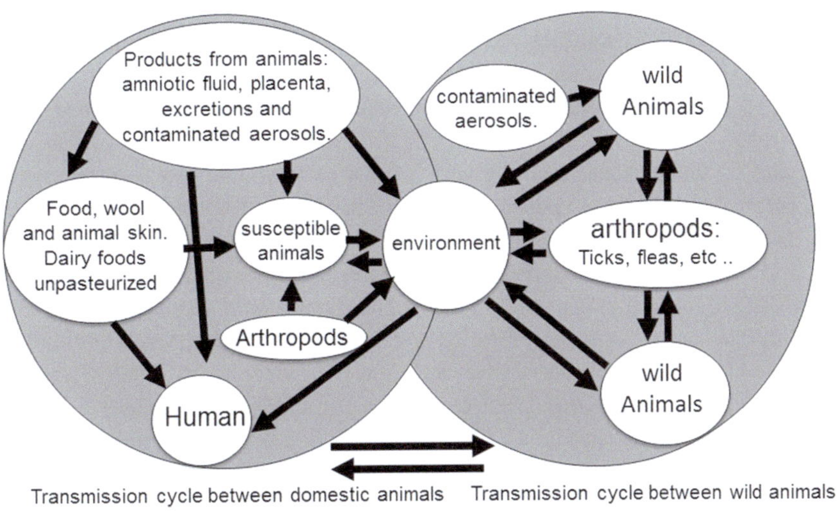

Fig. 12.4 Schematic showing the interrelationships of the domestic and wild animal infection cycles of *Coxiella burnetii*

autopsy, (iv) by intradermal inoculation, or (v) through blood transfusion. Although *C. burnetii* was isolated from arthropods, especially ticks, it is unlikely that the transmission of *C. burnetii* to humans by arthropods is significant.

12.3.3 Clinical Manifestations

Coxiella burnetii infection in humans can vary from asymptomatic infection to fatal disease, and symptomatic patients may present with a wide clinical spectrum, including pneumonia, hepatitis, encephalitis, myocarditis, and other manifestations. Four contributing factors can account for this variation in clinical presentation of Q fever: (i) the route of infection with *C. burnetii*, including aerosol or gastrointestinal tract; (ii) priming dose of *C. burnetii*; (iii) infectious variant of *C. burnetii*, which can have high potential virulence; and (iv) host factors, including the immune status of the infected patient.

The inoculation route of *C. burnetii* in humans can be partly determined by the dominant clinical manifestation. Pneumonia is most common when the source of transmission is via contaminated aerosols, while granulomatous hepatitis is the predominant clinical manifestations when transmission occurs through ingestion of raw milk. In humans, the severity of acute Q fever has also been associated with the dose of the infecting inoculum. Finally, host factors resulting in a state of immunosuppression or pregnancy may influence the course of infection, including developing chronic disease.

Infection with *C. burnetii* can present with acute or chronic clinical manifestations. However, nearly 60 % of Q fever cases are asymptomatic. Among the 40 % of symptomatic patients, most (38–40 %) will experience a mild disease without the need for hospitalization. Hospitalized patients account for only 2 % of infected individuals, and only one in ten (0.2 % of the total infected) develops chronic Q fever. These proportions correspond to the data obtained from analysis of confirmed cases in the south of France, where the incidences of Q fever and endocarditis are 50/100,000 and 0.1/100,000 inhabitants, respectively.

Symptomatic acute Q fever is primarily manifested as a self-limiting febrile illness, atypical pneumonia, or granulomatous hepatitis, while endocarditis is the most common form of chronic Q fever. However, considering the wide ranging and nonspecific clinical spectrum of Q fever, the disease should be systematically considered in febrile patients with recent contact with parturient animals.

12.3.3.1 Acute Q Fever

In acute Q fever cases, after a 2–3-week incubation period, the patient develops an acute infection characterized by high fever, fatigue, chills, and headache. The most common clinical manifestation of the acute form is a self-limiting febrile illness associated with severe headache. Atypical pneumonia is also a frequent clinical presentation, and the clinical picture may vary from clinically asymptomatic pneumonia diagnosed in chest X-ray to acute respiratory failure, although the latter continues to be extremely rare.

Hepatitis is another common manifestation of acute Q fever and is usually identified during laboratory investigation of patients with hepatic frame with increased levels of aspartate aminotransferase (AST), alanine aminotransferase (ALT), and alkaline phosphatase. Although jaundice is rare, hepatomegaly is sometimes clinically detectable.

In addition to the conditions described above and the typical complications of acute Q fever, infection with *C. burnetii* has also been reported to produce endocarditis, osteomyelitis, Q fever in childhood, neurological manifestations, rash, fever of unknown origin, thrombocytopenia, or reactive thrombocytosis.

12.3.3.2 Chronic Q Fever

The diagnosis of chronic Q fever can be established from the persistence of clinical manifestation for more than 6 months after the onset of symptoms. Chronic Q fever occurs in approximately 5 % of patients infected with *C. burnetii* and may develop insidiously over months to years after the acute disease. Endocarditis, vascular infections, osteoarticular infections, chronic hepatitis, chronic lung infections, chronic fatigue syndrome, chronicity favored by pregnancy and preterm delivery, and abortion are some of the clinical presentations that result from chronic infection.

12.3.4 Diagnosis

Given that clinical diagnosis is often difficult due to similarities with a number of infectious or non-infectious diseases, in most cases the diagnosis of Q fever is confirmed by serological testing. Although a variety of serological techniques are available, the indirect microimmunofluorescence test has become the reference technique (Fig. 12.5). Serologic diagnosis is easy to establish, although antibodies are usually only detected after 2–3 weeks of disease onset. Thus, serological tests should be matched with blood samples collected in the acute and convalescent phase, which are unavailable at early diagnosis. The serology test also allows the differentiation of acute and chronic infections of Q fever. Other methods that have been used include indirect immunofluorescence assay (IFA), agglutination, complement fixation, radioimmunoassay, indirect hemolysis test, enzyme-linked immunosorbent assay (ELISA), enzyme-linked immunofluorescent assay (ELIFA), dot immunoblotting, and Western blotting. The most commonly used techniques include IFA, ELISA, complement fixation, and agglutination. However, only the first two methods are commercially available.

The IFA remains the gold standard technique for the diagnosis of Q fever and has the advantage of requiring only small amounts of antigen—*C. burnetii* phase I and phase II with Nine Mile strain. During acute Q fever, seroconversion is usually detected 7–15 days after the onset of symptoms and the antibodies are detected by the third week in 90 % of cases. A titer of anti-IgG antibodies phase II of ≥200 and an IgM antibody titer antiphase II of ≥50 are considered to be significant, but the choice of the negative cutoff titers depends on the amount of antigen stimulation in the study population and can vary from one area to another. Chronic Q fever is characterized by the presence of antiphase I antibody, and an IgG antiphase I antibody titer of ≥800 is considered highly predictive of Q fever endocarditis.

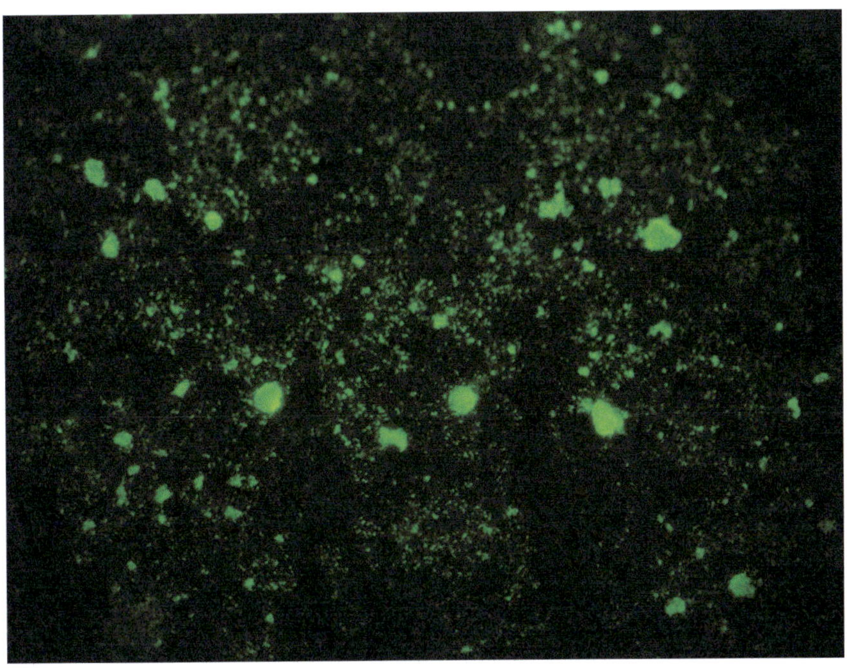

Fig. 12.5 Image of the result of indirect microimmunofluorescence test for *Coxiella burnetii*. 40× magnification (Laboratory of Hantaviruses and Rickettsiosis/Oswaldo Cruz Institute/ FIOCRUZ)

Cross-reactivity is a major source of confusion in the interpretation of serological results and this may vary according to the serological technique. Cross-reactions have been described between *C. burnetii*, *Legionella pneumophila*, *L. micdadei*, *B. henselae*, and *B. quintana*. Such cross-reactions should be considered for the etiology of atypical pneumonia and endocarditis with negative blood culture that can also be caused by *Legionella* and *Bartonella*, respectively. A differential diagnosis is easily established if quantitative data are determined for antibodies against both the *C. burnetii* antigens (i.e., antiphase I and antiphase II).

PCR has been successfully used to detect *C. burnetii* DNA in cell cultures and clinical specimens. The availability of oligonucleotides for amplifying specific *C. burnetii* genes has enabled a simple and reliable method for detecting these bacteria, even in tissue preserved in paraffin. In addition, PCR has been shown to be more sensitive than standard culture techniques for retrospective diagnosis of frozen samples and for monitoring patients treated for chronic Q fever. The detection of *C. burnetii* in tissues should also be performed, particularly in patients who are undergoing treatment for chronic Q fever. Samples can be tested fresh or after fixation in formalin and paraffin. Immunodetection is accomplished using immunoperoxidase or immunofluorescence techniques and specific polyclonal or monoclonal antibodies. Only the latter technique can be used on paraffin samples.

12.3.5 Treatment and Prevention

Doxycycline has the highest therapeutic efficacy against *C. burnetii* and is the treatment of choice for acute Q fever in adults and in children over 8 years of age and in children of all ages with severe disease. Doxycycline is to be administered orally for 14–21 days at a dosage of 100 mg twice daily for adults and 2.2 mg/kg body weight twice daily for children under 45 kg. If these antibiotics are contraindicated, other antibiotics, such as trimethoprim/sulfamethoxazole, macrolides, and fluoroquinolones, can be used (Table 12.2). The benefits of using doxycycline outweigh the potential risk of discoloration of the permanent teeth in children seriously ill or hospitalized less than 8 years of age with acute Q fever. Treatment should be given in the first 3 days of illness for maximum effectiveness and not be delayed pending the results of laboratory tests or because of negative initial laboratory results in the first week of illness.

Prophylactic treatment after a suspected exposure to *C. burnetii* is not recommended as it can

Table 12.2 Guidelines for Q fever treatment

Clinical manifestation	Patient cohort	Treatment	Time
Acute Q fever	Adults	Doxycycline (100 mg 2 × day)[a]	14 days
		Fluoroquinolones (200 mg 3 × day) or pefloxacin (400 mg)[b]	14–21 days
		Rifampin (1200 mg/day)[b]	21 days
	Pregnant	Trimethoprim (320 mg) and Sulfamethoxazole (1600 mg/day)[c]	>5 weeks
	Children (less than 45 kg)	Doxycycline (2.2 mg/kg 2 × day) max. 100 mg[a]	10–14 days
Chronic Q fever	Adults	Doxycycline (100 mg 2 × day) and Hydroxychloroquine (200 mg 3 × day)[a,c]	>18 months
	Children <12 years	Trimethoprim (4–6 mg/kg) and Sulfamethoxazole (20–30 mg/kg) 2 × day[a,d]	>18 months

[a]CDC, Centers for Disease Control and Prevention (2015)
[b]Angelakis and Raoult (2010)
[c]Carcopino et al. (2007)
[d]Lazzerini and Tickell (2011)

prolong the incubation period and does not prevent the infection from occurring. Likewise, treatment of asymptomatic infections or resolved cases is not routinely recommended, although it can be considered in patients with risk factors for developing chronic infection of Q fever.

Healthy patients without identified risk factor for chronic Q fever development should be clinically evaluated and subjected to serological testing by IFA 6 months after the acute infection. Patients with heart or other risk factors for the development of chronic disease should be evaluated clinically and serologically tested by IFA at 3, 6, 12, 18, and 24 months after acute infection. Pregnant women should be evaluated clinically and serologically, using the IFA at 3, 6, 12, 18, and 24 months after delivery.

Patients with chronic Q fever should be referred to a specialist in infectious diseases for tracking, since they require long-term treatment with antibiotics (using a combination of doxycycline and hydroxychloroquine), periodic diagnosis, and long-term monitoring.

Special care should be taken when introducing a new animal into a herd that is free of Q fever. Simply ensuring that the new animal is not infected is insufficient because Q fever is principally transmitted through the air. During outbreaks of Q fever, animal infections and contamination of the environment can be avoided or decreased by destroying placentas and fetuses (in order to avoid ingestion by wild or domestic carnivores). If possible, births must be confined to a specific location and disinfected without inducing aerosols. As in all zoonotic diseases, control of disease in animals will influence the level of disease observed in humans. Appropriate strategies of tick control and good hygiene practices can reduce environmental contamination. Fluids and infected fetal membranes, aborted fetuses, and contaminated bedding materials must be incinerated or buried after disinfection. In addition, the manure should be treated with lime or calcium cyanide (0.4 %) before spreading on fields, which should not be done in the wind (to prevent the spread of the organism over long distances). Although expensive, infected animals should be removed from herds or placed in separate confinement at delivery. Animal industry workers should be fully informed about the risk factors of contracting Q fever, and laboratories should be provided with appropriate safety devices and equipment (Fig. 12.6).

Leakage of potentially infectious materials must be decontaminated immediately by the use of 70 % ethyl alcohol, 5 % peroxide, or phenolic-based solutions. Biohazardous waste should be decontaminated by autoclaving and equipment/

Fig. 12.6 Personal protective equipment to be used indoors during collection of samples contaminated with *Coxiella burnetii*

Table 12.3 Inactivation of *Coxiella burnetii* in relation to various physical and chemical agents

	Agents	Inactivation
Physical agents	Temperature of 65 °C	30 min
	UV radiation	Resistant
Chemical agents	5 % Formaldehyde	Resistant
	Lysol (5 %)	Resistant
	70 % Ethyl alcohol	30 min
	Chloroform at 5 %	30 min
	Sodium hypochlorite 0.5 %	Resistant

instruments decontaminated by disinfectants, autoclaving, or boiling for 10 min. However, the spore-like shape of *C. burnetii* may be resistant to normal disinfectants (e.g., sodium hypochlorite and ultraviolet radiation), heat, drying, pressure, and oxidative and osmotic stress (Table 12.3).

Pasteurization at 72 °C for 15 s and sterilizing of milk of infected flocks are regularly recommended, even if an oral route is not considered a major mode of transmission. Worryingly, the suspension of *C. burnetii* in aqueous solutions of 0.5 % hypochlorite, 5 % Lysol, or 5 % formalin at does not completely inactivate the agent after 24 h at 24 °C.

Q-VAX is the only type of vaccine that provides human protection against Q fever and has been widely used in Australia. The vaccine causes only minor reactions, including edema and hyperemia and, more rarely, edema at the site of inoculation.

References

Angelakis E, Raoult D (2010) Review Q fever. Vet Microbiol 140:297–309

Arricau-Bouvery N, Rodolakis A (2005) Is Q fever an emerging or re-emerging zoonosis? Vet Res 36:327–349

Brenner DJ, O'Connor SP, Winkler HH et al (1993) Proposals to unify the genera *Bartonella* and *Rochalimaea*, with descriptions of *Bartonella quintana* comb. nov., *Bartonella vinsonii* comb. nov., *Bartonella henselae* comb. nov., *B. elizabethae* comb. nov., and to remove the Family Bartonellaceae from to order *Rickettsiales*. Int J Syst Microbiol 43:711–715

Breitschwerdt EB, Maggi RG, Chomel BB et al (2010) Bartonellosis: an emerging infectious disease of zoonotic importance to animals and human beings. J Vet Emerg Crit Care 20:8–30

Birtles R (2005) Bartonellae as elegant hemotropic parasites. Ann N Y Acad Sci 106:3270–3279

Billeter SA, Levy MG, Chomel BB et al (2008) Vector transmission of *Bartonella* species with emphasis on the potential for tick transmission. Med Vet Entomol 22:1–15

Carcopino X, Raoult D, Bretelle F et al (2007) Managing Q fever during pregnancy the benefits of long-term cotrimoxazole therapy. Clin Infect Dis 45:548–555

Cazorla C, Socolovschi C, Jensenius M et al (2008) Tick-borne diseases: tick-borne spotted fever rickettsioses in Africa. Infect Dis Clin North Am 22:531–544

CDC, Centers for Disease Control and Prevention (2015) Q fever. Access http://www.cdc.gov/qfever/symptoms/index.html

Chomel BB, Boulois HJ, Maruyama S et al (2006) *Bartonella* spp. in pets and effect on human health. Emerg Infect Dis 12:389–394

Ducan AW, Maggi RG, Breitschwerdt EB (2007) A combined approach for the enhanced detection and isolation of *Bartonella* species in dogs blood samples: pre-enrichment liquid culture followed by PCR and subculture onto agar plates. J Microbiol Methods 69:273–281

Dumler JS, Barbet AF, Bekker CP et al (2001) Reorganization of genera in the families Rickettsiaceae and Anaplasmataceae in the order Rickettsiales: unification of some species of *Ehrlichia* with *Anaplasma*, *Cowdria* with *Ehrlichia* and *Ehrlichia* with *Neorickettsia*, descriptions of six new species combinations and designation of *Ehrlichia equi* and 'HGE agent' as subjective synonyms of "*Ehrlichia phagocytophila*". Int J Syst Evol Microbiol 51:2145–2165

Eremeeva M, Dasch GA (2015) Challenges posed by tick-borne rickettsiae: eco-epidemiology and public health implications. Front Public Health 3:55

Georgiev M, Afonso A, Neubauer H et al (2013) Q fever in humans and farm animals in four European countries, 1982 to 2010. Euro Surveill 18:20407

Grisoli D, Million M, Edouard S et al (2014) Latent Q fever endocarditis in patients undergoing routine valve surgery. J Heart Valve Dis 23:735–743

Harms A, Dehio C (2012) Intruders below the radar: molecular pathogenesis of *Bartonella* spp. Clin Microbiol Rev 25:42–78

Hogerwerf L, van den Brom R, Roest HIJ et al (2011) Reduction of *Coxiella burnetii* prevalence by vaccination of goats and sheep, the Netherlands. Emerg Infect Dis 17:379–386

Kosoy M, Hayman DTS, Chan KS (2012) *Bartonella* bacteria in nature: where does population variability end and a species start? Infect Genet Evol 12:894–904

Labruna MB (2003) Brazilian spotted fever: the role of capybaras. In: Moreira JR, Ferraz KMPMB, Herrera EA et al (eds) Capybara: biology, use and conservation of an exceptional Neotropical species. Springer Science + Business Media, New York, pp 371–383

Labruna MB (2009) Ecology of *Rickettsia* in South America. Ann N Y Acad Sci 1166:156–166

Lamas CCA, Boia MN, Lemos ERS (2008) Human bartonellosis: seroepidemiological and clinical features with an emphasis on data from Brazil – a review. Mem Inst Oswaldo Cruz 103:221–235

Lazzerini M, Tickell D (2011) Antibiotics in severely malnourished children: systematic review of efficacy, safety and pharmacokinetics. Bull World Health Organ 89:593–606

Madariaga MG, Rezai K, Trenholme GM et al (2003) Q fever: a biological weapon in your backyard. Lancet Infect Dis 3:709–721

Maggi RG, Mascarelli PE, Pultorak EL et al (2011) *Bartonella* spp. bacteremia in high-risk immunocompetent patients. Diagn Microbiol Infect Dis 71:430–437

Maurin M, Raoult D (1999) Q fever. Clin Microbiol Rev 12:518–553

McClintic J, Srivastava S (2012) Imaging in the diagnosis and management of ocular cat scratch disease. Int Ophthalmol Clin 4:155–161

Miceli MH, Veryser AK, Anderson AD et al (2010) A case of person-to-person transmission of Q fever from an active duty serviceman to his spouse. Vector Borne Zoonotic Dis 10:539–541

Minick MF, Andreson BE, Lima A et al (2014) Oroya fever and verruga peruana: bartonelloses unique to South America. PLoS Negl Trop Dis 8:1–19

Nava S, Beati L, Labruna MB et al (2014) Reassessment of the taxonomic status of *Amblyomma cajennense* (Fabricius, 1787) with the description of three new species, *Amblyomma tonelliae* n. sp., *Amblyomma interandinum* n. sp. and *Amblyomma patinoi* n. sp., and reinstatement of *Amblyomma mixtum* Koch, 1844, and *Amblyomma sculptum* Berlese, 1888 (Ixodida: Ixodidae). Ticks Tick Borne Dis 5(3):252–276. doi:10.1016/j.ttbdis.2013.11.004

Nourse C, Allworth A, Jones A (2004) Three cases of Q fever osteomyelitis in children and a review of the literature. Clin Infect Dis 39:61–66

O'Neill TJ, Sargeant JM, Poljak Z (2014) The effectiveness of *Coxiella burnetii* vaccines in occupationally exposed populations: a systematic review and meta-analysis. Zoonoses Public Health 61:81–96

Paddock CD, Finley RW, Wright CS et al (2008) *Rickettsia parkeri* rickettsiosis and its clinical distinction from Rocky Mountain spotted fever. Clin Infect Dis 47(9):1188–1196

Parola P, Labruna MB, Raoult D (2009) Tick-borne rickettsioses in America: unanswered questions and emerging diseases. Curr Infect Dis Rep 11:40–50

Parola P, Paddock CD, Socolovschi C et al (2013) Update on tick-borne rickettsioses around the world: a geographic approach. Clin Microbiol Rev 26:657–702

Raoult D, Marrie T, Mege J (2005) Natural history and pathophysiology of Q fever. Lancet Infect Dis 5: 219–226

Rodolakis A (2009) Q Fever in dairy animals. Rickettsiology and rickettsial diseases-fifth international conference. Ann N Y Acad Sci 1166:90–93

Sander A, Berner R, Ruess M (2001) Serodiagnosis of cat-scratch disease: response to *Bartonella henselae* in children and a review of diagnostic methods. Eur J Clin Microbiol Infect Dis 20:392–401

Szabó MPJ, Pinter A, Labruna MB (2013) Ecology, biology and distribution of spotted-fever tick vectors in Brazil. Front Cell Infect Microbiol 3:1–9

Walker DH (2007) Rickettsiae and rickettsial infections: the current state of knowledge. Clin Infect Dis 45: S39–S44

Further Reading

Anderson AD, Kruszon-Moran D, Loftis AD et al (2015) Seroprevalence of Q fever in the United States, 2003–2004. Am J Trop Med Hyg 92:247–255

Angelakis E, Raoult D (2014) Pathogenicity and treatment of *Bartonella* infections. Intern J Antimicrob Agents 44:16–25

Breitschwerdt EB, Kordick DL (2000) *Bartonella* infection in animals: carriership, reservoir potential, pathogenicity, and zoonotic potential for human infection. Clin Microbiol Rev 13:428–438

Centers for Disease Control and Prevention (CDC) http://www.cdc.gov/travel/diseases/typhus.htm

Guptill L (2010) Bartonellosis. Veterinary Microbiol 140:347–359

National Association of State Public Health Veterinarians and National Assembly of State Animal Health Officials (2013) Prevention and control of *Coxiella burnetii* infection among humans and animals: guidance for a coordinated public health and animal health response. Available from http://www.nasphv.org/Documents/Q_Fever_2013.pdf

Oteo JA, Santiago N, Sousa R et al (2014) Guías Latinoamericanas dc la RIICER para el diagnóstico de las rickettsiosis transmitidas por garrapatas/Latinamerican guidelines of RIICER for diagnosis of tick-borne rickettsioses. Rev Chil Infect 31:54–65

13 Borreliosis

Cláudio Mafra and Carlos Emmanuel Montandon

Abstract

Bacteria of the genus *Borrelia* are responsible for causing numerous diseases that can affect humans and also animals. Among these diseases we have Lyme disease and relapsing fever. Lyme disease, whose etiological agents are bacteria of the *Borrelia burgdorferi* s.l., is considered a major emerging zoonosis, especially in the northern hemisphere. Very important in the past, but now with a reduced number of accounts, due to a decrease in its vector, the louse, relapsing fever is caused by at least 15 different *Borrelia* species. Several other borreliosis affect animals and deserve significant attention. Avian and bovine borreliosis can be cited as examples of these. Currently, the main method of control and prophylaxis of these diseases is the control of their vectors, since there are no vaccines for them. Epidemiological studies are also of great value for the control of borreliosis, since they allow monitoring the endemic areas for these ills.

Keywords

Borreliosis • *Borrelia* • Lyme disease • *Ixodes*

13.1 Introduction

Borreliosis are diseases caused by the *Borrelia* genus of microorganisms, which are bacteria of the Borreliaceae family. Among the infections brought by these organisms, Lyme disease is prominent, characterized by being an infectious multisystemic disorder affecting humans and also wild and domestic animals (Fonseca et al. 2005), with skin, joint, neurological, and heart problems. It is the main disease caused by the *Borrelia burgdorferi* s.l., which comprises a

C. Mafra (✉) • C.E. Montandon
Laboratory of Parasitology and Molecular Epidemiology, Department of Biochemistry and Molecular Biology, Federal University of Viçosa, Viçosa, MG, Brazil
e-mail: mafra@ufv.br; em.carlos1984@gmail.com

C.B. Marcondes (ed.), *Arthropod Borne Diseases*, DOI 10.1007/978-3-319-13884-8_13

group of at least 20 species (Clark et al. 2014). Lyme disease is considered an emerging zoonosis of great importance in the USA and Europe, being transmitted to humans by the bite of *Ixodes* ticks (Yoshinari 2009). *Borrelia* organisms are also responsible for many other diseases, among which can be cited: Brazilian Lyme-*simile* disease (Yoshinari et al. 2003), relapsing fever (El-Bahnsawy et al. 2012), avian borreliosis (Saif 2013), bovine borreliosis (Smith et al. 1978), and bovine enzootic abortion (Teglas et al. 2011).

13.2 History

Analyzing the history of borreliosis, when it comes to Lyme disease, although emerging, can be affirmed that it is certainly very old. The first borreliosis symptom reports were described by Buchwald in 1883 in Germany and by Marcus in 1910, demonstrating the relationship between these symptoms and the bite of *Ixodes* genus ticks. In 1922, the neurologists Garin, Bujadoux, and Bannwarth described a meningopolyradiculoneuritis frame, later known as Bujadoux-Garin-Bannwarth syndrome, now considered a typical form of neurologic manifestation of Lyme disease. Finally, in 1976, the disease was identified in the region of Lyme, Connecticut, USA, where several cases of acute arthritis in adolescents led to evidence of pathology, thereby leading it to be called Lyme disease[1]. Later, in 1981, Burgdorfer and colleagues isolated an unknown spirochete in *Ixodes scapularis* tick specimens collected in this region (Burgdorfer et al. 1982). Reviewing smears of *Ixodes ricinus* gut stained with Giemsa, listed in Switzerland, Burgdorfer showed similar spirochetes previously identified in *I. scapularis* ticks (Burgdorfer et al. 1983). These structures were also observed in skin lesions and blood samples obtained from patients, providing strong evidence that such a body was the causative agent of Lyme disease (Steere et al. 1983).

The demonstration about the success in growing spirochetes from material collected from the gut of the tick, designated as strain B31 (Barbour 1984) in Barbour-Stoenner-Kelly (BSK) medium, initiated intensive research on Lyme disease. After studies, the B31 strain was characterized and named *B. burgdorferi* s.s.

In the following years, several cases were reported in mainland Europe and Asia where *B. garinii* and *B. afzelii* spirochetes were described as etiological agents in these regions. In South America, positive serology results against bacteria of the *Borrelia* genus have been reported in humans at Argentina (Stanchi and Balague 1993) and Bolivia (Ciceroni et al. 1994).

In Brazil, the first reports of Brazilian Lyme-*simile* disease were through skin manifestations related to tick bites that occurred in 1987 in the city of Manaus, Amazonas State, during the XLII Brazilian Congress of Dermatology. In 1992, Yoshinari et al. described cases of patients with extracutaneous manifestations and positive serology using enzyme-linked immunosorbent assay (ELISA) and Western blotting targeting *B. burgdorferi* s.s. New cases of patients showing cutaneous involvement and positive serology results were also reported in São Paulo. The isolation and characterization of the causative agent were not possible for these cases (Yoshinari et al. 1993). More recently, in 2014, the presence of antibodies against *B. burgdorferi* s.s. was evidenced by ELISA in wild and domestic animals in the state of Minas Gerais, Brazil (Montandon et al. 2014).

A borreliosis described since the time of Hippocrates, relapsing fever, named after an outbreak in Edinburgh between the years 1843 and 1848, had its etiology described by Otto Obermeier in 1873. Since then many studies have been done. However, after the outbreak occurred in 1977 in the city of Lyme, an increasing number of studies on Lyme disease have been initiated, while relapsing fever has been largely forgotten and is currently regarded as a neglected tropical disease infection (Cutler 2009).

Avian borreliosis, in turn, was first described in an extensive outbreak among geese in Russia. A few years later in Brazil, Marchoux and Salimbeni incriminated *Argas* ticks as natural vectors of the disease, which was later confirmed

[1] See Chap. 3 for more information on the history of Lyme disease (CBM).

by several authors around the world. *Borrelia theileri*, the etiologic agent of bovine borreliosis, was identified over 100 years ago in South Africa and named in honor of its discoverer Arnold Theiler. This species is considered the least characterized *Borrelia* organism transmitted by ticks. Historically, these bacteria have been described in Africa, Australia, South America, and North America (McCoy et al. 2014). In the case of bovine enzootic abortion, it is believed that it was first reported in 1923 and later described in 1956. Subsequent occurrences of this disease were reported in parts of California (Hall et al. 2002).

13.3 Characterization

Borrelia organisms are bacteria belonging to the order Spirochaetales, Borreliaceae family, accounting for many different diseases that affect both men and animals (Fonseca et al. 2005). Among these diseases can be cited:

- Lyme disease
- Brazilian Lyme-*simile* disease, a variant of Lyme disease found in Brazil, transmitted by *Amblyomma sculptum* (*A. cajennense* complex) ticks (Yoshinari et al. 2003)
- Relapsing fever caused by *Borrelia recurrentis* when transmitted by lice and more than 20 species of the *Borrelia* genus transmitted by ticks (El-Bahnsawy et al. 2012)
- Avian borreliosis, whose etiologic agent is *Borrelia anserina*, causing anemia, fever, apathy, and high morbidity rates in birds (Saif 2013)
- Bovine borreliosis, caused by *B. theileri*, which determines a discrete process that generates anemia in ruminants and horses, being of low pathogenic action (Smith et al. 1978)
- Bovine enzootic abortion, a disease that can affect cattle, deer, and, occasionally, humans, caused by *Borrelia coriaceae* (Teglas et al. 2011)

Different species of the *Borrelia* genus can cause Lyme disease, constituting the *B. burgdorferi* s.l. group, with at least 20 species, including three with the greatest epidemiological importance, namely, *B. burgdorferi* s.s., *B. garinii*, and *Borrelia afzelii*. In the USA, *B. burgdorferi* s.s. is the main cause of Lyme disease, with *B. garinii* and *B. afzelii* incriminated as causative agents of disease in both Europe and Asia (Clark et al. 2014) (Fig. 13.1).

The *Borrelia* genus has a spiral shape without a rigid cell wall, with dimensions varying from 10 to 30 μm in length and 0.2–0.3 μm in diameter, having from 7 to 11 flagella on each end that confer mobility on these microorganisms. *Borrelia* organisms have a small linear chromosome, with up to 21 linear and circular plasmids. Its flagella

Fig. 13.1 A single *Borrelia* spirochete in vitro culture (Courtesy of Dr. Natalino Hajime Yoshinari)

have basically two types of flagellin (Fla), a smaller form, FlaA, 38 kDa, and another, larger form, FlaB, 41 kDa; both are conserved proteins in species of the *B. burgdorferi* s.l. complex. When cultured in vitro, borrelia organisms lose their flagella and spiral shape.

The composition of the *Borrelia* cell envelope is similar to that found in Gram-negative bacteria, highlighting as differences the absence of lipopolysaccharides and the abundance of lipoproteins in the outer membrane. Phosphatidylcholine, phosphatidylglycerol, and two atypical glycolipids: 1-*O*-palmitoyl-2-O-oleoyl-3-*O*-α-D-galactopiranosilglicerol and cholesteryl-6-O-palmitoyl-β-D-galactopyranoside, capable of stimulating the production of specific antibodies in rats, are included among the lipids found in *Borrelia* membranes. Cross-reactivity between these antibodies and the ganglioside component in neuron membranes is suggested as a possible cause for the neurological disorders observed in advanced cases of Lyme disease (Hossain et al. 2001).

Regarding genetic factors, take the example of the *B. burgdorferi* s.s. genome, a highly complex one with approximately 8 % of all open reading frames (ORFs) encoding lipoproteins and 6 % dedicated to motility and chemotaxis. Being fully sequenced, the *B. burgdorferi* B31 strain is composed of a small linear chromosome, with approximately 1 Mpb and 21 plasmids (12 circular and 9 linear) corresponding to 40 % of the total DNA bacteria. The main aspect of the *B. burgdorferi* s.s. genome is the large number of lipoprotein-coding sequences, including surface outer membrane proteins (outer membrane protein—Osp), which are subdivided into OspA to OSPF (Fraser et al. 1997). In contrast, this genome also encodes some enzymes required for few amino acid, fatty acids, nucleotide and biosynthetic pathways, and cofactors, explaining the dependency in relation to this spirochete on biomolecules being produced by the host cells. It was also found that the *B. burgdorferi* s.s. genome does not encode toxins to the host cells, with this extracellular pathogen causing infection due to its migration through the tissues, adhering to cells, and dodging the immune defenses (Steere et al. 2004).

With respect to molecular markers, the gene for the β-subunit of RNA polymerase (rpoB) is worth mentioning which is highly conserved in the *Borreliaceae* family, and the 16S-23S spacer of ribosomal DNA (rDNA) that, unlike rpoB, is quite divergent across *B. burgdorferi* s.l. complex species, allowing better differentiation between closely related species. Nevertheless, until now, using this gene for the detection of *Borrelia* genus bacteria in environmental samples has not generated satisfactory results, since this region has a size greater than 3 kb, which hampers its amplification by chain reaction polymerase (PCR), a limitation due to the low processivity of the Taq DNA polymerase enzyme (Lee et al. 2010).

13.4 Biology

Borrelia organisms are primarily transmitted by ticks although, experimentally and in rare cases, they can be transmitted by horse flies, mosquitoes, and fleas (Magnarelli et al. 1987). Within the *Spirochaetia* class, covering the *Borrelia*, *Leptospira*, and *Treponema* genera (Cavalier-Smith 2002), only borrelias are transmitted by blood-sucking ectoparasites.

In ticks, borrelia organisms usually develop as symbionts, acting as parasites when they infect humans and other animals (Hoogstraal 1985). Silva and Fikrig (1997) reported that in successive infections by *Borrelia* without involvement of the tick vector, there was a decrease in its pathogenicity, with the sample becoming apathogenic. A biochemical dependency was also found between *Borrelia* and their spirochete vectors, which occurs mainly in the intestinal tract of the tick during the development and multiplication of these organisms (Hoogstraal 1985). This phenomenon, as well as the chemotaxis of specific sites for spirochete in the digestive tract of the tick, is associated with the activation of genes responsible for certain phases of the *Borrelia* life cycle (Silva and Fikrig 1997).

The growth and multiplication of *Borrelia* in ticks are also influenced by the life cycle of this arthropod, with many spirochetes dying soon after the vector molts to the next stage and with the tick succumbing due to injuries to their internal organs caused by an excessive number of spirochetes (Smith et al. 1978).

In vector mode, *Borrelia* transmission by ticks may be transovarian or transstadial. In these *Borrelia* species, transmission occurs primarily by the transovarian route, being well characterized in *Argas* ticks with *B. anserine*. However, it is important to note that in these ticks, the transmission is intrinsically related to the strain, physiographic region, *Borrelia* species, and association with other pathogens (Hoogstraal 1985). Ixodid ticks can support both transmission modes (Smith et al. 1978), and this phenomenon is better understood in Lyme disease, with *I. scapularis* and other species of *Ixodes* vectors of *B. burgdorferi* s.l. (Burgdorfer et al. 1985) (Fig. 13.2).

For the host infection, the tick's fixing time is considered a highly relevant factor for *Borrelia* transmission efficiency, with some studies showing that a time of attachment longer than 48 h is necessary (Piesman et al. 1987). For ticks this is not a relevant parameter for transmission of the pathogen, since transmission to representatives of this family takes place both by saliva and coxal liquid, except at the larval stage that realizes a blood meal in a few minutes (Hoogstraal 1985).

Ticks have the potential to transmit almost all borrelias. In the five borrelia groups currently recognized, all are transmitted by ticks, with each one having a specific kind of spirochete and arthropod vector. In the Northern US and Europe, spirochete transmission of the *B. burgdorferi* s.l. group causing Lyme disease occurs mainly by the *Ixodes* tick genus. In warmer regions, such as in the Southern US, cases of Lyme disease are associated with infection by *B. lonestari*, which is transmitted by *Amblyomma americanum* ticks (Barbour et al. 1996). In South America, *Amblyomma* ticks are responsible for the transmission of *Borrelia* organisms (Yoshinari 2009).

As demonstrated in epidemiological studies, the principal natural reservoirs of *Borrelia* found in America, Europe, Asia, and South America are marsupials (opossum), rodents (bush rat, agouti, and capybara), insectivores (hedgehog), canids, equines, cattle, and deer (Costa et al. 2002).

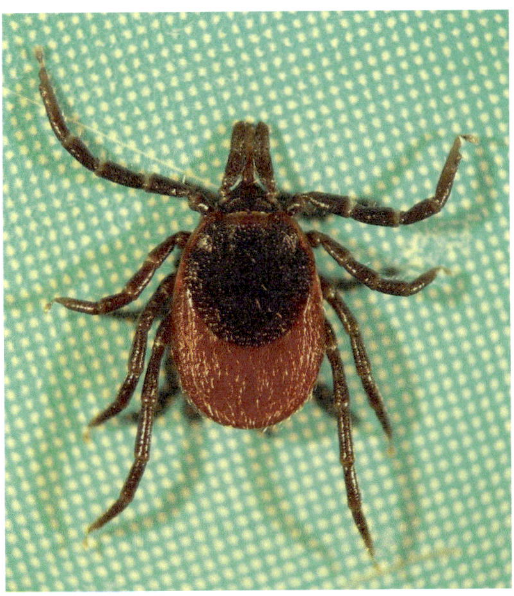

Fig. 13.2 *Ixodes scapularis* (female) (Courtesy of Dr. Jose M. Venzal)

13.5 Clinical signs

Clinical signs in animals with borreliosis, including humans, may vary according to the species of bacteria causing the disease. Among the first clinical signs of Lyme disease, skin lesions called erythema migrans, which expand from the site of the tick bite, appear 8–9 days after the bite in the classic cases (Fonseca et al. 2005). In Brazil, this lesion appears, on average, 30 days after exposure to ticks (Yoshinari et al. 2000). Although erythema is a unique feature of the disease, it is not present in all cases, manifesting in 70 % of American patients (Fonseca et al. 2005) (Fig. 13.3).

The classical erythema migrans lesion is characterized in its early stage by presenting a ring-shaped macula that is reddish, in various shades, measuring between 0.5 and 2 cm, which may arise in the tick bite site with a papule that

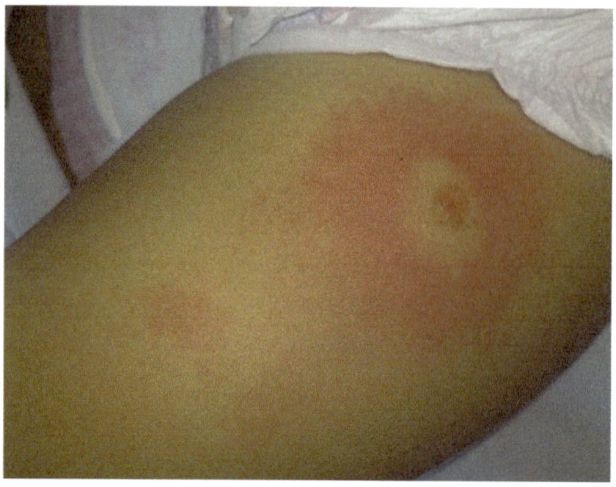

Fig. 13.3 Erythema migrans lesion showing typical lesion characteristics in syndrome of Baggio-Yoshinari (Courtesy of Dr. Natalino Hajime Yoshinari)

corresponds to the center of the lesion (Berger et al. 1985). It slowly progresses as centrifuge dispersion, rapidly expanding and assuming, in a short period of time, a plate appearance. Typically, about 14 days after the tick bite, the erythema can reach a diameter greater than 15 cm, together with other morphological changes, such as an oval or elongated shape (Steere et al. 1983). Generally the lesion is seen in isolation; however, various injuries can occur due to the dispersion of bacteria through the vascular and lymphatic system, receiving the designation of secondary annular lesions (Berger 1989). *Borrelia burgdorferi* s.l. has been found in biopsies of lesions, suggesting that expansion of the injury is also determined by migration of the spirochetes through the subcutaneous tissue (Malane et al. 1991).

In studies of lesion biopsies, the main histopathological findings are proliferation and dilation of blood vessels with vasculitis, displaying primary infiltrates of lymphocytes associated with plasma cells. With the evolution of the erythema migrans, there is a reduction in the inflammatory process, with pronounced atrophy of the epidermis and dermis.

In addition to lymphocytes observed in the infiltrate in the site of the tick bite, the presence of macrophages, mast cells, and neutrophils, as well as a small number of eosinophils, also is observed (Koning and Hoogkamp-Korstanje 1986). According to Berger (1989), in cases where only a single lesion like erythema migrans occurs, it is considered as a localized form with milder symptoms. When multiple injuries are involved including symptoms of greater intensity, erythema is considered widespread. The distinction between these two presentations is important because in individuals with a localized lesion the response to antibiotics is positive, demanding long therapy, sometimes requiring repetitions, with a poor prognosis in relation to ultimate cure in patients with widespread erythema migrans.

The main extracutaneous symptoms observed in patients with Lyme disease are similar to those of a common cold, among which are fever, fatigue, musculoskeletal discomfort, and headache (Berger 1989).

Secondary manifestations may come to pass, with involvement of the nervous system manifesting as diffuse neuropathy or local radiculopathy or, less frequently, by meningitis or encephalitis. These neurological complications are often associated with infections caused by *B. garinii*, predominately in Europe. Cranial neuritis may also be observed, resulting in hearing loss, facial paralysis, and optic neuritis (Halperin 1997), as well as intraocular inflammation caused by infection with *Borrelia* (Flach and Lavoie 1990). Among the neurological manifestations associated with Lyme disease, several authors have suggested the infection's association with *Borrelia* neurodegenerative disorders such as Alzheimer's disease (MacDonald 2006). These secondary manifestations may recur, lasting months, or become chronic (Yoshinari et al. 2000).

Another recurring symptom of Lyme disease is European atrophying chronic acrodermatitis, which, as its name implies, occurs in the chronic phase of the disease. The characteristic of elderly people, the formation of a red and blue edema, can be observed, located in the feet or lower legs, progressing to an atrophic wrinkled lesion, allowing it show to through subcutaneous vessels (Berger 1989).

Another situation that intrigues researchers and doctors is whether *B. burgdorferi* s.s. persists after conventional antimicrobial treatments. In recent studies of patients with consecutive episodes of erythema migrans, some strains of *B. burgdorferi* s.s. demonstrated that they are distinct based on their genotypes, indicating the occurrence of reinfection (Nadelman et al. 2012; Shapiro 2015). Recently, success was obtained in cultivating strains of *B. burgdorferi* s.s. found in ticks fed on patients treated with antibiotics, showing the presence of bacteria even after conventional treatment, demonstrating both the possibility of reinfection and the persistence of *B. burgdorferi* in patients who have already had or maintained Lyme disease (Shapiro 2015).

Regarding the symptoms, the Brazilian Lyme-*simile* disease displays the same symptoms of Lyme disease observed in the northern hemisphere, such as erythema migrans, cardiac involvement, and cases of neurological complications, including associations with meningitis, facial palsy, and bilateral sudden deafness (Costa et al. 1996), differing mainly from Lyme disease recurrence in terms the symptoms and development of autoimmune disorders. It is noteworthy that, unlike Lyme disease, where ticks the *I. ricinus* complex are seen as the main vectors in the northern hemisphere, Brazilian Lyme-*simile* disease cases are associated with the bite of *A. cajennense* ticks (Fonseca et al. 2005).

Symptoms of relapsing fever differ on some points when transmitted by ticks or lice. When transmitted by ticks, it is characterized by presenting an initial episode of incessant fever, which lasts 3–6 days. When the transmission is due to lice, multiple febrile episodes may take 1–3 days each. However, in both situations an incubation interval occurs between 4 and 14 days, with fever lasting from several hours to 4 days, accompanied by chills, headache, nausea, vomiting, sweating, abdominal pain, arthralgia, and cough. Acute respiratory distress syndrome and specific neurological symptoms (hemiplegia, facial paralysis, myelitis, and radiculopathy) may occur in the disease transmitted by ticks, being a point of distinction between these two forms (El-Bahnsawy et al. 2012).

The onset of borrelias in other animal species such as birds shows that the youngest animals apparently are more susceptible to *B. anserina*, depression with cyanosis in the head being the main symptom in the acute form, leading to mortality in up to 30 % of the herd. Birds affected by this disease also present body temperature beginning to rise just after the infection, with rapid weight loss, water consumption increase, and production of green excrement due to bile excess. In subacute and chronic cases, birds demonstrate weakness that ends in paralysis and death[2]. Pathological features include grossly enlarged spleen and stains due to subcapsular hemorrhage occurring at the site, being this the predominant lesion. In some cases focal necrotic hepatitis also occurs.

In cattle and other animals, *B. theileri* infection causes smooth and variable clinical signs, including fever and anemia (McCoy et al. 2014).

Bovine enzootic abortion lesions develop gradually over a period of 3 months. At the end of this period, the fetus may be aborted or born alive with serious complications, such as bleeding in the mucosa and petechial lesions in the conjunctiva and oral cavity, among other symptoms (Hall et al. 2002). The pathogenesis of this disease consists of a generalized inflammation with vascular lesions in most organs. In the brain vasculitis and granulomatous meningitis occur. The lung is affected by alveolar septal infiltration with mononuclear cells and interlobular septum, with granulomatous focal infiltration. Hepatic occurrence, although less specific, consists of a granulomatous infiltration with centrilobular congestion variable and atrophy of hepatic cord.

13.6 Diagnosis

The diagnosis of borreliosis in domestic animals, such as avian borreliosis and bovine enzootic abortion, in addition to serological and molecular tests, also is based on the characteristics of the macroscopic and histological lesions, as well as additional information from the herd history, such as exposure to tick vector.

[2]See Chap. 34 for more information on paralysis caused by ticks (CBM).

From a clinical point of view, diagnosis in the early stage of borreliosis, regardless of the etiological agent involved, is inaccurate due to the lack of specific symptoms. When it comes to Lyme disease, in the first 3 weeks of illness, patients have similar signs of a cold frame, associated with reddish cutaneous complications spreading throughout the body from the tick bite site. Only after a few months, when the disease moves to a chronic condition, the symptoms do become more characteristic, coming to show joint, heart, and neurological complications. Although the diagnosis of Lyme disease may be difficult and even unfeasible, it needs to be done as early as possible, including based on clinical aspects (Steere et al. 2004).

The initial laboratory diagnosis for borreliosis in animals, including man, is done on peripheral blood smears stained by Giemsa or Fontana methods, though causative agents can be detected only in cases of high spirochetemia. Due to this, the gold standard procedure for diagnosis of borreliosis in general is bacterium isolation by culture, followed by PCR confirmation or another specific test. This method is very costly, since the culture requires special materials such as the complex BSK medium and highly trained staff to perform the procedure, so it is not a routinely performed technique. In addition, these procedures are very time-consuming, taking between 2 and 6 weeks before they are useful to the clinical decision-making.

Another factor to be taken into consideration during the diagnosis of Lyme disease is the sensitivity of the described procedures, which are successful with between 40 and 70% of erythema migrans cases, in only 3–17% when using cerebrospinal fluid samples, and still further reduced in efficacy when using synovial fluid and other tissues. This is due to the requirement for a relatively complex culture medium for the cultivation of *Borrelia* in the laboratory, since these spirochetes do not possess genes encoding enzymes necessary for amino acid, fatty acid, and nucleotide biosynthetic pathways or nucleotide cofactors. This difficulty has also been found in *Borrelia* isolation from vector samples.

Therefore, applications that rely on immunological techniques such as ELISA or those based on molecular biology—PCR, nested PCR, and real-time PCR—are increasingly sought for early reliable diagnosis of borreliosis (Dunaj et al. 2013).

It is known, however, that serological confirmation of infection can be impaired by several factors, particularly where the possibility of coinfection by other spirochetes is possible, e.g., of the genus *Leptospira*, with the occurrence of cross-reactions. This occurs mainly when using whole-cell extracts of *Borrelia* as antigen for the reaction, resulting in false positives (Rogers et al. 1999). Another problem is the production of anti-*Borrelia* antibodies. Typically detected only in the first or second week after infection, specific IgM and IgG antibodies against *Borrelia* can persist for years, making it impossible to distinguish between past and recent infections in some patients (Steere et al. 2008).

The improvement of techniques based on molecular biology afforded the advent of more sensitive methods for the detection of DNA pathogens in environmental or human samples. In this field, PCR has proven a very promising technique due to its high sensitivity and specificity in the early diagnosis of borreliosis and also to confirm diagnoses obtained by other techniques. Introducing advantages over immunological techniques, PCR allows amplified sequences of the target DNA pathogen to be obtained, post-sequencing of which enables the performance of in silico analyses against other sequences stored in biological databases. Then, they can be compared at the molecular level with sequences from different species/strains of *Borrelia*, and phylogenetic relationships between them can be established. For the molecular detection of these bacteria, the choice of DNA extraction processes, primers, and reaction conditions should not differ from the standard, which should be aware when selecting the sequence to be amplified and the PCR method (Dunaj et al. 2013).

Different patterns of molecular marker sequences are accepted for phylogenetic analysis of *Borrelia* DNA sequences, among them some for Lyme disease such as fla, recA, 16S rDNA, p66, hbb, rpoB, intergenic spacers in the 23S and 5S genes, and plasmid carriers (OspA, OspB, and ospCVlsE) (Dunaj et al. 2013). In choosing the sequence to be amplified, one must also be assured that it is not homologous to genetic material from another organism, e.g., *Treponema*, *Leptospira*, or *Escherichia coli*, or the human or animal DNA being analyzed. Nested PCR has been shown to be a very sensitive tool for diagnosis through the detection of DNA pathogens, amplifying an internal sequence of a previously amplified fragment, which increases the sensitivity over traditional PCR.

13.7 Treatment

Symptoms and signs of clinical borreliosis are commonly treated with antibiotics to prevent manifestations of other symptoms. The classes of antibiotics that demonstrate greater efficacy against *Borrelia* are β-lactam, in particular cephalosporins, and, to a lesser degree, macrolides. Definition of the best treatment still generates much debate, since not all patients and species or strains of this pathogen respond significantly to the same base.

For the treatment of Lyme disease, based on the evidence available from randomized clinical trials, either more lenient or harsh recommendations have been published by the Infectious Diseases Society of America (IDSA) and the American Academy of Pediatrics, among other national and supranational medical associations, including some from Europe.

According to the IDSA, the therapy recommended against the main signal of Lyme disease, erythema migrans, and to stop evolution of the disease to more complex clinical situations is the following:

- Daily, during 14 days, two doses of 100 mg of doxycycline, orally, or 2 mg/kg of body weight for children; three daily doses of 500 mg of amoxycillin, for 14 days orally, or 50 mg/kg for children; two daily doses of 500 mg of amoxycillin, for 14 days orally, or 30 mg/kg of body weight for children.
- For the treatment of relapsing fever, experts recommend 500 mg of tetracycline, orally, four times/day, during 10 days, or 12.5 mg/kg of body weight for children. In cases where tetracycline is not effectively recommended, erythromycin can be used under the same conditions.

Borreliosis treatment in domestic animals is effective, being the use of oxytetracycline or chlortetracycline administered 1–2 mg/kg body weight recommended for avian borreliosis (Hampson and Swayne 2013). The treatment of bovine enzootic abortion consists in the administration of antimicrobial therapy to pregnant animal in order to save the calf, since the treatment does not have effect death of the newborn occurs within 48 h of birth.

13.8 Epidemiology

Among the borreliosis, Lyme disease is the one with the largest number of epidemiological studies due to its importance to public health. It is considered a bimodal disease, with a higher incidence in children aged 5–9 years and in adults older than 50 years, having no preference for gender, as in other immune-mediated diseases.

This disease has occurred almost exclusively in the northern hemisphere, with most of the confirmed cases from the USA and Europe, and to a lesser extent in Asia and North Africa. In the USA the situation is so critical that notification of Lyme disease has been obligatory since 1991, a growing and steady increase in the number of reported cases having been observed, ranging from 9908 cases in 1992 to a peak of 30,000 cases in 2009, coming to stabilize at a level of 25,000 cases. In the majority of European countries, Lyme disease is not a reportable disease, without until recently standardized case definitions.

Perhaps the incidence observed in the USA and Europe, restricting the majority of reported cases to the northern hemisphere, highlight the efficiency and scope of surveillance in these regions. In South America, there are many reports of this disease, with cases reported in Argentina (Stanchi and Balague 1993), Bolivia

(Ciceroni et al. 1994), and Brazil (Yoshinari 2009; Montandon et al. 2014).

Relapsing fever has been a global disease, remaining endemic in many parts of the world. Humans sporadically contract relapsing fever transmitted by ticks when entering natural environments infested with these vectors. Transmission by lice may generate an epidemic situation. Usually relapsing fever epidemics transmitted by lice occur in areas of overcrowding and poverty, e.g., as in situations caused by wars. Although the number of cases of relapsing fever seems to be falling, this could be due to inattention to this disease; reports of lice infestation resurgence among certain groups such as the homeless have been done, and it is the cause of the infection in a focus that occurred in Ethiopia (Cutler 2009).

Epidemiological studies of borreliosis in animals are still very meager, requiring more attention, since many of these cases have been reported, e.g., in poultry in Europe, North America, and Australia (Saif 2013).

13.9 Control and Prevention

Currently the most effective method available to prevent infection by bacteria of the genus *Borrelia* and other pathogens transmitted by ticks is to avoid exposure to this vector. However, in cases where exposure to these vectors is inevitable, procedures to mitigate the risk of infection are recommended, among which are included a thorough check of one's own body and the bodies of pets, with direct removal of posted ticks, and the use of protective clothing and repellents against ticks.

Antimicrobial prophylaxis for Lyme disease after a tick bite is generally not recommended due to the low rate of infection, even if the tick proves positive for *B. burgdorferi* s.s. (Shapiro et al. 1992). However, in randomized controlled trials, as performed by Nadelman et al. (2001), there was some effectiveness in preventing the development of erythema migrans with the use of a single dose of 200 mg of doxycycline within 72 h after removing *I. scapularis* ticks. IDSA recommends the preventive use of a single dose of doxycycline in the following situations: (I) the tick was fixed for a time equal to or greater than 36 h, being identified as *I. scapularis*; (II) the treatment can be administered within 72 h even after removal of the tick; (III) sites in which the rate of *Borrelia* infection in ticks is equal to or greater than 20 %; and (IV) the use of doxycycline is not a contraindicated drug for the concerned patient.

It needs to be considered that this drug is contraindicated for use in pregnant women and children up to 8 years of age. People who have removed ticks from themselves, even those who received antibiotic prophylaxis, should be carefully monitored for clinical signs and symptoms of the disease for up to 30 days, with attention mainly to the development of skin lesions due to tick bites, which may correspond to erythema migrans. Those who have signs of the disease should be advised to be examined by a doctor immediately, to have appropriate action taken regarding the diagnosis and possible treatment.

As verified with other diseases transmitted by ticks that have wild and domestic vertebrates as links in their transmission cycle, knowledge of the frequency of seropositives in wild and domestic animals is of great value in epidemiological studies of this zoonotical disease, considering the region, as some species animals may act as sentinels (Lemos et al. 2001). Data on the frequency of positive DNA *Borrelia* samples in ectoparasites thus become of extreme relevance to support the development of epidemiological surveillance strategies and the actions of health professionals in assessing geographical risk areas for the emergence of potential infections arising from agents carried by ticks (Montandon et al. 2014).

References

Barbour AG (1984) Isolation and cultivation of Lyme disease spirochetes. Yale J Biol Med 57(4):521–525

Barbour AG, Maupin GO, Teltow GJ et al (1996) Identification of an uncultivable *Borrelia* species in the hard tick *Amblyomma americanum*: possible agent of a Lyme disease-like illness. J Infect Dis 173:403–409

Berger BW (1989) Dermatologic manifestations of Lyme disease. Rev Infect Dis 11:S1475

Berger BW, Kaplan MH, Rothenberg IR, Barbour AG (1985) Isolation and characterization of Lyme disease spirochete from the skin of patients with erythema chronicum migrans. J Am Acad Dermatol 13:444–449

Burgdorfer W, Barbour AG, Hayes SF et al (1982) Lyme disease-a tick-borne spirochetosis? Science 216: 1317–1319

Burgdorfer W, Barbour AG, Hayes SF et al (1983) Erythema chronicum migrans: a tick borne spirochetosis. Acta Trop 40:79–83

Burgdorfer W, Lane RS, Barbour AG et al (1985) The western black-legged tick, *Ixodes* pacificus: a vector of *Borrelia burgdorferi*. Am J Trop Med Hyg 34: 925–930

Cavalier-Smith T (2002) The neomuran origin of archaebacteria, the negibacterial root of the universal tree and bacterial megaclassification. Int J Syst Evol Microbiol 52:7–76

Ciceroni L, Bartoloni A, Guglielmetti P et al (1994) Prevalence of antibodies to *Borrelia burgdorferi*, *Borrelia parkeri* and *Borrelia turicatae* in human settlements of the Cordillera Province, Bolivia. J Trop Med Hyg 97:13–17

Clark KL, Leydet BF, Threlkeld C (2014) Geographical and genospecies distribution of *Borrelia burgdorferi* sensu lato DNA detected in humans in the USA. J Med Microbiol 63:674–684

Costa IP, Yoshinari NH, Barros PJL et al (1996) Doença de Lyme em Mato Grosso do Sul: relato de três casos clínicos, incluindo o primeiro relato de meningite de Lyme no Brasil. Rev Hosp Clín Fac Med Sao Paulo 51:253–257

Costa IP, Bonoldi VL, Yoshinari NH (2002) Search for *Borrelia* sp. in ticks collected from potential reservoirs in an urban forest reserve in the State of Mato Grosso do Sul, Brazil: a short report. Mem Inst Oswaldo Cruz 97:631–635

Cutler SJ (2009) Relapsing fever—a forgotten disease revealed. J Appl Microbiol 108:1115–1122

Dunaj J, Moniuszko A, Zajkowska J et al (2013) The role of PCR in diagnostics of Lyme borreliosis. Przegl Epidemiol 67:35–39

El-Bahnsawy MM, Labib NA, Abdel-Fattah MA et al (2012) Louse and tick borne relapsing fevers. J Egypt Soc Parasitol 42:625–638

Flach AJ, Lavoie PE (1990) Episcleritis, conjunctivitis, and keratitis as ocular manifestations of Lyme disease. Ophthalmology 97:973–975

Fonseca AH, Salles RS, Salles SAN et al (2005) Borreliose de Lyme simile: uma doença emergente e relevante para a dermatologia no Brasil. An Bras Dermatol 80:171–178

Fraser CM, Casjens S, Huang WM et al (1997) Genomic sequence of a Lyme disease spirochaete, *Borrelia burgdorferi*. Nature 390:580–586

Hall MR, Hanks D, Kvasnicka W et al (2002) Diagnosis of epizootic bovine abortion in Nevada and identification of the vector. J Vet Diagn Invest 14:205–210

Halperin JJ (1997) Neuroborreliosis: central nervous system involvement. Semin Neurol 17:19–24

Hampson DJ, Swayne DE (2013) Avian intestinal spirochetosis. In: Saif YM (ed) Diseases of poultry, 13th edn. Wiley, New York, pp 922–940

Hoogstraal H (1985) Argasid and nuttalliellid ticks as parasites and vectors. Adv Parasitol 24:135–238

Hossain H, Wellensiek HJ, Geyer R et al (2001) Structural analysis of glycolipids from *Borrelia burgdorferi*. Biochimie 83:683–692

Koning J, Hoogkamp-Korstanje JA (1986) Diagnosis of Lyme disease by demonstration of spirochetes in tissue biopsies. Zbl Bakt Hyg A 263:179–188

Lee SH, Vigliotti VS, Vigliotti JS et al (2010) Increased sensitivity and specificity of *Borrelia burgdorferi* 16S ribosomal DNA detection. Am J Clin Pathol 133:569–576

Lemos ER, Alvarenga FB, Cintra ML et al (2001) Spotted fever in Brazil: a seroepidemiological study and description of clinical cases in an endemic area in the state of São Paulo. Am J Trop Med Hyg 65:329–334

MacDonald AB (2006) Plaques of Alzheimer's disease originate from cysts of *Borrelia burgdorferi*, the Lyme disease spirochete. Med Hypothesis 67:592–600

Magnarelli LA, Anderson JF, Schreier AB et al (1987) Clinical and serologic studies of canine borreliosis. J Am Vet Med Assoc 191:1089–1094

Malane MS, Grant-Kels JM, Feder H et al (1991) Diagnosis of Lyme disease based on dermatologic manifestations. Ann Intern Med 114:490–498

McCoy BN, Maïga O, Schwan TG (2014) Detection of *Borrelia theileri* in *Rhipicephalus geigyi* from Mali. Ticks Tick Borne Dis 5:401–403

Montandon CE, Yoshinari NH, Milagres BS et al (2014) Evidence of *Borrelia* in wild and domestic mammals from the state of Minas Gerais, Brazil. Rev Bras Parasitol Vet 23:287–290

Nadelman RB, Nowakowski J, Fish D et al (2001) Prophylaxis with single-dose doxycycline for the prevention of Lyme disease after an *Ixodes scapularis* tick bite. N Engl J Med 345:79–84

Nadelman RB, Hanincová K, Mukherjee P et al (2012) Differentiation of reinfection from relapse in recurrent Lyme disease. N Engl J Med 367:1883–1890

Piesman J, Mather TN, Sinsky RJ et al (1987) Duration of tick attachment and *Borrelia burgdorferi* transmission. J Clin Microbiol 25:557–558

Rogers AB, Smith RD, Kakoma I (1999) Serologic cross-reactivity of antibodies against *Borrelia theileri*, *Borrelia burgdorferi*, and *Borrelia coriaceae* in cattle. Am J Vet Res 60:694–697

Saif YM (ed) (2013) Diseases of poultry, 13th edn. Wiley, New York

Shapiro ED (2015) Repeat or persistent Lyme disease: persistence, recrudescence or reinfection with *Borrelia burgdorferi*? F1000 Prime Rep 7:11

Shapiro ED, Gerber MA, Holabird NB et al (1992) A controlled trial of antimicrobial prophylaxis for Lyme disease after deer-tick bites. N Engl J Med 327:1769–1773

Silva AM, Fikrig E (1997) *Borrelia burgdorferi* genes selectively expressed in ticks and mammals. Parasitol Today 13:267–270

Smith RD, Brener J, Osorno M, Ristic M (1978) Pathobiology of *Borrelia theileri* in the tropical cattle tick, *Boophilus microplus*. J Invertebr Pathol 32:182–190

Stanchi NO, Balague LJ (1993) Lyme disease: antibodies against *Borrelia burgdorferi* in farm workers in Argentina. Rev Saúde Públ 27:305–307

Steere AC, Coburn J, Glickstein L (2004) The emergence of Lyme disease. J Clin Invest 113:1093–1101

Steere AC, Grodzicki RI, Kornblatt AN et al (1983) The spirochetal etiology of Lyme disease. New Engl J Med 308:733–740

Steere AC, Mchugh G, Damle N, Sikand VK (2008) Prospective study of serologic tests for Lyme disease. Clin Infect Dis 47:188–195

Teglas MB, Mapes S, Hodzic E et al (2011) Co-infection of *Ornithodoros coriaceus* with the relapsing fever spirochete, *Borrelia coriaceae*, and the agent of epizootic bovine abortion. Med Vet Entomol 25:337–343

Yoshinari NH (2009) Uma longa jornada para entender a *Borrelia burgdorferi* no Brasil. Rev Bras Reumatol 49:483–486

Yoshinari NH, Oyafuso LK, Monteiro FGV et al (1993) Doença de Lyme. Relato de um caso observado no Brasil. Rev Hosp Clín Fac Med Sao Paulo 48:170–174

Yoshinari NH, Barros PJL, Gauditano G et al (2000) Report of 57 cases of Lyme-like disease (LLD) in Brazil. Arthritis Rheum 43:S188

Yoshinari NH, Abrão MG, Bonoldi VL et al (2003) Coexistence of antibodies to tick—borne agents of babesiosis and Lyme borreliosis in patients from Cotia county, State of São Paulo, Brazil. Mem Inst Oswaldo Cruz 98:311–318

Ehrlichiosis

14

Susan E. Little

Abstract

Ehrlichiosis is a term used to describe a group of tick-borne diseases caused by obligate intracellular bacteria of the genus *Ehrlichia* (*Rickettsiales*: *Anaplasmataceae*). A total of six species of *Ehrlichia* are recognized from humans and other animals: *E. canis*, *E. chaffeensis*, *E. ewingii*, *E. mineirensis*, *E. muris*, and *E. ruminantium*. Additional *Ehrlichia* spp. have been reported based on molecular and serologic evidence of infection, and new members of the genus are routinely discovered. Disease due to these agents occurs in people, dogs, and ruminants, with occasional reports from cats, horses, nonhuman primates, and other animals. Clinical signs of ehrlichiosis are best described in ruminants (*E. ruminantium* in Africa and the Caribbean) and dogs (*E. canis* worldwide and *E. ewingii* in the United States) and include severe, life-threatening febrile illness characterized by anorexia and lethargy; some agents are also associated with polyarthritis, neurologic disease, or diarrhea. *Ehrlichia* spp. also infect people; *E. chaffeensis*-induced human monocytic ehrlichiosis is the most commonly reported tick-borne infection in the southern United States. Ehrlichiosis is diagnosed through a combination of characteristic clinical signs, hematologic abnormalities, and specific serologic or molecular assays to detect evidence of infection. Doxycycline is considered the treatment of choice. Although research is promising, vaccines for many agents of ehrlichiosis are not widely available and prevention of disease focuses on tick control.

Keywords

Ehrlichia spp. • Ehrlichiosis • Tick • Zoonosis

14.1 Early History

In 1838, a Dutch settler in South Africa described a fatal disease of sheep following a severe tick infestation. The disease, which was also recognized

S.E. Little (✉)
Department of Veterinary Pathobiology, Center for Veterinary Health Sciences, Oklahoma State University, Stillwater, OK, USA
e-mail: susan.little@okstate.edu

C.B. Marcondes (ed.), *Arthropod Borne Diseases*, DOI 10.1007/978-3-319-13884-8_14

to affect cattle, goats, and wild ruminants, was eventually named heartwater due to the common presence of pericardial effusion and was ultimately attributed to infection with *Ehrlichia ruminantium* transmitted by *Amblyomma* spp. ticks (Bezuidenhout 2009). A second member of the genus, now referred to as *E. canis*, was first described from dogs in India in 1935 and later shown to be transmitted between dogs by *Rhipicephalus sanguineus* ticks (Donatien and Lestoquard 1937; reviewed in Little 2010). Subsequent species, including *E. chaffeensis* and *E. ewingii* in the southern United States, were initially thought to be atypical *E. canis* infections in people and dogs, respectively, but were later shown to be distinct organisms (Anderson et al. 1991; Anderson et al. 1992). Novel agents in this genus continue to be described in the literature.

The group of diseases referred to as ehrlichioses cause significant concerns for both public health and veterinary medicine. In naïve individuals, untreated infections are often fatal, and risk of infection is quite high, particularly in environments with intense tick populations.

14.2 Etiology

Ehrlichiosis is caused by bacteria of the genus *Ehrlichia* (*Rickettsiales*: *Anaplasmataceae*). Although currently comprised of six species (*E. canis*, *E. chaffeensis*, *E. ewingii*, *E. mineirensis*, *E. muris*, and *E. ruminantium*), the taxonomy of this genus and closely related organisms has been extensively revised in recent decades. For example, *E. ruminantium* was originally described as *Cowdria ruminantium* and the clinical disease referred to as "cowdriosis." Moreover, the genus formerly included additional agents such as *E. equi* (now *Anaplasma phagocytophilum*) and *E. risticii* (now *Neorickettsia risticii*) (Dumler et al. 2001). A number of novel organisms thought to be *Ehrlichia* spp. have been reported from molecular or serologic evidence of infection, but formal descriptions of these agents have not yet been published.

Ehrlichia spp. replicate inside cytoplasmic parasitophorous vacuoles within neutrophils, monocytes, macrophages, or endothelial cells, and infection can be confirmed by identifying intracellular inclusions within cells on stained blood smears or impression smears (Fig. 14.1). These inclusions, termed morulae, appear as characteristic densely staining granules within the cytoplasm of infected cells. The morulae are somewhat pleomorphic and density of appearance may vary with replication stage of the organisms (Rikihisa 1991). Table 14.1 summarizes *Ehrlichia* spp. described from animals and, where reported, tick vectors to date.

Human ehrlichiosis has been described from patients in the Americas, Africa, and Asia and attributed to infection with a variety of different *Ehrlichia* species. Most reported cases of human ehrlichiosis in the United States are caused by *E. chaffeensis* infection transmitted by *A. americanum* ticks, although *E. ewingii* (Fig. 14.2), also transmitted by *A. americanum*, and an *E. muris*-like agent transmitted by *I. scapularis* are increasingly recognized as causing disease (Salinas et al. 2010; Ismail et al. 2010; Pritt et al. 2011). In South America, the etiology and epidemiology of human ehrlichiosis are less clear. *Ehrlichia canis* has been documented in some cases of human disease, but *E. chaffeensis* and *E. ewingii* are also suspected to play a role in inducing a febrile, tick-borne disease in people; novel agents may also be involved (Vieira et al. 2011). Human infection with *E. ruminantium* has been reported from Africa, and *E. chaffeensis*-

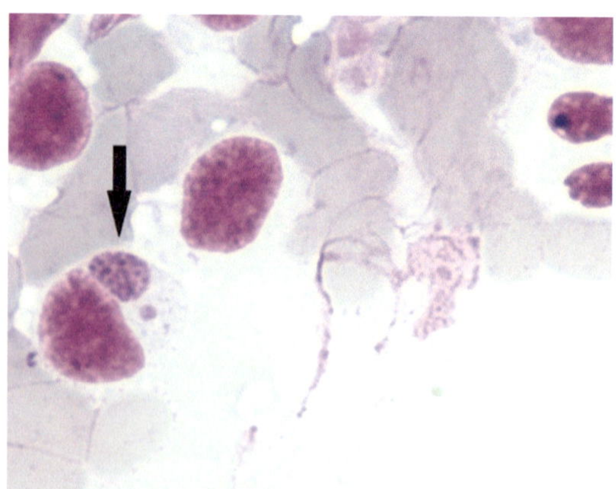

Fig. 14.1 *Ehrlichia canis* morula (*arrow*) in an infected monocyte

Table 14.1 *Ehrlichia* species described to date (reviewed in Little 2010; Ismail et al. 2010; Pritt et al. 2011; Vieira et al. 2011; Esemu et al. 2011)

Species	Principal vertebrate hosts	Primary tick vector(s)	Geographic distribution
E. canis[a]	Dogs	*Rhipicephalus sanguineus*	Worldwide
E. canis-like[b]	Cats	Unknown	North America (United States)
E. chaffeensis[c]	White-tailed deer	*Amblyomma americanum*	North America (United States), also reported elsewhere
E. ewingii	Dogs, white-tailed deer	*Amblyomma americanum*	North America (United States), also reported elsewhere
E. mineirensis	Cattle (presumed)	*Rhipicephalus microplus*	South America (Brazil)
E. muris	Mice	*Ixodes* spp.	Asia (Japan), North America (United States)
E. muris-like[b]	Mice	*Ixodes scapularis*	North America (United States)
E. ruminantium	Ruminants	*Amblyomma* spp. (several)	Africa, Caribbean
Panola Mountain *Ehrlichia* sp.[b]	White-tailed deer	*Amblyomma americanum*	North America (United States)
Ehrlichia sp.[b]	Cattle	Unknown	North America (Canada)
Ehrlichia sp.[b]	Horses	Unknown, *A. americanum* suspected	North America (United States)
Ehrlichia sp.[b]	Horses	Unknown	Central America (Nicaragua)

Not all organisms reported in the literature have been formally named

[a]Main species associated with canine ehrlichiosis worldwide. Canine ehrlichiosis is also caused by *E. ewingii*, the most common *Ehrlichia* sp. found infecting dogs in the United States; infections with *E. muris*, *E. chaffeensis*, *E. ruminantium*, and Panola Mountain *Ehrlichia* sp. have also been reported in dogs

[b]Infections described based on molecular sequence or serologic reactivity but responsible organisms not isolated in culture or formally named

[c]Main species associated with human ehrlichiosis, the most commonly reported tick-borne disease in the southern United States. The literature also contains reports of human ehrlichiosis caused by an *E. muris*-like agent, *E. ewingii*, *E. canis*, *E. ruminantium*, and Panola Mountain *Ehrlichia* sp. Human ehrlichioses due to both *E. canis* and *E. chaffeensis* have been reported in South America, but the primary causative agent is not yet clear

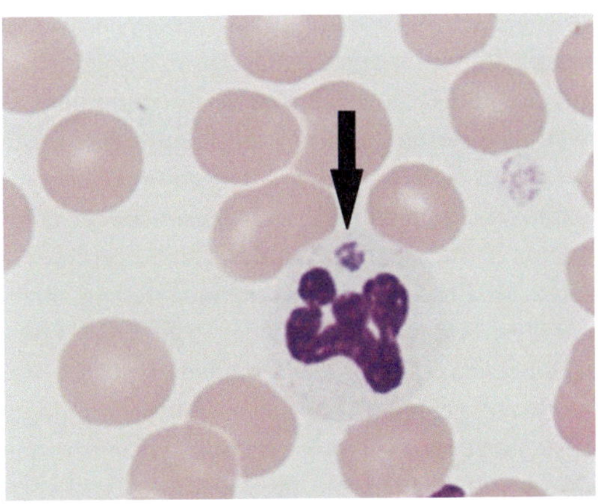

Fig. 14.2 *Ehrlichia ewingii* morula (*arrow*) in an infected neutrophil

related human ehrlichiosis has been suspected from Asia, but the full role of other *Ehrlichia* spp. in causing human disease in these regions is not clear (Esemu et al. 2011).

One reason for the confusion about the primary causative agents of ehrlichiosis is the diversity of *Ehrlichia* spp. in nature, widespread serologic cross-reactivity between the different organisms, and the apparent presence of a variety of novel agents in both tick vectors and wildlife reservoir hosts. For example, dogs in the United States may be infected with *E. canis*, *E. chaffeensis*, *E. ewingii*, Panola Mountain *Ehrlichia* sp., *E. muris*, or some combination of these organisms and likely other novel *Ehrlichia* spp. yet to be described, and all of them induce an antibody response that appears to cross-react on widely used immunofluorescent assays, making identification of the primary agent responsible for infection in a given patient difficult (Little 2010; Allison and Little 2013). Similar broad diversity of the *Ehrlichia* spp. infecting animals has been documented in other regions.

Molecular assays can identify the presence of specific agents. Targets for *Ehrlichia* sp. PCR assays include 16S rDNA, dsb, gltA (citrate synthase), groEL, p28, and VLPT (variable length PCR target). Because novel agents are common, sequence confirmation is recommended in epidemiologic surveys, especially when a new reservoir host, tick vector, or geographic location is reported for a given organism (Allison and Little 2013). Moreover, established molecular strategies may fail to detect the presence of novel agents. These issues can lead to misattribution of seroreactivity to a related etiologic agent, creating confusion. In 2009, an unexpected focus of canine seroreactivity to *Ehrlichia* sp. was identified in the northern United States. Because natural cycles of *Ehrlichia* sp. had not yet been documented in this area, a novel agent was suspected (Bowman et al. 2009). Subsequent work showed that a novel *E. muris*-like agent and *E. muris* were present in the region and infecting people and dogs (Pritt et al. 2011; Hegarty et al. 2012). Careful confirmation of the etiologic agents cycling in vertebrate hosts and tick vectors is necessary to insure accurate epidemiologic understanding of the public health and veterinary disease risks in a given geographic area.

14.3 Life Cycle

All *Ehrlichia* spp. described to date are transmitted between vertebrate hosts by feeding ixodid ticks (Table 14.1). Infections are not known to be maintained transovarially in tick populations but rather acquired from vertebrates when ticks feed as larvae or nymphs and then maintained transstadially to be transmitted at subsequent feeding of nymphs or adults (Nicholson et al. 2010). Because stages of ixodid ticks that serve as vectors for *Ehrlichia* spp. may survive in the environment, infection can be maintained in tick populations for several months. Intrastadial transmission has also been documented for *E. canis* and could be supported in nature by movement of male metastriate ticks between hosts (Bremer et al. 2005; Little et al. 2007). Transmission of *Ehrlichia* spp. has also been reported through blood transfusion and organ transplant (Starkey et al. 2014).

Transmission of *Ehrlichia* spp. does not appear to occur immediately upon attachment to the vertebrate host. Rather a brief period of activation—estimated to be 4–48 h—is observed before active transmission occurs (Nicholson et al. 2010). A recent study reported infection of dogs with *E. canis* after 3 or 6 h of feeding by *R. sanguineus sensu lato* ticks (Fourie et al. 2013a). *Ehrlichia canis* is thought to multiply within salivary glands, hemocytes, and midgut of the tick (Smith et al. 1976); the presence in the salivary gland prior to feeding may facilitate transmission soon after attachment. Moreover, individual species of *Ehrlichia* are usually only transmitted by specific vectors. Recent studies have shown that *R. sanguineus* is comprised of a group of morphologically similar species, and all lineages may not vector *E. canis* equally (Dantas-Torres et al. 2013; Cicuttin et al. 2015).

14.4 Clinical and Laboratory Findings

14.4.1 Canine Ehrlichiosis

Ehrlichial infection in dogs may lead to a broad spectrum of outcomes, from clinically inapparent to severe, fatal disease. This diversity in disease manifestation is considered due to variations in species and strains of *Ehrlichia* infecting dogs as well as differences in breed-related responses to infection, individual dog susceptibility, dose of pathogen transmitted, and presence of other agents (Little 2010). The high canine seroprevalence to *Ehrlichia* sp(p). documented in many surveys suggests subclinical infections commonly occur (Little 2010; Beall et al. 2012).

Severe canine disease is most commonly associated with *E. canis* infection and can present as an acute or chronic febrile illness. Lethargy, myalgia, anorexia, and bleeding diatheses often develop, and spleen and lymph nodes are enlarged in the majority of patients. In the severe, potentially fatal form of chronic canine ehrlichiosis, weight loss and anorexia are often seen together with myalgia, ocular lesions, and neurologic disease. Although any dog may be affected,

German shepherds are particularly susceptible to severe disease. Canine infections with *E. ewingii* and *E. chaffeensis* are considered to cause milder forms of disease and, more frequently, subclinical infections. Fever and neutrophilic polyarthritis may develop with *E. ewingii* infections in dogs (Little 2010).

Hematologic abnormalities characteristic of canine ehrlichiosis are most severe with *E. canis* infection and include pancytopenia, large granular lymphocytosis, and thrombocytopenia (Neer et al. 2002; reviewed in Little 2010). Abnormalities in platelet function are associated with bleeding tendencies. Dogs with ehrlichiosis also often have elevated liver enzymes and a decreased albumin to globuluin ratio due to hypoalbuminemia and concomitant hypergammaglobulinemia (Scorpio et al. 2008).

14.4.2 Ehrlichiosis (Heartwater) in Ruminants

Several forms of heartwater disease due to *E. ruminantium* infection have been described. Great variation apparently exists in the pathogenicity of different strains of *E. ruminantium*, and not all isolates are considered virulent (Spickler 2007). However, virulent strains are associated with high morbidity and mortality. Rarely, hyperacute disease has been described in non-African breeds of cattle, sheep, and goats and presents as a brief fever, severe diarrhea, and severe respiratory distress followed by sudden death. Acute disease is more common and presents with a sudden fever accompanied by anorexia, listlessness, diarrhea, and pulmonary edema. Neurologic signs develop as the disease progresses and affected animals usually die within a week. A prolonged subacute form and mild, subclinical infections have also been described.

Fatality rates in non-African breeds of cattle, sheep, and goats with virulent strains of *E. ruminantium* may exceed 80 %, while a majority of African breeds survive infection (Spickler 2007). Survival is also more commonly seen in young animals and some wild ruminants (Deem 1998). Recovery leads to complete homologous immunity.

At necropsy, the hydropericardium responsible for heartwater's sobriquet is most consistently present in sheep and goats. Pulmonary edema, hydrothorax, ascites, myocardial petechial hemorrhages, and meningeal congestion and edema are also commonly present (Spickler 2007).

14.4.3 Human Ehrlichiosis

Human ehrlichiosis is usually characterized by fever, severe headache, and general malaise. Myalgia and gastrointestinal distress are often present. Neurologic signs, including confusion, develop in approximately 20 % of patients, and in severe cases, difficulty breathing and bleeding diatheses are reported. Fatality rates are estimated to be 2–3 %.

Disease is more severe and fatalities more common in immunocompromised patients (Paddock and Childs 2003). A majority of children but a minority of adults with disease due to *E. chaffeensis* also develops a non-pruritic maculopapular to petechial rash or erythroderma. However, because skin lesions are inconsistent in *E. chaffeensis*-induced disease and not documented in other forms of human ehrlichiosis, the presence or absence of a rash should not be used in reaching a diagnosis (CDC 2015). Common hematologic and serum chemistry findings in human patients with ehrlichiosis include thrombocytopenia, leukopenia, and elevated liver enzymes (Paddock and Childs 2003).

14.5 Diagnosis

Diagnosis of ehrlichiosis relies upon a combination of characteristic clinical signs and hematologic findings along with confirmatory results on established laboratory assays. A history of exposure to infected ticks by virtue of residence or travel to an endemic area and the presence of hematologic abnormalities, particularly thrombocytopenia, also raise the index of suspicion. Direct diagnosis can be made by the identification of morulae in leukocytes on stained cytologic preparations of whole blood, buffy coat, or bone marrow; examining blood smears is usually more rewarding for identifying granulocytic agents. Morulae may also be identified in stained impression smears collected at necropsy (Allison and Little 2013).

Serologic and molecular assays are the most useful means of reaching a diagnosis of ehrlichiosis. To maximize the likelihood of confirming the infection in a clinically affected patient, both approaches should be used simultaneously (reviewed in Little 2010). Indirect fluorescent antibody tests using *E. chaffeensis*- and *E. canis*-infected cells as antigen have been available through diagnostic laboratories for decades and allow ready identification of antibodies reactive to *Ehrlichia* spp. in humans, dogs, and other species. Peptide-based assays are also widely available as commercial point-of-care tests for dogs. Both approaches are not specific for individual *Ehrlichia* sp. and results must be interpreted with caution (Allison and Little 2013). Laboratory-based peptide assays with greater specificity have been reported and aid interpretation for epidemiologic studies but are not available for individual patients. Antibody titers may not be present during acute infection.

Because many infections are fatal, antibiotic treatment should not be withheld from a clinically affected patient based on the absence of serologic confirmation. Moreover, antibodies often persist following resolution of disease, complicating interpretation of a single positive titer from a clinically normal patient.

PCR-based assays for *Ehrlichia* spp. are also widely available through diagnostic laboratories and can be used to detect the presence of organism in patients, vertebrate reservoir hosts, and tick vectors. Molecular assays for ehrlichial agents have been designed to maximize sensitivity of detection, and a positive result is confirmatory. However, patients may test negative if organisms are not present in circulation at detectable limits, as can occur during chronic, established infection or after antibiotic treatment has been instituted or when novel agents other than those for which the PCR has been optimized are responsible for disease (Allison and Little 2013). As with serologic assays, treatment of affected patients should not be delayed while awaiting results of PCR or deferred if results are negative when clinical signs are consistent.

Other coinfecting agents are commonly present in patients with ehrlichiosis and can be identified through testing for a panel of related vector-borne organisms known to be cycling in the region where the patient was infected.

14.6 Treatment

Doxycycline is the treatment of choice for ehrlichiosis; several protocols have been recommended. For dogs, the ACVIM consensus statement recommends 10 mg/kg q 24 h for 28 days (Neer et al. 2002). For people, the Centers for Disease Control and Prevention recommends adults be treated with 100 mg every 12 h for at least 3 days after resolution of fever and other clinical signs (usually 7–14 days) (CDC 2015). Prompt treatment with doxycycline, including in children of all ages and in young animals, is necessary to insure that clinical disease is resolved before the development of severe, adverse sequelae. Delaying treatment can result in fatalities. Concerns about tooth discoloration with doxycycline are not supported in the literature and should not limit the use of this potentially lifesaving antibiotic (Nicholson et al. 2010). Additional supportive therapy should be considered as indicated by clinical assessment, particularly in severely affected patients, and may include fluid administration, blood transfusion, or prednisone. Most patients respond to antibiotic treatment within 1–3 days. Lack of clinical response to doxycycline suggests other coinfecting agents may be present and causing disease or that illness is not due to ehrlichiosis (Little 2010; Allison and Little 2013; CDC 2015).

14.7 Epizootiology and Epidemiology

Risk of infection with *Ehrlichia* sp. in a given area is dependent upon tick exposure, which in turn is related to the density of ticks in the environment, a factor that may be heavily influenced by climate and habitat, as well as animal-specific characteristics such as breed, age, immune status, and lifestyle. In areas where heartwater is endemic, infection is almost assured and results in high morbidity and mortality in nonnative breeds of ruminants (Spickler 2007).

Similarly, in high tick challenge environments, ehrlichial infection in dogs as indicated by seroprevalence or molecular evidence of infection approaches 100 % (Otranto et al. 2008; Starkey et al. 2014). Seroprevalence in reservoir hosts is usually very high as well; virtually 100 % of white-tailed deer are seropositive for *E. chaffeensis* when the infection is present in a given area (Yabsley et al. 2003).

Infection is also common in people in areas where human ehrlichiosis is endemic. In one survey, 13 % of children in the southern United States, and as many as 22 % of those from certain endemic areas, had antibodies to *Ehrlichia* sp. although the great majority of these infections were not known to be associated with clinical disease (as reported in Paddock and Childs 2003). Human disease risk is associated with seroprevalence in both dogs and deer in a given area as well as habitat (Beall et al. 2012; Yabsley et al. 2003). Invasive honeysuckle has been shown to support high numbers of ticks infected with *E. chaffeensis*, presumably by creating ample browse for the deer reservoir hosts and protective habitat for the vector ticks (Allan et al. 2010).

Peak incidence of ehrlichiosis is closely tied to timing of tick activity. Although chronic canine ehrlichiosis may present at any time during the year, cases of acute disease in dogs are most commonly reported in the spring and summer months when activity of the vector ticks is high (Little 2010). Similarly, outbreaks of heartwater occur approximately 2–4 weeks after tick blooms and heavy tick infestations on ruminants (Spickler 2007). A majority (70 %) of cases of human ehrlichiosis due to *E. chaffeensis* are reported in May, June, and July when the *A. americanum* vector ticks are most active (as reported in Paddock and Childs 2003). As additional etiologic agents and tick vectors are identified and the disease is recognized in new geographic locations, phenology and timing of disease presentation are likely to shift.

14.8 Control and Prophylaxis

Preventing ehrlichiosis relies upon strict attention to tick control and limiting opportunities for tick infestations. In heavily infested environments, preventing ticks can be difficult. Routine use of acaricides has been shown to limit infection with *E. canis* in dogs using both experimental and natural challenge models. Treatment with fipronil, imidacloprid/permethrin, and imidacloprid/flumethrin has been shown to limit transmission of *E. canis* to dogs even in high natural challenge environments (Davoust et al. 2003; Otranto et al. 2008; Fourie et al. 2013b; Stanneck and Fourie 2013).

Tick control is also a mainstay of limiting disease due to heartwater in areas where *E. ruminantium* is endemic. However, removal of all ticks eliminates exposure of susceptible animals to small doses of the organisms which may serve an immunoprotective role against severe disease. Eradication of vectors for *E. ruminantium* has been pursued, primarily in the Caribbean, but the success of this approach is limited by the high fecundity of the *Amblyomma* vectors, the wide variety of hosts on which they feed, and the development of acaricide resistance (Spickler 2007).

Prophylactic administration of tetracycline antibiotics was recommended historically to prevent ehrlichiosis in dogs and ruminants in high-challenge environments when adequate tick control could not be achieved (Davidson et al. 1978; Spickler 2007). However, this approach is difficult to maintain and problematic in that it carries the risk of selection for resistance to the primary antibiotic considered highly effective for treating these infections. Although not yet documented, doxycycline resistance in *Ehrlichia* spp. could be particularly devastating should it occur with zoonotic agents.

Vaccination is available to prevent heartwater in domestic ruminants in areas where disease is endemic. Historically, vaccination against heartwater involved intentional infection with live *E. ruminantium* followed by antibiotic treatment; younger animals often survived infection without treatment (Spickler 2007). More recently, recombinant and inactivated vaccines have been developed. Efficacy of vaccination is limited by the antigenic diversity of *E. ruminantium* strains; vaccination results in limited protection against future infection with heterologous strains although inactivated vaccines may combine multiple strains to extend the spectrum of protection (Adakal et al. 2010). Commercial vaccines for canine ehrlichiosis are not yet available although recent research has identified a promising approach using an attenuated strain of *E. canis* (Rudoler et al. 2012). Human ehrlichiosis vaccines are not available.

References

Adakal H, Stachurski F, Konkobo M et al (2010) Efficiency of inactivated vaccines against heartwater in Burkina Faso: impact of *Ehrlichia ruminantium* genetic diversity. Vaccine 28:4573–4580

Allan BF, Dutra HP, Goessling LS et al (2010) Invasive honeysuckle eradication reduces tick-borne disease risk by altering host dynamics. Proc Natl Acad Sci U S A 107:18523–18527

Allison RW, Little SE (2013) Diagnosis of rickettsial diseases in dogs and cats. Vet Clin Pathol 42:127–144

Anderson BE, Dawson JE, Jones DC, Wilson KH (1991) *Ehrlichia chaffeensis*, a new species associated with human ehrlichiosis. J Clin Microbiol 29:2838–2842

Anderson BE, Greene CE, Jones DC, Dawson JE (1992) *Ehrlichia ewingii* sp. nov., the etiologic agent of canine granulocytic ehrlichiosis. Int J Syst Bacteriol 42:299–302

Beall MJ, Alleman AR, Breitschwerdt EB et al (2012) Seroprevalence of *Ehrlichia canis*, *Ehrlichia chaffeensis* and *Ehrlichia ewingii* in dogs in North America. Parasit Vectors 5:29

Bezuidenhout JD (2009) Heartwater: an abridged historical account. J S Afr Vet Assoc 80:208–209

Bremer WG, Schaefer JJ, Wagner ER et al (2005) Transstadial and intrastadial experimental transmission of *Ehrlichia canis* by male *Rhipicephalus sanguineus*. Vet Parasitol 131:95–105

Bowman D, Little SE, Lorentzen L et al (2009) Prevalence and geographic distribution of *Dirofilaria immitis*, *Borrelia burgdorferi*, *Ehrlichia canis*, and *Anaplasma phagocytophilum* in dogs in the United States: results of a national clinic-based serologic survey. Vet Parasitol 160:138–148

Centers for Disease Control and Prevention (2015) Ehrlichiosis. http://www.cdc.gov/ehrlichiosis/. Accessed 15 June 2015

Cicuttin GL, Tarragona EL, De Salvo MN et al (2015) Infection with *Ehrlichia canis* and *Anaplasma platys* (Rickettsiales: Anaplasmataceae) in two lineages of *Rhipicephalus sanguineus* sensu lato (Acari: Ixodidae) from Argentina. Ticks Tick Borne Dis 6:724–729

Dantas-Torres F, Latrofa MS, Annoscia G et al (2013) Morphological and genetic diversity of *Rhipicephalus sanguineus* sensu lato from the New and Old Worlds. Parasit Vectors 6:213

Davidson DE Jr, Dill GS Jr, Tingpalapong M et al (1978) Prophylactic and therapeutic use of tetracycline during an epizootic of ehrlichiosis among military dogs. J Am Vet Med Assoc 172:697

Davoust B, Marié JL, Mercier S et al (2003) Assay of fipronil efficacy to prevent canine monocytic ehrlichiosis in endemic areas. Vet Parasitol 112:91–100

Deem SL (1998) A review of heartwater and the threat of introduction of *Cowdria ruminantium* and *Amblyomma* spp. ticks to the American mainland. J Zoo Wildl Med 29:109–113

Donatien A, Lestoquard F (1937) State of the present knowledge concerning rickettsiosis of animals. Arch Inst Pasteur Alger 15:142–187

Dumler JS, Barbet AF, Bekker CP et al (2001) Reorganization of genera in the families Rickettsiaceae and Anaplasmataceae in the order Rickettsiales: unification of some species of *Ehrlichia* with *Anaplasma*, *Cowdria* with *Ehrlichia* and *Ehrlichia* with *Neorickettsia*, descriptions of six new species combinations and designation of *Ehrlichia equi* and 'HGE agent' as subjective synonyms of *Ehrlichia phagocytophila*. Int J Syst Evol Microbiol 51:2145–2165

Esemu SN, Ndip LM, Ndip RN (2011) *Ehrlichia* species, probable emerging human pathogens in sub-Saharan Africa: environmental exacerbation. Rev Environ Health 26:269–279

Fourie JJ, Stanneck D, Luus HG et al (2013a) Transmission of *Ehrlichia canis* by *Rhipicephalus sanguineus* ticks feeding on dogs and on artificial membranes. Vet Parasitol 197:595–603

Fourie JJ, Luus HG, Stanneck D et al (2013b) The efficacy of Advantix® to prevent transmission of *Ehrlichia canis* to dogs by *Rhipicephalus sanguineus* ticks. Parasite 20:36

Hegarty BC, Maggi RG, Koskinen P et al (2012) *Ehrlichia muris* infection in a dog from Minnesota. J Vet Intern Med 26:1217–1220

Ismail N, Bloch KC, McBride JW (2010) Human ehrlichiosis and anaplasmosis. Clin Lab Med 30:261–292

Little SE (2010) Ehrlichiosis and anaplasmosis in dogs and cats. Vet Clin Small Anim 40:1121–1140

Little SE, Hostetler J, Kocan KM (2007) Movement of *Rhipicephalus sanguineus* adults between co-housed dogs during active feeding. Vet Parasitol 150:139–145

Neer TM, Breitschwerdt EB, Greene RT et al (2002) Consensus statement on ehrlichial disease of small animals from the infectious disease study group of the ACVIM. American College of Veterinary Internal Medicine. J Vet Intern Med 16:309–315

Nicholson WL, Allen KE, McQuiston JH et al (2010) The increasing recognition of rickettsial pathogens in dogs and people. Trends Parasitol 26:205–212

Otranto D, Paradies P, Testini G et al (2008) Application of 10% imidacloprid/50% permethrin to prevent *Ehrlichia canis* exposure in dogs under natural conditions. Vet Parasitol 153:320–328

Paddock CD, Childs JE (2003) *Ehrlichia chaffeensis*: a prototypical emerging pathogen. Clin Microbiol Rev 16:37–64

Pritt BS, Sloan LM, Johnson DK et al (2011) Emergence of a new pathogenic *Ehrlichia* species, Wisconsin and Minnesota, 2009. N Engl J Med 365:422–429

Rikihisa Y (1991) The tribe Ehrlichieae and ehrlichial diseases. Clin Microbiol Rev 4:286–308

Rudoler N, Baneth G, Eyal O et al (2012) Evaluation of an attenuated strain of *Ehrlichia canis* as a vaccine for canine monocytic ehrlichiosis. Vaccine 31:226. doi:10.1016/j.vaccine.2012.10.003

Salinas LJ, Greenfield RA, Little SE et al (2010) Tickborne infections in the southern United States. Am J Med Sci 340:194–201

Scorpio DG, Wachtman LM, Tunin RS et al (2008) Retrospective clinical and molecular analysis of conditioned laboratory dogs (*Canis familiaris*) with serologic reactions to *Ehrlichia canis*, *Borrelia burgdorferi*, and *Rickettsia rickettsii*. JAALAS 47:23–28

Smith RD, Sells DM, Stephenson EH et al (1976) Development of *Ehrlichia canis*, causative agent of canine ehrlichiosis, in the tick *Rhipicephalus sanguineus* and its differentiation from a symbiotic *Rickettsia*. Am J Vet Res 37:119–126

Spickler AR (2007) Heartwater (Last Updated October 2007). http://www.cfsph.iastate.edu/DiseaseInfo/factsheets.php Accessed 15 June 2015

Stanneck D, Fourie JJ (2013) Imidacloprid 10%/flumethrin 4.5% collars (Seresto®, Bayer) successfully prevent long-term transmission of *Ehrlichia canis* by infected *Rhipicephalus sanguineus* ticks to dogs. Parasitol Res 112(Suppl 1):S21–S32

Starkey LA, Barrett AW, Chandrashekar R et al (2014) Development of antibodies to and PCR detection of *Ehrlichia* spp. in dogs following natural tick exposure. Vet Microbiol 173:379–384

Vieira RF, Biondo AW, Guimarães AM et al (2011) Ehrlichiosis in Brazil. Rev Bras Parasitol Vet 20:1–12

Yabsley MJ, Dugan VG, Stallknecht DE et al (2003) Evaluation of a prototype *Ehrlichia chaffeensis* surveillance system using white-tailed deer (*Odocoileus virginianus*) as natural sentinels. Vector Borne Zoonotic Dis 3:195–207

Anaplasmosis

15

Filipe Dantas-Torres and Domenico Otranto

Abstract

Anaplasmosis is a vector-borne disease caused by bacteria of the genus *Anaplasma*, which infect a wide range of wild and domestic animals. Currently, there are seven valid species in the genus *Anaplasma* (i.e., *Anaplasma bovis*, *A. caudatum*, *A. centrale*, *A. marginale*, *A. ovis*, *A. phagocytophilum*, and *A. platys*), some of which (*A. phagocytophilum*) are of great zoonotic concern. Biological vectors of *Anaplasma* species include ixodid ticks belonging to different genera. The prevalence of infection by *Anaplasma* in domestic animals such as cattle and dogs may vary widely from region to region and according to the level of exposure to tick vectors. The diagnosis of anaplasmosis is usually performed based on epidemiological data, on clinical signs, and on the retrieval of intracellular inclusions in stained blood smears upon cytological examination. Serological and molecular tools are available but most frequently restricted to research. The treatment of anaplasmosis is usually performed with tetracycline-class antibiotics, such as doxycycline, whereas the control of the disease requires a holistic approach whose ultimate aim is to reduce the degree of exposure to tick vectors.

Keywords

Anaplasma • Dogs • Livestock • Humans • Ticks • Mechanical transmission

15.1 Background

The first reports about anaplasmosis date from the late nineteenth century in South Africa. However, the identity of the causative agent of the disease was finally discovered only in 1910. That year, Max Theiler described the species *Anaplasma marginale*, which caused a serious disease in cattle, and

F. Dantas-Torres (✉)
Department of Immunology, Aggeu Magalhães Research Center, Recife, PE, Brazil

Department of Veterinary Medicine, University of Bari, Valenzano, Italy
e-mail: filipe.dantas@cpqam.fiocruz.br

D. Otranto
Department of Veterinary Medicine, University of Bari, Valenzano, Italy
e-mail: domenico.otranto@uniba.it

C.B. Marcondes (ed.), *Arthropod Borne Diseases*, DOI 10.1007/978-3-319-13884-8_15

after 1 year, he described a new species, *A. centrale*, which was responsible for causing a mild disease in cattle as well.

In the 1930s, Sanborn, Stiles, and Moc proved the mechanical transmission of *A. marginale* by horse flies (Tabanidae) and stable flies (*Stomoxys calcitrans*). In the late 1980s, the life cycle of *A. marginale* was demonstrated in ticks of the genus *Dermacentor* (Kocan et al. 1989).

The first cases of human anaplasmosis were reported in 1994 in the United States (Chen et al. 1994). On this occasion, inclusions (morulae) in the cytoplasm of granulocytes were observed in peripheral blood smears from six patients from Minnesota and Wisconsin with fever of unknown origin. Few years later, the first case of human anaplasmosis was described in Europe in a patient from Slovenia (Petrovec et al. 1997). Since the evidence of pathogenicity for humans, the interest of the scientific community on these rickettsiae arose.

The taxonomy of the genus *Anaplasma*, as well as other genera belonging to the order *Rickettsiales*, was greatly modified in recent years. In 2001, the taxonomy of the group was extensively reviewed and significant changes were proposed at generic and specific levels. Thereby, some species previously included in the genus *Ehrlichia* were transferred to *Anaplasma* (Table 15.1).

15.2 Etiology

Anaplasmosis is caused by bacteria of the genus *Anaplasma*, belonging to the family *Anaplasmataceae*, within the order *Rickettsiales*. The genus includes small (ca. 0.3 μm), gram-negative, pleomorphic bacteria that reside into the cells of the mononuclear phagocytic system, erythrocytes, and platelets of vertebrate hosts. For this reason, these bacteria can be observed in smears prepared from peripheral blood or aspirates of some organs such as the spleen, liver, and bone marrow (Dumler et al. 2001).

Currently, there are seven valid species of the genus *Anaplasma* (*A. bovis*, *A. caudatum*, *A. centrale*, *A. marginale*, *A. ovis*, *A. phagocytophilum*, and *A. platys*). Besides these, there is another species found in sheep (*A. mesaenterum*), whose validity has been discussed. Furthermore, another species (*A. odocoilei*) was recently described from white-tailed deer in North America.

Some *Anaplasma* species may display a certain degree of host specificity, although some of them (e.g., *A. phagocytophilum*) may infect more than one animal species, including human beings (Dumler et al. 2005). Indeed, *A. platys* has long been considered to be restricted to dogs and it has recently been reported in cats (Lima et al. 2010)

Table 15.1 Species of the genus *Anaplasma*, their hosts, vectors, and geographical distribution

Species	Main hosts	Vector genera	Distribution
A. bovis[a]	Cattle, rabbit	*Amblyomma*, *Haemaphysalis*, *Ixodes*, *Rhipicephalus* (*Boophilus*)	Africa, America, Asia
A. caudatum	Cattle	Unknown	North America
A. centrale	Cattle	*Haemaphysalis*	Europe, Africa, America, Asia
A. marginale	Cattle	*Dermacentor*, *Rhipicephalus* (*Boophilus*)	Cosmopolitan
A. mesaenterum[b]	Sheep	*Haemaphysalis*, *Ixodes*	Europe
A. odocoilei	White-tailed deer	Unknown	North America
A. ovis	Sheep, goat	*Dermacentor*, *Haemaphysalis*, *Hyalomma*, *Ixodes*, *Rhipicephalus*	Europe, Africa, North America, Asia
A. phagocytophilum[c]	Human, equine, sheep, goat, cattle, dog, cat, rodents	*Ixodes*, *Dermacentor* (?)	Cosmopolitan
A. platys[d]	Dog, cat, man	Unknown, *Rhipicephalus* (?)	Cosmopolitan

[a]Formerly *Eimeria bovis*
[b]Incertae sedis
[c]Formerly *E. equi*, *E. phagocytophila*, and *A. phagocytophila*
[d]Formerly *E. platys*

and in a man (Maggi et al. 2013), which suggests that its host range may have been underestimated.

Bacteria of the genus *Anaplasma* are difficult to isolate in culture medium (Chen et al. 1994) and some species (e.g., *A. bovis* and *A. platys*) have never been isolated. Generally, cell culture media are used, such as tick cells or cells of hematopoietic lineage (e.g., HL-60), where they can be maintained for long periods (>1 year) (Dumler et al. 2001). Complete genome sequences ($1.2–1.5 \times 10^6$ bp in size) have been determined for *A. phagocytophilum*, *A. centrale*, and *A. marginale* and are composed of a circular chromosome (Rar and Golovljova 2011).

15.3 Life Cycle

The life cycle of many species of *Anaplasma* is still not completely unraveled. So, most of the information illustrated herein refers to data about *A. marginale* whose life cycle in cattle and in *Rhipicephalus (Boophilus) microplus* ticks has been most intensively studied. Indeed, *Anaplasma* spp. are transmitted biologically by ticks and mechanically by various bloodsucking flies, such as horse flies (tabanids) and also eye flies (chloropids).

When an infected tick takes the blood meal, the rickettsiae localized in the salivary glands of the tick are inoculated into the dermis of the vertebrate host with the saliva. Depending on the *Anaplasma* species involved, inclusions may be observed inside many types of cells such as leukocytes, platelets, and erythrocytes according to the species considered. Inside these cells, intranuclear inclusion bodies are formed where the rickettsiae multiply before being released back into the bloodstream, infecting other cells.

When a noninfected tick takes a blood meal on a bacteremic host, the rickettsiae are ingested with the blood. After their release into the tick intestinal lumen, the rickettsiae invade the intestinal epithelium where the replication process starts. The rickettsiae then migrate to salivary glands via hemolymph, where a new replication process starts. This process culminates in the formation of infective rickettsiae during a new blood meal on a susceptible host.

To the best of our knowledge, there is no transovarial transmission of rickettsiae of the genus *Anaplasma*, which means that larvae of ticks originating from infected females are free of infection. Differently, transstadial transmission occurs, indicating that the presence of bacteremic hosts and affiliated tick vectors is a *conditio sine qua non* for the life cycle of *Anaplasma* species to be maintained in nature.

15.4 Clinical and Laboratorial Findings

Some clinical and laboratorial findings in anaplasmosis may be observed in different animal species. Anemia and thrombocytopenia, for example, are frequently reported in anaplasmosis, regardless of the animal species.

In dogs, arthritis, lethargy, fever, splenomegaly, lymph node enlargement, anorexia, vomiting, and weight loss are the main clinical findings observed in anaplasmosis caused by *A. phagocytophilum*. Similarly, it causes a mild disease in cattle and sheep, but it may increase the susceptibility of lambs to staphylococcal infection and louping ill virus. The mortality rate is negligible in cattle, and the main losses are associated to abortion when pregnant cows become infected. In sheep, *A. phagocytophilum* may persist in the blood for life, forming a major source of infection in an enzootic area.

The laboratorial findings include thrombocytopenia, leukopenia, anemia, and hyperproteinemia. Other findings such as hypoalbuminemia may also be present. In cases of anaplasmosis caused by *A. platys*, weight loss, weakness, apathy, anorexia, fever, neurological symptoms, and arthritis may be observed. Thrombocytopenia, usually associated with anemia and leukopenia, is the most common laboratorial findings. Nonetheless, dogs infected by *A. platys* are most frequently asymptomatic.

In cattle, anaplasmosis is more serious in animals with more than 2 years old, being potentially fatal mainly in infections by *A. marginale*. The main clinical signs are fever, jaundice, and anorexia. Decrease of milk production and abortion may also be observed. Generally, anemia is present but in variable degrees. *Anaplasma centrale* causes mild, inapparent disease in cattle, sheep, and goats. Nonetheless, a severe anemia may develop in goats suffering from concurrent diseases.

In equine anaplasmosis, the clinical signs include fever, lethargy, anorexia, limb edema, petechiae, jaundice, and ataxia. Some animals can become reluctant to move. The main laboratorial findings are leukopenia, thrombocytopenia, and anemia.

The clinical signs observed in most human patients are malaise, fever, myalgia, and headache. In some patients, the gastrointestinal and respiratory tracts, liver, and central nervous system may be involved. The most frequent laboratorial findings are thrombocytopenia, leukopenia, anemia, and increase in the level of transaminases (Dumler et al. 2005).

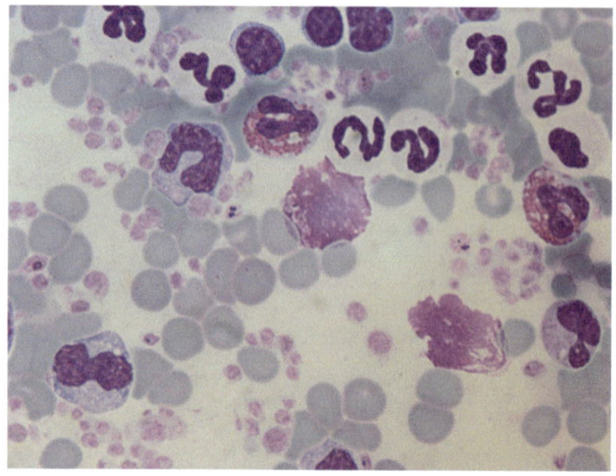

Fig. 15.1 Inclusions of *Anaplasma platys* in platelets of a buffy coat smear of a dog

15.5 Diagnosis

Taking into account the vast number of potentially clinical signs seen in the course of anaplasmosis, the diagnosis must be based on the detection (direct or indirect) of *Anaplasma* spp. in suspected animals. In endemic areas, the history of recent tick infestation associated with the presence of compatible clinical signs may be strongly suggestive of anaplasmosis.

In the acute phase, *Anaplasma* spp. inclusions can be easily found by cytological examination of peripheral blood smears stained with Giemsa stain or other ordinary staining method. According to the species, these inclusions may be found in erythrocytes (*A. caudatum*, *A. centrale*, *A. marginale,* and *A. ovis*), leukocytes (*A. bovis* and *A. phagocytophilum*), or platelets (*A. platys* and *A. odocoilei*).

For the diagnosis of *A. platys* infection, combining cytological examination of both blood and buffy coat smears in parallel (Fig. 15.1) enhanced the sensitivity from 61.5 and 67 %, respectively, to 87.3 % (Otranto et al. 2010). In addition, it has been suggested that during the early phase of *A. platys* infection in naïve young dogs, the cytological examination of blood or buffy coat smears allows the detection of the pathogen in approximately two-thirds of the infected dogs. Afterward, polymerase chain reaction (PCR) becomes more sensitive than cytology for diagnosing *A. platys* in naïve elder dogs (Otranto et al. 2010).

Similarly, in the chronic phase, the bacteremia is typically low and other diagnostic methods may be required. These methods include direct detection of *Anaplasma* spp. antigens using serological tests (e.g., enzyme-linked immunosorbent assay (ELISA)) or PCR-based tools (Vidotto and Marana 2001).

Many serological techniques have been used, including agglutination tests (and variants), the complement fixation, indirect immunofluorescence, ELISA (and its variants), and Western blot (Vidotto and Marana 2001). Recently, blood PCR has been employed in the characterization of *Anaplasma* species in many parts of the world, including Brazil (Ferreira et al. 2007). PCR has been a very useful tool for the diagnosis of *A. platys* infection in dogs, either in blood samples, spleen, and bone marrow (Eddlestone et al. 2007). Furthermore, PCR and DNA sequencing may also be useful to confirm *A. platys* infection in cats (Lima et al. 2010) and humans (Maggi et al. 2013; Breitschwerdt et al. 2014).

15.6 Treatment

Bacteria of the genus *Anaplasma* are sensitive to tetracyclines. Table 15.2 summarizes the protocols most frequently used for anaplasmosis treatment (Wormser et al. 2006).

Table 15.2 Protocols used in the bovine, equine, canine, and human anaplasmosis treatment

Anaplasmosis	Protocol
Bovine	Oxytetracycline (20 mg/kg), IM. Repeat when necessary after 7 days
Canine	Tetracycline (22 mg/kg) VO, t.i.d., during 14 days. Doxycycline (10 mg/kg), VO, s.i.d., during 28 days
Equine	Oxytetracycline (7 mg/kg), IV, s.i.d., during 7 days
Human	Doxycycline (100 mg), VO, b.i.d., during 10 days. For children 2 mg/kg/day b.i.d. are recommended (maximum of 100 mg per dose)

Cases of coinfection are common. For example, cases of coinfection with *A. marginale* and *Babesia bigemina* in cattle and with *A. platys*, *Hepatozoon canis*, *Leishmania infantum,* and *Babesia vogeli*, or *Ehrlichia canis* in dogs. Therefore, veterinary clinicians should be aware of the treatment of simultaneous infections, which may be present. Debilitated animals may require hospitalization and blood transfusion may be necessary in animals that present severe anemia.

15.7 Epizootiology

Anaplasmosis has a cosmopolitan distribution. Although some species of *Anaplasma* (e.g., *A. caudatum* and *A. mesaenterum*) have apparently a limited geographic distribution, others (e.g., *A. platys* and *A. marginale*) have been identified in all continents. The prevalence of infection varies widely according to the *Anaplasma* species, animal host, geographic region, and test used. Young animals are relatively resistant to *A. marginale* infection, but a hyperacute condition may develop in cattle older than 3 years with fatal exitus within 24 h. In addition, a mortality rate higher than 50 % may occur when susceptible animals move into an infected area or vector populations are introduced into a previously disease-free area. In endemic areas, the infection is maintained at herd level by recovered animals that will remain infective throughout their lives. For example, in Brazil, the prevalence of infection by *Anaplasma* spp. in cattle may reach 100 % based on serology.

In the United States, serological surveys indicate that the prevalence of infection by *A. phagocytophilum* in humans ranges between 15 and 36 %, with an annual incidence up to 58 cases per 100,000 inhabitants (Dumler et al. 2005). Although until now there are no reported cases of anaplasmosis in humans in some regions such as Brazil, cases of *A. phagocytophilum* infection have been reported in horses (Mesquita Neto et al. 1984) and dogs (Paiva Diniz et al. 2007). Therefore, the occurrence of human anaplasmosis should be considered where there is evidence of the infection in domestic or wild animal species. Indeed, the risk of infection by *Anaplasma* spp. is related to the exposure to infected vectors. Regarding the propensity to illness, risk factors are varied and generally related to immune status of the host. Indeed, young cattle are less affected than adults. European breeds are more susceptible to infection. Conversely, some animal species such as the African buffalo (*Syncerus caffer*) have been shown to be resistant to *A. marginale* infection.

Outbreaks of anaplasmosis, most frequently associated to high tick infestation or when animals from a non-endemic area are introduced in areas of active transmission, are generally associated with high morbidity and mortality.

Little is known about the risk factors associated with canine anaplasmosis. Many dogs infected by *A. platys* remain asymptomatic showing only a cyclic thrombocytopenia. In general, the risk of the disease is higher in dogs coinfected with other pathogens such as *B. vogeli*, *H. canis,* and *Ehrlichia canis*. For example, among a population of autochthonous dogs from Southern Italy exposed to many canine vector-borne pathogens that were PCR negative for *A. platys* at baseline, PCR positivity increased to 34.5 % and 57.7 % at 3 and 6 months follow-up, respectively, and remained high (30.4 %) at 1 year follow-up (Otranto et al. 2010). Interestingly, a recent study reported a prevalence of 13.19 % of *A. platys* infection in cats from Southeast Brazil as assessed by PCR.

Rickettsiae of the genus *Anaplasma* are transmitted by ticks, which are the only proven biological vectors (Samish et al. 1993; Stiller and Coan 1995; Dumler et al. 2005). Tick vector species are included in different genera (*Amblyomma*, *Dermacentor*, *Haemaphysalis*, *Ixodes*, and *Rhipicephalus*). As far as *A. platys*, evidence indicates the role of *R. sanguineus* sensu lato ticks in the transmission (Ramos et al. 2014), even though an experimental study has failed to demonstrate that (Simpson et al. 1991). Ticks of the genera *Dermacentor*, *Haemaphysalis,* and *Ixodes* have also been found naturally infected by *A. platys* (Parola et al. 2003; Kim et al. 2006).

Besides ticks, rickettsiae of the genus *Anaplasma* have been naturally found in a wide variety of hematophagous arthropods including flies, mosquitoes, and lice. The chewing louse *Heterodoxus spiniger*, which is commonly found infesting dogs in Brazil, has been found infected by *A. platys* in Australia. Nevertheless, the role of these arthropods in maintaining the transmission cycle of *Anaplasma* spp. is largely unknown. It is possible that some of them may act as mechanical vectors. Indeed, the mechanical transmission through hematophagous insects such as horse flies (*Tabanus*), deer flies (*Chrysops*), stable flies (*Stomoxys*), horn flies (*Haematobia*), eye flies (*Hippelates*), and mosquitoes (*Psorophora*) has been reported. Furthermore, the iatrogenic transmission by blood transfusion and fomites (e.g., surgical instruments and syringes) can also assume an important role in the spreading process of *A. marginale* in tick-free areas (Guglielmone et al. 1997). Transplacental transmission has also been usually reported (Ribeiro et al. 1995).

Anaplasma species infect a wide variety of animals, including humans. Domestic and wild ruminants have been incriminated as important reservoirs of *Anaplasma* spp. Although small rodents are important reservoirs of *A. phagocytophilum*, this rickettsia can infect a wide range of hosts (Stuen 2007).

15.8 Control and Prophylaxis

The prevention and control measures of anaplasmosis include vaccination (against bovine anaplasmosis), tick control, adoption of hygienic-sanitary measures, and occasionally chemoprophylaxis with tetracycline.

Nowadays, there are no vaccines for the prevention of anaplasmosis infection in animal other than bovines. Cattle vaccination has been effected in some countries, including Brazil. In spite of the efforts toward the development of new vaccines (Kocan et al. 2003), attenuated vaccines, originally developed in Australia in the 1960s and 1970s, are still currently used.

There are two types of attenuated vaccines commercially available in Brazil, the chilled and the frozen. Both of them are triple vaccines from attenuated *A. centrale*, *Babesia bovis,* and *B. bigemina* strains (Kessler et al. 2002). The frozen vaccines (ERITROVAC N2®, EMBRAVAC®-HEMOPAR) are stored in liquid nitrogen, allowing their transportation to any region of the country (Kessler et al. 1998). The vaccine may induce a level of protection of approximately 97 % (Kessler et al. 1998). Alternatively, a second dose of vaccine may be applied 60 days after the first. In this case, it is preferable to carry out a serological test and apply the second dose only in seronegative individuals (Kessler et al. 2002). The chilled vaccine (ERITROVAC) is also available in Brazil and has the same indication as the frozen vaccines but with a very short life span (7 days). A vaccine with live *A. centrale* is currently used in South Africa, Australia, Israel, and South America against *A. marginale*, although severe reactions are recorded in a proportion of vaccinated cattle. Vaccines with killed *A. marginale* and adjuvant have been used in the past where live vaccine cannot be employed (e.g., the United States), but they are not currently available. Although these vaccines elicited a certain degree of immunity against *A. marginale*, which prevented severe conditions, they did not protect vaccinated individuals against heterologous strains. Live vaccines based on *A. marginale* from tick cell cultures and subunit vaccines have also been investigated, but the control of the arthropod vectors remains the best control strategy. Indeed, the indication for cattle vaccination depends on the situation regarding the epizootiology of bovine babesiosis and anaplasmosis in the region. Cattle from free areas should not be vaccinated, unless if they originated from endemic areas. In areas with enzootic instability, the vaccination is recommended to all calves born outside the tick season period. In areas with enzootic stability, vaccination is recommended only for animals newly introduced from free areas or areas with enzootic instability. Vaccination of cattle from areas of enzootic stability should be performed only when an intensive tick control program is adopted.

The premunition is a control measure that has been widely used in the past and consists in the vaccination of cattle using blood of chronic carriers of the causative agents of bovine babesiosis and anaplasmosis. However, a veterinarian should follow premunized animals and treat them if clinical signs are present. In addition, this practice might be risky because of the transmission of other infectious agents and the difficulties inherent to the large-scale use. Besides these limitations, the premunition does not meet the requirements of safety due to the use of biological material and for these reasons, the premunition is no longer recommended in some regions (e.g., the United States).

In areas of low endemicity, the systematic control of ticks, the main vectors, is usually sufficient to reduce the risk of transmission of *Anaplasma* spp. However, considering that ticks are not the only sources of infection, hygienic and sanitary measures, such as control of flies, are important to reduce the risk of mechanical transmission of *A. marginale* to cattle. Where anaplasmosis is endemic, needles and syringes should not be used for more than one individual during handling cattle herds in order to avoid the risk of transmission by fomites.

Chemoprophylaxis with tetracycline or the use of acaricides of residual effect may be adopted in specific cases, such as in animals from non-endemic areas that will travel temporarily to areas of active transmission (Gonçalves 2000; Kocan et al. 2003). Before the adoption of this measure, it is important to consider many factors, including the value (affective, economic, or both) of the animal. It is important to emphasize that keeping the animals free of ticks during their temporary stay in areas of active transmission is the best strategy of preventing infection by *Anaplasma* spp. regardless of the animal species.

References

Breitschwerdt EB, Hegarty BC, Qurollo BA et al (2014) Intravascular persistence of *Anaplasma platys*, *Ehrlichia chaffeensis*, and *Ehrlichia ewingii* DNA in the blood of a dog and two family members. Parasit Vectors 7:298

Chen SM, Dumler JS, Bakken JS et al (1994) Identification of a granulocytotropic *Ehrlichia* species as the etiologic agent of human disease. J Clin Microbiol 32:589–595

Dumler JS, Barbet AF, Bekker CP et al (2001) Reorganization of genera in the families Rickettsiaceae and Anaplasmataceae in the order Rickettsiales: unification of some species of *Ehrlichia* with *Anaplasma*, *Cowdria* with *Ehrlichia* and *Ehrlichia* with *Neorickettsia*, descriptions of six new species combinations and designation of *Ehrlichia equi* and 'HGE agent' as subjective synonyms of *Ehrlichia phagocytophila*. Int J Syst Evol Microbiol 51:2145–2165

Dumler JS, Choi KS, Garcia-Garcia JC et al (2005) Human granulocytic anaplasmosis and *Anaplasma phagocytophilum*. Emerg Infect Dis 11:1828–1834

Eddlestone SM, Gaunt SD, Neer TM et al (2007) PCR detection of *Anaplasma platys* in blood and tissue of dogs during acute phase of experimental infection. Exp Parasitol 115:205–210

Ferreira RF, Cerqueira AMF, Pereira AM et al (2007) *Anaplasma platys* diagnosis in dogs: comparison between morphological and molecular tests. Int J Appl Res Vet Med 5:113–119

Gonçalves PM (2000) Epidemiologia e controle da tristeza parasitária bovina na região sudeste do Brasil. Cienc Rural 30:187–194

Guglielmone AA, Abdala AA, Anziani O et al (1997) Different seasonal occurrence of anaplasmosis outbreaks in beef and dairy cattle in an area of Argentina free of *Boophilus microplus* ticks. Vet Q 19:32–33

Kessler RH, Sacco MAS, Madruga CR et al (1998) Teste crítico de vacinas atenuadas de *Babesia bovis*, *B. bigemina* e *Anaplasma marginale* em novilhas da raça holandesa. Rev Bras Parasitol Vet 7:1–5

Kessler RH, Soares CO, Madruga CR et al (2002) Tristeza parasitária dos bovinos: quando vacinar é preciso. Embrapa Gado de Corte, Campo Grande

Kim CM, Yi YH, Yu DH et al (2006) Tick-borne rickettsial pathogens in ticks and small mammals in Korea. Appl Environ Microbiol 72:5766–5776

Kocan KM, Stiller D, Goff WL et al. (1989) The developmental cycle of *Anaplasma marginale* in *Dermacentor* spp. In: Proceedings of the 8th National Veterinary Hemoparasite Disease Conference, St. Louis, p 149–160

Kocan KM, de la Fuente J, Guglielmone AA et al (2003) Antigens and alternatives for control of *Anaplasma marginale* infection in cattle. Clin Microbiol Rev 16:698–712

Lima MLF, Soares PT, Ramos CAN et al (2010) Molecular detection of *Anaplasma platys* in a naturally-infected cat in Brazil. Braz J Microbiol 41:381–385

Maggi RG, Mascarelli PE, Havenga LN et al (2013) Co-infection with *Anaplasma platys*, *Bartonella henselae* and *Candidatus* Mycoplasma haematoparvum in a veterinarian. Parasit Vectors 6:103

Mesquita Neto FD, Araújo T, Vieira RM et al (1984) *Ehrlichia equi* em eqüinos no município de Betim - Minas Gerais. Arq Bras Med Vet Zootec 36:203–207

Otranto D, Testini G, Dantas-Torres F et al (2010) Diagnosis of canine vector-borne diseases in young dogs: a longitudinal study. J Clin Microbiol 48:3316–3324

Paiva Diniz PP, Schwartz DS, de Morais H et al (2007) Surveillance for zoonotic vector-borne infections using sick dogs from southeastern Brazil. Vector Borne Zoonotic Dis 7:689–697

Parola P, Cornet JP, Sanogo YO et al (2003) Detection of *Ehrlichia* spp., *Anaplasma* spp., *Rickettsia* spp., and other eubacteria in ticks from the Thai-Myanmar border and Vietnam. J Clin Microbiol 41:1600–1608

Petrovec M, LotricFurlan S, Zupanc TA et al (1997) Human disease in Europe caused by a granulocytic *Ehrlichia* species. J Clin Microbiol 35:1556–1559

Ramos RA, Latrofa MS, Giannelli A et al (2014) Detection of *Anaplasma platys* in dogs and *Rhipicephalus sanguineus* group ticks by a quantitative real-time PCR. Vet Parasitol 205:285–288

Rar V, Golovljova I (2011) *Anaplasma*, *Ehrlichia*, and "Candidatus Neoehrlichia" bacteria: pathogenicity, biodiversity, and molecular genetic characteristics, a review. Infect Genet Evol 11:1842–1861

Ribeiro MFB, Lima JD, Guimarães AM et al (1995) Transmissão congênita da anaplasmose bovina. Arq Bras Med Vet Zootec 47:297–304

Samish M, Pipano E, Hadani A (1993) Intrastadial and interstadial transmission of *Anaplasma marginale* by *Boophilus annulatus* ticks in cattle. Am J Vet Res 54:411–414

Simpson RM, Gaunt SD, Hair JA et al (1991) Evaluation of *Rhipicephalus sanguineus* as a potential biologic vector of *Ehrlichia platys*. Am J Vet Res 52: 1537–1541

Stiller D, Coan ME (1995) Recent developments in elucidating tick vector relationships for anaplasmosis and equine piroplasmosis. Vet Parasitol 57:97–108

Stuen S (2007) *Anaplasma phagocytophilum* - the most widespread tick-borne infection in animals in Europe. Vet Res Commun 31:79–84

Vidotto O, Marana ERM (2001) Diagnóstico em anaplasmose bovina. Cienc Rur 31:361–368

Wormser GP, Dattwyler RJ, Shapiro ED et al (2006) The clinical assessment, treatment, and prevention of Lyme disease, human granulocytic anaplasmosis, and babesiosis: clinical practice guidelines by the Infectious Diseases Society of America. Clin Infect Dis 43:1089–1134

Leishmaniasis 16

Carlos Brisola Marcondes, Domenico Otranto, and Filipe Dantas-Torres

Abstract

Leishmaniases are diseases caused by several species of *Leishmania* protozoa and occur in all continents, except Antarctica. Almost 20 species have been isolated from humans, and they may cause cutaneous, mucocutaneous, and visceral forms of the disease. The severity of the infection depends on the protozoon species and strain, host immune response, nutritional status, and, to a lesser extent, sand fly species involved in the transmission. Depending on the parasite tropism, amastigotes invade macrophages in the skin, mucosa, and internal organs, mostly the spleen, liver, and bone marrow, causing their depletion. Clinical presentations may vary from localized skin lesions to generalized disease, with fever, weight loss, enlargement of the spleen and liver, anemia, leucopenia, and thrombocytopenia. Transmission primarily occurs by the bites of sand flies, the only proven biological vectors of *Leishmania* parasites. Secondary transmission routes have been demonstrated in both humans and dogs, including congenital transmission. The epidemiology of leishmaniases is complex, as it involves different vector, host, and parasite species, living in a constantly changing environment. Disease diagnosis and treatment are still complicated, particularly in developing countries. More effective control strategies are urgently needed to reduce the burden of leishmaniases in endemic areas. Meanwhile, education and better living conditions for people residing in risk areas are key to controlling this group of diseases.

Keywords

Leishmaniasis • Kala-azar • Espundia • Visceral leishmaniasis • Cutaneous leishmaniasis

C.B. Marcondes (✉)
Department of Microbiology, Immunology and Parasitology, Center of Biological Sciences, Federal University of Santa Catarina,
88040-900 Florianópolis, SC, Brazil
e-mail: cbrisolamarcondes@gmail.com

D. Otranto
Department of Veterinary Medicine, University of Bari, Valenzano, Italy
e-mail: domenico.otranto@uniba.it

F. Dantas-Torres
Department of Immunology, Aggeu Magalhães Research Center, 50740465 Recife, PE, Brazil
e-mail: filipe.dantas@cpqam.fiocruz.br

C.B. Marcondes (ed.), *Arthropod Borne Diseases*, DOI 10.1007/978-3-319-13884-8_16

Leishmaniases are a group of diseases affecting humans and several other animal species, being widely distributed in all continents, except Antarctica. They cause a major burden in endemic areas, with approximately 1.5 million new cases each year and at least 350 million people at risk in the world. Their distribution is mostly tropical and subtropical, but they are expanding their distribution to new areas in Europe and in the Americas. Due to multiplicity of species of *Leishmania*, reservoir hosts, and vectors, their epidemiology is complex and should be studied locally to allow the planning of more effective control and prevention strategies.

The infection has been known for centuries, and cutaneous forms have been represented in potteries, "huacos," in pre-Colombian Andes and were well known in the Mediterranean region as "bouton de Aleppo." At the end of the nineteenth century, their etiological agents were described, and epidemics of visceral leishmaniasis (VL) occurred frequently associated to famine, for example, in the Indian continent.

Transmission of *Leishmania* spp. occurs through the bites of sand flies (Phlebotominae), the only proven biological vectors of the parasites, although biting midges infected by *Leishmania* sp. have been found in Australia. Sexual and congenital transmissions are considered secondary modalities of infection in dogs, as well as transmission via blood transfusion (Freitas et al. 2006). Amastigotes of *Leishmania infantum* have been found in 34–52 % of discarded syringes of drug addicts in a study in Madrid, which raised the question on possible mechanical transmission among drug users.

The origin of *Leishmania* was supposed to be South American, *ca.* 46–36 million of years ago. Then, *ca.* 14–24 million of years ago, the ancestor of *Leishmania donovani* and *L. infantum* would have diverged from other species in Central Asia before being imported from Europe with the conquistadores, about 500 years ago (Lukes et al. 2007).

16.1 Etiology

Several species of *Leishmania* can infect humans and other animals causing diseases with different degrees of severity. For example, *L. infantum* is usually associated with VL but can also cause small cutaneous lesions, whereas species that typically causes cutaneous leishmaniasis (CL) in humans but may eventually invade the viscera, as it is the case of *Leishmania amazonensis*.

Species of *Leishmania* have been classified according the localization of developmental stages in the digestive tract of their sand fly vectors and to their biochemical and molecular characters. Therefore, they have been distributed in two subgenera (*Leishmania* and *Viannia*) (Table 16.1), and *Leishmania* of lizards are included in the subgenus *Sauroleishmania*. Besides the occurrence of *Leishmania* in Eurasia, Africa, and the Americas, a *Leishmania* infecting and causing cutaneous lesions on macropods (kangaroos, wallaroos, and wallabies) has been recently characterized in the north of Australia.

Correct identification of species of *Leishmania* causing human disease is crucial for a better evaluation of the prognosis of the disease and for planning control strategies. The etiological identification is particularly important in the New World, where several *Leishmania* spp. may simultaneously occur. Also in the Old World, the etiological diagnosis may be very important for planning strategies to follow. For example, in addition to *L. infantum*, *L. tropica* has been implicated in some cases of VL in the south of Iran and Iraq. The utilization of insecticide impregnated collars in dogs in Iran reduced the incidence of *L. infantum* in both dogs and children, but this strategy would probably have no impact on *L. tropica* incidence, which emphasizes the need for a precise parasite identification to elaborate proper control strategies.

The etiological diagnosis of leishmaniasis should ideally rely not only on cytological examination of stained smears prepared with material obtained by puncture of cutaneous lesions or viscera but also on the isolation of parasite in culture and characterization by multilocus enzyme electrophoresis or other reference method. In this perspective, molecular techniques are frequently necessary for the correct identification at species level.

Monoclonal antibodies also are useful and can be obtained free of charge from WHO. Nonetheless, they may present cross-reactions

Table 16.1 Principal causative agents of human leishmaniasis (modified from Cantacessi et al. 2015)

Leishmania species	Principal tropism	Geographical distribution	Notes on the infection in dogs[a]
Leishmania aethiopica	C	Old World: Ethiopia and Kenya	–
L. amazonensis	C	New World: Argentina, Bolivia, Brazil, Colombia, Ecuador, French Guiana, Peru, Suriname, and Venezuela	VL cases in Brazil
L. archibaldi[b]	V	Old World: Ethiopia, Kenya, Lebanon, and Sudan	VL cases in Sudan
L. braziliensis	C, MC	New World: Argentina, Belize, Bolivia, Brazil, Colombia, Costa Rica, Ecuador, Guatemala, French Guiana, Honduras, Mexico, Nicaragua, Panama, Paraguay, Peru, and Venezuela	CL cases in Argentina, Bolivia, Brazil, Colombia, Peru, and Venezuela
L. colombiensis	C	New World: Colombia, Panama, and Venezuela	VL in a dog in Venezuela
L. donovani	V	Old World: Bangladesh, Bhutan, China, Cyprus, Djibouti, Ethiopia, India, Iraq, Israel, Kenya, Nepal, Saudi Arabia, Somalia, Sri Lanka, Sudan, Ukraine, Uganda, and Yemen	Dogs are commonly infected in some countries (e.g., Sudan), but their role as reservoirs is unknown
L. garnhami[b]	C	New World: Costa Rica and Venezuela	–
L. guyanensis	C	New World: Argentina, Bolivia, Brazil, Colombia, Ecuador, French Guiana, Guyana, Peru, Suriname, and Venezuela	CL cases in Colombia
L. infantum	V, C	Old World: Afghanistan, Albania, Algeria, Armenia, Azerbaijan, Bosnia and Herzegovina, Bulgaria, Central African Republic, China, Cyprus, Croatia, Egypt, France, Gambia, Georgia, Greece, Iraq, Iran, Israel, Italy, Libyan Arab Jamahiriya, Jordan, Kazakhstan, Kirgizstan, Lebanon, Macedonia, Malta, Morocco, Mauritania, Monaco, Montenegro, Oman, Pakistan, Palestine, Portugal, Syria, Romania, Senegal, Saudi Arabia, Slovenia, Spain, Sudan, Tunisia, Turkmenistan, Turkey, Ukraine, Uzbekistan, and Yemen New World: Argentina, Bolivia, Brazil, Colombia, Costa Rica, El Salvador, Guatemala, Honduras, Mexico, Nicaragua, Paraguay, and Venezuela	VL cases usually found in areas where human cases are reported. Autochthonous cases reported in dogs in the USA (no human case reported so far)
L. killicki	C	Old World: Algeria, Libyan Arab Jamahiriya, and Tunisia	–
L. lainsoni	C	New World: Bolivia, Brazil, French Guiana, Peru, and Suriname	–
L. lindenbergi	C	New World: Brazil	–
L. major	C	Old World: Afghanistan, Algeria, Azerbaijan, Burkina Faso, Cameron, Chad, Egypt, Ethiopia, Georgia, Ghana, Guinea, Guinea-Bissau, India, Iraq, Israel, Libyan Arab Jamahiriya, Jordan, Kazakhstan, Kenya, Kuwait, Mali, Morocco, Mauritania, Mongolia, Niger, Nigeria, Oman, Pakistan, Palestine, Saudi Arabia, Syria, Iran, Senegal, Sudan, Tunisia, Turkmenistan, Uzbekistan, and Yemen	CL in Egypt and Saudi Arabia
L. mexicana	C	New World: Belize, Colombia, Costa Rica, Ecuador, Guatemala, Mexico, and the USA	CL in Ecuador and the USA
Leishmania naiffi	C	New World: Brazil and French Guiana	–

(continued)

Table 16.1 (continued)

Leishmania species	Principal tropism	Geographical distribution	Notes on the infection in dogs[a]
L. panamensis	C, MC	New World: Colombia, Costa Rica, Ecuador, Guatemala, Honduras, Nicaragua, and Panama	CL in Ecuador and Colombia
L. peruviana	C	New World: Peru	CL in Peru
L. pifanoi[b]	C	New World: Venezuela	CL in Ecuador
L. shawi	C	New World: Brazil	–
L. tropica	C	Old World: Afghanistan, Azerbaijan, Egypt, Ethiopia, Greece, India, Iraq, Israel, Iran, Jordan, Kenya, Morocco, Namibia, Pakistan, Palestine, Saudi Arabia, Syria, Turkmenistan, Turkey, Uzbekistan, and Yemen	CL cases in India, Iran, Israel, Morocco, and Syria
L. venezuelensis	C	New World: Venezuela	–

Abbreviations: *C* dermotropic, *V* viscerotropic, *MC* mucotropic

[a]In addition, *L. arabica* has been reported in dogs in Saudi Arabia. Moreover, other *L.* species (e.g., *L. equatorensis* and *L. utingensis*) have been described from wildlife and/or sand flies but not have not till now been detected in humans or dogs

[b]Species status is under discussion

among closely related species and are not able to discriminate strain-level differences. Indeed, different *L. braziliensis* genotypes have been associated to different forms of CL (Schriefer et al. 2004), and, in this perspective, molecular techniques are accelerating and refining the etiological diagnosis of leishmaniasis at both species and strain level.

16.2 Life Cycle of *Leishmania*

Mammals are usually infected when metacyclic promastigotes are inoculated during the bite of sand flies. Promastigotes are phagocytized by macrophages, in which they become amastigotes (Fig. 16.1), multiplying and eventually destroying them, before invading other macrophages. When amastigotes are ingested by a sand fly (Fig. 16.2) during its blood feeding, they can differentiate in its digestive tract to different promastigote forms (Fig. 16.3). The cycle in sand flies involves at least six forms (amastigotes, procyclic promastigotes, nectomones, leptomones, haptomones, and metacyclic promastigotes) (Kamhawi 2006), but for didactic purposes, only promastigotes are usually referred as occurring in these insects.

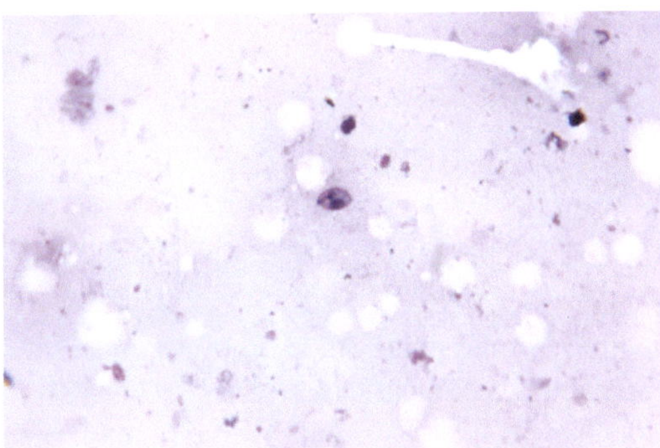

Fig. 16.1 Amastigote of *Leishmania*. Kindly authorized by M. Steindel

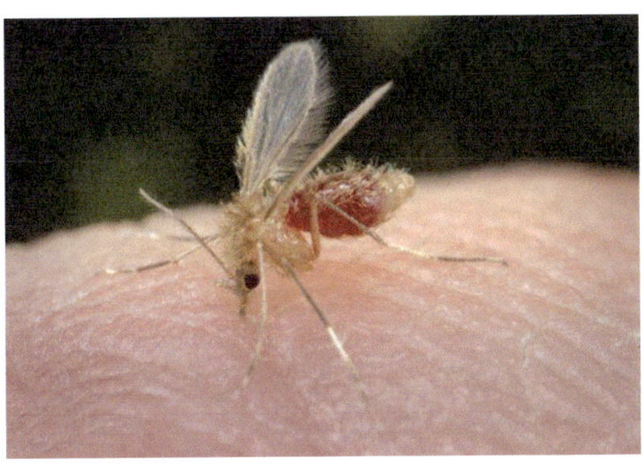

Fig. 16.2 Phlebotomine sand fly biting skin. From CDC

The destruction of macrophages in several parts of the organism will cause several consequences, and the skin, mucosa, and several viscera can be affected, sometimes reducing the host defense against the invaders. The effort of the organism to produce macrophages can reduce the production of erythrocytes and other blood cells, causing anemia with tendency to hemorrhage. If amastigotes are concentrated in the skin and mucosa or abnormal reactions occur, variable lesions can be found. The development of VL or CL will depend on the species of *Leishmania* involved and several other factors. For example, dogs remain infected for long periods of time and frequently have a high rate of infection with no clinical manifestations even if seropositive. The latter animals are less infectious than dogs with multiple clinical signs. Also, the sand fly species involved in transmission of *L. infantum* seem to affect the clinical presentation in humans as it does the individual susceptibility and immune response.

16.3 Clinical Forms of Leishmaniasis

16.3.1 Visceral Leishmaniasis

16.3.1.1 Agents

VL is caused by *L. donovani* (in Africa and Indian subcontinent, where the disease is also known as kala-azar) and *L. infantum* (in the Mediterranean area, Middle East, China, and Southwest Asia and American continent). In addition, *L. amazonensis* has been incriminated in some cases in Brazil. Visceralization is usually influenced by several factors, including the biology of the parasite, the host immunological status, and the vector strain.

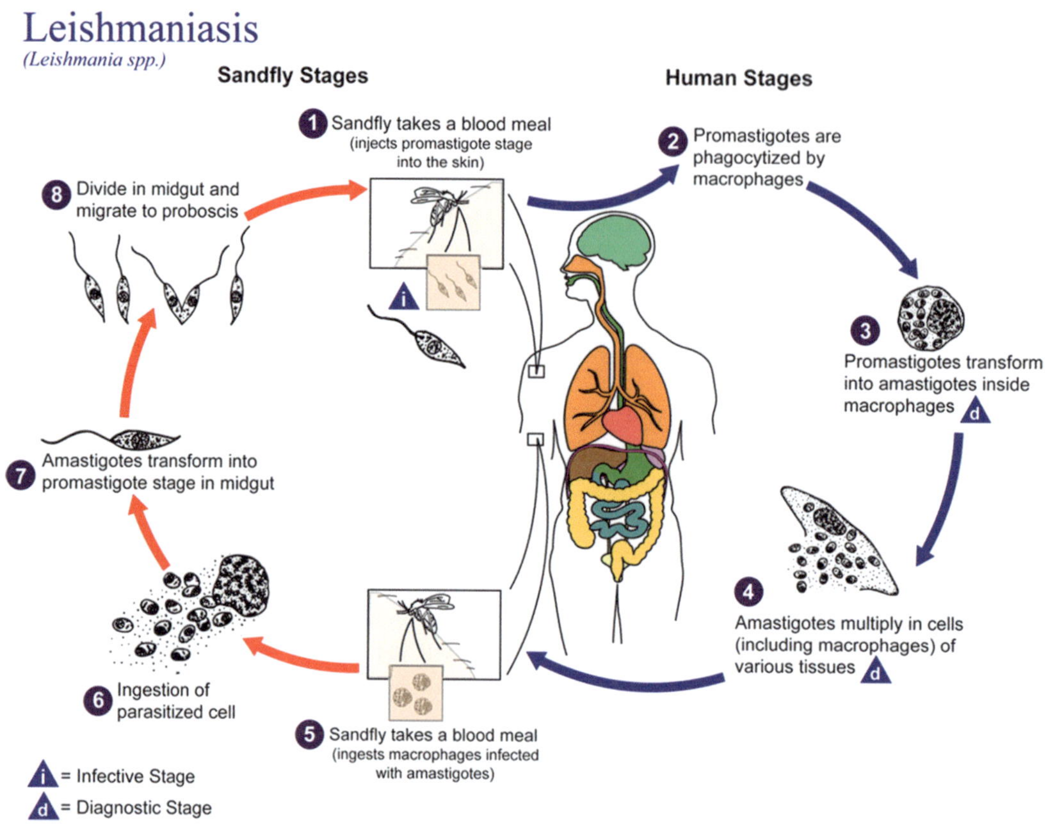

Fig. 16.3 Cycle of *Leishmania*. From CDC

16.3.1.2 Disease in Humans and Domestic Animals

The disease is characterized by the development of amastigotes in visceral macrophages, mostly in the spleen, liver, lymph nodes, and intestinal mucosa. After inoculation by the vector, VL evolves for several weeks, with mild symptoms, including fever and enlarged lymph nodes, spleen, and liver, weakening worsened by anemia, immunosuppression due to the destruction of macrophages and deviation of function of hematopoietic tissues, and weight loss. Undernourished children, after a period from 1 month to 1 year, can die, mostly of concurrent infections, like pneumonia, diarrhea, and tuberculosis, which can often confuse the diagnosis.

Although the clinical presentation is similar in endemic areas, enlarged lymph nodes are rarely observed in the Indian continent, while hyperpigmentation, previously common in this region, is today rare. The ratio of asymptomatic infections to clinical cases varies from 1:2.6 in Sudan to 50:1 in Spain and is strongly influenced by species of parasites and nutrition.

In dogs the disease is characterized by usually long incubation periods (from 2 months to years), a systemic evolution due to the infection of macrophages in all tissues and organs and alterations due to an exacerbated immune response with a hyperproduction of antibodies and immunocomplex deposition (e.g., uveitis, vasculitis, and glomerulonephritis).

Although it is difficult or even impossible to identify specific clinical signs for the infection, infected dogs appear in poor body conditions, with enlarged lymph nodes and other nonspecific clinical signs. Dog owners may refer a sudden aging of the animal in a few months. In addition, cutaneous lesions, when present, are characterized by non-pruritic exfoliative dermatitis (large, abundant, and gray scales), alopecia, and skin ulcers at the bony prominences, ear pinnae, and footpad. However, nodular or pustular dermatitis, nose bleeding, colitis, as well as neurological disorders may also occur.

16.3.1.3 Leishmaniasis-HIV Coinfection

Leishmaniasis can affect the presentation, diagnosis, and course of malignant disorders, sometimes complicating treatment (Kopterides et al. 2007). HIV infection, due to low cell-mediated immune response, mostly if associated to malnutrition, increases risk of developing clinical disease (Chappuis et al. 2007). For example, low CD4+ counts (<100 cells/ml) and the absence of an increase of these cells at follow-up are predictors of VL relapse, when associated to HIV (Cota et al. 2011). Coinfection by HIV can cause different clinical consequences (Table 16.2), and special care with diagnosis and follow-up is necessary. *Leishmania*-HIV coinfected patients tend to be more frequently malnourished, to have less fever and hepatosplenomegaly, compared to non-HIV-infected ones, possibly complicating initial diagnosis (Cota et al. 2014).

16.3.1.4 PKDL

In East Africa and the Indian subcontinent, after symptomatic or asymptomatic VL by *L. donovani*, a skin infection, called post-kala-azar dermal leishmaniasis (PKDL), develops respectively in 50–60 % and 5–10 % of patients, during or after the treatment. It is characterized by maculopapular, papular, and nodular rash, starting around the mouth and affecting most of the face, trunk, arms, and legs (Zijlstra et al. 2003) (Fig. 16.4). In Indian subcontinent, the nodules tend to evolve to plaques, while in Africa ulceration is frequent (Singh et al. 2011). Eyes can be affected, causing blindness (Khalil et al. 2011).

Some cases of PKDL caused by *L. infantum* were reported from northwestern Iran (Badirzadeh et al. 2013). In Sudan, most cases are spontaneously cured, but in the Indian subcontinent, most need treatment (Zijlstra et al. 2003). Discrepancies

Table 16.2 Clinical conditions under which patients with leishmaniasis should be offered an HIV test (from Rabello et al. 2003, authorized by Maney Publishing)

Visceral leishmaniasis	Cutaneous leishmaniasis
Development of any clinical form of leishmaniasis more than a year after the patient was in an area where human infection with *Leishmania* occurs	Development of any clinical form of leishmaniasis more than a year after the patient was in an area where human infection with *Leishmania* occurs
Patient is an intravenous drug user	Cutaneous or mucosal clinical forms with parasites observed in the viscera
The absence of antileishmanial antibodies in a patient with the typical symptoms of VL	Diffuse clinical form
The presence of circulating amastigotes	Any clinical form in a patient giving a negative result in a (Montenegro) skin test
Involvement of systems/organs rarely affected in VL[a]	The presence of amastigotes in a smear of a mucosal lesion
Treatment failure after therapy with pentavalent antimonial[b]	Therapeutic failure with a pentavalent antimonial drug[c]
Late relapse[d]	Late relapse (>6 months after clinical recovery)
Development of other infections indicative of immunodeficiency, during or after treatment[e]	Cutaneous lesions appearing after active mucosal lesion
Isolation of a normally dermotropic species of *Leishmania*	Isolation of *L. infantum* (or other normally viscerotropic *Leishmania* sp.) from skin or mucosal lesions

[a]Such as the respiratory tract, digestive tract, or skin

[b]The absence of clinical cure after two attempts at treatment with meglumine antimoniate (10–20 mg SbV/kg/day for 30 days), with a 3-month follow-up after each period of treatment

[c]The absence of clinical cure after two attempts at treatment with meglumine antimoniate (10–20 mg SbV/kg/day for 20–30 days), with a 3-month follow-up after each period of treatment

[d]After at least 12 months of clinical recovery

[e]Such as herpes zoster or tuberculosis

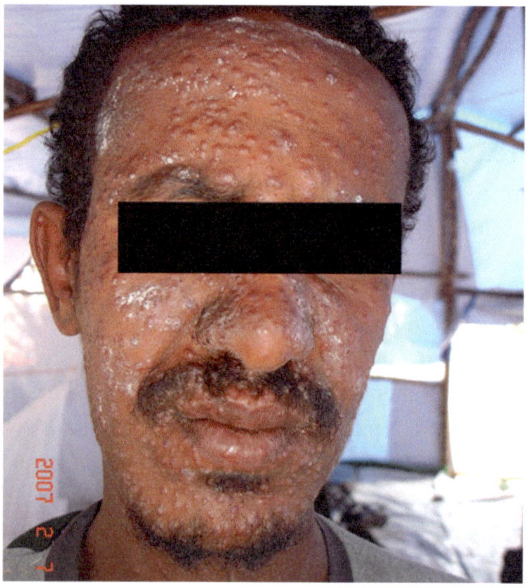

Fig. 16.4 Post-kala-azar dermal leishmaniasis (PKDL). From WHO

between immunohistopathological findings in Indian and Sudanese patients were reported (Zijlstra et al. 2003). Four cases of PKDL in the south of the Brazilian state of Bahia were associated to *L. amazonensis*.

PKDL is frequently more severe at a younger age and a shorter interval between VL treatments, and has been associated to more severe disease. Differential diagnosis with leprosy, in both areas, must be done; neuritis in small cutaneous nerves in some cases of PKDL may cause some confusion in the differential diagnosis. Besides the clinical importance of PKDL, the patients have been supposed to constitute sources of amastigotes for sand flies in the long term and between epidemics.

Infection by *L. infantum* in dogs may span from the absence of clinical signs (in even more than 50 % of the infected animals) to skin and ocular lesions, lymph node enlargement, severe weight loss, splenomegaly, onychogryphosis, and renal failure. Other *Leishmania* species have been implicated in cases of canine VL, including *L. colombiensis* and *L. amazonensis*, but the actual clinical and epidemiological relevance of these cases remains uncertain.

16.3.2 Cutaneous Leishmaniasis

16.3.2.1 Agents

CL is caused by several species of *Leishmania*, which varying epidemiology and clinical presentations. *L. braziliensis* is the most common species in the American continent, but at least 12 other species (*L. colombiensis*, *L. amazonensis*, *L. garnhami*, *L. guyanensis*, *L. infantum*, *L. lainsoni*, *L. lindenbergi*, *L. mexicana*, *L. naiffi*, *L. panamensis*, *L. peruviana*, *L. pifanoi*) can cause cutaneous lesions. In the Old World, *L. aethiopica*, *L. infantum*, *L. killicki*, *L. major*, and *L. tropica* besides *L. donovani* (in PKDL) can cause CL (Table 16.1).

16.3.2.2 Disease in Humans

After transmission of promastigotes by sand fly bites, usually a small red papule develops, rupturing after some weeks, producing a scaly plate (Fig. 16.5) or more frequently a slowly growing ulcer (Fig. 16.6), which may spontaneously heal. Lesions usually occur in exposed areas, but multiple lesions can be produced by hematogenous or lymphatic dispersion to districts other than those of original bites. In *L. guyanensis*, transmitted by sand flies associated to large trees, a massive attack by these insects can cause dozens of lesions. Simultaneous to cutaneous lesions or even 20 years after their healing, the nasopharyngeal mucosa can be affected. Nasal lesions are always present, with nodules and infiltration of cartilaginous septum, obstruction of nostril, perforation of the septum, and even total destruction of the nose (Fig. 16.7). In part of the patients, the pharynx, palate, trachea, and upper lip can be affected, with lymphadenopathy. Mucosal lesions are associated to poorly modulated T-cell response, with high levels of IFN-γ and TNF-α and a decreased ability of IL-10 and TGF-β to modulate this response (Bacellar et al. 2002).

Diffuse CL, with several non-ulcerated nodules in all parts of the body, is mostly caused by *L. amazonensis* in the Amazon region and also by *L. braziliensis*. Indeed, the latter species may cause disseminated CL, with several papulous, acneiform, and rarely ulcerated lesions, associated to chills and fever, which indicates hematog-

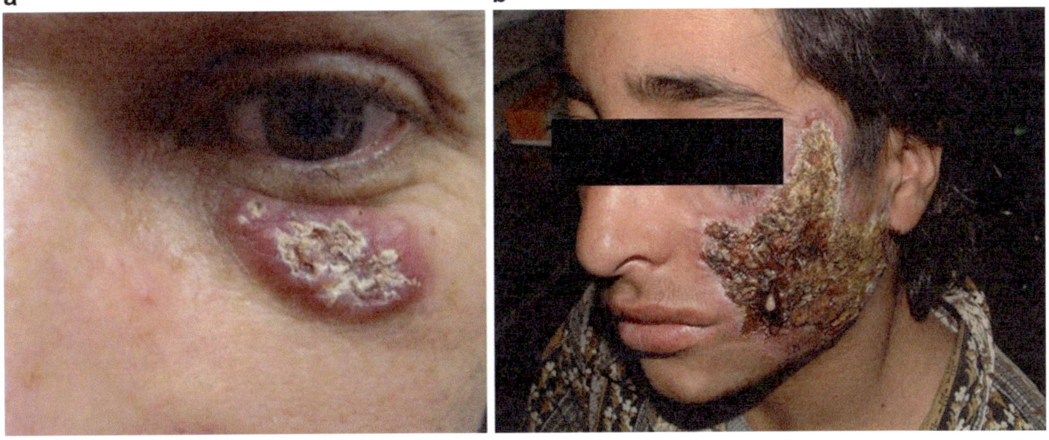

Fig. 16.5 Cutaneous lesions caused by *L. major*. (**a**) Negev desert. (**b**) Iran. Kindly authorized by O. Katz and R. Yaghoob

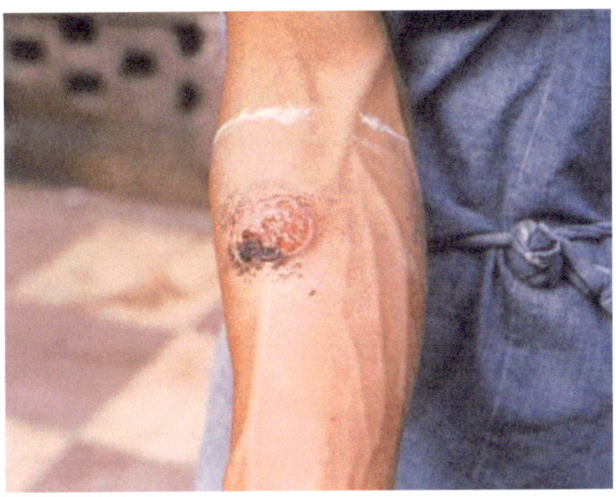

Fig. 16.6 Ulcer caused by *L. braziliensis*. From CDC

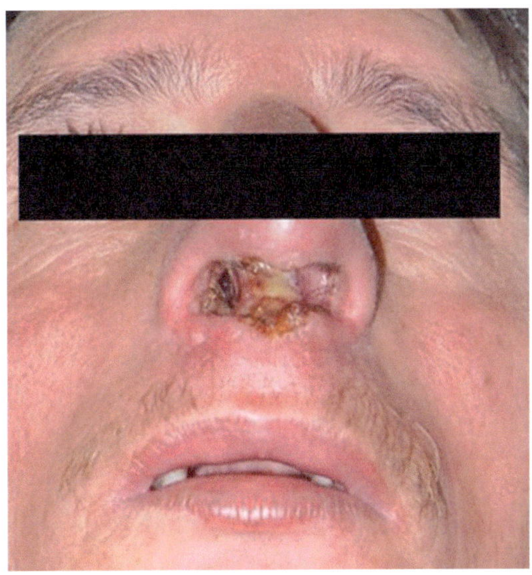

Fig. 16.7 Mucocutaneous leishmaniasis (*L. braziliensis*) from southern Brazil. Kindly authorized by M. Steindel

enous dispersion (Guimarães et al. 2005). Diffused CL is characterized by persistent T-cell anergy. Patients usually present negative results in the leishmanin test and do not respond to therapy with antimonials and amphotericin B, only to pentamidine. In turn, the disseminated CL (Fig. 16.8) is frequently associated with mucosal lesions, with patients being positive in the leishmanin test and responding to amphotericin B therapy but not to antimonials.

Most cases of mucocutaneous leishmaniasis (MCL) have been associated to *L. braziliensis* and *L. panamensis* in the New World and to *L. aethiopica* in the Old World. *Leishmania donovani* may cause mucosal lesions in Sudan, usually not associated to cutaneous lesions (El-Hassan and Zijlstra 2001). The association to HIV infection favors the onset of diffuse or mucosal forms (Table 16.2).

Cutaneous lesions can become infected by bacteria, being painful and wet, restricting amastigotes mostly to ulcer borders. Antibiotics are frequently the first treatment for many cutaneous lesions, and, although the failure of treatment is not an adequate method of diagnosis, the nonhealing of ulcers after elimination of bacteria by this treatment may raise suspicion of CL. As examples,

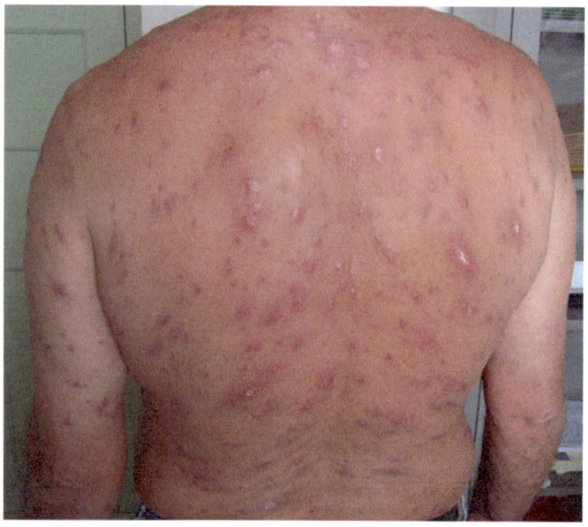

Fig. 16.8 Disseminated cutaneous leishmaniasis caused by *L. braziliensis*, south of Brazilian state of Bahia. Kindly authorized by L. H. Guimarães

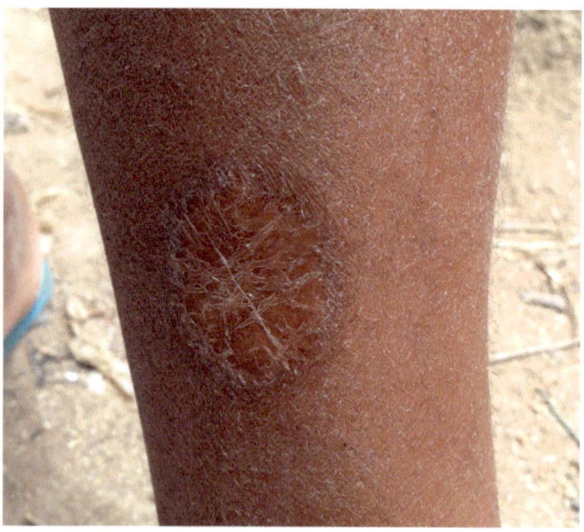

Fig. 16.9 Scar of dermal leishmaniasis from Brazilian state of Pernambuco

in Alagoa Grande, in northeastern Brazil, many CL patients referred previous treatment by "a painful injection with a white liquid" (penicillin), before the correct diagnosis,[1] and a girl with a lesion in the ear in Ecuador had previously been treated by oral ciprofloxacin and clindamycin for 10 days and fusidic acid as a topical ointment, with no improvement (Hinojosa et al. 2013).

Scars derived from CL are atrophic and whiter than the surrounding skin (Fig. 16.9), in consequence more evident in dark skin patients, and amastigotes can be present in the scars, even several years after healing, potentially dispersing through the organism and causing new lesions.

[1]Personal observation (CBM)

16.3.2.3 Disease in Domestic and Sylvatic Animals

Several species of *Leishmania* can cause lesions similar to those observed in humans, mostly in domestic animals (Fig. 16.10), but amastigotes can be present in intact skin, mostly in sylvatic animals, facilitating the infection of the vectors. Besides cutaneous lesions observed in VL by *L. infantum* (e.g., non-pruritic exfoliative dermatitis with large, abundant, and gray scales, alopecia, skin ulcers at the bony prominences, ear pinnae, footpad), *L. braziliensis*, *L. peruviana*, and other species which cause CL in humans may also induce localized skin lesions in dogs (Dantas-Torres 2009). Most of the dogs infected by *L. braziliensis* live in rural areas, and they may present single cutaneous or mucosal lesions. Their role as spreaders of the infection to humans has been suspected even if it is actually only minor.

16.4 Diagnosis

The diagnosis of visceral and CL in humans and dogs is a complex task, which relies on several aspects (e.g., anamnesis, epidemiology, clinical presentation, and laboratory findings), including laboratory evidence of infection, which ultimately drives the clinician from a suspicion to the direct or indirect confirmation of the infection. Indeed, before clinical examination laboratory testing, data on the provenience, age, sex, attitude, and lifestyle of the patient as well as the occurrence of other concomitant infections needs to be carefully considered in order to appreciate the likely risk of infection. After this preliminary assessment and a complete clinical examination, parasitological, serological, and/or molecular tests should be requested to confirm the diagnosis.

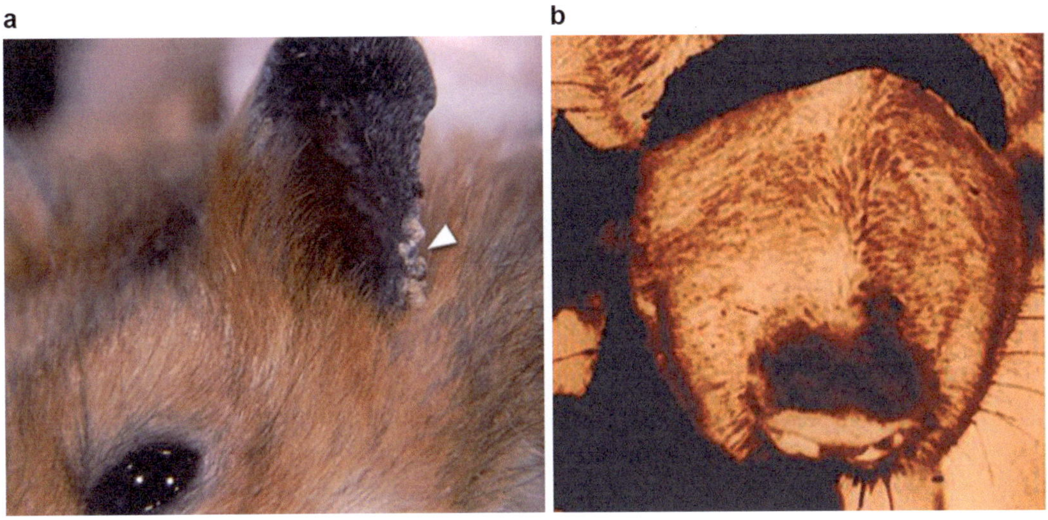

Fig. 16.10 Cutaneous lesions in dogs: (**a**) *L. braziliensis* (American continent); (**b**) *L. major* (Iran). Kindly authorized by CDC and R. Yaghoob

16.4.1 Clinical Diagnosis

In human medicine, long-term low fever, associated to hepatosplenomegaly, lymph node enlargement, and anemia, mostly in known endemic areas, raises the suspicion of VL. An adequate comparative diagnosis has to be made with other diseases, like schistosomiasis, mononucleosis, AIDS, malaria, and other diseases with similar presentation.

CL must be suspected for cutaneous lesions, mostly localized in regions of the body not covered by clothes and/or in patients potentially exposed to sand fly bites, with more than 1 month of development. The appearance of lesions, if an ulcer has already developed, may lead to suspect CL. However, lesions characterized by several nodules following the trajectory of lymphatic vessels and simulating sporotrichosis may finally conduct to the diagnosis of an infection by *L. guyanensis* (Hinojosa et al. 2013). Atypical manifestations (erysipeloid, *recidiva cutis*, and disseminated CL) were related to *L. panamensis* in Ecuador.

In MCL caused by *L. braziliensis* and *L. panamensis* in New World and *L. donovani* in Sudan, mucosa and cartilaginous parts of the nose and nearby areas are largely compromised. The PKML is differentiated from several other diseases by histopathology and appropriate laboratory tests. Less often, mucosa of other regions of the face, like tongue and lips, is affected (Mignogna et al. 2014), and dentists can collaborate on the diagnosis. Allergic rhinitis, histoplasmosis, *cancrum oris*, lymphoma, and other neoplasia, leprosy, and sarcoidosis may mimic this form of leishmaniasis, and parasitological, immunological, and molecular techniques should drive physicians to the correct etiological diagnosis. The presence of concomitant or previous cutaneous lesions may be suggestive for MCL.

Clinical signs in canine leishmaniasis (CanL) are often inconsistent being primarily characterized by emaciation, enlargement of peripheral lymph nodes, non-pruritic exfoliative dermatitis with alopecia and also nodular or pustular dermatitis, lesions of mucous membranes (nodules and ulcers), and ocular lesions (blepharitis, keratoconjunctivitis, uveitis). Also laboratory abnormalities, such as hyperglobulinemia (polyclonal beta and/or gammaglobulinemia), hypoalbuminemia, and decreased albumin/globulin ratio, may be suggestive of CanL. Clinical CanL is characterized by a wide spectrum of clinical signs, which may occur alone or in combination, and severity (e.g., weight loss, lethargy, muscular atrophy, anemia, lymphadenomegaly, splenomegaly, epistaxis, diarrhea, renal disorders, ocular lesions, polyarthritis, onychogryphosis, and skin lesions). Importantly, in endemic areas, a large number of subclinical infections have been recorded in dogs and humans. Despite the conflicting results of different studies regarding the infectiousness of asymptomatic versus sick dogs, the former may eventually serve as a source of infection to sand fly vectors, albeit to a lesser extent in comparison with sick dogs. The diverse presentation of disease in infected animals is primarily linked to the many pathogenic mechanisms involved and the immune response elicited in the host, which ultimately make the diagnosis of infection in clinically healthy but infected animals challenging.

16.4.2 Parasitological Diagnosis

Cytological examination of Giemsa-stained smears prepared with bone marrow aspirates, preferentially from sternum, is usually useful for the diagnosis of VL. Liver and spleen aspirates may be productive but are too risky. HIV patients, mostly if severely immunosuppressed, may present other opportunistic infections, which may complicate the diagnosis. In these patients, the parasite load is higher, sometimes in unusual sites. All patients with unusual presentations of leishmaniasis need to be tested to HIV.

For the parasitological diagnosis of CL, samples can be obtained, with adequate local anesthesia, from skin lesions by slide impression, biopsy, or aspiration from the border of the ulcer, which are richer in protozoa and less contaminated by bacteria. The material obtained from skin lesions should be smeared on a glass slide, stained by Giemsa or similar. Other preparations with staining adequate for fungi and culture media can be utilized for diagnosis of concomitant or alternative infections. It can also be inoculated in the snout, feet, or peritoneal cavity of laboratory animals, preferably the golden hamster, for in vivo isolation and subsequent identification, mostly in reference laboratories. Nonetheless, with the advent of molecular techniques, the biological isolation is not any more commonly employed, also due to ethical reasons. In vitro culture, although reliable, is prone to microbiological contamination, especially if skin samples or samples collected under field conditions are used.

In MCL, parasites are usually rare, and detection of parasite DNA by PCR is a sensitive method for diagnosis. Amastigotes are particularly easy to find in cutaneous lesions in PKDL (Singh et al. 2011) and in those caused by *L. amazonensis*, but monoclonal antibodies and molecular techniques should be utilized in case of negative results.

The observation of amastigotes by cytology or histology in lymph node or bone marrow samples is often used for diagnosing CanL. Cytological examination is a straightforward method, though time-consuming, but may lack in sensitivity, particularly in asymptomatic dogs. Thus, a negative cytological examination does not exclude the possibility of infection in a dog with suspicious CanL. As far as histology, it may present high sensitivity if immunohistochemistry is performed.

16.4.3 Immunological Diagnosis

Indirect immune fluorescence test (IFAT) and mostly enzyme-linked immunosorbent assay (ELISA) displayed good performance in terms of sensitivity and specificity for the diagnosis of VL but require equipments that are not adapted to field conditions. Conversely rapid immunochromatographic tests (ICT) are well adapted for field utilization. Persistent positive reactions are a limitation of all serological tests, and these tests must always be utilized in conjunction with adequate clinical evaluation.

Leishmanin skin test (Montenegro reaction) (injection of a suspension of killed *Leishmania* promastigotes sonicated and kept in merthiolate solution at 1:10,000, with a volume of 0.1 mL of the antigen with 40 μg/mL of nitrogen) is usually positive from 1 month after infection, for most cases of CL, except those of anergic diffuse CL, mostly caused by *L. amazonensis*, and some cases of disseminated CL, and continues to be positive even after cure.

IFAT can detect both symptomatic and asymptomatic dogs, and it is widely used in large screening surveys. Among serological tests, IFAT still represents one of the most widely used serological techniques for diagnosing CanL, even if this technique has recently been replaced by Immunochromatographic Strip Test (ICT) in Brazil, for example. However, the definitive diagnosis of *L. infantum* infection in asymptomatic dogs is troublesome since both serological and parasitological methods have inherent limitations, especially in terms of sensitivity in this particular group of animals. Indeed, serology may not be a good indicator of infection in cross-sectional studies due to the varying time spanning between infection and seroconversion (i.e., from 1 to several months).

For HIV-VL patients, different serological tests (e.g., IFAT and ELISA) have shown variable, often low, sensitivities and specificities in the Mediterranean region (Monge-Maillo et al. 2014) and in Latin America (Lindoso et al. 2014). Nonetheless, tests such as immunoblotting and direct agglutination test (DAT) have shown better results for the diagnosis of VL in this group of patients (Monge-Maillo et al. 2014; Lindoso et al. 2014)..

16.4.4 Molecular Diagnosis

A number of PCR assays such as conventional and real-time PCR on various tissues, including blood, lymph node, bone marrow, and spleen, have improved the sensitivity of the diagnosis of *Leishmania* infection in both animals and humans. PCR methods have been standardized for the amplification of several target genes including genomic DNA of the internal transcribed spacer 1 (ITS1) region of the ribosomal RNA (rRNA) and kinetoplast DNA (kDNA) minicircle sequence. For sure the sensitivity of the PCR is correlated with the number of copies of target genes and with the type of samples used. For example, PCR tests of material obtained from bone marrow, lymph nodes, and the skin are preferred samples in relation to blood. Refined techniques, such as real-time PCR and loop-mediated isothermal amplification (LAMP), further increased the sensitivity and specificity of the molecular diagnosis of leishmaniases.

Although the use of molecular diagnostic techniques has become widespread in recent years, they are still mostly restricted to urban centers and to reference laboratories. The lack of standardization of molecular assays is a problem in both private and public laboratories. Scientists should endeavor to standardize the protocols, including target genes, primers, amplification conditions, and samples to be tested in order to make studies comparable and to standardize the diagnosis of the disease in both developing and developed countries.

For HIV-VL patients, real-time PCR (qPCR) is indicated to predict relapse after treatment; parasite loads >0.03 and 0.9 parasites/mL, respectively, at 3 and 12 months after treatment indicate 100 % of probability of a relapse (Lindoso et al. 2015).

16.5 Treatment

Treatment should be started soon after the diagnostic confirmation, and, in areas where different *Leishmania* species may occur, the diagnosis should ideally be at species level. National and regional guidelines provided by public health authorities should be followed. Early diagnosis and treatment are essential to prevent mucosal lesions and parasite dissemination.

There are several drugs and protocols for treatment, and the choice will depend on the clinical form, species of *Leishmania*, and other factors, like concomitant diseases and malnutrition. More detailed schemes must be easily chosen in WHO's (2010) review. Some more usual drugs and schemes are referred below.

16.5.1 Pentavalent Antimonials

Meglumine antimoniate (8.1 % of Sb^{5+}) and sodium stibogluconate (10 % of Sb^{5+}) selectively inhibit leishmanial glycolysis and fatty acid oxidation. Their toxicity is similar, and they must be injected intravenously or intramuscularly, by infusion (5–10 min) or through a fine needle (23–25 gauge) to avoid the risk of thrombosis. Anorexia, abdominal pains, nausea, vomiting, headache, and other symptoms are usual. Electrocardiographic changes, depending on the doses and duration of treatment, including T-wave inversion, prolonged Q-T interval, and arrhythmia, are the most common, and, rarely, cardiotoxicity and sudden death can occur. The prolongation of Q-T interval signals the onset of serious and fatal arrhythmia. The rise of pancreatic enzymes, but not clinical pancreatitis, is common. Elevated liver enzyme concentration, leucopenia, anemia, and thrombocytopenia are not uncommon. Patients must be monitored for serum chemistry, complete blood counts, and electrocardiography. If serious side effects arise, mostly hepato- and cardiotoxicity, the drug must be changed. Mortality on VL-HIV patients is particularly high when treated by antimonial, and WHO recommends utilizing liposomal amphotericin B for these individuals.

Association of immunotherapy by the injection of dead promastigotes of *L. amazonensis* to antimonial reduced the doses and length of treatment of infections by *L. braziliensis* in Brazil (Mayrink et al. 2006). Antimonial association with pentoxifylline, an inhibitor of TNF-α, decreased the inflammatory reaction responsible for the tissue damage in mucocutaneous lesions.

16.5.2 Amphotericin B

It acts by binding to ergosterol in the parasite's cell membrane and altering its permeability. However, their use is limited by the nephrotoxicity and might endanger dogs with CanL that already have existing renal pathology. Treatment must always be given in hospital, for continuous monitoring of patients. Amphotericin B deoxycholate may be replaced by lipidic formulations, equally efficacious but less toxic.

16.5.3 Pentamidine

This drug has been utilized in French Guiana for CL caused by *L. guyanensis*, usually not very responsive to treatment by antimonials, but the high doses (>2000 mg) necessary for the treatment of MCL can cause diabetes mellitus. Even one dose of 200 mg can cause glicosuria and hyperglycemia, and blood sugar and glycemia must be carefully checked.

16.5.4 Milfetosine

Milfetosine was originally an oral anticancer drug, shown to have good antileishmanial action. Gastrointestinal side effects can force a temporary interruption of treatment, and it can have some hepatotoxicity and is potentially teratogenic.

16.5.5 Intralesional Treatment

It seems to be more efficient in curing cutaneous lesions caused by *L. major* than by *L. tropica* (Minodier et al. 2005). However, an efficacy of 75.2% and 97.2% in curing *L. tropica* infection was obtained in Afghanistan and Turkey, respectively. The injections are painful and the quantity of material can attain 1 L (Minodier et al. 2005). In a study in state of Rio de Janeiro performed on patients infected by *L. braziliensis* (Oliveira-Neto et al. 2008), intralesional treatment achieved good results (80% of healing after 12 weeks), without relapses or mucosal lesions in 10-year observation; nevertheless such treatment, due to the risk of mucosal lesions caused by *L. braziliensis* and *L. panamensis*, would not be recommended for leishmaniasis treatment in the New World, except for *L. mexicana* (Blum et al. 2004).

16.5.6 Others

Cryotherapy and heat for cutaneous lesions and natural products have been tested and deserve additional studies.

As far as CanL treatment, there are different standardized therapeutic protocols in Europe, with first- and second-line drugs (reviewed in Solano-Gallego et al. 2011). The most frequently used drug combinations are meglumine antimoniate plus allopurinol and miltefosine plus allopurinol. Both meglumine antimoniate and miltefosine are registered for veterinary use in most European countries. The prognosis depends mostly on the clinical staging and on the protocol used. Whenever possible, medicines primarily used for treating humans should be avoided in dogs. For example, amphotericin B is rarely used for dogs in Europe, as this is the main drug used for humans. The Ministry of Health in Brazil is against the treatment of CanL with drugs currently used for humans, to prevent the development of resistance, but there is an intense debate on this issue currently in Brazil.

16.6 Epidemiology

Detailed reviews of distribution of all human-infecting species of *Leishmania*, their vectors, and reservoirs were recently published by WHO (2010) and Ready (2013).

VL is the disease of the poorest of the poor, mostly in Indian subcontinent and eastern Africa. The density of *Ph.*[2] *argentipes* in houses in the Indian subcontinent has been shown to be related to the quality of houses, dampness of floor, presence of animals in and around houses, and socioeconomic conditions, and these factors are also important for endophilic vectors. Poverty, including poor housing and peridomestic sanitary conditions and difficult access to healthcare, may help increasing vector populations and their contact with humans, causing more disease and hampering treatment. VL is more frequent in populations living in poor suburban areas, where vectors and reservoirs may be abundant.

Leishmania transmission is influenced by several meteorological and behavioral factors related to complex epidemiology, contact between vectors and reservoirs, and human biting. Topography and vegetation coverage are very important risk indicators. For example, human-sand fly contact is increased in villages situated in rock hills and river banks in East Africa, in areas of transmission of *L. aethiopica* (WHO 2010), and houses situated in the periphery of a village in French Guiana, near primary forest, had a higher incidence of infection by *L. guyanensis*.

A recent comprehensive review listed 42 species of *Phlebotomus* in the Old World and 56 of *Lutzomyia*[3] in the New World as vectors of *Leishmania* spp. (Maroli et al. 2013). Nonetheless, different criteria have recently been proposed for the incrimination of sand fly species as vectors of *Leishmania* spp., and currently there is no consensus (Maroli et al. 2013; Ready 2013). In any case, the diversity of potential vectors of *Leishmania* is remarkable.

16.6.1 Visceral Leishmaniasis

Although VL is known from 74 countries, 90 % of cases occur in six countries: India, Bangladesh, Sudan, South Sudan, Ethiopia, and Brazil. VL is mostly caused by *L. infantum* (synonymous of *Leishmania chagasi* or *Leishmania infantum chagasi*) and *L. donovani*, but other species such as *L. tropica* and *L. amazonensis* can also invade viscera and cause VL. The disease caused by *L. infantum* is a zoonosis, whereas that caused by *L. donovani* is a zoonosis in East Africa and an anthroponosis in the Indian continent.

VL causes an estimated 50,000 deaths annually, a rate only surpassed by malaria among protozoal diseases, with 2,357,000 disability-adjusted life years (DALY). Since a great proportion of cases is asymptomatic or undiagnosed, the estimate of 0.5 million new cases annually (WHO 2010) may be an underestimate.

In Europe, VL by *L. infantum* occurs in the Mediterranean region. Dogs are considered to be the main reservoirs of the parasite, but other ani-

[2] Special abbreviations (Marcondes 2007), similar to those for mosquitoes, were utilized for sand fly genera.

[3] *Lutzomyia* lato sensu, as proposed by D. Lewis and revised by Young and Duncan, and not s.s. as utilized in this chapter

mals including foxes (*Vulpes vulpes*), black rats (*Rattus rattus*), and hares (*Lepus europaeus* and *Lepus granatensis*) have also been found naturally infected and suspected to play a role in the epidemiology of VL. Humans and cats may also infect sand flies. VL in this region is mostly restricted to children, but adults became half of the total number of infected patients, when HIV positive. *Leishmania infantum*, besides causing VL, has also been incriminated as the agent of CL in several countries around the Mediterranean Sea and in Kyrgyzstan and Mauritania. Due to the dog and human traveling and dog relocation from southern to central Europe, the disease has the potential to spread to northern temperate areas, mostly in Austria,[4] Switzerland, and Germany, where potential vectors may occur. Parasites are usually transmitted by *Ph. ariasi* and *Ph. Perniciosus* in France, these plus *Ph. perfiliewi* and *Ph. neglectus* in Italy, and some others have also been suspected as vectors in other countries. VL caused by *L. infantum* is widely distributed in Central Asia and China, and several species of *Phlebotomus* (*Larroussius*) are suspected as vectors (Ready 2013).

In the Middle East (a poorly defined region from Turkey to Afghanistan[5]), several species of *Leishmania* may cause VL and CL, and many vectors and reservoirs for *L. infantum* (e.g., dogs and rodents) are known to occur. For example, *L. tropica* and mostly *L. infantum* have been isolated from human VL patients, being the latter isolated from wild hamsters along with *L. donovani*.

Circa 4500–6800 cases per year of VL occur in the American continent (Alvar et al. 2012), of which 96 % occur in Brazil. The disease, previously found mostly in the northeastern Brazil, is nowadays present in all regions, including in center-west and southeast, and it expanded into several medium and large cities, such as Campo Grande and Belo Horizonte. VL is also spreading in northern Argentina and Paraguay, Colombia, Venezuela, and some Central American countries. *Lutzomyia longipalpis* is the primary vector in most VL foci in Latin America. However, other species such as *Lutzomyia cruzi* (in Brazil), *Pifanomyia evansi* (in Colombia), and *Migonemyia migonei* (in Argentina and Brazil) may also be involved in some foci, mostly in areas without *L. longipalpis*.

Dogs are important reservoirs and wild dogs and opossums have sometimes been found infected; infectivity of wild dogs for sand flies is considered negligible for the introduction of parasites in new areas in Amazon region. The role of chickens in the epidemiology of VL is considered ambiguous, because they may contribute for increasing domestic populations of *L. longipalpis*, but they may also potentially reduce the number of sand fly bites on potential reservoirs, thus acting as zooprophylatic agents.

VL caused by *L. donovani* is distributed through northeastern India, Bangladesh, Bhutan, and Nepal, where *Ph. argentipes*, a sand fly mostly associated to cattle, is the primary vector. Antibodies against *Leishmania* and sand fly saliva have been detected in cattle but no parasite DNA was found. Similarly, 3.1 % of tested goats in Bihar (total number 867) reacted to rK39 antigen, and 20 of them had *Leishmania* DNA; the recent finding of goats infected by *L. donovani* (Singh et al. 2013) and the lack of scientific evidence (Singh 2014, pers. commun.) enhance the need to more thorough studies on the occurrence of the infection in this animal species. Cases of VL and/or CL attributed to *L. donovani* have also been found in some Middle East countries (Israel, Iraq, Saudi Arabia, Yemen), in China, and in Europe (Ukraine) (WHO 2010). In the East of Africa, the disease, also caused by *L. donovani*, is widely distributed, causing many cases and mortality, mostly in Sudan; in *Acacia balanites* forest, it is transmitted by *Ph. orientalis,* while *Ph. martini* s.l. is the vector in savanna with termite mounds (Ready 2013). Epidemics are related to mass migrations, malnutrition, and associated diseases. Infection of 6.9 % *Ph. orientalis* was observed, and mongooses were suspected as

[4] *Phlebotomus mascitti*, considered a poor vector, has been found in Austria.

[5] This concept of the Middle East is as arbitrary as the more usual, proposed by the American naval officer and historian Alfred Thayer Mahan in 1902, which included lands from Arabia to India.

reservoirs in Dinder National Park, which is currently inhabited. Although dogs are very attractive to *Ph. orientalis* and may become infected by *L. donovani*, they are not considered as important reservoirs of this parasite.

HIV pandemic influenced the occurrence of leishmaniasis, mostly of VL. In Europe, HIV patients living in rural areas are highly exposed to the risk of infection and disease development (Desjeux and Alvar 2003). In Brazil, the dissemination of HIV from big cities to smaller ones and rural areas overlapped with the distribution of leishmaniasis (Rabello et al. 2003). As a consequence, from 2001 to 2012, the incidence of VL-HIV increased tenfold in Brazil, possibly also influenced by the improvement of HIV diagnosis (Lindoso et al. 2014).

16.6.1.1 Other Modes of Transmission

The role of ticks and fleas as possible vectors of *L. infantum* has been suspected, mostly in areas in the absence of the proven sand fly vectors. However, caution is needed before considering any of these arthropods as actual vectors of this parasite. Although *L. infantum* may remain viable in other arthropods, there is no proof that they may act as vectors.

Several observations of congenital transmission of *L. donovani* and *L. infantum* in humans (Figueiró-Filho et al. 2004) and of *L. infantum* in dogs in the USA have been reported (Petersen and Barr 2009). In an early study, no infection was confirmed in 63 puppies from 18 naturally *L. infantum*-infected female dogs from Belo Horizonte, Brazil. Later on, a study reported a vertical transmission in Brazil, as confirmed by PCR and immunohistochemistry techniques in samples from the spleen and liver of two stillborn pups from a female dog naturally infected with *L. infantum*. For VL cases in dogs in the east of the USA and Canada, congenital and venereal transmission should also be suspected, even if *Psathyromyia shannoni* is susceptible to infection and widely distributed in these areas. Lesions in the genital tract of male dogs positive for *L. infantum* and the presence of *Leishmania* in 13/15 (86.6%) dogs indicate the possibility of venereal transmission.

The recent finding in Finland of three autochthonous cases of VL in dogs, probably infected through contact to a dog that visited Spain, emphasizes the possibility of direct transmission of the parasite and the need of attention to the possibility of occurrence of disease in areas without sand flies.

High prevalence of VL was observed in hemodialysis patients and blood donors in Natal (Luz et al. 1997), a low-endemicity area in Brazil, while several cases of transmission of *Leishmania* through blood transfusion in humans have been reported (Cardo 2006). Intravenous drug users are at greater risk of infection by HIV and by *Leishmania*/HIV in southwestern European countries (Desjeux and Alvar 2003). Transmission by blood transfusion among dogs was reported in Brazil and in the USA. Overall, these reports indicate the occurrence of secondary modes of transmission, but their actual epidemiological significance is still poorly understood.

16.6.2 Cutaneous Leishmaniasis

CL is widely distributed and in expansion in all continents, possibly influenced by human activities, like agricultural projects, deforestation, and migration, and climate changes. Several species of *Leishmania* can cause CL, and new species are likely circulating among sylvatic mammals, possibly infecting humans in the future. For example, a new species, *L. siamensis*, has been associated to VL, CL, and disseminated CL in HIV-infected humans in Thailand, and a similar species was found infecting horses in central Europe. The epidemiology of CL is extremely variable and is briefly described below.

In the New World, *L. braziliensis* causes most cases of CL and is broadly distributed. It is transmitted by several species of sand flies, both by sylvatic species like *Psychodopygus wellcomei* and *Ps. complexus* and by those well adapted to domiciles, like *Nyssomyia intermedia*, *Ny. neivai*, and *Ny. whitmani*. Rodents are considered important reservoirs, but opossums and other mammals have been found infected; dogs may be infected but do not seem to constitute good reservoirs

(Dantas-Torres 2007); cats and donkeys also have been found infected.

Leishmania mexicana is found in Central America and has been mostly associated to *Ny. olmeca*. The closely related *L. amazonensis*, in the Amazon region, is associated to floodable areas ("igapós") and transmitted mostly by *Bichromomyia flaviscutellata*, whose preference to bite rodents may explain its low incidence in humans; it has also been found in other regions of Brazil and nearby countries, with unknown vectors and reservoirs.

Leishmania guyanensis occurs in forested areas in Argentina, Bolivia, Brazil, Colombia, Ecuador, French Guiana, Guyana, Peru, Suriname, and Venezuela. Sloths (Choloepus didactylus) and anteaters (Tamandua tetradactyla) are the main reservoirs. The most important vector is Ny. umbratilis, which can bite in great numbers when humans invade their microhabitat (around trunks of large trees). *Leishmania panamensis*, distributed in Central America and eastern Ecuador, is mostly associated to preserved forests, with sloths as reservoirs and sand flies of *Nyssomyia* and *Psychodopygus* as major vectors. Transmission has also been confirmed in peridomestic environments of Andean valleys with coffee plantations in Colombia. Other species, like *L. peruviana*, occurring in high Andean valleys in Peru, and *L. colombiensis*, occurring in forests in Colombia, Panama, and Venezuela, have a more restricted distribution.

In the Old World, *L. tropica* causes anthroponotic CL (ACL) in India, Central Asia, the Middle East (from Afghanistan to Israel and Syria), and Morocco; cases are found predominantly in densely populated settlements, transmitted among humans by *Ph. sergenti*, and patients with leishmaniasis *recidivans* can be infective for sand flies for several years. In the same areas, sporadic cases of CL caused by *L. tropica* have been associated to other reservoirs, mostly hyraxes. In the focus at Rajasthan desert, in the northwest of India, cases of CL are caused by *L. tropica*, but some cases in the state of Himachal Pradesh are caused by *L. donovani*, and cases of CL caused by this species have been reported from Kenya, Nepal, and Sri Lanka.

Leishmania major causes zoonotic CL (ZCL), mostly associated to arid regions, with gerbils (*Rhombomys opimus* and *Psammomys obesus*) as reservoirs in Central Asia, North Africa, and Palestine, besides sub-Saharan areas; in most regions, *Ph. papatasi* has been incriminated as the major vector, but *Ph. duboscqui* in sub-Saharan Africa and probably *Ph. salehi* in southeast Iran, Pakistan, and northwest India may be vectors (Ready 2013).

16.7 Control and Prophylaxis

16.7.1 Visceral Leishmaniasis

Control of VL requires variable strategies according to parasites involved and to disease epidemiology. In all foci, active detection of human cases and treatment, besides improvement of nutrition, is highly recommended to reduce mortality and morbidity.

In *L. donovani*-caused VL at Indian subcontinent, treatment of human patients and control of *Ph. argentipes* by the utilization of DDT in houses and animal sheds reduced the incidence, during and after the control of malaria. However, restriction of DDT utilization and the appearance of resistance in sand flies, besides operational difficulties and resistance to drugs, complicated VL control in this subcontinent. The presence of cattle near houses has a doubtful effect on control, because they can deviate biting from humans, but the production of dung can increase populations of sand flies. The utilization of impregnated bed nets has had a partial effect on quantity of bites (Gidwani et al. 2011), causing variable reduction of incidence. Since PKDL and asymptomatic patients constitute a good source of parasites, they also need to be adequately managed. Although infection of goats and oxen by *L. donovani* has been found in some Indian foci, their role as reservoirs needs to be established.

In the African foci of VL caused by *L. donovani*, control depends on the biology of vectors and reservoirs. Since this parasite is transmitted by exophilic vectors, application of insecticides on termite mounds (for *Ph. martini*) needs to be adequately evaluated. The use of insecticide-treated bed nets reduced the number of bites of *Ph. martini* and the incidence in Sudan, and a good effect was also obtained with permethrin-impregnated curtains in Kenya. Passive diagnosis and treatment are important to reduce mortality and morbidity, but the influence on transmission is unknown. In densely populated areas, mostly with nonimmune migrants (or refugees), spraying of houses may be necessary; though the actual effect in the long term has not been fully evaluated.

In the American continent, the association of elimination of positive dogs; spraying walls of houses, henhouses, and kennels with insecticides; and treatment of human cases was proposed in the 1950s and has been extensively applied. However, difficulty on the elimination of dogs, restriction of the utilization of DDT, and poor effect of pyrethroids on *L. longipalpis* has jeopardized control, stimulating the discussion on the use of DDT for the control of VL in Brazil (Marcondes and Costa 2014).

The application of insecticides in the environment may have momentary effect, but they are usually unsustainable in the long term for technical and economic reasons. Importantly, environmental and human health hazard concerns around the employment of organochlorines and other chemical groups (e.g., organophosphates and carbamates) have spurred their substitution by synthetic pyrethroids (e.g., α-cypermethrin, cypermethrin, deltamethrin, and λ-cyhalothrin), which are currently used by public health authorities, usually when human cases are reported and a high density of sand flies is found nearby human habitations. Other compounds have shown promising results under laboratory conditions, but their use under field conditions need to be better assessed. For example, azadirachtin, a secondary metabolite present in neem seeds, was demonstrated to be a potent growth inhibitor (for larvae) and a potent sterilizer (for females) that could be used against *L. longipalpis*. Nonetheless, the limited information on the breeding sites of this vector currently impairs its utilization, at least, to control pre-imago stages. In any case, the insecticidal spraying should also be associated with the elimination of microhabitats favorable to sand fly development (e.g., crevices and cracks on the walls and humid soil in shaded areas).

In foci of VL caused by *L. infantum* with dogs as known or assumed reservoirs, infection in dogs should be monitored. The World Health Organization recommends that the management of infected dogs should be treatment or elimination. In the same way, stray and feral dog populations should be controlled, even if these actions may not be enough to control the disease, also due to the presence of other reservoirs (WHO 2010). On the other hand, the use of topical insecticides on dogs may reduce the infection in both dogs and humans (Otranto and Dantas-Torres 2013). Both laboratory and field studies have demonstrated the efficacy of different products currently available in the veterinary market. For instance, a spot-on formulation of permethrin plus imidacloprid[6] has been highly effective in preventing *L. infantum* infection in dogs in Italy under field conditions for 1 month. Similarly, different studies have proven the usefulness of collars containing deltamethrin or flumethrin plus imidacloprid in protecting dogs from *L. infantum* infection for up to 8 months (Killick-Kendrick 1999; Otranto et al. 2013). Laboratory studies showed an anti-feeding effect on sand flies ranging from 84 to 96 % and an insecticidal activity of near 100 %, in the few fed females. Field studies demonstrated that a collar containing 4 % deltamethrin (Maroli et al. 2010) and a spot-on formulation with 10 % imidacloprid and 50 % permethrin reduced the risk of *L. infantum* infection in dogs. Furthermore, the field efficacy of a new collar containing 10 % imidacloprid and 4.5 % flumethrin was also demonstrated in different studies in southern Italy. Altogether, these results indicate the usefulness of topical insecti-

[6]When applied in the general environment, imidacloprid is very toxic for bees and bumblebees, (Van der Sluijs et al. 2013), but since it has very high LD_{50} for mammals (2000 mg/kg for dermal exposition in rats and 450 and 131 mg/kg for ingestion, respectively, for rats and mice) (Hovda and Hooser 2002), the application on dogs would not probably be dangerous (CBM).

cides to prevent *L. infantum* infection in dogs in endemic areas. Nonetheless, the large-scale use of these collars may be impaired by several factors, including operational, economic, and political reasons, particularly in developing countries.

As aforementioned, the treatment of dogs suffering from VL is routinely performed in Europe, where different drugs have been marketed for use in dogs, including meglumine antimoniate and miltefosine. The main objective of the treatment of dogs is to promote clinical cure, prevent relapses, and reduce parasite burden. All dogs, under treatment or not, must also be treated with repellents to reduce the risk of reinfection and of transmission to sand flies. While the reduction of transmissibility is not the primary objective of the treatment, studies have demonstrated that some drug combinations were effective in blocking the transmission of parasites to sand flies (Miró et al. 2011).

Removing positive dogs has been practiced for over 50 years in Brazil, but both practical and scientific evidence indicates that this strategy should be replaced by more effective ones, such as the use of topical repellents. Several factors may explain the inefficacy of dog culling for the control of VL, including delay between sample collection and dog removal, rapid replacement of removed dogs by susceptible puppies, high number of dogs daily exposed to infected sand fly bites, presence of other reservoirs, lack of infrastructure and human resources, among others. Mathematical models have indicated that the control of insect vectors and reduction of infectivity of reservoirs are more efficient than dog culling for controlling human VL (Dye 1992).

Vaccines for dogs may be very useful for reducing the risk of disease development but has limited impact on the risk of infection, as vaccinated dogs remain exposed to sand fly bites. To date, three vaccines have been registered for use in dogs, two in Brazil (Leishmune® and Leish-Tec®), and one in Europe (CaniLeish®); the license for one of these vaccines (Leishmune®) has recently been suspended in Brazil. When exposed to infection, vaccinated dogs usually are less prone to disease development, and even if the infection is established, they are less infectious to sand flies. Indeed, these vaccines represent an additional tool for the prevention of CanL, but it is important to emphasize that the use of repellents and/or fast-killing insecticides is the simplest and most effective way to reduce the risk of infection in dogs. While the use of repellents and vaccines can undoubtedly reduce the risk of infection and disease development in dogs, further studies are necessary to better quantify their impact on the incidence of the disease in humans in different epidemiological contexts.

16.7.2 Cutaneous Leishmaniasis

Control of CL depends on the epidemiology of disease in each area. Personal protection, including the utilization of impregnated bed nets and repellents, is useful for reducing the exposure to sand fly bites and the risk of infection by most species of *Leishmania*.

In the anthroponotic cycle of *L. tropica*, diagnosis and treatment of infected humans, associated to insecticide use on houses and annexes, and reduction of potential breeding sites for the vectors, by asphalting streets and covering the soil of backyards with bricks, are considered efficient strategies. Control must be carefully planned to effectively reduce the global number of cases, and not only temporarily reducing the risk of infection in children (WHO 2010).

In foci of *L. major* with *Rhombomys* as reservoirs, the control of rodent colonies by plowing or poisoning and by preventing colonization or reinvasion by artificial (irrigation canals) or natural barriers has led to very good results in the former USSR. However, for foci with *Psammomys* in the Middle East and Tunisia, this strategy has been difficult to implement. Control of CL caused by *L. aethiopica* in the East of Africa has been obtained by the control of hyraxes around houses, and breeding rabbits (which are very attractive to *Ph. papatasi* but not susceptible to *L. major* infection) around houses was proposed for the control of *L. major* in rural areas in Tunisia.

In the New World, indoor and peridomestic transmission of *L. peruviana* and *L. braziliensis* has been controlled by the utilization of insecticides on houses and annexes, but for the later and for other species present in the forests, personal protection by repellents and clothes is the only available prophylactic measure. Spraying insecticides on large trees caused a temporary reduction of the quantity of vectors of *L. panamensis* and *L. guyanensis*, respectively, in Panama and Brazil, but this method is obviously not viable for large-scale use. Although deforestation around villages in Panama and French Guiana was efficient to reduce infection by *L. panamensis* and *L. guyanensis*, respectively, such modification has a doubtful effect on the transmission of *L. braziliensis*, due to replacement of vectors (sylvatic → intermediate → domestic). Although the reduction of humidity and shadow around houses and removal of pigsties and henhouses reduced the quantity of potential vectors in the Brazilian state of Paraná (Teodoro et al. 2004), the direct association between sand flies and garbage is doubtful.

16.8 Conclusion

Due to the multifaceted epidemiology and local nature of leishmaniases, control strategies must be adequate to local situations, being very difficult to propose general rules. Knowledge on the *Leishmania* species circulating in humans, reservoirs, and sand flies in a given area is necessary to attain efficient control. Information on the local ecology of vector species, seasonal fluctuation, and frequency of contact with humans and reservoirs and blood feeding patterns is primordial before any hope of control.

As a matter of fact, leishmaniases remain neglected diseases that affect disproportionally the poorest of the poor. Therefore, better living conditions, including adequate housing and food supply, are a priority for affected populations living in endemic areas in Latin America, Africa, and Asia. Last but not least, education is fundamental to provide affected populations with information on disease prevention and control. Without access to healthcare and information diseases such as leishmaniases will remain one of the main Achilles' hells of public health in developing countries.

References

Alvar J, Vélez ID, Bern C et al (2012) Leishmaniasis worldwide and global estimates of its incidence. Plos One 7, e35671

Bacellar O, Lessa H, Schriefer A et al (2002) Up-regulation of Th1-type responses in mucosal leishmaniasis patients. Infect Immun 70:6734. doi:10.1128/IAI.70.12.6734-6740.2002

Badirzadeh A, Moheball M, Ghasemian M et al (2013) Cutaneous and post kala-azar dermal leishmaniasis caused by *Leishmania infantum* in endemic areas of visceral leishmaniasis, northwestern Iran 2002–2011: a case series. Pathog Global Hlth 107:104–197

Blum J, Desjeux P, Schwartz E et al (2004) Treatment of cutaneous leishmaniasis among travellers. J Antimicrob Chemother 53:158–166

Cantacessi C, Dantas-Torres F, Nolan MJ et al (2015) The past, present, and future of *Leishmania* genomics and transcriptomics. Trends Parasitol 31:100–108

Cardo LJ (2006) *Leishmania*: risk to the blood supply. Transfusion 46:1641–1645

Chappuis F, Sundar S, Hailu A et al (2007) Visceral leishmaniasis: what are the needs for diagnosis, treatment and control? Nat Rev Microbiol 5:873–882

Cota GF, Sousa MR, Mendonça ALP et al (2014) *Leishmania*-HIV co-infection: clinical presentation and outcomes in an urban area in Brazil. PLoS Negl Trop Dis 8, e2816. doi:10.1371/journal.pntd.0002816

Cota GC, Sousa MR, Rabello A (2011) Predictors of visceral leishmaniasis relapse in HIV infected patients: a systematic review. PLoS Negl Trop Dis 5, e1153. doi:10.1371/journal.pntd.0001153.g001

Dantas-Torres F (2007) The role of dogs as reservoirs of *Leishmania* parasites, with emphasis on *Leishmania* (*Leishmania*) *infantum* and *Leishmania* (*Viannia*) *braziliensis*. Vet Parasitol 149:139–146

Dantas-Torres F (2009) Canine leishmaniosis in South America. Paras Vect 2 (Suppl I). doi:10.1186/1756-3305-2-S1-S1

Desjeux P, Alvar J (2003) *Leishmania*/HIV co-infections: epidemiology in Europe. Ann Trop Med Parasitol 97(Suppl 1):S3–S15

Dye C (1992) Leishmaniasis epidemiology: the theory catches up. Parasitology 104:S7–S18

El-Hassan AM, Zijlstra EE (2001) Leishmaniasis in Sudan: 2. Mucosal leishmaniasis. Trans R Soc Trop Med Hyg 95(Suppl 1):S1/19–S1/26

Figueiró-Filho EA, Duarte G, El-Beitume P et al (2004) Visceral leishmaniasis (kala-azar) and pregnancy. Infect Dis Obstet Gynecol 12:31–40

Freitas E, Melo MN, Costa-Val AP et al (2006) Transmission of *Leishmania infantum* via blood transfusion in dogs: potential for infection and importance of clinical factors. Vet Parasitol 137:159–167

Gidwani K, Picado A, Rijal S et al (2011) Serological markers of sand fly exposure to evaluate insecticidal nets against visceral leishmaniasis in India and Nepal: a cluster-randomized trial. PLoS Negl Trop Dis 5, e1296

Guimarães LH, Machado PRL, Lessa HA et al (2005) Clinical aspects of tegumentary leishmaniasis. Gaz Med Bahia 75:66–74

Hinojosa MC, Alvarez DR, Kato H et al (2013) Cutaneous sporotrichoid lesion in a patient from a subtropical region of Ecuador. Enferm Infecc Microbiol Clin. doi:10.1016/j.eimc.2014.02.007

Hovda LR, Hooser SB (2002) Toxicology of newer insecticides for use in dogs and cats. Vet Clin North Am Small Anim Pract 23:455–467

Kamhawi S (2006) Phlebotomine sand flies and *Leishmania* parasites: friends or foes? Trends Parasitol 22:439–445

Killick-Kendrick R (1999) The biology and control of phlebotomine sand flies. Clin Dermatol 17:279–289

Kopterides P, Mortzoukou EG, Skopelitis E et al (2007) Aspects of the association between leishmaniasis and malignant disorders. Trans R Soc Trop Med Hyg 101:1181–1189

Lindoso JA, Cota GF, Cruz AM et al (2014) Visceral leishmaniasis and HIV coinfection in Latin America. PLoS Negl Trop Dis 8, e3136

Lindoso JAL, Lima ACS, Cunha MA et al (2015) Diagnosing neglected tropical diseases in HIV coinfection. Hum Parasit Dis 7:11–18. doi:10.4137/HPD.S19569

Lukes J, Mauricio IL, Schönian G et al (2007) Evolutionary and geographical history of the *Leishmania donovani* complex with a revision of current taxonomy. PNAS 104:9375—9380. doi:10.1073_pnas.0703678104

Luz KG, Silva VO, Gomes EM et al (1997) Prevalence of anti-*Leishmania donovani* antibody among Brazilian blood donors and multiply transfused hemodialysis patients. Am J Trop Med Hyg 57:168–171

Marcondes CB (2007) A proposal of generic and subgeneric abbreviations for Phlebotomine sandflies (Diptera: Psychodidae: Phlebotominae) of the world. Entomol News 118:351–356

Marcondes CB, Costa CHN (2014) Visceral leishmaniasis control in Brazil: time to re-evaluate DDT? Terr Arthr Rev 7:21–39

Maroli M, Gradoni L, Oliva G et al (2010) Guidelines for prevention of leishmaniasis in dogs. J Am Vet Med Assoc 236:1200–1206

Maroli M, Feliciangeli MD, Bichaud L et al (2013) Phlebotomine sand flies and the spreading of leishmaniases and other diseases of public health concern. Med Vet Entomol 27:123–147

Mayrink W, Botelho ACC, Magalhães PA et al (2006) Immunotherapy, immunochemotherapy and chemotherapy for American cutaneous leishmaniasis treatment. Rev Soc Bras Med Trop 39:14–21

Mignogna MD, Celentano A, Leuci S et al (2014) Mucosal leishmaniasis with primary oral involvement: a case series and a review of the literature. Oral Dis. doi:10.1111/odi.12268

Minodier P, Noël G, Blanc P et al (2005) Traitement des leishmanioses cutanées de l'adulte e de l'infant. Med Trop 65:487–495

Miró G, Gálvez R, Fraile C et al (2011) Infectivity to *Phlebotomus perniciosus* of dogs naturally parasitized with *Leishmania infantum* after different treatments. Paras Vectors 4:529

Monge-Maillo B, Norman FF, Cruz I et al (2014) Visceral leishmaniasis and HIV coinfection in the Mediterranean region. PLoS Negl Trop Dis. 8(8), e3021. doi:10.1371/journal.pntd.0003021

Oliveira-Neto MP, Schubach A, Mattos M et al (2008) Intralesional therapy of American cutaneous leishmaniasis with pentavalent antimony in Rio de Janeiro, Brazil — an area of *Leishmania (V.) braziliensis*. Int J Dermatol 36:463–468

Otranto D, Dantas-Torres F (2013) The prevention of canine leishmaniasis and its impact on public health. Trends Parasitol 29:339–345

Otranto D, Dantas-Torres F, de Caprariis D et al (2013) Prevention of canine leishmaniosis in a hyper-endemic area using a combination of 10% imidacloprid/4.5% flumethrin. PLoS One 8, e56374

Petersen CA, Barr SC (2009) Canine leishmaniasis in North America: emerging or newly recognized? Vet Clin North Am Small Anim Pract 39:1065–1074

Rabello A, Orsini M, Disch J (2003) *Leishmania*/HIV co-infection in Brazil: an appraisal. Ann Trop Med Parasitol 97(Suppl 1):S18–S28

Ready PD (2013) Biology of Phlebotomine sand flies as vectors of disease agents. Annu Rev Entomol 58:227–250

Schriefer A, Schriefer ALF, Góes-Neto A et al (2004) Multiclonal *Leishmania braziliensis* population structure and its clinical implication in a region of endemicity for American tegumentary leishmaniasis. Infect Immun 72:508–514

Singh S, Sharma U, Mishra J (2011) Post-kala-azar dermal leishmaniasis: recent developments. Int J Dermatol 50:1099–1108

Singh N, Mishra J, Singh R et al (2013) Animal reservoirs of visceral leishmaniasis in India. J Parasitol 99:64–67

Solano-Gallego L, Miró G, Koutinas A et al (2011) LeishVet guidelines for the practical management of canine leishmaniosis. Parasit Vectors 4:86, http://www.parasitesandvectors.com/content/4/1/86

Teodoro U, Silveira TG, dos Santos DR et al (2003) Influência da reorganização, da limpeza do peridomicílio e a da desinsetização de edificações na densidade populacional de flebotomíneos no Município de Doutor Camargo, Estado do Paraná, Brasil. Cad Saúde Publ 19:1801–1813

Van der Sluijs JP, Simon-Delso N, Goulson D et al (2013) Neonicotinoids, bee disorders and the sustainability of pollinator services. Curr Opin Environ Sustain 5: 293–305

WHO (2010) Towards universal access: scaling up priority HIV/AIDS interventions in the health sector - progress report. WHO, Geneva. Available at: http://www.whoint/hiv/pub/2010progressreport/full_report_enpdf. Accessed in 31 May 2013

Zijlstra EE, Musa AM, Khalil EAG et al (2003) Post-kala-azar dermal leishmaniasis. Lancet Infect Dis 3:87–98

Chagas Disease (American Trypanosomiasis)

17

João Carlos Pinto Dias

Abstract

American trypanosomiasis was previously an enzooty of wild mammals and invertebrate vectors in American continent. Human infection is endemic in Latin America, progressively being spread to non-endemic countries throughout human migration. Today, a yearly incidence of 20,000 new cases and a prevalence of around 8 million individuals are estimated. About 15 or 20 species of the vector (Hemiptera, Reduviidae) are relevant to human disease. Congenital, blood transfusion, organ transplantation, accidents, and oral routes are other possibilities for parasite transmission. Wild reservoirs belong to several orders, being dogs, cats, rats, rabbits, and guinea pigs, the principal domestic reservoirs. Acute human infection (few weeks, high parasitemia and low circulating antibodies) is followed by a chronic phase (low parasitemia, high level of antibodies). Acute cases are usually oligosymptomatic, but severe myocarditis and meningoencephalitis can occur, mainly in low age and immunodepressed individuals. An asymptomatic indeterminate chronic form is used to follow the acute period, but up to 40 % of the cases will develop a cardiac and/or digestive (megaesophagus, megacolon) involvement. Usual diagnosis requires parasitological procedures in acute and serological techniques in the chronic phase. Specific treatment is more effective in congenital, acute, recently transmitted, and low age cases. Supportive medical care is required to manage arrhythmias, heart failure, and digestive "megas." Preventive measures include elimination of domiciliated triatomines, selection of noninfected blood and organ donors, and the detection and treatment of congenital cases. Acute cases (especially of oral transmission) must be investigated, in view of an effective epidemiological surveillance.

Keywords

Chagas disease • Epidemiology • Clinical management • Prevention

J.C.P. Dias (✉)
René Rachou Center of Research, Oswaldo Cruz Foundation, 30190-002 Belo Horizonte, MG, Brazil
e-mail: jcpdias@cpqrr.fiocruz.br

C.B. Marcondes (ed.), *Arthropod Borne Diseases*, DOI 10.1007/978-3-319-13884-8_17

17.1 Introduction and General Epidemiological Aspects

Chagas disease, a metaxenic protozoonosis caused by *Trypanosoma* (*Schizotrypanum*) *cruzi* (*T. cruzi*), was described in 1909 by the physician Carlos Chagas, in Minas Gerais, Brazil. This brilliant researcher discovered the parasite and its vectors, began the study of clinical aspects, and found the first reservoirs. Originally, the American trypanosomiasis existed only as an enzooty, circulating the parasite among wild mammals and invertebrate vectors. Much later, human beings approached this chain; thus, the human infection emerged in the domestic environment, with a close relationship with socioeconomic and ecological situations (settlements, migration, living standards, landscape, altitude, temperature, humidity, etc.), therefore emerging a typical zoonosis or zooanthroponosis. Human Chagas disease (HCD) is endemic throughout much of Mexico, Central America, and South America, having also spread to non-endemic countries all over the world by means of the migration of thousands of infected individuals (Fig. 17.1) (Miles et al. 2004; Dias et al. 2013).

In endemic areas, HCD is mainly transmitted by bloodsucking triatomine bugs of the genera *Triatoma*, *Rhodnius*, and *Panstrongylus*. Bugs get infected with *T. cruzi* parasites when taking blood meals from infected mammalian hosts. Transmission to humans occurs when the conjunctiva, mucosal tissue, or bite wound becomes contaminated with infected feces from the bugs. Triatomine bugs thrive under poor housing conditions (such as mud walls and thatched roofs); thus, people living in rural areas are at the greatest risk of being infected. Since the second half of the twentieth century, stringent measures of control have led to large reductions in vector transmission over a wide endemic region (Carlier et al. 2002; Coura and Dias 2009; WHO 2014).

HCD may also be transmitted by blood transfusion, organ transplantation, oral and congenital routes (primary and secondary routes), or very rarely by exceptional alternative routes such as sexual and by non-triatomine vectors. In immunosuppressed circumstances, HCD becomes a very worrying situation, mainly in those chronic individuals who acquire HIV/AIDS, frequently with a recrudescence of *T. cruzi* infection (WHO 2002; Dias et al. 2011).

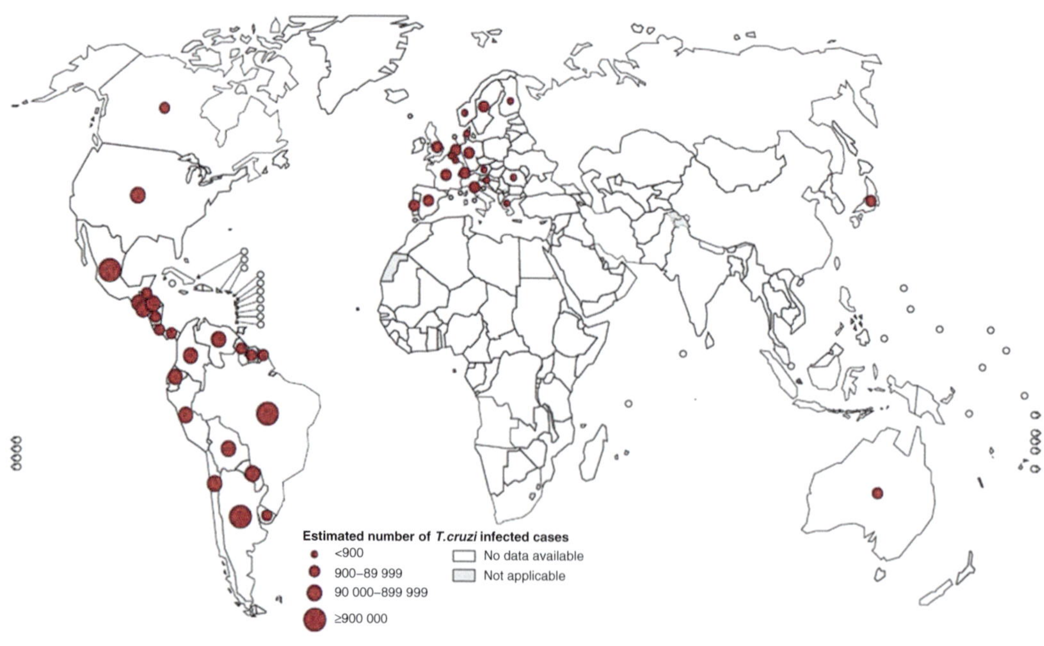

Fig. 17.1 Global distribution of HCD based on official estimates between 2006 and 2010 (From WHO 2014)

In vertebrate hosts, the infection develops in two clinical phases, the acute one (initial, short lasting, with a high number of circulating parasites and a few antibodies) and the chronic one (long term, with a few circulating parasites and a high level of antibodies). The morbidity in humans is mainly due to cardiac, digestive, and neurological disorders in both phases, in spite of most cases being asymptomatic or oligosymptomatic. The most serious situation depends on the advanced chronic heart disease, with difficult clinical management, employment disability, high medical costs, and high mortality (30–40 % of cardiac patients). Mortality in the acute phase may reach 10 % in low age children (Miles et al. 2004; Dias et al. 2013).

In this chapter, the intention is to present a general overview of American trypanosomiasis, centered in the human infection. To those involved in parasitology, a relevant role is reserved in the study and research on this protozoosis, aiming at its definitive control.

17.2 The Etiologic Agent

Trypanosoma cruzi belongs to the order Kinetoplastida, family Trypanosomatidae. It is the single human trypanosome known to be transmitted by the excreta of an invertebrate vector (systematic section *Stercoraria*), in contrast with trypanosomes transmitted by the bite of vectors, the section *Salivaria*, e.g., *T. brucei*, the agent of sleeping sickness and other diseases (Miles et al. 2004).

17.2.1 Evolution of Parasitic Forms

In order to adapt to different bioecological situations, *T. cruzi* assumes different evolutionary forms: trypomastigotes, amastigotes, spheromastigotes, and epimastigotes (Carlier et al. 2002; Zingales et al. 2009; Souza 2000; Andrade 2013).

Trypomastigotes are the infective forms of the parasite, with a fusiform nucleus, and a terminal positioned large corpuscle, with a high density of DNA, the kinetoplast. The flagellum is too long, emerging from the kinetoplast and scrolling through the cell body in the longitudinal direction, thus elevating the cell membrane, the "undulating membrane." The anterior pole of the parasite leaves the free portion of the flagellum. Longitudinally, the trypomastigotes measure between 20 and 25 μm (Fig. 17.2). They are extracellular forms, which may be found in peripheral blood of the vertebrate host and in the end portions of the digestive tract and in Malpighian tubules of the insect vector (here called *metacyclic* trypomastigotes).

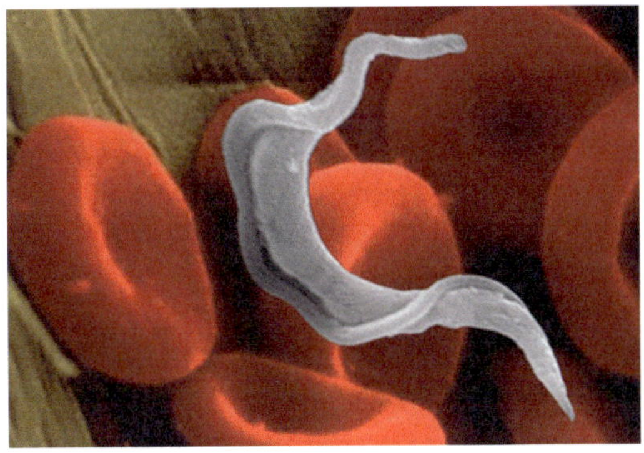

Fig. 17.2 *Trypanosoma cruzi* trypomastigote in peripheral mammalian blood (WHO 2014)

Amastigotes are rounded forms without an exteriorized flagellum, measuring between 2 and 4 μm in diameter. They are responsible for binary multiplications of the parasite in the vertebrate host, where they remain inside different cell types. Epimastigote forms are a very flexible transition between trypomastigotes and amastigotes, found basically in the digestive tract of the insect vector, in the anal glands of certain marsupials, and in artificial means of cultivation. Spheromastigotes are also forms of multiplication, detected in the stomach of the vector (Souza 2000; Brener 1997).

17.2.2 Different Groups of *T. cruzi* in Nature

Modern studies employing biochemistry, molecular, and phylogenetic analyses have led to recognize six basic groups of *T. cruzi* such as that more connected to the sylvatic cycle and to the northern hemisphere (called group I) and another one that is more connected to domestic situations (group IIb). Such a biodiversity of the parasite fauna has a practical importance, involving not only regional differences in morbidity, but other epidemiological implications such as the behavior of subpopulations in distinct species of vector, the variation in susceptibility to chemotherapy, the histotrophic preferences in vertebrate hosts, etc. Phylogenetic studies based on DNA sequence data indicate that the genetic exchange has contributed to the evolution of the parasite, enabling the natural generation of hybrid individuals and clones (Souza 2000; Andrade 2013). The following scheme resumes the present *T. cruzi* classification in six groups and their respective geographical distribution (Table 17.1) (Zingales et al. 2009).

In a large extent of South and Central America, another flagellate, *Trypanosoma rangeli*, is found parasitizing triatomine bugs (mainly of the genus *Rhodnius*), found in the hemolymph and in the salivary glands of the insect (see Chap. 19). With antigenicity next to *T. cruzi*, often *T. rangeli* may cause difficulties for the immunological diagnosis of Chagas disease, generating false-positive results in conventional serological tests.

17.2.3 Life Cycles of *T. cruzi*

Characterized by their restricted hematophagy, invertebrate vectors ingest trypomastigote forms of the parasite when feeding on infected mammals. The first transformations take place in the stomach, starting a cycle of multiplication and permanence of *T. cruzi* in the insect. Spheromastigotes and epimastigotes are involved, being more abundant in the midgut, where their replication is extremely active. In general, a population of epimastigotes usually remains along the digestive tract throughout the life of the infected insect, always in multiplication, with forms adhered to the wall and others that will move to the terminal bowel and the Malpighian tubules, where the differentiation for trypomastigotes occurs. Triatomine dejections are usually complex, merging feces and urine, both containing metacyclic trypomastigotes. The complete cycle of *T. cruzi* in the insect generally lasts between 2 and 4 weeks, being shorter in cases of massive ingestion of blood trypomastigotes (e.g., acute disease) and when the strain is better adapted to the vector. It is well established that *T. cruzi* infection does not cause impairments or disease in the invertebrate vector (Carlier et al. 2002; Souza 2000; Brener 1997).

Only small- and medium-sized mammals, including humans, are vertebrate hosts of *T. cruzi*. A large number of cells and tissues are involved in the initial phase of the vertebrate cycle, particularly smooth and striated muscle fibers, macrophages, fibroblasts, and epithelial cells.

Since the cell penetration, trypomastigotes differentiate themselves in amastigotes that multiply and break the cell, when a new transformation to trypomastigotes occurs. They fall into the bloodstream and invade neighboring cells, for a new cycle. Remaining undifferentiated amastigotes will be killed and degenerate, a fact of the greatest importance in the genesis and dynamics of local inflammatory processes (Andrade 2013).

In the acute phase, the number of circulating trypomastigotes and parasitized cells increases exponentially with the succession of cycles of intracellular parasite. Progressively, an immune response is built by the host and able to suppress the acute process, thus gradually decreasing the parasitemia. This situation marks the beginning of the chronic phase (Dias et al. 2013).

The histotrophic preferences of *T. cruzi* in vertebrates are quite variable. At the penetration site, a typical invasion of macrophage-phagocytic cells occurs, being followed by the parasite dissemination to several organs and tissues, mainly to cardiac cells, smooth muscles, and nervous system. In some hosts, the acute process is not contained and death supervenes (Souza 2000; Andrade 2013).

Table 17.1 Classification and distribution of different groups of *T. cruzi* (Zingales et al. 2009)

Trypanosoma Cruzi classification									
Group	Zymodeme	Biodemes or types	Clones	Lineage	DTU	Lineage	Zyroodeme	Geographical areas	Environments
I	Z1	III	1–25	1	1	2	ZIV	Endemic in northern South America and Central America (and whole endemic area)	Exclusive nearby sylvatic in *Diddphis marsupialis* (and domestic in Venezuela)
IIa	Z3		26–29	2	2a	1	–	More restricted. Amazon basin and North up to the US	Almost exclusively sylvatic
IIb	Z2 (Brazilian Z2)	II	30–34	2	2b	1	Z I ZA?	More restricted. Endemic in Southern cone countires of South America	Almost domestic (humans and triatomines)
IIc	Z3/Z1 ASAT		35–37	2	2c	1	–	More restricted. Amazon basin and Paraguay	Almost exclusively sylvatic
IId#	Z 2b? (Brazilian Z2)	I?	38, 39	2	2d	1, 3	–	More restricted. Endemic in Southern cone countires of South America	Almost domestic (humans and triatomines)
IIe#			40–43	2	2e	1	Z III	More restricted. Endemic in Southern cone countires of South America	Almost domestic (humans and triatomines)

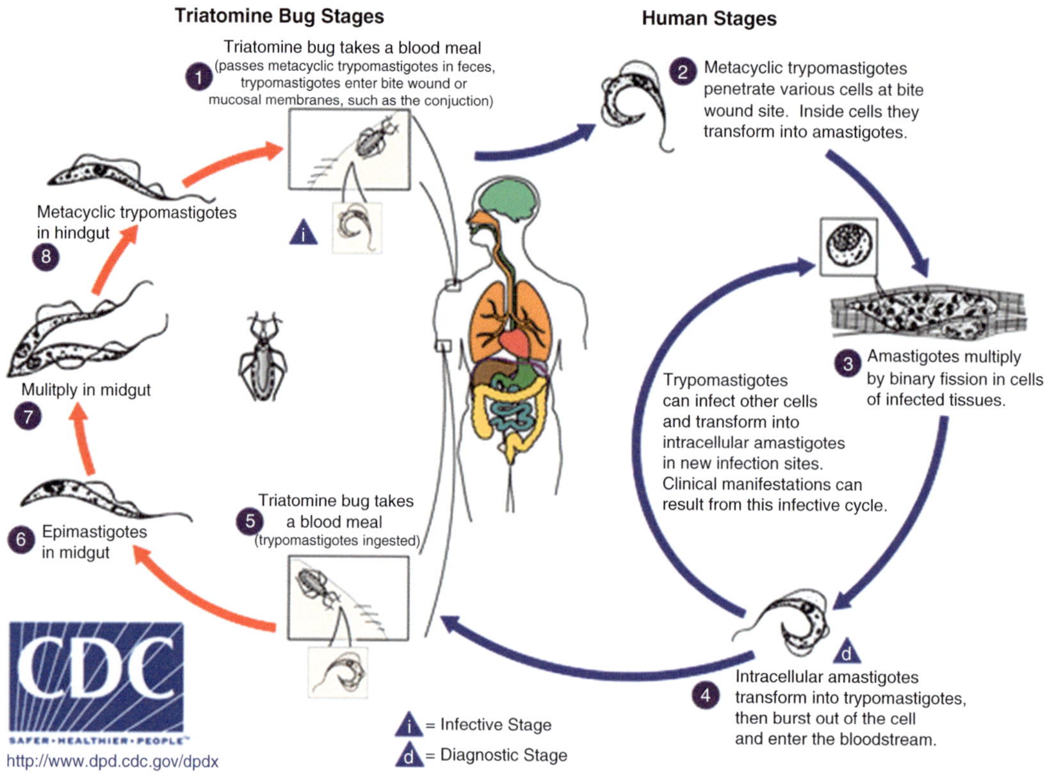

Fig. 17.3 Life cycles of *Trypanosoma cruzi* in vector and human beings. An infected triatomine insect vector (or kissing bug) takes a blood meal and releases trypomastigotes in its feces near the site of the bite wound. Trypomastigotes enter the host through the wound or through intact mucosal membranes, such as the conjunctiva ❶. Common triatomine vector species for trypanosomiasis belong to the genera *Triatoma*, *Rhodnius*, and *Panstrongylus*. Inside the host, trypomastigotes invade cells near the site of inoculation, where they differentiate into intracellular amastigotes ❷. Amastigotes multiply by binary fission ❸ and differentiate into trypomastigotes and then are released into the circulation as bloodstream trypomastigotes ❹. Trypomastigotes infect cells from a variety of tissues and transform into intracellular amastigotes in new infection sites. Clinical manifestations may result from this infective cycle. The bloodstream trypomastigotes do not replicate (different from the African trypanosomes). Replication resumes only when parasites enter another cell or are ingested by another vector. The bug becomes infected by feeding on human or animal blood that contains circulating parasites ❺. The ingested trypomastigotes transform into epimastigotes in the vector's midgut ❻. Parasites multiply and differentiate in the midgut ❼ and differentiate into infective metacyclic trypomastigotes in the hindgut ❽ (CDC 2013)

A global view of *T. cruzi* cycles in vector and human individuals may be seen in Fig. 17.3.

A particular cycle of *T. cruzi* was described inside the anal glands (odorants) of some marsupials (*Didelphis* sp.), very similar to what was observed in triatomines, with the presence of trypomastigotes, epimastigotes, and spheromastigotes. As the product of these glands may be launched by the animal in the external environment in order to repel the enemies, another form of parasite transmission by contamination of mucosa becomes possible (Deane et al. 1984).

17.3 Pathology, Clinical Aspects, Diagnosis, and Management of HCD

17.3.1 Routes of Parasite Transmission (Miles et al. 2004; Brener 1997; CDC 2013; Dias et al. 2011)

Different routes of *T. cruzi* transmission to humans have been described. Accordingly, their numerical grade of occurrence may be classified as primary, secondary, and tertiary routes, a situation that may change according to the historical period and to different geographic and sociopolitical circumstances.

17.3.1.1 Main Routes

- *Vector transmission*. It is the main original route of HCD transmission, occurring from the Southern USA to Patagonia, where several triatomine species dwell. A great positive impact of vector control on the incidence of HCD has been observed in all endemic areas where sustainable control programs were implemented.
- *Transfusion transmitted HCD*. It was the second most important route from 1960 to 2000, when blood transfusion was spread all over the world and blood donors were not universally screened. Its risk was estimated between 12 and 25 % in a routine 500 ml blood bag collected from chronic infected donors, naturally being higher in the cases of acute donors, total blood, and repeated transfusions.
- *Congenital route*. It is important in endemic and non-endemic areas, depending on the contamination of children from infected women, during pregnancy or birth. The risk of transmission occurs between 0.5 and 8.0 %, in different studies, apparently being higher in areas of an existing active vector transmission.

17.3.1.2 Secondary Routes

- *Oral route*. It is very common in the sylvatic cycle, in which several mammals eat infected triatomines and smaller mammals. Human ingestion of certain uncooked foods, such as sugarcane and açai juices, has caused family or community oral outbreaks of Chagas disease. Individually, cases of oral transmission have also been described in certain primitive cultures that have the habit of eating raw animals (rabbits, opossums, armadillos, etc).
- *Organ transplantation*. Transmission of *T. cruzi* by organ transplantation (kidney, liver, heart, and bone marrow transplants) has been observed in the case of an infected donor and a vulnerable recipient.
- *Accidents in laboratory*. Several possibilities have been detected in laboratories and in the healthcare system, in situations involving *T. cruzi* manipulation.

17.3.1.3 Other Eventual or Possible Routes

Other forms of transmission involving different arthropods (flies, mosquitoes, fleas, ticks, and lice), sexual contamination, etc. seem to be extremely rare and never proven. The specific case of transmission by breast-feeding in infected mothers should be considered as an extremely rare form of oral transmission, involving not only the milk itself but also the contamination delivered from bleeding mammillary injuries.

17.3.2 The Natural History of HCD

The human infection with *T. cruzi* involves two different clinical and anatomic-pathological phases, as follows: the *acute* one (initial, short duration (few weeks), with apparent or unapparent clinical forms) and the *chronic* one (long duration, with indeterminate, cardiac, digestive, or mixed clinical form). In Fig. 17.4, the natural history of HCD is summarized, taking into account the possible perspectives for the evolution and the impact of a specific treatment (Dias et al. 2013).

The main aspects involving the pathology and pathogenesis of HCD may be summarized as follows (Andrade and Andrade 1999; Kierszenbaum 2007; Rassi et al. 2009b).

In the *acute phase*, there is a general infection with the parasite, with high levels of circulating trypomastigotes and intense replication of parasites in different tissues, where a parasite-related inflammation occurs, usually followed by cell degeneration and also cell necrosis. The main defense cells involved are monocytes, macrophages, and lymphocytes. Intensity and preferential localization of these lesions may vary according to the parasite strain, the age and the immunological competence of the host, the inoculum, etc. Heavily parasitized sites in ACD are the heart, several skeletal and smooth muscles, and the central nervous system. An acute

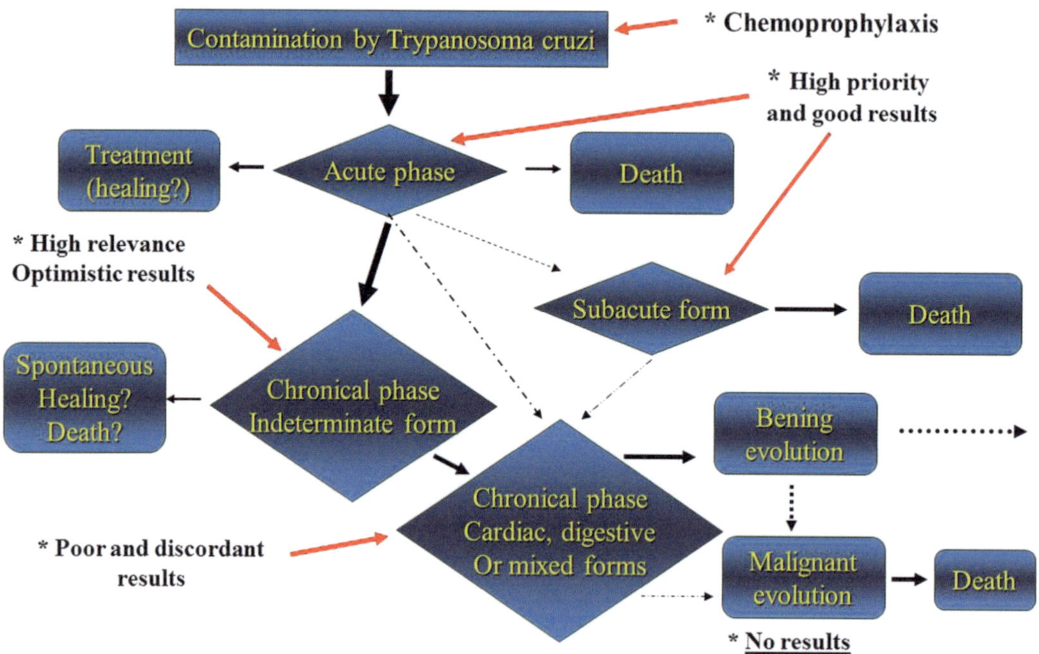

Fig. 17.4 A general schedule of the natural history of human Chagas disease, regarding the specific treatment available in 2014 (Adapted from Dias 1990)

myocarditis frequently occurs and represents, particularly in younger age and/or immunocompromised individuals, the main cause of death in ACD. Cell infiltrates are seen around parasite cells dissociating cardiac fibers and giving rise to edema, vascular changes, and regressive changes. Natural killer cells and cytotoxic lymphocytes adhere to the capillary endothelium, occasioning endothelial swelling, platelet aggregation, and fibrin thrombosis. In addition to lesions in the contractile myocardium, pathological changes may involve the conduction system and intracardiac nerves. In particular, the autonomic nervous system is highly affected in ACD, with the destruction of many autonomic neurons (especially parasympathetic ones) occurring near parasite nests.

In the *chronic phase*, circulating parasites are scarce and intracellular parasitism remains in the heart or in other foci. In the heart, chronic microscopic lesions are composed of diffuse and/or focal infiltration of mononuclear cells, hypertrophy of myocardial fibers, lytic myocardial necrosis, thin-walled congested small blood vessels, the presence of fibroblasts, and interstitial and perimuscular fibrosis. Intracellular amastigotes are rarely found in conventional microscopy, but they are always present near the infiltrates, detected by immunocytochemical methods. Anatomically, more severe cases present cardiomegaly, apical and/or lateral aneurisms, parietal thrombosis, and a peculiar disruption of myocardial architecture, caused by advanced fibrosis and cell destruction. Large coronary vessels are preserved, but endotheliitis and microthrombosis occur in lower vessels. A physiological elongation of remaining cardiomyocytes results in cardiomegaly, but, in the medium term, this compensatory mechanism becomes insufficient and heart failure supervenes. In parallel (and characteristically), the inflammatory process substantially affects the intracardiac nervous system of formation and conduction of the electrical stimulus, leading to several degrees of atrioventricular and intraventricular blocks, which contributes to cardiac failure and sudden deaths. At the walls, inflammatory and fibrotic lesions lead to extrasystolic arrhythmias as well as atrium and/or ventricle fibrillation. In digestive chronic lesions, in addition to myositis, a very characteristic picture of ganglionitis and periganglionitis occurs, depending on the presence of the parasite and occasioning a typical extensive loss of intramural autonomic neurons. Destruction and elongation of smooth myocytes, in addition to interstitial fibrosis, result in megaviscera, at medium and long term. Lesions in the central nervous system are more evident in acute and subacute HCD and are particularly severe in immunodepressed individuals. In such cases, an acute meningoencephalitis supervenes with a high amount of parasites being present in liquor and nervous structures. In severe cases of HIV/AIDS coinfection, big pseudotumoral inflammatory formations are seen in the brain.

17.3.3 Acute Chagas Disease (ACD)

In human beings, the parasite penetrates through damaged skin (due to triatomine biting) and other possibilities (scratching) or by intact mucous membranes or is injected into the bloodstream, via blood transfusion. In vector transmission, typical lesions at the site of penetration ("portal of entry") start between 10 and 15 days after inoculation, as it is a cell-mediated hypersensitivity reaction. As the parasite spreads in the body, it multiplies in several organs. The high blood concentration of parasites represents the fundamental characteristic of ACD, being detectable by direct parasitological tests. In this phase, the parasite level will be modulated by the presence and amount of IgG antibodies that begin to appear at the end of the second week after the contamination. The maximum level of these antibodies will occur between the fourth and eighth week, and after that a very stable "plateau" will remain for the infected individuals (both non-treated and cured) (Andrade and Andrade 1999; Dutra and Gollob 2008).

The balance of parasite/antibodies is summarized in Fig. 17.5, in which the curve of parasitemia is higher in the acute phase (with higher IgM and lower IgG antibody concentration), in contrast with the chronic phase (Dias et al. 2013; CDC 2013).

It must to be remembered that the so-called congenital Chagas disease is basically an acute situation in which the parasite is transmitted from an infected woman to her baby, during pregnancy or childbirth. Infected pregnant women in Latin America have a prevalence ranging from 1 to 51 % in some urban centers and from 1 to 81 % in some rural settings. The risk of transmission may

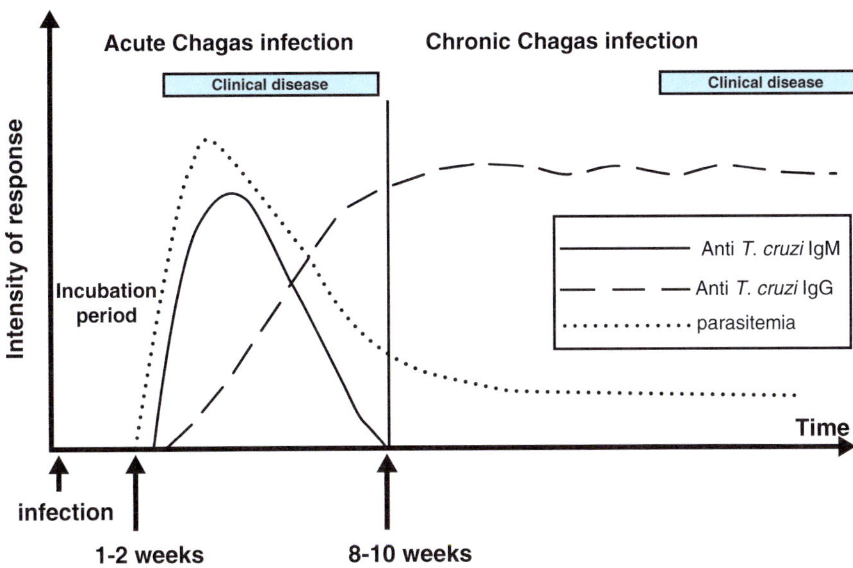

Fig. 17.5 Evolution of parasite and antibody levels in human Chagas disease (From Dias et al. 2013)

vary from <1 and 29 % in chronic infected women of childbearing age, with an estimate of around 15,000 congenital *T. cruzi* infections occurring per year (OPAS 2006; Torrico 2014).

In addition to the risk of miscarriage or premature birth, neonatal infection with *T. cruzi* causes an acute form of Chagas disease in the baby, mostly asymptomatic (Torrico 2014). This form of the disease is a major public health priority because it is frequent, severe, identifiable, and curable.

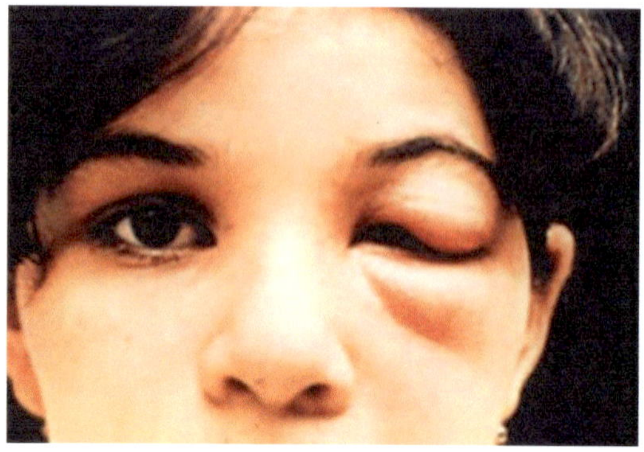

Fig. 17.6 Romaña's sign (From CDC 2013)

17.3.3.1 Clinical Aspects

In general, the acute disease is usually asymptomatic and most of acute cases remain undetected in endemic areas. The clinical scope of this phase has a pleomorphic pattern, with more apparent and severe cases occurring in younger individuals (0–3 years of age). In practically all apparent or unapparent acute cases, a febrile picture is present, more specifically a long-duration moderate fever (up to 4 weeks, oscillating between 37.5 and 38.5 °C). In most detected acute cases, a moderate depletion of the clinical general state is described, with patients usually presenting pallor, adynamia, and irritability (Laranja et al. 1956; Maguire 1999; Dias et al. 2013).

In cases of vector transmission, some *signs of portal of entry* may be detected, corresponding to the local inflammatory reaction to parasite penetration ("chagomas"). Among them, the most striking and characteristic is the so-called Romaña's sign, an ophthalmic and ganglionic complex, which corresponds to a unilateral eyelid edema accompanied by dacryoadenitis and regional ganglia infarction (Fig. 17.6). The clinical expression of ACD is more intensive among children, generally involving fever, tachycardia micropolyadenitis, general edema, hepatosplenomegaly, and signs of portal of entry. Nonspecific prostration, diarrhea, vomiting, anorexia, headache, and myalgias have been described in most apparent cases (Laranja et al. 1956; Dias et al. 2013). Signs and/or symptoms of acute cardiomyopathy, of greater or lesser intensity, are detected in most cases and also in experimental models (Carlier et al. 2002; Andrade and Andrade 1999).

Esophagopathy in ACD is little studied. It is known that, particularly in central and southeastern Brazil, the esophagus is precociously affected, by means of intense intramural denervation in the autonomic nervous system. On the other hand, colopathy in ACD apparently has never been reported (Carlier et al. 2002; Dias et al. 2013).

Mortality in ACD usually varies between 5 and 10 % of cases, occurring especially in children under 2 years of age (Laranja et al. 1956; CDC 2013).

In congenital transmission, most cases are asymptomatic. Some clinical signs in exceptionally severe infections are similar to those produced by other pathogens of the ToRCH complex [toxoplasmosis, other (syphilis), rubella, CMV, HSV]. They are marked by stillbirths or have variable clinical manifestations, such as prematurity, fever, hepatosplenomegaly, and signs of acute myocarditis or CNS involvement (meningoencephalitis). In some cases, it is possible to detect cerebral calcifications, whose clinical significance is still undetermined (Torrico 2014).

In transfusion transmitted HCD, the signs of entry do not exist, and the incubation period is usually larger than in the vector transmission (up to 2–3 months), generally occurring among adult individuals. Likewise, in oral transmitted HCD, signs of entry do not exist either, with the affected age groups varying a lot (Dias et al. 2013).

The anatomic substrate of ACD is basically the inflammatory process installed in affected organs around the parasite cells. Focal and generalized reaction depends mainly on the parasite antigens delivered by degenerated *T. cruzi* forms when the parasite cell is disrupted. The infiltration is composed mainly of monocytes, lymphocytes, and macrophages, which attack the parasites and also normal cells, producing local edemas and initial repairing fibrosis (Andrade and Andrade 1999).

In Table 17.2, the main clinical findings of ACD related to symptomatic cases are summarized (Carlier et al. 2002; Maguire 1999).

Table 17.2 Main clinical findings in apparent acute Chagas disease

Signs and symptoms	General characteristics	Practical notes
Portal of entry (inoculation "chagomas")	Indurated and erythematous skin lesions with exfoliative desquamation in the end of the evolution. Painless or little pain, violet color, usually present in limbs or face. Very frequent satellite adenopathy	Amastigote forms may be found in local biopsies
Romaña's sign	Eyelid edema with satellite adenopathy and dacryoadenitis. Reduced palpebral fissure. Itching, tearing, and pain, mild and local, may occur. This signal is mentioned in more than 50 % of the cases described	Differential diagnosis: insect bites, myiasis, conjunctivitis and styes, trauma, orbital cellulitis, angioneurotic edema, and cavernous sinus thrombosis
Other "chagomas"	Less frequent: metastatic (distant from the primary inoculation, often by hematogenous or lymphatic propagation), sometimes located in the cheek (lipogenic chagoma)	Described in Argentina
Fever	Generally moderate (±38 °C) and continuous, lasting between 7 and 30 days. May have high peaks in the afternoon. Even in the unapparent cases, it is present, with a short duration and lower temperatures	It does not usually respond to common antipyretic drugs
Adenopathy	There are usually small and multiple lymph nodes in various plexuses, hardened, are not coalescing, and do not suppurate; they also present downstream inoculation *chagomas*	Biopsy may show parasites. Lymphocyte hyperplasia is usual. Swollen ganglia may persist for months after the acute phase
Hepatosplenomegaly	From about 20–40 % of the cases; usually detected in younger ages, with a small increase in volume; both viscera hardened and somewhat painful to palpation; concomitance of passive congestion and tissue degeneration	Pay attention to other febrile hepatosplenomegalies in endemic areas (visceral leishmaniasis, acute toxoplasmosis, schistosomiasis, etc.)
Generalized edema	Hardened, snap, diffuse, and cold, not leaving the "godet" depression. Very early. More visible on the face, extremities, and scrotum	May be superimposed by an edema due to heart failure
Local edema	Situated at the point of parasite penetration, having a reddish or red-violet color, presenting induration and slight soreness	Inflammatory origin. It is part of the *chagomas* (inoculation or metastatic)
Impaired general health status	Asthenia, adynamia, continuous crying, and suffering	Especially detected in younger children
Signs of acute myocarditis	Detection varying between 5 and 50 % of cases, on average. Abnormal ECG in 30 % or more of the cases published suggesting an acute myocarditis (inversion of T and increased PR). Possible presence of heart failure (poor prognosis): easy fatigue, orthopnea, gallop rhythm, and increased venous pressure. X-ray shows that a global cardiomegaly is present from 15 to 60 % of cases, with generally clear lung fields. A pericardial effusion is shown in the most severe cases	Differential diagnosis with other acute myocarditis (rheumatic fever, toxoplasmosis, diphtheria, toxic, syphilitic, etc.) and with endocarditis. Histologically, mononuclear inflammation, usually diffuse and predominantly subendocardial. Myocytolysis and intercellular edema. The parasite is easily detected in cardiac myocyte cells
Signs of meningoencephalitis	Especially in children younger than 2 years (1–10 %), generally associated with heart disease manifestations. Opisthotonos-like posture, neck stiffness, and other traditional signs of meningism are commonly present. Other symptoms are frequent vomiting, headache, agitation, strabismus, obnubilation, prostration, convulsions, abdominal pain, etc. Clean cerebrospinal fluid, harboring parasites	Very bad prognosis, with autopsy findings showing severe inflammatory changes in the brain and meninges

17.3.3.2 Evolution, Prognosis, and Mortality in ACD

In general, non-treated cases of ACD evolve for the remission of their clinical symptoms in periods varying between 2 and 12 weeks after diagnosis, most of them between 5 and 8 weeks. The natural remission of clinical symptoms is slow, and progressively, however, cardiomegaly, diffuse lymphadenopathy, and hepatosplenomegaly last longer (Laranja et al. 1956).

The specific treatment usually shortens this time of development, rapidly reducing the parasitemia and making fever and tachycardia stop. The vast majority of patients evolve into a long chronic indeterminate form. In a few patients, electrocardiographic changes remain, such as stretching of PR and repolarization abnormalities, leading directly to a chronic cardiac form (Dias et al. 2013).

Death in acute cases usually occurs in the first 3 weeks of disease manifestation, basically due to heart failure and/or meningoencephalitis, which have a poor prognosis with manifestations of meningoencephalitis, large cardiomegaly, heart failure, and some ECG abnormalities, mainly extrasystolic arrhythmias, intraventricular blockings, and low QRS voltage (CDC 2013; Dias et al. 2013).

17.3.3.3 Medical Management of ACD

When properly treated, ACD may heal in proportions that vary generally between 30 and 90%, according to different authors. As far as cases were specifically treated and cured of acute infection, late manifestations of visceral involvement (heart, esophagus, and colon) due to *T. cruzi* (Viotti and Vigliano 2014) were not made evident in the long term. See details of specific treatment in the section of specific treatment of Chagas disease. It is very important to detect and report ACD, both because of its vulnerability to treatment and also to strengthen the epidemiological surveillance on the trypanosomiasis (Rassi et al. 2009b; Viotti and Alvarez 2014). Each new case indicates the occurrence of active transmission and may mean that other acute cases of infection are occurring at the same time and place, which may be avoided from the required notification.

17.3.4 Indeterminate Chronic Chagas Disease (ICD)

This is the most common clinical form of HCD in the general population of endemic areas, presenting an enormous clinical and social interest.

Following the acute phase, most infected people enter into a prolonged asymptomatic form of the disease (called "indeterminate chronic") during which few or no parasites are found in the blood. It is characterized by positive serological tests or the detection of the parasite or its fractions by indirect parasitological tests (xenodiagnosis, blood culture, and polymerase chain reaction (PCR)). They are clinically asymptomatic, with a normal ECG and a normal X-ray (heart, esophagus, and colon). However, more sophisticated tests, such as dynamic electrocardiography, stress and pharmacological tests, radionuclide scanning, etc., may detect some degree of variation, which does not invalidate the concept above and nor does it mean, necessarily, the disease is evolving (Maguire 1999; Brasil 2005).

During ICD, most people are unaware of their infection. Many of them may remain asymptomatic for life and never develop Chagas-related symptoms. They are always healthy, the majority corresponding to young and middle-aged adults. ICD may evolve to a definite clinical form, in a proportion of 2–3% of cases/year, or remain asymptomatic indefinitely (Maguire 1999). Management of these individuals is simple and its objective is basically to detect any early evidence of evolution of the disease.

A clinical review of the ICD carrier should be conducted annually in their own basic health network or in mixed healthcare units that have ECG and X-ray equipment. If possible, a Doppler echocardiogram will be useful, as it is able to detect regional dyskinesia and even initial lesions in asymptomatic individuals with a normal ECG (Rassi et al. 2009a).

The prognosis is very good in the medium term. In general, no deaths have been reported due to HCD in "indeterminate" individuals, over the 10 years following their serological diagnosis. In Brazilian endemic areas, the evolution to an early cardiomyopathy or esophagopathy is usually around 2–3 % of the cases, yearly (Dias et al. 2013). In ICD, the physical capacity is always great; thus, individuals are able to practice any profession; the only existing labor restrictions are to operators of heavy machinery, of collective vehicles and aircrafts (WHO 2002; CDC 2013).

The specific treatment for ICD is more and more recommended by several researchers. Its rationality is based on the idea that the elimination (or strong reduction) of the existing parasites will stop or at least delay the clinical evolution (Viotti and Vigliano 2014).

17.3.5 The Chronic Chagas Heart Disease (CCHD)

This is the most important clinical form of HCD, precisely because of its impact on morbidity and mortality: it is the form which limits labor productivity, causes absenteeism, decreases the quality of life, and kills patients (Laranja et al. 1956). Being more severe and more precocious in males, several pathogenic factors will affect the heart, leading to a progressive impairment of its contractile function and/or rhythm. In its late evolution, thromboembolic phenomena may occur as a result of stasis, endocardial changes, and a dilation of the cardiac chambers. CCHD is also characterized as a thromboembolic syndrome, which can compromise the lungs, brain, kidneys, spleen, and other sectors, enabling other immediate causes of death from the disease. Once installed, CCHD is usually progressive, and its main consequences (arrhythmias, heart failure, and thromboembolic features) may be associated with each other and boost in a reciprocal way. In summary, CCHD is characterized by parasite persistence and immunological unbalance favoring systemic inflammatory profile (Andrade and Andrade 1999; Rassi et al. 2009b). For instance, cytokine TNF seems to play an important role in *T. cruzi*-induced immune deregulation and cardiac abnormalities (Vilar-Pereira et al. 2014).

In summary, to the extent that CCHD progresses anatomically, the heart performance also deteriorates, in terms of rhythm disturbs and cardiac output. The progressive degrees of heart failure usually follow the levels of dilation of the cardiac chambers. The symptomatology of CCHD, almost absent in the early stages, becomes increasingly more evident, to the extent that the electrical disturbances and the heart failure will already be in progress (Rassi et al. 2009b).

ECG works as the strongest diagnostic element of CCHD, typical of the intraventricular blockades (particularly of right bundle branch associated with left anterior hemiblockade), the multifocal extrasystoles, the changes of T wave, and the bradyarrhythmias. In CCHD, the conventional ECG is very helpful to establish disease prognosis. The most severe cases usually present findings such as frequent and polymorphic extrasystoles, paroxystic ventricular tachycardia, atrial fibrillation, extensive inactive areas, left bundle branch blockade, low voltage of QRS, and advanced AV blocks. On the other hand, X-ray highly concerns with CCHD prognosis, always being the cardiomegaly associated with advanced CCHD and heart failure. Images of the lungs are usually normal, only reaching the present signs of congestion and liquid spilling in the final stages of CCHD (Laranja et al. 1956; Maguire 1999).

In practice, most patients with HCD may be assisted by a general practitioner in the basic health network. In order to do so, a relatively elementary propaedeutic tool is required. For the initial screening of chronic patients, the minimum propaedeutic arsenal requires: a good physician, a serological diagnosis, an ECG, and a set of X-ray apparatus capable of taking a simple chest X-ray (heart area and contrasted esophagus) and a simplified barium enema (Dias et al. 2013).

Cardiomegaly on CCHD has a direct relationship with the heart failure, while the functional capacity is significantly reduced in dilated myocardial fibers. Transparency of the pleuropulmonary fields is very characteristic, but an effusion is only detected when the pulmonary venous congestion appears, in terminal stages of heart failure.

In other less conventional exams, CCHD also has important changes, with more sophisticated procedures available for its assessment. Among the most useful ones are dynamic electrocardiography (Holter), Doppler echocardiography (M-mode and two dimensional), treadmill test, scintillography, etc., which are indicated both for the evaluation of arrhythmic events and heart failure and for obtaining a better clinical monitoring and management.

In more advanced stages of CCHD, especially when heart failure is present, a syndrome of thromboembolism may be transient, on account of displaced thrombi particularly the tip of the left ventricle and right atrium, detectable in the echocardiogram and ventriculography. The resulting framework consists of secondary infarctions of several organs, predominantly the lungs, brain, kidneys, spleen, and mesenteric venous system. It may be asymptomatic or present the usual clinical characteristics (pain, dyspnea, hematuria, etc.) and central neurological changes (aphasia, dyslalia, loss of consciousness, motor abnormalities). Consequences are generally severe as death may be precipitated in a patient already depleted and affected by important hemodynamic problems (Andrade and Andrade 1999; Rassi et al. 2009b).

The main elements of risk for thromboembolic events are (a) severe myocardial dysfunction (functional classes III and IV, chronic visceral venous congestion, dilation of the heart area, decreased ventricular ejection fraction); (b) ECG showing advanced AV or IV blocks, electrically inactive areas, and atrial fibrillation; (c) apical lesion in LV; (d) the presence of intracardiac thrombosis; and (e) a previous thromboembolic occurrence (Rassi et al. 2009b).

In people who have been immunosuppressed (e.g., due to AIDS or chemotherapy), heart Chagas disease may be triggered with intense myocarditis and several parasites found in the circulating blood. This occurrence may potentially cause severe subacute cardiomyopathy, resulting in heart failure and death when not adequately treated. A practical classification of CCHD in progressive degrees and in relation to medical actions is shown in Table 17.3.

The clinical and surgical management of the CCHD is of utmost importance in the context of the life of the infected person, provided that the earlier and the more properly it is installed, the greater benefits it will bring to the patients (Laranja et al. 1956; WHO 2002; Kierszenbaum 2007). In principle, any infected chronic individual should be considered as having a cardiac potential, with the objective of considerably reducing risk factors of CCHD evolution. The basic goal is to avoid the installation and/or the progression of arrhythmic situations and heart failure, consequently improving the quality of life and its duration (Rassi et al. 2009b; Viotti and Vigliano 2014).

An efficient therapeutic arsenal is currently available for CCHD management, including antiarrhythmic compounds (amiodarone, sotalol, propafenone), angiotensin-converting enzyme inhibitors, vasodilators, selective beta-blockers (carvedilol), diuretics (furosemide, spironolactone), and cardiotonics (digoxin).

In parallel, in the last few decades, surgical resources and electronic tools have been incorporated for CCHD management, such as cardiac pacemakers, implantable defibrillators, techniques for ablation of arrhythmogenic foci, aneurysmectomies and withdrawal of thrombi, heart transplant, etc. (Carlier et al. 2002; Maguire 1999; Brasil 2005).

Unfortunately, ventriculectomy and cardiomyoplasty have not succeeded in advanced CCHD. There are expectations regarding the deployment of stem cells in injured hearts, to recover cardiomyocytes and remodel the myocardium. The management of the thromboembolism aims at reducing precipitating factors and prevention routine in high-risk situations with aspirin and other anticoagulants, as well as the employment of these drugs and even surgeries in patent thromboembolism (Dias et al. 2013; Rassi et al. 2009b).

17.3.6 Digestive Chronic Forms

The main changes are in the esophagus and colon, resulting in organ dilatation commonly named "megas." In Central Brazil, esophagopathy occurs between 7 and 10 % of chronically infected individuals and colonopathy between 5 and 10 %.

Table 17.3 Summarized classification of the human chronic Chagas heart disease (Adapted from Dias et al. 2013)

Grade	Clinical criteria and NYHA classification	Chest X-ray and echocardiography	ECG	Prognosis	Labor medical decision
I	No signs or symptoms (NYHA I)	Normal	Normal (basal)	Very good	Able to all professions
II	Absence or minimal symptoms and signals (NYHA II)	Normal X-ray *Echo*: normal or presenting minimal and isolated dyskinesia, EjF > 60 %	Minimal alterations such as rare and isolated extrasystoles	Very good	If functional assessment is not possible, consider inability to harder activities, indicating functional rehabilitation
III	Absence of symptoms and signals under major physical efforts (NYHA II)	Heart shadow: borderline or small increase *Echo*: dyskinesia, diastolic dysfunction, increased oro-valvular, EjF < 50 %	CRBBB, T wave alterations, sinoatrial block Junction rhythm A-V block II (Mobitz I), inactive septal area	Regular	Inability to perform activities that require physical effort of moderate intensity. Evaluate the possibility of vocational rehabilitation
IV	Presence of symptoms and signals under moderate efforts (NYHA II or III)	Slight or moderate increase of cardiac shadow *Echo*: dyskinesia, systolic and diastolic dysfunction, EjF < 50 %	CRBBB + LAHB, LBBB Isolated monomorphic extrasystoles, A-V block (III), inactive areas	Reserved	Permanent omni-professional inability. It is accepted that some light physical activities may be carried out
V	Overt heart failure Orthopnea (NYHA III to IV)	Slight or moderate increase of cardiac shadow. Possible lung congestion and or pleural spilling *Echo*: dyskinesia, systolic and diastolic dysfunction, enlargement of the left atrium, EjF < 40 %	Frequent polymorphic extrasystoles, extensive inactive areas, advanced A-V and bi(tri)fascicular blocks, R/T phenomena Atrium flutter or fibrillation	Extremely reserved	Permanent omni-professional inability, restricting any physical activity

NYHA New York Heart Association, *ECG* electrocardiogram, *EjF* ejection fraction, *CRBB* complete right bundle branch block, *A-V* atrioventricular, *LAHB* left anterior bundle branch block

17.3.6.1 Esophagopathy

It affects mainly male individuals aging 20 years or more, featuring four stages or degrees of evolution, according to the functional status and the radiological morphology of the organ. The fundamental symptom is dysphagia, which leads people to ingest large volumes of water to facilitate swallowing. Other symptoms are pain when swallowing (odynophagia) and hiccups. The main complications are pneumonia, food regurgitation, and esophagitis. There is a higher incidence of esophageal cancer in patients with chagasic megaesophagus, in comparison with noninfected individuals (Rezende and Moreira 2004).

The diagnosis of chagasic megaesophagus is clinical and radiological, eventually being complemented by per-oral endoscopy, pharmacological tests, and records of motor activity. A very simple radiological procedure consists of a radiograph carried out 1 min after a barium contrast meal: retention of contrast for as long as 1 min still indicates a functional disturb. Organ dilatation and elongation indicate higher degrees of esophagopathy (Rezende and Moreira 2004; Maguire 1999; Dias et al. 2013).

In initial levels, general measures and noninvasive procedures should be tried such as an adequate dietary routine (dried, cold, and irritating foods are contraindicated), patients should chew well and eat peacefully, etc. For cardiac dilation, isosorbide mononitrate may be used (one tablet 30 min before a meal) as well as the infiltration of botulinum toxin in the lower esophagus. Probes and balloons tires may be useful in the early stages, in spite of possible recurrences in the medium term. The extra mucosa myotomy (Heller's esophageal cardiomyotomy) has been successfully indicated for grades 1–3 of esophageal achalasia, with best results when performed by video laparoscopy. The main purpose of the surgery is to correct dysphagia. Plastics in the sphincter are directed toward the prevention of esophageal-gastric reflux. For advanced dolichomega-esophagus, the replacement of the lower part of the esophagus by jejunal loop is indicated (Rezende and Moreira 2004).

17.3.6.2 Colopathy

This occurs especially among older age groups of patients with Chagas disease, appearing in several countries of the Southern Cone, with prevalence between 2 and 8 %, mainly after 40 years of age. There is no difference regarding sex (Carlier et al. 2002; Rezende and Moreira 2004).

Motor disorders are important, resulting in colon dilatation, and long megacolon may be severe, leading to situations of extreme discomfort and even death (in the case of sigmoid volvulus). The main symptom is constipation. Other symptoms are flatulence and dyschezia (difficulty in defecating). The main complications are fecaliths, fecal impaction, and sigmoid volvulus.

Constipation is progressive; the patient generally remains constipated for 3–7 days, sometimes reaching 2–3 months. Constipation is usually accompanied by bloating, abdominal pain, flatulence, and fecalith. Diagnosis is made by a contrasted colon X-ray (barium enema) that shows a predominant dilatation in the rectum and sigmoid.

The differential diagnosis is with the congenital megacolon, pregnancy, and different causes of chronic constipation (neoplastic diseases, intestinal parasitic infections, constipation due to inadequate diets, etc.) (Maguire 1999; Rezende and Moreira 2004).

A similar clinical picture to chagasic megacolon has been reported in the higher regions in the Andes mountains, affecting adult individuals with negative serological tests for *T. cruzi*. This is the so-called Andes megacolon, whose etiology remains obscure (Dias et al. 2013).

The objective is to treat constipation and prevent complications. In the early stages, the dietary adequacy and the use of mild laxatives may be useful. In the presence of a fecalith, this must be eliminated through washes and/or manual draining via anorectal. In some cases, surgical procedure may be necessary by using laparotomy and direct manipulation. For advanced megacolon, surgery is required, mainly through the Duhamel-Haddad technique, which consists of the resection of the dilated segment and lower colon remaining until the rectal stump (Silva 1997). In cases of volvulus of the sigmoid, its reduction by rectum sigmoidoscopy should be tried before greater surgical interventions (Rezende and Moreira 2004; Brasil 2005).

17.3.6.3 Other Manifestations in the Digestive Tract

The same physical and pathogenic bases of *T. cruzi* infection may affect other parts of the digestive tract, in lower incidence than in the colon and esophagus. Some rare and oligosymptomatic "megas" may occur in the duodenum, stomach, small intestine, and parotids (Carlier et al. 2002; Maguire 1999).

17.3.7 Other Chronic Forms

Theoretically, *T. cruzi* may lodge and cause damage in virtually all components of the vertebrate host. In HDC, parasitism is predominant in the heart and in the digestive tract, as seen above. There is also an involvement of the nervous system (central, peripheral, and autonomous), especially in some situations of immunodepression (particularly in chagasic patients with AIDS); there may be severe meningoencephalitis, characterized by tumors in the brain filled with parasites. Psychological changes (irritability, exacerbated emotional instability) may also occur in cases of autonomic denervation, having as substrate a hyper-sympathetic reactivity to environmental and emotional stimuli.

Secondary locations of the parasite are described in the liver, biliary ducts, pancreas, urinary tract, bronchi, and salivary and sweat glands, eventually causing mild and nonspecific symptoms. In all of these situations, the presence of the parasite, the inflammatory response, and the autonomous denervation are usually the main physiopathogenic factors of the disease (Köberle 1968; Dias et al. 2013).

17.3.8 Special Clinical Situations

In practice, the study and the management of HCD should be conducted in the context of everyday life of infected people. Of greatest interest, the following situations are summarized, each one constituting a wide field for research (Carlier et al. 2002; Oliveira 1997; Andrade and Andrade 1999; Dias et al. 2013; Ferreira 2013).

17.3.8.1 Pregnancy

Most pregnant women are usually in the indeterminate chronic form, which does not compromise the development of pregnancy or childbirth. Those few cases harboring severe chagasic cardiomyopathy deserve special care during the pregnancy and birth, in order to prevent an evolution to heart failure and sudden death. Specific treatment is not recommended in pregnancy, due to possible teratogenic effects of the available drugs. In extreme cases of mother cardiomyopathy, therapeutic abortions may be cogitated. Eventual cases of congenital transmitted disease must be managed through early diagnosis and etiological treatment of infected children. In (rare) cases of pregnant women in the acute phase, the risk of congenital transmission to the baby is naturally greater.

17.3.8.2 Surgical Interventions for Chagas Disease

In general, chronic patients can withstand the most common surgeries, including under general anesthesia. Nevertheless, those with severe arrhythmias and decompensated heart failure should have their surgeries postponed as much as possible. It is preferable to wait for an improvement of their clinical situation, thus reducing the risk of cardiac arrest, thromboembolisms, and ventricular fibrillation, naturally increased by the surgical stress.

17.3.8.3 Chagas Disease in Elderly Individuals

In recent years, a tendency to reach the chagasic patients in more advanced ages occurs all around the world, in contract with the situation 50 years ago, when the chronic disease was predominant among young people. Some factors such as the earlier and better management of patients, better healthcare coverage, and possibly a reduction of exogenous reinfection may explain this. Today, the infected elderly assume complex clinical patterns, based on the natural aging process itself and on the superposition of other morbid aging factors such as arterial hypertension, cardioangiosclerosis, chronic neurological processes, etc. The clinical management of such cases is complicated, given the several physiopathogenic mechanisms involved and even the potentiating of a factor on the other.

17.3.8.4 Chagas Disease in Immunocompromised Patients

Acute and subacute trypanosomiasis may arise in immunodepressed chagasic individuals, being most notable the cases of *T. cruzi* and HIV coinfection when AIDS is being developed. This is a situation of great severity, almost always expressing very intense myocarditis and/or meningoencephalitis, with high lethality. The clinical manifestation usually occurs; the rate of circulating CD4+ lymphocytes is reduced to less than 200 cells per mm^3.

17.4 Diagnosis of HCD

In practice, the etiological diagnosis is necessary for individual cases, blood bank control, prenatal routine, and population studies. It should be complemented with epidemiological, clinical, and pathological studies. The main elements of laboratory diagnosis are different in acute or chronic disease: in the acute phase, the detection of the circulating parasites has priority, performed by means of direct and indirect parasitological methods; in the chronic phase, circulating antibodies will be detected based on serological tests. In summary, for both acute and chronic HCD, a series of parasitological, serological, and immunochromatographic (rapid) tests are available now and may be complemented by molecular methods (CDC 2013; Viotti and Vigliano 2014).

Although a number of methods are available, current diagnostic approaches are insufficient to respond to aspects of the disease such as congenital Chagas, infection in immunocompromised patients, and monitoring treatment. Therefore, there is a growing need to develop and market new diagnostic tests that may be used in the remote areas where the disease is endemic and with a performance good enough to guide clinical care. The main methods and available tools have been summarized (Dias et al. 2013; Maguire 1999; Carlier et al. 2002; Brasil 2005; Viotti and Vigliano 2014).

17.4.1 Acute Phase

By definition, the basic finding is the high level of circulating parasites in peripheral blood, which may be detected by direct parasitological tests (see Fig. 17.2). Indirect methods help in the diagnosis, but do not properly characterize the acute infection.

In case of suspicion, in addition to the clinical and epidemiological approaches, nonspecific tests may be of great value to assist in etiologic evidence and differential diagnosis such as Schilling's hemogram, cerebrospinal fluid analysis, and specific tests for other acute diseases such as typhoid fever, visceral leishmaniasis, bacterial meningitis, toxoplasmosis, endocarditis, mononucleosis, etc.

17.4.1.1 Parasitological Methods

The simplest technique consists of direct microscopy on fresh blood drop, examined between the slide and coverslip. The positive test shows the parasite (usually in serpentine movement between red cells and leukocytes) with its elongated shape, large kinetoplast, and very mobile flagella (Fig. 17.2). Under clinical suspicion, being negative in the first attempt, the examination should be repeated three or four times a day, for several days, a strategy which significantly increases the chance of parasite detection.

The technique of thick smear (as used for malaria) may also be employed, although it is less sensitive than the fresh exam. Nowadays, the direct parasitological diagnosis has been improved by means of enrichment methods such as *microhematocrit* (centrifugation and examination of leukocyte cream) and the *Strout* (centrifugation of plasma after clot removal). Molecular techniques, such as PCR, allow to detect (with great sensitivity) fractions of parasite DNA but do not characterize the acute disease.

To increase the sensitivity, indirect parasitological methods may be used, by increasing the amount of parasites, in passages by triatomine bugs ("xenodiagnosis"), through blood cultures or through subinoculation in sensitive animals. In xenodiagnosis, 40 noninfected triatomine nymphs are recommended, preferably of the most common species in the region. As a rule, this amplification has better results between 30 and 60 days of incubation, mainly in xenodiagnosis and in blood cultures. A technique of artificial xenodiagnosis has been used: the patient's blood is collected with a syringe and kept warm and heparinized, inside a glass apparatus provided with a thin membrane, through which insects are feed, to prevent allergic reactions to bite.

For blood culture, 30 ml of heparinized blood are submitted to centrifugation for 30 min, following the immediate discarding of the plasma. The material should be washed a second time with the culture medium itself (usually LIT, liver infusion tryptose), sowing the solid part in six tubes of the medium that will be incubated at 28 °C.

17.4.1.2 Immunological Diagnosis

Serological tests do not define for themselves the acute disease, but may help indirectly in its diagnosis, being very useful in individual cases and in the epidemiological surveillance system.

Antibodies of the IgG class will appear in the 3rd or 4th week of acute infection. IgM antibodies are suggestive of acute disease, but may appear sporadically in the chronic phase. Their use is limited by low availability and easy deterioration of the reagents.

For practical purpose, it is suggested that in suspicion of ACD with negative parasitological tests, a first conventional serology for IgG. If negative, it should be repeated about 30 days later. A serum conversion, in this case, indicates an acute disease in progress or regression.

17.4.1.3 Molecular Methods

Their aim is to amplify and detect the nuclear or kinetoplastic DNA of the parasite possibly present in a blood sample from the patient, through molecular appropriate sounding.

The most used is PCR (polymerase chain reaction), due to its relative simplicity and high specificity. It is as sensitive as the xenodiagnosis and blood culture in the acute phase, being higher in chronic phase. Because of its unavailability in endemic areas, it has a limited use. It is particularly suitable for solving problems of discordant serology in suspected cases of chronic infection.

It may be negative in infected individuals, as it is a detection of particles in *suspension*. These may not be found in the aliquot examined. Therefore, PCR is less sensitive than the conventional serological tests, which detect the antibodies that are present in the form of a *solution*.

17.4.1.4 Other Diagnostic Methods in the Acute Phase

Amastigote nests may be detected in biopsies of the lymph nodes, skin, muscles, and heart (biopsies monitoring in cardiac transplants). The parasite is well stained by eosin-hematoxylin technique in optical microscopy, but the method is gaining much greater sensitivity and specificity with the modern techniques of immunohistochemistry as the peroxidase-antiperoxidase, radioimmunoprecipitation assay, etc.

17.4.2 Laboratory Diagnosis of the Chronic Phase

17.4.2.1 Immunological Diagnosis

In usual practice, diagnosis is based on the detection of circulating IgG antibodies by conventional techniques such as hemagglutination (IHAT), immunofluorescence (IFAT), and ELISA (enzyme-linked immunosorbent assay). They use crude or purified antigens, available in commercial kits, with good reproducibility, simplicity, and high levels of sensitivity and specificity.

In sophisticated reference laboratories, modern techniques such as Western blot, radioimmunoprecipitation assay, and recombinant immunoblot are being used. The classical CFT (Guerreiro and Machado) technique was abandoned in the 1960s. For a conclusive diagnosis, WHO recommends the use of two different techniques in the analysis of each serum, being advised against examinations in "pool."

The IHAT is quite simple, with a step of the serum dilution and another for the addition of antigen. Its "cutoff" is between titles 1:16 and 1:20, according to the kit used. The cutoff is located around 1:20 in IFAT, being frankly negative titles below 1:10 and positive equal or larger than 1:40. In ELISA, results are expressed in degrees of absorbance and were calculated according to different parameters. A ratio of 1.0 or less suggests negativity, while above 1.2 it is considered a positive serum. Nowadays, several "rapid tests" are being developed by different laboratories, aiming to optimize the diagnosis in remote places and emergent situations.

For discordant sera, a simple repetition of the exam with the same techniques may define the outcome in 95 % of patients. Persisting disagreement, the employment of nonconventional tests in reference laboratories is advised. In large-scale surveys, a simple system is the collection of blood on filter paper (Whatman no. 1), with digital puncture, for processing after elution by IHAT, ELISA, or IFAT.

The repetition of the test is recommended in all positive sera and in part (5–10 %) of negative sera. The titles of conventional tests have no correlation either with the clinical form of the patient or with the severity or prognosis. After a specific treatment, the patient that is shown to be completely and persistently negative with conventional techniques will be healed.

Up to the moment, a "gold standard" that is able to achieve absolute sensitivity and specificity is not available. False-positive results in cases of other infections and injuries may occur, as in visceral leishmaniasis, L-leprosy, in some collagenoses and in the infection with *Trypanosoma rangeli*.

Modern and alternative techniques employ recombinant antigens, purified, synthetic peptides, and mixtures able to identify multiple epitopes of the parasite, being usually qualitative and having a high specificity. Some examples are the "PaGIA" (agglutination of particles sensitized polymers with three synthetic peptide), the "Tuners," the "INNO-LIA" (that uses strips of nitrocellulose coated with recombinant antigens), the "Chembio" (rapid test by immunochromatography), etc. A new ELISA test with three recombinant antigens is very promising in terms of high sensitivity and specificity. Western blot is a very specific technique, used in reference laboratories. It consists of parasitic electrophoresis in the presence of the patient's serum, forming characteristics of antigen-antibody reaction bands in positive cases.

17.4.2.2 Parasitological Diagnosis

Parasitological methods are indicated in research of drugs and in rare cases of diagnostic clarification. Due to the shortage of circulating parasites, only indirect methods are employed, with sensitivity between 30 (xenodiagnosis) and 80 % (blood cultures, PCR). A combination of xenodiagnosis and blood cultures may increase the sensitivity and facilitate the isolation of the parasite. As for PCR, its specificity is virtually absolute, but its sensitivity never reaches 100 %, mainly in the chronic infection phase.

17.4.2.3 Other Procedures for Laboratory Diagnosis

In immunosuppressed and reactivated cases, cutaneous affections and cerebrospinal fluid may contain several parasites, which are detectable by biopsy and culture. In individuals who died with suspicion of HCD, the conventional serological tests in pericardial fluid are feasible and may be complemented by the parasite finding in lesions on the myocardium and other organs, especially through immunohistochemical techniques.

The World Health Organization (WHO 2014) recommends that individual diagnosis of Chagas disease should be done using two tests based on different principles and different antigens. This is due to the suboptimal sensitivity and/or specificity reported for the existing tests. Diagnosis of acute, congenital, and reactivated forms of Chagas disease is based on direct detection of parasites, while diagnosis of chronic disease is generally based on serological testing. Although a number of methods are available, current diagnostic approaches are insufficient to respond to aspects of the disease such as congenital Chagas, infection in immunocompromised patients, and monitoring the treatment. Therefore, there is a growing need to develop and market new diagnostic tests that may be used in the remote areas where the disease is endemic and which perform well enough to guide clinical care.

17.5 Specific Treatment of Chagas Disease

For infected individuals, in addition to the symptomatic and social management, the specific treatment aiming to *T. cruzi* elimination has assumed increasing interest and importance in recent decades. The main practical aspects are presented right away (Maguire 1999; Carlier et al. 2002; WHO 2002; Brasil 2005; Rassi et al. 2009b; Dias et al. 2013; Viotti and Vigliano 2014; Coura et al. 2014; Vilar-Pereira et al. 2014).

The treatment should be considered in most existing cases of HCD, since it is estimated that the current available drugs might eliminate or reduce at least 43.7 % of the parasite load, possibly decreasing morbidity and mortality. This would lead to a decrease in the burden of the disease by reducing public health costs and preventing the economic and social impact produced by the deterioration of the quality of life of infected people; the expenses in heart disease drugs, pacemakers, defibrillators, transplants, etc. caused by the symptomatic treatment of the chronic phase would also decrease.

Treatment of Chagas disease currently relies on only two drugs: benznidazole and nifurtimox. It has been estimated that both drugs achieve parasitological cure in 60–85 % of acute cases. Treatment is mandatory during the acute phase of infection (including congenital and reactivated infections) and for all chronically infected patients without advanced cardiomyopathy.

Nifurtimox (3-methyl-4-(5'-nitrofurylideneamino)-tetrahydro-4H-1,4-thiazine-1,1-dioxide) and *benznidazole* (N-benzyl-2-nitro-1-imidazoleacetamide) are active in animals and in humans, mainly in acute phase and in recent chronic situations. Both drugs are administered orally, having a mean life of 12 h in peripheral blood. The parasiticidal activity basically consists of inhibiting protein and RNA synthesis in peripheral blood and inside infected cells. Both products have hepatic metabolism and are eliminated in the urine.

17.5.1 Drug Presentation, Dosage, and Indications

- *Benznidazole*: tablets of 100, 50, and 12.5 mg, administered in daily doses of 5 mg/kg body weight (adults) and 7–10 mg/kg/day for children, twice a day (every 12 h), for 60 days. The mean half-life of elimination varies between 10.5 and 13.6 h.
- *Nifurtimox*: tablets of 120 mg, for daily doses of 7–10 mg/kg (adults) and 12–20 mg/kg (children), in two or three daily doses for 60 days.

Current indications for the use of these drugs may be summarized as follows:

- All acute and subacute cases, including congenital
- Chemoprophylaxis (in laboratory accidents and transplants)
- Episodes of reactivation in immunosuppressed individuals
- Low age and recent chronic cases
- Chronic indeterminate and initial clinical forms (experimental)

Notes: (a) chronic clinically advanced cases should not be treated; (b) in case of laboratory accident, a 10-day prophylactic treatment must start immediately, thus preventing parasitic invasion; (c) eventual cross-drug resistance of some strains of *T. cruzi* may occur, depending on these products.

Management of both drugs is relatively simple and should be administered on ambulatory situations under medical supervision, in most cases. The contraindications basically apply for pregnant women and people with hepatic or renal failure and severe neurological problems. During the treatment, the *ingestion of alcohol is prohibited*. Medical attention is needed on side effects, febrile episodes, and emergence of leukopenia. A simple leukocyte count is required every 15 or 20 days.

17.5.2 Treatment Effectiveness and Clinical Benefits

Taking into account that the effectiveness of available drugs may vary according to the region and the patient profile, it oscillates as follows:

- Acute cases: 50–90 %
- Congenital: 90–100 %
- Chemoprophylaxis (accidents and transplants): 100 %
- Recent chronic cases: 50–75 %
- Old chronic cases: 8 % or more, if evaluated between 15 and 20 years after treatment

The clinical benefit of specific clinical treatment is substantial in acute and congenital cases, possibly also in recent chronic and in indeterminate form. There are no data on tangible long-term benefits in chronic initial forms. In cases of reactivation, the effects on the parasitism and the clinical picture are favorable in the short term, but the incidence of recurrence is usually high.

In addition to its importance as a form of primary prevention to reduce or prevent new cases of congenital Chagas disease, the recommendation of indicating trypanocidal treatment would also serve as a secondary prevention mechanism due to the protective effect observed during the clinical course. By eliminating or reducing at least 43.7 % of the parasite load, the risks of morbidity and mortality of the infection would decrease significantly. This would lead to a decrease in the burden of the disease by reducing public health costs and preventing the economic and social impact produced by the deterioration of the quality of life of infected people; the expense in heart disease drugs, pacemakers, defibrillators, transplants, etc. caused by the symptomatic treatment of the chronic phase would decrease too.

17.5.3 Adverse Effects

The side effects of both drugs affect approximately 30 % of patients, being more frequent and intense in adults. As for benznidazole, the predominant effects are dermatological (urticarial scaly plaques) and gastrointestinal, having also clinically relevant (although rare) peripheral neuropathy and a depletion of leukopoiesis. All these effects are temporary and reversible with the suspension of the drug.

Dermatopathy usually affects the end of the first week, with a peripheral polyneuropathy in the last few weeks of treatment. In the case of an important leukopenia, drug suspension and corticoid administration are suggested. For nifurtimox, anorexia, gastrointestinal, dermatological, and neuropsychological problems may occur.

17.5.4 Cure Certification

In practice, a single final criterion is consensus: the *complete and permanent negative conventional serological tests*. Parasitological tests and PCR will only have value if positive, when they indicate treatment failure. Clinical improvement has some value in acute cases or reactivation, but it does not confirm infection cure.

The dosage of antibodies able to detect live trypomastigotes via complement-mediated lysis and flow cytometry is a promising method, and its negative results precede the disappearance of conventional antibodies.

17.5.5 New Drugs

Several compounds have been investigated, such as some antifungal drugs (posaconazole, ravuconazole, itraconazole) that prevent the synthesis of ergosterol by the parasite. Other targets in test concern with metabolic enzymatic pathways essential to the parasite, such as trypanothione reductase, DNA topoisomerase, cysteine proteinase, dihydrophosphate reductase, inhibitors, etc. New therapeutic alternatives for *T. cruzi* elimination have also been evaluated. Some unspecific drugs might be highly valuable as adjuvants in ST, acting in the chain of inflammatory response. For instance, cytokine TNF (found in high levels among patients with advanced cardiomyopathy) plays an important role stimulating the inflammatory process. Some drugs used in autoimmune diseases, such as *infliximab G*, are able to inhibit TNF activity, thus reducing the intensity of inflammatory process. Benefits in infected rats have been stimulating, opening up new possibilities in HCD treatment.

17.6 Epidemiology

17.6.1 Distribution

HCD is endemic in 18 Latin American countries, with approximately 40–50 million people still exposed to the risk of infection (OPAS 2006; Rassi et al. 2009b). Table 17.4 summarizes in general groups the countries where the American trypanosomiasis occurs naturally, according to some basic epidemiological characteristics.

HCD is originally from Latin America, but in the recent decades, it has been increasingly

Table 17.4 General epidemiological characteristics of American trypanosomiasis in Latin America (Modified from Carlier et al. 2002)

Groups	Countries	Epidemiological characteristics	Observations
I	Argentina, Bolivia, Brazil, Chile, Ecuador, Honduras, Paraguay, Peru, Uruguay, and Venezuela	Domestic cycles with areas of high prevalence of human infection. Important prevalence of chronic chagasic cardiomyopathy; digestive forms existing in the south of the equatorial line. Wild cycles important in various natural environments. Sylvatic populations of *T. infestans* detected in restricted areas of Bolivia and Chile	Vector and transfusion control programs exist, under different conditions of coverage and sustainability. The elimination of domestic *T. infestans* and *R. prolixus* is an attainable goal
II	Colombia, Costa Rica, and Mexico	Domestic and sylvatic cycles exist. There are areas with significant HCD prevalence and infected blood donors. Chronic cardiomyopathy is detected	Control programs are starting or absent
III	El Salvador, Guatemala, Nicaragua, and Panama	Moderate HCD prevalence. Lack of detailed clinical information, in spite of detected cases of heart disease	The program started in Guatemala and Honduras in 2009
IV	Antilles, Bahamas, Belize, Cuba, USA, Guyana, Haiti, Jamaica, Suriname	Sylvatic cycles exist. Rare human cases are detected. Very little clinical information	Control programs are absent

detected in the United States of America, Canada, and several European and some Western Pacific countries (Fig. 17.4). This is due mainly to the migration of infected individuals from endemic to non-endemic countries, incurring medical costs and the possibility of *T. cruzi* transmission by congenital, blood transfusion and organ transplantation routes (Carlier et al. 2002; Rassi et al. 2009b; Coura et al. 2014; WHO 2014).

Currently, there is a general trend of concentrating chronic cases in higher age groups. Nevertheless, in some regions of Bolivia, Paraguay, and Central America, active vector transmission still occurs, prevailing mainly among young people. In older patients it is common to observe the superposition of other chronic degenerative diseases. Regarding the coinfection of HCD with HIV, myocarditis and encephalitis are usually frequent and very severe (Dias et al. 2013; Carlier et al. 2002).

Since the second half of the twentieth century, stringent measures of control have led to large reductions in the transmission over a wide endemic region. Nevertheless, the still existing possibilities of *T. cruzi* transmission and the existence of 8–10 million infected people make it essential to prevent and manage this trypanosomiasis and the dissemination of the knowledge of it.

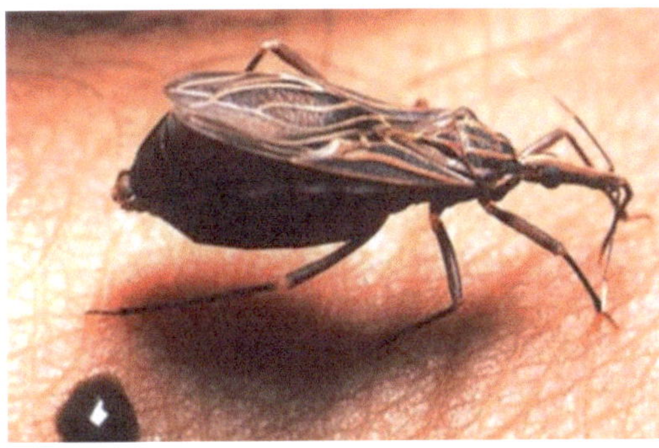

Fig. 17.7 Bloodsucking adult *Rhodnius prolixus* and its possible contaminant feces on human skin (WHO 2014)

17.6.2 Natural Vectors and Reservoirs of *T. cruzi*

In its original sylvatic cycle, the parasite has been circulating in sylvatic conditions for millions of years among invertebrate vectors and several species of susceptible mammals.

17.6.2.1 Vectors

Vectors of American trypanosomiasis are insects of the order Hemiptera, family Reduviidae, and subfamily Triatominae, with a wide distribution in different ecotopes and natural regions of tropical and subtropical regions in the American continent, chiefly between latitudes 40° N and 45° S, at altitudes of up to 1500 m (Fig. 17.7).

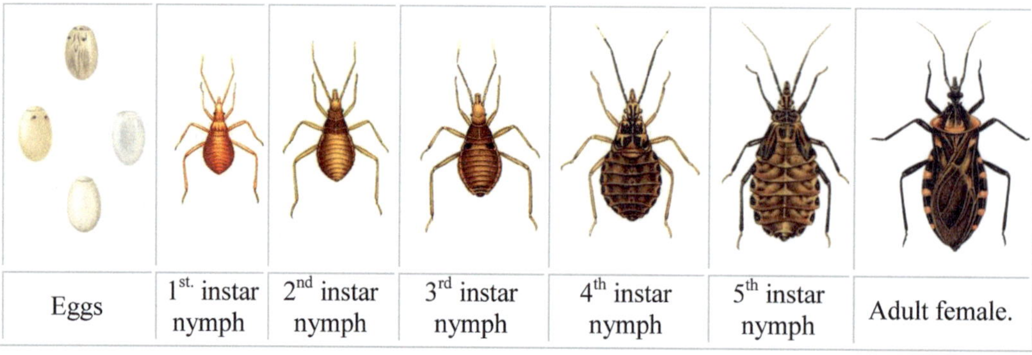

Fig. 17.8 Evolutive cycle of *Panstrongylus megistus* (From Carcavallo et al. 2000)

Among people, they are commonly known as *kissing bugs* (USA); *barbeiros*, *bicudos*, or *chupões* (Brazil); *vinchucas*, *chipos*, *chirimachas*, or *chinches* (Hispanic America), etc. (Schofield 1994).

More than 140 species of triatomine bugs have been described, most of them living in association with birds and several other wild vertebrates. At the medical and social standpoints, less than twenty species are able to invade and colonize artificial ecotopes and transmit the parasite to human beings. They have a relatively large size and are strictly hematophagous, eventually performing cannibalism and coprophagy. Their life span is relatively long (1–2 years) and with a paurometabolous cycle[1] of five nymphal instars and an adult stage. The adult stage is characterized by the emergence of the wings and the complete sexual differentiation. The egg-egg cycle may last from a few months to 2 years, depending on the species and environmental conditions, particularly temperature, humidity, and food availability. Figure 17.8 shows the complete cycle for *Panstrongylus megistus*, according to the Atlas of Chagas disease vectors in the Americas (Carcavallo et al. 2000).

Most species live in sylvatic environments, having nocturnal habits and reduced ability to fly.

[1]Immature stage (nymphs) of paurometabolous insects (bugs, lice, etc.) lives in the same environment of adults, while those of hemimetabolous ones (dragonflies, stoneflies, etc.) live in different environments and in both cases have similar feeding habits (CBM).

Female adults are generally more active than the males, with greater dispersion capacity and longevity. Females usually have a period of 3–4 months of egg laying and a final production of 100–200 eggs per year. Eggs are hatched on average between 18 and 25 days, according to the species and the environmental conditions.

Regarding the best ability to transmit HCD, the following characteristics of triatomine species are particularly important (Schofield 1994):

- Capacity of hosting and good reproduction in artificial ecotopes
- High degree of anthropophily and ingestion of significant amounts of blood
- Short time for meal and a short interval between the meal and the defecation
- High susceptibility to infection by *T. cruzi* and high capacity of metacyclogenesis

Thus, the main species involved in the transmission of HCD would be *Triatoma infestans* (in the South American Cone) and *Rhodnius prolixus* (Venezuela, Colombia, and Central America), followed by *T. dimidiata* (Andean countries, Central America), *Panstrongylus megistus* (mainly in Brazil), *T. brasiliensis* and *T. pseudomaculata* (northeastern Brazil), *T. maculata* [Venezuela and Roraima State (Brazil)], and *T. barberi*, *T. pallidipennis*, and *T. longipennis* (Mexico) (Fig. 17.9). Other species have a sylvatic origin, but present a small potential of housing invasion and colonization such as *T. sordida*, *T. tibiamaculata*, *T. vitticeps*, *R. neglectus*, *R.*

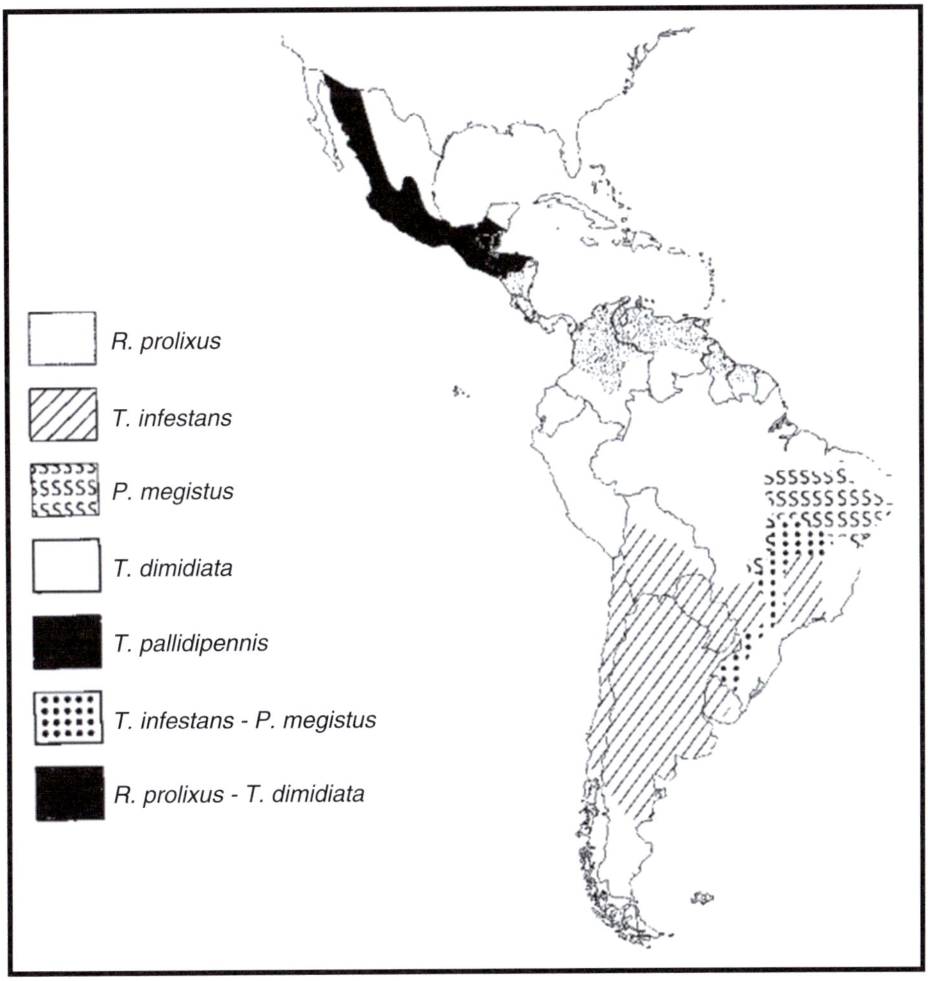

Fig. 17.9 Main Chagas disease vectors in Latin America (From WHO 2014)

ecuadoriensis, *R. nasutus*, other species of the *phyllosoma* complex, etc. (Carcavallo et al. 2000; Carlier et al. 2002). On the other hand, there are exclusively sylvatic species, which never invade the human environment, such as *Cavernicola* spp., *Psammolestes* spp., *Belminus* spp., *Dipetalogaster maximus*, *Microtriatoma* spp., etc. (Carcavallo et al. 2000; Dias et al. 2013).

In its dynamics, vector transmission of *T. cruzi* is considered a slow and difficult process, from the differentiation to metacyclic trypomastigotes in the insect until the penetration and reproduction in cells of the vertebrate host. Dryness of vector dejections, thickness and nonspecific defense systems of the vertebrate skin, heterophilic antibodies, and specific/nonspecific defense systems, such as properdins, are considered concrete barriers to parasite transmission (Brener 1997; Souza 2000). In general, the highest rates of vector transmission occur in large concentrations of vectors and parasites. In the domestic cycle, situations involving the presence of *T. infestans* and *R. prolixus* are the most remarkable, mainly in the case of long-standing dwelling infestation. A similar phenomenon is observed in the sylvatic cycle, when reduced, natural spaces provide a high concentration of triatomines and trypanosomes, inducing a strong pressure of transmission to susceptible vertebrates (Schofield 1994; Dias and Schofield 2004).

17.6.2.2 Vertebrate Hosts

In general, wild reservoirs and several domestic ones become infected with the contamination by excretions of infected triatomine or, orally, by the ingestion of infected triatomines and/or other mammals. There is a large variety of infection rates of natural reservoirs, due to different factors such as resistance/susceptibility, environmental vector, density, etc. (Antunes 1999; Dias et al. 2013).

Sylvatic reservoirs, which under bio-eco-ethological circumstances approach the human dwelling, play an important role in the natural chain. Classified as *synanthropic*, the most important ones are marsupials and rodents, followed by some edentulous and chiropteran.

Populations of *T. cruzi* have been more frequently detected in groups I and III in wild reservoirs. Nevertheless, group II has also been isolated in synanthropic reservoirs, which characterizes the interaction among different cycles of the parasite.

Among the natural reservoirs, parasitemia is usually high in young and newly infected animals, sometimes being detected by direct examination. In summary, the main orders and species of mammals, which are carriers of the parasite in the sylvatic cycle, are as follows (Carlier et al. 2002; WHO 2002):

- Marsupialia (*Didelphis albiventris*, *D. azarae*, *D. marsupialis*, etc.)
- Edentata (*Dasypus novencintus*, *D. mexicanus*, *Bradypus infuscatus*, etc.)
- Chiroptera (*Phyllostomus hastatus*, *Carollia perspicillata*, *Desmodus rotundus*, etc.)
- Carnivora (*Cerdocyon thous*, *Dusycon griseus*, *D. vetulus*, *Urocyon cinereoargenteus*, *Nasua* sp., etc.)
- Rodentia (*Sciurus* spp., *Akodon* spp., *Neotoma* spp., etc.)
- Primates (*Alouatta* spp., *Ateles* spp., *Callicebus* spp., etc.)
- Lagomorpha (*Sylvilagus* spp.)

In this list, the main reservoirs of HCD are marsupials and rodents, being found infected in almost all Latin American countries. Several of them, although having a wild origin, are very close to humans and come to live indoors, particularly in times of harvest and grain storage. Between 5 and 20 % of some American opossums (*Didelphis* spp.) harbor in their anal glands a particular cycle of *T. cruzi* (Deane et al. 1984).

From a practical point of view, the elimination of vertebrate reservoirs is not considered effective in the control of HCD, although this environmental control, from the perspective of their detachment from human dwellings, has a relative interest (Dias and Schofield 2004).

In the domestic cycle, the most important reservoirs are usually human beings and the small- and medium-sized mammals that live nearby humans, mainly dogs and cats. There are also several domestic rodents, particularly *Rattus rattus* (common house rat), *Rattus norvegicus* (sewer rat), *Mus musculus* (mouse), and *Cavia porcellus* (guinea pig), which may directly infect bigger mammals when eaten by them. Rabbits and guinea pigs raised by Andean people have particular importance, due to the hundreds of triatomines captured in their nurseries. Pigs and goats, although possibly infected, are not important reservoirs, as their parasitemia is extremely low and transient (Antunes 1999; Dias et al. 2013).

Among domestic reservoirs, a gradual reduction of their rates of natural infection with *T. cruzi* has been observed after the control of domiciliated triatomines. For example, rates in cats and dogs from endemic areas of Minas Gerais and Sao Paulo (Brazil) were between 5 and 30 % in the 1950s, but today they rarely exceed 0.5 % (Coura and Dias 2009). These data show the fundamental importance of triatomine infestation in the maintenance and expansion of the domestic cycle, and they also show the minor importance of congenital transmission among these reservoirs.

Considering pathogenicity, domestic animals have variable degrees of damage in experimental studies, developing acute forms, arrhythmias, congestive heart, esophagus and colon dysfunction, dysautonomia, etc., in dogs, monkeys, rabbits, and rodents (Andrade 2013; Dias et al. 2013).

17.7 Prophylaxis

Prophylaxis of HCD must have high priority in Latin American endemic countries because of its high social and economic impact and of its technical feasibility regarding the transmission interruption. Since the 1950s, based on the works of Emmanuel Dias, Pedreira de Freitas, Cecílio Romaña, Victor Nussenzweig, and other pioneer authors, preventive measures have been considered the first strategy to face HCD, mainly by means of vector and blood bank control (Dias and Schofield 1999). In recent years, epidemiological surveillance and specific treatment have been incorporated (Miles et al. 2004; Coura and Dias 2009).

Overall, the classic levels of primary, secondary, and tertiary prevention should be considered in the prophylaxis of HCD. At primary level, as a safe and effective vaccine is not available, vector control and blood banks are the most impactful procedures. At the primary level, interventions are designed to prevent parasite transmission. At the secondary level, they aim to prevent or inhibit the disease in cases where infection has occurred, whereas in the tertiary level they seek to limit human incapacity and reintegrate infected people into the community. On the other hand, several observations and mathematical models indicate that social and political development of the endemic regions might be sufficient to control HCD. Yet this route is complex in less developed regions; specific technical interventions must be made against the most vulnerable levels in the cycle of the parasite.

Current preventive measures focus on the following main approaches (Dias and Schofield 2004):

1. Elimination of vectors in domestic settings by means of insecticides, home improvement, and educational campaigns
2. Screening blood donors
3. Maternal screening and treatment of infected newborns
4. Active acute case finding, notification, and treatment, with active and immediate epidemiological prospection in family and contacts (particularly important in oral outbreaks)
5. Chemoprophylaxis in the case of laboratory accidents and organ transplants with infected donors
6. Specific massive treatment of young infected individuals

Considering briefly each one, it might be summarized as follows.

17.7.1 Vector Control

The objective is to eliminate contact between domestic vectors and susceptible people. The ideal strategy involves regular systemic chemical vector control in parallel with physical housing improvement, to be performed continuously in contiguous endemic areas and followed by epidemiological surveillance (Dias and Schofield 2004). The elective insecticides are synthetic alpha-cyano-substituted pyrethroids generally applied as wettable powder or flowable suspension concentrate formulations, which have an effective residual action. Some currently employed pyrethroids are shown in Table 17.5.

Resistance to pyrethroids has been detected in certain areas of Bolivia and Argentina, a situation

Table 17.5 Formulations, LD_{50}, and field concentration of the commonly used pyrethroids for indoor spraying against Chagas disease vectors

Insecticide	Formulation	LD_{50}[a]	Field concentration[b]
Deltamethrin	Suspension concentrate	1.54	25.00
Lambda-cyhalothrin	Wettable powder	0.11	30.00
Cypermethrin	Wettable powder	2.86	125.00
Beta-cyfluthrin	Suspension concentrate	0.32	25.00
Beta-cypermethrin	Suspension concentrate	1.56	50.00

[a]μg/g body weight
[b]mg of active ingredient/m^2 of treated surface

in which their substitution for other chemical groups (mainly carbamates and organophosphate products) is recommended. To improve residual action, modern microencapsulated formulations are being tested (Coura and Dias 2009).

Chemical vector control methods are based on three main phases of intervention: preparatory, attack, and surveillance. After the geographical and entomological mapping of the region, the attack phase generally involves large-scale spraying of all existing premises, a work that must be complemented by following regular inspections and selective spraying of residual foci (systemic approach). In lower degrees of infestation (5 % or less), the surveillance phase must be implemented, based on effective community participation in focal detection and good housing maintenance, under decentralized (municipal) organization. This is considered the most critical phase of the HCD control program, because of several problems concerning its continuity and sustainability (Dias and Schofield 2004).

Following this general schedule, results have been very good in the Southern Cone of America, Venezuela, Guatemala, and Honduras. In places where vector control has been satisfactorily implemented, acute cases of HCD are no longer seen, and serology screening tests of young children (born since the launch of the program) show no or very low rates of infection. At short/medium term, vector control will also have a positive impact on other transmission routes, by progressive reduction of infected blood donors and women of childbearing age (Antunes 1999; Coura and Dias 2009; Moraes-Sousa and Silva 2009; Dias et al. 2013).

Since 1991 a new political and technical approach to improve national control programs in endemic areas has been stimulated by PAHO/WHO: the multinational initiatives for vector and blood bank control. The cooperation of efforts, techniques, epidemiological data, etc. among well-defined macro-regions (Southern Cone, Andean countries, Central America and Mexico, Amazonian countries) has been officially established by the respective countries, with PAHO technical assistance.

17.7.2 Control of Transfusion Transmission

In endemic countries, the most effective approach to avoid blood-transmitted HCD consists of screening all blood donors using at least two serological tests performed using different techniques. Seropositive candidates must be excluded as blood donors and delivered to the local or regional health system to receive adequate medical and social assistance. In some situations of emergence or in the particular need to use blood from an infected individual, chemoprophylaxis is indicated, with the addition of trypanocidal drugs, such as gentian violet to stored blood (Brasil 2005; Moraes-Sousa and Silva 2009; CDC 2013).

17.7.3 Preventing Congenital HCD

To this day, prevention during pregnancy has not been possible because the available drugs are toxic and teratogenic and also because preventive abortion is unacceptable. The best procedure consists of an early diagnosis of the newborn followed by a specific treatment with benznidazole or nifurtimox. The ideal schedule is to implement routine serological screening in pregnant women in endemic areas (or coming from endemic areas), following the newborns of positive mothers by means of clinical and laboratorial procedures. Clinical indicators are prematurity, fever, persistent tachycardia, and hepatosplenomegaly, but in most cases, babies are asymptomatic. To suspected cases, parasitological procedures are indicated, chiefly in the first weeks of life, by means of microhematocrit technique and/or indirect tests (xenodiagnoses and hemocultures). Positive cases must be immediately treated. Conventional serology is not a diagnostic test for newborns because they will remain seropositive for 6 or 7 months due to the passive transference of IgG antibodies from the infected mother. The best option, in the case of negative parasitological tests, is to use conventional serology after the baby is 7 or 8 months old (when maternal antibodies will have disappeared), immediately treating those with positive results. In recent studies, thc specific treatment of young chagasic women, before pregnancy, is being strongly recommended to reduce the chances of congenital transmission in future pregnancies (Brasil 2005; Torrico 2014).

17.7.4 The Prevention in Organ Transplantation

Serology in endemic areas is mandatory for organ donors and recipients. In the case of a positive donor and a susceptible recipient, the best option is to discharge the donor. Nevertheless, in urgent or specific cases of histocompatibility, the transplant must proceed, and the best strategy is the specific treatment of the donor (ideally 10 days before the surgery) and of the recipient the day before the surgery and for 9 days afterward (Dias and Schofield 2004; WHO 2014).

17.7.5 Preventing Laboratory Accidents

Laboratories dealing with *T. cruzi* should incorporate a very severe routine of personal training and security rules, under periodic supervision of reference centers. They should have a restricted entrance, and the use of gloves, protective goggles, masks, closed shoes, and long-sleeved gowns should be obligatory for technicians working closely with the parasite. The periodic serological checks of the staff are also highly recommendable. In case of accidents involving the skin or mucosa, immediate local sterilization must be provided. In case of eye contamination, silver nitrate washes are indicated. If a person is likely to have been infected, an immediate expected specific treatment should be given for at least 10 days (Dias et al. 2013).

17.7.6 Massive Treatment of Young Infected Individuals

To this day, evidences of the positive benefit of the treatment in low age groups have been accumulated, considering the possibility of cure as much as the interruption (or delaying) of the disease evolution. Taking into account these premises, in parallel with the relatively little problems of side effects, some countries such as Argentina and Bolivia are including the treatment of young people in their national Chagas disease programs. In WHO and in multinational initiatives for HCD prophylaxis, this idea is also being considered to be implemented worldwide soon, with strong possibilities to expand the treatment to individuals in the chronic indeterminate form and even to cases of initial cardiomyopathy (see The Natural History of HCD above).

17.8 Final Considerations and Expected Future

Many accounts of HCD still present a fatalistic picture of an incurable infection consigning millions of Latin American rural, ignorant, and extremely poor individuals, a neglected disease with no or very little possibilities to be treated or prevented. However, this picture has changed in recent years. Advances in prophylaxis, diagnosis, and clinical attention have led to a much better disease management and to an increasing awareness that transmission may be interrupted and the clinical course of infection attenuated. A subtle change from the classical paradigm of a neglected disease to a disease of neglected populations is arising all over the world, stimulating hopes and providing dignity to the victims of *T. cruzi* as well as a series of political and scientific challenges to governmental and academic institutions. New possibilities for HCD prophylaxis and treatment are available, and more and more depend on political will and appropriate public health organization. Transfusion and congenital transmission are being highly minimized and will disappear or be eliminated in the next 20 or 30 years. Vector should be controlled in endemic areas, still largely dependent on regional and local decentralized health systems by means of an appropriate epidemiological surveillance. Oral transmission probably will continue to appear in eventual outbreaks, but under an expected trend in focus and reduction, depending chiefly on social advances and a better environmental management. In the future, infected individuals will remain more and more in higher age groups in parallel with the progressive reduction of the universal prevalence of infection. Optimistic epidemiological data together with other social priorities, such as underemployment, violence, global market, dengue fever, AIDS/HIV, degenerative diseases, etc., pose a considerable risk to minimize the importance of American trypanosomiasis. Thus, the reduction of human and financial resources to face HCD is predictable. As a whole, the great challenge now is to maintain HCD in the governmental and academic agendas of affected countries, for the next two or three decades.

References

Andrade SG (2013) Biodemas, zimodemas e esquizodemas: sua relação com a patologia da doença de Chagas. In: Coura JR (ed) Dinâmica das doenças infecciosas e parasitárias, 2nd edn. Guanabara Koogan, Rio de Janeiro, pp 669–686

Andrade AA, Andrade SG (1999) Pathology and pathogenesis. In: Gilles HM (ed) Protozoal diseases. Arnold, London, pp 323–336

Antunes CMF (1999) The epidemiology of Chagas disease. In: Gilles HM (ed) Protozoal diseases. Arnold, London, pp 351–369

Brasil (2005) Consenso Brasileiro sobre a doença de Chagas. Rev Soc Bras Med Trop 38 (Suppll III):1–39

Brener Z (1997) *Trypanosoma cruzi*: morfologia e ciclo evolutivo. In: Dias JCP, Coura JR (eds) Clinica e Terapêutica da doença de Chagas. Uma abordagem prática para o clínico geral. Fiocruz Ed, Rio de Janeiro, pp 25–32

Carcavallo RU, Girón IG, Jurberg J et al (2000) Atlas of vectors of Chagas' disease, vol 1. Fiocruz Editora, Rio de Janeiro

Carlier Y, Dias JCP, Luquetti AO et al (2002) Trypanosomiase américaine ou maladie de Chagas. Enciclop Médico-Chirurg 8:505–520

CDC (Centers for Disease Control and Prevention) (2013) Chagas disease. www.cdc.gov/parasites/chagas/

Coura JR, Dias JCP (2009) Epidemiology, control and surveillance of Chagas disease: 100 years after its discovery. Mem Inst Oswaldo Cruz 104(Suppl I):31–40

Coura JR, Albajar-Viñas P, Junqueira ACV (2014) Ecoepidemiology, short history and control of Chagas disease in the endemic countries and the new challenge for non-endemic countries. Mem Inst Oswaldo Cruz 109:856–862

Deane MP, Lenzi HL, Jansen AM (1984) *Trypanosoma cruzi*: vertebrate and invertebrate cycles in the same mammal host, the opossum *Didelphis marsupialis*. Mem Inst Oswaldo Cruz 79(Suppl 1):513–515

Dias JCP (1990) Doença de Chagas: clínica e terapêutica. Ministério da Saúde (SUCAM), Brasília

Dias JCP, Schofield CJ (1999) The evolution of Chagas disease (American Trypanosomiasis) control after 90 years since Carlos Chagas discovery. Mem Inst Oswaldo Cruz 94 (Suppl I):103–121

Dias JCP, Schofield CJ (2004) Control of Chagas disease. In: Maudlin I, Holmes PH, Miles MA (eds) The trypanosomes. CABI, London, pp 547–564

Dias JCP, Amato Neto V, Luna EJA (2011) Mecanismos alternativos de transmissão do *Trypanosoma cruzi* no Brasil e sugestões para sua prevenção. Rev Soc Bras Med Trop 44:375–379

Dias JCP, Pereira JB, Macedo VO (2013) Doença de Chagas. In: Coura JR (ed) Dinâmica das doenças infecciosas e parasitárias, 2nd edn. Guanabara Koogan, Rio de Janeiro, pp 606–641

Dutra WO, Gollob KJ (2008) Current concepts in immunoregulation and pathology of human Chagas disease. Infect Dis 21:287–292

Ferreira MS (2013) Infecções no hospedeiro imunocomprometido. In: Coura JR (ed) Dinâmica das Doenças Infecciosas e Parasitárias, 2nd edn. Guanabara Koogan, Rio de Janeiro, pp 209–219

Kierszenbaum F (2007) Mechanisms of pathogenesis in Chagas disease. Acta Parasitol 52:1–12

Köberle F (1968) Chagas' heart disease and Chagas' syndromes: the pathology of American trypanosomiasis. Adv Parasitol 6:63–116

Laranja FS, Dias E, Nóbrega GC et al (1956) Chagas' disease. A clinical, epidemiologic and pathologic study. Circulation 14:1035–1060

Maguire JH (1999) Clinical features, diagnosis and treatment. In: Gilles HM (ed) Protozoal diseases. Arnold, London, pp 336–350

Miles MA, Yeo M, Gaunt MW (2004) Epidemiology of American Trypanosomiasis. In: Maudlin I, Holmes PH, Miles MA (eds) The Trypanosomiasis. CAB International, Oxfordshire, pp 243–252

Moraes-Sousa H, Silva MMF (2009) Control of transfusion transmission of Chagas disease in different

Brazilian regions. Rev Soc Bras Med Trop 42(Suppl II):103–105
Oliveira W Jr (1997) O cardiopata chagásico em situações especiais. In: Dias JCP, Coura JR (eds) Clinica e Terapêutica da doença de Chagas. Uma abordagem prática para o clínico geral. Fiocruz Editora, Rio de Janeiro, pp 297–322
OPAS (Organización Panamericana de la Salud) (2006) Estimacion cuantitativa de la enfermedad de Chagas en las Americas. OPAS, Montevideo
Rassi A Jr, Dias JC, Marin-Neto JAA et al (2009a) Challenges and opportunities for primary, secondary, and tertiary prevention of Chagas' disease. Heart 95:524–534
Rassi A Jr, Rassi A, Marin-Neto J (2009b) Chagas heart disease: pathophysiologic mechanisms, prognostic factors and risk stratification. Mem Inst Oswaldo Cruz 104(Suppl 1):152–158
Rezende JM, Moreira H (2004) Forma digestivas da doença de Chagas. In: Castro LP, Coelho LGV (eds) Gastroenterologia. Medsi, Rio de Janeiro, pp 325–392
Schofield CJ (1994) Triatominae: biology and control. Eurocommunica, Sussex
Silva AL (1997) Formas digestivas: tratamento cirúrgico. In: Dias JCP, Coura JR (eds) Clinica e Terapêutica da doença de Chagas. Uma abordagem prática para o clínico geral. Fiocruz Editora, Rio de Janeiro, pp 177–200
Souza W (2000) O parasito e sua interação com os hospedeiros. In: Brener Z, Andrade ZA, Barral Neto M (eds) *Trypanosoma cruzi* e doença de Chagas, 2nd edn. Guanabara Koogan, Rio de Janeiro, pp 88–126
Torrico F (2014) Chagas congénito. In: Viotti RJ, Vigliano CA (eds) Enfermedad de Chagas. Un enfoque práctico basado en la investigación médica. Editorial Panamericana, Buenos Aires, pp 117–128
Vilar-Pereira G, Silva AA et al (2014) Tumor necrosis factor is a therapeutic target for immunological unbalance and cardiac abnormalities in chronic experimental Chagas-Heart disease. Mediators Inflamm 2014:1–16. doi:10.1155/2014/798078
Viotti RJ, Alvarez MG (2014) Métodos diagnósticos. In: Viotti RJ, Vigliano CA (eds) Enfermedad de Chagas. Un enfoque práctico basado en la investigación médica. Editorial Panamericana, Buenos Aires, pp 155–172
Viotti RJ, Vigliano CA (2014) Tratamiento etiológico. In: Viotti RJ, Vigliano CA (eds) Enfermedad de Chagas. Un enfoque práctico basado en la investigación médica. Editorial Panamericana, Buenos Aires, pp 253–262
WHO (2002) Control of Chagas disease. Second report of the WHO Expert Committee. WHO Technical Report Series 905, Geneva
WHO (World Health Organization) (2014) Chagas disease. http://www.who.int/chagas/en/
Zingales B, Andrade SG, Briones MRS et al (2009) A new consensus for *Trypanosoma cruzi* intraspecific nomenclature: second revision meeting recommends TcI to TcVI. Mem Inst Oswaldo Cruz 107:1051–1054

Sleeping Sickness and Nagana Disease Caused by *Trypanosoma brucei*

18

Dietmar Steverding

Abstract

The hemoflagellate *Trypanosoma brucei* is the causative agent of human and animal African trypanosomiasis, also known as sleeping sickness and nagana disease, respectively. The infectious disease is transmitted by the bite of infected tsetse flies and afflicts mainly rural populations in sub-Saharan Africa. The subspecies *T. b. gambiense* and *T. b. rhodesiense* are responsible for the two forms of human African trypanosomiasis, the West and East African sleeping sickness, respectively. A third subspecies, *T. b. brucei*, is only infective for animals. The disease progresses in two stages. In the first stage, the parasites are restricted to the blood and lymphatic system, while in the second stage they invade the central nervous system. The presence of the parasites in the brain of humans is associated with disturbance of the sleep-wake cycle which is the main characteristic symptom of second-stage sleeping sickness and in many languages gives the disease its name. Without treatment, infected humans and animals will die within months or years. Only few drugs are available for treatment of sleeping sickness and nagana disease, but most of them are difficult to administer and show serious side effects. Although *T. brucei* has been extensively studied and many unique molecular and cellular features have been discovered for this organism, no new drugs have been developed for treatment of sleeping sickness and nagana disease since the 1970s. New methods for diagnosis, therapy, and vector control are needed if eradication of the devastating disease is to be achieved.

Keywords

Human African trypanosomiasis • Sleeping sickness • *Trypanosoma brucei* • Nagana • Cattle

D. Steverding (✉)
Norwich Medical School, University of East Anglia,
Bob Champion Research and Education Building,
James Watson Road, Norwich NR4 7UQ, UK
e-mail: dsteverding@hotmail.com

C.B. Marcondes (ed.), *Arthropod Borne Diseases*, DOI 10.1007/978-3-319-13884-8_18

18.1 Introduction

Trypanosoma brucei ssp. is the etiological agent of sleeping sickness in humans (human African trypanosomiasis) and nagana disease in livestock (animal African trypanosomiasis). The subspecies *T. b. gambiense* and *T. b. rhodesiense* are infectious to both humans and animals, whereas the subspecies *T. b. brucei* is only infective to animals. The protozoan parasites are transmitted by the bite from infected tsetse flies (*Glossina* spp.) and live and multiply extracellularly in blood and tissue fluids of their mammalian host. Both diseases are exclusively found in sub-Saharan Africa and are restricted to the distribution of tsetse flies. African trypanosomiasis is one of the most neglected tropical diseases, affecting mainly poor people living in remote areas of Africa. Sleeping sickness is a major cause for both morbidity and mortality, while nagana disease is of economic importance making livestock breeding very difficult in the concerned regions.

18.2 Morphology and Physiology of *T. brucei*

Trypanosoma brucei is a zooflagellate distinguished by the presence of a single flagellum (Fig. 18.1). Another characteristic feature of *T. brucei* is that it undergoes morphological changes during the course of its life cycle (polymorphism). In the insect vector, the parasite occurs in an epimastigote form and in a trypomastigote form, while the latter is the only found in the mammalian host. Both forms have a C- or S-shape and are between 20 and 50 μm long and 1–3 μm wide (Fig. 18.2). In the trypomastigote form, the flagellum emerges at the posterior end, whereas in the epimastigote form, the flagellum exits in the middle of the cell. In both forms the flagellum is attached along its length to the cell body by a network of cytoskeletal and membranous connections that collectively make up the flagellum attachment zone (Ralston et al. 2009). In cross-section, the flagellum has the typical $9\times2+2$ microtubule pattern. Near the basal body that anchors the flagellum, a DNA-containing, disklike structure can be observed, the kinetoplast. This organelle defines flagellates of the class of Kinetoplastida to which *T. brucei* belongs. The shape of the cell body is stabilized by microtubules orientated parallel to the long axis of the cell forming a helical array linked to the membrane. The cell surface of *T. brucei* is covered with a coat of glycoprotein attached to the cell membrane via GPI anchors. In the case of the mammalian bloodstream forms, this coat is composed of densely packed variant surface glycoprotein (VSG) which undergoes antigenic variation in order to escape the immune response of the mammalian host (see below) (Borst and Rudenko 1994). The coat of insect life cycle stages is formed of an invariant glycoprotein termed procyclin (Roditi et al. 1998). In other respects, *T. brucei* has all the typical organelles found in other eukaryotic cells. However, the parasite has only one elongated mitochondrion. In addition, *T. brucei* possesses a special organelle called glycosome, peroxisome-related microbodies found only in trypanosomatids.

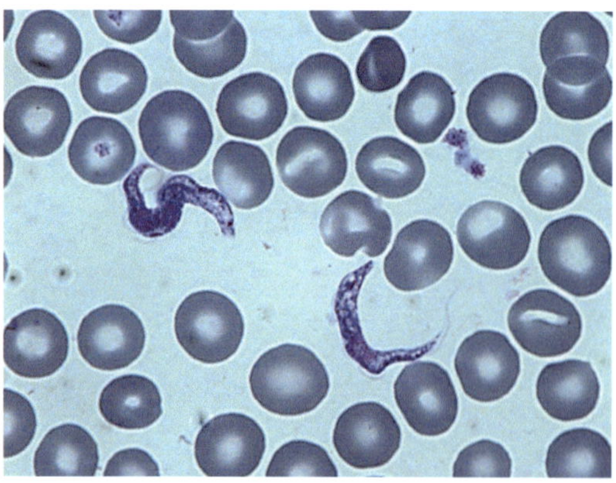

Fig. 18.1 *Trypanosoma brucei* in human blood. Giemsa-stained blood smear

Asexual proliferation of *T. brucei* occurs by asymmetric fission. First, the basal body is duplicated, and then the kinetoplast followed by the outgrowth of a new flagellum, and finally the nucleus. Sexual reproduction of *T. brucei* has been shown to occur in the insect vector (Gibson 2015). Meiosis happens in the salivary glands and promastigote-like gametes interact with each

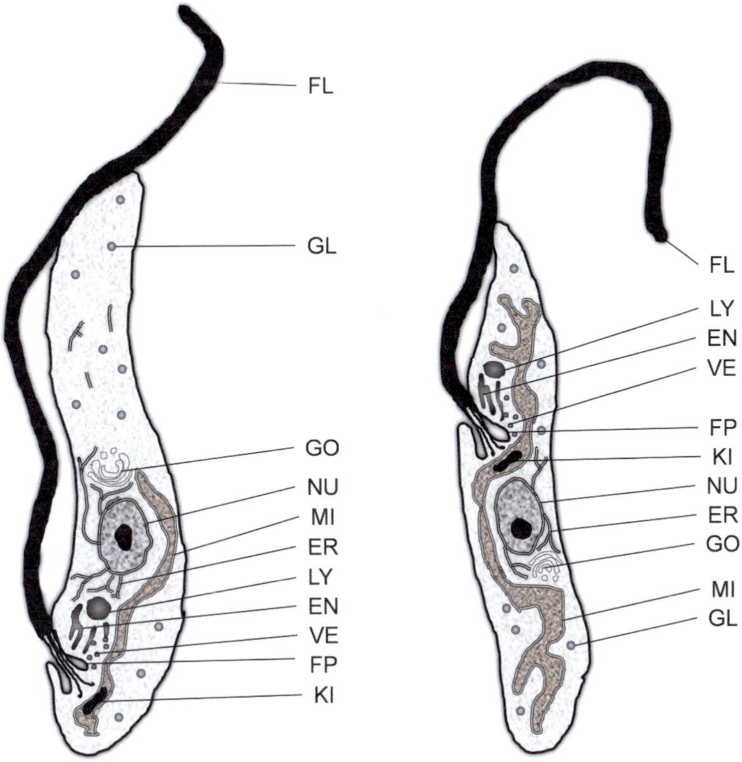

Fig. 18.2 Ultrastructure of *T. brucei*. Shown are the bloodstream trypomastigote form of the mammalian host (*left*) and salivary gland epimastigote form of the insect vector (*right*). *EN* endosome, *ER* endoplasmic reticulum, *FL* flagellum, *FP* flagellar pocket, *GL* glycosome, *GO* Golgi apparatus, *KI* kinetoplast, *LY* lysosome, *MI* mitochondrion, *NU* nucleus, *VE* vesicle

other via their flagella followed by cell fusion (Gibson 2015).

Trypanosoma brucei lives only on soluble nutrients which are abundantly available in tissue fluids and intestinal contents of their hosts. The uptake of nutrients occurs via endocytosis at the membrane of the flagellar pocket, an invagination of the plasma membrane where the flagellum exits the cytoplasm. Mammalian bloodstream forms have a glycolytic metabolism and rely entirely on glucose as energy source. Glucose is broken down to pyruvate which is then excreted. They consume about 50–100% of their dry weight in glucose per hour, and as they do not store polysaccharides, they need a continuous supply of the sugar. Mammalian bloodstream forms do not have a functional tricarboxylic acid cycle and are deficient of respiratory chain enzymes. Insect forms have an oxidative metabolism but mainly burn up amino acids, especially proline.

18.3 Molecular and Cellular Features of *T. brucei*

Although *T. brucei* is a eukaryotic cell, the parasite has developed special molecular and cellular features which are unique to this organism.

18.3.1 Antigenic Variation

As extracellular blood parasite *T. brucei* is constantly exposed to the immune defense of its mammalian host. However, the parasite found a way to circumvent the host's immune response: it changes regularly its surface antigens (Vickerman 1978; Horn 2014). Each bloodstream form trypanosome is covered by a 15 nm thick surface coat (Fig. 18.3) composed of densely packed variable protein molecules known as variant surface glycoproteins (VSGs). During an infection with *T. brucei*, the host's immune system produces antibodies against the VSG molecules of the first intruders. Most parasites of this first *T. brucei* population are killed by the antibodies, but a few cells survive because they produce a new, different VSG molecule (Fig. 18.4). These cells multiply and give rise to a new population of parasites until the immune system has produced new antibodies against this second generation of VSG molecules (Fig. 18.4). Once more most parasites are killed yet a few have replaced their surface coat with a new VSG molecule. Again, these cells escape the attack of the antibodies and expand to a new parasite population (Fig. 18.4). This process of antigenic variation recurs numerous times during the course of an infection because *T. brucei* has about 2000 genes for different VSG molecules (Berriman et al. 2005). The switching between VSG genes can occur by two major mechanisms, duplicative transposition and in situ switch (Fig. 18.5) (Horn 2014). In order to be expressed, a VSG gene has to be in an expression site, a specialized locus at the end of chromosomes. Although *T. brucei* has approximately 20 expression sites, only one is fully active at a time in bloodstream forms to transcribe its VSG gene. In the case of duplicative transposition, the VSG gene of the active expression site is replaced by a copy of a non-telomeric, archived VSG gene. In the second mechanism, the in situ switch, the active expression is switched off, while an inactive expression site is switched on with the result that another VSG gene is transcribed.

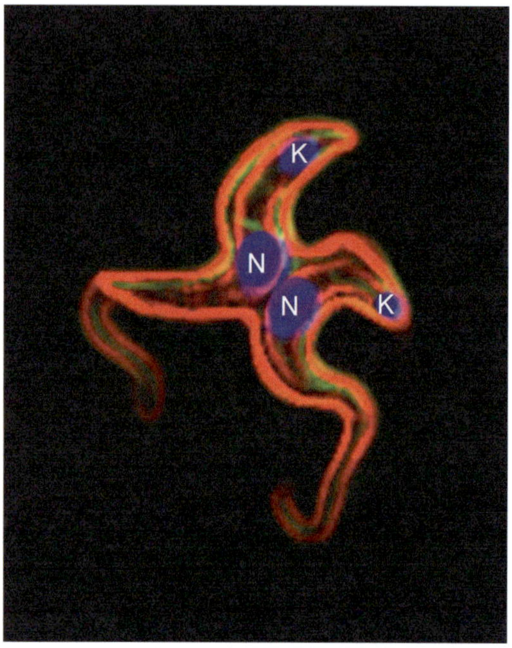

Fig. 18.3 Surface coat of bloodstream forms of *T. brucei*. The surface coat (*red*) has been labeled with a fluorescent antibody specific for the variant surface glycoprotein (VSG). Nucleus (N) and kinetoplast (K) have been labeled with the DNA stain 4′,6-diamidino-2-phenylindole (DAPI)

18.3.2 *Trans*-splicing

In contrast to most other eukaryotic cells, the protein-coding genes of *T. brucei* do not have introns. Instead they are organized in polycistronic

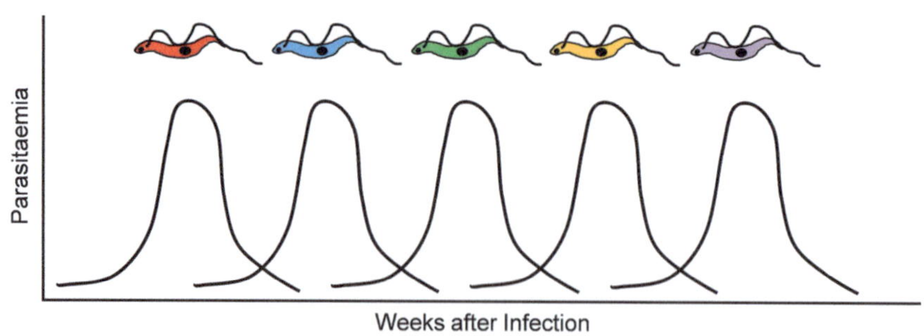

Fig. 18.4 Antigenic variation of *T. brucei*. During the course of an infection with *T. brucei*, different clones with different surface coats (indicated by different colors) outgrow in succession. Although the immune system produces each time specific antibodies which kill most parasites, some few survive because they have changed their surface coat and thus evade the immune response of their host

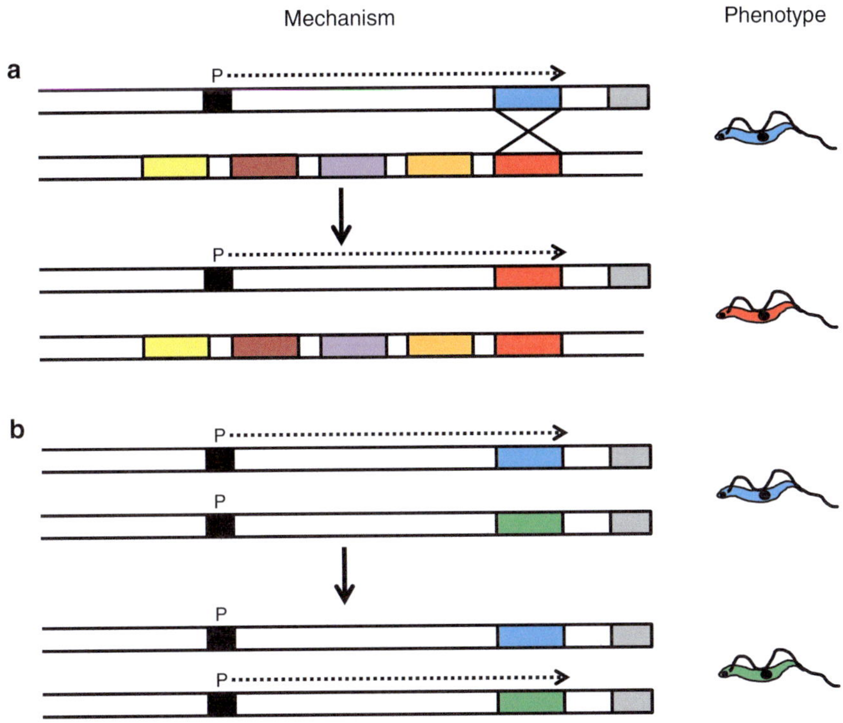

Fig. 18.5 Molecular mechanisms for VSG switching in *T. brucei*. Only VSG genes situated in an expression site located at the end of chromosomes (telomeres, indicated by a *gray box*) are expressed. *Trypanosoma brucei* has about 20 expression sites, and only one of which is active at a given time. In addition to VSG genes located in expression sites, the *T. brucei* genome contains 2000 different VSG genes (indicated by *colored boxes*) most of which are located in subtelomeric regions of different chromosomes. (**a**) Duplicative transposition: the transcription of the "blue" VSG gene is driven by the promoter (*black box*) of the active expression site and the corresponding VSG surface coat is "blue" (phenotype). During the switching process, the "blue" VSG gene is replace by a copy of a non-telomeric "red" VSG gene and the VSG surface coat changes to "red." (**b**) In situ switch: the active expression site containing the "blue" VSG gene is switched off and the inactive expression site harboring the "green" VSG gene is switched on. As a result, the VSG surface coat changes from "blue" to "green"

transcription units. These units are transcribed by an upstream located promoter as a whole and a polycistronic pre-mRNA is produced, i.e., a transcript that contains the information for several proteins (similar to bacterial transcription). Mature monocistronic mRNAs are created in a concerted action of *trans*-splicing and polyadenylation (Fig. 18.6) (Liang et al. 2003). This process involves the addition of a 39-nucleotide-long spliced leader sequence (mini-exon) from a donor RNA at the 5′-terminus of a protein-encoding RNA and 3′-end cleavage/polyadenylation. Thus, through the process of *trans*-splicing, exons of two different RNAs are linked to each other.

18.3.3 RNA Editing

RNA editing is the posttranscriptional correction of mRNAs through insertion and/or deletion of one or several nucleotides. This unusual process was first discovered in *T. brucei* and concerns the protein-coding genes of the mitochondrial DNA (Stuart et al. 2005; Aphasizheva and Aphasizhev 2016). As most of the mitochondrial genes are encrypted, RNA editing is necessary to change illegible sequences into translatable sequences. Through insertion and to a minor degree deletion of uridine residues, correct open reading frames and an initiation site are created (Fig. 18.7).

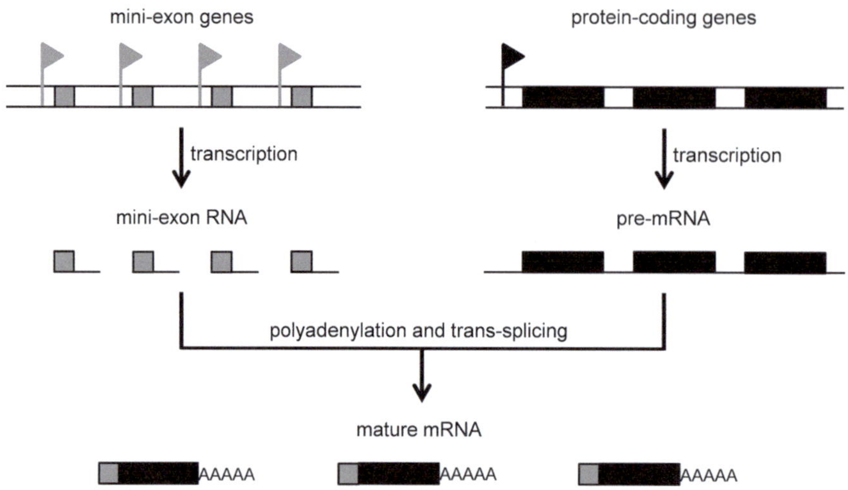

Fig. 18.6 Trans-splicing and polyadenylation in *T. brucei*. Tandem repeats of mini-exon genes (*left*, *gray boxes*) are individually transcribed from promoters (*gray flags*) located directly in front of each mini-exon gene generating 140 nt long mini-exon RNAs. Polycistronic arranged protein-coding genes (*right*, *black boxes*) are transcribed together from a single promoter (*black flag*), producing a polycistronic pre-mRNA. Pre-mRNAs are matured by cleavage for poly-A addition and by trans-splicing for joining with mini-exon RNAs to generate mature mRNA molecules (*bottom*)

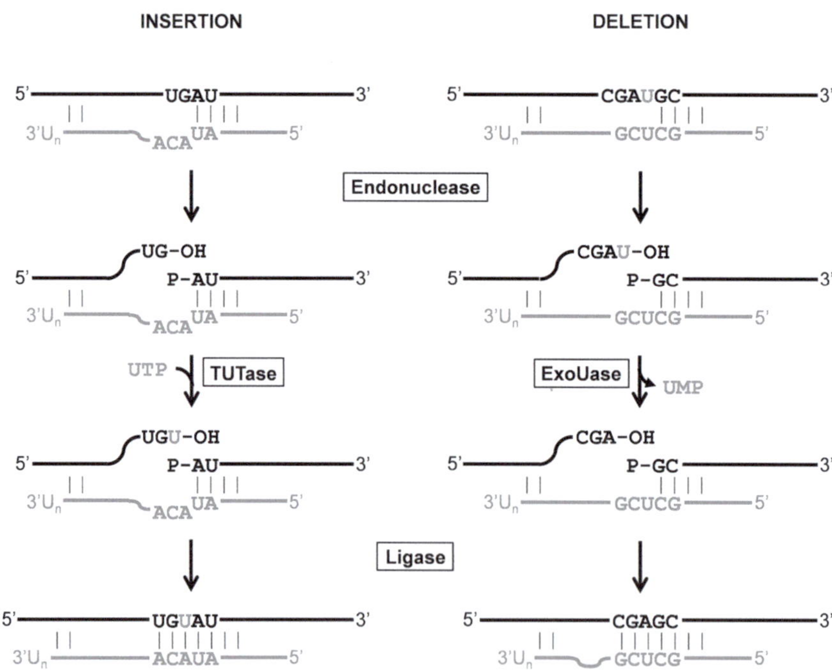

Fig. 18.7 General mechanism of RNA editing in *T. brucei*. Shown are the catalytic events in insertion (*left*) and deletion (*right*) of uridine residues. After the 5′ region of a gRNA (*gray*) has formed an anchor duplex with its cognate mRNA (*black*) 3′ region at the editing site, the pre-mRNA is cleaved by an endonuclease. Uridine residues (*gray*) are either added to the 5′ cleavage fragment by a TUTase (using UTP) in the insertion editing or removed from the 5′ cleavage fragment by an ExoUase (released as UMP) in the deletion editing, as specified by the sequence of the gRNA. The 5′ and 3′ fragments are then religated by an RNA ligase. Several cycles of editing occur until all of the sites specified by a gRNA have been processed

This can be quite extensive and up to 50 % of the nucleotides in an mRNA molecule may have been added in course of the editing. This process is regulated by small RNA molecules called guide RNAs (gRNAs), which encode complementary sequence information for the insertion and deletion of uridine residues. In *T. brucei*, RNA editing is developmentally regulated and occurs only in the insect forms.

18.3.4 Kinetoplast

The kinetoplast contains the mitochondrial DNA and consists of two classes of circular DNA molecules: ~50 identical 22-kb maxicircles and ~10,000 heterogeneous1-kb minicircles (Povelones 2014). The DNA molecules are catenated together similar to a chain armour and form an extraordinary disklike structure located within the matrix of the single mitochondrion (Figs. 18.2 and 18.3). The maxicircles code for 18 cryptogenes and gRNAs, while the minicircles code for more than 1200 different gRNAs.

18.3.5 Glycosome

The glycosome is a special organelle in trypanosomes which is related to peroxisomes of higher eukaryotes (Haanstra et al. 2016). In mammalian bloodstream forms of *T. brucei*, glycosomes contain the first seven enzymes of the glycolytic pathway (Fig. 18.8) which made up 90 % of their

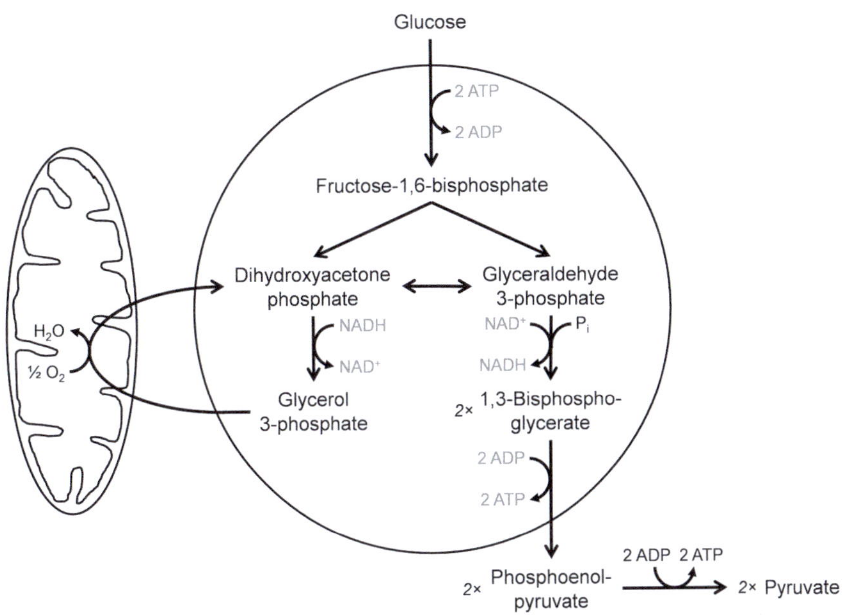

Fig. 18.8 Schematic representation of the catalytic pathways in the glycosome of bloodstream forms of *T. brucei*. The microbody-like organelle comprises the first seven enzymes involved in glycolysis (hexokinase, phosphoglucose isomerase, phosphofructokinase, aldolase, triose-phosphate isomerase, glyceraldehyde-3-phosphate dehydrogenase, and phosphoglycerate kinase), while the last three glycolytic enzymes (phosphoglycerate mutase, enolase, and pyruvate kinase) are located in the cytosol. The consumption and production of ATP and NADH is balanced within the glycosome (shown in *gray*): the consumption of ATP by hexokinase and phosphofructokinase is compensated by the ATP production by phosphoglycerate kinase, while the NADH generated by glyceraldehyde-3-phosphate dehydrogenase reaction is reoxidized by glycerol-3-phosphate dehydrogenase using dihydroxyacetone phosphate. The resulting glycerol-3-phosphate is then reoxidized back to dihydroxyacetone phosphate by a mitochondrial glycerol-3-phosphate oxidase which comprises an FAD-dependent glycerol-3-phosphate dehydrogenase, ubiquinone, and a cyanide-insensitive alternative oxidase using molecular oxygen as final electron acceptor. This respiratory process is not involved in free-energy transduction. The glycosome itself is not involved in the net production of ATP which occurs in the cytosol where phosphoenolpyruvate is converted to pyruvate

protein content. This compartmentation of glycolysis is unique among eukaryotes. It appears to be essential for the regulation of the glycolytic pathway and enables bloodstream forms of *T. brucei* to overcome short periods of anoxia.

18.3.6 Alternative Oxidase

As mammalian forms of *T. brucei* rely on blood glucose as their sole source of energy and do not express respiratory chain enzymes, they need an alternative way to reoxidize the NADH produced during glycolysis. A cyanide-insensitive alternative oxidase located in the mitochondrion is key to this process (Chaudhuri et al. 2006). NADH is reoxidized through a glycerol-3-phosphate/dihydroxyacetone phosphate shuttle between glycosomes and the mitochondrion, where an alternative oxidase transfers the electrons to oxygen to form water (Fig. 18.8). The trypanosome alternative oxidase is essential for the survival of bloodstream forms of *T. brucei*. As mammals do not have an alternative oxidase, this enzyme is a promising chemotherapeutic drug target for African trypanosomiasis.

18.3.7 Trypanothione

In *T. brucei* the cellular thiol redox homeostasis is maintained by trypanothione, a thiol molecule composed of two molecules of glutathione joined by a spermidine linker (Fig. 18.9) (Krauth-Siegel and Comini 2008). The major function of trypanothione is to donate electrons for the synthesis of DNA and for the reduction of hydroperoxides. Trypanothione is kept reduced by a parasite-specific trypanothione reductase which links the NADPH- and the thiol-based redox metabolisms (Fig. 18.9). As mammals use the glutathione/glutathione reductase system for maintenance of their cellular thiol redox homeostasis, trypanothione reductase has been considered as a target for the development of drugs against *T. brucei*.

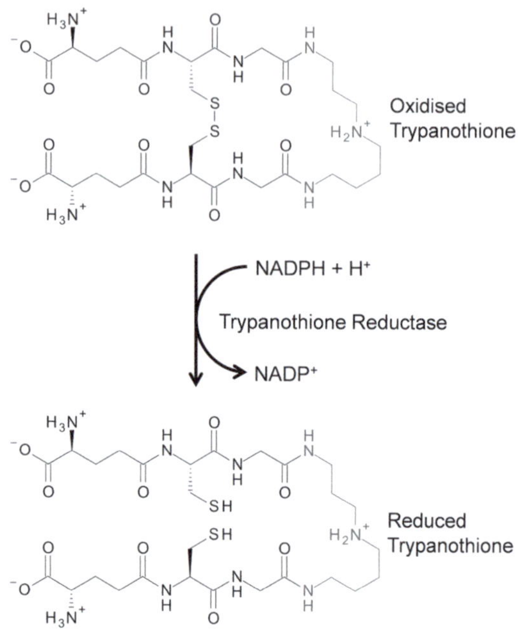

Fig. 18.9 Structure of the oxidized and reduced forms of trypanothione. The molecule consists of two glutathione moieties (*black*) linked via a spermidine bridge (*gray*). Reduction of oxidized trypanothione into its dithiol form is catalyzed by the NADPH-dependent flavoenzyme trypanothione reductase

18.4 Life Cycle of *T. brucei*

Trypanosoma brucei has a complex digenetic life cycle between a mammalian host and an insect vector (Fig. 18.10). During a blood meal of an infected tsetse fly, about 20,000 infectious metacyclic trypomastigote forms of *T. brucei* are injected with saliva into the connective tissue of the skin of a human host. Within the next 2–4 weeks, the parasites transform into actively dividing long slender trypomastigotes and are spread via the lymph and blood throughout the body. The parasites continuously replicate by binary fission until a certain cell density is reach. At the peak of parasitemia, most long slender forms change into nondividing short stumpy trypomastigotes (pleomorphism). The short stumpy forms are eliminated by the immune response of the human host and subsequently the number of parasites in the blood

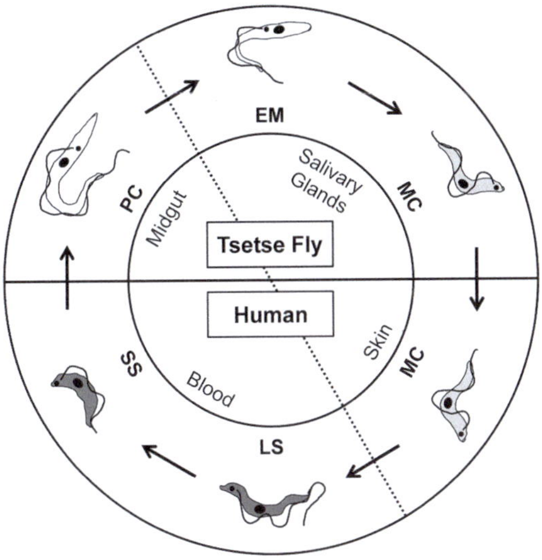

Fig. 18.10 The life cycle of *T. brucei*. With the bite of an infected tsetse fly, metacyclic forms (MC) are injected into the skin of a human host. After transforming into long slender forms (LS), the parasite enters the lymphatic system and the bloodstream. A tsetse fly gets infected when it takes up short stumpy forms (SS) with a blood meal. In the insect midgut, the parasite differentiates into procyclic forms (PC). After migration into the salivary glands, the parasite develops into epimastigote forms (EM). Finally, the epimastigote forms transform into human-infective metacyclic forms (MC). Life cycle stages with a variant surface coat are shown in *gray*

falls considerable. However, some of the remaining long slender forms escape the immune response by antigenic variation (see above) and give rise to another parasitemic wave. This interplay between long slender and short stumpy forms continues almost indefinitely in the human host and is the reason for the chronic nature of sleeping sickness.

Tsetse flies become infected with trypanosomes when taking a blood meal containing short stumpy forms. The short stumpy forms are preadapted to the life in the insect vector and transform in actively dividing procyclic trypomastigotes in the fly's midgut. Eventually the procyclic forms leave the midgut and migrate to the salivary glands where they change into epimastigote forms. At first, the epimastigotes multiply by asymmetric fission and then transform into nondividing, human infectious metacyclic forms. The cycle in the tsetse flies takes about 2–3 weeks.

18.5 Human Infectivity of *T. brucei*

Generally, humans are resistant to infection by African trypanosomes. This innate immunity is due to a trypanosome lytic factor (TLF) found in normal human serum and consists of apolipoprotein L-I (apoL1) and haptoglobin-related protein (Hpr) (Vanhollebeke and Pays 2010). These proteins are associated with two serum complexes, a high density lipoprotein subfraction (HDL_3) termed TLF1 and an IgM/apolipoprotein A-I complex called TLF2 (Vanhollebeke and Pays 2010). Uptake of these factors leads to killing of trypanosomes through lysis induced by the pore-forming activity of apoL1. Only the two subspecies *T. b. gambiense* and *T. b. rhodesiense* can resist this defense mechanism and thus can infect humans causing sleeping sickness.

In *T. b. rhodesiense* the factor inactivating apoL1 is a VSG-like protein termed serum resistant associated (SRA) protein (Xong et al. 1998). The N-terminal helical domain of the SRA protein interacts with the C-terminal helical domain of apoL1 in the lysosome preventing the release of the pore-forming protein and subsequently the lysis of the parasite (Vanhamme et al. 2003). *Trypanosoma brucei gambiense* does not contain the SRA protein and has developed independently a different, multifactorial mechanism to resist the lysis by TLF. First, a specific truncated *T. b. gambiense* VSG (TgsGP) confers resistance to apoL1 by stiffening and increasing the curvature of endosomal membranes which probably impairs the pore-forming activity of apoL1 (Uzureau et al. 2013). Second, apoL1 is degraded faster in *T. b. gambiense* due to increased cysteine protease activity in early endosomes (Uzureau et al. 2013). Third, TLF-1 is inefficiently taken up by *T. b. gambiense* due to a single mutation in the TLF-1 receptor (DeJesus et al. 2013).

18.6 Clinical Presentation of Sleeping Sickness and Nagana Disease

18.6.1 Sleeping Sickness

Human African trypanosomiasis occurs in two forms depending on the *T. brucei* subspecies involved (Table 18.1). *Trypanosoma brucei gambiense* is responsible for West African sleeping sickness (Fig. 18.11) which accounts for 98 % of all reported sleeping sickness cases and causes a chronic infection (WHO 2016). A patient can be infected for months or even years before any major symptoms of the disease present. *Trypanosoma brucei rhodesiense* is the etiological agent of East African sleeping sickness (Fig. 18.11) which currently accounts for only 2 % of all notified sleeping sickness infections and gives rise to an acute illness (WHO 2016). First symptoms of the disease are already observed a few weeks after infection. Without treatment, both forms of sleeping sickness are normally fatal. The mean survival time for West African sleeping sickness patients is 3 years, while that of East African sleeping sickness patients is 6 months.

The symptoms of sleeping sickness are generally the same for both forms of the disease but differ in their severity and frequency of appearance

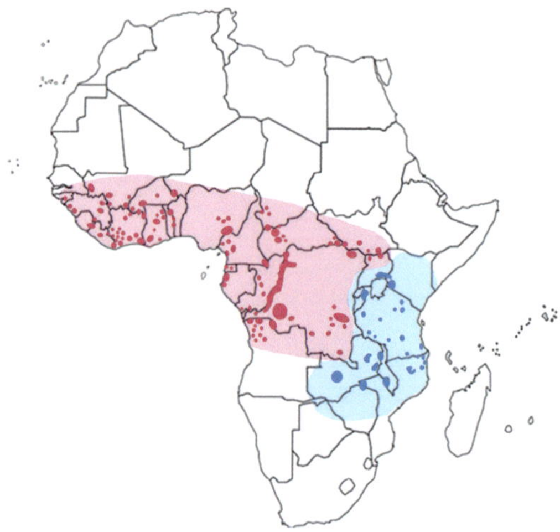

Fig. 18.11 Distribution of *T. brucei*, sleeping sickness, and nagana disease. The approximate geographical distributions of the two human pathogenic subspecies *T. b. gambiense* and *T. b. rhodesiense* are indicated by the *red and blue shaded areas*, respectively. The geographical extension of *T. b. brucei* and nagana disease encompasses both *shaded areas*. The distribution of the two forms of sleeping sickness is indicated by *red* (West African sleeping sickness) and *blue* (East African sleeping sickness) spots

Table 18.1 The two forms of human African trypanosomiasis

	West African sleeping sickness	East African sleeping sickness
Pathogen	*Trypanosoma brucei gambiense*	*Trypanosoma brucei rhodesiense*
Endemic countries	Angola, Benin[a], Burkina Faso[a], Cameroon, Chad, Central African Republic, Congo, Côte d'Ivoire, Democratic Republic of the Congo, Equatorial Guinea, Gabon, Gambia[a], Ghana, Guinea, Guinea Bissau[a], Liberia[a], Mali[a], Niger[a], Nigeria, Senegal[a], Sierra Leone[a], South Sudan, Togo[a], Uganda	Botswana[a], Burundi[a], Ethiopia[a], Kenya, Malawi, Mozambique[a], Namibia[a], Rwanda[a], Swaziland[a], United Republic of Tanzania, Uganda, Zambia, Zimbabwe
Vectors	Palpalis group:	Morsitans group:
	Glossina palpalis gambiensis	*Glossina morsitans morsitans*
	G. p. palpalis	*G. m. centralis*
	G. fuscipes fuscipes	*G. pallidipes*
	G. f. quanzensis	
Natural habitat	Forested rivers and shores	Savannah
Animal reservoir	Predominantly anthroponosis: animal reservoir is considered as epidemiologically unimportant	Zoonosis: large natural host reservoir consisting of both wild and domestic animals (ungulates)
Clinical picture	Chronic disease	Acute disease
	CNS involvement after months	CNS involvement after weeks
	Death within years	Death within months

[a]No reported new cases since 2000

(WHO 2013). The disorder occurs in two stages, the first or early stage, also known as the hemolymphatic stage, defined by the restriction of the parasites to the lymphatic and blood system, and the second or late stage, also known as the neurological stage, characterized by the active invasion of the central nervous system by the parasites. Around 5 days after the bite of an infected tsetse fly, a local inflammatory skin reaction called trypanosome chancre may develop at the inoculation site. This trypanosome chancre is observed in about 50% of all *T. b. rhodesiense* infections but rarely seen in *T. b. gambiense* cases.

18.6.1.1 First-Stage Symptoms

Early stage symptoms appear about 2–4 weeks after infection when the parasites have spread into the lymph and blood. These include chronic and intermittent fever, headache, pruritus, lymphadenopathy, weakness, anemia, musculoskeletal pain, hepatosplenomegaly, cardiac involvement, and endocrine dysfunctions (WHO 2013). The fever episodes can last from 1 day up to 1 week and are separated by afebrile intervals of a few days to a month. They occur when the parasitemia drops and are triggered by cell debris and metabolites from disintegrating parasites. Fevers are more pronounced and more frequent in the early stage than in the late stage of the disease, while the reverse is true for headaches.

Itching is a more subjective complaint and occurs more frequently with the duration and severity of the disorder. Generalized lymphadenopathy is a typical sign for human African trypanosomiasis. In West African sleeping sickness, enlargement of the posterior cervical lymph nodes, known as Winterbottom's sign, is characteristic, whereas in East African sleeping sickness, submandibular, axillary, and inguinal lymph nodes tend to be swollen. Musculoskeletal pain, hepatomegaly, and splenomegaly are other common nonspecific signs often observed in sleeping sickness patients.

Electrocardiographic alterations (Q-Tc prolongation, repolarization changes, and low voltage) are frequently observed in both stages of West African sleeping sickness patients but rarely lead to clinically relevant heart failure. However, in East African sleeping sickness, myocarditis, pericarditis, and congestive cardiac failure can develop early and lead to death during the first stage. Endocrine disorders may include a permanent feeling of being cold, reduced or increased appetite, excessive thirst, loss of libido, impotence, and amenorrhea.

18.6.1.2 Second-Stage Symptoms

Late-stage symptoms appear after weeks in East African sleeping sickness and after months is West African sleeping sickness. This stage of the disease is associated with severe headache, sleep abnormalities, motor system disturbances, and mental changes (WHO 2013). The sleep disorder is the main characteristic sign of late-stage sleeping sickness and in many languages gives the disease its name. It consists of a reversal of the circadian rhythm of the sleep-wake cycle with night-time insomnia and day-time somnolence and frequent uncontrollable sleep episodes of short duration. The reason for the sleep disturbance lies in a dysregulation of the normal sleep-wake cycle and a fragmentation of the sleep pattern. In addition, the structure of the sleep is altered with an early onset of rapid-eye movement sleep. Motor abnormalities include motor weakness, tremor, bradykinesia, walking difficulties, abnormal movements, and speech disturbances (i.e., slurred speech). Another characteristic feature of the late stage of sleeping sickness is the impairment of mental functions and personality and behavioral changes. These include lassitude, apathy, loss of concentration, confusion, mood changes, anxiety, irritability, hallucinations, delirium, and the inability to cope with the tasks of everyday life. Finally the patient enters a terminal somnolent and an advanced cachexial state. Eventually an untreated patient will succumb to starvation or to secondary infections (mostly pneumonia).

18.6.2 Nagana Disease

Trypanosoma brucei brucei causes relative mild infections in wild animals, whereas it can cause severe, often fatal disease in domestic animals. In cattle and pigs the course of the disease is relatively mild, while in sheep and goats more moderate and in horses, donkeys, and dogs rather severe (Taylor and Authié 2004). The symptoms of the disease are fever, listlessness, emaciation, drop in milk production, hair loss, discharge from eyes, edema, anemia, and paralysis (Taylor and Authié 2004). As the disease progresses, the animals become weaker and weaker and eventually unfit for work. The poor condition of infected animals gave the disease its name from the Zulu word "N'gana" which means powerless/useless.

18.7 Diagnosis of Sleeping Sickness and Nagana Disease

18.7.1 Sleeping Sickness

The presence of clinical symptoms (fever, swollen lymph nodes, or neurological signs) can raise suspicion of sleeping sickness in a patient but is insufficient in order to start chemotherapeutic treatment (WHO 2013). Diagnosis of sleeping sickness is generally laboratory-based and follows a three-step pathway: screening, parasitological confirmation, and staging (Fig. 18.12). However, as sleeping sickness occurs in rural areas, only those diagnostic methods that are cheap, quick, simple, and reliable should be used (WHO 2013).

18.7.1.1 Screening

Serological Tests

The standard serological assay for the detection of *T. b. gambiense* is the long-established card agglutination test for trypanosomiasis (CATT) (WHO 2013). The test was developed in 1978 and is provided by the Institute for Tropical Medicine in Antwerp. The CATT employs fixed and stained bloodstream forms of *T. b. gambiense* of the variable antigen type LiTat1.3[1] and is based on direct agglutination of specific antibodies with antigens exposed on the surface of the trypanosomes. The assay can be performed on undiluted whole blood or on diluted blood, serum, or plasma. One drop of the test sample is put on the test area of a plastified card and then mixed with one drop of the CATT antigen. The card is rocked on an orbital rotator for 5 minutes. Visible agglutination indicates a positive reaction. The sensitivity of the CATT ranges between 78 and 99 % with a negative predicted value of 99–100 %. However, the low positive predictive value of the CATT is the reason why all positive tested individuals require parasitological confirmation. Because of the much higher antigenic variation of *T. b. rhodesiense*, no equivalent of the CATT is available for the detection of this subspecies.

Immunofluorescence assay (IFA) is another serological technique for diagnosis of sleeping sickness (WHO 2013). The IFA can be used for serodiagnosis of both *T. b. gambiense* and *T. b. rhodesiense* infections. In this test, whole trypanosomes coated on slides are incubated with diluted test serum, then with fluorescein-labeled anti-human IgG and finally inspected with a fluorescence microscope. The sensitivity of the test with serum has been reported to be 75–95 % for West African sleeping sickness and 71–92 % for East African sleeping sickness.

Numerous enzyme-linked immunosorbent assays (ELISAs) have been developed for the detection of antibodies in sleeping sickness patients (WHO 2013). For diagnosis of *T. b. rhodesiense* infections, crude trypanosome extracts are usually used as antigens, while for *T. b. gambiense* infections purified LiTat 1.3 or 1.5 VSG antigens are employed. The various ELISAs have sensitivities and specificities of 95–100 % and 97–100 %, respectively.

More recently, two rapid diagnostic tests (RDTs) have been developed for *T. b. gambiense* sleeping sickness, the SD Bioline HAT marketed

[1] The identity of a clone of *T. brucei* is determined by the variable antigen expressed at the surface. Different variable antigen types are designated by a code composed of letters indicating the laboratory that originally characterized the antigen type, a letter indicating the subgenus, and two letters as an abbreviation of the words "antigenic type," followed by a number. For example, LiTat 1.3 stands for Lille *Trypanozoon* antigen type 1.3. The molecule responsible for a variable antigen type is the VSG.

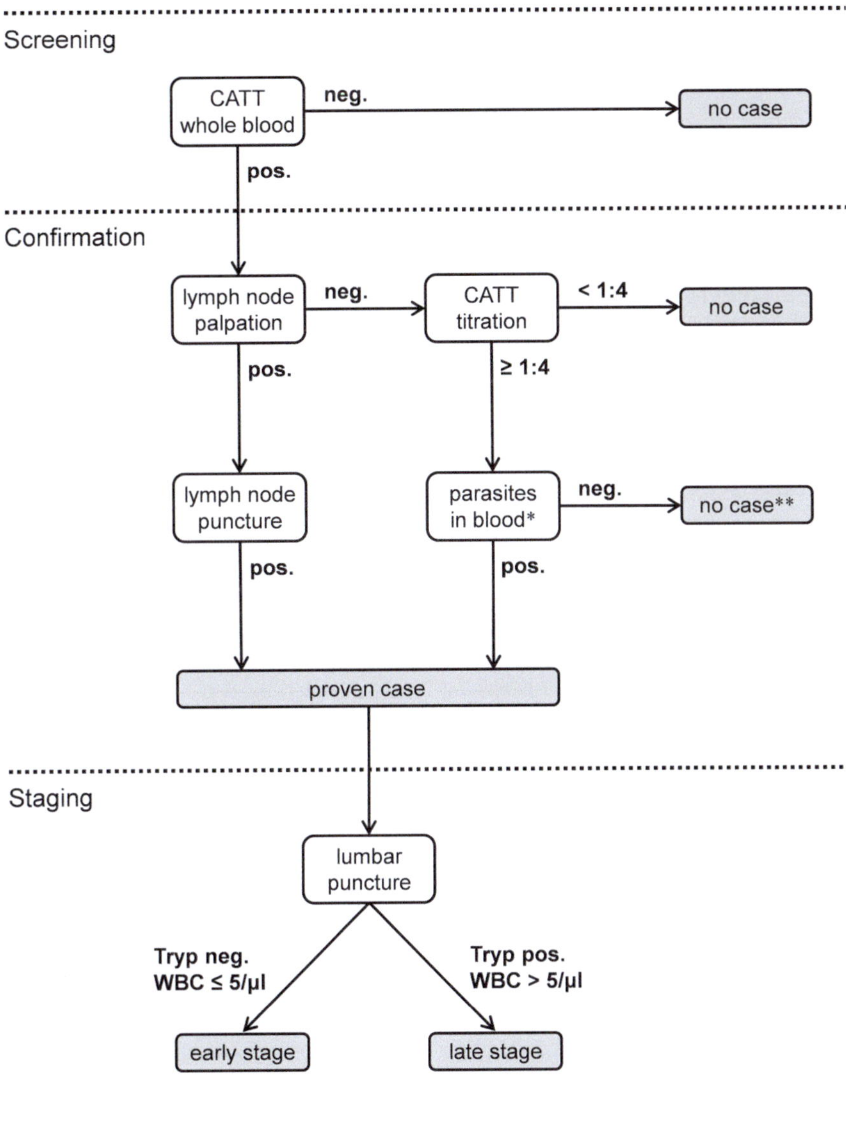

Fig. 18.12 Example of a diagnostic algorithm for *T. b. gambiense* sleeping sickness. As the prevalence of sleeping sickness varies greatly between foci, the diagnostic algorithm applied should be adapted according to the disease prevalence. The algorithm shown is for populations with rather high disease prevalence (above 2 %) (Chappuis et al. 2005). *Asterisk*: For the detection of trypanosomes in the blood, concentration methods such as microhematocrit centrifugation technique (mHCT), quantitative buffy coat test (QBC), or mini-anion-exchange centrifugation technique (mAECT) may be needed to employ. *Double asterisk*: In high prevalence areas, serologically suspected patients (positive CATT of 1:4) with negative parasitological examination should be monitored when their CATT plasma end-dilution titer is ≥1:16. *CATT* card agglutination test for trypanosomiasis, *neg.* negative, *pos.* positive, *Tryp neg.* trypanosome negative, *Tryp pos.* trypanosomes positive, *WBC* white blood cells

by Standard Diagnostic, Inc. and the HAT Sero-*K*-SeT introduced by CorisBioConcept. Both RDTs are lateral flow tests (immunochromatographic strip tests) and use equimolar mixtures of the *T. b. gambiense* VSGs LiTat 1.3 and 1.5 as antigens (Büscher et al. 2014). On whole blood samples, the HAT Sero-*K*-SeT test performs better than the SD Bioline HAT (sensitivity 98.5 % vs. 89.3 %, specificity 98.6 % vs. 94.6 %).

Molecular Tests

Molecular methods to detect trypanosome DNA or RNA for diagnosis of sleeping sickness have been developed over the years. Although molecular techniques such as polymerase chain reaction (PCR) or nucleic acid sequence-based amplification (NASBA) show generally high sensitivity and specificity and can be performed on various specimens (e.g., whole blood, frozen blood, blood stored on filter paper, cerebrospinal fluid (CSF)), their requirements of laboratory instruments for nucleic acid extraction, amplification, and visualization limit their use to reference centers. In contrast to conventional molecular methods, loop-mediated isothermal amplification (LAMP) is a much simpler technique that does not require nucleic acid extraction and amplifies DNA at constant temperature.

A LAMP test has recently been developed into a commercially available kit, the Loopamp™ *Trypanosoma brucei* kit, which is marketed by Eiken Chemical Co., Ltd. This field-applicable test uses the repetitive insertion mobile element (RIME) as targeted DNA and can be used for the detection of both *T. b. gambiense* and *T. b. rhodesiense*. The test shows good diagnostic accuracy and excellent reproducibility with a sensitivity of 87.3–93.0% and a specificity of 92.8–96.4% (Mitashi et al. 2013).

18.7.1.2 Parasite Detection

The demonstration of the presence of parasites in *T. b. gambiense* sleeping sickness is notoriously difficult because of the cyclical parasitemia which makes it necessary to repeat examinations on a daily basis. In *T. b. rhodesiense* sleeping sickness, the detection of parasites is easier because of the much higher density of circulating trypanosomes. Parasitological confirmation relies usually on microscopic search for trypanosomes in aspirates collected from enlarged cervical lymph nodes or in blood films (WHO 2013). Because of the low parasitemia in West African sleeping sickness, concentration methods such as microhematocrit centrifugation technique (mHCT), quantitative buffy coat test (QBC), or mini-anion-exchange centrifugation technique (mAECT) may be necessary in order to increase the density of trypanosomes in test samples (WHO 2013).

18.7.1.3 Staging

Once sleeping sickness has been confirmed in a patient, the determination of the disease stage is mandatory because of the toxicity of the drugs used for treatment of the late stage of the disease. For stage determination, CSF taken by lumbar puncture is examined for the presence of trypanosomes and for an increased white blood cell (WBC) count (WHO 2013). The CSF is examined under a microscope using a counting chamber. Patients with no trypanosomes and <5 WBC/ml in the CSF are considered to be in the first stage, while those with trypanosomes and/or ≥5 WBC/ml in the CSF are regarded to be in the late stage (WHO 2013). As the number of WBCs in normal CSF is usually low and an accurate WBC count is imperative for correct patient management, the CSF should not be diluted for counting. As the sensitivity of direct detection of trypanosomes in CSF is usually low, parasites can be concentrated by centrifugation.

18.7.2 Nagana Disease

A presumptive diagnosis of nagana disease can be based on clinical signs such as anemia in combination with poor condition. However, as for sleeping sickness, parasitological demonstration of the presence of trypanosomes in the blood of an animal is mandatory before commencing any treatment. The techniques for parasite detection are similar to those describe for sleeping sickness and include direct microscopy of blood films, concentration techniques, and animal inoculation (e.g., mice or rats) (Molyneux 1975).

18.8 Treatment of Sleeping Sickness and Nagana Disease

18.8.1 Sleeping Sickness

Therapy for sleeping sickness is currently unsatisfactory as only a five drugs, pentamidine, suramin, eflornithine, nifurtimox, and melarsoprol (Fig. 18.13), are available. All of these drugs were developed over 40 years ago (Steverding

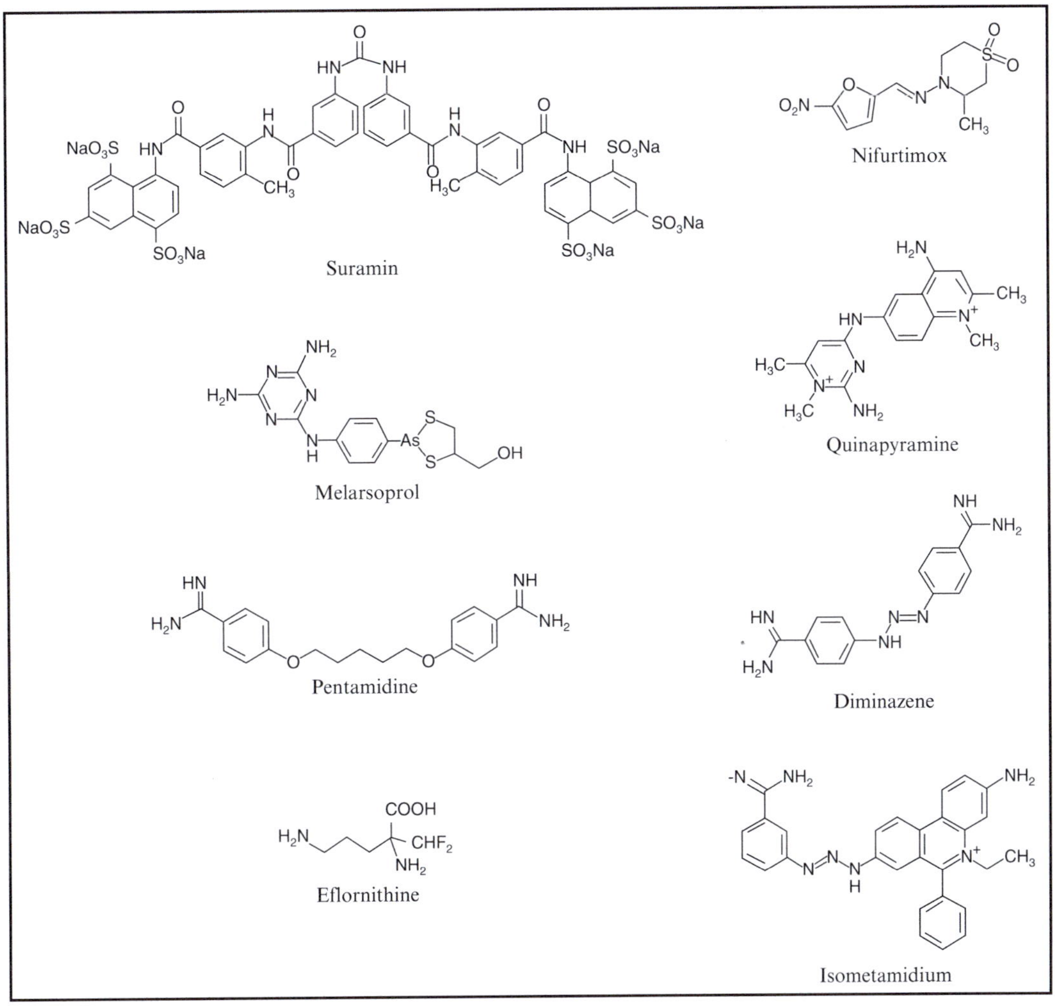

Fig. 18.13 Structures of drugs currently used for treatment of African trypanosomiasis. Suramin, melarsoprol, pentamidine, and eflornithine are used as monotherapies, while nifurtimox is used together with eflornithine in combination for therapy of sleeping sickness. Quinapyramine, diminazene, and isometamidium are used for treatment of nagana disease caused by *T. brucei*

2010). The main drugs used for treatment of early- and late-stage sleeping sickness have to be administered parenterally, display toxic side effects, and are sometimes ineffective. All anti-sleeping sickness drugs are donated to the World Health Organization by Bayer and Sanofi and the companies have agreed to secure the supply of the drugs until at least 2017 (WHO 2013).

Pentamidine is used as first-line treatment for early-stage *T. b. gambiense* sleeping sickness (Table 18.2). Pentamidine is a synthetic aromatic diamidine, chemically related to the antidiabetic drug phenformin. It was introduced in 1937 and is nowadays given as the isethionate salt (WHO 2013). The most common adverse side effects are hypoglycemia and hypotension (WHO 2013). Because of the hypoglycemic effect of pentamidine, patients are given sweet food or drinks before the administration of the drug. Pentamidine can also be used as alternative treatment for early-stage *T. b. rhodesiense* sleeping sickness (Table 18.2).

Although *suramin* is effective against the first stage of both forms of sleeping sickness, today its use is restricted to first-line treatment for early-stage *T. b. rhodesiense* disease (Table 18.2). This

Table 18.2 Treatment for early and late-stage sleeping sickness

Disease	Stage	First-line treatment	Dosage	Alternative treatment(s)
T. b. gambiense sleeping sickness	Early stage	Pentamidine isethionate	4 mg/kg/day i.m. or i.v. (diluted in saline and given in 2-h infusion) for 7 days	
	Late stage	Nifurtimox plus eflornithine (NECT)	Nifurtimox: 15 mg/kg/day p.o. in three doses for 10 days Eflornithine: 400 mg/kg/day i.v. in two 2-h infusions (each dose diluted in 250 ml water) for 7 days	Eflornithine: 400 mg/kg/day i.v. in four 2-h infusions (each dose diluted in 100 ml water) for 10 days Melarsoprol (second-line treatment in case of relapse after NECT): 2.2 mg/kg/day i.v. (dissolved in propylene glycol) for 10 days
T. b. rhodesiense sleeping sickness	Early stage	Suramin	Test dose of 4–5 mg/kg i.v. (day 1), then 20 mg/kg i.v. (dissolved in water) weekly for 5 weeks (e.g., days 3, 10, 17, 24, 31) (maximal dose/injection: 1 g)	Pentamidine isethionate: 4 mg/kg/day i.m. or i.v. (diluted in saline and given in 2-h infusion) for 7 days
	Late stage	Melarsoprol	2.2 mg/kg/day i.v. (dissolved in propylene glycol) for 10 days	

is because suramin is also active against onchocerciasis which is prevalent in West African sleeping sickness areas and its treatment with suramin could give rise to severe immunological reactions in coinfected patients (WHO 2013). Suramin is a polysulfonated naphthyl urea introduced in 1922 (WHO 2013). Frequent side effects are renal failure, peripheral neuropathy, skin lesions, and bone marrow toxicity (WHO 2013). Because hypersensitivity reactions (anaphylactic shock) can occasionally occur, the drug needs to be tested at a low dose before commencing treatment.

The standard first-line treatment for late-stage *T. b. gambiense* sleeping sickness is now the *nifurtimox-eflornithine* combination therapy (NECT) (Table 18.2). *Nifurtimox* is a 5-nitrofuran derivative and was introduced for treatment of Chagas disease in the late 1960s (Steverding 2008). In the past, nifurtimox was used on compassionate grounds for treatment of *T. b. gambiense* sleeping sickness patients who did not respond to melarsoprol (see below). Adverse effects associated with nifurtimox are anorexia, weight loss, psychological changes, excitability, sleepiness, and gastrointestinal symptoms. *Eflornithine* (DL-α-difluoromethylornithine, DFMO) is a specific and irreversible suicide inhibitor of ornithine decarboxylase. It was initially developed in the 1970s as a potential anticancer drug and was approved in 1990 for treatment of late-stage sleeping sickness caused by *T. b. gambiense* (Steverding 2008). The much shorter half-life of ornithine decarboxylase in *T. b. rhodesiense* is the reason why this drug is ineffective against this subspecies. The major drawbacks of eflornithine are its high costs and its requirement for large doses. The main side effects of the drug include bone marrow toxicity, alopecia, seizures, and gastrointestinal symptoms (WHO 2013). Compared with eflornithine monotherapy, NECT shows less adverse events and has a lower relapse rate (WHO 2013). Eflornithine monotherapy remains an alternative to NECT when nifurtimox is not available or is contraindicated (Table 18.2) (WHO 2013).

Melarsoprol is the only treatment for late-stage *T. b. rhodesiense* sleeping sickness (Table 18.2). Melarsoprol is a trivalent organic arsenical and was introduced in 1949 for treatment of the late stage of both West and East African sleeping sickness (Steverding 2008). The drug has a wide range of adverse effects including hepatic toxicity, peripheral neuropathy, skin rash, acute phlebitis, and vein sclerosis (WHO 2013). However, the most serious complication is a reactive encephalopathy which occurs in 5–18 % of patients, 10–70 % of whom die, leading to an average mortality for the treatment of about 5 % (WHO 2013).

Melarsoprol is nowadays only indicated in the treatment of *T. b. gambiense* sleeping sickness in the case of a relapse after NETC (Table 18.2) (WHO 2013).

After successful treatment, patients should be regularly monitored every 6 months for 2 years. Patients are considered cured if no trypanosomes can be found in blood, lymph, and CSF, and the WBC count in the CSF remains normal at the end of the follow-up period.

18.8.2 Nagana Disease

For treatment and prophylaxis of nagana disease caused by *T. brucei*, only three drugs, quinapyramine, diminazene, and isometamidium (Fig. 18.13), are available (Holmes et al. 2004). All these drugs have been developed and introduced between 1950 and 1961. Quinapyramine is a bis-quaternary compound and can be used as therapeutic as well as prophylactic drug (Table 18.3). It is effective against *T. brucei* infection in cattle, sheep, goats, and pigs. Due to the development of drug resistance, quinapyramine was withdrawn from the market in many parts of Africa in the 1970s. Quinapyramine is nowadays mainly used as prophylactic in pigs to prevent infections with *T. brucei*. Diminazene is like pentamidine an aromatic diamidine. Because diminazene is very quickly eliminated, it can be only used for therapy of nagana disease (Table 18.3). As it is not very toxic for cattle, it is mainly employed for treatment of bovine trypanosomiasis. It is also used for therapy of *T. brucei* infection in dogs. Isometamidium is a phenanthrinium derivative and effective against *T. brucei* infections in equids (Table 18.3). The drugs can be used as therapeutic and prophylactic agent (Table 18.3).

18.9 Epidemiology of Sleeping Sickness and Nagana Disease

18.9.1 Geographical Distribution of *T. brucei*, Sleeping Sickness, and Nagana Disease

The distribution of *T. brucei* is restricted to that of tsetse flies which are exclusively found in a belt that stretches south of the Sahara Desert and north of the Kalahari Desert between 14° North and 20° South latitude (Fig. 18.11) (Franco et al. 2014; Welburn et al. 2016). Whereas nagana disease occurs throughout this area, sleeping sickness has a spatially discrete distribution and currently occurs in some 250 rural foci scattered over 36 African countries (Fig. 18.11) (Franco et al. 2014; WHO 2016). In addition, the two human pathogenic subspecies of *T. brucei* have distinct allopatric distribution which coincides with the geographical extension of their respective tsetse fly vectors. *Trypanosoma brucei gambiense* has a much broader range and is found in 24 countries in West and Central Africa, while *T. b. rhodesiense* occurs in 13 countries in Eastern and Southern Africa (Fig. 18.11 and Table 18.1) (Franco et al. 2014; Welburn et al. 2016; WHO 2016). Uganda is the only country in which both *T. b. gambiense* and *T. b. rhodesiense* co-occur but in different regions without overlapping (Franco et al. 2014; WHO 2016).

18.9.2 Vectors

Tsetse fly populations are one of the most important factors in the epidemiology of sleeping sickness and nagana disease. There are 31 species and subspecies of tsetse flies but only members of the

Table 18.3 Treatment for *T. brucei* nagana disease

Drug	Animal	Use	Dosage
Quinapyramine	Cattle Sheep Goats Pigs	Therapy Prophylaxis	5 mg/kg s.c. 3–5 mg/kg s.c.
Diminazene	Cattle Dogs	Therapy	3.5–7 mg/kg i.m.
Isometamidium	Equids	Therapy Prophylaxis	0.25–0.5 mg/kg i.m. 0.5–1 mg/kg i.m.

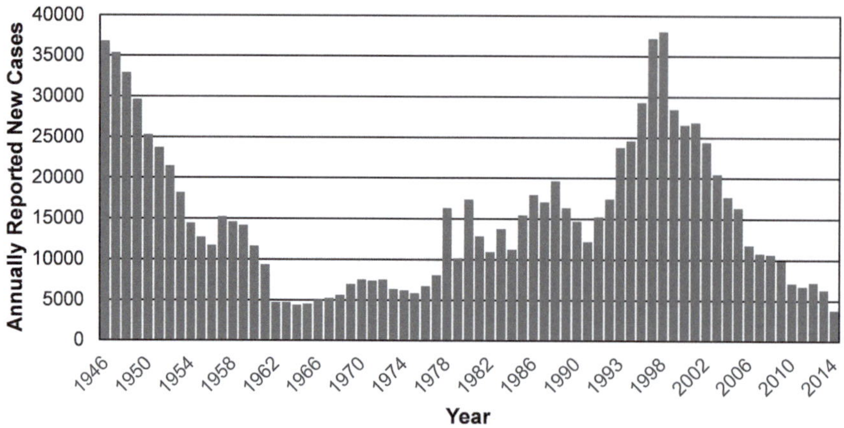

Fig. 18.14 Number of annually reported new cases of sleeping sickness since the end of the Second World War up to 2014. Data according the World Health Organization (WHO 2000, 2013, 2016; Franco et al. 2014)

palpalis and the morsitans groups are important vectors for the transmission of *T. brucei* (Franco et al. 2014). Whereas *T. b. brucei* is transmitted by both groups of tsetse flies, *T. b. gambiense* and *T. b. rhodesiense* are transmitted by members of the palpalis and morsitans group, respectively (Table 18.1) (Franco et al. 2014). The reason for the focal distribution of sleeping sickness (Fig. 18.11) is unknown and there are areas where tsetse flies are found but no cases of sleeping sickness. Transmission of the disease usually occurs when people enter the natural habitat of tsetse flies (Table 18.1) to carry out activities such as farming, fishing, and hunting. Humans are the epidemiologically important reservoir for West African sleeping sickness, while animals for East African sleeping sickness (Table 18.1). Livestock usually contract nagana disease when farmers take their animals into tsetse fly-infested areas for grazing.

18.9.3 Epidemics

There are some historical accounts of sleeping sickness epidemics from as long ago as medieval times (Steverding 2008). In the twentieth century, three major sleeping sickness epidemics are documented (Steverding 2008; Welburn et al. 2016; WHO 2016). The first began in 1896 and lasted until 1906 with 300,000 to 500,000 people estimated to have died. It affected mainly the Busoga focus in Uganda and Kenya and the Congo Basin. The second started in 1920 in a number of African countries which eventually began to die down in the late 1940s. The availability of chemotherapy in combination with systematic case detection and vector control led to a dramatic reduction in the incidence of sleeping sickness in the mid-1960s (Fig. 18.14). Since the end of the 1970s, sleeping sickness had reemerged in many African countries. The resurgence of the disease was closely linked to the collapse of the surveillance and control programs in most endemic countries after decolonization. This was the beginning of the third sleeping sickness epidemic in the twentieth century (Fig. 18.14). By the end of the 1990s, the scale of the epidemic had almost reached pre-World War II levels. Due to reinforced surveillance, the number of reported new cases has steadily dropped ever since (Fig. 18.14).

18.9.4 Current Situation

Since 2009, the number of registered cases of sleeping sickness fell below 10,000 for the first time in 50 years and in 2014 there were 3796 cases recorded (Fig. 18.14) (Franco et al. 2014; WHO 2016). At present the actual number of human African trypanosomiasis cases is estimated to be below 20,000 and the population at risk to be 65–70 million people (Franco et al. 2014; Welburn et al. 2016; WHO 2016). Most sleeping sickness cases are reported from the Democratic Republic of Congo accounting for 90 % (5642) of all new cases registered in 2013 (Franco et al. 2014). Seventeen endemic countries seem to be free of sleeping sickness as they have not reported any new cases for over a decade (Table 18.1) (WHO 2013, 2016).

Because of nagana disease, livestock production is difficult in many parts of sub-Saharan Africa. The affected area comprises currently about 10 million km^2 of arable land. The situation is exacerbated by the increasing occurrence of drug resistant trypanosomes.

18.9.5 Sleeping Sickness in Non-endemic Countries[2]

Although sleeping sickness is uncommon among patients from non-endemic countries, 244 cases of the disease have been documented in Europe and the USA between 1902 and 2012 (Neuberger et al. 2014). Most of these sleeping sickness cases were reported before 1920 and after 2000. However, there has been a change in the epidemiology of sleeping sickness among patients from non-endemic countries over the past century. In the colonial era (1902–1966), 74.3 % of cases were diagnosed in expatriates, missionaries, and soldiers, while in the post-colonial era (1967–2012), 72.8 % of the patients were tourists visiting safari parks in eastern and southern Africa. This change in patients affected is reflected by a shift in the *T. brucei* subspecies involved. Whereas before 1966 the majority of cases were due to infections with *T. b. gambiense* (76.2 %), after 1966 most cases were the result from infections with *T. b. rhodesiense* (76.6 %).

[2]Dr. Manuel Augusto Pirajá da Silva (1939), a pioneer in the study of intestinal schistosomiasis, suggested that many cases of "banzo" (sadness) of African slaves in Brazil were due to sleeping sickness (CBM).

18.10 Vector Control

As sleeping sickness and nagana disease are transmitted by tsetse flies, vector control is an important strategy for protecting humans and animals from acquiring trypanosome infections: without tsetse flies, there can be no African trypanosomiasis. The aim of control measures is to lower the number of tsetse flies or to eradicate tsetse fly populations in order to reduce the spread of disease or eliminate transmission entirely. Several control strategies have been tried since the discovery of tsetse flies as the vector for African trypanosomiasis.

One of the oldest measures to reduce tsetse fly populations is bush clearing in order to destroy resting and breeding habitats of the insects (Steverding 2008; WHO 2013). This strategy was introduced in 1910 in many parts of Africa, although with varying degree of success. Elimination of wild animal hosts is another effective measure in the control of tsetse flies, which has been employed since the 1920s (Steverding 2008; WHO 2013). The reduction of game animals on which tsetse flies feed reduces the number of tsetse flies rapidly, thus lessening trypanosomiasis transmission. However, habitat and game destruction are not anymore justifiable because of conservation concerns.

With the discovery of insecticides, ground and aerial spraying was the method of choice for tsetse fly vector control during the 1950s to the 1970s (Steverding 2008; WHO 2013). As most insecticides have the potential to interfere with ecosystems by affecting nontarget species, their use is highly controversial nowadays. In addition, there also exists the risk of the development of resistance to insecticides, although this has not yet been reported for tsetse flies (WHO 2013). Nevertheless, insecticide spraying still remains an important part of an integrated tsetse fly vector control management.

Trapping is another useful measure to reduce quickly tsetse fly populations, owing to their low reproduction rate (WHO 2013). The traps deployed have different sizes and shapes and usually consist of black and blue panels of cotton cloth. The trapping efficiency can be increased when the traps are combined with appropriately designed dispensers for attractants (odor-baited traps) such as acetone, a tsetse fly-attracting chemical found in the breath of cattle. Additional impregnation of traps with insecticides ensures that caught tsetse flies will be killed. Odor-baited and insecticide impregnated traps have been used in many African countries and have shown to suppress tsetse fly populations by 99 % (WHO 2013). Cattle coated with insecticides can serve as mobile bait to catch and destroy tsetse flies (WHO 2013).

Sterile male technique is the key method when it comes to eradicating tsetse flies (WHO 2013). The basis for this method is the fact that female tsetse flies usually mate only once in their lifetimes. Females fertilized by sterile male will produce no viable offspring resulting in a decline in the wild tsetse fly population. Male tsetse flies are sterilized by gamma radiation and then released into the wild in overwhelming numbers, approximately at ten sterile males to one female. This method works best when the density of tsetse fly population is low. Therefore, tsetse fly populations need to be reduced first by over 95 % using other methods (WHO 2013). The sterile male technique can only be used effectively if the targeted area is isolated; otherwise, reinvasion from neighboring tsetse fly-infested regions will occur. This measure was successfully used in the eradication of *G. austeni* from the island of Zanzibar off the Tanzanian coast.

References

Aphasizheva I, Aphasizhev R (2016) U-insertion/deletion mRNA-editing holoenzyme: definition in sight. Trends Parasitol 32:144–150

Berriman M, Ghedin E, Hertz-Fowler C et al (2005) The genome of the African trypanosome *Trypanosoma brucei*. Science 309:416–422

Borst P, Rudenko G (1994) Antigenic variation in African trypanosomes. Science 264:1872–1873

Büscher P, Mertens P, Leclipteux T et al (2014) Sensitivity and specificity of HAT Sero-*K*-SeT, a rapid diagnostic test for serodiagnosis of sleeping sickness caused by *Trypanosoma brucei gambiense*: a case–control study. Lancet Glob Health 2:e359–e363

Chappuis F, Loutan L, Simarro P et al (2005) Options for field diagnosis of human African trypanosomiasis. Clin Microbiol Rev 18:133–146

Chaudhuri M, Ott RD, Hill GC (2006) Trypanosome alternative oxidase: from molecule to function. Trends Parasitol 22:484–491

DeJesus E, Kieft R, Albright B et al (2013) A single amino acid substitution in the group 1 *Trypanosoma brucei gambiense* haptoglobin-hemoglobin receptor abolishes TLF-1 binding. PLoS Pathog 9, e1003317

Franco JR, Simarro PP, Diarra A et al (2014) Epidemiology of human African trypanosomiasis. Clin Epidemiol 6:257–275

Gibson W (2015) Liaisons dangereuses: sexual recombination among pathogenic trypanosomes. Res Microbiol 166:459–466

Haanstra JR, González-Marcano EB, Gualdrón-López M et al (2016) Biogenesis, maintenance and dynamics of glycosomes in trypanosomatid parasites. Biochim Biophys Acta 1863:1038–1048

Holmes PH, Eisler MC, Geerts S (2004) Current chemotherapy of animal trypanosomiasis. In: Maudlin I, Holmes PH, Miles MA (eds) The trypanosomiases. CABI, Wallingford, pp 431–444

Horn D (2014) Antigenic variation in African trypanosomes. Mol Biohcem Parasitol 195:123–129

Krauth-Siegel RL, Comini MA (2008) Redox control in trypanosomatids, parasitic protozoa with trypanothione-based thiol metabolism. Biochim Biophys Acta 1780:1236–1248

Liang XH, Haritan A, Uliel S et al (2003) *Trans* and *cis* splicing in trypanosomatids: mechanism, factors, and regulation. Eukaryot Cell 2:830–840

Mitashi P, Hasker E, Ngoyi DM et al (2013) Diagnostic accuracy of loopamp *Trypanosoma brucei* detection kit for diagnosis of human African trypanosomiasis in clinical samples. PLoS Negl Trop Dis 7, e2504

Molyneux DH (1975) Diagnostic methods in animal trypanosomiasis. Vet Parasitol 1:5–17

Neuberger A, Meltzer E, Leshem E et al (2014) The changing epidemiology of human African trypanosomiasis among patients from nonendemic countries – 1902–2012. PLoS One 9, e88647

Roditi I, Furger A, Ruepp S et al (1998) Unravelling the procyclin coat of *Trypanosoma brucei*. Mol Biochem Parasitol 91:117–130

Pirajá da Silva MA (1939) Inércia das funções vitais alimentação. In: von Martius CF (ed) Natureza, doenças, medicina e remédios dos índios brasileiros. Companhia Editora Nacional, São Paulo, p 31

Povelones ML (2014) Beyond replication: division and segregation of mitochondrial DNA in kinetoplastids. Mol Biochem Parasitol 196:53–60

Ralston KS, Kabututu ZP, Melehani JH et al (2009) The *Trypanosoma brucei* flagellum: moving parasites in new directions. Annu Rev Microbiol 63:335–362

Steverding D (2008) The history of African trypanosomiasis. Parasit Vectors 1:3

Steverding D (2010) The development of drugs for treatment of sleeping sickness: a historical review. Parasit Vectors 3:15

Stuart KD, Schnaufer A, Ernst NL et al (2005) Complex management: RNA editing in trypanosomes. Trends Biochem Sci 30:97–105

Taylor K, Authié EML (2004) Pathogenesis of animal trypanosomiasis. In: Maudlin I, Holmes PH, Miles MA (eds) The trypanosomiases. CABI, Wallingford, pp 331–353

Uzureau P, Uzureau S, Lecordier L et al (2013) Mechanism of *Trypanosoma brucei gambiense* resistance to human serum. Nature 501:430–434

Vanhamme L, Paturiaux-Hanocq F, Poelvoorde P et al (2003) Apolipoprotein L-1 is the trypanosome lytic factor of human serum. Nature 422:83–87

Vanhollebeke B, Pays E (2010) The trypanolytic factor of human serum: many ways to enter the parasite, a single way to kill. Mol Microbiol 76:806–814

Vickerman K (1978) Antigenic variation in trypanosomes. Nature 273:613–617

Welburn SC, Molyneux DH, Maudlin I (2016) Beyond tsetse—implications for research and control of human African trypanosomiasis epidemics. Trends Parasitol 32:230–241

WHO (2000) African trypanosomiasis. In: WHO report on global surveillance of epidemic-prone infectious diseases. World Health Organization, Geneva, pp 95–106

WHO (2013) Control and surveillance of human African trypanosomiasis. World Health Organ Tech Rep Ser 984:1–237

WHO (2016) Trypanosomiasis, human African (sleeping sickness).World Health Org Fact Sheet 259. http://www.who.int/mediacentre/factsheets/fs259/en/. Accessed 7 Apr 2016

Xong HV, Vanhamme L, Chamekh M et al (1998) A VSG expression site-associated gene confers resistance to human serum in *Trypanosoma rhodesiense*. Cell 95:839–846

Other Major Trypanosomiasis 19

Patricia H. Stoco, Luiz C. Miletti, Kim Picozzi, Mário Steindel, and Edmundo C. Grisard

Abstract

In this chapter we will address *Trypanosoma* species of medical (*Trypanosoma rangeli*) or veterinary (*T. vivax*, *T. evansi*, *T. equiperdum*, *T. congolense*, and *T. simiae*) importance. *Trypanosoma rangeli* is a non-pathogenic trypanosome of the Latin American continent, which infects several mammalian species, including humans, and is transmitted through the bite of triatomine bugs (Chap. 18 has more details on the biology of triatomine bugs) mainly those belonging to the genus *Rhodnius*. Along with the controversial taxonomic position and the unknown life cycle in the vertebrate host, the sharing of antigenic determinants with *T. cruzi* (Chagas disease) and their sympatric distribution in an extensive geographical area are of major importance for diagnosis and epidemiology. *Trypanosoma equiperdum* may cause "dourine," a sexually transmitted trypanosomiasis of equids. *Trypanosoma evansi* is the etiological agent of "Surra," a disease of veterinary importance of horses, camels, and several livestock species worldwide, causing significant morbidity and mortality. The parasite is transmitted noncyclically by hematophagous flies in Africa, Asia, and South America and also by the common American vampire bat, *Desmodus rotundus*. *Trypanosoma vivax* is also transmitted mechanically to a variety of wild and domestic ungulate species in both American and African continents; in the last it is more commonly vectored by the tsetse flies, having a major impact on livestock production. Two other tsetse-transmitted trypanosomes are *Trypanosoma congolense*, the agent of

P.H. Stoco (✉) • M. Steindel • E.C. Grisard
Department of Microbiology, Immunology and Parasitology, Center of Biological Sciences, Federal University of Santa Catarina, Florianópolis 88040-900, SC, Brazil
e-mail: patricia.stoco@ufsc.br; mario.steindel@ufsc.br; edmundo.grisard@ufsc.br

L.C. Miletti
Department of Animal and Foods Production, Center of Agroveterinary Sciences, State University of Santa Catarina, Lages, SC, Brazil
e-mail: luiz.miletti@udesc.br

K. Picozzi
School of Biomedical Sciences, Centre for Infectious Diseases, College of Medicine and Veterinary Medicine, University of Edinburgh, Edinburgh, UK
e-mail: Kim.Picozzi@ed.ac.uk

C.B. Marcondes (ed.), *Arthropod Borne Diseases*, DOI 10.1007/978-3-319-13884-8_19

"Nagana," and *Trypanosoma simiae*, the agent of acute swine trypanosomiasis. In this chapter we address historical aspects of these parasites and their biology, focusing on the diseases and their relevance to humans and livestock.

Keywords
Dourine • Nagana • *Trypanosoma rangeli* • Tsetse flies • Surra

19.1 *Trypanosoma (Herpetosoma) rangeli*

19.1.1 Historical Aspects

The Venezuelan physician Enrique Tejera described *Trypanosoma rangeli* in 1920 while observing flagellate protozoans in the intestinal content of *Rhodnius prolixus* (Hemiptera: Reduviidae). Since then, emphasis has been placed upon the biology of *T. rangeli* such that the parasite life cycle within the mammalian host is still unknown. The geographical distribution in Latin America is related to the presence of natural reservoirs and triatomine vectors especially those belonging to the genus *Rhodnius*. Although considered as nonpathogenic for mammals, human infection by *T. rangeli* induces a humoral immune response whose antibodies also recognize antigens from *Trypanosoma cruzi*, the etiological agent of Chagas disease. Such cross-reactivity allows false-positive results on serological assays confounding diagnosis on regions where these two parasites overlap (Grisard et al. 1999; Guhl and Vallejo 2003).

19.1.2 Biology

Trypanosoma rangeli is pleomorphic, a characteristic shared with *T. cruzi*, presenting as short to very long forms in the blood of mammalian, gut, and hemolymph of triatomines or in culture in vitro (Fig. 19.1). Blood trypomastigotes are 26–38 µm long, including flagella, and a well-developed undulating membrane is easily observed. A single nucleus is located on the anterior part of the cell and a small, sub-terminal kinetoplast is observed (Fig. 19.2a).

The parasite is cultivated in LIT (*liver infusion tryptose*) alone or NNN (*Novy-MacNeal-Nicolle*) overlaid with LIT, both supplemented with 10–20 % fetal calf serum (FCS). In culture, short and long epimastigotes and/or trypomastigotes with variable lengths (10–50 µm) are observed and the doubling time takes 26.1 h (E. C. Grisard, personal information). Within the intestinal tracts of infected triatomines, most of the parasite forms are either short or long dividing epimastigotes and more rarely trypomastigotes. Some epimastigotes cross from the lumen of the gut to the hemolymph, and all of them transform into long epimastigote forms that are able to invade the hemocytes as well as

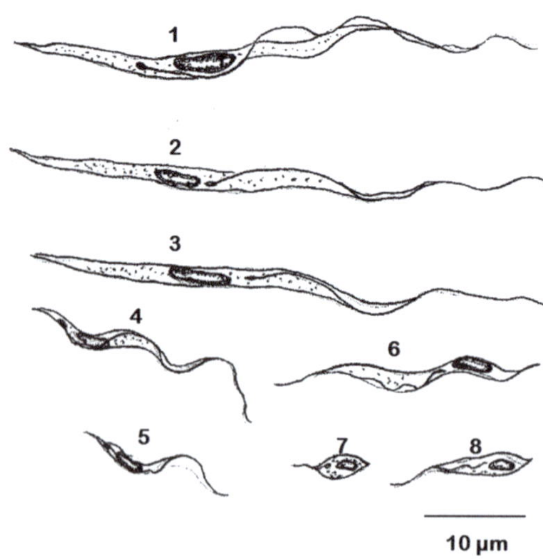

Fig. 19.1 General morphology of *Trypanosoma rangeli* in axenic culture. 1 = long trypomastigote, 2 and 3 = long epimastigotes, 4 and 5 = short trypomastigotes, 6–8 = short epimastigotes

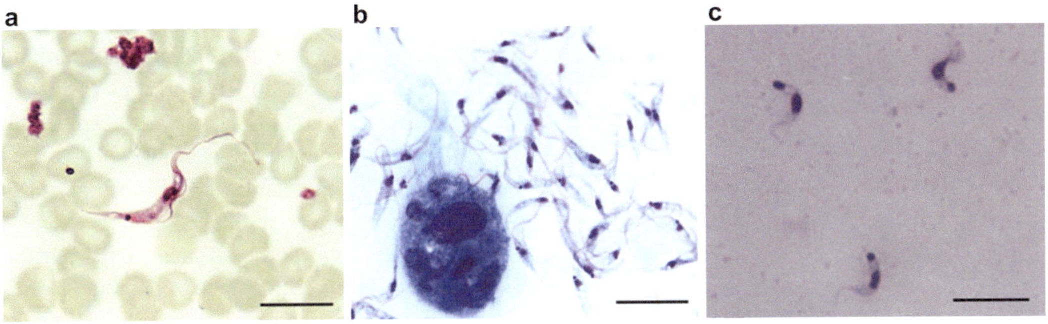

Fig. 19.2 *Trypanosoma rangeli* blood trypomastigotes (**a**), epimastigotes in hemolymph with hemocyte interaction (**b**), and metacyclic trypomastigote (**c**). Giemsa staining. Black bars = 10 μm

the salivary glands (Fig. 19.2b) (Meirelles et al. 2005). Round-shaped forms resembling amastigotes (amastigote-like), but showing no signs of division, can also be observed within hemocytes. The epimastigotes in the hemolymph migrate and penetrate the salivary glands of the insect where the parasite performs metacyclogenesis in the gland lumen to turn into infective metacyclic trypomastigote stage. The infective trypomastigotes are very short (10–12 μm long) presenting reduced free flagella if compared to blood trypomastigotes and the sub-terminal kinetoplast (Fig. 19.2c).

19.1.3 Molecular Characterization

Trypanosoma rangeli is a hemoflagellate protozoan parasite of the Order Kinetoplastida, Family Trypanosomatidae. The positioning of this taxon has been controversial due to the mixture of biological characteristics shared with American and African trypanosomes. While *T. rangeli* is primarily transmitted via bite of infected triatomine bugs (anterior transmission), which is a characteristic of the African trypanosomes (*T. brucei*), the blood trypomastigotes are biologically compatible with other species from the *Herpetosoma* subgenus, being the only form of the parasite detectable in mammalian hosts. Due to these mixed phenotypic characteristics, inclusion of *T. rangeli* in a new subgenus named *Tejeraia* was formerly proposed but still not widely adopted (Grisard et al. 1999).

The use of biological, biochemical, immunological, and molecular markers has demonstrated remarkable genotypic and phenotypic differences among *T. rangeli* strains (Silva et al. 2004). Thus, the intraspecific genetic variability of this taxon seems to be as variable as *T. cruzi*, despite the reduced number of genes on multi-copy families such as trans-sialidases, mucins, and mucin-associated proteins (MASP) and the shorter telomeric regions. In this sense, molecular characterization of *T. rangeli* strains isolated from distinct geographical regions and hosts was carried out using a variety of methods and markers such as kinetoplast minicircles DNA (kDNA), mini-exon or spliced-leader gene, 24SαrDNA, and the internal transcribed spacers (ITS) flanking the 5.8S subunit of the ribosomal RNA genes (rDNA). This analysis has revealed the existence of four distinct genetic groups (A–D) circulating in Central and South America (Maia Da Silva et al. 2004) (Fig. 19.3).

The *T. rangeli* kinetoplast DNA network contains three types of minicircles named KP1, KP2, and KP3. These circular DNA molecules can be distinguished by their distinct sizes due to the presence of four (KP3), two (KP2), or a single (KP1) conserved region. Based on the observation that some *T. rangeli* strains lack KP1 minicircles, parasite strains were typed based on this characteristic and grouped according to the presence (KP1+) and absence of (KP1−) of this particular molecule.

Further studies on *T. rangeli* strains from both genetic groups showed that KP1(+) Colombian

Fig. 19.3 Distribution of the *Trypanosoma rangeli* genotypes in Central and South America. Genotypes A to E are according to Maia da Silva et al. (2004) and KP1+ (*white letters*)/KP1− (*red letters*) according to Urrea et al. (2005). Genotype "B" (*yellow letter*) does not correlate with any KP1 group. Figure originally drawn by Dr. Gustavo A. Vallejo (reprinted with permission)

strains are highly associated to *R. prolixus* and the domestic cycle of the infection, whereas KP1(−) is more commonly aligned with *R. colombiensis* and the sylvatic transmission cycle, revealing a close parasite-vector coevolution (Vallejo et al. 2009). So far, typing of all *T. rangeli* strains isolated in Brazil are KP1(−) strains.

Considering the nuclear DNA, the *T. rangeli* karyotype obtained by pulsed-field gel electrophoresis (PFGE) revealed 16 chromosomal bands from 0.40 to 3.44 Mb and the absence of minichromosomes as observed in *T. cruzi*. A chromosomal size polymorphism was observed between KP1(+) and KP1(−) strains and, thus, may indicate that the actual number of chromosomes may be higher due to the co-migration of chromosomes of similar sizes (Stoco et al. 2014).

The *T. rangeli* genome is of ~24 Mb, being the smallest and least repetitive trypanosomatid genome if compared to the TriTryps (*T. cruzi*, *T. brucei*, *L. major*), *T. cruzi marinkellei*, and *Leishmania* spp. genomes available. A phylogenomic comparison positioned *T. rangeli* closer to *T. cruzi* than *T. brucei* at the genomic level.

A deep comparison was made between the *T. rangeli* genome and a range of pathogenic parasites. A total of 7613 coding sequences (CDS) were described, among which, a high percentage (~65.6 %) encode for hypothetical proteins. 7101 genes (93 %) are shared with other trypanosomatids that infect humans, *T. cruzi*, *T. brucei*, or *L. major*. Based on the genome sequencing and analysis, several new features were described for *T. rangeli* such as the absence of functional RNAi machinery, a higher sensitivity to oxidative stress in comparison to *T. cruzi*, and a reduced number and diversity of multigene virulence factor families such as MASPs, trans-sialidases, and mucins associated with infection of the mammalian host.

19.1.4 Transmission

Trypanosoma rangeli is primarily transmitted via bite of an infected vector (inoculative or anterior transmission) during blood meal, being different from *T. cruzi*, whose posterior or contaminative transmission occurs via contact of infected triatomine feces with mucosal membranes or through a lesion on the mammalian host skin. Although possible under experimental conditions, posterior transmission of *T. rangeli* is considered as having no epidemiological impact under natural conditions. Other transmission routes to humans such as oral or congenital infection are well described for *T. cruzi*, but there are no reports on nonvectorial transmission of *T. rangeli*.

The *T. rangeli* life cycle on mammalian hosts is still poorly understood, having scarce and conflicting reports in the literature (Vallejo et al. 2003). It is known that just 24 h after the infective bite, *T. rangeli* trypomastigotes are seen in the mammal bloodstream and a low level parasitemia is usually detectable for a week, decreasing thereafter to a level undetectable on the 15th day of infection. After this short acute phase with low level or undetectable parasitemia, parasites are only detected by indirect methods such as hemoculture and xenodiagnosis. Despite this subpatent parasitemia, experimental infection in distinct mammalian species has indicated that infection might persist for up to 3 years, although to date no signs of intracellular forms have been observed in any of the tissues assessed by histopathology. In vitro studies on the interaction of well-characterized *T. rangeli* strains with distinct cell lines have shown that this parasite is able to infect different cell lines at a low level; however, no signs of parasite division were found during a 120 h period of observation.

There are a few reports describing long-term blood parasitemia and the presence of amastigote nests in several tissues of mice experimentally infected with the *T. rangeli* strains Dog-82 and C-23. There are no reports of these experiments having been successfully repeated since their original publication. Considering the occurrence of natural mixed infections (*T. rangeli/T. cruzi*) in both mammals and triatomine vectors and the possibility of cross-contamination in laboratories that manipulate these parasites, the results obtained with Dog-82 and C-23 strains should now be further confirmed by molecular techniques.

The biological cycle of the parasite initiates when a triatomine bug ingests blood trypomastigotes from an infected mammal during the blood meal. Once in the insect, these trypomastigotes transform to epimastigotes in the stomach and midgut and start multiplying by binary fission, colonizing the whole intestinal tract. It is possible to observe both epimastigotes and trypomastigotes on the triatomines feces, but as mentioned above, the contaminative (posterior) transmission is considered of low epidemiological relevance (Fig. 19.4).

Trypanosoma rangeli epimastigotes have the unique characteristic to evade the intestinal tract and reach the insect hemocoel 20–40 days after the infective blood meal. Once in the hemolymph, parasites multiply intensively either freely or in clusters; after a variable period of time (usually 2–6 weeks) depending on the strain and the triatomine species, epimastigotes reach the salivary glands where they penetrate through the cells via an active process (Meirelles et al. 2005). On the salivary gland lumen, differentiation of epimastigotes to infective metacyclic trypomastigotes takes place (metacyclogenesis), and these metatrypanosomes are injected into the mammalian host bloodstream during the insect's blood meal.

However, a vector species-parasite strain correlation is well established for *T. rangeli* genetic groups (KP1+ and KP1-) having major influence on the parasite development within distinct triatomine species. Thus, distinct degrees of susceptibility are observed among triatomine species, especially those belonging to genus *Rhodnius*, to *T. rangeli* strains. Although variable and depending on the triatomine species, a single fully infected salivary gland may have up to 500,000 parasites.

Experimental infections by *T. rangeli* strains in distinct triatomine species revealed interesting and intriguing results (Urrea et al. 2005). A vector species-parasite strain coevolution has been proposed; thus, the finding of this parasite in a given triatomine species hemolymph does not imply that this species is a *T. rangeli* vector. Although *T. rangeli* infection is harmless for the mammalian hosts, the significant pathogenic effects that *T. rangeli* may cause to triatomines are well established by promoting defective or lethal effects in both molting and feeding processes.

19.1.5 Disease, Pathology, and Symptomatology

Infection by *T. rangeli* in mammals is considered nonpathogenic; however, such infections represent a serious problem for Chagas disease diagnosis due to the serological cross-reactivity, leading to misdiagnosis and inaccurate epidemiological inferences. Different authors have pointed out distinct levels of serological cross-reactivity between *T. rangeli* and *T. cruzi* in humans and experimental models. The well-described genomic variability of these parasites may influence the cross-reactivity levels, depending on the parasite strain and the host species.

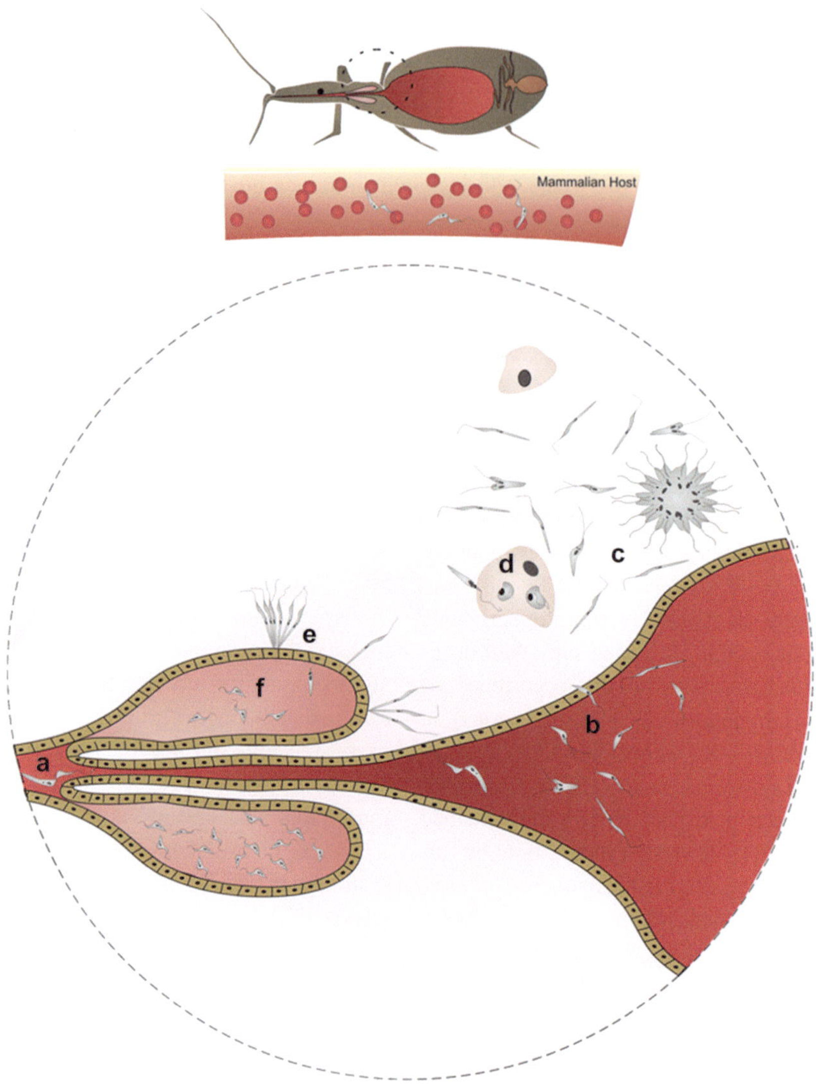

Fig. 19.4 Schematic representation of the *Trypanosoma rangeli* life cycle in the triatomine vector. A triatomine acquires *T. rangeli* infection through probing an infected mammalian host. Once ingested, blood trypomastigotes (**a**) develop into epimastigote forms, which are predominant in the midgut (**b**). Short epimastigotes multiply, colonize the insect's midgut, and invade the hemocoel (**c**), where they freely divide as long epimastigotes. Some parasites are observed inside hemocytes (**d**), but no evident signs of intracellular division are present. Long epimastigotes adhere to and penetrate the triatomines salivary glands (**e**), where they differentiate into short infective metacyclic trypomastigotes (**f**), which are injected along with the saliva during the triatomine blood meal. The life cycle and replication sites within the mammalian host are still unknown

Due to the absence of a pathogenic effect, most human infections by *T. rangeli* have been diagnosed as a result of serological surveys for Chagas disease whenever the diagnostic methods allow specific detection of each species (see Sect. 19.1.6). Although a few symptoms like fever or discrete lymphadenopathy may be present during the acute phase on the infection by *T. rangeli*, there are no reports of clinical symptoms on patients exclusively infected by this parasite.

Considering that almost 60 % of the patients infected with *T. cruzi* are within the indeterminate phase of the disease (positive serology and no clinical symptoms), and that there are a number of common (diagnostic) antigens shared with *T. rangeli*, it is noteworthy to mention that *T. rangeli*-infected patients may be underdetected, especially in areas where these parasites coexist and *Rhodnius* species are the local vectors.

19.1.6 Diagnosis

Diagnosis of *T. rangeli* infection in humans and other mammalian species can be achieved via direct demonstration of the parasite by fresh blood examination, in Giemsa-stained blood smears, hemoculture in LIT, NNN or blood-agar+LIT, or xenodiagnosis using triatomines from *Rhodnius* genus. However, as commonly observed for *T. cruzi*, these methods are of low sensitivity due to the very low blood parasitemia of *T. rangeli* in mammalian hosts.

The classical method for identification of *T. rangeli* in triatomines is the direct observation of flagellates in the hemolymph and salivary glands; this can be easily achieved by light microscopy.

There are a few serological tests based on *T. cruzi*-specific recombinant antigens that can be used for specific diagnosis of *T. rangeli* infection. However, due to the parasite polymorphism, the coexistence with *T. cruzi* and the sharing of soluble antigens between these parasite species, the occurrence of misdiagnosis, especially false positive, may occur. Several molecular techniques not used in routine diagnosis allow specific detection of *T. rangeli*. The polymerase chain reaction (PCR) followed by hybridization with species-specific probes directed to the kDNA minicircles is of high sensitivity and specificity, allowing precise differentiation from *T. cruzi*.

Based on the sequencing of the *T. rangeli* genome, comparative proteomics studies have been carried out using both epimastigotes and trypomastigotes from *T. cruzi* and *T. rangeli* using sera from chagasic patients presenting distinct clinical forms which are pointing out target proteins for the development of new and specific diagnostic methods.

19.1.7 Epidemiology

Since *T. rangeli* is transmitted among mammals, especially by triatomines of the genus *Rhodnius*, and has been found infecting the same reservoirs as described for *T. cruzi*, there is a clear superimposed geographical distribution in Central and South Americas. There is no precise estimation of the number of *T. rangeli*-infected humans in these areas, but over 3000 cases have been described in seven Central and South American countries (Fig. 19.5). In Brazil, *T. rangeli* infection in humans was described at the Barcelos Municipality, Amazonas State (northern Brazil).

However, natural infection of mammalian reservoirs and triatomine bugs has been reported in the Brazilian States of Amazonas, Pará, Rondônia, Tocantins, Ceará, Alagoas, Bahia, Minas Gerais,

Fig. 19.5 Distribution of *Trypanosoma rangeli* findings in humans, triatomines, or mammalian reservoirs in Central and South America. *Shaded countries* are areas of human Chagas disease transmission

Table 19.1 Natural or experimental infection of salivary glands from distinct triatomine species by *Trypanosoma rangeli*

Species	Infection	
	Natural	Experimental
Rhodnius prolixus	X	X
R. pallescens	X	X
R. ecuadoriensis	X	X
R. dalessandroi	X	
R. pictipes	X	
R. robustus	X	X
R. neglectus		X
R. neivai		X
R. domesticus		X
R. colombiensis	X	X
R. nasutus	X	X
R. brethesi		X
R. montenegrensis	X	
Triatoma dimidiata		X
T. patagonica		X
T. protracta		X
T. infestans		X
T. vitticeps		X

Modified from D'Alessandro and Saravia (1992)

and Santa Catarina. In these areas, dozens of distinct mammalian species belonging to five orders (Carnivora, Edentata, Marsupialia, Primates, and Rodentia) and six triatomine species have been found naturally infected. Thirteen out of the 19 triatomine species of the genus *Rhodnius* are proved *T. rangeli* vectors under natural or experimental conditions (Table 19.1). Also, some species of the genus *Triatoma* have revealed their ability to transmit *T. rangeli* under experimental conditions, being considered as putative vectors (Table 19.1).

19.1.8 Prophylaxis and Treatment

Considered as a nonpathogenic parasite of mammals, human infection by *T. rangeli* is benign and asymptomatic. Having no reports of any pathology inflicted by the parasite, there is no indication for treatment.

Common epidemiological features between *T. rangeli* and *T. cruzi*, such as transmission by triatomine bugs and a wide range of mammalian hosts within a superimposed geographical area, indicate that most of the prophylactic measures recommended for Chagas disease are also applicable to prevent infection by *T. rangeli*. In fact, efforts on combating *T. cruzi* vectors carried out during many years on the southern cone of Latin America should have affected *T. rangeli* epidemiology and transmission, which was not assessed at that time.

19.2 *Trypanosoma (Trypanozoon) evansi*

19.2.1 Historical Aspects

Trypanosoma evansi (Steel 1885) was originally observed by Griffith Evans in the bloodstream of infected horses and dromedary camels in India, being the first pathogenic trypanosome described in the literature (Desquesnes 2004). On his original description, Griffith describes motile microbes that once transferred from sick to healthy horses induced pathogenesis. Convinced that parasites were the cause of the disease, he proposed that infection occurs via ingestion of contaminated water (Fallis 1986). Later, Steel (1885) described spirochetes in the blood of sick horses, having named the organism as *Spirochaete evansi*, which was redescribed by Balbiani (1888). In 1896 the parasite was placed in genus *Trypanosoma*, and Leonard Roger described its transmission by tabanid flies in 1899 (Roger 1901; Hoare 1972). While first described in Brazil in the nineteenth century (Desquesnes 2004), it is believed that *Trypanosoma evansi* was introduced to the South American continent in the late sixteenth century as an infection of horses from Spanish conquerors.

19.2.2 Biology

Trypanosoma evansi belongs to the subgenus *Trypanozoon* (section Salivaria), Family Trypanosomatidae, and Order Kinetoplastida.

Rarely and occasionally infecting humans (Joshi et al. 2005), *Trypanosoma evansi* is biologically similar to *T. equiperdum*, the causative agent of dourine (Brun et al. 1998), and it morphologically resembles the long, slender forms of *T. brucei brucei, T. b. gambiense*, and *T. b. rhodesiense* found in the tsetse fly. *Trypanosoma evansi* varies between 15 and 34 μm length, showing a long free flagellum and thin posterior extremity with a small, sub-terminal kinetoplast. However, it is not unusual for akinetoplastic strains to be described when directly observed, these were considered a distinct species in Latin America and named as *T. equinum*, and this is no longer considered a valid species but a synonym of *T. evansi* (Uilenberg 1998). Along the intraspecific variability observed among strains, long and shorter forms may be observed in distinct hosts (Fig. 19.6).

Molecular characterization of *T. evansi* strains isolated in Asia, Africa, and South America by isoenzymes and DNA minicircles analysis showed a high genetic homogeneity indicating a single origin (Ventura et al. 2002).

However, some reports suggest that *T. evansi* may have derived from *T. brucei* in several instances. Despite such genetic similarity, *T. evansi* is not able to multiply in the vector, and due to loss of the maxicircles of kinetoplastic mitochondrial DNA, its cyclical transmission by flies of the genus *Glossina* does not occur (Lun and Desser 1995).

Trypanosoma evansi outer membrane is coated by a dense variable surface glycoprotein (VSG) layer, which acts as major immunogenic antigens eliciting specific antibodies. *Trypanosoma evansi* is able to evade the host immune response by switching the VSG, a phenomenon known as antigenic variation; this immunological avoidance strategy is common among the African trypanosomes.

19.2.3 Transmission

Trypanosoma evansi is mechanically transmitted by biting flies. The dissociation between the parasite and a specific insect vector has resulted in *T. evansi* having the largest distribution among the pathogenic parasites of the genus *Trypanosoma*.

Mechanical transmission of *T. evansi* in America occurs via the bite of bloodsucking horse flies of the genus *Tabanus* (Diptera: Tabanidae). Due to the lack of cyclical development of the parasite in these American vectors, transmission occurs via parasites present in the proboscis of the horse flies after feeding on infected animals. *Tabanus* is the main vector of *T. evansi* in the

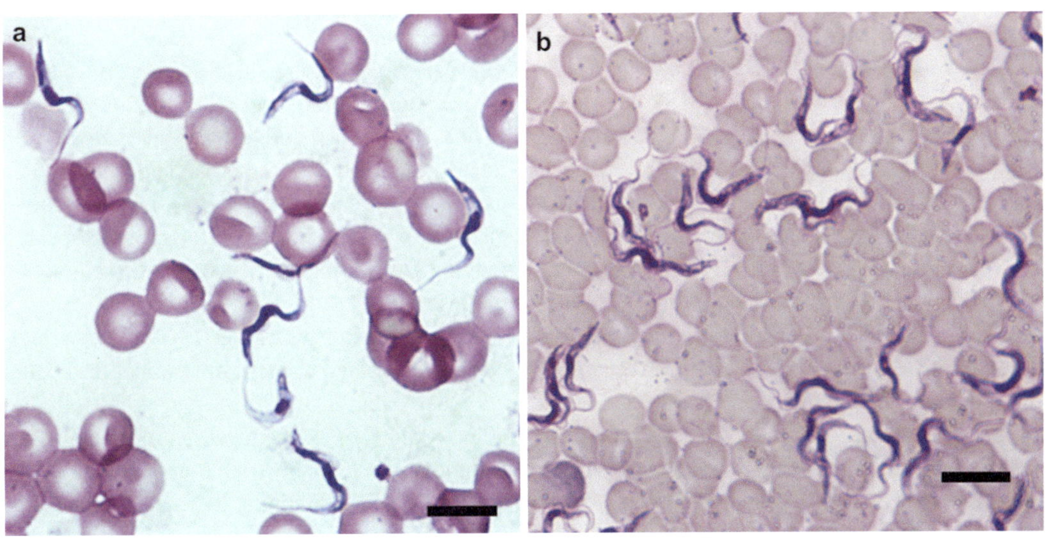

Fig. 19.6 *Trypanosoma evansi* blood trypomastigotes. Parasitemia observed after 24 (**a**) and 72 h of infection of Swiss mice (**b**). Giemsa staining. Black bar = 10 μm

Brazilian Pantanal region and *T. nebulosus* is the suspected vector in Colombia. However, flies belonging to the genera *Stomoxys* and *Haematobia (=Lyperosia)* are also considered as putative vectors in other continents. Besides these flies, the vampire bat *Desmodus rotundus* is also able to transmit the parasite in South America.

A variety of mammalian species (camels, horses, donkeys, cattle, zebu cattle, goats, pigs, dogs, buffaloes, elephants, capybaras, coatis, tapirs, deer, and several wild rodents) are known as hosts of the parasite. Experimental infection in mice results in a fast-growing parasitemia that can be detected hours after infection, reaching a countless number of parasites in 3–4 days (Fig. 19.6). These trypomastigote forms can be stored in liquid nitrogen upon addition of 10% glycerol or DMSO.

Although *T. evansi* can be maintained in vivo by successive blood passages in laboratory rodents, in vitro axenic cultivation is difficult but sometimes is possible after adaptation of parasites in cultures containing defined cell lines, for example, murine fibroblasts (Hirumi et al. 1997).

19.2.4 Disease

The disease caused by *T. evansi* is referred by a number of distinct denominations around the world, commonly known as "Surra" in Africa and Asia and "Mal das cadeiras," "Derrengadera," or "Murrina" in Latin America (Desquesnes et al. 2013).

In Africa, *T. evansi* is mainly found infecting camels, known as the main hosts, but can also be found in horses and dogs, to whom the infection is often fatal. In Asia, *T. evansi* is a major parasite for water buffaloes (*Bubalus bubalis*), representing a major economic issue, while in Latin America the parasite is found infecting domesticated animals like horses, cattle, buffaloes, sheep, and goats, also incurring in major economic losses (Desquesnes et al. 2013). In Brazil, the main natural reservoirs are capybaras (*Hydrochaeris hydrochaerus*) and coatis (*Nasua nasua*) (Silva et al. 1995).

The first case of human infection by *T. evansi* was reported in India in 2005. Later, it was found that the human infection was only possible because the patient presented a lack of trypanolytic activity due to frame shift mutations in both apoL-I alleles.

19.2.5 Pathogenesis and Clinical Signs

The pathogenic effects of *T. evansi* to the susceptible hosts are classical and distinct from any other mammalian pathogenic trypanosomes. The observation of *T. evansi*-infected animals characteristically occurs during outbreaks of the disease. During these outbreaks, the initial phase is characterized by elevated morbidity and mortality of the animals. Over time, the outbreaks are characterized by a reduction of the incidence of disease, of the infection rates, and of the severity of the disease.

In horses, the disease manifestation is most severe, often reaching mortality rates over 90%. After 1–2 weeks of incubation, the most frequent symptoms are recurrent fever with transient hyperthermia, anemia, weakness, ophthalmic infection with lacrimation and conjunctivitis, cutaneous rash, lymphadenopathy, weight loss, and, classically, edema of the intermandibular region, the limbs, and the abdomen (Silva et al. 1995). Except for the progressive anemia, most of these symptoms may disappear during the course of infection. The parasite multiplies at the horse fly bite site where sometimes causes a cutaneous reaction. In the acute phase, other signs as petechiae on mucous membranes, urticaria, and keratitis are usually observed. However, such symptoms are of variable intensity in the different host species, ranging from a total absence to a dissemination of these signs.

The chronic phase of the *T. evansi* infection is closely associated with the invasion of the central nervous system (CNS) by the parasite (Rodrigues et al. 2009). The consequent neurological manifestations are highly variable and dependent on factors that are inherent to the host (species, susceptibility) and to the parasite (virulence).

The invasion of the CNS by *T. evansi* may occur within 2 weeks after infection, which is a frequent complication in infected horses, leading to the development of meningitis. Moreover, intraocular hemorrhage due to vitreous humor invasion by the parasite has also been reported. After the invasion of the CNS, symptoms of posterior paralysis are progressively observed, reaching a stage where the infected animal cannot stand still. Parasitemia is variable during this period, being relatively high in the late stages of the disease progression.

19.2.6 Diagnosis

Clinical signs of *T. evansi* infection are usually evident, but since these symptoms do not have a unique etiology, they should not be used as the sole form of diagnosis. Despite the variable parasitemia, direct demonstration of the parasite on fresh blood examinations, stained blood smears, or microhematocrit is usually possible and allows the determination of the etiologic agent. However, in areas where *T. evansi* and *T. brucei* occurs in sympatry such as in Africa, microscopic observation can lead to misdiagnosis. Likewise, the coexistence of *T. evansi* and *T. vivax* in the Americas may lead to misdiagnosis of these etiologic agents (Silva et al. 2002). Thus, a variety of techniques are currently being used for laboratory diagnosis of *T. evansi* infection.

The subinoculation in animals susceptible to infection (rats and mice) is often effective for parasitic propagation. However, due to the variable parasite infectivity to such models and the susceptibility of these laboratory animals, this method has been discontinued, as it does not offer diagnostic discrimination.

A range of tools, based on the detection of an immune response against *T. evansi*, such as ELISA and indirect or direct latex agglutination and hemagglutination, have been used with variable specificity. In addition to these assays, agglutination card to trypanosomiasis (CATT) also shown promising results, although the reproducibility of the CATT test/*T. evansi*® is as yet unsatisfactory and cross-reactions with *T. vivax* have been observed.

However, the immunodiagnostics of *T. evansi* infection is limited by the antigenic variability related to the VSGs. In addition the presence of an immune response does not differentiate between current and historic infections as antibodies directed against the parasite persist for more than 1 year. In this sense, the search and use of monoclonal antibodies directed against a diagnostic antigen has been encouraged. Among the various studies, a monoclonal antibody specific for the *T. brucei* group has been demonstrated to aid the detection of infection by capture ELISA in a wide variety of hosts; this approach has proven to be very sensitive allowing the detection of antigens in parasitological negative animals.

The use of PCR in diagnosis of infection has shown variable reproducibility, sensitivity, and specificity. To date, the molecular detection for *Trypanozoon* subgenus parasites targets a multicopy element originally identified by Sloof et al. (1983). The longevity of the TBR primers is testament to their role as the gold standard, at least in the first instance. If necessary, subspecies confirmation can be obtained with more specific primers such as TE2249/TE2250, which target a mini-circle sequence, or even the use of specific probes. Other primers specific for RoTat or non-RoTat strains can be used. Another widely used molecular marker for the specific detection and characterization of isolates of *T. evansi* is the internal transcribed spacers (ITS) of the ribosomal RNA gene (Desquesnes et al. 2013).

During the last few years, a variety of PCR assays, including quantitative PCR (qPCR) assays, have been described in the literature. The qPCR assay targeting the ITS region was tested with *T. evansi* strains isolated from different hosts and geographic regions, revealing to be a sensitive and highly specific assay, allowing early diagnosis and differentiation of *T. evansi* from *T. brucei* and other related species that may also infect the same hosts.

19.2.7 Epidemiology

In Africa *T. evansi* is widely distributed, occurring from northern border of the Sahara desert to the east, including Sudan and northern Kenya. These areas correspond to the north of tsetse belt distribution, where this trypanosome is responsible for the "Surra" in camels, horses, and dogs.

In Asia the infection is widely distributed; *T. evansi* is mainly a disease of horses and buffaloes, and the strains are more pathogenic than those of Africa and America. For example, in India the parasite is present all over the country affecting many kinds of hosts (cattle, horses, donkeys, camels, and dogs). Elsewhere, in countries such as Cambodia, Laos, Vietnam, and Malaysia, *T. evansi* is given little attention, with few epidemiological surveys in recent history despite the endemic nature of the parasite (Uilenberg 1998).

In Central and South American countries (Panama, Venezuela, Colombia, Brazil, Bolivia, Paraguay, Argentina, and Peru), "Surra" is predominantly a disease that affects horses and sporadically buffaloes. Different from other continents, in Latin America *T. evansi* is found in a wide range of wild reservoirs like capybaras (*Hydrochoerus hydrochaeris*), coatis (*Nasua nasua*), tail deers (*Odocoileus virginianus chiriquensis*), brocket deers (*Mazama temama*), wild pigs (*Tayassu tajacu*), wild dogs (*Canis azarae*), and ocelots (*Felis pardalis*) (Seidl et al. 1998).

Considering the variety of susceptible wildlife species which includes the role of the vampire bat as a biological vector, the extensive use of horses, the use of herding dogs, the sympatric occurrence of bloodsucking Diptera, and the migration of herds over the period of flooding in the wetlands of Pantanal region, *T. evansi* transmission in Brazil and Venezuela has a complex epidemiological scenario.

With regard to the epidemiology of this infection, it is important to also mention the role of trade in the spread of infection; for example, in Europe, mainly in Spain and France, infection by *T. evansi* has been reported in horses and camels, probably due to the import of infected animals (Desquesnes 2004). The reports of *T. evansi* recognizing endemic areas highlight the necessity of appropriate surveillance systems to minimize the risk of onward infection and consequent economic losses.

19.2.8 Prophylaxis and Treatment

There are a variety of drugs that may be used for the treatment of *T. evansi* infection. These drugs can be divided into two groups, drugs solely used as curative and drugs used in mixed curative/preventive schemes. A curative drug is the one able to eliminate the parasite and normally disrupt or block one or more vital process, which are essential to the invading microorganism (Uilenberg 1998; Desquesnes et al. 2013). The curative/preventive drug scheme is also known as chemoprophylaxis, mixing the treatment of the infected animal and prevention against new infections.

Among the available drugs, the most widely used curative drug is diminazene aceturate, used mainly in cattle and small ruminants. Other drugs include isometamidium chloride used in cattle, Cymelarsan®, which is registered only to use for curative treatment for camels, and suramin and quinapyramine, for camels and horses, respectively (Silva et al. 2002).

Since the use of some of these drugs has been performed for more than three decades in different geographic regions, different degrees of parasitic resistance have been reported in the literature. An important factor is that the parasites exhibit a high degree of cross-resistance, since the available drugs have related chemistries. As the treatment (prophylactic or curative) of an animal or an entire herd is based primarily on cost, other strategies for prophylaxis and control can be adopted with relative efficiency.

There is limited information on the economic impact of *T. evansi* infection in endemic or non-endemic countries. A detailed study of the cost-effectiveness of treatment of horses in the Brazilian Pantanal region found that, regardless of the size of property and number of animals, an outbreak of *T. evansi* in the Pantanal would cost around US$ 2.4 million (Seidl et al. 1998). The only other study using bioeconomic infectious disease models has estimated losses up to U$158,000/year on livestock in the Philippines (Dargantes et al. 2009).

19.3 *Trypanosoma (Dutonella) vivax*

19.3.1 Historical Aspects

The German physician Hans Ziemann originally described *Trypanosoma vivax* in 1905 from blood of livestock animals in Cameroon (Steverding 2008); following identification the parasite is now known to be endemic across the tsetse belt of sub-Saharan Africa.

Introduction in South America probably occurred in the nineteenth century due to import of cattle from Senegal. Since then, livestock infection was reported in Venezuela (1920), Guadalupe (1926), Martinique (1929), Colombia (1931), Suriname (1938), Panama (1941), and Guyana (1953) with more recent reports from Brazil (1972), Peru (1978), and Bolivia (1996) as a result of trade cross the Pantanal region (Osório et al. 2008).

19.3.2 Biology

Trypanosoma vivax belongs to the genus *Trypanosoma*, subgenus *Dutonella* (Salivarian section), Family Trypanosomatidae, and Order Kinetoplastida. *Trypanosoma vivax* is a typical kinetoplastid protozoan parasite, having the extracellular trypomastigote and epimastigote forms with a single anterior flagellum. It differs from other trypanosomes by its reduced length (21.0–25.4 μm), a large and terminal kinetoplast which appears cup shaped in electron micrographs (Gardiner 1989); this motile parasite also has rapid vibrational movement in wet blood films making it easy to discern with direct observation. *Trypanosoma vivax* is morphologically related to *T. uniforme*, a much smaller trypanosome (~18.0 μm in length), and rarely reported, the most recent reports of which were from a giraffe in Tanzania in 1967 and a *T. uniforme*-like infection of cattle observed in Ontario, Canada.

While in tsetse flies (*Glossina* spp.), the vector of *T. vivax* in the African continent, both trypomastigote and epimastigote forms are observed, only trypomastigotes are found within the bloodsucking insects that mechanically transmit the parasite in South America. Despite the importance of the parasite morphology for diagnosis, comparative studies using parasites from naturally or experimentally infected rodents showed the existence of a longer, more granular bloodstream forms with sub-terminal kinetoplast, indicating the plasticity of this taxon and that morphology cannot be used as unique marker for diagnosis and specific identification (Gardiner 1989).

In general, it is accepted that in the mammalian host, the slender forms of the parasite (trypomastigotes), observed in the early stage of infection, evolve into stumpy forms with a variable surface glycoprotein coat. *Trypanosoma vivax* can infect a variety of hosts, especially ungulates, but has also been found infecting wild animals, such as deer, that are considered reservoirs of the parasite.

As the *T. vivax* genome sequencing at the Sanger Institute is undergoing, the availability of these data will soon allow the understanding of important biological aspects as well as the identification and characterization of novel diagnostic and/or prognostic/virulence markers.

19.3.3 Transmission

In Africa, *T. vivax* is transmitted by tsetse flies, genus *Glossina*. The parasite develops in the proboscis of this invertebrate host, where trypomastigotes evolve to epimastigotes. *Glossina* spp. are the only known vectors in which *T. vivax* is able to multiply and remain in the infective phase throughout the insect's life (Osório et al. 2008). However, predominance of *T. vivax* in *Glossina*-free areas, and its persistence in areas where *Glossina* spp. have been controlled, indicates the possibility of mechanical transmission of this parasite in Africa. This has been demonstrated for *Stomoxys* spp., where under experimental condition success rates of 11.5% for *T. brucei*, 3.4% for *T. vivax*, and 0.9% for *T. evansi* were achieved (Mihok et al. 1995). This has also been reproduced for *T. vivax* under seminatural conditions.

In Latin America, *T. vivax* is mainly transmitted by horse flies and *S. calcitrans*. However, biting midges (*Culicoides* spp., Ceratopogonidae), *Haematobia* spp., and certain Culicidae among other bloodsucking insects might also be involved in parasite transmission. Considering the geographical distribution of *T. vivax*, which includes South America, Asia, and Africa, it is estimated that more than half billion cattle and over 100 million buffaloes are at risk of acquiring the infection in these continents.

19.3.4 Disease, Pathology, and Symptomatology

In Africa, the term "Nagana" refers to animal trypanosomiasis that can be due to *T. vivax, T. congolense*, or *T. brucei brucei*[1] in cattle and small ruminants or to *T. brucei, T. congolense, T. evansi*, and *T. equiperdum* in horses. In Latin America the *T. vivax* infection is commonly known as "secadera," "huquera," or "cachohueco" (Desquesnes 2004). In South America *T. vivax* infects the same farm-raised animals as in Africa and natural infection was also described in llamas (*Lama glama*).

Experimental infections of different species of domestic ruminants (cattle, sheep, and goats) with *T. vivax* revealed major differences on the parasitemia levels, including variation in the prepatent period and the oscillation of parasitemia. In addition experimentally infected animals appear to have no hepatic or renal dysfunction (Desquesnes 2004). The postmortem analysis in experimentally infected animals reveals lymph node swelling, thickening of the cerebral meninges, meningitis, meningoencephalitis, and meningoencephalomyelitis.

Such variability impairs the parasitological diagnosis, leading to false-negative results, suggesting that *T. vivax* infection has three distinct phases: acute, subacute, and chronic.

Acute infection with *T. vivax* can be fatal to cattle and is characterized by an elevated, long-term parasitemia. The main clinical signs observed in the acute phase are progressive anemia associated with immunosuppression accompanied by a weight loss, sublingual and dewlap edema, lethargy, and lacrimation. In the most severe cases, fatal paresis of the hindquarters often occurs, and if untreated, animals may die within 4–5 weeks (Desquesnes 2004).

The commonly observed mucous discoloration is associated with anemia and can be monitored by assessment of hematocrit and hemoglobin levels, as well as a decrease on the number of white blood cells and neutrophils.

Infection by *T. vivax* also has effects on fertility of bovines and ovines. Multiple abnormalities have been identified in the sperm morphology of experimentally inoculated animals, and abortion has been reported in experimentally infected sheep. However in cases of abortion due to *T. vivax*, the parasite has not been found in fetal blood or in amniotic fluid.

The "hemorrhagic syndrome" (high and persistent parasitemia, fever, pernicious anemia, and generalized hemorrhages of viscera and mucosa) described in East Africa (Gardiner 1989) and in West Africa (Vohradsky 1971) has not been reported in South America.

However, some studies indicate the involvement of the nervous system, characterized by the lack of coordination, muscular spasms, strabismus, and blindness, the latter being reversible by administration of specific treatment (Gardiner 1989).

19.3.5 Diagnosis

Trypanosoma vivax infection can be diagnosed by parasitological, immunological, and molecular methods (Desquesnes 2004). Despite the reduced sensitivity and the morphological plasticity, morphology and the measurement of blood forms remain an important diagnostic method.

Trypanosoma vivax is not as easily cultured in vitro as other trypanosomes such as *T. cruzi* and *T. rangeli*. Although time consuming and expensive, establishment of an axenic culture is often possible aiming parasite isolation and/or bulk parasite producing for downstream assays, having no applicability for diagnostic purposes (Gumm 1991).

[1] See Chap. 18 for nagana caused by *Trypanosoma brucei brucei* (CBM).

Variations of classical parasitological techniques have been suggested, revealing distinct positive and negative predictive values. Along stained blood smears and lymph nodes puncture, the microhematocrit technique (Woo technique) is one of the most widely used methods for diagnosis of *T. vivax* infection (Woo 1970). This technique allows early detection of motile trypanosomes between the leukocyte layer and the plasma, usually 6–10 days before detection in fresh or stained blood smears (Molyneux 1975).

Most serological tests available lack specificity due to cross-reaction with other trypanosomatids, being particularly problematic in endemic areas where *T. evansi* infection also occurs. A promising ELISA test directed to a *T. vivax* soluble antigen was used for the detection of active infections; however, in Latin America this approach failed to specifically detect *T. vivax* infections (Desquesnes 1995).

Serological techniques such as immunofluorescence (IF) and ELISA can be used for direct and indirect diagnosis. Both indirect IF (IFA) and indirect ELISA utilize parasitic proteins to assess for the presence of an immune response through prior exposure; these are useful techniques for epidemiological investigations, especially for the determination of *T. vivax* distribution within the mammalian host. Indirect hemagglutination (IHA) has also been described for with the original methods for *T. evansi* since adapted for *T. vivax* using soluble antigens to sensitized red blood cells. In all these tests, standardization is still a problem due to antigenic variation of the parasite. Many problems are also associated with cross-recognition between the pathogenic trypanosomal species, and work is ongoing to develop appropriate criteria for immunological screening.

The PCR technique has been used for detection of *T. vivax* infection in different hosts. A variety of assays based on the use of species-specific sequences is available in the literature, presenting variable levels of specificity and sensitivity. In 1997, species-specific oligonucleotides were designed to amplify tandemly reiterated sequence of the *T. vivax*. More recently, the internal transcribed spacer 1 (ITS-1) of the ribosomal RNA (rRNA) has been described, although agreement between this pan-trypanosomal target and the species-specific primers is poor and subject to ongoing investigation (Ahmed et al. 2013).

19.3.6 Treatment and Prophylaxis

In general, *T. vivax* prophylactic treatment is effective, differing, however, in respect of drug used and the dosage. The use of chemoprophylaxis should consider several factors such as (1) the availability of drugs, (2) the half-life of the drug of choice, (3) the prevalence of infection in a herd or given area, (4) the sensitivity of pathogens to the drug, (5) timing of higher incidence of vectors, as well as (6) the cost of treatment considering the number of animals.

The use of diminazene aceturate (DA) for treatment of trypanosomiasis is effective, especially when it is caused by *T. vivax*, but having no prophylactic effect. Infections of cattle and sheep by *T. vivax* resistant to DA are currently treated with isometamidium chloride being the therapy of choice for prevention and treatment. Other drugs such as homidium chloride and homidium bromide that are closely related derivatives of isometamidium chloride are also used. These drugs present good index of cure as well as prophylactic effects that can last for 1–3 months as reported. The indication for treating cattle, sheep, and goats against trypanosomiasis due to *T. vivax* and *T. congolense* is homidium chloride (Uilenberg 1998). It should be noted that suramin and melarsomine are ineffective against *T. vivax*.

Although treatment of livestock should be carried out, it is relevant to avoid the consumption of meat and milk, so as preparation of dairy products, from treated animals during a 30-day period post-treatment.

19.3.7 Epidemiology

In tropical Africa, *T. vivax* is found in an extensive area that is a habitat for flies of the genus *Glossina*. Outside the tsetse area, the parasite is also found in most of the western, eastern, and central regions in Africa like Burkina Faso,

Democratic Republic of Congo, Ethiopia, Ghana, Nigeria, Sudan, Tanzania, and Zambia (Uilenberg 1998).

Studies revealed immunological evidence for *T. vivax* exposure in many South and Central American countries, including El Salvador, Costa Rica, Colombia, Ecuador, Peru, Paraguay, and Brazil. However, when interpreting this information, it is important to consider the specificity of these diagnostic tools.

In Brazil, the distribution of *T. vivax* occurs primarily in regions of extensive farming, where climatic conditions allow the sympatric occurrence of horse flies. For example, in the Pantanal of Mato Grosso and Mato Grosso do Sul States in Brazil, the epidemiology of *T. vivax* is directly associated with periods of drought and flooding of these wetlands. In 3–5 months of flooding that occurs every year, there is a decrease in pasture area, consequently increasing animal density within those areas that remain above water. These factors cause nutritional problems placing the animal under stress, predisposing the animal to present anomalous higher levels of parasitemia, and allowing the parasite to be transmitted by the abundant horse flies population during this period.

19.4 *Trypanosoma (Trypanozoon) equiperdum*

19.4.1 Historical Aspects

Trypanosoma equiperdum is a venereal infection of Equidae; the condition is referred to as dourine or covering sickness.

The historical aspects of the disease and the parasite occurrence are well described in Claes et al. (2005) and Brun et al. (1998). Briefly, the early reports of dourine-like symptoms are mentioned on Arabian texts, with the first description of the disease in an Algerian horse reported by Ammon and Dirkhausen in 1796. Rouget described the parasite in 1894 from blood smears from an infected horse, and Doflein postulated the name *T. equiperdum* in 1901 (Gardiner and Mahmoud 1992).

Since then, sporadic infections by this parasite have been observed in the European and Western European countries (France, Hungary, Germany, and Spain) but also in Canada, Russia, Turkey, Ethiopia, and Mongolia. A search for dourine cases from 1996 to 2004 on the Handistatus II database revealed that outbreaks and/or cases of dourine were consistently reported in Botswana, Namibia, and South Africa, but sporadically in other countries such as Senegal. In 2015, cases of dourine became notifiable to the World Organisation for Animal Health—the OIE—according to whom dourine is nowadays restricted to the African continent with outbreaks and isolated cases mainly occurring in Botswana and Ethiopia (Fig. 19.7).

In Ethiopia control measures for dourine are nowadays solely based on general surveillance. In Botswana dourine is monitored for borders protection and by general surveillance. Also, Botswana adopts treatment as one of the disease control measures, including treatment of the infected animals.

19.4.2 Biology and Transmission

A major problem on the study of dourine is the isolation and in vitro or in vivo maintenance of *T. equiperdum* isolates. After more than a century from the discovery of the parasite, only a few strains are available under laboratory conditions (Claes et al. 2005).

While some studies propose that *T. equiperdum* and *T. evansi* may belong to the same species, others speculate that *T. equiperdum* evolved from an equine strain of *T. evansi* that has since adapted to develop a tissue parasitism (Hoare 1972). However, it has been hypothesized that *T. equiperdum* and *T. evansi* are sister taxa descending from *T. brucei* as their common (Silva et al. 2002). More recently, analysis using RAPD, genotyping, and DNA fingerprinting has confirmed *T. equiperdum* within the *Trypanozoon* subgenus, proposing the hypothesis that *T. evansi* has originated from *T. equiperdum* variant and the influence of a genetic makeup on the evolution of the subgenus *Trypanozoon* (Li et al. 2005).

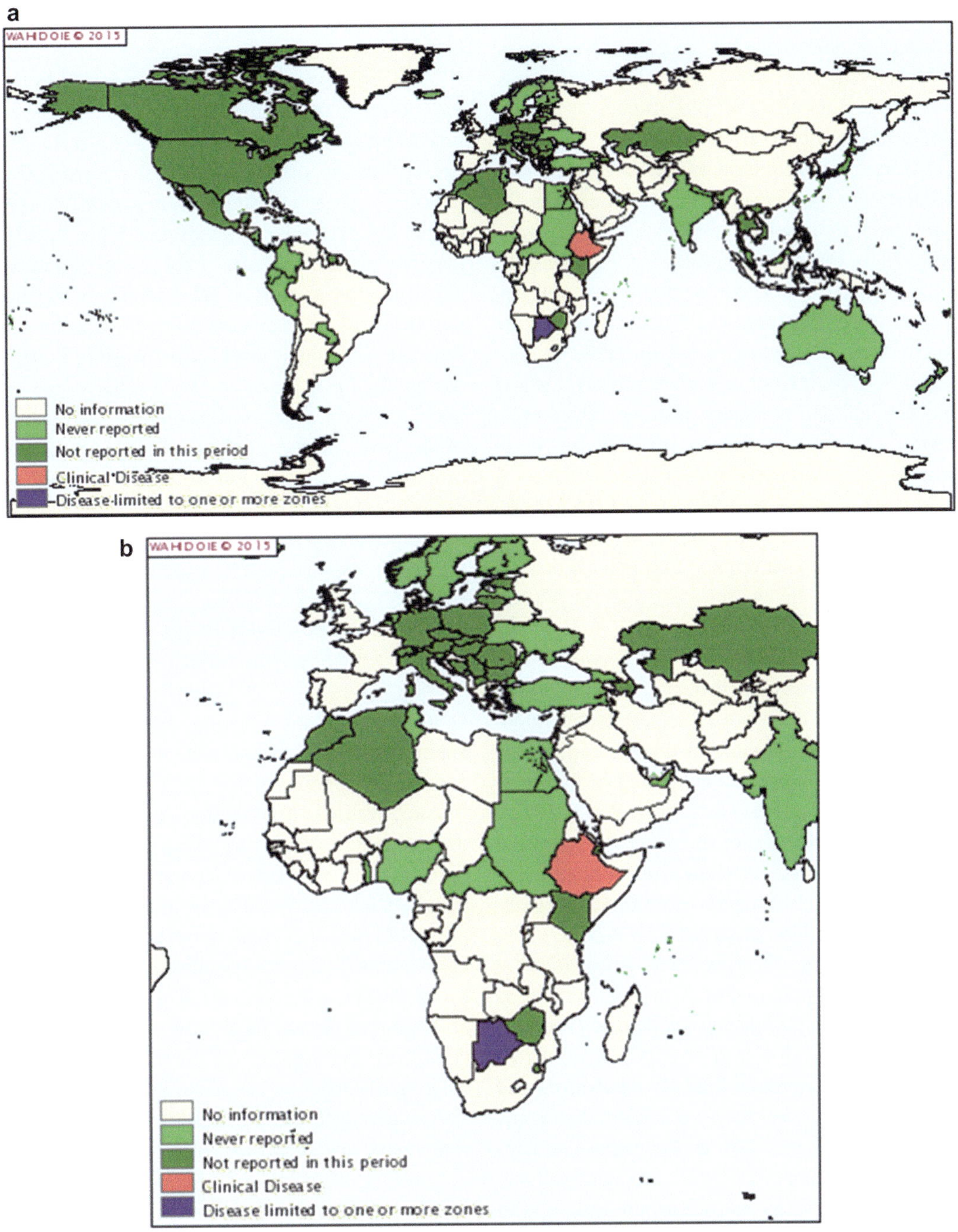

Fig. 19.7 Global distribution of dourine reports and cases (**a**) and details of the disease distribution in Africa (**b**) during January–July, 2014. Maps were generated by the World Animal Health Information database—WAHID (http://www.oie.int/wahis_2/public/wahid.php) on January 16, 2015

Infected equids such as horses, mules, and donkeys are the reservoirs of the parasite, and no other natural or sylvatic reservoirs are known. Although difficult, experimental infection of *T. equiperdum* can be achieved in laboratory models where the parasite has adapted to infect rodents, rabbits, and dogs; onward transmission to domestic animals such as goats and sheep can be achieved with these previously adapted parasites. Such infections demonstrate clinical symptoms of dourine, as observed in horses.

Although observation of the *T. equiperdum* bloodstream form is rare in natural infections, the parasite is morphologically related to *T. evansi* (Fig. 19.6), presenting with the same dimensions as *T. evansi*, it is difficult to phenotypically distinguish between these two species under light microscopy. *Trypanosoma equiperdum* is also taxonomically related to *T. evansi* as revealed by biochemical and molecular analysis (Brun et al. 1998). These same analyses revealed that these two species are closer to each other than to *T. brucei*. However, the presence of kDNA maxicircles in *T. equiperdum*, which are absent in *T. evansi*, is a major difference that reflects distinct evolutionary pathways for these species (Frasch et al. 1980).

Pleomorphism is unusual to *T. equiperdum* such that it is considered to be a monomorphic species. However, when in experimental infections in rodents, the parasite usually exhibits an increasing pleomorphism according to the number of passages (Blacklock and Yorke 1913). Also, cyclic maintenance in rodents or in culture in vitro induces the occurrence of dyskinetoplastic parasites, a phenomenon of multiple and independent origins also observed in *T. evansi*.

In nature, the infection by *T. equiperdum* is restricted to equids, among which the parasite is sexually transmitted during the coitus. The parasite can be found within vaginal secretions of infected mares, seminal fluid of the stallion, and mucous membranes and exudates of the penis and vagina. It is proposed that parasites can penetrate the intact mucous membranes and establish the infection. The trypanosomes are observed in the mucosal membrane capillaries of the urogenital tract, being rarely observed circulating within the infected animal's bloodstream. Also, newborn or lactant foals may become infected via mucosa during parturition and nursing or even while drinking milk from an infected dam. Symptomless infections by *T. equiperdum* are reported in the literature and may have influence on the clinical diagnosis (Brun et al. 1998; Claes et al. 2005).

Although hypothesized in the past due to the rare presence of bloodstream forms, there are no reports on the transmission of *T. equiperdum* by tsetse flies or other mechanical vectors. The absence of vectorial transmission is a unique feature of this particular trypanosome. Thus, the dispersion of the parasite may have primarily occurred via transport of infected and/or undiagnosed horses to many areas worldwide. In this sense, dourine has spread across Western Europe during the WW2 due to the extensive use of horses, mainly by the German army. By the end of the WW2, a systematic screening and control program, in combination with enforcement of sanitary control, allowed control of transmission of *T. equiperdum*.

19.4.3 Disease, Pathology, and Symptomatology

Dourine is marked by symptoms such as fever, anemia, plaques and/or skin eruption, genital edema, and paralysis, resulting in death of the infected animal. Among the equids, horses are the most susceptible to infection, which is potentially fatal if not treated. The infection of other equids such as donkeys or mules is usually asymptomatic and, thus, may be underreported. Like the other animal trypanosomiasis (such as Surra and Nagana), dourine also presents a variable clinical picture depending on the host susceptibility and the strain of the infecting parasite (Stephen 1986).

The pathognomonic clinical symptoms of dourine are the appearance of skin lesions, which are coetaneous or edematous urticarial skin plaques known as "silver dollar plaques." Although occurring anywhere on the animal body, these lesions are usually located over the ribs, having 5–8 cm in diameter and 1 cm in thickness, persisting for over a week. Despite the presence of these plaques, accurate diagnosis cannot exclusively rely on this clinical aspect due to the possible occurrence of these urticarial plaques also on *T. evansi*-infected animals. Among the diseases that must be considered for differential diagnosis, we must cite Surra, Nagana, coital exanthema, and contagious equine metritis.

19.4.4 Diagnosis, Treatment, and Prophylaxis

Due to the absence of bloodstream forms during the entire infection, diagnosis of *T. equiperdum* infection mostly relies on the same indirect methods used for diagnosis of *T. evansi* infection above described. Most of the serological tests used for *T. evansi* (indirect fluorescence assay or IFA, ELISA, complement fixation or CF, card agglutination test for trypanosomiasis or CATT) are capable to detect *T. equiperdum* infection with variable sensitivity and specificity (Gardiner and Mahmoud 1992). Nevertheless, these studies must be updated toward the development of new, specific, and sensitive serological test for *T. equiperdum* infection.

Due to the scarce number of nucleotide (166) and protein (89) sequences available to date (January 2015), molecular diagnosis cannot be performed as for *T. evansi* and other trypanosomes. The reduced number of available strains and the uncertainties concerning the origins of these strains (Claes et al. 2005) demands efforts toward isolation of well-characterized strains in order to sequence the parasite genome aiming new molecular diagnostic tools as well as a comprehension of the evolution of this taxon.

Animals presenting suggestive clinical symptoms of dourine such as swelling and/or edema of the genitalia should have the lymph and edematous fluids of the genital organs and genital washes examined for the presence of trypanosomes under light microscopy. Giemsa-stained blood smears should also be prepared using peripheral blood collected with EDTA. Examination of the pellet from double-centrifuged plasma may also reveal trypanosomes.

From the veterinary point of view and in accordance to the OIE guidelines, dourine control is based on animal slaughtering (or castration) since there are no vaccines available and treatment is not recommended since parasitological cure is not achieved with efficacy, despite the disappearance of the clinical signs. Despite such guidelines, it is postulated that treatment of *T. equiperdum* infection is based on the same drugs used for *T. evansi* infection. Although not tested for their clinical efficacy in vivo, former in vitro studies pointed out that suramin, diminazene, and quinapyramine are as effective against *T. equiperdum* as they are against *T. evansi* (Brun and Lun 1994). Diminasan® and Cymelarsan® were also tested in vivo in rodents and horses where 0.25–0.5 mg/kg body weight of Cymelarsan® cleared parasitemia 24 h posttreatment in horses infected with the *T. equiperdum* 834/940 Dodola strain. However, authors point out that such activity may be a strain-dependant phenomenon and deserve further studies (Hagos et al. 2010).

Sanitary measures are effective and consist of effective control of the occurrence of the disease (notification, monitoring, surveillance, and compulsory testing), slaughtering of the infected animals, control of animal movements (borders protection), and good sanitary practices on animal facilities, especially during mating on horse farms. Prophylaxis and control of dourine thus requires a multidisciplinary approach, relying on veterinary and sanitary actions; disease control can be achieved if the endemic/affected countries adopt a multinational, long-term, and well-defined approach.

Both the WAHID and Handistatus II databases reveal that distinct African countries adopt measures such as border control, use of quarantine for imported/exported animals, testing and stamping, and treatment of infected animals. However, these measures appear to be independently taken by distinct countries, discontinued or not reported, not allowing a clear comprehension of the dourine status and transmission in this continent.

19.5 *Trypanosoma (Nannomonas) congolense*

19.5.1 Historical Aspects, Biology, and Transmission

Trypanosoma congolense Broden, 1904, is an important pathogen of livestock contributing to the disease referred to as "Nagana" that affects over 40 countries in the African continent. Infecting around 100–120 million animals, mostly small ruminants and cattle, *T. congolense* is the most important agent of animal trypanosomiasis in Africa, determining a severe economic impact, losses in productivity, and death of the infected animals (Peacock et al. 2012).

Although revealing an extensive morphological variability, the parasite is among the smallest trypanosomes in length, and its morphology is typical of a *Nannomonas* species presenting a mid-sized kinetoplast of marginal positioning. *Trypanosoma congolense* is not a host-specific pathogen, infecting a variety of mammalian species such bovines, equines, sheep, goats, pigs, camels, and dogs.

Transmitted by tsetse flies, this parasite presents an interesting but quite a complex life cycle, it proliferates in the infected host bloodstream, and these forms are infective to the tsetse fly. After the infective meal, the bloodstream forms differentiate into procyclic forms within the *Glossina* spp. midgut, from established infection parasites which will migrate to the insect's proboscis where they differentiate into the mammalian-infective metacyclic forms. It is noteworthy to mention that *T. congolense* infective forms (metacyclics) are derived from epimastigotes that develop within the *Glossina* spp. mouthparts. Isolation of these epimastigotes is possible and the parasite, unlike most of the *Nannomonas*, can be maintained in vivo in rodents under laboratory conditions (Peacock et al. 2012).

Trypanosoma congolense can also be maintained in vitro in MEM media supplemented with distinct serum and supplements as detailed in Coustou et al. (2010). The distinct media types allow the cultivation of all life cycle stages of the parasite, including the infective metacyclic forms that are obtained via metacyclogenesis.

Along with the morphological plasticity, *T. congolense* reveals an interesting genotypic variability despite being a species of the *Nannomonas* subgenus that is considered a monophyletic group. Characterization of *T. congolense* strains by biochemical and molecular tools pointed out the occurrence of three distinct genetic groups that are named as savanna, riverine/forest, and Kenya coast (also referred as Kilifi) (Stevens and Brisse 2004).

The intraspecific variability was initially revealed by isoenzyme electrophoresis, then by DNA probing, RAPD, and finally by comparative analysis of the 18S rDNA gene (Stevens and Brisse 2004). Currently, the reference *T. congolense* genome (IL3000 strain) is available at the GenBank and TriTrypDB, contributing an important tool for comparative genomics toward the development of new diagnostic and therapeutic methods. This particular strain was originally isolated from an infected cow in the Transmara District, Rift Valley Province, Kenya, and can be maintained in vivo by cyclic passages among laboratory rodents via tsetse flies as well as in vitro.

19.5.2 Disease, Pathology, and Symptomatology

Trypanosoma congolense is primarily an intravascular parasite. Although not fully understood, acute Nagana in cattle is generally characterized by anemia, pronounced loss of appetite and weight, pyrexia, apathy, reduced reproductive capacity, and immunosuppression of the infected animal; the host may recover depending on the infecting strain (Coustou et al. 2010).

Strains from the savanna group are commonly observed infecting cattle and reveal a variable degree of pathogenicity, while strains from the riverine/forest and Kenya coast groups are less pathogenic and are not specifically associated to any host species (Uilenberg 1998; Stevens and Brisse 2004). Nevertheless, mixed infection by *T. congolense* strains from different genetic groups in the same host and in the same vector has already been detected in South Africa and may be present in distinct areas.

Infection by *T. congolense* tends to be less acute and less pathogenic than *T. vivax* infections, although lethality seems to be the same (4–7 weeks after infection) for those animals with progressive disease determined by savanna-type strains (Stephen 1986; Coustou et al. 2010).

A detailed study on the virulence of each *T. congolense* genetic group in zebu cattle pointed out that infection by savanna-type strains consistently revealed higher parasitemia levels and resulted in animal death if compared to the riverine/forest-type and Kilifi-type strain infection, from which animals have recovered within 3–4 months (Bengaly et al. 2002).

A *postmortem* pathological examination of the zebu cattle experimentally infected with a savanna-type strain revealed a general edematous disorder with enlarged, often hemorrhagic lymph nodes, heart and liver congestion, and the later revealing signs of necrosis (Bengaly et al. 2002). An investigation of 62 *T. congolense* savanna-type strains isolated from the domestic and from the sylvatic transmission cycles revealed an uneven distribution of the virulent strains over these transmission cycles and the high proportion of highly virulent strains on the sylvatic cycle (Van Den Bossche et al. 2011). In this same study, virulence tests revealed reproducible results in mice and cattle, reinforcing the usefulness of this model.

19.5.3 Diagnosis, Treatment, and Prophylaxis

Due to the vascular localization and the high parasitemia in cattle, *T. congolense* infection can be primarily diagnosed via parasitological tests. A comparison of distinct methods revealed that analysis of the buffy coat using phase-contrast microscopy (Murray technique) (Murray et al. 1977) was the most sensitive method for *T. congolense* infection when compared to direct examination of thick or thin blood smears, wet blood films, and the Woo technique (microhematocrit technique) (Woo 1970), allowing simultaneous detection of anemia as one of the pathognomonic signs of the disease.

Intraperitoneal inoculation of blood (0.2–0.4 mL) from naturally infected cattle to mice has also been used as a diagnostic test, allowing isolation of the parasite strain aiming downstream characterization by biological, biochemical, and molecular methods (Van Den Bossche et al. 2011).

Indirect methods can also be employed for early diagnosis of *T. congolense* infection or subclinical or sub-patent infections. In this sense, the complement fixation test (CFT), the indirect fluorescent antibody test (IFAT), the card agglutination test (CATT), the enzyme-linked immunosorbent assay (ELISA), and the polymerase chain reaction (PCR) methods have been used to detect animal trypanosomiasis (Ahmed et al. 2013).

Aiming the improvement of the sensitivity and specificity, a method coupling the PCR and the ELISA assays was developed for specific detection of *T. congolense* savanna-type using biotinylated probes. Using the buffy coats from naturally infected cattle, the method proved to be more sensitive than the Murray method, showing less cross-reaction (Sow et al. 2006).

The first molecular methods used for *T. congolense* detection were based on repetitive DNA sequences detected by comparative studies using pulsed-field gel electrophoresis and restriction enzyme profiling. The use of species-specific probes directed to repetitive DNA sequences of *T. congolense* (probe pgNRE-372) and *T. simiae* (probe pgNS-600) allowed specific detection of these morphologically related parasites on Southern blot assays (Uilenberg 1998). Since then, species-specific DNA targets for diagnosis of *T. congolense*, *T. vivax*, or *T. brucei*, the etiological agents of trypanosomiasis of cattle, have been developed. Nowadays, a commercial *T. congolense* DNA detection kit named VetPCR™ is available from Bioingentech biotechnologies, Inc. (www.bioingentech.com), amplifying a 315bp species-specific amplicon plus an internal control of the reaction.

A comparative study of distinct PCR-based methods targeting the internal transcribed spacer (ITS) sequence of the rRNA gene used for diagnosis of *T. congolense* infection showed that genetic variability of this sequence of trypanosomes from the *Nannomonas* and *Trypanozoon* subgenus infecting cattle may impair the detection. A detailed description of the primer sets used is available on the study of Ahmed et al. (2013). In conclusion, the sensitivity of the molecular diagnosis relies on biological and genetic aspects of the infecting parasite.

The availability of the *T. congolense* genome draft allows further comparative studies aiming the development of new and accurate molecular diagnosis as well as functional characterization of genes related to virulence and/or pathogenesis toward the understanding of Nagana (Coustou et al. 2010). Also, *T. congolense* is an interesting biological model due to the availability of all life cycle stages under laboratory conditions, allowing functional studies to address the mechanisms of pathogenesis and the search for putative therapeutic targets.

Comparative biological and molecular characterization of *Nannomonas* (*T. congolense* and *T. godfreyi*) and *Pycnomonas* (*T. suis*) trypanosomes isolated from both wild and domestic suids in Africa has clearly separated *T. congolense* in a distinct branch and suggests that *T. suis* and *T. simiae* may be a single species.

Due to the biological similarities, treatment and prophylaxis of *T. congolense* infection in cattle is based on the same measure preconized for *T. vivax* but may vary according to the host species.

19.5.4 Epidemiology

Trypanosoma congolense is probably the most prevalent and widespread pathogenic trypanosome in tropical Africa in terms of both geographical and host range across sub-Saharan Africa.

The parasite is found infecting ruminants, pigs, dogs, and other domestic animals throughout the tsetse belt. Camels are highly susceptible to *T. congolense* but do not usually penetrate into the tsetse region. N'Dama cattle and the West African shorthorn breeds are more resistant to development of disease after *T. congolense* infections than Zebu cattle or European.

Geographically, *T. congolense* infections have been described in countries within the tsetse belt such as Zambia, Burkina Faso, Nigeria, Kenya, Ghana, Uganda, Mali, and Cameroon and sporadically in countries outside tsetse belt such as South Africa.

An interesting fact was observed in experiments carried out in Burkina Faso on the differential ability of infection between the *T. congolense* groups and species of *Glossina*. These experiments have shown higher infection rates with *T. congolense* savanna type in *G. morsitans morsitans* and *G. morsitans submorsitans* (both belonging to the savanna group tsetse) than in the *Glossina* species of palpalis and riverine group (*G. tachinoides* and *G. palpalis gambiensis*).

Since *T. congolense* is divided into three types, studies have already shown that there is a clear difference in pathogenicity between the three types of *T. congolense*; the savanna type was virulent, while the forest type was of low pathogenicity and the Kilifi type was nonpathogenic. Among the various strains of savanna type, strains from sylvatic cycle (in which trypanotolerant wildlife species are the reservoir of the trypanosomes) are more virulent than those of domestic cycle (strains isolated from cattle) (Van Den Bossche et al. 2011).

Mixed infection with different types of *T. congolense* and other trypanosomatids has also been described (Bengaly et al. 2002).

19.6 *Trypanosoma (Nannomonas) simiae*

19.6.1 Historical Aspects, Biology, and Transmission

Sir David Bruce reported the original description of *T. simiae* in 1912 while observing Giemsa-stained blood trypomastigotes of the parasite taken from naturally infected monkeys and goats (Bruce et al. 1912). At that time, the parasite was revealed to be monomorphic and to have an average length of 17.5 μm (14–24 μm). Recent studies, however, revealed that *T. simiae* presents the same pleomorphic characteristics of other *Nannomonas* species, appearing as three distinct blood forms within the infected host. The elongated-stout form (well-developed recurrent flagellum) comprises around 90% of the forms found in the blood stream, while the long-slender form with a poorly developed recurrent flagellum and the short forms, which are quite indistinguishable from *T. congolense*, comprise around 7 and

3% of the parasites in the bloodstream (Bruce et al. 1912). Compared to other tsetse-transmitted trypanosomes, the medium-sized kinetoplast is located marginally and subterminally (Eisler et al. 2004).

Also transmitted by tsetse flies, *T. simiae* infects wild and domestic suids determining a hyperacute or acute disease usually leading to death of the infected animal within 3–6 days. Although isolated reports indicate that *T. simiae* is also able to infect distinct species (camel, horse, ox, monkey, goat) (Bruce et al. 1912; Hoare 1972), the domestic pig is considered to be the natural and primary host of this parasite, thus predominantly being a problem of pig farming in the tsetse-infested area of the African continent.

Trypanosoma simiae does not infect laboratory rodent models but can be cultivated in vitro on monoaxenic cultures or even axenically under the same conditions above described for *T. congolense*. It is noteworthy to mention that *T. simiae* trypomastigotes can be directly isolated from infected tsetse flies and cultivated on MEM media over a monolayer of bovine aortic endothelial cells (BAE) as described by Zweygarth et al. (1992).

It is well established that *Glossina* species present a variable susceptibility to distinct *T. simiae* strains and that male specimens are more susceptible to *T. simiae* infection than females. While in Kenya hyperacute infections in pigs seem to be related to *T. simiae* transmission by *Glossina brevipalpis* and chronic infections are related to transmission via *G. pallidipes*, others propose that virulence of the parasite in pigs is not influenced by the *Glossina* vector species (Moloo et al. 1994). Nevertheless, a number of *Glossina* species are incriminated as *T. simiae* vectors along with *G. brevipalpis* and *G. pallidipes*: *G. morsitans centralis*, *G. palpalis gambiensis*, *G. fusca*, and *G. austeni* (Moloo et al. 1994), resulting in a wide geographical area of transmission in Africa. The possibility of mechanical transmission of *T. simiae* by other blood sucking arthropods is also considered.

Despite its importance for pig farming, *T. simiae* is among the least studied trypanosomes. On the Pubmed database (January 2015), only 96 records on scientific or medical abstracts/citations, among which, 86 are full-text journal articles, were found. Few studies on isoenzymes and schyzodeme variability, cross-DNA probing, RFLP, and RAPD were carried out using a limited number of isolates. These studies, however, allowed a clear differentiation of *T. simiae* from *T. congolense*, pointing out the occurrence of distinct *T. simiae* lineages in Africa (Coustou et al. 2010), probably related to the distinct tsetse species and/or transmission areas.

19.6.2 Disease, Pathology, and Symptomatology

Trypanosoma simiae causes a hyperacute or acute trypanosomiasis in domestic pigs, which usually present depression, polypnea, innapetence, and fever, often leading to animal death in a couple of hours to several days.

Few descriptions of the pathological effects of *T. simiae* infection in pigs are available. The infected animals reveal a disseminated intravascular coagulation and hemorrhage on the lungs, heart, and meninges. Studies revealed the presence of *T. simiae* in the cerebrospinal fluid of experimentally infected domestic pigs, indicating that this parasite is able to invade the central nervous system, which is probably the cause of the acuteness of the disease (Zweygarth and Röttcher 1987).

19.6.3 Diagnosis, Treatment, and Prophylaxis

Diagnosis of *T. simiae* infection can be performed via direct observation of parasites in the blood stream (wet films or Giemsa-stained smears) that are abundant in the acute phase. However, due to the short-term hyperacute aspect of the infection, the polymorphism of the parasite, the absence of pathognomonic signs, and the occurrence of other trypanosomes that can also infect pigs in Africa, a morphology-based diagnosis is difficult, and clinical or laboratorial diagnosis, if performed, should often be achieved *postmortem*.

Due to the characteristic acuteness of the *T. simiae* infection, the use of concentration methods such as the microhematocrit centrifugation (Woo technique) improves the sensitivity of detection of *T. simiae* in the bloodstream and/or cerebrospinal fluid (CSF).

Treatment is often administered too late to influence the outcome of this hyperacute infection, allowing the parasite to invade the central nervous system. Treatment of experimentally infected pigs using a combination of diminazene aceturate and a 5-substituted 2-nitroimidazole compound named Ro 15-0216 was successful, while treatment using each compound in separate was ineffective (Majiwa and Webster 1987).

The *T. simiae* genome is among the least studied trypanosomatid genomes, having only 29 nucleotide and seven protein sequences available at the GenBank (January 21, 2015). However, a repetitive DNA element was described in the *T. simiae* genome allowing distinction from *T. congolense* (savanna and Kilifi cell types) in hybridization assays, which is of utmost interest as a diagnostic tool. Also, the probes were able to specifically detect *T. simiae* in mouthparts of tsetse flies, not showing cross-reaction with other trypanosome species occurring in these vectors (*T. congolense, T. brucei*) (Hamill et al. 2013).

Due to the transmission of *T. simiae* by tsetse flies, the same vector control measures adopted for other human or animal trypanosomiasis transmitted by *Glossina* species will also work on the control of this parasite.

19.6.4 Epidemiology

Despite their biological differences, animal trypanosomiasis in Africa is caused by several species of trypanosomes, usually separated as the tsetse or cyclically transmitted trypanosomes (*T. brucei* subspecies, *T. vivax*, *T. congolense*, and *T. simiae*), the trypanosomes mechanically transmitted by Tabanidae and *Stomoxys* sp. (*T. vivax* and *T. evansi*) and the sexually transmitted *T. equiperdum* (Masake et al. 1997).

In the case of *T. simiae*, little is known about the epidemiology of the disease as a whole in the African continent. Considering the *Glossina* spp.-infested area of approximately 10 million km^2, the existence of asymptomatic wild hogs infected by the parasite, and the growing domestic pig farming industry, we may expect a vast area of distribution of this parasite. The fact that wild suids such as the bushpig (*Potamochoerus porcus*) or the warthog (*Phacocoreus aethiopicus*) are also infected by *T. simiae* in Africa, where they act as asymptomatic hosts, is a complicating factor for the disease epidemiology. Despite this, infection by *T. simiae* in pigs seems to be restricted to focal outbreaks associated to high mortality rates.

Although contested, it has been proposed that severity of *T. simiae* infection in pigs is related to the *Glossina* species transmitting the parasite since susceptibility of these vectors is variable depending on the parasite strain. Considering such hypothesis and the fact that domesticated pigs can also harbor the human infective trypanosome *T. b. rhodesiense*, as revealed in a study in Tanzania (Majekodunmi et al. 2013), animal trypanosomiasis has been studied as a group in Africa, consisting in a major medical and veterinary problem due to difficult diagnosis.

Acknowledgments To Dr. Gustavo A. Vallejo from Universidad del Tolima, Ibagué, Colombia, for the Fig. 19.3.

References

Ahmed HA, Picozzi K, Welburn SC et al (2013) A comparative evaluation of PCR-based methods for species-specific determination of African animal trypanosomes in Ugandan cattle. Parasit Vectors 6:316

Bengaly Z, Sidibe I, Ganaba R et al (2002) Comparative pathogenicity of three genetically distinct types of *Trypanosoma congolense* in cattle: clinical observations and haematological changes. Vet Parasitol 108:1–19

Blacklock B, Yorke W (1913) The trypanosome causing Dourine (Mal de Coit or Beschalseuche). Proc R Soc B B87:89

Bruce D, Harvey D, Hamerton AE et al (1912) The morphology of *Trypanosoma simiae*, sp. nov. Proc R Soc Lond B 83:477–481

Brun R, Hecker H, Lun ZR (1998) *Trypanosoma evansi* and *T. equiperdum*, distribution, biology, treatment and phylogenetic relationship (a review). Vet Parasitol 79:95–107

Brun R, Lun ZR (1994) Drug sensitivity of Chinese *Trypanosoma evansi* and *Trypanosoma equiperdum* isolates. Vet Parasitol 52:37–46

Claes F, Büscher P, Touratier L et al (2005) *Trypanosoma equiperdum*: master of disguise or historical mistake? Trends Parasitol 21:316–321

Coustou V, Guegan F, Plazolles N et al (2010) Complete in vitro life cycle of *Trypanosoma congolense*: development of genetic tools. PLoS Negl Trop Dis 4, e618

D'Alessandro A, Gore-Saraiva N (1992) *Tryoanosoma rangeli*. In: Kreier JP (ed) Parasitic Protozoa. CA: Academic Press, San Diego, v2, pp 1–54

Dargantes AP, Mercado RT, Dobson RJ et al (2009) Estimating the impact of *Trypanosoma evansi* infection (surra) on buffalo population dynamics in southern Philippines using data from cross-sectional surveys. Int J Parasitol 39:1109–1114

Desquesnes M (1995) Evaluation of three antigen detection tests (monoclonal trapping ELISA) for African trypanosomes, with an isolate of *T. vivax* from French Guyana. Ann N Y Acad Sci 791:172–184

Desquesnes M (2004) Livestock trypanosomiases and their vectors in Latin America. OIE, Paris

Desquesnes M, Holzmuller P, Lai DH et al (2013) *Trypanosoma evansi* and surra: a review and perspectives on origin, history, distribution, taxonomy, morphology, hosts, and pathogenic effects. Biomed Res Int. doi:10.1155/2013/321237

Eisler MC, Dwinger RH, Majiwa PAO et al (2004) Diagnosis and epidemiology of African animal trypanosomiasis. In: Maudlin I, Holmes PH, Miles MA (eds) The trypanosomiasis. CABI Publishing, Oxford, pp 253–267

Fallis AM (1986) Griffith Evans 1835–1935 discover of the first pathogenic trypanosome. Can Vet J 27:336–338

Frasch ACC, Hajduk SL, Hoeijmakers JHJ et al (1980) The kinetoplast DNA of *Trypanosoma equiperdum*. Biochim Biophys Acta 607:397–410

Gardiner PR (1989) Recent studies of the biology of *Trypanosoma vivax*. Adv Parasitol 28:229–317

Gardiner PR, Mahmoud MM (1992) Salivarian trypanosomes causing disease in livestock. In: Kreier JP, Baker JR (eds) Parasitic protozoa, vol 2, 2nd edn. Academic, San Diego, pp 295–302

Grisard EC, Steindel M, Guarneri AA et al (1999) Characterization of *Trypanosoma rangeli* strains isolated in Central and South America: an overview. Mem Inst Oswaldo Cruz 94:203–209

Guhl F, Vallejo GA (2003) *Trypanosoma (Herpetosoma) rangeli* Tejera, 1920: an updated review. Mem Inst Oswaldo Cruz 98:435–442

Gumm I (1991) The axenic cultivation of insect forms of *Trypanosoma* (*Duttonella*) *vivax* and development to the infective metacyclic stage. J Protozool 38:163–171

Hagos A, Goddeeris BM, Yilkal K et al (2010) Efficacy of Cymelarsan and Diminasan against *Trypanosoma equiperdum* infections in mice and horses. Vet Parasitol 171:200–206

Hamill LC, Kaare MT, Welburn SC et al (2013) Domestic pigs as potential reservoirs of human and animal trypanosomiasis in Northern Tanzania. Parasit Vectors 6:322

Hirumi H, Martin S, Hirumi K et al (1997) Cultivation of bloodstream forms of *Trypanosoma brucei* and *T. evansi* in a serum-free medium. Trop Med Int Health 2:240–244

Hoare CA (1972) The trypanosomes of mammals: a zoological monograph. Blackwell Scientific Publications, Oxford

Joshi PP, Shegokar VR, Powar RM (2005) Human trypanosomiasis caused by *Trypanosoma evansi* in India: the first case report. Am J Trop Med Hyg 73:491–495

Li FJ, Gasser RB, Zheng JY et al (2005) Application of multiple DNA fingerprinting techniques to study the genetic relationships among three members of the subgenus *Trypanozoon* (Protozoa: Trypanosomatidae). Mol Cell Probes 19:400–407

Lun Z, Desser SS (1995) Is the broad range of hosts and geographical distribution of *Trypanosoma evansi* attributable to the loss of maxicircle kinetoplast DNA? Parasitol Today 11:131–133

Majekodunmi AO, Fajinmi A, Dongkum C et al (2013) A longitudinal survey of African animal trypanosomiasis in domestic cattle on the Jos Plateau, Nigeria: prevalence, distribution and risk factors. Parasit Vectors 6:239

Majiwa PAO, Webster P (1987) A repetitive deoxyribonucleic acid sequence distinguishes *Trypanosoma simiae* from *T. congolense*. Parasitology 95:543–558

Masake RA, Majiwa PA, Moloo SK et al (1997) Sensitive and specific detection of *Trypanosoma vivax* using the polymerase chain reaction. Exp Parasitol 85:193–205

Meirelles RMS, Henriques-Pons A, Soares MJ et al (2005) Penetration of the salivary glands of *Rhodnius domesticus* Neiva & Pinto, 1923 (Hemiptera: Reduviidae) by *Trypanosoma rangeli* Tejera, 1920 (Protozoa: Kinetoplastida). Parasitol Res 97:259–269

Mihok S, Kangethe LJ, Kamau GK (1995) Trials of traps and attractant for *Stomoxys* spp. (Diptera; Muscidae). J Med Entomol 32:283–289

Moloo SK, Zweygarth E, Sabwa CL (1994) Comparative study on the susceptibility of different laboratory strains of *Glossina* species to *Trypanosoma simiae*. Med Vet Entomol 8:225–230

Molyneux DH (1975) Diagnostic methods in animal trypanosomiases. Vet Parasitol 1:5–17

Murray M, Murray PK, McIntyre WIM (1977) An improved parasitological technique for the diagnostic of African trypanosomiasis. Trans R Soc Trop Med Hyg 71:325–326

Osório A, Madruga CR, Desquesnes M et al (2008) *Trypanosoma* (*Duttonella*) *vivax*: its biology, epidemiology, pathogenesis, and introduction in the New World—A review. Mem Inst Oswaldo Cruz 103:1–13

Peacock L, Cook S, Ferris V et al (2012) The life cycle of *Trypanosoma (Nannomonas) congolense* in the tsetse fly. Parasit Vectors 5:109

Rodrigues A, Fighera RA, Souza TM et al (2009) Neuropathology of naturally occurring *Trypanosoma evansi* infection of horses. Vet Pathol 46:251–258

Roger L (1901) The transmission of the *Trypanosoma evansi* by horse flies and other experiments pointing to the probable identity of Surra in India and Nagana or tsetse fly disease in Africa. Proc Roy Soc B 68:163

Seidl A, Moraes AS, Silva RAMS (1998) A financial analysis of treatment strategies for *Trypanosoma evansi* in the Brazilian Pantanal. Prev Vet Med 33:219–234

Silva FM, Noyes H, Campaner M et al (2004) Phylogeny, taxonomy and grouping of *Trypanosoma rangeli* isolates from man, triatomines and sylvatic mammals from widespread geographical origin based on SSU and ITS ribosomal sequences. Parasitol 129:549–561

Silva RAMS, Barros ATM, Herera HM (1995) Trypanosomiasis outbreaks due to *Trypanosoma evansi* in the Pantanal, Brazil. A preliminary approach on risk factors. Rev Elev Med Vet Pays Trop 48:315–319

Silva RAMS, Seidl A, Ramirez L et al (2002) *Trypanosoma evansi* e *Trypanosoma vivax*: biologia, diagnóstico e controle. EMBRAPA Pantanal, Campo Grande. http://www.cpap.embrapa.br/publicacoes/online/Livro015.pdf

Sloof P, Bos JL, Konings AFJ et al (1983) Characterization of satellite DNA in *Trypanosoma brucei* and *Trypanosoma cruzi*. J Mol Biol 167:1–21

Sow A, Sidibé I, Desquesnes M et al (2006) The application of PCR-ELISA to the detection of *Trypanosoma congolense* type savannah (TCS) in bovine blood samples. Trop Biomed 23:123–129

Stephen LE (1986) Trypanosomiasis, a veterinary perspective. Pergamon Press, Oxford 572 pp

Stevens JR, Brisse S (2004) Trypanosomes of medical and veterinary importance: systematics. In: Maudlin I, Holmes PH, Miles MA (eds) The trypanosomiasis. CABI Publishing, Oxford, pp 1–23

Steverding D (2008) The history of African trypanosomiasis. Parasit Vectors 1:3

Stoco PH, Wagner G, Gerber A et al (2014) Genome of the a virulent human-infective trypanosome—*Trypanosoma rangeli*. PLoS Negl Trop Dis 8, e3176

Uilenberg G (1998) A field guide for the diagnosis, treatment and prevention of African animal trypanosomiasis. Food and Agriculture Organization of the United Nations. Rome © FAO 1998 M-27. Available on line at ftp://ftp.fao.org/docrep/fao/006/x0413e/x0413e00.pdf. Accessed 21 Jan 2015

Urrea DA, Carranza JC, Cuba CA et al (2005) Molecular characterisation of *Trypanosoma rangeli* strains isolated from *Rhodnius ecuadoriensis* in Peru, *R. colombiensis* in Colombia and *R. pallescens* in Panama, supports a co-evolutionary association between parasites and vectors. Infect Genet Evol 5:123–129

Vallejo GA, Guhl F, Carranza JC et al (2003) Parity between kinetoplast DNA and mini-exon gene sequences supports either clonal evolution or speciation in *Trypanosoma rangeli* strains isolated from *Rhodnius colombiensis, R. pallescens* and *R. prolixus* in Colombia. Infect Genet Evol 3:39–45

Vallejo GA, Guhl F, Schaub GA (2009) Triatominae-*Trypanosoma cruzi/T. rangeli*: vector-parasite interactions. Acta Trop 110:137–147

Van Den Bossche P, Chitanga S, Masumu J et al (2011) Virulence in *Trypanosoma congolense* Savannah subgroup. A comparison between strains and transmission cycles. Parasit Immunol 33:456–460

Ventura RM, Takeda GF, Silva RAMS et al (2002) Genetic relatedness among *Trypanosoma evansi* stocks by random amplification of polymorphic DNA and evaluation of a synapomorphic DNA fragment for species-specific diagnosis. Int J Parasitol 32:53–63

Vohradsky F (1971) Clinical signs, daily rate of infection, physical changes of the blood and pathomorphological changes in cattle artificially infected by *Trypanosoma vivax*. Rev Elev Med Vet Pays Trop 24:251–263

Woo PTK (1970) The haematocrit centrifuge technique for the diagnosis of African trypanosomiasis. Acta Trop 27:384–386

Zweygarth E, Moloo SK, Kaminsky R et al (1992) Axenic in vitro cultivation of *Trypanosoma simiae* bloodstream trypomastigotes. Acta Trop 52:79–81

Zweygarth E, Röttcher D (1987) The occurrence of *Trypanosoma (Nannomonas) simiae* in the cerebrospinal fluid of domestic pigs. Parasitol Res 73:479–480

Complementary On-line Resources

http://www.rangeli.lncc.br/—The *Trypanosoma rangeli* Genome Resource

http://tritrypdb.org/—TriTrypDB—Kinetoplastid Genomics Resource

http://www.who.int/tdr/—World Health Organization

http://www.oie.int—World Organization of Animal Health

http://www.oie.int/wahis_2/public/wahid.php—World Animal Health Information database—WAHID

http://web.oie.int/hs2/report.asp—Handistatus II Database (1996–2004)

http://www.sanger.ac.uk/resources/downloads/protozoa/trypanosoma-vivax.html—*Trypanosoma vivax* Genome Resource

http://web.oie.int/eng/maladies/Technical%20disease%20cards/DOURINE_FINAL.pdf—Dourine: aetiology, epidemiology, diagnosis, prevention and control references

ftp://ftp.fao.org/docrep/fao/006/x0413e/x0413e00.pdf—A field guide for the diagnosis, treatment and prevention of African animal trypanosomiasis—FAO

Malaria

20

Keith H. Carter, Rainier P. Escalada,
and Prabhjot Singh

Abstract

Malaria is a preventable and curable but life-threatening disease caused by *Plasmodium* parasites that are transmitted to people through the bites of infected *Anopheles* mosquitoes. Approximately half of the world's population is at risk although most malaria cases and deaths occur in sub-Saharan Africa. In 2013, 97 countries and territories had ongoing malaria transmission, including 21 countries from the region of the Americas. Early diagnosis and treatment reduces disease and transmission and prevents deaths. All cases of suspected malaria must be confirmed using parasite-based diagnostic testing before administering treatment. There are currently no licensed vaccines against malaria, and the risk of resistance to treatment remains a global threat, but effective antimalarial treatment combination options are available. In areas receptive and vulnerable to malaria, international travelers and migrants are recommended to take chemoprophylaxis. Vector control is a primary intervention that can reduce malaria transmission from very high levels. With the significant decline in malaria cases globally in recent years, and renewed global interest on effectively interrupting transmission, a number of countries with low and unstable transmission are currently working toward malaria elimination.

Keywords

Malaria • Paludism • Anemia • Fever • Splenomegaly • Culicidae

20.1 Introduction

Malaria is an acute and chronic disease caused by *Plasmodium* parasites, a large genus of parasitic protozoa that currently has approximately 200 known species. The four species of *Plasmodium* that cause disease in humans are *P. vivax*, *P. falciparum*, *P. malariae*, and *P. ovale*. *Plasmodium*

K.H. Carter (✉) • R.P. Escalada • P. Singh
Department of Communicable Disease and Health Analysis, Pan American Health Organization/World Health Organization, Washington, DC, USA
e-mail: carterke@paho.org; escaladr@paho.org; singhpra@paho.org

C.B. Marcondes (ed.), *Arthropod Borne Diseases*, DOI 10.1007/978-3-319-13884-8_20

vivax and *P. falciparum* are the most common, with the latter also considered the most fatal. In recent years, simian parasites like *P. knowlesi* or *P. cynomolgi*, which are found in some forested areas of Southeast Asia, have been reported to likewise cause human cases of malaria.

The disease often manifests as a febrile illness which is caused by the asexual reproduction of parasites in erythrocytes. Resulting symptoms are variable, but the disease is usually characterized by paroxysms of fever and chills, anemia, sweating, and splenomegaly. *Plasmodium falciparum* is the most pathogenic parasite, can cause severe and complicated malaria, and can result in death. Humans infect female mosquitoes of the genus *Anopheles* which are the vectors that transmit parasites to other susceptible humans.

20.2 Life Cycle of the Parasite

The life cycle consists of phases in both the human host and in the mosquito of the genus *Anopheles*, the disease vector (Fig. 20.1).

20.2.1 Phase in the Vector

When a female anopheline feeds on the blood of an infected human being, it can ingest gametocytes, the sexual stages of the parasite. The presence of both developing microgametocytes or male forms and the female macrogametocytes in the mosquito stomach results in the subsequent union of gametes in the stomach of the mosquito and the formation of motile ookinetes. The ookinetes penetrate the stomach wall and form oocysts in the outer wall of the stomach. Each oocyst that ruptures releases thousands of motile sporozoites in the mosquito's body cavity that migrate to the mosquito salivary glands. Based on this, it can be noted that from one pair of gametocytes, several thousands of sporozoites may actually emerge. This process of sporozoite development, which begins with the binding of micro- and macrogametes up to the presence of sporozoites in the salivary glands, is known as *sporogony* and varies in terms of time between parasite species (Table 20.1).

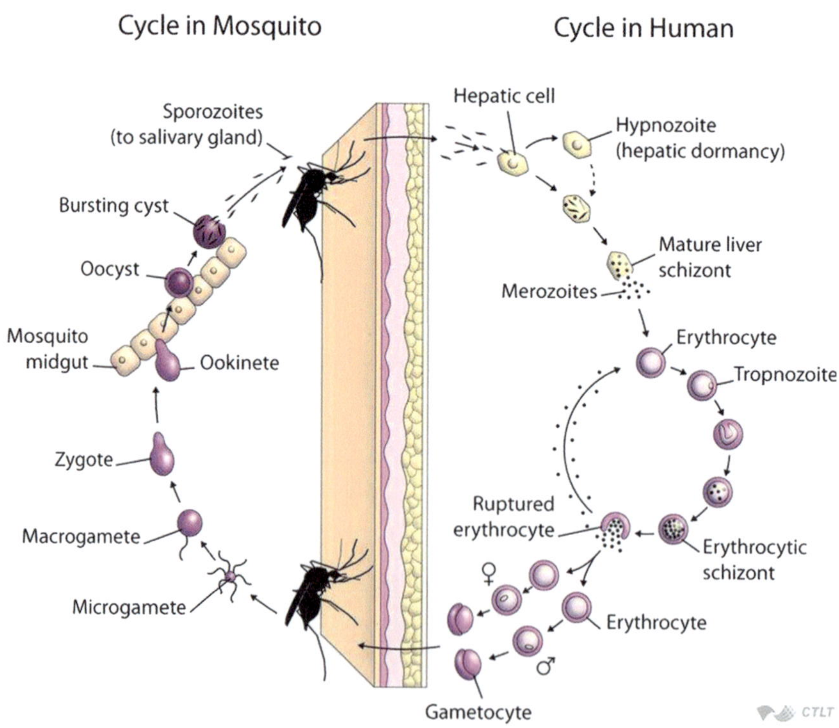

Fig. 20.1 Life cycle of malaria parasites (*Source*: Johns Hopkins Bloomberg School of Public Health)

Female anopheline mosquitoes with sporozoites in their salivary glands are infectious. Sporozoites accumulate in salivary glands in the course of their maturation and are injected with the mosquito's saliva. The sporozoites may be injected even during probing so the ingestion of blood is not critical for a human to become infected.

20.2.2 Phases in the Human Being

20.2.2.1 Exoerythrocytic Phase

When an infected female *Anopheles* feeds on a human being, the sporozoites in its saliva are injected intradermally where some are destroyed by the local macrophages or may enter the lymphatic system, while others are able to access the subcutaneous blood vessels. The sporozoites which are able to enter the human circulatory system can reach the liver within 30 min to a few hours. Inside the liver cells, hepatocytes, each sporozoite develops into a schizont containing thousands of merozoites. The circumsporozoite protein of the parasite facilitates the environment for the growth and development of the parasite in the liver cells (Singh et al. 2007). This process, which includes the entry of the sporozoites, to the rupture of hepatocytes, and merozoite entry into erythrocytes, is known as the exoerythrocytic phase which may last 6–16 days, depending on the parasite species (Table 20.1).

The duration of the exoerythrocytic cycle between *P. vivax* and *P. falciparum* is similar and relatively longer compared with *P. ovale* and *P. malariae*. Also, the exoerythrocytic cycles of *P. vivax* and *P. ovale* usually have two cycles, one in which the merozoites develop and break for 6–9 days and another in which some merozoites, known as hypnozoites, go into a state of dormancy in the liver and may be undeveloped for a long time until they develop and rupture leading to erythrocyte infection and relapse. This does not occur with *P. falciparum* and *P. malariae* and therefore these parasites do not cause relapses (Table 20.1).

20.2.2.2 Erythrocytic Phase

When the merozoites are released after rupture of the hepatocyte, there is an exchange between substances in the merozoite surface and the erythrocytes, allowing its entry into the erythrocyte. This process takes place very quickly which minimizes the exposure of the antigens on the surface of the parasite, affording parasites protection from the host immune response (Greenwood et al. 2008). Molecular interactions between distinct ligands on the merozoite and host receptors on the erythrocyte membrane play a key role in the invasion of the merozoites into the red blood cells. *Plasmodium vivax* invades only the red blood cells that are Duffy blood group positive, using the Duffy-binding protein

Table 20.1 Some characteristics of malaria parasites. Adapted from Giles and Warrell (1993)

	P. vivax	*P. falciparum*	*P. malariae*	*P. ovale*
Erythrocytic cycle (hours)	42–48	48	12–16	9
Exoerythrocytic cycle (days)	6–8	5.5–7	72	49–50
Sporogony in mosquito (minimum and range in days)	8 (8–16)	9 (9–22)	16 (16–35)	12 (12–14)
Appearance of gametocytes in blood (days)	3	10	…	…
Occurrence of secondary erythrocytic cycle	Yes	No	Yes	No
Merozoites in each hepatocyte	10,000	40,000	2000	15,000
Parasites in each erythrocyte	12–24	38	11	8
Incubation period (range in days)	12–17 or more	9–14 or more	18–40 or more	16–18 or more
Duration of untreated infection (years)	1.5–4	1–2	3–50	1.5–4
Severity of first febrile attack	Mild to severe	Severe—among nonimmune	Slight	Slight
Duration of febrile attack (hours)	8–12	16–36 or more	8–10	8–12
Recurrence of febrile attack (hours)	48	No	72	48

and the reticulocyte homology protein, found mostly on the reticulocytes. *Plasmodium falciparum* uses several different receptor families and alternate invasion pathways that are highly redundant, allowing it to invade any red cell (Weatherall et al. 2002).

The young forms of the parasite that develop in the erythrocytes subsequent to its invasion are known as rings and feed on hemoglobin. Within the erythrocytes, the parasite numbers expand rapidly. Despite the immunologic advantage provided by the erythrocytes to the growing parasite, the lack of standard biosynthetic pathways and intracellular organelles in the red cells tend to create obstacles for the fast-growing intracellular parasites.

As a result of the growth within the erythrocyte, they rupture and release merozoites that invade other erythrocytes, and the cycle is repeated each time with the invasion of previously uninfected erythrocytes. Each erythrocytic cycle occurs between about 2 and 3 days (Table 20.1). The cycles are repeated until the immunity of the person and/or antimalarial drugs slows the process. These merozoites are asexual but there are some which undergo differentiation and become sexual forms known as (macro- and micro)gametocytes. When the gametocytes are ingested from the human blood by female *Anopheles* mosquitoes, it becomes infected and capable of transmitting the disease.

20.3 Etiology and Pathophysiology

The pathophysiological changes observed among infected individuals are mainly caused by parasites in the erythrocytes and involve many organ systems. No pathophysiological changes are induced by sporozoites, gametocytes, or the exoerythrocytic phase. The presence of parasites in the erythrocytes results in erythrocyte rupture and sequestration, biochemical and electrolyte imbalances, and disorders of the hematological, gastrointestinal, pulmonary, neurological, and renal systems.

During the life cycle of the parasite in the blood, the destruction of erythrocytes results in anemia in conjunction with leukopenia and thrombocytopenia which are acute manifestations of malaria infections. When erythrocytes rupture, malaria pigment, other elements of parasites, and the erythrocyte membrane and intracellular contents enter the peripheral circulation. The circulating malaria pigments are taken up by macrophages leading to the production of pro-inflammatory cytokines, especially tumor necrosis factor and other pyrogenic cytokines, causing fever. Some of these factors may promote the coagulation cascade and result in disseminated intravascular coagulation.

Plasmodium falciparum is the most pathogenic of the parasites and is associated with severe morbidity and mortality resulting from other pathophysiological changes. When *P. falciparum* infects erythrocytes, it leads to expression of many surface proteins including *P. falciparum* erythrocyte membrane protein-1 (PfEMP-1) on the outer membrane of these infected erythrocytes making them more rigid and inflexible. When passing through microvessels in the tissues, they get stuck owing to adhesion to the endothelial cells, a process called cytoadherence. This cytoadhesion is largely due to the interaction of PfEMP-1 on infected erythrocytes with intercellular adhesion molecule 1 (ICAM-1) and CD36 acting as receptors on the endothelial membranes except in placental malaria. In the latter chondroitin sulfate A (CSA) is considered to be the cytoadhesion receptor for PfEMP-1 proteins. Accumulation of infected erythrocytes can lead to blockage of small vessels, limiting the oxygen delivery to tissues. However, this phenomenon of cytoadherence is apparently not uniform, and postmortem analyses of tissues in people who have succumbed to cerebral *P. falciparum* malaria have shown that the brain is preferentially the most parasite-infiltrated organ. Failure of the central nervous system is characteristic of cerebral malaria and may present with focal symptoms, such as seizures, deviated fixed gaze, or coma. Hypoglycemia, metabolic acidosis, and hyponatremia occur frequently especially in infections due to *P. falciparum.* Renal changes may also occur, resulting in the increase of urea, creatinine, and proteinuria. Acute renal failure is a complication of severe malaria that occurs more in adults. Pulmonary changes are also more common in adults than children and can result in adult respiratory distress syndrome and pulmonary edema.

20.3.1 Pathological Anatomy

The mortality associated with malaria is almost always due to infection with *P. falciparum.* However, malaria deaths caused by *P. vivax*, *P. ovale*, and *P. malariae* do occur, and those due to *P. vivax* are of particular importance due to the large number of infections of that species in some geographic regions of the world. The brain of patients who die from malaria typically presents with congestion, edema, and petechiae in the white matter. The veins and capillaries are usually full of parasitized erythrocytes and bleeding is observed. Bleeding and stroke are also possible conditions. The spleen is enlarged and has a gray color, with tensed vessels containing parasitized erythrocytes. The liver is also enlarged and has a gray color. The kidneys are slightly increased in size and congested. It may have bleeding, pigmentation in the glomeruli, and parasitized erythrocytes in the capillaries. In patients dying of hemoglobinuria, the kidneys are enlarged and exhibit edema and tubular necrosis. Often, the lungs are congested and with edema and with parasitized erythrocytes present in the lung capillaries. Minimal changes are observed in the heart with parasitized erythrocytes in the congested capillaries. Congestion and edema can also be observed in the gastrointestinal tract particularly in the intestinal mucosa with parasitized erythrocytes in vessels. In the placenta, parasitized erythrocytes are often seen in the maternal portion but almost never in the fetal portion.

20.3.2 Clinical Features and Manifestations

The host or the patient's previous exposure or immunity to malaria affects the symptomatology and incubation period, but individuals with malaria typically become symptomatic a few weeks after infection. The typical incubation period also varies with each *Plasmodium* species (Table 20.1). Clinical symptoms may include the following: fever, chills, sweating, headache, malaise, back pain, muscle pain, dizziness, abdominal pain, nausea, vomiting, diarrhea, poor appetite, cough, and dyspnea. At the same time, the patient may also have postural hypotension, anemia, hepatosplenomegaly, jaundice, hypoglycemia, leukopenia, and thrombocytopenia.

The paroxysms of fever, chills, and sweats are also variable depending on species. Typically, they begin with a period of shivering and chills, lasting anywhere from 1 to 2 h and followed by a high fever. The patient then experiences excessive diaphoresis, and subsequently, the body temperature drops to normal or below normal. Several small fever spikes a day instead of the classic paroxysms may be reported by many patients early in the infection. Patients with long-standing infections are more likely to present with classic fever patterns. However, periodicity in febrile attacks is often not observed in *P. falciparum* infections so in general, the occurrence of periodicity of fever is not a reliable clue to the diagnosis of malaria.

As malaria symptoms are related to the maturation and rupture of infected erythrocytes, the clinical picture of infections with *P. vivax* and *P. ovale* parasitemias may last from days to a few months after the initial attack. The persistence of parasitemia from the initial infection can result to the recrudescence of symptoms within a short period of time. Also, without appropriate treatment, exoerythrocytic parasites that persist in the parenchymal cells of the liver or hypnozoites can cause relapses ranging from a few weeks up to 9 months. Relapses of *P. vivax* can last up to 3 or 4 years. Without treatment, *P. malariae* can persist up to 50 years (Table 20.1).

Infections caused by *P. falciparum* in patients without immunity to infection are very dangerous and can result in death. As the parasite has developed resistance to most commonly used antimalarials in almost all parts of the world, the patient's condition may worsen and result in death unless an early diagnosis is made and the patient is treated in a timely and appropriate manner. Patients with complicated and severe

Table 20.2 Signs and symptoms of severe malaria in adults and in children. From World Health Organization (2012)

Sign or symptom	Adults	Children
Duration of illness	5–7 days	Shorter (1–2 days)
Respiratory distress/deep breathing (acidosis)	Common	Common
Convulsions	Common (12 %)	Very common (30 %)
Posturing (decorticate/decerebrate and opisthotonic rigidity)	Uncommon	Common
Prostration/obtundation	Common	Common
Resolution of coma	2–4 days	Faster (1–2 days)
Neurological sequelae after cerebral malaria	Uncommon (1 %)	Common (5–30 %)
Jaundice	Common	Uncommon
Hypoglycemia	Less common	Common
Metabolic acidosis	Common	Common
Pulmonary edema	Uncommon	Rare
Renal failure	Common	Rare
CSF opening pressure	Usually normal	Usually raised
Bleeding/clotting disturbances	Up to 110 %	Rare
Invasive bacterial infection (coinfection)	Uncommon (<5 %)	Common (10 %)

malaria infections associated with *P. falciparum* can have very high fever, convulsions, delirium, coma, acute pulmonary distress syndrome, acute renal failure, liver failure, abdominal pain, nausea, vomiting, and diarrhea. Likewise, they may have severe anemia, hypoglycemia, and metabolic acidosis. Fortunately, *P. falciparum* infections do not result in the presence of hypnozoites in the liver and therefore do not relapse if the parasites are effectively eliminated in the blood during treatment. However, if the treatment was not successful in eliminating viable parasites in the blood, patients may experience recrudescence (renewed presence of parasites in circulation) within a short period of time.

Among children, congenital malaria may result from any of the four species of parasites during the first few months of life. In areas where the disease is endemic, infants receive maternal antibodies and may only have fever without other symptoms. Likewise, in these areas, it is quite difficult to make the differential diagnosis between congenital malaria and malaria transmitted by mosquitoes after the first 2 weeks of birth. In non-endemic areas, congenital malaria may present with fever, anemia, hepatomegaly, and splenomegaly, and patients may be anorexic and appear lethargic. Infection can also cause death in the first days of life.

20.3.3 Severe Malaria

Plasmodium falciparum accounts for the vast majority of severe malaria cases although *P. vivax* and *P. knowlesi* (Kantele and Jokiranta 2011; Daneshvar et al. 2009) can also cause severe disease. Uncomplicated malaria caused by these parasites can also result to severe form if effective treatment is delayed. Among children, severe *P. falciparum* malaria may develop so rapidly that early treatment may not even be possible.

As previously described, uncomplicated *P. falciparum* malaria is highly variable in clinical presentation among adults and children (Table 20.2) and mimics other diseases. Fever is common and often intermittent but may be absent in some cases. Young children may manifest with irritability, refusal to eat, and vomiting; and on physical examination, fever may be the only sign. If early diagnosis and management is missed, patients with falciparum malaria may deteriorate rapidly.

20.4 Diagnostic Approach

Many people with malaria infections do not present the typical symptoms of fever and chills with a well-defined periodicity. Persons living in non-endemic areas and returning from a trip to an endemic area can present with symptoms which can mimic any other febrile illness such as hepatitis, yellow fever, dengue, typhoid fever, leptospirosis, tuberculosis, trypanosomiasis, visceral leishmaniasis, or viral infections. Persons living in endemic areas have some degree of immunity and may have very mild symptoms or may even be asymptomatic. Malaria infection may also result from the transfusion of unscreened blood of an affected person. For these reasons, conducting a thorough *clinical history* is very important because malaria should be considered as a differential diagnosis for people who reside or come from an area where malaria transmission is possible, particularly for patients presenting clinical symptoms that may include fever, chills, sweating, headache, muscle pain, lack of appetite, malaise, anemia, drowsiness, or complications associated with the disease.

In many non-endemic areas, the disease is often overlooked, and in the absence of facilities for good quality laboratory diagnosis, the clinical picture may deteriorate rapidly resulting in serious consequences. As realities and contexts in various countries vary, national policies on the diagnosis and treatment of suspected and confirmed malaria cases should be taken into account. All cases of suspected malaria must be confirmed using parasite-based diagnostic testing before administering treatment, which suspicion should only be considered solely on the basis of clinical when a parasitological diagnosis is not accessible (World Health Organization 2010). In many countries, patients with confirmed *P. vivax* infection, potential glucose-6-phosphate dehydrogenase (G6PD) deficiency status, must also be borne in mind, as this condition has important implications in terms of treatment.

20.4.1 Microscopic Diagnosis

Microscopic examination of a blood smear is the gold standard for the diagnosis of the disease, and its use is particularly promoted and encouraged in the region of the Americas. The test may be extended to include a thick blood smear where detection and identification of parasites are easier and in which the erythrocytes are grouped and visualized more easily, but thin blood smears permit more accurate species identification. In general, blood smears are stained with Giemsa, but other dyes which have been used include Field, Leishman, Romanowsky, and Wright. Treatment regimens may vary according to parasite species, and it is therefore very important to identify and confirm the causative parasites of disease.

A large variation in parasite morphology can be observed in microscopic diagnosis for malaria. The different *Plasmodium* species exhibit characteristic morphological features in the various stages of their life cycle in the erythrocytes (Table 20.3). Most malaria infections (microscopic and submicroscopic) should be considered as potentially infectious and able to contribute to ongoing transmission. There is no need for routine detection of gametocytes using sensitive nucleic acid amplification methods in malaria surveys or clinical settings (World Health Organization 2014a, b, c, d, e).

20.4.2 Rapid Diagnostic Tests

In endemic and non-endemic areas where there is difficult or nonexistent access to the microscopic diagnosis, the use of rapid diagnostic tests (RDTs) is recommended. Rapid tests detect parasite antigens in the peripheral blood. In general, RDTs detect the histidine-rich protein 2 which is specific to *P. falciparum* (HRP2), the protein lactate dehydrogenase (pLDH), and aldolase. These rapid tests allow the differentiation between *P. falciparum* and *P. vivax* infections.

As RDTs have varying indications, sensitivity, and specificity in terms of use, users are strongly advised to obtain guidance from the World Health Organization (WHO) RDT procurement recommendations and the results of the periodic RDT product testing (World Health Organization 2014a, b, c, d, e).

Table 20.3 Species identification of malaria parasites in Giemsa-stained thick blood films (World Health Organization 2010)

Species		Trophozoite	Schizont	Gametocyte
Plasmodium falciparum	Young, growing trophozoites and/or mature gametocytes usually seen	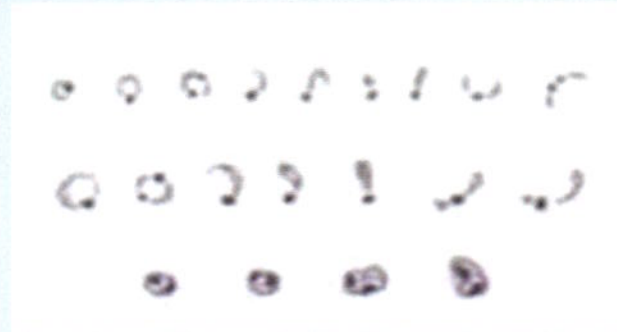 *Size*, small to medium; *number*, often numerous; *shape*, ring and comma forms, common; *chromatin*, often two dots; *cytoplasm*, regular, fine to fleshy; *mature forms*, sometimes present in severe malaria, compact with *pigment* as few coarse grains or a mass	 Usually associated with many young ring forms. *Size*, small, compact; *number*, few, uncommon, usually in severe malaria; *mature forms*, 12–30 or more merozoites in compact cluster; *pigment*, single dark mass	 Immature pointed-end forms, uncommon. *Mature forms*, banana shaped or rounded; *chromatin*, single, well defined; *pigment*, scattered, coarse, rice grain-like; pink extrusion body sometimes present. Eroded forms with only chromatin and pigment often seen
Plasmodium vivax	All stages seen; Schüffner stippling in "ghost" of host red cells, especially at film edge	 *Size*, small to large; *number*, few to moderate; *shape*, broken ring to irregular forms, common; *chromatin*, single, occasionally two; *cytoplasm*, irregular or fragmented; *mature forms*, compact, dense; *pigment*, scattered, fine	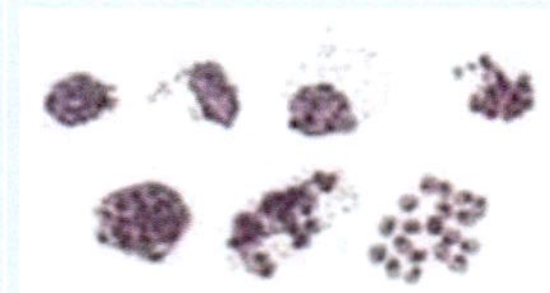 *Size*, large; *number*, few to moderate; *mature forms*, 12–24 merozoites, usually 16, in irregular cluster; *pigment*, loose mass	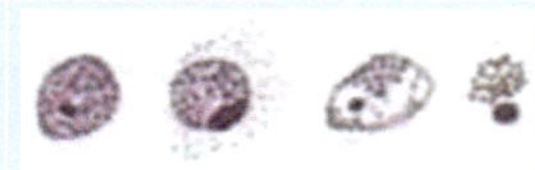 Immature forms difficult to distinguish from mature trophozoites; *mature forms*, round, large; *chromatin*, single, well defined; *pigment*, scattered, fine. Eroded forms with scanty or no cytoplasm and only chromatin and pigment present

Plasmodium ovale	All stages seen; prominent Schüffner stippling in "ghost" of host red cells, especially at film edge	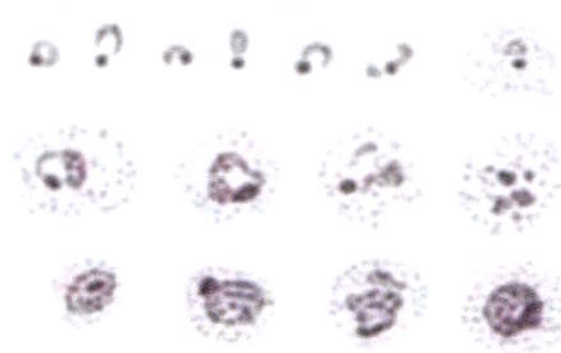 *Size*, may be smaller than *P. vivax*; *number*, usually few; *shape*, ring to rounded, compact forms; *chromatin*, single, prominent; *cytoplasm*, fairly regular, fleshy; *pigment*, scattered, coarse	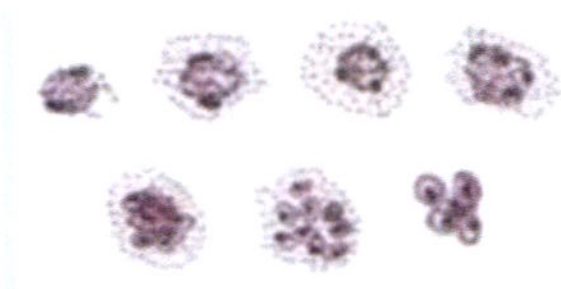 Immature forms difficult to distinguish from mature trophozoites; *mature forms*, round, may be smaller than *P. vivax*; *chromatin*, single, well defined; *pigment*, scattered, coarse. Eroded forms with only chromatin and pigment present	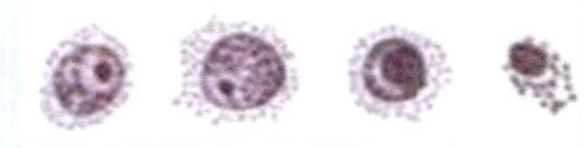 *Size*, rather like *P. malariae*; *number*, few; *mature forms*, 4–12 merozoites, usually 8, in loose cluster; *pigment*, concentrated mass
Plasmodium malariae	All stages seen	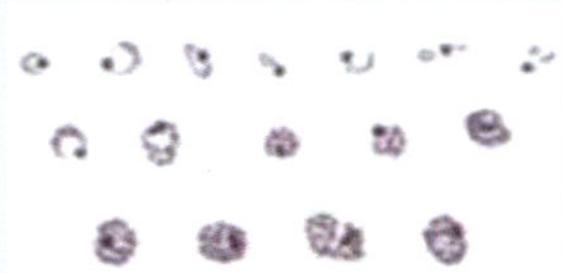 *Size*, small; *number*, usually few; *shape*, ring to rounded, compact forms; *chromatin*, single, large; *cytoplasm*, regular, dense; *pigment*, scattered, abundant, with yellow tinge in older forms	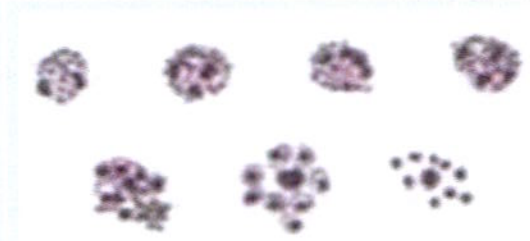 *Size*, small, compact; *number*, usually few; *mature forms*, 6–12 merozoites, usually 8, in loose cluster, some apparently without cytoplasm; *pigment*, concentrated	 Immature and certain *mature forms* difficult to distinguish from mature trophozoites. *Mature forms*, round, compact; *chromatin*, single, well defined; *pigment*, scattered, coarse, may be peripherally distributed. Eroded forms with only chromatin and pigment present

20.4.3 Other Diagnostic Methods

Nucleic acid amplification techniques are available and are more sensitive in detection of malaria compared to RDTs and microscopy. As these tests are expensive and require significant expertise, their use should be considered only in low transmission settings where there is already widespread implementation of malaria diagnostic testing and treatment and low parasite prevalence rates (e.g., <10 %). The use of nucleic acid amplification (NAA)-based methods should not divert resources away from malaria prevention and control interventions and strengthening of the healthcare services, including the surveillance system (World Health Organization 2014a, b, c, d, e).

20.5 Treatment

The treatment of malaria has different objectives, including the treatment of uncomplicated malaria to cure the infection, reduce the transmission, and prevent the emergence and spread of resistance to antimalarials. In people with severe malaria, the goal of treatment is to prevent death and prevention of neurological sequelae (World Health Organization 2015).

Different types of antimalarials act at different stages of the life cycle of the parasite. Those that destroy the asexual parasites (trophozoites) in erythrocytes are known as blood schizonticides and are used in patients with symptoms. Tissue schizonticides act against the exoerythrocytic stages (hypnozoites) caused by *P. vivax* and *P. ovale* in the liver. Gametocytocides destroy sexual stages (gametocytes), while sporontocides inhibit parasite development of oocysts in the anopheline vector.

According to structure, most antimalarials can be divided into the following groups:

Group 1. Quinine, quinidine, and mefloquine
Group 2. Chloroquine and amodiaquine
Group 3. Pyrimethamine and proguanil
Group 4. Sulphadoxine
Group 5. Primaquine
Group 6. Artemisinin and its derivatives
Group 7. Atovaquone
Group 8. Antibiotics with antimalarial effect as clindamycin, doxycycline, and tetracycline

Though they may have different modes of action, antimalarials of groups 1 and 2 have similar effects as both blood schizonticides and gametocytocides for *Plasmodium* parasites other than those of *P. falciparum.*

In infections with *P. falciparum*, a large variation in the number and size of trophozoites can be found about 10–12 days after infection in a primary attack; gametocytes appear 8–10 days after the primary attack which means that early diagnosis and timely and appropriate treatment will interrupt the development of gametocytes and the ability to infect the anopheline mosquito.

20.5.1 Treatment of Uncomplicated *P. falciparum*

WHO recommendations for the treatment of uncomplicated *P. falciparum* malaria are:

- The treatment of choice for uncomplicated *P. falciparum* malaria is a combination of two or more antimalarial drugs with different mechanisms of action. The use of combination instead of single-drug therapies provides the benefit that combinations are more effective and prevents selection of mutant parasites resistant to a single drug as the partner drug kills it.
- Artemisinin-based combination therapies (ACTs) are the recommended treatments for uncomplicated *P. falciparum* malaria. Some areas in the world continue to use chloroquine as the parasite species has still not developed resistance to this antimalarial. However, this is an exception worldwide rather than a rule and unless otherwise known, ACTs are preferred and recommended.
- Artemisinin is a potent antimalarial but has a short half-life owing to which it is combined with other antimalarials with longer half-lives.

Thus, the following ACTs are currently recommended for use:

- Artemether plus lumefantrine
- Artesunate plus amodiaquine
- Artesunate plus mefloquine
- Artesunate plus sulfadoxine-pyrimethamine
- Dihydroartemisinin plus piperaquine

The choice of ACT in a country or region is based on the resistance level of artemisinin partner drug in the combination:

- In areas of multidrug resistance (Southeast Asia): artesunate + mefloquine, artemether-lumefantrine, dihydroartemisinin + piperaquine.
- In areas without multidrug resistance (Africa and other areas): artemether-lumefantrine, artesunate + amodiaquine, artesunate + sulfadoxine-pyrimethamine.
- The artemisinin component of an ACT should be given at least 3 days for optimal effect, and artemether-lumefantrine should be used with a regimen of six doses.
- Amodiaquine + sulfadoxine-pyrimethamine may be considered as an interim option in situations where ACTs may not be available.
- Although artemisinin has a gametocytocidal effect, the current recommendation is to add a single dose of primaquine (0.25 mg/kg; 0.75 mg/kg is also acceptable and used in some countries) as a gametocytocide to the treatment regimen, usually given on the first day of treatment and irrespective of the presence or absence of gametocytes in the patient's peripheral blood.

20.5.2 Treatment Regimens with Recommended ACTs

20.5.2.1 Artemether Plus Lumefantrine

Currently available as a fixed-dose co-formulated tablet containing 20 mg artemether and 120 mg lumefantrine, the recommended treatment regimen is artemether-lumefantrine tablets twice daily given for 3 days. A total therapeutic dose includes 1.4–4 mg/kg of artemether, 10–16 mg/kg of lumefantrine, and 0.75 mg/kg of primaquine. It should be given after a meal rich in fat as fat improves the absorption of lumefantrine (Table 20.4).

Table 20.4 Treatment regimen using artemether plus lumefantrine combination

Body weight in kg/(age in years)		Time (hours)					
		0	8	24	36	48	60
5–14	(<3)	1	1	1	1	1	1
15–24	(≥3–8)	2	2	2	2	2	2
25–34	(≥9–14)	3	3	3	3	3	3
>34	(>14)	4	4	4	4	4	4

20.5.2.2 Artesunate Plus Amodiaquine

Fixed-dose formulations of the drug with tablets are available with 25/67.5 mg, 50/135 mg, or 100/270 mg of artesunate and amodiaquine, respectively. Full recommended treatment regimen is of 4 mg/kg/day of artesunate and 10 mg/kg/day of amodiaquine administered once daily for 3 days along with 0.75 mg/kg of primaquine given on the first day (Table 20.5).

Table 20.5 Treatment regimen using artesunate plus amodiaquine combination

Age	Dose in mg (no. of tablets)					
	Artesunate (50 mg)			Amodiaquine (153 mg)		
	Day 1	Day 2	Day 3	Day 1	Day 2	Day 3
5–11 months	25 (1/2)	25	25	76 (1/2)	76	76
≥1–6 years	50 (1)	50	50	153 (1)	153	153
≥7–13 years	100 (2)	100	100	306 (2)	306	306
>13 years	200 (4)	200	200	612 (4)	612	612

20.5.2.3 Artesunate Plus Mefloquine

Separate tablets containing 50 mg of artesunate and 250 mg of mefloquine available, but not promoted, in order to reduce potential use of either as monotherapy. The recommended treatment is 4 mg/kg/day of artesunate administered once daily for 3 days, 25 mg/kg of total dose of mefloquine generally divided over 2 days (10 and 15 mg/kg/day) or 3 days (8.3 mg/kg/day), and 0.75 mg/kg of primaquine given on the first day. Mefloquine may cause nausea and vomiting among other side effects, and splitting the dose over a longer period (2–3 days) generally reduces the likelihood and severity of side effects. The use of mefloquine within 60 days of initial treatment could lead to increased risk of neuropsychiatric side effects, and thus its concomitant use within 60 days should be avoided. If two separate *P. falciparum* infections occur within the same patient within 60 days, the second episode should be treated with an ACT other than artesunate plus mefloquine if the first episode was also treated with it (Table 20.6).

20.5.2.4 Artesunate Plus Sulfadoxine-Pyrimethamine

The recommended treatment is a target dose of 4 mg/kg/day of artesunate given once a day for 3 days and sulfadoxine-pyrimethamine in a dose of 25/1.25 mg/kg on day 1. A 0.75 mg/kg dose of primaquine on day 1 should be added to this (Table 20.7).

Table 20.6 Treatment regimen using artesunate plus mefloquine combination

Age	Dose in mg (no. of tablets)					
	Artesunate (50 mg)			Mefloquine (250 mg)		
	Day 1	Day 2	Day 3	Day 1	Day 2	Day 3
5–11 months	25 (1/2)	25	25	–	125 (1/2)	–
≥1–6 years	50 (1)	50	50	–	250 (1)	–
≥7–13 years	100 (2)	100	100	–	500 (2)	250 (1)
>13 years	200 (4)	200	200	–	1000 (4)	500 (2)

20.5.2.5 Dihydroartemisinin Plus Piperaquine

A total dose of 4 mg/kg of dihydroartemisinin and 18 mg/kg/day of piperaquine once a day each for 3 days is recommended. Additionally primaquine (0.75 mg/kg) should be added on the first day of treatment.

20.5.3 Recommendations for Second-Line Antimalarial Treatment

It is possible that the same person may get infected with *P. falciparum* malaria more than once within the same month; however, when presenting fever and malaria parasites in peripheral blood after 14 days of start of treatment with ACT, the patient should be treated as new infection and treated accordingly with recommended ACTs. However, rarely some patients with uncomplicated *P. falciparum* malaria treated with ACTs might resolve, but parasites may reappear in peripheral blood within 14 days of start of treatment. In these cases it is recommended to use the following:

- Another ACT known to be effective in the region.
- Artesunate (2 mg/kg of body weight once a day) + tetracycline (4 mg/kg body weight four times a day) or doxycycline (3.5 mg/g body weight once daily) or clindamycin (10 mg/kg body weight twice daily). Any of these combinations should be given for at least 7 days.

Table 20.7 Treatment regimen using artesunate plus sulfadoxine-pyrimethamine combination

Age	Dose in mg (no. of tablets)					
	Artesunate (50 mg)			Sulfadoxine-pyrimethamine (500/25)		
	Day 1	Day 2	Day 3	Day 1	Day 2	Day 3
5–11 months	25 (1/2)	25	25	250/12.5 (1/2)	–	–
≥1–6 years	50 (1)	50	50	500/25 (1)	–	–
≥7–13 years	100 (2)	100	100	1000/50 (2)	–	–
>13 years	200 (4)	200	200	1500–1575 (3)	–	–

- Quinine (10 mg/kg body weight three times a day) + tetracycline or doxycycline or clindamycin can also be used. Any of these combinations should be given for at least 7 days.

20.5.4 Treatment in Specific Populations and Situations

20.5.4.1 Pregnant Women

Pregnant women with acute symptoms of malaria are a high-risk group and should receive efficacious antimalarial treatment. Malaria in pregnancy is associated with low birth weight, increased anemia, and, in areas of low transmission, an increased risk of severe malaria. The use of primaquine in pregnant women should be avoided as it may cause harm to the fetus. Recommendations for the treatment of uncomplicated *P. falciparum* malaria in pregnancy are:

- First trimester: quinine and clindamycin for 7 days should be given. Quinine monotherapy can be given if clindamycin is unavailable. If this fails, artesunate and clindamycin should be used. Evidence on the effect of artemisinin and its partner drug in ACTs in first trimester pregnancy is presently limited, but no adverse effects have been seen. Thus ACTs are only recommended for use if they are the only treatment immediately available or treatment failure with previous regimens is seen.
- Second and third trimesters: Any ACT known to be effective in the country/region except dihydroartemisinin and piperaquine, or artesunate plus clindamycin for 7 days, or quinine and clindamycin for 7 days.

If neither one of the above options nor clindamycin is available, monotherapy can be given.

20.5.4.2 Lactating Women

The use of primaquine in lactating women should be avoided unless the glucose-6-phosphate dehydrogenase (G6PD) deficiency status of the infant is known. Primaquine is known to cause hemolysis in G6PD-deficient individuals. Tetracycline is also contraindicated as it affects the bones and teeth of children. Any other antimalarial, preferably an ACT, should be used for the treatment of uncomplicated *P. falciparum* malaria in breastfeeding women.

20.5.4.3 Infants

Infants should receive standard antimalarial treatment (including ACTs) except for primaquine to be avoided in the first 6 months of infancy and tetracyclines at least until 8 years of age. As pediatric formulations are not available for most ACTs, appropriate care should be taken to provide a correct dose according to the body weight of the child. Infants are more likely to vomit or regurgitate antimalarial treatment than older children or adults, requiring repeat doses if necessary. Taste, volume, consistency, and gastrointestinal tolerability are important and crucial to treatment of infants. Artemisinin appears to be safe and well tolerated by young children, and so the choice of the ACT will be determined largely by the safety and tolerability of the partner drug. The limited information available does not indicate the specific difficulties ACT currently recommended in childhood.

20.5.4.4 Travelers

Travelers to malaria-endemic areas are often nonimmune adults, either from cities with little or no transmission in endemic countries, or visitors from non-endemic countries. Both tend to be at higher risk of contracting malaria and its consequences, as they have no immunity to malaria. When diagnosed with malaria in an endemic area, they should be treated as any other patient and according to malaria guidelines prevalent. If malaria is diagnosed on return to the patient's home country (non-endemic), the following treatment is recommended:

- Atovaquone-proguanil (15/6 mg/kg, the usual dose for adults, four tablets once daily for 3 days)
- Artemether-lumefantrine (adult dose, four tablets twice daily for 3 days)
- Quinine (10 mg salt/kg body weight every 8 h) + doxycycline (3.5 mg/kg body weight

once daily) or clindamycin (10 mg/kg body weight twice a day), all drugs approved for 7 days

20.5.4.5 Coexisting Morbidities

HIV Infection

A growing number of people in malaria-endemic areas are living with HIV. It is increasingly clear that as HIV progresses and immunosuppression worsens, the manifestations of malaria also worsen. In pregnant women, adverse effects on birth weight increase. In patients with partial immunity to malaria, the severity of infection is greater. The treatment with sulfadoxine-pyrimethamine should not be administered to patients with co-trimoxazole, as they have probably a higher risk of sulfonamide-related adverse effects, and, in any case, since both drugs have antimalarial activity, malaria infection is likely to be resistant to sulfadoxine-pyrimethamine. Depending on the setting of the transmission of malaria, HIV-infected individuals have an increased risk of asymptomatic parasitemia, clinical malaria, or severe and complicated malaria. Therefore they have a greater need for control measures of malaria than people not infected with HIV. The following are the recommendations for treatment of uncomplicated *P. falciparum* malaria in patients with HIV infection:

- Patients with HIV infection who develop malaria should receive standard ACT.
- ACTs or intermittent preventive therapy with sulfadoxine-pyrimethamine should not be administered to HIV-infected patient receiving co-trimoxazole (trimethoprim and sulfamethoxazole).
- ACTs with amodiaquine should be avoided in HIV-infected patients receiving zidovudine or efavirenz.

Severe Malnutrition

Malaria and malnutrition frequently coexist. Malnutrition in absorption may be reduced due to diarrhea, vomiting, intestinal transit, and rapid atrophy of the intestinal mucosa. The drug absorption by intramuscular and possibly rectal routes may be slower, and the decline in muscle mass may make it difficult to administer repeated intramuscular injections. Although there are many reasons for drug kinetics to be different in malnourished patients compared to others, changes in the current recommendations of dosage in mg/kg of body weight are not recommended.

20.5.4.6 Treatment of Uncomplicated *P. vivax* Malaria

The recommendations for treating uncomplicated *P. vivax* malaria are:

- A total dose of 25 mg/kg of chloroquine divided over 3 days in combination with 0.25 mg/kg/day of primaquine taken once daily with food for 14 days is the treatment of choice for the chloroquine-sensitive infections. In Oceania and Southeast Asia, the dose of primaquine should be 0.5 mg/kg/day for 14 days.
- ACTs with mefloquine, amodiaquine, or piperaquine along with primaquine should be used in *P. vivax* malaria resistant to chloroquine.
- In moderate G6PD deficiency, primaquine 0.75 mg/kg should be administered once a week for 8 weeks. In severe G6PD deficiency, primaquine should not be given.
- In pregnant or lactating women in whom primaquine is contraindicated, weekly chemoprophylaxis with chloroquine can be considered until such time that primaquine can be safely given.

20.5.4.7 Treatment of Complicated *P. falciparum* Malaria

Severe malaria is a medical emergency. After rapid clinical assessment and confirmation of the diagnosis, full doses of parenteral antimalarial treatment should be started without delay with any effective antimalarial first available:

- For adults, artesunate 2.4 mg/kg BW IV or IM given on admission (time=0), then at 12 and 24 h, and then once a day is the recommended treatment. Artemether, or quinine, is an acceptable alternative if parenteral artesunate is not available: artemether 3.2 mg/kg BW IM given on admission and then 1.6 mg/kg BW per day or quinine 20 mg salt/kg BW on admission (IV infusion or divided IM injection) and then 10 mg/kg BW every 8 h; infusion rate should not exceed 5 mg salt/kg BW per hour.

- For children less than 20 kg in weight, artesunate 3 mg/kg BW IV or IM given on admission (time = 0), then at 12 and 24 h, and then once a day is the recommended treatment. Artemether, or quinine, is an acceptable alternative if parenteral artesunate is not available: artemether 3.2 mg/kg BW IM given on admission and then 1.6 mg/kg BW per day or quinine 20 mg salt/kg BW on admission (IV infusion or divided IM injection) and then 10 mg/kg BW every 8 h; infusion rate should not exceed 5 mg salt/kg BW per hour.
- Whenever parenteral treatment is started, it should be continued for at least 24 h or till the patient can take treatment orally, whichever is later chosen.
- Thereafter, treatment should be continued by giving a complete course of a recommended ACT:
  - Artemether plus lumefantrine
  - Artesunate plus amodiaquine
  - Dihydroartemisinin plus piperaquine
  - Artesunate plus sulfadoxine-pyrimethamine
  - Artesunate plus clindamycin or doxycycline (pregnancy)
  - Quinine plus clindamycin or doxycycline (pregnancy)

20.5.4.8 Treatment of Complicated *P. vivax* Malaria

Plasmodium vivax malaria may also cause a severe and debilitating febrile illness. It can lead to severe manifestations like cerebral malaria, severe anemia, severe thrombocytopenia and pancytopenia, jaundice, splenic rupture, acute renal failure, and acute respiratory distress syndrome. Treatment of severe *P. vivax* malaria is similar to that of severe *P. falciparum* malaria.

20.5.4.9 Treatment of Malaria Caused by *P. ovale*, *P. malariae*, and *P. knowlesi*

Resistance of *P. ovale* and *P. malariae* to antimalarials is not well characterized, and infections caused by these two species are generally considered sensitive to chloroquine.

The recommended treatment of malaria caused by *P. ovale* is the same as that for *P. vivax*, namely, chloroquine and primaquine. *P. malariae* and *P. knowlesi* should be treated with the standard regimen of chloroquine for malaria by *P. vivax*, but they do not require primaquine as hypnozoites are not formed in infections by these species.

Those with severe *P. knowlesi* malaria are treated as a severe *P. falciparum* malaria infection, except that the thresholds for severe malaria are changed to parasite density higher than 100,000 asexual forms/μL of whole blood or jaundice and parasite density higher than 20,000/μL.

20.5.5 Antimalarial Drug Resistance

Antimalarial drug resistance has been documented in *P. falciparum*, *P. vivax*, and *P. malariae*. Parasite resistance results in a delayed or incomplete clearance of parasites from the patient's blood when the person is being treated with an antimalarial. The problem is compounded by cross-resistance, in which resistance to one drug confers resistance to other drugs that belong to the same chemical family or which have similar modes of action. Several antimalarials had to be removed from markets after the spread of parasite resistance in previous decades.

Enhanced surveillance of antimalarial drug efficacy is important to detect and respond to emerging drug resistance of treatment combinations or the individual partner drugs. Countries should monitor the efficacy of first-line malaria therapies, and treatment failure should prompt a change in national antimalarial treatment policy.

ACTs remain effective provided that the partner drugs remain efficacious. Unfortunately,

P. falciparum resistance to artemisinin has emerged in five countries of the Greater Mekong Subregion—Cambodia, the Lao People's Democratic Republic, Myanmar, Thailand, and Viet Nam. This is a serious public health concern that threatens the sustainability of the global effort to reduce and eliminate malaria. The global plan for artemisinin resistance containment (GPARC) emphasizes among others that a key priority for endemic countries and the global malaria community is protecting the efficacy of ACTs and developing new combinations.

The scale-up of all basic malaria interventions, including vector control, diagnostic testing, and quality-assured treatment, contributes to reducing the potential emergence of resistance. National regulatory authorities should regulate production, marketing authorization, export, import, and use of oral artemisinin-based monotherapies. Countries should also take regulatory action to remove fake and substandard drugs from health facilities to preserve the efficacy of ACTs (World Health Organization 2011).

20.5.6 Vaccines

As in other complex parasites, the development of a vaccine against the species of *Plasmodium* pathogenic to humans is a very challenging task. Currently, there is no licensed and commercially available malaria vaccine, although they are expected to be an important addition to the arsenal of tools against malaria in the coming years. A number of vaccine candidates are currently in various stages of development, with different modes of action, to prevent *P. falciparum* and *P. vivax* infections.

20.5.6.1 *P. falciparum* Vaccines

The most advanced *P. falciparum* malaria vaccine candidate is the RTS,S/AS01. The need for preventive measures such as the use of long-lasting insecticidal nets, confirmatory diagnosis either by microscopy or RDT, and treatment with ACTs will continue, even if RTS,S/AS01 becomes available and is used (World Health Organization 2013a, b, c).

20.5.6.2 *P. vivax* Vaccines

Funding for *P. vivax* vaccine candidates is relatively smaller compared with *P. falciparum* vaccine candidates. Much less is known about this species of the malaria parasite, but there is evolving interest in the light of its significance in malaria elimination efforts. In 1999, a *P. vivax* network to create, improve, and expand research on the parasite, cGMP6 production of candidate vaccines for human clinical trials, clinical testing site development, and funding was established in Asia (World Health Organization 1999). Currently, there are at least two ongoing *P. vivax* vaccine-related projects: ChAd63/MVA PvDBP and the experimental transmission of *P. vivax* to *Anopheles* (EVITA) which is enabling technology, not a vaccine (World Health Organization 2014a, b, c, d, e).

20.5.6.3 The Global Portfolio of Vaccines Against Malaria

WHO keeps track of the various malaria vaccine projects at advanced preclinical and clinical stages globally. Below is a table pertaining to ongoing projects from various research and pharmaceutical institutions that are currently contributing to the global malaria vaccine development efforts (Table 20.8).

20.6 Epidemiology

Malaria transmission involves the interaction between the malaria parasite, the human host, the *Anopheles* vector, and physical, biological, and socioeconomic factors that shape the environment. The intensity of transmission depends on the prevalence and incidence of different parasites in the human population; characteristics of the mosquito vectors such as its efficiency as a vector, density, feeding, and resting habits; susceptibility to infection and longevity that allows the development of sporozoites; human susceptibility to infection; and environmental factors that favor the presence of vector breeding population. Given the interaction of all these parameters, the disease is much focused, and in many areas where it is endemic, transmission

Table 20.8 Global malaria vaccine pipeline (World Health Organization 2014a, b, c, d, e)

TRANSLATIONAL PROJECTS				*VACCINE CANDIDATES*	
Phase 1a	**Phase 2a**		**Phase 1b**	**Phase 2b**	**Phase 3**
ChAd63/MVA ME-TRAP + Matrix M™	Ad35.CS/ RTS,S-AS01	M3V.Ad.*Pf*CA	Ad35.CS	ChAd63/MVA ME-TRAP	RTS,S-AS01
Polyepitope DNA EP 1300	Ad35.CS/ Ad26.CS	M3V.D/ Ad.*Pf*CA	EBA 175.R2	GMZ2	
PfCelTOS FMP012	ChAd63/ MVA (CS, TRAP, AMA)	ChAd63/MVA MSP 1	SE36	MSP3 [181-276]	
CSVAC		ChAd63.AMA1/ MVA.AMA1	*Pf*SPZ		
ChAd63.AMA/ MVA.AMA1 +Al/CPG7909		FMP2.1-AS01B (AMA1 3D7)			
SR11.1					
ChAd63/MVA PvDBP					
*Pfs*25-EPA					
*Pfs*25-VLP					

P. falciparum **vaccines:** Pre-erythrocytic; Blood-stage; Transmission-blocking
P. vivax **vaccines:** Pre-erythrocytic; Blood-stage; Transmission-blocking

Data source: http://www.who.int/vaccine_research/links/Rainbow/en/index.html

varies with temperature, rainfall, and humidity. Lack of access to health services affecting diagnosis and treatment of the disease, poor housing conditions, and lack of measures to reduce the contact between man and vector as a direct result of poverty are major conditions associated with malaria transmission.

After the discovery of the residual activity of dichlorodiphenyltrichloroethane (DDT) and successes with its use for interruption of transmission over large areas, among others, in Italy and Greece in Europe, Venezuela and Guyana in the Americas, the World Health Organization (WHO) launched the global strategy of malaria eradication in 1955. This led to development and availability of tools and interventions and the scaling up of efforts to combat the disease. Interruption of malaria transmission was achieved in the former Soviet Union, most countries in Europe, Australia, many countries in Asia, North America, and all the Caribbean islands with the exception of the island of Hispaniola, shared by Haiti and the Dominican Republic.

However, the disease has persisted in many countries worldwide and even reemerged or has been reintroduced in many countries where it was greatly reduced and/or eliminated. Acknowledging that eradication would be difficult to achieve globally and given its high burden in Africa, where efforts at eradication had not been attempted, in 1992 the WHO decided to change the strategy to focus to reduce burden through prevention and control of the disease, an extremely important cause of childhood morbidity and mortality. Technically, the components of the focus were on early diagnosis and timely and appropriate treatment of cases; design and implementation of selective and sustainable preventive measures, including vector control; early detection, containment, or prevention of epidemics; and strengthening local

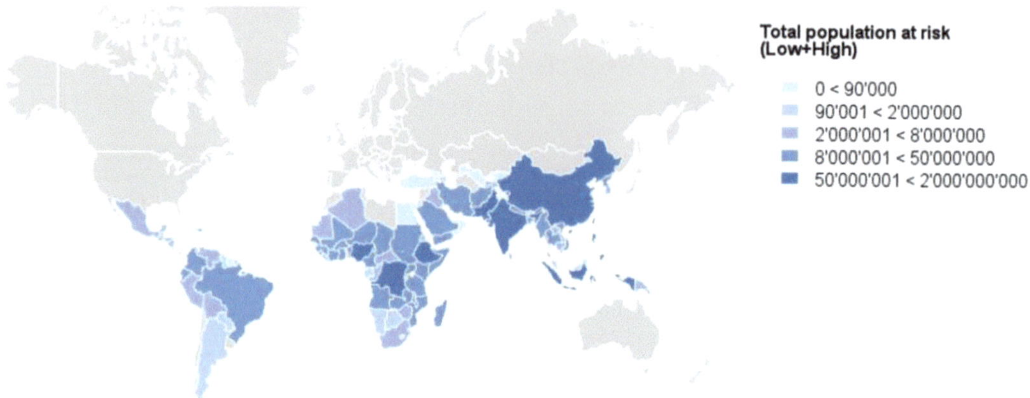

Fig. 20.2 Global malaria map, based on total populations at risk (World Health Organization and Medicines for Malaria Venture 2014)

capacities to enable and to promote regular analysis of the malaria situation in a country, including the analysis of the ecological, social, and economic determinants of disease research.

As of 2013, the World Health Organization (WHO) reported that 106 countries have ongoing malaria transmission (Fig. 20.2). Approximately 3.3 billion people are at risk of the disease, of which 1.2 billion are at high risk or living in areas where more than one malaria case occurs per 1000 population (World Health Organization 2014a, b, c, d, e). Estimates indicate that there were 198 million cases of malaria and 584,000 deaths in 2013 worldwide, with 90 % of all malaria deaths occurring in sub-Saharan Africa. Of the estimated deaths, 78 % were among children under 5 years of age. Outside of Africa, most of the malaria occurs in South and Southeast Asian and South American countries.

Most of the malaria is caused by *P. falciparum* especially in Africa, Papua New Guinea, and Hispaniola where it is almost the exclusive malaria parasite species. *Plasmodium vivax* predominates in Central American, South American, Central Asian, and Middle Eastern countries. Almost equal distribution of the two species is seen in South and Southeast Asian countries. *Plasmodium ovale* is restricted in its distribution to African countries and rarely found elsewhere, while *P. malariae* infections have been reported from tropical countries across the world. Recently, *P. knowlesi* human infections have been reported from areas adjacent to forests in Southeast Asian countries.

20.7 Vectors and Vector Control

About 60–70 of the approximate 490 species of *Anopheles* mosquitoes and sibling species worldwide can transmit malaria. Throughout the world, 30 of these vectors are of major importance (World Health Organization 2014a, b, c, d, e) in terms of transmitting malaria to humans (Fig. 20.3).

Vector control is an essential component of all malaria control strategies and contributes in accelerating efforts toward malaria elimination. It has two core interventions—long-lasting insecticidal nets (LLINs) and indoor residual spraying (IRS)—which should be implemented on the basis of epidemiological and entomological surveillance data. Supplementary methods such as larval source management, improved housing including screens, and use of repellents and mosquito coils may also be appropriate in specific settings.

20.7.1 Long-Lasting Insecticidal Nets (LLINs)

LLINs provide both a physical barrier and an insecticidal effect that reduce human-mosquito contact. They last longer than conventional

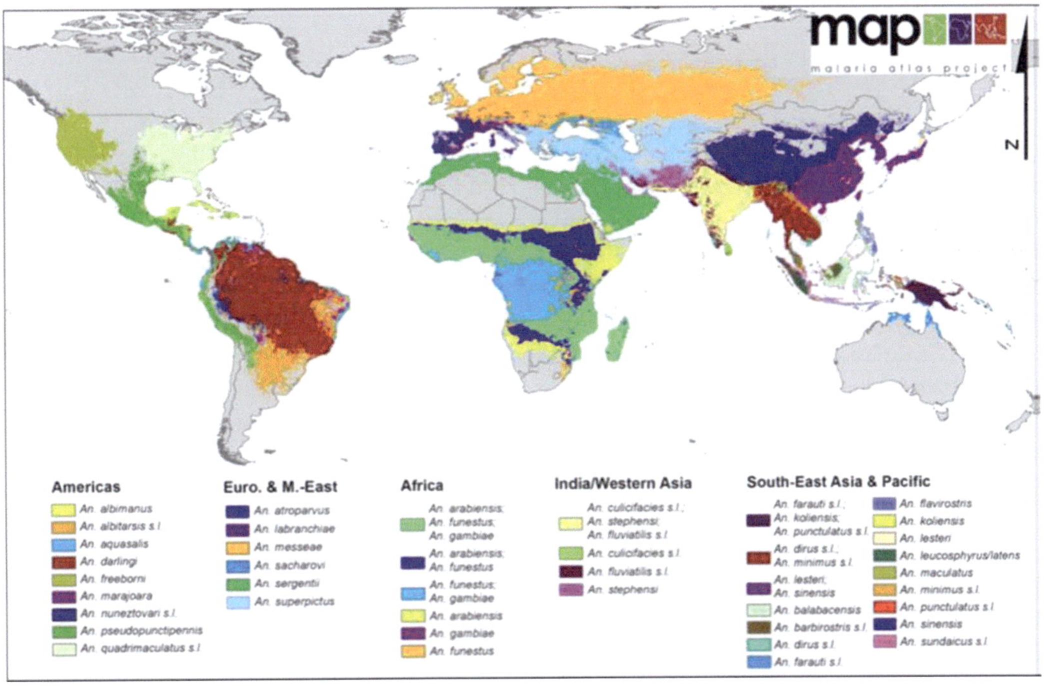

Fig. 20.3 Global distribution of dominant or potentially important malaria vectors (Sinka et al. 2012)

insecticide-treated bed nets, but still need to be replaced regularly. They also have an important community effect that leads to large-scale killing of mosquitoes so they should be provided in sufficient numbers to cover everyone exposed to transmission in target communities. This can be achieved through free LLIN mass campaigns and continuous distribution. Protection of risk groups, especially young children and pregnant women in high transmission areas, should be given preference in situations when there are temporary gaps in LLIN coverage. This can minimize deaths from malaria and should not replace the goal of universal coverage (World Health Organization 2013a, b, c).

20.7.2 Indoor Residual Spraying (IRS)

IRS involves spraying an effective dose of insecticide, once or twice per year, on indoor surfaces where malaria vectors are likely to rest after biting. It reduces malaria transmission by reducing the survival of mosquitoes that enter houses or sleeping units. IRS requires a high level of coverage in space and time to achieve its full effect as a method for community protection.

There are currently 12 insecticides recommended for IRS use by the WHO Pesticide Evaluation Scheme (WHOPES) (WHOPES 2013). They come from only four classes of insecticide: pyrethroids, organochlorines (dichlorodiphenyltrichloroethane, DDT), organophosphates, and carbamates. When implemented properly, IRS is highly effective and can provide protection to communities through a rapid mass effect on vector populations, reducing densities and longevity of vectors and their capacity to transmit malaria parasites. The quality of the spraying operation is also an important variable in determining IRS effectiveness, and at least 80% of premises in target communities must be properly sprayed. IRS is effective usually for 3–6 months, but occasionally up to 9 months, depending on the insecticide used, the type of surface sprayed, and the seasonality of transmission.

20.7.3 Monitoring and Management of Insecticide Resistance

In 2012, the global plan for insecticide resistance management (GPIRM) in malaria vectors was launched which encouraged affected countries to develop and implement comprehensive insecticide resistance management strategies and have preemptive plans in place. To preserve the susceptibility of malaria vectors to currently available insecticides, the recommendations include: periodic changes in the class of insecticides used for indoor residual spraying and the use of multiple interventions in combination. The strategic pillars of GPIRM are the following: plan and implement insecticide resistance management strategies in malaria-endemic countries; ensure proper, timely entomological and resistance monitoring and effective data management; develop new, innovative vector control tools; fill gaps in knowledge on mechanisms of insecticide resistance and the impact of current insecticide resistance management strategies; and ensure that enabling mechanisms (advocacy, human and financial resources) are in place (World Health Organization 2012). Mosquito resistance is currently reported in 64 countries affecting all major vector species and all classes of insecticides (World Health Organization 2013a, b, c).

20.7.4 Malaria Elimination

Malaria elimination is the interruption of local mosquito-borne malaria transmission or the reduction to zero of the incidence of malaria infection in a defined geographical area. After elimination, continued measures are required to prevent reestablishment of transmission. Malaria eradication is defined as the permanent reduction to zero of the worldwide incidence of malaria infection caused by a particular malaria parasite species (World Health Organization 2014a, b, c, d, e).

In recent years, there have been significant reductions in malaria burden in the Americas and other regions including Africa particularly in some countries in south of the Sahara, where malaria incidence and mortality among children and adults have declined. These marked reductions in the levels of transmission are among the important driving forces behind the current global interest on malaria elimination and subsequent global eradication (Pan American Health Organization 2010).

While continuing to gain momentum and support for malaria elimination efforts is generally increasing, the path toward malaria-free status is clearly challenging and resource intensive. The malaria elimination process or continuum is characterized by four distinct program phases: control, pre-elimination, elimination, and prevention of reintroduction. A set of specific program interventions pertaining to prevention, treatment, surveillance, monitoring and evaluation, and health systems strengthening characterizes each of the program phases (Table 20.9).

In 2014, the WHO Global Malaria Programme initiated the process of consolidating the Post-2015 Global Technical Strategy (GTS) for Malaria which is the primary theme of accelerating malaria elimination. Efforts to reorient control programs toward elimination require an updating of strategies (PAHO 2010). Among malaria control programs, the overall objective is reducing malaria cases and deaths by providing access to preventive methods, diagnostic testing, and treatment to the entire population at risk. In the elimination phase, malaria cases have diminished and have become relatively focal. Programs at this stage should focus interventions on detecting all malaria cases, preventing onward transmission, managing malaria foci, and managing imported malaria cases. In order for a country to be certified and declared as malaria-free, it must demonstrate an effective malaria surveillance system and that there have been no cases of locally transmitted malaria for the past 3 years (Mendis et al. 2009; World Health Organization 2014a, b, c, d, e).

Table 20.9 Criteria for classifying countries according to malaria program phase (World Health Organization 2012)

	Pre-elimination	Elimination	Prevention of reintroduction
Malaria situation in areas with most intense transmission			(1) Recently endemic country with zero local transmission for at least 3 years (2) Country on the register or supplementary list that has ongoing local transmission[a]
Test positivity rate	≤5 % among suspected malaria patients (PCD) throughout the year		
API in the district with the highest number of cases/1000 population/year (ACD and PCD)[b], averaged over the last 2 years	<5 (less than 5 cases/1000 population)	<1 (less than 1 case/1000 population)	
Total number of reported malaria cases nationwide		A manageable number, e.g., <1000 cases nationwide (local and imported)	
Case management			Imported malaria. Maintain capacity to detect malaria infection and manage clinical disease
All cases detected in the private sector are microscopically confirmed	National policy being rolled out	Yes	Yes
All cases detected in the public sector are microscopically confirmed	National policy being rolled out	Yes	Yes
Nationwide microscopy quality assurance system covers public and private sector	Initiated	Yes	Yes
Radical treatment with primaquine for *P. vivax*	National policy being updated	National policy fully implemented	Yes
Treatment with ACT plus single-dose primaquine for *P. falciparum*	National policy being updated	National policy fully implemented	Yes
Surveillance			Vigilance by the general health services
Malaria is a notifiable disease nationwide (<24–48 h)	Laws and systems being put in place	Yes	Yes
Centralized register on cases, foci, and vectors	Initiated	Yes	Yes
Malaria elimination database	Initiated	Yes	Certification process (optional)
Active case detection in groups at high risk or with poor access to services ("proactive" case detection)	Initiated	Yes	In residual and cleared-up foci, among high-risk population groups
Case and foci investigation and classification (including "reactive" case detection and entomological investigation)	Initiated	Yes	Yes

[a]Ongoing local transmission = two consecutive years of local *P. falciparum* malaria transmission or three consecutive years of local *P. vivax* malaria transmission in the same locality or otherwise epidemiologically linked

[b]The API has to be evaluated against the diagnostic activity in the risk area (measured as the ABER). Low values of ABER in a district raise the possibility that more cases would be found with improved diagnostic efforts

References

Daneshvar C, Davis TME, Cox-Singh J et al (2009) Clinical and laboratory features of human *Plasmodium knowlesi* infection. Clin Infect Dis 49:852–860

Giles HM, Warrell DA (1993) Bruce-Chwatt's essential malariology, 3rd edn. Little Brown, Boston, MA

Greenwood BM, Fidock DA, Kyle DE et al (2008) Malaria: progress, perils, and prospects for eradication. J Clin Invest 118:1266–1276. doi:10.1172/JCI33996FullTextat, http://www.jci.org/articles/view/33996/files/pdf

Johns Hopkins Bloomberg School of Public Health. Life cycle of the malaria parasite. In: Epidemiology of infectious diseases. Johns Hopkins Bloomberg School of Public Health, Creative Commons BY-NC-SA. http://ocw.jhsph.edu

Kantele A, Jokiranta S (2011) Review of cases with the emerging fifth human malaria parasite, *Plasmodium knowlesi*. Clin Infect Dis 52:1356–1362

Mendis K, Rietveld A, Warsame M et al (2009) From control to malaria eradication: the WHO perspective. Trop Med Int Health 14(7):1–7

Pan American Health Organization (2010) Guide for the reorientation of malaria control programs with a view toward elimination of the disease. PAHO, Washington, DC, http://www2.paho.org/hq/dmdocuments/2011/Guide_Reor_Malaria%20Control.pdf

Pan American Health Organization (2014) Interactive malaria statistics. PAHO, Washington, DC, http://www.paho.org/hq/index.php?option=com_content&view=article&id=2632&Itemid=2130&lang=en

Singh AP, Buscaglia CA, Wang Q et al (2007) *Plasmodium* circumsporozoite protein promotes the development of the liver stages of the parasite. Cell 131:492–504

Sinka ME, Bangs MJ, Manguin S et al (2012) A global map of dominant malaria vectors. Parasit Vectors 5:69

Weatherall DJ, Miller LH, Baruch DI et al (2002) Malaria and the red cell. Haematology 1:35–57, http://asheducationbook.hematologylibrary.org/cgi/reprint/2002/1/35

WHO Pesticide Evaluation Scheme (2013) WHO recommended insecticides for indoor residual spraying against malaria vectors. WHO, Geneva, http://www.who.int/whopes/Insecticides_IRS_Malaria_25_Oct_2013.pdf

World Health Organization (1999) Meetings on *Plasmodium vivax* and *Schistosoma japonicum* in Asia. TDR News, WHO, Geneva

World Health Organization (2010) Basic malaria microscopy, 2nd edn. WHO, Geneva, http://www.who.int/entity/malaria/publications/atoz/9241547820/en/index.html

World Health Organization (2011) Global plan for artemisinin resistance containment. WHO, Geneva, http://apps.who.int/iris/bitstream/10665/44482/1/9789241500838_eng.pdf?ua=1

World Health Organization (2012) Management of severe malaria: a practical handbook, 3rd edn. WHO, Geneva, http://www.who.int/iris/bitstream/10665/79317/1/9789241548526_eng.pdf?ua=1

World Health Organization (2013a) Malaria entomology and vector control. WHO, Geneva, http://apps.who.int/iris/bitstream/10665/85890/1/9789241505819_eng.pdf

World Health Organization (2013b) Questions and answers on malaria vaccines. WHO, Geneva, http://www.who.int/immunization/research/development/WHO_malaria_vaccine_q_and_a_Oct2013.pdf?ua=1

World Health Organization (2013c) Recommendations for achieving universal coverage with long-lasting insecticidal nets in malaria control. WHO, Geneva, http://www.who.int/entity/malaria/publications/atoz/who_recommendations_universal_coverage_llins.pdf

World Health Organization (2014a) World malaria report 2014. WHO, Geneva, http://www.who.int/malaria/publications/world_malaria_report_2014/en/

World Health Organization (2014b) Malaria rapid diagnostic test performance. Results of WHO product testing of malaria RDTs: round 5 (2013). WHO, Geneva, http://apps.who.int/iris/bitstream/10665/128678/1/9789241507554_eng.pdf?ua=1&ua=1

World Health Organization (2014c) Policy recommendation on malaria diagnostics in low transmission settings. WHO, Geneva, http://www.who.int/malaria/publications/atoz/who-recommendation-diagnostics-low-transmission-settings-mar2014.pdf?ua=1

World Health Organization (2014d) Procedures for certification of malaria elimination. Wkly Epidemiol Rec 89(29):321–336, http://www.who.int/wer/2014/wer8929.pdf?ua=1

World Health Organization (2014e) Tables of malaria vaccine projects globally. WHO, Geneva, http://www.who.int/immunization/research/development/Rainbow_tables/en/

World Health Organization (2015) Guidelines for the treatment of malaria, 3rd edn. WHO, Geneva, http://www.who.int/malaria/publications/atoz/9789241549127/en/

World Health Organization and Medicines for Malaria Venture (2014) Global malaria mapper. WHO and MMV, Geneva, http://www.who.int/malaria/publications/world_malaria_report/global_malaria_mapper/en/

Babesiosis 21

Filipe Dantas-Torres, Leucio Camara Alves, and Gerrit Uilenberg

Abstract

Babesiosis is a tick-borne disease caused by protozoa of the genus *Babesia*. Several species of *Babesia* have been associated with disease in domestic animals, including *Babesia bovis* and *B. bigemina* in cattle, *B. caballi* in horses, and *B. canis*, *B. vogeli*, and *B. rossi* in dogs. Some species (e.g., *B. divergens*, *B. microti*, *B. venatorum*, and *B. duncani*) may also infect and cause diseases in humans. The infectivity and pathogenicity of each *Babesia* species may vary widely. For instance, *B. vogeli* is less pathogenic than other species that affect dogs, such as *B. canis* and *B. rossi*. The diagnosis of babesiosis relies on suggestive clinical signs (e.g., fever, lethargy, and jaundice) and laboratory abnormalities (e.g., anemia and thrombocytopenia) and should be ideally confirmed by cytology or molecular biology techniques. The treatment of animal babesiosis is usually carried out with drugs like imidocarb dipropionate and diminazene diaceturate, whereas for humans, different drug combinations may be used. While vaccines have been licensed for preventing babesiosis in some animal species (e.g., cattle and dogs), the control and prevention of the disease is primarily based on tick control.

Keywords

Babesia • Dogs • Cats • Livestock • Humans • Ticks • Anemia • Fever • Zoonosis

21.1 Early History

In 1888, Victor Babeş, trying to identify the cause of a severe febrile illness responsible for the deaths of 30,000–50,000 cattle in Romania, identified small coccoid microorganisms inside the erythrocytes of sick cattle (Babeş 1888), which he called "haematococcus" in 1889 (Babeş 1889). In 1893, Starcovici created the genus *Babesia* in honor of Babeş and proposed the names *Babesia*

F. Dantas-Torres (✉)
Department of Immunology, Aggeu Magalhães Research Center, Oswaldo Cruz Foundation, Recife, Brazil

Department of Veterinary Medicine, University of Bari, Valenzano, Italy
e-mail: filipe.dantas@cpqam.fiocruz.br

L.C. Alves
Department of Veterinary Medicine, Federal Rural University of Pernambuco, Recife, Brazil
e-mail: leucioalves@gmail.com

G. Uilenberg
"A Surgente", Route du Port, 20130 Cargèse, France
e-mail: uilenber@club-internet.fr

C.B. Marcondes (ed.), *Arthropod Borne Diseases*, DOI 10.1007/978-3-319-13884-8_21

bovis (for the parasite found in Romania) and *Babesia ovis* (the parasite of a disease in sheep found by Babeş in 1892) (Starcovici 1893).[1] In the same year, Smith and Kilbourne (1893) identified the agent of Texas cattle fever in the United States, naming it *Pyrosoma bigeminum*. The genus name *Pyrosoma* was preoccupied; therefore, *Babesia* is the correct genus name.

Nowadays, babesiosis is a group of tick-borne diseases of both medical and veterinary concerns. Different *Babesia* species may infect domestic animals and some of them are responsible for severe losses in the livestock industry worldwide.

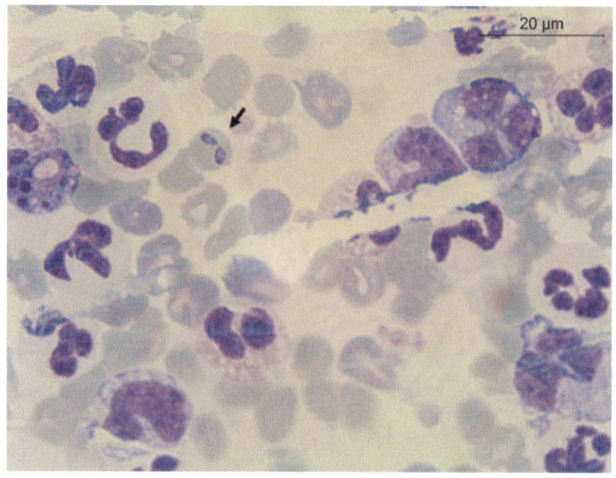

Fig. 21.1 Large *Babesia* merozoites within a red blood cell in a blood film of a dog

21.2 Etiology

Babesiosis is caused by a protozoan parasite of the genus *Babesia* (Piroplasmida: Babesiidae), which currently includes more than 100 species. Morphologically, babesiae are pleomorphic but typically seen as piriform intra-erythrocyte inclusions. Depending on the size of intra-erythrocytic forms, babesiae are classified as "large babesiae" (2.5–5.0 µm) or "small babesiae" (1.0–2.5 µm) (Fig. 21.1). This classification is important to determine the possible species involved during cytological examination of stained blood smears, though it does not allow a diagnosis to the species level, particularly in animals such as dogs that may be infected by different *Babesia* species. Table 21.1 summarizes the main species of *Babesia* of veterinary and medical relevance.

Human babesiosis is mostly caused by *B. microti* and *B. duncani* in the United States, whereas most cases in Europe are caused by *B. divergens* and *B. venatorum*. Many of the human (and even some of the bovine) cases ascribed to *B. divergens* before the description of *B. venatorum* might in fact be due to the latter, but this is impossible to know retrospectively. Although there are reports of human babesiosis in Brazil (Alecrim et al. 1983), the species involved has not been determined. A case of human babesiosis supposedly caused by *B. bovis* has also been reported in Colombia (Ríos et al. 2003), which emphasizes the need for further studies related to human babesiosis in South America.

The genera *Babesia* and *Theileria* used to be defined by life cycle and morphological studies, but nowadays, phylogenetic data (mainly based on 18S rRNA gene sequences) are also providing valuable information on the specific identity of piroplasmid parasites. Nonetheless, due to the lack of life cycle and genetic data for several *Babesia* species, the validity of some species listed above needs further confirmation. For instance, *B. beliceri* and *B. occultans* are morphologically similar, infect cattle, and have *Hyalomma* ticks as vectors. The validity of *B. foliata* and *B. taylori* needs confirmation. The validity of *B. jakimovi* should be assessed through its comparison with other *Babesia* spp. of Cervidae transmitted by *Ixodes* ticks. There may well be more than one species under the name *B. motasi*.

The inclusion of *B. microti* in the genus *Babesia* is questionable, not only because schizonts have been observed, but also based on genetic data (Uilenberg 2006). In the same way, *Rangelia vitalii* is the causative agent of a disease called canine rangeliosis in South America (Argentina, Brazil, and Uruguay). Canine rangeliosis is a severe, hemolytic, and hemorrhagic disease popularly referred to as *nambiuvú* (=bleeding ears) in Tupi-Guarani and *peste de sangue* (=bleeding plague) and *febre amarela dos*

[1]An interesting review of the first discoveries on *Babesia* was published (Köhler and Köhler 2003) (C.B. Marcondes, personal communication).

Table 21.1 *Babesia* species of medical and veterinary concern (Uilenberg 2006; Solano-Gallego and Baneth 2011; Schnittger et al. 2012; Wong et al. 2012; Bosman et al. 2013)

Species	Principal hosts	Vector genus	Geographical distribution
B. beliceri	Cattle	*Hyalomma*	Russia
B. bigemina	Cattle, buffalo	*Rhipicephalus*	Africa, America, Asia, Australia, Europe
B. bovis	Cattle, buffalo, human?	*Rhipicephalus*	Africa, America, Asia, Australia, Europe
B. caballi	Horse, donkey, mule	*Dermacentor*, *Hyalomma*, *Rhipicephalus*	Africa, America, Asia, Europe
B. canis	Dog, cat?	*Dermacentor*	Europe
B. cati[a]	Cat?	Unknown	Asia (India)
B. conradae	Dog	Unknown	America (USA)
B. crassa	Sheep, goat	Unknown	Asia
B. divergens[b]	Cattle, human	*Ixodes*	Europe
B. duncani[c]	Human	Unknown	USA
B. felis	Cat	Unknown	Africa
B. foliata[d]	Sheep	Unknown	Asia (India)
B. taylori	Goat	Unknown	Asia (India)
B. gibsoni	Dog	*Haemaphysalis*	Africa, America, Asia, Europe
B. hongkongensis	Cat	Unknown	Asia (Hong Kong)
B. jakimovi	Cattle, roe deer, elk, reindeer	*Ixodes*	Siberia
B. lengau	Cheetah, cat	Unknown	South Africa
B. major	Cattle	*Haemaphysalis*	Africa, Asia, Europe
B. microti[c]	Rodent, human	*Ixodes*	Africa, America, Asia, Australia, Europe
B. motasi	Sheep, goat	*Haemaphysalis*, *Rhipicephalus*	Africa, Asia, Europe
B. occultans	Cattle	*Hyalomma*	Africa
B. orientalis	Buffalo	*Rhipicephalus*	Asia
B. ovata	Cattle	*Haemaphysalis*	Asia
B. ovis	Sheep, goat	*Rhipicephalus*	Africa, Asia, Europe
B. perroncitoi	Pig	Unknown	Africa, Asia (China), Europe (Italy)
B. presentii	Cat	Unknown	Asia (Israel)
B. rossi	Dog	*Haemaphysalis*	Africa
B. trautmanni	Pig	*Rhipicephalus*	Africa, Asia (former USSR), Europe
B. venatorum[b]	Roe deer, human	*Ixodes*	Europe
B. vogeli	Dog, cat	*Rhipicephalus*	Africa, America, Asia, Australia, Europe

The taxonomic status of some species may be arguable/doubtful

[a]This species was described from a wild cat in India, but the authors were not able to transmit the infection to domestic cats (Mudaliar et al. 1950)

[b]Main species associated with human babesiosis in Europe. Cases of human infection with other species (e.g., *B. bovis*) have also been reported, but some have not been properly documented

[c]Main species associated with human babesiosis in the United States

[d]The name of this species is often given as '*B. foliate*' (a common autocorrect typo)

? = reported, but confirmation is needed

cães (=yellow fever of dogs) in Portuguese (França et al. 2014). The taxonomic status of its causative agent is currently under discussion. *Rangelia vitalii*, which is transmitted by *Amblyomma aureolatum*, is able to invade endothelial cells, erythrocytes, and leukocytes (Silva et al. 2011), and differs from most *Babesia* sensu stricto species that typically infect erythrocytes only. Nonetheless, its inclusion in the genus *Babesia* has been proposed in the past and more recently by some authors (Schnittger et al. 2012; Soares et al. 2014). However, the inclusion of this parasite in the genus *Babesia* would require a redefinition of this genus.

Another enigmatic member of the Piroplasmida infecting dogs in Europe is the so-called *Theileria annae* (also referred to as *Babesia* Spanish dog isolate, *Babesia microti*-like, *Babesia* (*Theileria*) *annae*, and *Babesia* cf. *microti*). This parasite was originally classified as a *Theileria* sp. by Zahler et al. (2000), but other authors (e.g., Schoeman 2009; Schnittger et al. 2012) have placed it in the genus *Babesia*. Recently, Baneth et al. (2015) also considered it as a *Babesia* sp. and proposed the name *Babesia vulpes* but gave no valid reason for changing the species epithet from *annae* to *vulpes*. According to the International Code of Zoological Nomenclature, name-bearing types must be explicitly fixed for new species names published after 1999 (Article 16.4), but no types have been fixed and/or deposited for either *Theileria annae* or *Babesia vulpes*. In the same way, every new name published after 1930 must be accompanied by a description or definition that states in words characters that are purported to differentiate the taxon (Article 13.1.1), but no descriptions or illustrations have been provided for either *Theileria annae* or *Babesia vulpes*. For these reasons, *Theileria annae* and *Babesia vulpes* are nomina nuda. Until a complete, formal description of this species of the Piroplasmida is available in the literature, we suggest the use of *Theileria annae* isolate. So far, dogs and foxes have been suggested as principal hosts of this parasite, whose tick vector remains unknown.

21.3 Life Cycle

The life cycle of *Babesia* spp. is complex and involves at least three stages of reproduction: gamogony (formation and fusion of gametes within the tick gut), sporogony (asexual reproduction in the salivary glands of the tick), and merogony (asexual reproduction in the vertebrate host).

When an infected tick feeds on a susceptible host, sporozoites present in the salivary glands are deposited in the dermis of the vertebrate host. The sporozoites invade erythrocytes and differentiate into trophozoites. They divide asexually by binary fission (merogony) yielding two, four, eight, or more merozoites. After intense multiplication, erythrocyte rupture occurs and free merozoites invade other erythrocytes (Mehlhorn and Schein 1984).

Some merozoites may assume an ovoid aspect, being referred to as gamonts or pre-gametocytes. When ingested by a tick during its blood feeding, merozoites that did not differentiate into gamonts do not develop in the tick gut. Meanwhile, the gamonts differentiate into gametes, also known as ray bodies due to their cytoplasmic projections, which give them a streaked appearance. These uninucleated forms are fused, forming a diploid zygote (gamogony). These zygotes undergo meiosis and give rise to elongated, motile kinetes, which invade the intestinal epithelium, where they multiply by sporogony, reaching the hemocele and migrating to various tick tissues such as salivary glands, ovaries, and eggs. In the salivary glands, the kinetes form a large multinucleate sporoblasts that give rise to thousands of sporozoites that will infect the vertebrate host when the tick has molted to the next stage (transstadial transmission: larva to nymph or nymph to adult). When the tick takes a new blood meal, sporozoites are inoculated into the dermis of the vertebrate host, restarting the cycle. In typical cases, kinetes invade the tick ovaries and eggs, and infective sporozoites are formed in the salivary gland of the next generation of larvae or nymphs (transovarial transmission) (Mehlhorn and Schein 1984).

21.4 Clinical and Laboratorial Findings

The clinical signs resulting from *Babesia* spp. infection are generally similar in different animal species. Clinical signs usually range from mild to severe, although infected animals may remain asymptomatic. Hyperacute cases usually occur in animals that have been recently introduced to endemic areas. This may lead to death, especially in young animals.

The parasitism and the destruction of erythrocytes lead to anemia, which is usually the first laboratory abnormality to appear. It is accompanied by hyperthermia. Intravascular and extravascular hemolyses occur, causing hemoglobinemia with subsequent hemoglobinuria and, in some cases, jaundice. Other clinical signs such as lethargy, anorexia, weight loss, diarrhea, lymphadenomegaly, and splenomegaly may also be observed in animals with babesiosis (Solano-Gallego and Baneth 2011).

Cerebral babesiosis may occur in the course of *B. bovis* and *B. rossi* infections in cattle and dogs, respectively, due to the sequestration of infected erythrocytes in cerebral capillaries (Jacobson 2006; Sondgeroth et al. 2013). Clinically, cerebral babesiosis is characterized by incoordination, convulsions, paddling movements, posterior paralysis, and coma. Mortality rate in complicated cases of canine babesiosis by *B. rossi* is around 15 % (Jacobson 2006).

Human babesiosis is characterized by a clinical condition similar to that observed in malaria. This involves nonspecific clinical signs such as fatigue, lethargy, anorexia, myalgia, nausea, headache, night sweats, tremors, abdominal pain, emotional liability, depression, hematuria, hepatomegaly, and splenomegaly (Setty et al. 2003; Hildebrandt et al. 2013). Occasionally, the renal function may be affected (Setty et al. 2003). The case fatality rate of human babesiosis can range from 5 % in the United States up to 42 % in Europe, particularly when diagnosis and treatment are not properly performed. Human cases caused by *B. divergens* are most often fatal in splenectomized persons, while this species appears to be normally apathogenic for humans.

Laboratorial findings observed in the course of canine babesiosis are nonspecific and have a limited diagnostic value. However, the complete blood count (CBC) is an important tool in the clinical assessment of patients. In severe cases, hematocrit may be quite low, indicating the need for blood transfusion. Coagulation disorders have also been reported in severe cases in dogs (Ruiz de Gopegui et al. 2007). In general, animals infected with *Babesia* spp. exhibit regenerative anemia, reticulocytosis, abnormal leukocyte count, increased serum protein, aspartate aminotransferase (AST), alanine aminotransferase (ALT), urea, creatinine, and alkaline phosphatase. Anemia is initially of the normocytic normochromic variety, progressing to macrocytic hypochromic anemia over the course of the infection.

In human babesiosis, anemia (hemolytic) ranges from mild to moderately severe, with a high reticulocyte count, thrombocytopenia, high levels of urea and creatinine in serum, and proteinuria.

21.5 Diagnosis

A clinical diagnosis of babesiosis is based on the association of clinical and epidemiological data (e.g., history of tick bites, living in or visiting endemic areas). A definitive diagnosis of *Babesia* spp. infection is performed using direct methods such as cytological examination of stained-blood smears (Fig. 21.1). Ideally, thick and thin blood smears should be prepared from capillaries in the ear or tail tip, which often contains a higher concentration of babesial parasites.

A number of serologic tests have been developed for detecting antibodies to *Babesia* in different animals. The most widely used serological tests are the indirect immunofluorescence assay (Fig. 21.2) and an enzyme-linked immunosorbent assay. Different rapid immunochromatographic tests have been developed for the detection of antibodies to *B. bovis*, *B. bigemina*, *B. microti*, and *B. gibsoni*. A recently developed rapid immunochromatographic test was successfully applied for the simultaneous serodiagnosis of bovine babesioses caused by *B. bovis* and *B. bigemina* (Kim et al. 2008). In general, serologic tests have been used both on an individual basis and in population surveys.

Different molecular assays, such as polymerase chain reactions (PCR), real time PCR, and loop-mediated isothermal amplification (LAMP) assays have been developed for the specific detection of *Babesia* spp. in vertebrate hosts and tick vectors. Indeed, molecular tools are very useful to confirm the infection, especially in chronic cases when the level of parasitemia may be low. It is also important to confirm the species

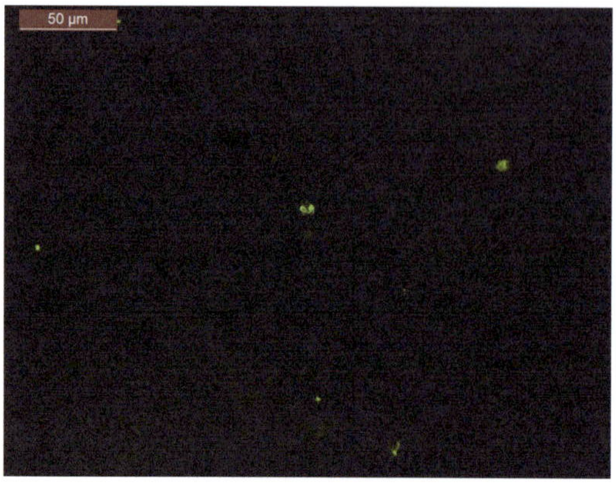

Fig. 21.2 Positive indirect immunofluorescence assay with *Babesia canis* antigen. In the center, two brightly fluorescent pear-shaped forms can be seen witihn a red blood cell

Table 21.2 Protocols used in the treatment of bovine, canine, equine and human babesiosis

Babesiosis	Protocol (s)
Bovine	Diminazene diaceturate (3–5 mg/kg)
	Imidocarb dipropionate (1–3 mg/kg)
	Amicarbalide (5–10 mg/kg)
Canine	Imidocarb dipropionate (5–6.6 mg/kg), SC/ IM, two doses, with an interval of 2–3 weeks
	Diminazene diaceturate (5 mg/kg), IM (may be toxic)
Equine	Imidocarb dipropionate (2–3 mg/kg), two doses, with an interval of 24 h between doses
Human	Quinine (600 mg, PO, t.i.d.) + clindamycin (600 mg, PO, t.i.d. or 1.2 mg, parenterally, b.i.d.), for 7–10 days
	Atovaquone (750 mg, b.i.d.) + azithromycin (500–1000 mg, s.i.d.)

of *Babesia* involved and to assess the therapeutic response of treated individuals.

When dealing with a suspected case of babesiosis, it is also important to assess the occurrence of concurrent infections with other tick-borne pathogens, such as *Ehrlichia canis* in dogs and *Anaplasma marginale* in cattle.

21.6 Treatment

Several drugs and protocols have been used for treating animals and humans suffering from babesiosis. The choice of the drug to be used depends on its availability in the market and the species of *Babesia* (Vial and Gorenflot 2006). The main therapeutic protocols used to treat babesiosis in different animal species are summarized in Table 21.2. Blood transfusion and supporting therapy should also be considered on an individual basis.

21.7 Epidemiology

The risk and the prevalence of *Babesia* infection in different animal populations are related to several factors, including the presence and abundance of tick vectors. The abundance of tick vectors may vary widely from region to region according to the host availability, landscape features, and climatic conditions, particularly temperature, humidity, and photoperiod, which are important for their development. From an epizootiological point of view, an area of enzootic stability is defined as an area with bovine herds with a high prevalence of antibodies to *Babesia* spp. and rare cases of clinical babesiosis. On the other hand, an area of enzootic instability is usually defined as an area where bovine herds present a low prevalence of antibodies to *Babesia* spp. and frequent outbreaks of clinical babesiosis. A severe, potentially fatal form of the disease may also occur in animals from non-endemic areas that are introduced to endemic areas. Risk factors for bovines include breed, age, immune status, *Babesia* strain, climate, and herd management.

Babesiosis incidence in endemic areas is high during the tick season. For instance, cases of human babesiosis usually occur between May and October in Europe. The risk of *B. divergens* infection in humans is higher in rural regions with a high concentration of bovines. In the United States, transmission normally occurs between May and September. The proximity of areas coinhabited by rodents (primary hosts of

B. microti) infested by ticks, as well as advanced age and immunological status, plays an important role in the genesis of the disease in humans. Other factors such as coinfection with human immunodeficiency virus (HIV), (Froberg et al. 2004), stress, and the use of immunosuppressive drugs may increase the risk of clinical disease. Importantly, splenectomized patients are more susceptible than patients with spleens and effective immune systems (Hunfeld et al. 2008). In these patients, the parasitemia is usually high, and the disease is often fatal, particularly when the causal agent is *B. divergens*.

The tick species involved in the transmission vary in accordance with the species of *Babesia* and the region. Several ixodid species have been found naturally infected by different *Babesia* species, which may infect a wide range of domestic and wild animals. Various wild animals have been found naturally infected by *Babesia* species worldwide, including in Brazil. However, until now, the role of these animals in the maintenance of the enzootic or zoonotic cycle of the parasite is still not completely clear. For example, the tick vectors and natural hosts of *B. duncani* (Conrad et al. 2006), an emerging pathogen of humans in the United States, are still unknown. Indeed, the reservoirs and/or tick vectors for many zoonotic babesial parasites are still unknown (reviewed in Yabsley and Shock 2012).

Transmission of *B. gibsoni* in dogs via blood transfusion, contaminated equipment, and transplacental transmission has been demonstrated, and evidence indicates possible dog to dog transmission via bite wounds, saliva, or ingested blood (reviewed in Solano-Gallego and Baneth 2011). Transmission through blood transfusions and a few cases of transplacental/perinatal transmission have been demonstrated in humans as well (reviewed in Vannier et al. 2008). Moreover, several cases of human babesiosis have been reported after blood transfusion between patients, where the donor had previously traveled to endemic areas.

21.8 Control and Prophylaxis

As for other tick-borne pathogens, the best way to prevent *Babesia* spp. infection is to avoid contact between the animals and tick vectors. However, this measure is difficult to apply particularly in some situations where animals and humans live in close contact with tick-infested environments. Systematic chemical control is thus recommended in areas where animals are exposed to ticks. Several products (e.g., permethrin, flumethrin, amitraz, fipronil, and other chemicals) for tick control in different animal species are available in the veterinary market. The use of repellents and/or fast killing acaricides may help reduce tick infestation and also the transmission of *Babesia* spp. and other tick-borne pathogens in dogs (e.g., Dantas-Torres et al. 2013).

Eventually, the control of ticks in the environment may be required. During the application of the acaricide products, the use of personal protective equipment is essential and should be carried out by trained and authorized personnel. Other strategies such as pasture rotation may also help reduce tick levels and the risk of infection.

Several vaccines against different *Babesia* species have been developed, most of which use live attenuated organisms, soluble antigens from culture, and recombinant antigens. Although their success has been limited, two vaccines are currently available in Europe, one containing soluble antigens derived from the culture of *B. canis* and the other also contains culture-derived soluble antigens of both *B. canis* and *B. rossi*. Chemoprophylaxis with imidocarb dipropionate or doxycycline has shown variable results and is not recommended currently for the routine prevention of babesiosis in dogs (Solano-Gallego and Baneth 2011).

Blood donors, whether animal or human, should be periodically screened for *Babesia* infection in order to prevent transmission via blood transfusion, which appears to be more common than previously recognized (Bloch et al. 2012; Lobo et al. 2013).

References

Alecrim I, Pinto B, Avila T et al (1983) Registro do primeiro caso de infecção humana por *Babesia* spp. no Brasil. Rev Pat Trop 12:11–29

Babeş V (1888) Sur l'hémoglobinurie bactérienne du boeuf. CR Acad Sci Paris 107:692–694

Babeş V (1889) Die Aetiologie der seuchenhaften Hämoglobinurie des Rindes. Virchows Arch 115(Folge 11, Band 5):81–108

Baneth G, Florin-Christensen M, Cardoso L et al (2015) Reclassification of *Theileria annae* as *Babesia vulpes* sp. nov. Parasit Vectors 8:207

Bloch EM, Herwaldt BL, Leiby DA et al (2012) The third described case of transfusion-transmitted *Babesia duncani*. Transfusion 52:1517–1522

Bosman AM, Oosthuizen MC, Venter EH et al (2013) *Babesia lengau* associated with cerebral and haemolytic babesiosis in two domestic cats. Parasit Vectors 6:128

Conrad PA, Kjemtrup AM, Carreno RA et al (2006) Description of *Babesia duncani* n. sp. (Apicomplexa: Babesiidae) from humans and its differentiation from other piroplasms. Int J Parasitol 36:779–789

Dantas-Torres F, Capelli G, Giannelli A et al (2013) Efficacy of an imidacloprid/flumethrin collar against fleas, ticks and tick-borne pathogens in dogs. Parasit Vectors 6:245

França RT, Da Silva AS, Loretti AP et al (2014) Canine rangeliosis due to *Rangelia vitalii*: from first report in Brazil in 1910 to current day—a review. Ticks Tick Borne Dis 5:466–474

Froberg MK, Dannen D, Bakken JS (2004) Babesiosis and HIV. Lancet 363:704

Hildebrandt A, Gray JS, Hunfeld KP (2013) Human babesiosis in Europe: what clinicians need to know. Infection 41:1057–1072

Hunfeld KP, Hildebrandt A, Gray JS (2008) Babesiosis: recent insights into an ancient disease. Int J Parasitol 38:1219–1237

Jacobson LS (2006) The South African form of severe and complicated canine babesiosis: clinical advances 1994–2004. Vet Parasitol 138:126–139

Kim CM, Blanco LB, Alhassan A et al (2008) Development of a rapid immunochromatographic test for simultaneous serodiagnosis of bovine babesioses caused by *Babesia bovis* and *Babesia bigemina*. Am J Trop Med Hyg 78:117–121

Köhler M, Köhler W (2003) Zentralblatt für Bakteriologie—100 years ago Victor Babeş and enzootic haemoglobinuria of cattle. Int J Med Microbiol 293:233–239

Lobo CA, Cursino-Santos JR, Alhassan A et al (2013) *Babesia*: an emerging infectious threat in transfusion medicine. PLoS Pathog 9, e1003387

Mehlhorn H, Schein E (1984) The piroplasms: life cycle and sexual stages. Adv Parasitol 23:37–103

Mudaliar SV, Achary GR, Alwar VS (1950) On a species of *Babesia* in an Indian wild cat (*Felis catus*). Indian Vet J 26:392–395

Ríos L, Alvarez G, Blair S (2003) Serological and parasitological study and report of the first case of human babesiosis in Colombia. Rev Soc Bras Med Trop 36:493–498

Ruiz de Gopegui R, Peñalba B et al (2007) Clinicopathological findings and coagulation disorders in 45 cases of canine babesiosis in Spain. Vet J 174:129–132

Schnittger L, Rodriguez AE, Florin-Christensen M et al (2012) *Babesia*: a world emerging. Infect Genet Evol 12:1788–1809

Schoeman JP (2009) Canine babesiosis. Onderstepoort J Vet Res 76:59–66

Setty S, Khalil Z, Schori P et al (2003) Babesiosis. Two atypical cases from Minnesota and a review. Am J Clin Pathol 120:554–559

Silva AS, França RT, Costa MM et al (2011) Experimental infection with *Rangelia vitalii* in dogs: acute phase, parasitemia, biological cycle, clinical-pathological aspects and treatment. Exp Parasitol 128:347–352

Smith T, Kilbourne FL (1893) Investigation into the nature, causation and prevention of Texas fever or southern cattle fever. US Dept. Agric., Bur. Anim. Ind., 8th and 9th Ann. Repts, pp 177–301

Soares JF, Dall'Agnol B, Costa FB (2014) Natural infection of the wild canid, *Cerdocyon thous*, with the piroplasmid *Rangelia vitalii* in Brazil. Vet Parasitol 202:156–163

Solano-Gallego L, Baneth G (2011) Babesiosis in dogs and cats—expanding parasitological and clinical spectra. Vet Parasitol 181:48–60

Sondgeroth KS, McElwain TF, Allen AJ et al (2013) Loss of neurovirulence is associated with reduction of cerebral capillary sequestration during acute *Babesia bovis* infection. Parasit Vectors 6:181

Starcovici C (1893) Bemerkungen über den durch Babes entdeckten Blutparasiten und die durch denselben hervorgebrachten Krankheiten, die seuchenhafte Hämoglobinurie des Rindes (Babes), das Texasfieber (Th. Smith) und der Carceag der Schafe (Babes). Zbl Bakt I Abt 14:1–8

Uilenberg G (2006) *Babesia*—a historical overview. Vet Parasitol 138:3–10

Vannier E, Gewurz BE, Krause PJ (2008) Human babesiosis. Infect Dis Clin North Am 22:469–488

Vial HJ, Gorenflot A (2006) Chemotherapy against babesiosis. Vet Parasitol 138:147–160

Wong SS, Poon RW, Hui JJ et al (2012) Detection of *Babesia hongkongensis* sp. nov. in a free-roaming *Felis catus* cat in Hong Kong. J Clin Microbiol 50:2799–803

Yabsley MJ, Shock BC (2012) Natural history of zoonotic *Babesia*: role of wildlife reservoirs. Int J Parasitol Parasites Wildl 2:18–31

Zahler M, Rinder H, Schein E et al (2000) Detection of a new pathogenic *Babesia microti*-like species in dogs. Vet Parasitol 89:241–248

Theileriosis

22

Filipe Dantas-Torres and Domenico Otranto

Abstract

Theileriosis is a tick-borne disease caused by protozoa of the genus *Theileria* (Piroplasmida: Theileriidae). Several species of *Theileria* have been associated with the disease in domestic animals, including *Theileria parva* and *Theileria annulata* in cattle, *Theileria equi* in horses, and "*Theileria annae*" in dogs. The infectivity and pathogenicity of each *Theileria* species may vary widely. For instance, *Theileria buffeli* is considered of minor significance in cattle as compared with *T. parva*, the agent of East Coast fever found in cattle in East and Central Africa. The diagnosis of theileriosis is based on suggestive clinical signs and laboratory abnormalities, and it is confirmed by cytological or molecular detection of the parasite in blood samples. The treatment of the disease is usually carried out with the same drugs used for treating babesiosis, and the control of *Theileria* infection is based on the control of ticks.

Keywords

Theileria • Theileriosis • East Coast fever • Ticks

22.1 Background

Theileriosis is a group of diseases caused by intracellular protozoa of the genus *Theileria* (Piroplasmida: Theileriidae), which affects primarily ruminants, but also horses and dogs. Theileriosis is responsible for important economic losses to the livestock industry worldwide, particularly in Africa and Asia.

The earliest records of theileriosis in cattle date from the missionary chronicles of Fr. Monclaro from 1569, but it was in the early twentieth century that the disease (nowadays

F. Dantas-Torres (✉)
Department of Immunology, Aggeu Magalhães Research Center, Recife, PE, Brazil

Department of Veterinary Medicine, University of Bari, Valenzano, Italy
e-mail: filipe.dantas@cpqam.fiocruz.br

D. Otranto
Department of Veterinary Medicine, University of Bari, Valenzano, Italy
e-mail: domenico.otranto@uniba.it

C.B. Marcondes (ed.), *Arthropod Borne Diseases*, DOI 10.1007/978-3-319-13884-8_22

known as East Coast fever) gained more attention from the scientific community. The disease was recorded in Southern Rhodesia in 1902 and initially mistaken for redwater by Theiler. In 1903, Lounsbury showed that the tick *Rhipicephalus appendiculatus* was involved in the transmission of this protozoan, and, in 1904, Theiler named the protozoan as *Piroplasma parvum*. A few years later (1907), Bettencourt, França, and Borges found a new species of protozoa (namely, *Theileria cervi*), which had the same characteristics of that one described by Theiler. The aforementioned Portuguese researchers described the genus *Theileria* in honor to Theiler, including the species they discovered and the one described by the renowned African veterinary bacteriologist. Today, *T. parva* is well known as the etiological agent of East Coast fever, a life-threatening disease, which affects cattle in East and Central Africa. However, several species of *Theileria* may actually cause disease in cattle and other domestic animals as discussed below in this chapter.

22.2 Etiology

Protozoa of the genus *Theileria* are genetically, morphologically, and biologically related to those within the genus *Babesia* (Sivakumar et al. 2014). Differently from *Babesia*, no *Theileria* species has been reported in humans so far. The species of veterinary concern are listed in Table 22.1.

With some rare exceptions, cases of *Theileria* infections in dogs and horses are associated with "*T. annae*" (for details on the current taxonomic status of this parasite, see Chap. 21) and *T. equi*, respectively. "*Theileria annae*" has been predominately reported in dogs in southern Europe, but it was recently found in foxes in Portugal and Germany (Cardoso et al. 2013; Najm et al. 2014). This parasite has also been reported in other animals, such as cat, horse, and donkey. Furthermore, dogs have also been found infected by *T. annulata* and *T. equi*.

The horse parasite *T. equi* was redescribed in the late 1990s based on its biological, morphological, biochemical, and molecular characteris-

Table 22.1 *Theileria* species of veterinary concern

Species	Principal hosts	Proven/putative vectors	Geographical distribution
T. annae[a]	Dogs, foxes	*Ixodes*, *Rhipicephalus*	Europe
T. annulata	Cattle, Asiatic buffalo, bison	*Hyalomma*	Africa, Europe, Asia
T. buffeli	Cattle, African buffalo, Asiatic buffalo	*Haemaphysalis*	Asia, Australia, Europe, America, Africa
T. equi[b]	Horse, donkey, mule	*Amblyomma*, *Dermacentor*, *Hyalomma*, *Rhipicephalus*	Asia, Australia, Europe, America, Africa
T. lestoquardi[c]	Sheep and goats	*Hyalomma*	Europe, Asia, Africa
T. luwenshuni	Sheep and goats	*Haemaphysalis*	Asia
T. mutans	Cattle, African buffalo	*Amblyomma*	Africa
T. orientalis	Cattle	*Haemaphysalis*	Asia, Australia, Europe, America, Africa
T. ovis[d]	Sheep and goats	*Rhipicephalus*	Europe, Asia, Africa
T. parva	Cattle, African buffalo, Asiatic buffalo, waterbuck	*Rhipicephalus*	Africa
T. separata	Sheep and goats	*Rhipicephalus*	Africa
T. sergenti	Cattle, Asiatic buffalo	*Haemaphysalis*	Asia
T. taurotragi	Cattle, sheep, goats, eland (*Taurotragus oryx*)	*Rhipicephalus*	Africa
T. uilenbergi	Sheep and goats	*Haemaphysalis*	Asia
T. velifera	Cattle, African buffalo	*Amblyomma*	Africa

[a]*Nomen nudum*. For more information, see Chap. 21
[b]Formerly *Babesia equi*
[c]Formerly *T. hirci*
[d]Used synonymously with *T. recondita*

tics, being transferred to the genus *Theileria*. In fact, protozoa of the genus *Babesia* and *Theileria* are very similar morphologically, but are genetically distinct and present different biological life cycles. With the progress of the studies on the biology and phylogeny of these protozoa, new taxonomic changes within the genera *Theileria* and *Babesia* are expected to occur.

In contrast with dogs and horses, large and small ruminants may be infected by a broad range of *Theileria* species (Ahmed et al. 2006; Sivakumar et al. 2014). Indeed, at least eight *Theileria* species infect cattle and five infect small ruminants around the world, causing disease with varying degrees of severity. From global perspective, *T. parva* and *T. annulata* are the most important species infecting cattle, whereas *T. lestoquardi* is the most clinically relevant agent of theileriosis in small ruminants in a number of countries. However, the importance of each species may vary locally. For instance, *T. luwenshuni* and *T. uilenbergi* are important agents of theileriosis in small ruminants in China. It results evident that the molecular characterization of *Theileria* species occurring in each endemic area is pivotal for understanding the disease epizootiology and transmission dynamics.

22.3 Life Cycle

The life cycle of protozoa belonging to the genus *Theileria* is very similar to that of *Babesia* (see Chap. 21) and therefore will not be discussed in detail in this chapter. During the blood meal, sporozoites are inoculated in the vertebrate host with the tick saliva, thereby invading lymphoid cells in which they form schizonts. The presence of these schizonts stimulates the division of host cells that give rise to new infected cells. These spherical schizonts (=macro-schizonts) start to produce merozoites, which further differentiate into micro-schizonts. After the disruption of the lymphoid cells, merozoites are released and invade erythrocytes. During the intraerythrocytic phase, elongated comma-shaped, spherical, or ovoid parasites can be observed. During the process of division by binary fission within the erythrocyte, the protozoa may form a group of four parasites (tetrads), which occasionally resembles a Maltese cross.

When a tick takes a blood meal on an infected host, the ovoid-shaped protozoa (=gamonts) differentiate in the so-called ray bodies, which display flagella-like protrusions and four nuclei. These ray bodies divide and give rise to four mononuclear gametes. Two gametes fuse originating a zygote that will give rise to a motile ookinete, which migrates to tick hemolymph. When the tick molts (e.g., from larva to nymph) and takes a new blood meal, the ookinete invades the salivary glands differentiating into sporonts, which will generate thousands of sporozoites. Then, during a subsequent blood meal, the tick inoculates the sporozoites (infective stage) into the new vertebrate host.

22.4 Clinical and Laboratorial Findings

Most animals infected by *Theileria* protozoa remain asymptomatic. For instance, this is frequently the case of horses infected by *T. equi*. In acute cases, however, clinical signs such as fever, dyspnea, pale mucous membranes, icterus, and hemoglobinuria are frequently observed. Acute cases are usually associated with high mortality rates. Hepatomegaly and splenomegaly may be present. Cases of abortion and neonatal death have been linked to *T. equi* infections.

On cattle, the pathogenicity of the infection varies according to the *Theileria* species involved. Most species are usually mildly pathogenic (e.g., *T. buffeli*, *T. mutans*, and *T. sergenti*) or nonpathogenic (e.g., *T. velifera*). On the other hand, other species may cause severe production losses and deaths in dairy and beef herds. For instance, a recent study carried out in eastern Australia demonstrated that clinical theileriosis by *T. orientalis* in dairy cattle can cause significant milk production losses (Perera et al. 2014). From a global perspective, *T. parva* and *T. annulata* are the most pathogenic species infecting cattle. Indeed, the mortality in cases of *T. parva* infection may reach

100 %. On the other hand, *T. buffeli* is generally a benign parasite, also associated with mild disease in cattle. Clinical signs of *Theileria* infection in cattle are fever, tachycardia, dyspnea, pale mucous membranes, icterus, hemoglobinuria, enlarged lymph nodes, diarrhea (blood may be present), and dehydration. Deaths are frequent in the first 3 weeks after initial infection by *T. parva*. The principal laboratorial abnormalities are anemia, low-packed cell volume, and leukocytosis. Furthermore, urea, creatinine, liver enzymes, and total protein may be increased.

Theileria lestoquardi is the most virulent species in sheep and goats, which causes a disease called malignant theileriosis featured by fever, anorexia and weight loss, listlessness, lymphadenopathy, edema of the throat, difficulty in breathing, anemia, and icterus. Morbidity rate can approach 100 % and the disease can often be fatal.

"*Theileria annae*" infection in dogs may result in severe disease, and dogs may present with weakness, fever, lethargy, hemoglobinuria, tachycardia, and tachypnea. Laboratory findings may include moderate to severe regenerative anemia and thrombocytopenia. Azotemia has also been recorded in some cases and has been associated with a poor prognosis in canine theileriosis in Spain (Camacho et al. 2004).

22.5 Diagnosis

The diagnosis is based on the observation of parasites in blood smears stained with Giemsa or similar. Serologic tests have also been used, such as indirect immunofluorescence, complement fixation, and enzyme-linked immunosorbent assay. Several molecular tools (i.e., PCR and its variants) have been standardized for detecting *Theileria* spp. DNA in blood samples from different animal species (Heim et al. 2007; Kim et al. 2008; Zhang et al. 2014). For instance, a duplex real-time PCR was developed for the simultaneous detection of *T. equi* and *Babesia caballi* infections in horses (Heim et al. 2007) as well as a multiplex real-time PCR assay, for the diagnosis of *T. uilenbergi*, *T. luwenshuni,* and *T. ovis* in small ruminants (Zhang et al. 2014). Nonetheless, molecular tools are generally restricted to research institutes and universities, particularly in developing countries.

22.6 Treatment

Drugs such as parvaquone (two injections of 10–20 mg/kg, IM, at 48 h intervals), buparvaquone (two injections of 2.5 mg/kg, IM, at 48 h intervals), and halofuginone (two injections of 1.2 mg/kg, PO, at 48 h intervals) are available to treat *T. parva* and *T. annulata* infections in cattle. A combination of parvaquone-plus-furosemide proved useful in the management of pulmonary signs in cattle with *T. parva* theileriosis (Mbwambo et al. 2002). While relatively ineffective against *T. parva* theileriosis, oxytetracycline (single injection of 20 mg/kg, IM) is reported to be effective against *T. annulata*, as well as a single injection of buparvaquone at 2.5 or 5 mg/kg.

Imidocarb dipropionate is the drug most widely used for the treatment of equine theileriosis. While a previous study indicated that five consecutive injections of imidocarb dipropionate (4.7 mg/kg, IM, at 72 h intervals) were not able to eliminate *T. equi* and *B. caballi* from naturally infected horses (Butler et al. 2008), a recent study demonstrated this drug (four injections of 4 mg/kg, IM, at 72 h intervals) successfully cleaned *T. equi* infection in five out of six experimentally infected horses (Grause et al. 2013).

There is limited information on the treatment of canine theileriosis, but two SC injections of 6 mg/kg of imidocarb dipropionate, 14 days apart, are reputed to be effective, even in cases of *T. equi* infection (Rosa et al. 2014).

22.7 Epidemiology

The prevalence of equine infection by *T. equi* can vary widely according to region. In Brazil, studies have demonstrated high seroprevalence rates, varying from 58 to 100 % (Heuchert et al. 1999; Ribeiro et al. 1999; Xuan et al. 2001; Heim et al. 2007). Once infected by *T. equi*, equines become asymptomatic carriers of the infection for the whole life, being an important source of infection to tick vector. In hyperendemic areas clinical cases are rare, since the foals acquire maternal immunity (=passive immunity) after ingestion of colostrum, whereas animals from free-living area recently introduced in endemic areas may develop a severe, life-threatening disease. The prevalence of *T. equi* increases with age (Rüegg et al. 2007) and is even higher in equine that had contact with bovines as a most likely effect of the increases in the risk of exposure to *Rhipicephalus* (*Boophilus*) *microplus* ticks, a proven vector of *T. equi* (Guimarães et al. 1998). Other transmission forms for *T. equi* have been described including mechanic and vertical transmission (Allsopp et al. 2007). Interestingly, vertical transmission has been demonstrated for other *Theileria* species, including *T. sergenti* in cows (Baek et al. 2003) and *T. lestoquardi* in sheep (Zakian et al. 2014). Vertical transmission and transmission by bites have also been speculated for "*T. annae*" in dogs. Nonetheless, the epidemiological significance of this transmission routes is uncertain.

Two different epidemiological situations regarding East Coast fever (*T. parva*) have been described in cattle: endemic stability (high seroprevalence rates and low disease incidence) and endemic instability (low seroprevalence rates and high disease incidence). In the first case, zebu cattle exposed to *T. parva* for generations and subjected to little or no tick control usually become immune to infection, rarely developing overt disease. In the second scenario, the level of exposure is lower (lower tick pressure and low prevalence of infection) and periodic epidemic outbreaks may occur. For instance, a population-based study in Uganda concluded that endemic stability for theileriosis caused by *T. parva* existed in the study area, and the risk of outbreaks of East Coast fever in indigenous cattle was regarded as minimal (Kivaria et al. 2004). In general, taurine breeds of cattle are more susceptible to both *T. parva* and *T. annulata* than indigenous zebu cattle or Sanga breed. However, mortality may be high in zebu cattle under 1 year (Thumbi et al. 2013). Similarly, cattle introduced into endemic areas are more susceptible to theileriosis than cattle from endemic areas, regardless of the breed. Regarding the vector, *T. parva* is transmitted by *Rhipicephalus* ticks, while *T. annulata* is transmitted by *Hyalomma* ticks. Moreover, African buffalo (*Syncerus caffer*) and waterbuck (*Kobus* spp.) are reservoirs for *T. parva*, whereas water buffalo (*Bubalus bubalis*) plays this role for *T. annulata*.

The closely related parasites *T. orientalis*, *T. buffeli*, and *T. sergenti* are worldwide distributed and transmitted by *Haemaphysalis* ticks. Mortality rate by these parasites is usually low in indigenous cattle. Nonetheless, a study conducted in the state of Victoria, Australia, revealed that oriental theileriosis (*T. orientalis*) was associated with abortion and cow deaths, affecting mainly beef cows of more than 2 years of age (Perera et al. 2013).

Theileria buffeli and another infection have been reported in cattle in the USA (e.g., Texas, North Carolina, Missouri, and Michigan). *Amblyomma americanum* and *Dermacentor variabilis* ticks have been found naturally infected by an undesignated *Theileria* sp. (type D SSU rRNA gene) in the USA (Chae et al. 1999) and may be involved in the transmission of this parasite to cattle in this country.

Theileria lestoquardi is diffused in northern Africa, southern Europe, and the Middle East in relationship to the presence of its tick vector, *Hyalomma anatolicum*. This *Theileria* causes severe losses in small ruminants including abortion (Zakian et al. 2014). The disease may be mild in young lambs and kids due to maternal immunity, whereas acute forms resembling the East Cost fever may occur in elder animals. However, it seems that goats are more resistant than sheep to this protozoan. Information about the distribution of *T. uilenbergi* and *T. luwenshuni* is scant although they have been described as pathogenic species of sheep and goats in China (Yin et al. 2007). This species occurs where *Haemaphysalis qinghaiensis* is prevalent since both nymphs and adults transmit both species of *Theileria*. No difference has been demonstrated in the susceptibility of goats and sheep to the infection (Yin et al. 2007).

Information on the epidemiology of "*T. annae*" infection in dogs is scarce. Most cases are observed in autumn and winter. The distribution of "*T. annae*" is uncertain, but *Babesia microti*-like piroplasms have been described in dogs in Spain, Croatia, Sweden, and the USA and in foxes in Spain, Portugal, Germany, and North America (the USA and Canada) (Solano-Gallego and Baneth 2011). "*Theileria annae*" has been reported in other animal species (e.g., cats), but foxes and dogs are probably the principal hosts of this parasite, which seems to be primarily transmitted by *Ixodes* ticks (e.g., *I. canisuga*, *I. hexagonus*, and *I. ricinus*).

22.8 Control and Prophylaxis

For over 30 years, live attenuated vaccines have been used to prevent bovine theileriosis in Africa (McKeever 2007). Despite their proven efficacy, such vaccines have not been used on a large scale, mainly due to the lack of infrastructure for its production and distribution (Morrison and McKeever 2006). The concern about the risk of introducing new strains on local populations of ticks has inhibited the widespread use of these vaccines in areas of endemic stability for bovine theileriosis (McKeever 2007). Recent advances in the knowledge of molecular biology and immunological aspects involved in the *Theileria*–host interactions have opened new perspectives in the search for new vaccines against theileriosis (Morrison and McKeever 2006). For instance, a study demonstrated the efficacy of a live attenuated candidate vaccine against malignant ovine theileriosis by *T. lestoquardi* (Ahmed et al. 2013). Despite the progress, further studies are still needed to obtain a more effective and easy-to-deliver vaccine against bovine theileriosis.

In the absence of an effective vaccine, the best way to prevent theileriosis in cattle, small ruminants, horses, and dogs is by means of integrated tick control programs, which necessarily need to include also the systematic use of acaricides. At the same time, pasture sharing with horses and cattle should be avoided, so as to reduce the risk of exposure to tick vectors. Certainly, the prevalence of *Theileria* infection tends to be significantly lower in areas where control measures against ticks are adopted systematically.

References

Ahmed JS, Luo J, Schnittger L et al (2006) Phylogenetic position of small-ruminant infecting piroplasms. Ann N Y Acad Sci 1081:498–504

Ahmed BM, Taha KM, Enan KA et al (2013) Attenuation of *Theileria lestoquardi* infected cells and immunization of sheep against malignant ovine theileriosis. Vaccine 31:4775–4781

Allsopp MT, Lewis BD, Penzhorn BL (2007) Molecular evidence for transplacental transmission of *Theileria equi* from carrier mares to their apparently healthy foals. Vet Parasitol 148:130–136

Baek BK, Soo KB, Kim JH et al (2003) Verification by polymerase chain reaction of vertical transmission of *Theileria sergenti* in cows. Can J Vet Res 67:278–282

Butler CM, Nijhof AM, van der Kolk JH et al (2008) Repeated high dose imidocarb dipropionate treatment did not eliminate *Babesia caballi* from naturally infected horses as determined by PCR-reverse line blot hybridization. Vet Parasitol 151:320–322

Camacho AT, Guitian EJ, Pallas E et al (2004) Azotemia and mortality among *Babesia microti*-like infected dogs. J Vet Intern Med 18:141–146

Cardoso L, Cortes HC, Reis A et al (2013) Prevalence of *Babesia microti*-like infection in red foxes (*Vulpes vulpes*) from Portugal. Vet Parasitol 196:90–95

Chae JS, Levy M, Hunt J Jr et al (1999) *Theileria* sp. infections associated with bovine fatalities in the United States confirmed by small-subunit rRNA gene analyses of blood and tick samples. J Clin Microbiol 37:3037–3040

Grause JF, Ueti MW, Nelson JT et al (2013) Efficacy of imidocarb dipropionate in eliminating *Theileria equi* from experimentally infected horses. Vet J 196:541–546

Guimarães AM, Lima JD, Ribeiro MF (1998) Sporogony and experimental transmission of *Babesia equi* by *Boophilus microplus*. Parasitol Res 84:323–327

Heim A, Passos LM, Ribeiro MF et al (2007) Detection and molecular characterization of *Babesia caballi* and *Theileria equi* isolates from endemic areas of Brazil. Parasitol Res 102:63–68

Heuchert CM, Giulli V Jr, Athaide DF et al (1999) Seroepidemiologic studies on *Babesia equi* and *Babesia caballi* infections in Brazil. Vet Parasitol 85:1–11

Kim CM, Blanco LB, Alhassan A et al (2008) Diagnostic real-time PCR assay for the quantitative detection of

Theileria equi from equine blood samples. Vet Parasitol 151:158–163

Kivaria FM, Heuer C, Jongejan F et al (2004) Endemic stability for *Theileria parva* infections in Ankole calves of the Ankole ranching scheme, Uganda. Onderstepoort J Vet Res 71:189–195

Mbwambo HA, Sudi FF, Mkonyi PA et al (2002) Comparative studies of the efficacy of parvaquone and parvaquone-plus-frusemide in the treatment of *Theileria parva* infection (East Coast fever) in cattle. Vet Parasitol 108:195–205

McKeever DJ (2007) Live immunisation against *Theileria parva*: containing or spreading the disease? Trends Parasitol 23:565–568

Morrison WI, McKeever DJ (2006) Current status of vaccine development against *Theileria* parasites. Parasitology 133:S169–S187

Najm NA, Meyer-Kayser E, Hoffmann L et al (2014) A molecular survey of *Babesia* spp. and *Theileria* spp. in red foxes (*Vulpes vulpes*) and their ticks from Thuringia, Germany. Ticks Tick Borne Dis 5:386–391

Perera PK, Gasser RB, Anderson GA et al (2013) Epidemiological survey following oriental theileriosis outbreaks in Victoria, Australia, on selected cattle farms. Vet Parasitol 197:509–521

Perera PK, Gasser RB, Firestone SM et al (2014) Oriental theileriosis in dairy cows causes a significant milk production loss. Parasit Vectors 7:73

Ribeiro MF, Costa JO, Guimarães AM (1999) Epidemiological aspects of *Babesia equi* in horses in Minas Gerais, Brazil. Vet Res Commun 23:385–390

Rosa CT, Pazzi P, Nagel S et al (2014) Theileriosis in six dogs in South Africa and its potential clinical significance. J S Afr Vet Assoc 85:1–7

Rüegg SR, Torgerson P, Deplazes P et al (2007) Age-dependent dynamics of *Theileria equi* and *Babesia caballi* infections in southwest Mongolia based on IFAT and/or PCR prevalence data from domestic horses and ticks. Parasitology 134:939–947

Sivakumar T, Hayashida K, Sugimoto C et al (2014) Evolution and genetic diversity of *Theileria*. Infect Genet Evol 27C:250–263

Solano-Gallego L, Baneth G (2011) Babesiosis in dogs and cats—expanding parasitological and clinical spectra. Vet Parasitol 181:48–60

Thumbi SM, Bronsvoort MB, Kiara H et al (2013) Mortality in East African shorthorn zebu cattle under one year: predictors of infectious-disease mortality. BMC Vet Res 9:175

Xuan X, Nagai A, Battsetseg B et al (2001) Diagnosis of equine piroplasmosis in Brazil by serodiagnostic methods with recombinant antigens. J Vet Med Sci 63:1159–1160

Yin H, Schnittger L, Luo J et al (2007) Ovine theileriosis in China: a new look at an old story. Parasitol Res 101(Suppl 2):S191–S195

Zakian A, Nouri M, Barati F et al (2014) Vertical transmission of *Theileria lestoquardi* in sheep. Vet Parasitol 203:322–325

Zhang X, Liu Z, Yang J et al (2014) Multiplex PCR for diagnosis of *Theileria uilenbergi*, *Theileria luwenshuni*, and *Theileria ovis* in small ruminants. Parasitol Res 113:527–531

Hepatozoonosis

23

Filipe Dantas-Torres and Domenico Otranto

Abstract

Canine hepatozoonosis is a vector-borne disease caused by apicomplexan protozoa of the genus *Hepatozoon*. Dogs may be infected by two *Hepatozoon* species, namely, *Hepatozoon americanum* and *Hepatozoon canis*. The first species is considered to be restricted to North America, whereas the second has a cosmopolitan distribution. Recently, feline infection by *Hepatozoon felis* showed to be prevalent in some cat populations worldwide. *Hepatozoon* infection typically occurs when a susceptible animal ingests a tick vector harboring mature oocysts with sporozoites, even if intrauterine transmission has been described for both *H. canis* and *H. americanum* for which also carnivorism of the intermediate host was suggested. While *H. canis* meronts develop in the lymphatic organs (e.g., spleen and lymph nodes), bone marrow, and liver, mostly causing mild disease, *H. americanum* may cause debilitating conditions, associated with musculoskeletal pain. The diagnosis of canine hepatozoonosis is mainly based on the presence of compatible clinical signs and the cytological or molecular detection of gamonts in blood samples. Several therapeutic protocols have been proposed, but data on the treatment of the disease are still incipient. Currently, no vaccine is available for the prevention of canine hepatozoonosis, and therefore the control of this infection should rely on the control of ticks.

Keywords

Hepatozoon • Dogs • Cats • Ticks • Neutrophils • Muscle

F. Dantas-Torres (✉)
Department of Immunology, Aggeu Magalhães Research Center, Recife, PE, Brazil

Department of Veterinary Medicine, University of Bari, Valenzano, Italy
e-mail: filipe.dantas@cpqam.fiocruz.br

D. Otranto
Department of Veterinary Medicine, University of Bari, Valenzano, Italy
e-mail: domenico.otranto@uniba.it

23.1 Early History

The first cases of canine hepatozoonosis were reported in India, in 1905. At that time, the causative agent of the disease was described as a parasite found inside leukocytes of dogs, therefore named *Leukocytozoon canis*. Some years later,

C.B. Marcondes (ed.), *Arthropod Borne Diseases*, DOI 10.1007/978-3-319-13884-8_23

the causative agent of canine hepatozoonosis was transferred to the genus *Hepatozoon*, which included parasites found in neutrophils of mice, and therefore renamed as *Hepatozoon canis*.

In North America, the first reported cases of canine hepatozoonosis date back to 1978. Initially, these cases were assigned to *H. canis*; however, in the late 1990s, researchers determined that canine hepatozoonosis in the United States was caused by another species, more pathogenic, formally described in 1997 as *Hepatozoon americanum*. In this region, mixed infections by *H. americanum* and *H. canis* have been molecularly recognized (Allen et al. 2008; Li et al. 2008).

23.2 Etiology

More than 340 species of *Hepatozoon* (Adeleorina: Hepatozoidae) infect a wide variety of animals, especially reptiles and mammals but also amphibians and birds (Baneth 2011). Canine hepatozoonosis is caused by *H. canis* and *H. americanum*, and therefore they are the species of interest in the present chapter (Baneth et al. 2007). In addition, *Hepatozoon felis* is the primary cause of *Hepatozoon* infection in cats, in which the protozoan has tropism for muscular tissues. While *H. americanum* is limited in distribution to the New World, *H. canis* has been reported from different regions of southern Europe, America, Africa, and Asia (Götsch et al. 2009; Abd Rani et al. 2011; Baneth 2011; Otranto et al. 2011; Dantas-Torres et al. 2012). Nonetheless, though in some geographical areas (e.g., Brazil) canine hepatozoonosis is mostly associated with *H. canis* infection (Rubini et al. 2006), the involvement of other species cannot be ruled out (Paludo et al. 2005).

A recent study conducted in Israel provided histopathologic, hematologic, clinical, epidemiological, and genetic evidence indicating that feline infection is primarily caused by *H. felis* (Baneth et al. 2013). Similarly to *H. americanum*, this species infects muscular tissues and may be highly prevalent in certain feline population.

In the canine host, meronts develop in the lymphatic organs (e.g., spleen and lymph nodes), bone marrow, and liver (Baneth et al. 2001). They contain micromerozoites aligned around a central structure ("wheel spoke" shape), which develops into a larger macromerozoite. Micromerozoites penetrate neutrophils and monocytes, in which they mature to the gamont stage (Fig. 23.1) that is found in circulating leukocytes of parasitemic dogs and is featured by an elongated shape measuring about 11.42 μm in length and 5.39 μm in width. In ticks, oocysts consist of a membrane that envelops hundreds of sporocysts in which the infective sporozoites are found (Fig. 23.2).

23.3 Life Cycle

The life cycle of *Hepatozoon* spp. on dogs is quite complex and involves a cascade of events which occur in vertebrate and invertebrate hosts and, under controlled laboratory condition, it may take 81 days (Baneth et al. 2007).

When a dog ingests infected ticks or their parts (i.e., while grooming or biting areas where the vector is attached), the sporozoites are released from sporocysts into the intestinal lumen. Once released, the sporozoites reach the intestinal epithelium and, via bloodstream and lymph, migrate to various organs, including the spleen, liver, bone marrow, lungs, lymph nodes, and muscles. In these sites, especially in the parenchymal organs (liver and spleen) and in the bone marrow, the sporozoites give rise to macromeronts and micromeronts (containing macro- and micromerozoites). When released, the micromerozoites invade leukocytes, particularly neutrophils and monocytes, and mature into gamonts, which may be retrieved in the blood 28 days post-infection. On the other hand, it is thought that macromerozoites undergo a secondary merogony (=schizogony), originating new meronts.

During the blood meal, the tick ingests *H. canis* gamonts. The gamonts are released from leukocytes in the tick gut and associate together in syzygy to form a zygote, which will give rise to an ookinete. The mobile ookinete penetrates the

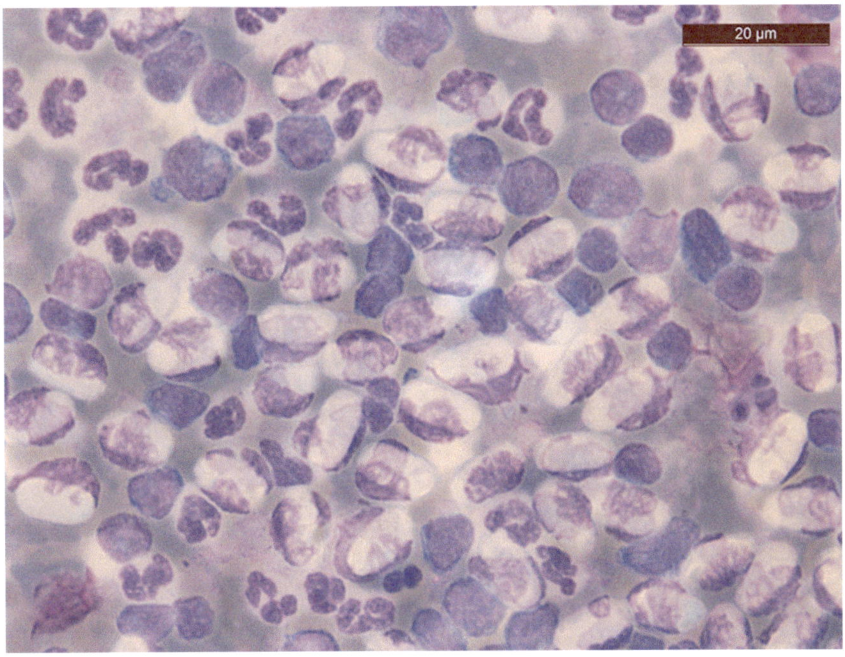

Fig. 23.1 Blood buffy coat Giemsa-stained smear, a typical ellipsoidal-shaped gamont of *Hepatozoon canis* in a leukocyte

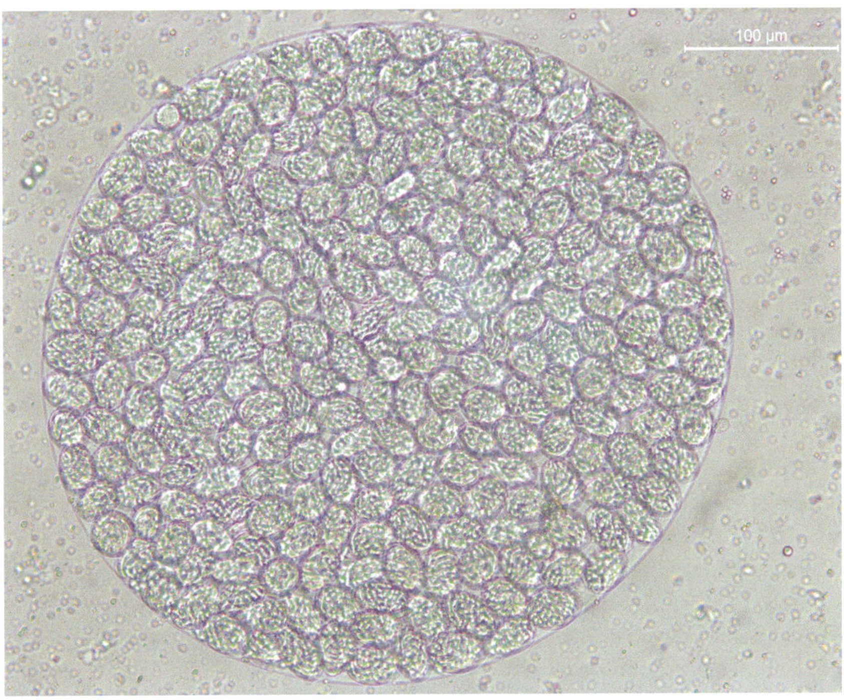

Fig. 23.2 *Hepatozoon canis* oocyst, containing numerous sporocysts and sporozoites

intestinal epithelium and reaches the tick hemocoel where it differentiates into an oocyst. The sporogony lasts according to the species and tick stage involved. In *Rhipicephalus sanguineus* sensu lato, mature oocysts (Fig. 23.2), containing various sporocysts and elongated sporozoites, may be detected in 30-day-old nymphs. Interestingly, an infected vector may contain several mature oocysts, which remain in the hemocoel until their ingestion by a receptive animal.

Nonetheless, monozoic cysts of *H. canis* have also been described in tissues of naturally or experimentally infected dogs, thus suggesting that the infection might be transmitted by carnivorism (Baneth and Shkap 2003). Finally, *H. canis* infection has also been associated, in the absence of ticks, with intrauterine transmission (Murata et al. 1993).

23.4 Clinical and Laboratorial Findings

Clinical signs of hepatozoonosis in dogs may vary according to the *Hepatozoon* species involved and to the nutritional and immunological status of each infected individual. Indeed, *H. canis* protozoan usually causes chronic infection with mild or no clinical alterations to its host (Baneth et al. 2003) or induces severe clinical manifestations (e.g., lethargy, fever, anorexia, weight loss, lymphadenomegaly, and anemia), when associated with a high parasite load (Baneth and Weigler 1997; Otranto et al. 2011; Dantas-Torres et al. 2012). Severe clinical manifestations of hepatozoonosis have also been recorded in cases of coinfection with multiple canine vector-borne pathogens (Sasanelli et al. 2009; Otranto et al. 2012).

Differently from *H. canis*, the infection by *H. americanum* is associated with fever, muscular pain, myositis, and ocular discharge that may lead to debilitation and death (Baneth et al. 2003; Vincent-Johnson 2003). Muscular pain may be generalized with gait abnormalities and even complete recumbence and inability to rise.

The most frequent laboratory finding is leukocytosis, characterized by the presence of mature neutrophils, with or without a left shift. Some dogs have a non-regenerative, normochromic-normocytic anemia. Thrombocytopenia has also been often observed. Moreover, an increase of alkaline phosphatase and creatine kinase activities is observed, with or without the presence of hypoalbuminemia and hypoglycemia. Necropsy of dogs with high parasitemia revealed hepatitis, pneumonia, and glomerulonephritis associated with the presence of *H. canis* meronts (Baneth and Weigler 1997).

23.5 Diagnosis

The diagnosis of canine hepatozoonosis is based on the association of clinical and epidemiological findings with the presence of intracytoplasmatic ellipsoidal-shaped gamonts in neutrophils and monocytes in blood cytology (Fig. 23.1) and of meronts and/or monozoic cysts in tissues at the histopathological visualization. Parasitemia is commonly observed in dogs infected with *H. canis*. On the other hand, gamonts of *H. americanum* species are rarely seen in blood smears. For this reason, the biopsy of the skeletal muscle is currently considered the gold standard for the diagnosis of the infection by *H. americanum*. Biopsy samples can be processed by conventional histological techniques and stained with hematoxylin-eosin. Immunohistochemical techniques may increase the sensitivity of the biopsy for detecting *H. americanum*.

Serological tests, particularly an indirect immunofluorescence assay (IFA) and an enzyme-linked immunosorbent assay (ELISA), have been used mainly in epidemiological studies showing a high sensitivity to dogs with chronic infections (Baneth et al. 1998).

During the last decade, a number of molecular methods based on conventional (Baneth et al. 2000) and real-time polymerase chain reaction (PCR) (Li et al. 2008) have been used not only for the characterization of *Hepatozoon* species involved in cases of canine hepatozoonosis but also to better understand the spread of this protozoan in canine populations (Otranto et al. 2011; Dantas-Torres et al. 2012). These methods used on blood samples have been shown to be more sensitive and specific for the diagnosis of *H. canis* than the other tools (Karagenc et al. 2006).

23.6 Treatment

Currently, there is no specific drug effective for the treatment of canine hepatozoonosis. Despite numerous attempts, even using different drugs in association, the results generally are contradictory and the effectiveness of some protocols has been questioned. The use of imidocarb dipropionate (6 mg/kg, SC, two doses at 14 days interval) associated with tetracycline (22 mg/kg, t.i.d., VO during 14 days) resulted in a clinical improvement and reduced parasitemia. However, a longitudinal study using imidocarb dipropionate treatment and evaluation of parasitemia by buffy coat smears and PCR, at different follow-up points, indicated that the parasite is often not completely eliminated (Sasanelli et al. 2010). Other treatment protocols may include doxycycline (10 mg/kg, s.i.d. for 21 days). Additionally, the combination of sulfadiazine-trimethoprim (15 mg/kg, b.i.d.), clindamycin (10 mg/kg, t.i.d.), and pyrimethamine (0.25 mg/kg, s.i.d.) for 14 days or until the remission of clinical signs, has also been recommended, but relapses may occur 6 months after end of the treatment (De Tommasi et al. 2014).

23.7 Epidemiology

Canine hepatozoonosis has a worldwide distribution. Cases associated with *H. americanum* are exclusively found in the New World, whereas those associated with *H. canis* infection have been reported in many countries in Africa, Asia, Europe, and the Americas.

The transmission of *H. canis* and *H. americanum* occurs when a dog ingests an infected tick containing mature oocysts of the protozoa in the hemocoel. The brown dog tick (*Rhipicephalus sanguineus* sensu lato) (Ixodida: Ixodidae) is the main vector of *H. canis* (Dantas-Torres 2008, Baneth 2011), although oocysts of this protozoan have also been detected in other ixodid species feeding on dogs, including *Haemaphysalis longicornis* and *H. flava* in Japan (Murata et al. 1995) and *Amblyomma ovale* in Brazil (Forlano et al. 2005; Rubini et al. 2009). Furthermore, oocysts of *H. canis* have been found in *Rhipicephalus (Boophilus) microplus* collected from an infected dog in Brazil (Miranda et al. 2011). A recent study confirmed that *H. canis* is not transmitted transovarially, but it is transmitted transstadially from larvae to nymphs of *R. sanguineus* sensu lato, developing sporozoites in oocysts that may infect dogs (Giannelli et al. 2013). In the United States, *H. americanum* is transmitted by the Gulf Coast tick *Amblyomma maculatum*.

Other carnivores (e.g., wild felids, wolves, and jackals) have been found naturally infected by *H. canis* and *H. americanum*. For instance, *H. canis* has been reported in red foxes from Croatia, Italy, Germany, Portugal, Hungary and Spain, with positivity rates up to 75.6 % in Portugal (reviewed in Otranto et al. 2015). Nonetheless, the role of wildlife in maintaining the enzootic life cycles of *H. canis* and *H. americanum* has not yet been completely understood, as well as the role of other ixodids in transmitting the infection.

23.8 Prophylaxis

Currently, there are no commercial vaccines available for the prevention of canine hepatozoonosis. Therefore, the only effective way to prevent the infection by *Hepatozoon* spp. in dogs is through the systematic control of ticks on animals and in the environment.

References

Abd Rani PA, Irwin PJ, Coleman GT et al (2011) A survey of canine tick-borne diseases in India. Parasit Vector 4:141

Allen KE, Li Y, Kaltenboeck B, Johnson EM et al (2008) Diversity of *Hepatozoon* species in naturally infected dogs in the southern United States. Vet Parasitol 154:220–225

Baneth G (2011) Perspectives on canine and feline hepatozoonosis. Vet Parasitol 181:3–11

Baneth G, Shkap V (2003) Monozoic cysts of *Hepatozoon canis*. J Parasitol 89:379–381

Baneth G, Weigler B (1997) Retrospective case–control study of hepatozoonosis in dogs in Israel. J Vet Intern Med 11:365–376

Baneth G, Aroch Tal N, Harrus S (1998) *Hepatozoon* species infection in domestic cats: a retrospective study. Vet Parasitol 79:123–133

Baneth G, Barta JR, Shkap V et al (2000) Genetic and antigenic evidence supports the separation of *Hepatozoon canis* and *Hepatozoon americanum* at the species level. J Clin Microbiol 38:1298–1301

Baneth G, Samish M, Alekseev E et al (2001) Transmission of *Hepatozoon canis* to dogs by naturally-fed or percutaneously-injected *Rhipicephalus sanguineus* ticks. J Parasitol 87:606–611

Baneth G, Mathew JS, Shkap V et al (2003) Canine hepatozoonosis: two disease syndromes caused by separate *Hepatozoon* species. Trends Parasitol 19:27–31

Baneth G, Samish M, Shkap V (2007) Life cycle of *Hepatozooncanis* (Apicomplexa: Adeleorina: Hepatozoidae) in the tick *Rhipicephalus sanguineus* and domestic dog (*Canis familiaris*). J Parasitol 93:283–299

Baneth G, Sheiner A, Eyal O et al (2013) Redescription of *Hepatozoon felis* (Apicomplexa: Hepatozoidae) based on phylogenetic analysis, tissue and blood form morphology, and possible transplacental transmission. Parasit Vectors 6:102

Dantas-Torres F (2008) The brown dog tick, *Rhipicephalus sanguineus* (Latreille, 1806) (Acari: Ixodidae): from taxonomy to control. Vet Parasitol 152:173–185

Dantas-Torres F, Latrofa MS, Weigl S et al (2012) *Hepatozoon canis* infection in ticks during spring and summer in Italy. Parasitol Res 110:695–698

De Tommasi AS, Giannelli A, de Caprariis D et al (2014) Failure of imidocarb dipropionate and toltrazuril/emodepside plus clindamycin in treating *Hepatozoon canis* infection. Vet Parasitol 200:242–245

Forlano M, Scofield A, Elisei C et al (2005) Diagnosis of *Hepatozoon* spp. in *Amblyomma ovale* and its experimental transmission in domestic dogs in Brazil. Vet Parasitol 134:1–7

Giannelli A, Ramos RAN, Di Paola G et al (2013) Transstadial transmission of *Hepatozoon canis* in *Rhipicephalus sanguineus* nymph stage. Vet Parasitol 4:542–546

Götsch S, Leschnik M, Duscher G et al (2009) Ticks and haemoparasites of dogs from Praia, Cape Verde. Vet Parasitol 166:171–174

Karagenc TI, Pasa S, Kirli G et al (2006) A parasitological, molecular and serological survey of *Hepatozooncanis* infection in dogs around the Aegean coast of Turkey. Vet Parasitol 135:113–119

Li Y, Wang C, Allen KE, Little SE et al (2008) Diagnosis of canine *Hepatozoon* spp. infection by quantitative PCR. Vet Parasitol 157:50–58

Miranda RL, Castro JR, Olegário MM et al (2011) Oocysts of *Hepatozoon canis* in *Rhipicephalus* (*Boophilus*) *microplus* collected from a naturally infected dog. Vet Parasitol 177:392–396

Murata T, Inoue M, Tateyama S et al (1993) Vertical transmission of *Hepatozoon canis* in dogs. J Vet Med Sci 55:867–868

Murata T, Inoue M, Taura Y et al (1995) Detection of *Hepatozoon canis* oocyst from ticks collected from the infected dogs. J Vet Med Sci 57:111–112

Otranto D, Dantas-Torres F, Weigl S et al (2011) Diagnosis of *Hepatozoon canis* in young dogs by cytology and PCR. Parasit Vector 4:55

Otranto D, Dantas-Torres F, Tarallo VD et al (2012) Apparent tick paralysis by *Rhipicephalus sanguineus* (Acari: Ixodidae) in dogs. Vet Parasitol 188:325–329

Otranto D, Cantacessi C, Pfeffer M et al (2015) The role of wild canids and felids in spreading parasites to dogs and cats in Europe. Part I: Protozoa and tick-borne agents. Vet Parasitol 213:12–23

Paludo GR, Friedmann H, Dell'Porto A et al (2005) *Hepatozoon* spp.: pathological and partial 18S rRNA sequence analysis from three Brazilian dogs. Parasitol Res 97:167–170

Rubini AS, Santos Paduan K, Perez RR et al (2006) Molecular characterization of feline *Hepatozoon* species from Brazil. Vet Parasitol 137:168–171

Rubini AS, Paduan KS, Martins TF et al (2009) Acquisition and transmission of *Hepatozoon canis* (Apicomplexa: Hepatozoidae) by the tick *Amblyomma ovale* (Acari: Ixodidae). Vet Parasitol 164:324–327

Sasanelli M, Paradies P, Lubas G et al (2009) Atypical clinical presentation of coinfection with *Ehrlichia*, *Babesia* and *Hepatozoon* species in a dog. Vet Rec 164:22–23

Sasanelli M, Paradies P, Greco B et al (2010) Failure of imidocarbdipropionate to eliminate *Hepatozoon canis* in naturally infected dogs based on parasitological and molecular evaluation methods. Vet Parasitol 171:194–199

Vincent-Johnson NA (2003) American canine hepatozoonosis. Vet Clin North Am Small Anim Pract 33:905–920

Lymphatic Filariasis

24

Eliana Maria Mauricio da Rocha, Gilberto Fontes, and John Patrick Ehrenberg

Abstract

Lymphatic filariasis (LF), frequently called elephantiasis, is a debilitating disease with serious social and economic impact. The infection is caused by worms of the species *Wuchereria bancrofti*, *Brugia malayi*, and *B. timori*. Parasites are transmitted by infected blood-feeding mosquitoes of different species of *Culex*, *Aedes*, *Anopheles*, and *Mansonia*. LF is endemic in 73 countries in Asia, Africa, and the Americas, with an estimated 120 million people affected, being 108 million of them infected with *Wuchereria bancrofti* and 12 million with *Brugia malayi* or *B. timori*. Filariasis is characterized by a broad spectrum of clinical disease, ranging from asymptomatic microfilaremic to acute and chronic lymphatic manifestations. In symptomatic patients the most common manifestations are hydrocele, lymphedema, and elephantiasis. The female worm produces microfilariae found in the bloodstream exhibiting a circadian rhythm (diurnal or nocturnal periodicity). Diagnosis can be performed by parasitological methods based on the detection of microfilariae by microscopic examination. Blood sample should be taken depending on parasite periodicity, when microfilariae concentration in the blood is higher. Serological techniques based on antigen detection or DNA amplification by polymerase chain reaction (PCR) can also be used. Diethylcarbamazine citrate (DEC) is considered the standard drug for filariasis therapy. However, chronic patients may not benefit from antifilarial drug requiring specific

E.M.M. da Rocha • G. Fontes (✉)
Federal University of São João del Rei, Campus Centro Oeste, Rua Sebastião Gonçalves Coelho, 400, 35501-296 Divinópolis, Minas Gerais, Brazil
e-mail: eliana.rocha@pq.cnpq.br; gilberto.fontes@pq.cnpq.br

J.P. Ehrenberg
Western Pacific Regional Office of the World Health Organization, Avenida Cedro 303, Cholul, 97305, Merida, Yucatan, Mexico
e-mail: ehrenbej@gmail.com

C.B. Marcondes (ed.), *Arthropod Borne Diseases*, DOI 10.1007/978-3-319-13884-8_24

care to restore function of the affected area. Elimination programs strategies are mainly based on annual mass drug administration, environmental sanitation, vector control, treatment of infected individuals, and morbidity management to alleviate the physical disability and psychological suffering of chronically affected patients.

Keywords
Lymphatic filariasis • *Wuchereria bancrofti* • *Brugia* • *Culex quinquefasciatus* • Neglected tropical disease

24.1 Introduction

Filariases are infectious diseases caused by filarial worm parasites. They have the following common characteristics: fine and delicate white worms and their females produce embryos or microfilariae which can be found in lymphatic or blood vessels, subcutaneous tissues, peritoneal cavities, or mesentery. Many of the species are capable of infecting vertebrates. However, only nine of these species are classified as nematodes of the Onchocercidae family (*Wuchereria bancrofti*, *Brugia malayi*, *B. timori*, *Onchocerca volvulus*, *Mansonella ozzardi*, *M. perstans*, *M. streptocerca*, *Loa loa, and Dirofilaria immitis*), and one belonging to the family Drancunculidae (*Dracunculus medinensis*) infects human beings (Fontes and Rocha 2011). The natural hosts of *Dirofilaria immitis* are dogs, wolves, and felines. However, human cases have been identified in some countries as accidental hosts in which the worms are not capable of completing their development leading to the formation of pulmonary nodules.

Human lymphatic filariasis can result in elephantiasis and is considered by the World Health Organization as a neglected disease. It is caused by *Wuchereria bancrofti*, *Brugia malayi*, *and B. timori* transmitted from person to person through the bite of mosquitoes infected with a larval stage of the parasite. *Wuchereria bancrofti* is the only etiological agent of lymphatic filariasis in Africa and the Americas (WHO 2008, 2011). *Brugia malayi* causes disease in the South and Southeast Asia and in the Eastern Pacific, while *B. timori* is reported in Timor and other Indonesian Lesser Sunda Islands. Lymphatic filariasis is prevalent in 73 countries of Asia, Africa, and the Americas exposing about 1.4 billion people to the risk of infection (WHO 2011, 2014). Highest prevalence rates are encountered among the poorest tropical and subtropical areas in the world where they pose a serious public health problem. It is currently estimated that approximately 120 million people may be infected throughout the world, 90% of whom are infected by *W. bancrofti*, the remainder by *B. malayi* and *B. timori. W. bancrofti and B. timori* infect exclusively human beings, the only source of infection for mosquito vectors. *Brugia malayi* can parasitize humans, primates, domestic animals (cats), and wild animals.

24.2 *Wuchereria bancrofti*

24.2.1 Morphology

Developmental stages of *W. bancrofti* in humans include adult worms (males and females) and embryos known as microfilariae. The larval stages develop in the mosquito vector. Adult worms inhabit lymphatic vessels and are long white and threadlike roundworms. The male measures 3.5–4 cm in length and 0.1 mm in diameter presenting a sharp anterior side and a ventrally enrolled posterior side, whereas the female measures 7–10 cm in length and 0.3 mm in diameter. Microfilariae which are eliminated by the pregnant females are enclosed by a flexible sheath and move actively in the bloodstream of the host. They range in size 250–300 μm in length and 7–9 μm in width. The sheath is included among the morphological criteria to identify different filarial parasites and establish a differential diagnosis since some microfilariae species found in blood lack this structure (Fig. 24.1) (Fontes and Rocha 2011).

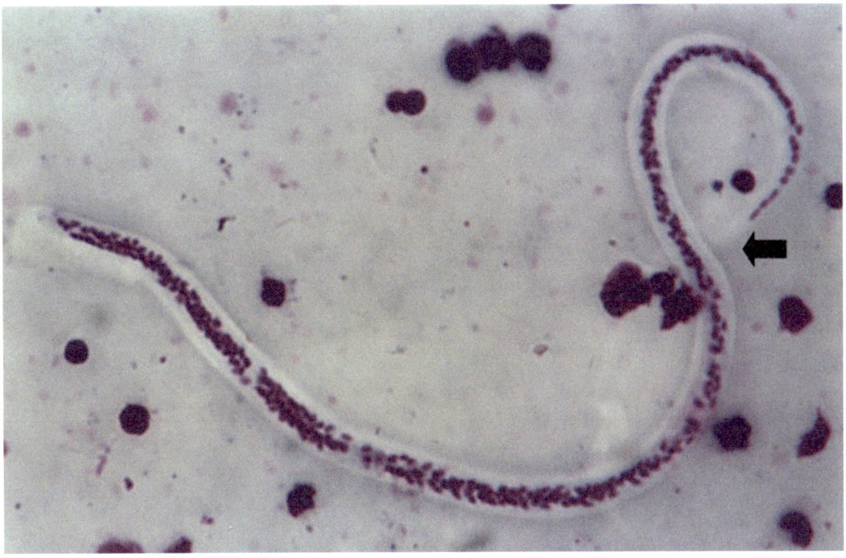

Fig. 24.1 *Wuchereria bancrofti* microfilariae in thick blood smear stained by Giemsa method (*arrow*—observe the characteristic coating sheath). Original-Eliana M. Mauricio da Rocha and Gilberto Fontes

The microfilariae ingested by the mosquito vector loose their sheaths and develop into the first-stage larvae (L_1). This larval stage is sausage shaped and measures around 300 μm in length. L_1 larva grows in length and width and develops into the second-stage larva (L_2). Subsequently L_2 molts into the third-stage larva (L_3) measuring 1.5–2 mm in length. The third-stage infective larva migrates to the mosquito's proboscis and can be transmitted to human hosts (Fig. 24.2).

24.2.2 Habitat

The female and male adult worms' lifespan ranges from 8 to 10 years. Their habitats are the human lymphatic vessels and ganglia of the spermatic cord, legs, scrotum, and occasionally breasts and arms. The microfilariae produced by the female leave the lymphatic ducts to reach the blood circulatory system of the host. The larval stages develop within the invertebrate hosts (mosquitoes).

24.2.3 Life Cycle

Wuchereria bancrofti life cycle is heteroxenic. This means it alternates between different host species throughout development, humans and mosquitoes. The female mosquitoes are infected upon feeding human microfilariae carriers. In mosquitoes midgut, microfilariae lose their sheaths, cross the stomach wall into the cavity, and reach the thoracic muscles where they develop into sausage stage or L_1. The first molt takes place 6–10 days later leading to the development of the L_2 stage. This stage undergoes further morphological changes into a second molt 10–15 days later transforming in L_3 (Paily et al. 2009). The third-stage larva is approximately 2 mm in length and is the infectious form of the parasite which migrates toward the proboscis of the vector and is transmitted to the new host during blood feeding. The life cycle within the invertebrate hosts from microfilariae to L_3 takes between 14 and 20 days at temperatures of 20–25 °C, but in higher temperatures may occur in a shorter period.

L_3 larvae are transmitted during a mosquito blood meal into the human host. The larvae emerge from the mosquito mouthparts and penetrate the skin of the human host through insect bite wound (larvae are not inoculated by the mosquitoes). Once in the human host, the larvae move through the lymphatic vessels to regional lymph nodes. Upon reaching the lymphatic system,

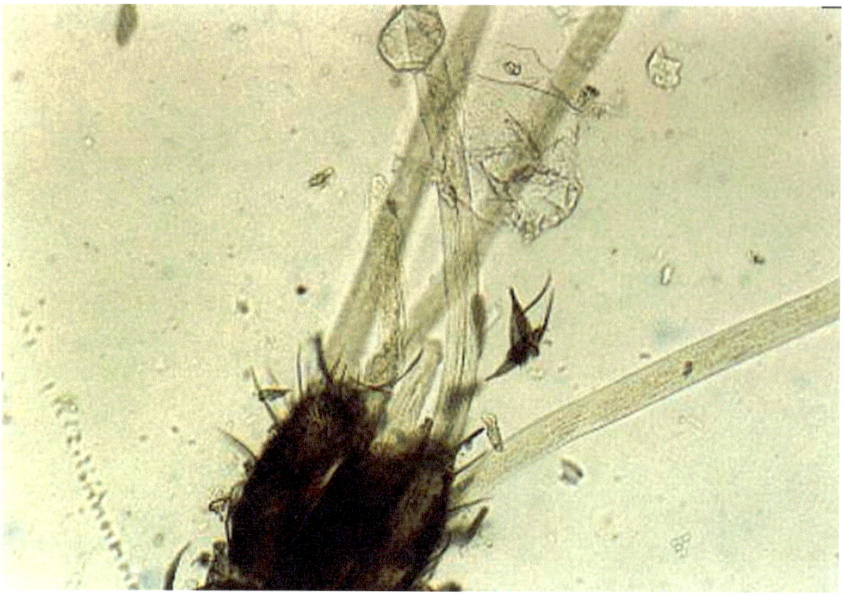

Fig. 24.2 Infective larvae (L_3) of *Wuchereria bancrofti* emerging from the proboscis of female *Culex quinquefasciatus*. Original-Eliana M. Mauricio da Rocha and Gilberto Fontes

they mature and develop into adult worms (males or females), and 7–9 months later pregnant females start to release microfilariae.

24.2.4 Transmission and Vectors

Filariasis is transmitted among humans when female mosquitoes' blood feed and infective larvae are deposited on the surface of the skin. The larvae penetrate by the wound produced by mosquito bite. Apparently the stimulus that provokes the release of the L_3 larvae from the mosquitoes' proboscis is the heat emanating from the human body. Skin humidity (sweat combined with high air humidity) allows larval movement and penetration. Mosquito lifespan is around 1 month and the development of the filarial parasite in the mosquito (from microfilaria to L_3 larva) takes place within approximately 20 days. Thus, the length of time the vector can transmit the infection to human host is of short duration. Different species of mosquitoes are capable of transmitting *W. bancrofti.* This depends on the geographic region. Mosquitoes belonging to *Culex quinquefasciatus* are the main vectors of this parasite in the Americas. *Anopheles* species, particularly *A. gambiae* and *A. funestus*, are the most common vectors in Africa, while *Aedes*, *Mansonia*, and *Ochlerotatus*[1] mosquitoes transmit the infection in Asia and the Pacific Islands (WHO 2013).

24.2.5 Microfilariae Periodicity

Microfilariae of *W. bancrofti* undergo nocturnal periodicity in the bloodstream of the human host in most of the endemic regions of the world (WHO 2013). During the day microfilariae live in deep capillary blood vessels, particularly in the lungs, and circulate at night through the peripheral blood vessels. The peak microfilaremia

[1]Although several taxonomic modifications have been proposed on Culicidae genera, mostly splitting Aedes by Reinert et al. (2009) (and accepted by CBM), this taxonomy has been used in the chapters according to authors' preference. To facilitate utilization by health personnel, all new aedine genera can be considered Aedes (CBM).

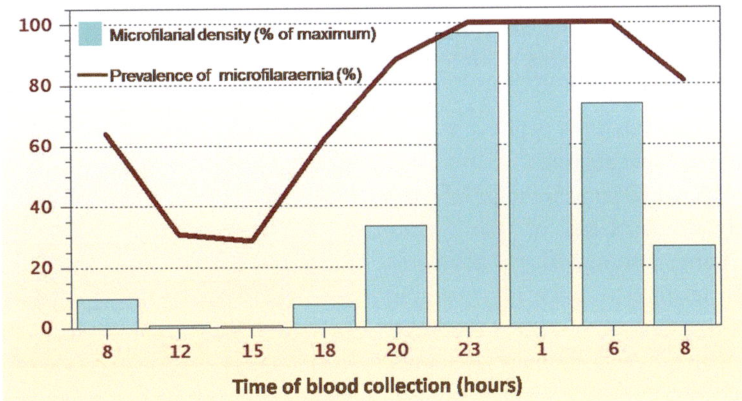

Fig. 24.3 Periodicity of *Wuchereria bancrofti* microfilariae in human peripheral blood: Percentage of patients with positive tests and percentage of average microfilaremic from analysis of drops of 60 mm^3 of peripheral blood stained by Giemsa method and harvested at different times to over 24 h (pool of 45 microfilariae carriers from Maceió, Alagoas State, Northeast Brazil) (Profs. Eliana M. Mauricio da Rocha and Gilberto Fontes, 1994)

occurs around midnight, decreasing the overall density of parasites during the day (Fig. 24.3) (Fontes et al. 2000). This variation in the concentration of blood microfilariae within a 24 h period is independent of the parasite load and gender of the host (Fontes et al. 2000). The peak of peripheral microfilariae coincides in most endemic regions with the period of highest blood-feeding activity of the principal insect vector, *Culex quinquefasciatus*. In the South Pacific and Southeast Asia where the vectors belong to the genus *Aedes* (blood feeding occurs during the day), microfilariae can be detected in peripheral blood at any time with the peak densities seen in the late afternoon and early evening hours. In these filarial infections, the microfilariae are considered non-periodical or diurnal sub-periodical (Harinasuta et al. 1970; Hawking et al. 1981).

24.2.6 Pathogenicity and Clinical Manifestations

Lymphatic filariasis or Bancroftian filariasis is characterized by a broad spectrum of clinical disease, ranging from asymptomatic microfilaremic carriers to chronic lymphatic irreversible manifestations such as elephantiasis. Infection outcomes can be caused by the adult worms living in the lymphatic system or can be the result of the hosts' immune response against the microfilariae and the parasite antigens. The prepatent period fluctuates between 7 and 9 months. Incubation period varies from few months to various years.

Clinical manifestations of lymphatic filariasis are classified into: subclinical or asymptomatic forms, acute forms, chronic irreversible lymphedema, and tropical pulmonary eosinophilia (TPE) (Dreyer et al. 2000).

Asymptomatic individuals are those with microfilariae in their blood but without apparent symptomatology. Lymphoscintigraphy and ultrasonography have allowed to demonstrate that the asymptomatic individuals actually present damage to lymph vessels (dilation and endothelium proliferation) or renal abnormalities (microscopic hematuria) (Shenoy 2008). These individuals deserve early medical attention.

The majority of the acute manifestations are retrograde lymphangitis primarily of the limbs associated with fever and general malaise, funiculitis, and epididymo-orchitis. Acute lymphangitis is of short duration and usually evolves from the root of the limbs toward the extremities. Enlarged lymph nodes usually appear around the inguinal and axillary regions as well as around the epitrochlear region.

Studies show the filarial parasites to be infected with endosymbiotic bacteria of the genus *Wolbachia.* According to some authors, an inflammatory response leading to acute lymphangitis and damage to the lymphatic system is produced by the filarial antigens and lipopolysaccharide (LPS) of the bacteria (Pfarr et al. 2009). Both of these are released as the result of the death and degeneration of the parasite within the human tissues.

Only a small proportion of the infected individuals develop chronic manifestations characterized by lymphedema, hydrocele, chyluria, and elephantiasis. The onset of chronic filarial disease tends to occur few years after the initial acute attacks of those affected in the endemic areas.

Hydrocele is the most common chronic clinical manifestation and usually occurs in the absence of previous inflammatory reactions. Patients with hydrocele may have microfilaremia in peripheral blood and in the scrotal sac fluid. Elephantiasis usually affects the lower limbs and scrotum. It tends to be associated with recurrent inflammatory episodes. The seriousness of the clinical manifestations increases with age. Chronic lesions such as elephantiasis (Fig. 24.4) can be irreversible leading to disabilities and social stigma. Some patients experience renal involvement characterized by chyluria which is the excretion of milky urine due to excretion of chyle (mixture of lymphatic fluid and chylomicrons).

Tropical pulmonary eosinophilia (TPE) is a syndrome characterized by the occurrence of bronchial asthma symptoms, a relatively rare manifestation.

Almost 25 million men out of the 120 million estimated cases worldwide suffer with genital disease (more commonly hydrocele), and close to 25 million women are affected by lymphedema or elephantiasis of the lower limbs (WHO 2014).

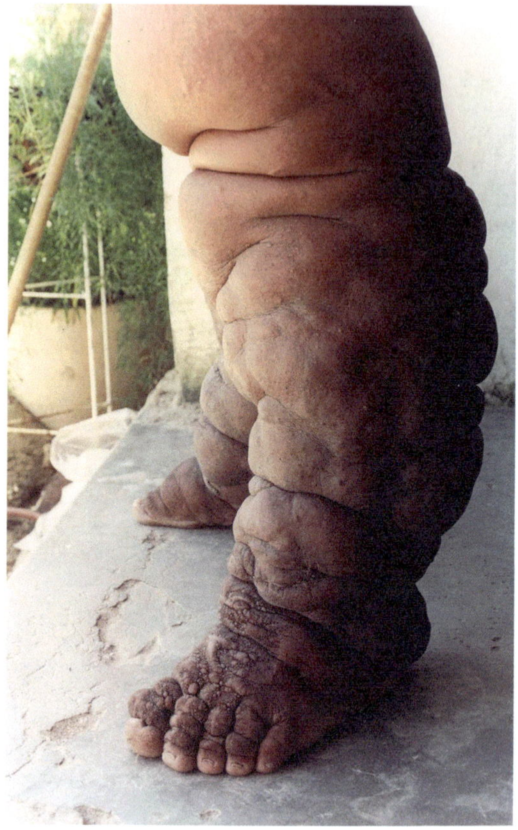

Fig. 24.4 Patient presenting elephantiasis of the lower limbs. Original-Eliana M. Mauricio da Rocha and Gilberto Fontes

24.2.7 Diagnosis

24.2.7.1 Clinical Diagnosis

Clinical diagnosis is difficult because most infected people are asymptomatic and lack abnormal physical examination findings. Furthermore, lymphatic filariasis patients may experience signs or symptoms similar to other types of diseases. A clinical history of recurrent fever in endemic areas associated with adenolymphangitis could indicate filarial infection. However, it can be hard to distinguish clinically adenolymphangitis from erysipelas or cellulitis, skin infections caused by bacteria.

Tropical pulmonary eosinophilia can be suspected in patients showing respiratory symptoms such as cough, breathlessness, wheezing, and chest pain.

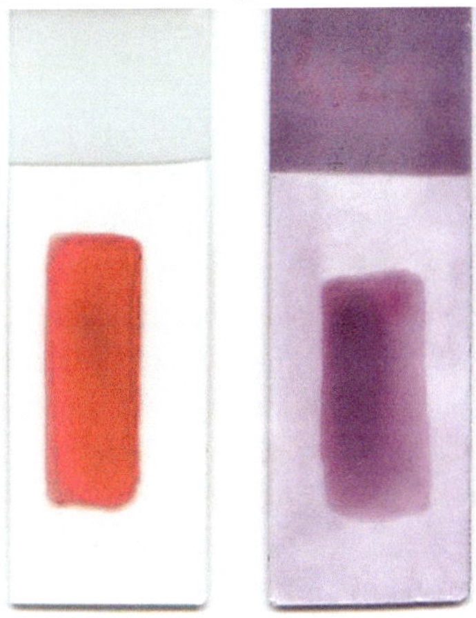

Fig. 24.5 (**a**) Thick blood smear. (**b**) Thick blood smear dehemoglobinized in water, fixed with methanol, and Giemsa stained. Original-Eliana M. Mauricio da Rocha and Gilberto Fontes

Lymphatic damage associated with elephantiasis has other causes than filariasis, such as leprosy, congenital malformations in the lymphatic system, among others.

24.2.7.2 Laboratory Diagnosis

Examination of Blood for Microfilariae

Detection of microfilariae in peripheral blood can be performed by various parasitological methods. The most common technique is based on the thick smear prepared with 20–80 μl of blood taken from a finger prick (Fig. 24.5). This should be done at night, between 10 p.m. and midnight, because of the nocturnal periodicity of microfilariae in peripheral blood of the human host throughout most of the endemic regions of the world (Fontes et al. 2000). After the smears are allowed to dry thoroughly at room temperature for 12 h, they are dehemoglobinized in water, fixed in methanol,[2] and stained with hematoxylin-eosin or Giemsa. The slides are then microscopically examined for the detection of microfilariae. The thick blood technique is still used in epidemiological surveys because it is fast, practical, easy to perform, and is an inexpensive method. However, due to its low sensitivity may fail to identify patients with low microfilariae levels (<10 microfilariae/ml of blood). In order to overcome this limitation and increase the sensitivity of this technique, the preparation of more than one slide per sample is recommended. It has also been demonstrated that over 70 % of the tests could be false negatives if the blood sample is collected at 3 pm (Fig. 24.3) (Fontes et al. 2000).

Although not routinely used, concentration techniques such as membrane filtration method increase diagnosis sensitivity. This technique consists of blood filtration through a polycarbonate membrane with 3–5 μm pore size and is an alternative to detect patients with low microfilariae densities. Samples up to 10 ml of blood can be analyzed. This procedure is quite sensitive and is usually used for the diagnosis of individual cases and in post treatment evaluation.

If membrane technique is not available, it is possible to use the technique described by Knott. This method consists of diluting blood samples 1:10 in 2 % formalin followed by centrifugation. The pellet is examined unstained or fixed and stained under the microscope for microfilariae counting. This technique is less sensitive (but less expensive) than the filtration technique. The sediment must be examined carefully since the parasites can be missed because of the viscosity of the preparation.

Microfilariae may be absent in blood but present in urine (chyluria and hematuria) or hydrocele liquid analyzed using a concentration technique.

Immunodiagnosis: Antibody Detection

Antibody-detecting immunoassays are not suitable for the diagnosis of lymphatic filariasis as they are complex to interpret. The principal limitations lie in the difficulty to distinguish among the past exposure from the present infection, as well as between those who are not infected but may have been continuously exposed to parasite antigens in the endemic area. Another problem are the cross-reactivities between filarial from non-filarial worms in lymphatic filariasis endemic areas.

[2]Methanol is very toxic for humans and primates by ingestion, skin contact, or breathing. In rhesus monkeys, minimal lethal dose by oral route, which causes symptoms similar to those in humans (irreversible blindness, acidosis, and death), is 3 g/kg.

Promising results have been obtained through IgG4 antibody assays in blood samples collected on filter paper, against recombinant antigens. One of them, Bm14, reacts against sera of patients infected with *Brugia* or *Wuchereria* with a good sensitivity (above 90 %) but low specificity considering samples from patients with other filarial infections had positive antibody tests (Weil et al. 2011). Standardization for its use in large-scale surveys is still required.

Immunodiagnosis: Antigen Detection Detection of *Wuchereria*-soluble antigens using monoclonal antibodies can be carried out by enzyme-linked immunosorbent assay (ELISA) (semiquantitative results) or by rapid immunochromatographic test (ICT card test) yielding qualitative results (positive/negative) using patient blood or serum. The advantage of antigen detection assays is that they may be present in the circulation at any time allowing for blood collection during the day. In addition, a positive result indicates active infection, as monoclonal antibodies used in this test will recognize adult worm antigens even in the absence of microfilariae.

The ELISA technique was developed using a monoclonal antibody against *Onchocerca gibsoni*, a filarial parasite of the cattle. This antibody is known as Og4C3 and is used for the detection of *W. bancrofti* circulating antigens in the human serum. This test does not have good sensitivity in low microfilaremia carriers or amicrofilaremic individuals (70–75 %) but it does have high specificity (98–100 %).

The ICT card test uses the monoclonal antibody AD12 to detect soluble antigens released by living adult *W. bancrofti* worms. The test is simple and can be used in the laboratory and clinical settings or under field conditions (Weil and Ramzy 2007). It is particularly rapid, can be performed on daytime capillary fingertip blood samples, and has a high specificity (98–100 %). The disadvantage is its high cost.

DNA Amplification by Polymerase Chain Reaction The polymerase chain reaction (PCR) is very sensitive in amplification of DNA sequences of *W. bancrofti* using species-specific primers (Vasuki et al. 2003). Various biological specimens can be used for PCR testing, including blood, urine, and even saliva of the patients. PCR assay allows diagnosis in various fluid samples collected during the day, and it is especially useful to discriminate *W. bancrofti* infections that may coexist with other filarial parasites. Primers most frequently used in the detection of *W. bancrofti* DNA were obtained by cloning and characterizing a noncoding repeat DNA sequence (Sspl) of the parasite's genome with 195pb. PCR can be used for detection of parasite DNA in biological materials from infected patients and in mosquitoes collected in the endemic areas.

Detection of Adult Wuchereria bancrofti Worms Ultrasonography is used to reveal the presence and location of the live adult worm, especially in the lymphatic vessels of the scrotal area of asymptomatic and amicrofilaremic patients. This is a noninvasive technique capable of detecting infections before the appearance of clinical manifestations. It also enables the monitoring of therapeutic efficacy through the loss of mobility of the adult worms.

In patients with lymphadenopathy, biopsy can detect adult worms. However, this procedure is rarely used for this purpose. Lymphoscintigraphy can be helpful to detect anatomical and functional changes of lymphatic vessels, especially in patients with lymphedema. This technique is not a useful tool for diagnosis but is recommended to identify lymphatic disease of the limbs.

Filarial Parasite Detection in the Mosquito Vectors

Determination of filarial infection in vector population is an important complementary tool, along with determination of human prevalence rates, in order to assess the impact of disease control strategies.

Wuchereria bancrofti larvae detection in mosquitoes can be performed through conventional method based on mosquito dissection and microscopic identification of the parasite. This technique allows specific identification of the larvae as well as the quantification of its various stages.

This procedure can be used to calculate the infection index (percentage of mosquitoes infected with any of the larval stages) and the infectivity index (percentage of mosquitoes harboring infective larvae—L_3). However, this is a laborious and a time-consuming technique, and sometimes it might be difficult to differentiate the larvae through its morphological characteristics, especially in areas where there are several species of filarial nematodes including animal parasites.

PCR is an alternative to dissection and it is a highly specific and sensitive assay being able to detect *W. bancrofti* infection in pools of mosquitoes. This method, termed molecular xenomonitoring, is suitable for testing large numbers of mosquitoes in a relatively short period of time (Bockarie 2007).

24.2.8 Treatment

Lymphatic filariasis treatment is recommended for all individuals with active infection regardless of whether they have or not have clinical manifestations related to the parasite infection. The overall goals of filariasis treatment are to prevent morbidity, reverse some of the manifestations caused by the parasite (lymphangitis, lymphedema, and hydrocele), and prevent transmission to new hosts. The treatment of choice for individual patients is diethylcarbamazine citrate (DEC). According to the World Health Organization (WHO) recommendations, DEC should be given at 6 mg/kg of body weight per day, orally for 12 days (WHO 1992). This treatment can be repeated if necessary until disappearance of parasitemia. For mass drug administration in high-endemic areas, the WHO recommends the use of single dose of DEC (6 mg/kg of body weight) plus 400 mg albendazole once a year for at least 5 years (WHO 2013). DEC is rapidly absorbed through the digestive tract reaching peak serum levels within 1–2 h after drug intake. Its mechanism of action is not known.

During the treatment, especially in the first 3 days, patients can experience adverse reactions due to parasite death (fever, headache, body and joint pain, transient hematuria, and malaise) and local reactions such as lymphangitis, funiculitis, and orchitis. These manifestations tend to disappear spontaneously and do not require interruption of treatment.

DEC causes a rapid and significant disappearance of microfilariae in blood. Parasite clearance can be observed within hours following initiation of therapy. DEC has also a considerable action against mature worms. The direct evidence of adult filaria killing (or macrofilaricidal effect) is the formation of inflammatory nodules along the lymphatic vessels after treatment (filarial granulomas). Granulomatous reactions are characterized by accumulation of eosinophils and fibrosis around degenerate and dead parasites. Indirect evidence of the adulticidal action of the antifilarial drug is the disappearance of microfilariae after treatment. Therapeutic failures can occur and are apparently caused by drug resistance despite repeated cycles of drug administration. DEC is also the drug of choice for tropical pulmonary eosinophilia, but therapy might be extended for at least 3 weeks. Patients improve clinically within a few days of starting the medication and pulmonary function returns to normal as long as the lung is not extensively damaged.

Treatment during the acute phase can prevent the development of chronic infection. In cases of early detection of lymphedema or hydrocele, treatment can reverse lymphatic pathology. However, patients with intense hydrocele or elephantiasis do not improve clinically following treatment with DEC.

Alternative drug regimen options include combinations of DEC with ivermectin or albendazole. Ivermectin is a semisynthetic broad-spectrum antibiotic effective in reducing microfilaremia at a single dose of 200–300 μg/kg of body weight. Ivermectin has a potent microfilaricidal action but no effect on adult worms, being useful to reduce the infection of mosquitoes. Thus, microfilaremia can reappear in a large proportion of patients. Albendazole does not have a short-term effect on the microfilariae if administered in a single dose. But, its combined use with DEC is recommended because of its broad anthelmintic effect in patients with *W. bancrofti* and intestinal parasite coinfection. This is usually the case in low socioeconomic areas lacking environmental sanitation.

Fig. 24.6 Geographical distribution of lymphatic filariasis, according to the World Health Organization (*Source:* http://www.asnom.org/oh/fr/0544_filarioses.php?PHPSESSID=575c0fab93319ef01b0da3f811702e7f. Accessed 10 Jan 2015)

Lymphedema management consists of intensive local hygiene using water, soap, and, if necessary, antibiotics to control secondary bacterial colonization that can worsen the clinical manifestations. Hygiene of the affected limb, active physiotherapy, postural drainage, and the use of compresses reduce lymphatic edema decreasing the chance of elephantiasis development. The affected limb must be exercised regularly in order to activate the circulation of the lymph. Elastic bandaging of the altered limb is also recommended to reduce swelling.

Surgery can be performed to alleviate hydrocele and chylocele and is also indicated in advanced cases of elephantiasis of the limbs, scrotum, and breasts. However, surgical procedures to reduce elephantiasis are usually esthetically not satisfactory.

24.2.9 Epidemiology

Lymphatic filariasis is after malaria the most important parasitic vector-borne disease in the world, due to physical disability and economic impact. The estimated number of people infected with *W. bancrofti* in 73 countries is approximately 112 million (WHO 2014). The most important endemic areas are in Asia, Africa, and the Pacific Islands (Fig. 24.6). The highest prevalence rates of infections are reported in Southeast Asia. This region is responsible for approximately 65 % of the cases followed by Africa, the South Pacific, and the Americas (WHO 2010). In India, 42 % of the population is at the risk of *W. bancrofti* infection (Ramaiah and Ottesen 2014). It is estimated that the economic loss related to lymphatic filariasis in India alone could reach up to one billion US dollars a year (WHO 2010). In Africa the disease occurs in the sub-Saharan region, to the south reaching Angola and Mozambique and the coast of Madagascar.

In the Americas active foci of transmission occur in Haiti, Dominican Republic, Guyana, and Brazil. In Haiti lymphatic filariasis is widely spread across the country, while in the rest of the Americas, the distribution of the infection is highly focal. Transmission interruption has been confirmed in Costa Rica, Suriname, and Trinidad and Tobago, all of which considered no longer endemic (WHO 2010).

The majority of microfilaremic people remains asymptomatic and can act as sources of infection for mosquitoes. From an epidemiological point of view, these individuals need to be treated to avoid the spread of the parasite. Longevity of adult worms is long with a high fertility. This means they can produce microfilariae for years even in the absence of reinfection.

Infection rates are higher among young adult males ages 18 and 25 years (Rocha et al. 2000). Detection of parasites in children is a strong indicator of active transmission in the area. If only people over 30 years are found positive, it may represent that transmission is no longer active and those carriers were exposed to the parasite prior to the elimination of transmission.

Multiple factors that affect the epidemiology and transmission of *W. bancrofti*:

- Presence of the domestic mosquitoes *Culex quinquefasciatus* in most of the endemic areas of the world
- High densities of mosquito vectors
- Infected microfilariae carriers (absence of animal reservoirs)
- History of long-term residence in filariasis endemic areas (>10 years) as a risk factor for infection (Rocha et al. 2000)
- Combination of high temperature and high relative humidity environment (important determinants in the successful development and survival of the filarial parasites in the mosquitoes and in the migration of the third-stage larvae into the human host skin through the bite wound)
- High infection rates of Bancroftian filariasis vectors

24.2.10 Prophylaxis and Control

Lymphatic filariasis is one of the nine human infectious diseases targeted for global elimination. The strategies of elimination filariasis programs aim to achieve two main objectives: interruption of transmission and morbidity control (Ichimori et al. 2014). There are currently neither prophylactic drugs available nor vaccines for this disease. Interruption of transmission is primarily based on three components: environmental sanitation, vector control, and treatment of infected individuals (WHO 2013).

Populations at risk of lymphatic filariasis greatly benefit from environmental sanitation management. Such actions depend on the improvement of the sanitation infrastructure such as drainage projects, recovery of environmentally degraded areas, cover and seal cesspools, among other measures to eliminate potential mosquitoes breeding sites decreasing the risk of infection.

An alternative way to control mosquitoes is to eliminate their larval habitat through regular applications of chemical larvicides.

Biological control is another option against insect larvae by the use of entomopathogenic bacteria (*Bacillus sphaericus* and *B. thuringiensis israelensis*). These bacteria release a d-endotoxin which is toxic upon ingestion and damages the insect larval midgut leading to its death. Biological larvicides have the advantages of exhibiting a selective mode of action, minimal residual activity, and no side effects either on humans or on the environment. Unfortunately, Bti is inactivated in polluted water, where *Cx. quinquefasciatus* immatures are found, and resistance to Bs is very frequent.

The use of floating layers of expanded polystyrene beads in mosquitoes' breeding sites, such as pit latrines, soakage pits, and overhead tanks, is also a viable alternative to inhibit egg laying. This measure also kills the immature stages by forming a thick blanket on the water surface.

Insecticides such as malathion and pyrethroids are used against adult vector species.

Individual and family protection measures are also important to avoid exposure to mosquitoes, including the use of protective clothing, topical repellent lotions, and bed nets with or without impregnated insecticide.

A comprehensive control strategy includes selective treatment of infected individuals as well as mass drug administration (MDA) and morbidity management. Annual MDA for 4–6 years is recommended in areas where lymphatic filariasis represents a public health problem. According to WHO, required levels of coverage must include at least 65 % of the inhabitants at risk in endemic areas. Combination of two drugs is delivered to the target population (WHO 2014). DEC and albendazole are administered in *W. bancrofti* transmission areas. In regions where onchocerciasis is co-endemic with lymphatic filariasis, DEC must be replaced by ivermectin since it may cause serious adverse reactions in patients infected by *O. volvulus*. Instead, a combination of albendazole and ivermectin is distributed in MDA programs in areas with both worms.

Cooking salt fortified with DEC is a substitute to mass drug administration. It is used as an option in places where tablet delivery is likely to be complicated because of logistical difficulties and costs (Lammie et al. 2007).

Over 40 million people are currently incapacitated by filarial disease, including 25 million men with genital disease and 15 million people with lymphedema. Morbidity management aims to prevent, to care, and to reduce the suffering and disability of patients suffering with chronic manifestations of the disease.

WHO expects to eliminate lymphatic filariasis as a public health problem by the year 2020 based on regional and multi-sector programs that have been developed in endemic countries (WHO 2010).

24.3 Brugia malayi and B. timori

These parasites are also etiological agents of human lymphatic filariasis. *Brugia malayi* is found in South and Southeast Asia, the Western Pacific, India, Indonesia, Malaysia, the Philippines, and Thailand. *Brugia timori* occurs only in some islands of Eastern Indonesia (Fischer et al. 2004). Brugian filariasis infects approximately 13 million people.

Brugia malayi hosts include humans, primates, domestic animals (primarily cats), and wild animals, whereas *Brugia timori* has only been found infecting humans.

Adult worms of both species can be found in the lymphatic vessels of the hosts. These filarial parasites look like those of *W. bancrofti* but are smaller: males measure around 2 cm in length and the females approximately 5 cm. As the microfilariae also have a sheath, they must be differentiated from other species according to the size of their cephalic space and caudal nuclei. *Brugia malayi* microfilariae are 230 μm in length and are found in the peripheral blood of the hosts. There are two circadian variations patterns based on the periodicity of microfilariae: the nocturnal periodical (more frequent) and the sub-periodic or aperiodic form (less common) depending on the endemic area.

Both *Brugia* species can be found in the same *W. bancrofti* endemic areas and may coinfect the same human host, but they are transmitted by different vector species. *Brugia malayi* is transmitted primarily by *Mansonia*, *Aedes*, and *Anopheles* species, although *Culex* mosquitoes can also act as vectors. The nocturnal periodic form is transmitted by *Anopheles* and *Mansonia* mosquitoes. The sub-periodic or the periodic form main vectors are *Mansonia* species. Microfilariae of *B. timori* exhibit a strictly nocturnal periodicity in peripheral blood and are transmitted by *Anopheles barbirostris* mosquitoes. The life cycle of these parasites is biologically similar to that of *W. bancrofti*.

Clinical manifestations in humans due to *Brugia* closely resemble those of *W. bancrofti* but genital involvement is rare. Signs and symptoms caused by *Brugia* usually appear earlier than those caused by *W. bancrofti*.

Diagnosis and treatment are the same as described for Bancroftian filariasis. A rapid immune-chromatographic test using a recombinant antigen BmR1 for the detection of IgG4 was developed for the diagnosis of *Brugia malayi and B. timori* infections (*Brugia* Rapid test) (Rahmah et al. 2003).

Brugian filariasis, along with Bancroftian filariasis, is also targeted by the Global Program to Eliminate Lymphatic Filariasis launched by the WHO.

References

Bockarie MJ (2007) Molecular xenomonitoring of lymphatic filariasis. Am J Trop Med Hyg 77:591–592

Dreyer G, Norões J, Figueiredo-Silva J et al (2000) Pathogenesis of lymphatic disease in Bancroftian filariasis: a clinical perspective. Parasitol Today 16: 544–548

Fischer P, Supali T, Maizels RM (2004) Lymphatic filariasis and *Brugia timori*: prospects for elimination. Trends Parasitol 20:351–355

Fontes G, Rocha EMM (2011) *Wuchereria bancrofti*—Filariose linfática. In: Neves DP, Melo AL, Linardi PM et al (eds) Parasitologia humana, 12th edn. Atheneu, Rio de Janeiro, pp 323–333

Fontes G, Rocha EMM, Brito AC et al (2000) The microfilarial periodicity of *Wuchereria bancrofti* in northeastern Brazil. Ann Trop Med Parasitol 94:373–379

Harinasuta C, Sucharit S, Deesin T et al (1970) Studies on the nocturnally subperiodic strain of *Wuchereria bancrofti* from West Thailand. Southeast Asian J Trop Med Public Health 1:152–158

Hawking F, Jennings T, Louis FJ et al (1981) The mechanisms which affect the periodic cycle of Pacific *Wuchereria bancrofti* microfilariae. J Helminthol 55:95–100

Ichimori K, King JD, Engels D et al (2014) Global programme to eliminate lymphatic filariasis: the processes underlying programme success. PLoS Negl Trop Dis 8, e3328

Lammie P, Milner T, Houston R (2007) Unfulfilled potential: using diethylcarbamazine-fortified salt to eliminate lymphatic filariasis. Bull World Health Organ 85:545–549

Paily KP, Hoti SL, Das PK (2009) A review of the complexity of biology of lymphatic filarial parasites. J Parasit Dis 33:3–12

Pfarr KM, Debrah AY, Specht S et al (2009) Filariasis and lymphoedema. Parasite Immunol 31:664–672

Rahmah N, Shenoy RK, Nutman TB et al (2003) Multicentre laboratory evaluation of *Brugia* rapid dipstick test for detection of brugian filariasis. Trop Med Int Health 8:895–900

Ramaiah KD, Ottesen EA (2014) Progress and impact of 13 years of the global programme to eliminate lymphatic filariasis on reducing the burden of filarial disease. PLoS Negl Trop Dis 8, e3319

Rocha EMM, Fontes G, Brito AC et al (2000) Bancroftian filariasis in urban areas of Alagoas State, Northeast Brazil: study in general population. Rev Soc Bras Med Trop 33:545–551

Shenoy RK (2008) Clinical and pathological aspects of filarial lymphedema and its management. Korean J Parasitol 46:119–125

Vasuki V, Hoti SL, Sadanandane C et al (2003) A simple and rapid DNA extraction method for the detection of *Wuchereria bancrofti* infection in the vector mosquito *Culex quinquefasciatus* by SspI PCR assay. Acta Trop 86:109–114

Weil GJ, Ramzy RM (2007) Diagnostic tools for filariasis elimination programs. Trends Parasitol 23:78–82

Weil GJ, Curtis KC, Fischer PU et al (2011) A multicenter evaluation of a new antibody test kit for lymphatic filariasis employing recombinant *Brugia malayi* antigen Bm-14. Acta Trop 120(Suppl 1):S19–S22

WHO – World Health Organization (1992) Lymphatic filariasis: the disease and its control. Fifth report of the WHO Expert Committee on Filariasis, Geneva Technical reports series, 821, 75 pp

WHO – World Health Organization (2008) Global programme to eliminate lymphatic filariasis. Progress report on mass drug administrations in 2007. Wkly Epidemiol Rec 83:333–348

WHO – World Health Organization (2010) Progress report 2000–2009 and strategic plan 2010–2020 of the global programme to eliminate lymphatic filariasis: halfway towards eliminating lymphatic filariasis, Geneva. Available: http://whqlibdoc.who.int/publications/2010/9789241500722_eng.pdf

WHO – World Health Organization (2011) Global programme to eliminate lymphatic filariasis: progress report on mass drug administration in 2010. Wkly Epidemiol Rec 86(35):377–388

WHO – World Health Organization (2013) Lymphatic filariasis: a handbook of practical entomology for national lymphatic filariasis elimination programmes, Geneva, WHO/HTM/NTD/PCT/2013.10. Available: http://apps.who.int/iris/bitstream/10665/87989/1/9789241505642_eng.pdf

WHO – World Health Organization (2014) Lymphatic filariasis. Fact sheet No. 102. Available: http://www.who.int/mediacentre/factsheets/fs102/en/ Accessed 10 Jan 2015

Onchocerciasis 25

James Lee Crainey, Jansen Fernandes Medeiros, Felipe Arley Costa Pessoa, and Sérgio Luis Bessa Luz

Abstract

Onchocerciasis is a neglected tropical disease that has left hundreds of thousands of the world's poorest people irreversibly blinded and has caused skin disease-induced misery for many more. The symptoms of the disease are caused by the death of skin-dwelling *Onchocerca volvulus* microfilariae. It is thought that the release of the filarial parasites' intracellular bacteria *Wolbachia* is the principal cause of an immune response that can lead to a range of progressively worsening symptoms including intense itching, various forms and degrees of skin disease, and visual impairment. The level of exposure an individual has to *O. volvulus*-infected simuliid vector bites correlates with the level of microfilariae found in their skin, and this, microfilarial load, correlates with the severity of symptoms an individual suffers from. The oldest people in an onchocerciasis focus typically have the most severe symptoms, and the most hyperendemic foci tend to have the most severely debilitated individuals. For more than 50 years, the WHO has led international efforts to control the disease. To reduce transmission, the WHO initially targeted the juvenile stages of the disease's simuliid vectors with the river application of the insecticide temephos. In the 1980s, however, the WHO's control policy shifted to community-directed treatment with ivermectin, which both reduced the parasites' transmission and benefited those treated by retarding the disease's clinical progression. Following the success of past control programs, policy makers are now targeting the elimination of

Due to the intimate relationship among lymphatic filariasis, onchocercosis, mansonelliasis, and loiasis on diagnosis, distribution, and treatment, all four chapters need to be checked when solving problems related to any of them (CBM).

J.L. Crainey • F.A.C. Pessoa • S.L.B. Luz (✉)
Leônidas e Maria Deane Institute (Fiocruz Amazônia), Rua Teresina, 476, Adrianópolis, Manaus, Amazonas 69027-070, Brazil
e-mail: james.lee@fiocruz.br; felipe.pessoa@fiocruz.br; sergio.luz@fiocruz.br

J.F. Medeiros
Fundação Oswaldo Cruz - Fiocruz Rondônia, Rua da Beira, 7671, Lagoa, Porto Velho, Rondônia 76812-245, Brazil
e-mail: jmedeiro@gmail.com

C.B. Marcondes (ed.), *Arthropod Borne Diseases*, DOI 10.1007/978-3-319-13884-8_25

new *O. volvulus* infections and the eventual global elimination of the parasite. Novel onchocerciasis treatment programs (OTPs) are presently being developed and evaluated to meet the challenges posed by OTPs expanding into *Loa loa* endemic areas and suboptimal ivermectin patient responses.

Keywords
Onchocerciasis • *Onchocerca volvulus* • Vectors • Black flies

25.1 Introduction

Onchocerciasis is a filarial disease that is vectored by black flies from the genus *Simulium*. The disease is exclusively caused by *Onchocerca volvulus* parasites and most commonly manifests clinically as an itchy skin disease but can also cause a wide spectrum of visual impairments which can progress to complete and irreversible blindness if left untreated (Crosskey 1990; Hoerauf et al. 2003). A recent assessment of the disease burden caused by onchocerciasis estimated that in 2015 there were over 21 million people infected with the *O. volvulus* parasite and around 1 million people living with *O. volvulus*-induced blindness or visual impairment and more than twice as many people suffering from the severe and debilitating itch that the parasite can also induce (Coffeng et al. 2014). Overall, however, the parasite's prevalence and its corresponding disease burden appear to be in decline, with control activities of the WHO's African Program of Onchocerciasis Control (APOC) alone estimated to have averted over 1.9 million disability-adjusted life years (DALYs) (Coffeng et al. 2014).

Also commonly known as river blindness because of the blinding pathology it can cause and because of the epidemiological link it has with the river breeding sites of its simuliid vectors, onchocerciasis is most prevalent in West Africa. The *O. volvulus* parasite, however, also occurs in East Africa, Latin America, and Yemen; reports of the parasite occurring in India and elsewhere have been made but are most likely erroneous (Bradley et al. 2005; Simonsen et al. 2014). Based on the present African distribution of both *O. volvulus* and *O. onchengi* (a very closely related cattle parasite) and the fact both parasites are vectored by the same species of black fly, it is presently thought that *O. volvulus* evolved from a domestic cattle-infecting ancestor not much earlier than ~5000 years ago (Morales-Hojas et al. 2006; Morales-Hojas 2009). Latin American *O. volvulus* parasites were once thought to cause a different disease (Robles disease) and indeed to be a different species, called *O. caecutiens*. However, *O. volvulus* parasites are today thought to have begun infecting humans in Africa and then later spread to Latin America with the slave trade (Blacklock 1927; Morales-Hojas et al. 2006; Crump et al. 2012). This perspective was recently supported by the complete sequencing of a Brazilian *O. volvulus* parasite's mitogenome, which, when compared with African references, revealed that parasite's radiation throughout Africa most likely predates its arrival in Latin America (Crainey et al. 2016). Although, thus, onchocerciasis is most likely a relatively new form of human parasitism that arrived only very recently in Latin America, it has probably, nevertheless, still been causing human misery in many of the world's poorest areas for hundreds if not thousands of years (Morales-Hojas et al. 2006; Morales-Hojas 2009; Crainey et al. 2016).

With the aim of reducing transmission, onchocerciasis control in Africa in the 1930s and 1940s was initially focused on the removal of vector breeding sites. This was achieved by riverbank clearance of vegetation, which the simuliid vector larvae cling to in order to filter feed on passing organic matter. Although this proved effective, the approach quickly shifted to chemical larviciding. Initially this was done with the river application of organochlorine insecticide dichlorodiphenyltrichloroethane (DDT), but the choice of insecticide quickly changed to the highly effective organophosphate, temephos (sold under the trade name Abate) (Davies 1994). Despite the unquestioned success of larviciding at reducing transmission, in the 1980s onchocerciasis disease control efforts again shifted focus and began to directly target the parasite in order to simultaneously suppress transmission and reduce the disease burden on those already infected (Davies 1994; Bradley et al. 2005). This shift came after the pharmaceutical company Merck Sharp & Dohme decided to donate the highly effective antifilarial drug ivermectin to the WHO for onchocerciasis control. Although vector targeting continued to play a part in onchocerciasis disease control, it was limited to areas where such efforts were deemed to be especially effective, such as Bioko (an island off the west coast of Africa), where experts believed (and then showed) vector elimination was feasible (Bradley et al. 2005; Traoré et al. 2009).

The global decline of onchocerciasis and its associated disease burden is in no small part attributable to the WHO's control and elimination efforts which operated in Africa through the internationally coordinated Onchocerciasis Control Program (OCP) from 1974 until 2002 and the African Program for Onchocerciasis Control (APOC) from 1995 to 2014 (Bradley et al. 2005; PENDA 2013). In a survey evaluating the success of the WHO's mass drug treatment program in three formally hyperendemic foci in Africa, *O. volvulus* was completely undetected in almost half a million assayed vectors and only found in a handful of >5000 individuals whose skin was assayed for the parasite (Traoré et al. 2012). The results of this survey, which was completed 3 years after the cessation of mass drug administration, were interpreted as supporting the notion that onchocerciasis elimination is achievable with the ivermectin-only approach—even in some of Africa's most challenging settings (Traoré et al. 2012).

A similar picture of successful community-directed treatment with ivermectin (CDTi) disease control has also emerged in Latin America where the WHO launched its more ambitiously named "Onchocerciasis Elimination Program for the Americas" (OEPA) in 1992 (Rodríguez-Pérez et al. 2011). While transmission appears still to be ongoing in the Brazilian side of the Amazonia focus, it is reported to have stopped on the Venezuelan side of this focus and in all the other 12 Latin American onchocerciasis foci where the WHO regarded the parasite as once endemic (Suwa et al. 2015; Botto et al. 2016). The phenomenal success of the CDTi programs for onchocerciasis control in Africa and elimination in Latin America was recently recognized by the Nobel Prize awarding committee who awarded Satoshi Ōmura and William Campbell with a 1/3 share each in the 2015 Nobel Prize for Medicine and Physiology (Molyneux and Ward 2015). Satoshi Ōmura was recognized for his discovery and culturing of the bacteria *Streptomyces avermitilis* and William Campbell for his involvement in the development of ivermectin from *S. avermitilis* avermectin fermentation products and for his proposal that the drug could be used in the treatment of onchocerciasis (Molyneux and Ward 2015). The success of CDTi also led the WHO and other important research fund donors, like the Bill & Melinda Gates Foundation, to begin targeting the elimination of new onchocerciasis infections in Africa and indeed the global elimination of the parasite (PENDA 2013; Turner et al. 2015a).

CDTi has, however, not been successful everywhere, and there are now several reports in Africa of continuing *O. volvulus* transmission occurring at sites where CDTi has been deployed for far longer than the expected adult life span of *O. volvulus*, and there are also growing concerns about patients' suboptimal ivermectin treatment responses (Lamberton et al. 2015; Eisenbarth et al.2016). To meet the challenge of new elimination targets and to safeguard hard-won gains in disease burden reductions, novel onchocerciasis reduction strategies, including the launching of a novel vaccine, are presently been proposed, evaluated, and tested (Molyneux et al. 2014; Kelly-Hope et al. 2015; Turner et al. 2015a). As the severe adverse events (SAEs) associated with ivermectin treatment occur significantly more frequently in people that are heavily infected with *Loa loa* microfilariae, onchocerciasis treatment regimes in such co-endemic areas need to be selected with great care (Boussinesq 2006; Kelly-Hope et al. 2014). Elimination programs that are expanding into poorly defined hypoendemic areas and areas where *L. loa* infections are also endemic, therefore, need to accurately map and delineate onchocerciasis endemic and co-endemic areas as carefully as possible (Molyneux et al. 2014; Kelly-Hope et al. 2015).

To meet these emergent mapping needs and also to help with the selection of treatment programs, new remote sensing-based disease mapping and filarial diagnostic tools are being developed and evaluated (Kelly-Hope et al. 2015; Molyneux et al. 2014; Vlamink et al. 2015). Diagnostic tools to assist with the increased and evolving needs of recrudesce epidemiological monitoring, as well as for the detection of light-filarial infections, are also in the process of being developed and evaluated and are also likely to be important if disease elimination targets are to be met (Vlamink et al. 2015; Drame et al. 2016; Golden et al. 2016; Thiele et al. 2016). In addition to this, onchocerciasis researchers are also investigating the importance of zooprophylaxis for onchocerciasis transmission and discussing the possible scaling up of vector control efforts in certain onchocerciasis endemic areas (Lamberton et al. 2015, 2016; Eisenbarth et al. 2016). Even investigations into optimal onchocerciasis drug treatments and drug treatment regimens for nascent elimination programs are still very much ongoing (Turner et al. 2015b; Frempong et al. 2016). Although, thus, many key public health policy makers now agree that the complete global elimination of *O. volvulus* parasites is a realistic medium-term target and most are now aiming to achieve this, how and when exactly this target can be realistically met is far from clear (Molyneux 2015; Turner et al. 2015a).

25.2 Parasite Morphology

Adults can be found in subcutaneous fibrous nodules, which vary in size from about that of a pea to that of a hen's egg. Reports of up to 50 filariae in a single nodule exist, but represent exceptional cases rather than the norm. On average nodules contain around five adult worms; males and females are typically found at a ratio of around 1:1.6, and thus a typical nodule might be expected to harbor two males and three females. It is generally believed that females are found in greater abundance within nodules because of the migratory nature of males, although many other factors could also contribute to this. Mature females are not thought to move between nodules; however, it is clear that they don't always inhabit them as the microfilariae they release can be detected in the skin of people who do not have visible nodules (Greene et al. 1985; Kläger et al. 1996; Muller 2002; Bradley et al. 2005).

Onchocerca volvulus parasites are long, white, and slender and tapered at both ends. Worms are recorded to yellow with age, with young worms reportedly more of a pure white color (Muller 2002; Bradley et al. 2005). Both male and female adult *O. volvulus* parasites have conspicuous close-set annular ridges marking their cuticles and running almost the whole course of their bodies from their heads to their tails (Fig. 25.1a, b). Muller described adult worms as having two circles of four papillae and a large lateral pair of

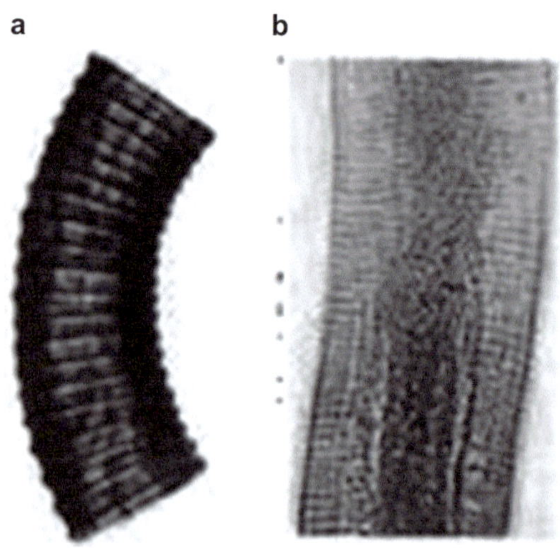

Fig. 25.1 Adult *Onchocerca volvulus* cuticular ridges. (**a**) A drawing made from an adult female *O. volvulus* fragment illustrating the close-set annual cuticular ridges of an *O. volvulus* worm, taken from Fülleborn (1908). (**b**) A photograph of an adult male *O. volvulus* worm showing the close-set cuticular ridges of an *O. volvulus* worm, taken from Fülleborn (1908)

papillae at their anterior end (Muller 2002). Reports on the overall size of adult male and female *O. volvulus* parasites vary. Markedly conflicting accounts of adult parasite length can be found in standard reference texts and technical review papers and sometimes even within the same review or textbook (see, for example, Cook and Zumla 2009, pp. 1492, 1689; Crump et al. 2002, p. 281).

The radically varying and conflicting accounts concerning the basic morphology of *O. volvulus* that are commonplace in modern scientific literature seem likely to be explained by two key factors. The first factor is that very few firsthand accounts of adult parasites have ever been published, probably because the brittle adult specimens are often obtained from highly coiled knots of adult worms and thus often break into pieces that cannot be reliably pieced back together (Prout 1901; Muller 2002; Bradley et al. 2005). The second factor that seems to be responsible for this is that phenomenon is the great influence that Sir Patrick Manson's erroneous accounts of the parasite's length have had on modern scientific literature (see Manson 1893 and the first six editions of Manson's Tropical Diseases: A Manual of the Diseases of Warm Climates). In Manson's and the world's first account of adult *O. volvulus* parasites, he described females (reasonably accurately) as 60–70 cm in length and males (erroneously) as about half their size (Manson 1893). Subsequently, in the first six editions of his highly influential tropical disease manual (that is now in its 23rd edition), Manson went on to give wildly varying and contradictory reports of the parasite species' length which, to the best of our knowledge, have gone unnoted until this day. In attempt to avoid further confusion, the parasite length measurement ranges quoted here are from firsthand primary accounts only and, therefore, exclude Manson's accounts (Prout 1901; Robles 1919). All other parasite measurements quoted here are based on either firsthand descriptions or sources that have not provided measurement lengths in serious conflict with those of Prout and Robles (i.e., Crosskey 1990; Muller 2002; Bradley et al. 2005; Simonsen et al. 2014).

Males are between about 24 and 42 mm in length and between 0.13 and 0.2 mm in breadth. The male nerve ring is noted to be approximately 140 μm for the tip of the head, which is rounded (Fig. 25.2a). The alimentary canal is long and straight and ends at the anogenital pore (cloaca) between approximately 0.05 and 0.07 mm from the tip of the tail, which is bulbous and characteristically forms a spiral (Fig. 25.2b). The male reproductive tract, which also feeds into the cloaca, is recorded to also span much of the length of the worm. Two unequally sized copulatory

a b c

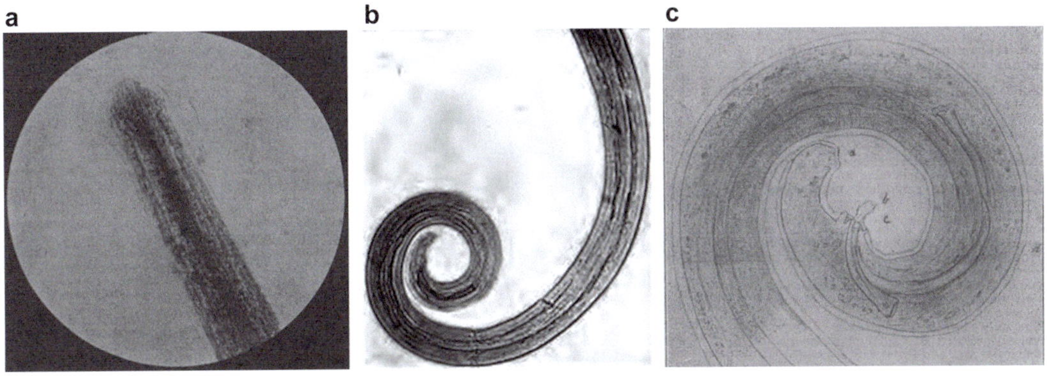

Fig. 25.2 Adult *Onchocerca volvulus* male parasites. (**a**) A photograph of the rounded anterior end (head) of an adult male *O. volvulus* parasite (taken from Prout 1901). (**b**) A photograph of the posterior tail end of an adult male *O. volvulus* parasite showing its characteristic curled tail end (taken from Fülleborn 1908). (**c**) A drawing of the posterior end of an adult male *O. volvulus* parasite showing its unequally sized spicules and paired papillae (taken from Prout 1901)

spicules can be seen stemming from the cloaca: the smaller measures between 0.08 and 0.082 mm and ends in a club shape and can be seen protruding ~0.018 mm from the cloaca; the larger is reported to measure between 0.166 and 0.28 mm and ends in a trumpet shape (Fig. 25.2c). Parasites are reported to have a variable number of terminal papillae. Usually they are recorded having two sets of paired preanal and two sets of paired postanal terminal papillae (Prout 1901; Robles 1919; Kläger et al. 1996; Muller 2002; Bradley et al. 2005).

Female adult parasites, which are unusually long for filariae, have been recorded measuring between ~400 and 500 mm in length and between 0.3 and 0.36 mm in breadth. The female nerve ring is reported to be localized around 170 μm from the worms' anterior end where the head is round and blunt; the tail is described as slightly curled, but not spiraled as they are seen in males (Fig. 25.3a, b). The female alimentary tract runs almost the entire length of the body from the mouth to the anus, which is reported to be around 210 μm from the tip of its tail. The female reproductive tract runs parallel to the alimentary canal, which is much thinner (Fig. 25.3c). The female genital pore has been measured occurring between 0.55 and 0.76 mm from the anterior end of the worm. Unsheathed microfilariae and eggs in varying stages of development can be seen within fertilized females' bicornuate uteri from between about 1.5 cm from a worm's anterior end and about 4.35 cm from its posterior end (Fig. 25.3c) (Prout 1901; Robles 1919; Bradley et al. 2005; Kläger et al. 1996; Muller 2002; Bradley et al. 2005).

Onchocerca volvulus microfilariae develop from eggs in the female uteri. The eggs have striated shells and measure approximately 30–50 μm in diameter; the unsheathed microfilariae can also be seen in the female uteri, but are most commonly found in the skin (see Figs. 25.3c and 25.4). *Onchocerca volvulus* microfilariae are recorded measuring between 186 and 360 μm and between 5 and 9 μm in width. Like the adult specimens, they have tough cuticles, which are textured with annular ridges. The head of the microfilariae are typically swollen and crowned with a disk-shaped cephalic cap, which is about 2.4 μm in diameter. A hook protrudes (~0.6 μm) from one side of the cephalic cap, which also has a central mouth and amphidial openings: one of which is located below the hook and one of which is at the extreme anterior tip. The cephalic head-space, which is free from nuclei, usually extends between 5.8 and 9.6 μm from the anterior tip of the worm. The microfilariae nerve rings are located in the anterior end of the filariae and are between 4.8 and 6.7 μm in width and are either completely or almost completely nuclei-free (Fig. 25.5). The tail is usually described as tapered to a sharp point and is also nuclei-free. This caudal tail space measures between about 8.3 and 16.5 μm in length (Franz and Schulz-Key 1981; WHO 1997).

The larvae of *O. volvulus* develop through three stages (L_1–L_3) within the parasites' simuliid

a b c

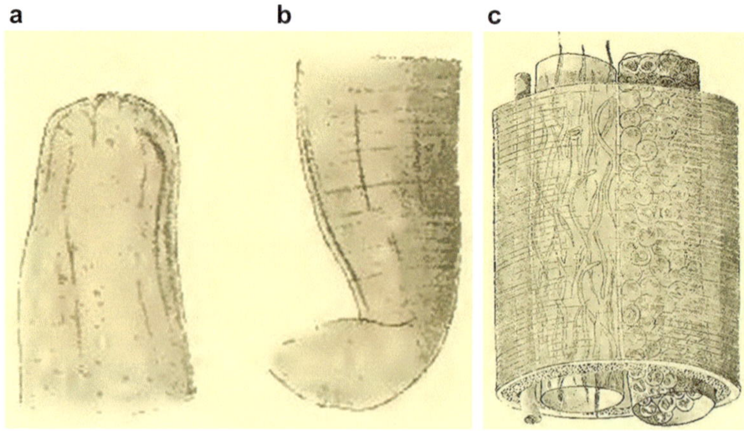

Fig. 25.3 Adult *Onchocerca volvulus* female parasites. (**a**) An anterior end drawing showing the bluntly rounded head of an adult female *O. volvulus* parasite (taken from Prout 1901). (**b**) A posterior end drawing of an adult female *O. volvulus* parasite (take from Prout 1901). (**c**) A drawing of a fertilized adult female *O. volvulus* parasite containing eggs and embryos at varying stages of development (taken from Prout 1901)

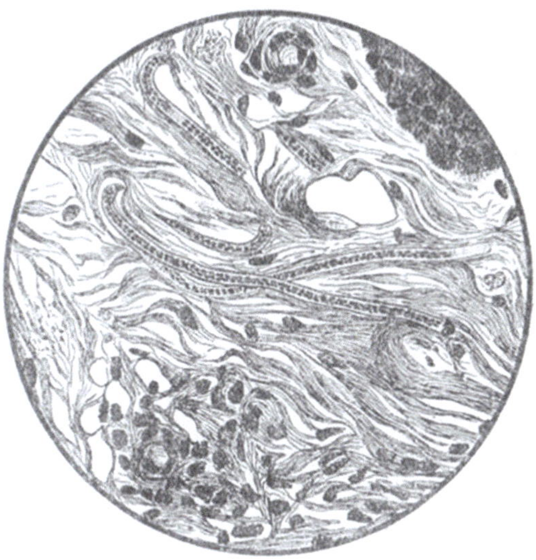

Fig. 25.4 A drawing taken from skin section showing *Onchocerca volvulus* microfilariae in situ (originally from Sharp 1926)

Fig. 25.5 An *Onchocerca volvulus* microfilariae (taken from Sharp 1926)

hosts. Sausage stage (L_1) larvae retain the tail and the hooked cephalic cap of the microfilariae stage, but are considerably thicker and shorter, typically measuring around 160 μm by 30 μm. The L_1 larvae still have annular ridges at either end; however, the rest of the cuticle appears smooth. The anal opening of L_1 larvae is found about 9 μm from the posterior tip on the lateral side. L_2 larvae measure between around 260 μm and around 370 μm in length. At the anterior end of L_2 larvae, there is a large central stoma and no cephalic cap; L_2 larvae, however, retain their tails. The much larger L_3 larvae measure between about 400 and 700 μm in length and 20 μm in width. The cephalic head region of L_3 larvae has two circles of four papillae, which ring the central mouth opening. Paired amphids lie on opposite sides of the mouth, close to the inner ring of four papillae. The anal opening is ringed by a rumpled cuticular doughnut-shaped swelling approximately 5.8 μm in diameter and about 18 μm from the posterior tip of the of L_3 larvae. The L_3 larvae have blunt posterior ends with three papillae, which vary in length from 1.1 to 1.9 μm (Franz and Schulz-Key 1981; Trpis et al. 2001).

25.3 Life Cycle

Onchocerca volvulus eggs develop into microfilariae in the adult female uterus, following an adult female's fertilization. Adult females from the nodule-forming *O. onchengi* cattle parasite simultaneously produce progeny sired by several different males. Fertilized *O. volvulus* females incased in nodules probably do the same and release around 500–1000 microfilariae a day from their vulvas. These microfilariae take around 4–6 weeks to migrate into the dermal layer where they can reportedly exist for up to 2 years and can reach densities of up to 50 microfilariae per milligram of the skin (see Fig. 25.4). The vast majority of these microfilariae represent a dead end in the parasite's life cycle; only the few parasites that are consumed by an appropriate simuliid vector species (see Table 25.1) can continue it. The simuliid vectors of *O. volvulus* use their sharp-toothed maxillae to tear open the dermal tissue, which the parasites inhabit, and then pool feed on the blood that floods the wound from the host's ruptured capillaries. During the course of a blood meal, which can last as long as 15 minutes, *O. volvulus* microfilariae migrate to the blood pool from which a vector is feeding on and are sucked into its gut with its blood meal

Table 25.1 The known distribution of probable and known *Simulium* vectors of *O. volvulus* past and present

The known distribution of probable and known *Simulium* vectors of *O. volvulus* past and present
Africa and Yemen
S. albivirgulatum (Democratic Republic of the Congo, Angola, Botswana, Congo, Gabon, South Africa (KZN), Zambia, Zimbabwe)
S. damnosum complex
S. damnosum s.l. strain (Benin, Burkina Faso, Cameroon, Central African Republic, Ghana, Guinea, Guinea-Bissau, Ivory Coast, Liberia, Mali, Niger, Nigeria, Senegal, Sierra Leone, South Sudan, Sudan, Togo, Uganda)
S. kilibanum (Burundi, Democratic Republic of the Congo, Kenya, Malawi, South Africa, Swaziland, Tanzania, Uganda)
S. konkourense (Ethiopia, Guinea, Guinea-Bissau, Liberia, Sierra Leone)
S. leonense (Guinea, Liberia, Sierra Leone)
S. mengense (Cameroon, Central African Republic)
S. rasyani (Yemen)
S. sanctipauli (Ghana, Guinea, Ivory Coast, Liberia, Mali, Nigeria)
S. sirbanum (Benin, Burkina Faso, Cameroon, Central African Republic, Ethiopia, Ghana, Guinea-Bissau, Ivory Coast, Mali, Niger, Nigeria, Sierra Leone, Sudan, Togo, Uganda)
S. soubrense (Burkina Faso, Benin, Ghana, Guinea, Ivory Coast, Liberia, Mali, Nigeria, Sierra Leone, Togo)
S. squamosum (Cameroon, Benin, Central African Republic, Democratic Republic of the Congo, Ivory Coast, Ghana, Guinea, Liberia, Nigeria, Sierra Leone, Togo)
S. thyolense (Malawi, Tanzania)
S. yahense (Benin, Cameroon, Equatorial Guinea [Bioko], Ghana, Guinea, Ivory Coast, Nigeria, Liberia, Sierra Leone, Togo)
S. neavei (Democratic Republic of the Congo, Kenya, Uganda)
S. ethiopiense (Ethiopia)
S. woodi (Malawi, Tanzania)
Latin America
S. callidum (Mexico, Belize, Costa Rica, El Salvador, Guatemala, Panama)
S. exiguum (complex) (Argentina, Bolivia, Brazil, Colombia, Ecuador, Guatemala, Guyana, Mexico, Panama, Peru, Venezuela)
S. guianense s.l. (Guyana, Brazil, French Guiana, Suriname, Venezuela)
S. horacioi (Guatemala, Mexico, Panama, Venezuela)
S. ochraceum s.l. (Belize, Colombia, Costa Rica, Cuba, Dominican Republic, Ecuador, French, Guiana, Guadeloupe, Guatemala, Jamaica, Mexico, Montserrat, Panama, Peru, Puerto Rico, Trinidad, Venezuela, Virgin Islands)
S. incrustatum (Brazil, Colombia, Ecuador, Panama, Paraguay, Trinidad, Venezuela)
S. limbatum (Brazil, Guyana, Venezuela)
S. metallicum (Belize, Brazil, Colombia, Costa Rica, Ecuador, El Salvador, French Guiana, Guatemala, Honduras, Jamaica, Mexico Nicaragua, Panama, Peru, Trinidad, Venezuela)
S. oyapockense s.l. (Argentina, Bolivia, Brazil, Colombia, Ecuador, French Guiana, Guyana, Paraguay, Uruguay, Venezuela)
S. quadrivittatum (Belize, Colombia, Costa Rica, Cuba, Dominican Republic, Ecuador, Guatemala, Jamaica, Mexico, Nicaragua, Panama, Puerto Rico, Venezuela)

Species and cytospecies names given in italics

(Crosskey 1990; Hildebrandt et al. 2012; Muller 2002; Taylor et al. 2010).

The microfilariae pass through two molts before reaching the infectious L_3 larval stage, which they achieve in a period of around 6–9 days, although the process and its details vary with vector species and can reportedly be slowed with low temperatures and even halted with temperatures below 18 °C. Following ingestion and before migrating to a vector's thorax, the microfilariae need to pierce through the vector's peritrophic membrane, which is secreted from its stomach lining following a blood meal. *Onchocerca volvulus* microfilariae morph into L_1 stage larvae without molting. Approximately 24 hours after a *S. damnosum s.l.* vector has taken a blood meal, most of the surviving parasites are in the L_1 stage. In the period between 5 and 6 days after a *S. damnosum s.l.* blood meal, almost all of the *O. volvulus* parasites can be expected to have molted at least once (shedding their tails) and to be found in the thorax. The parasites develop in the thorax lying parallel to the vector's flight muscle fibers. The second molt of *O. volvulus* parasites also takes place in the thorax, from about 6 days after the blood meal. Seven days after the blood meal, most of the larvae have molted twice and are in the L_3 phase. Between 7 and 9 days after the blood meal, L_3 larvae can typically be found in a vector's head and mouthparts (Franz and Schulz-Key 1981; Crosskey 1990; Trpis et al. 2001; Muller 2002; Bradley et al. 2005).

The infectious L_3 stage larvae escape their simuliid hosts and enter the dermis of a new human host after their vector has opened a fresh dermal wound. The L_3 larvae molt into L_4 stage larvae in the dermis of their new human host close to this wound site and then molt once more before reaching the adult phase. The microfilariae of fertilized adult females can be detected in skin snips taken from artificially infected chimpanzees from around 7 months after an infection; typically, however, *O. volvulus* infections are thought to take about 12–18 months to become patent. Adult parasites are reportedly very long-lived in humans and thought to survive up to 13–15 years, a view supported by data from the WHO's mass drug treatment programs (Soboslay et al. 1992; Bradley et al. 2005; Taylor et al. 2010; Traoré et al. 2012).

25.4 Clinical Symptoms

The clinical manifestation of *O. volvulus* seems to be largely related to the parasitic load and the duration of infection and seems to be caused mostly or entirely by the human inflammatory immune response to the death of microfilariae, as many as 100,000 of which have been estimated to die every day within a heavily infected person. The most obvious clinical signs of onchocerciasis, like skin nodules and blindness, are strongly

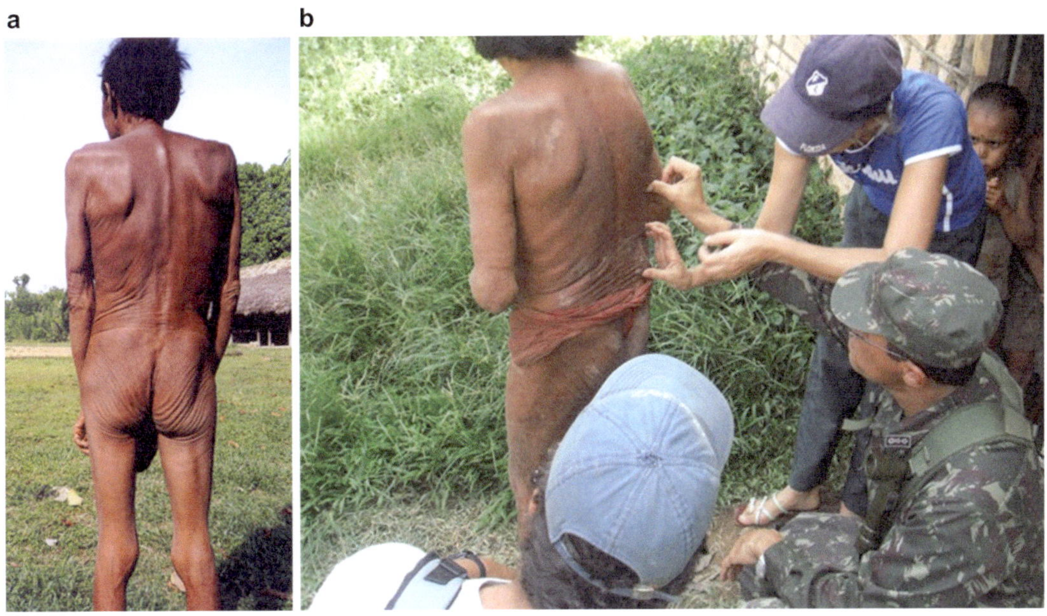

Fig. 25.6 Yanomami tribesman from the Amazonia onchocerciasis focus displaying classic symptoms of onchocerciasis skin disease including skin atrophy (**a**) and depigmentation (**b**)

correlated with heavy microfilarial loads and with exposure to infection and long-standing infections (Muller 2002; Bradley et al. 2005; Taylor et al. 2010).

Approximately 28 % of people who test positive for the existence of *O. volvulus* microfilariae, with microscopic examination of skin snips, appear to be completely asymptomatic (Blacklock 1927). Most of those displaying clinical symptoms have some form of skin condition (see Fig. 25.6) with only a small proportion presenting with severe visual impairments or blindness. With the exception of skin nodules, which usually cause no significant discomfort, all other forms of the onchocerciasis skin complaints are commonly associated with intense itching which can lead to insomnia and substantial reductions in an individual's workplace productivity and secondary dermal infections (Muller 2002; Bradley et al. 2005; Taylor et al. 2010; Simonsen et al. 2014). Murdoch et al. (1993) classified the typical skin complaints associated with onchocerciasis into five broad categories: acute papular onchodermatitis (APOD), chronic papular onchodermatitis (CPOD), lichenified onchodermatitis, atrophy, and depigmentation.

APOD and CPOD are both commonly associated with intense itching and very often affect the shoulder region. APOD, which is characterized by small widely dispersed pruritic papules, erythema, and edema, is also commonly found on the arms and trunk (Murdoch et al. 1993; Bradley et al. 2005; Taylor et al. 2010). CPOD, which is often found accompanying other symptoms like APOD and skin atrophy, is characterized by dispersed elevated flat-topped papules, which vary in size from about 3–9 mm in diameter (Murdoch et al. 1993; Muller 2002; Bradley et al. 2005; Taylor et al. 2010). In addition to the shoulder areas, CPOD is also often found on the buttocks and surrounding areas. Skin atrophy caused by onchocerciasis is also commonly seen in the buttocks and presents like prematurely aged skin, with thin skin appearing wrinkled and without elasticity. Skin atrophy, in severe cases, can result in hanging folds of excess skin, which often create a clinical presentation referred to as hanging groin (Muller 2002; Bradley et al. 2005; Simonsen et al. 2014). Depigmentation of the skin, especially of the shins, is also a very common presentation of long-standing onchocerciasis infections throughout Africa. This sort of depigmentation, which is also sometimes known as "leopard skin," is characterized by skin mottled by depigmentation (Bradley et al. 2005; Taylor et al. 2010; Simonsen et al. 2014). All the abovementioned skin pathologies can be found in combination with one another and in varying severity (Murdoch et al. 1993; Bradley et al. 2005; Taylor et al. 2010).

In Latin America, where the vast majority of those infected are light skinned, a clinical presentation which Robles refers to as "erisipela de la costa" can often be observed (Robles 1919; Rodríguez-Pérez et al. 2011). This presentation is thought to be an acute immune response to infection and is characterized by a visible reddening of papule-afflicted skin (Rodríguez-Pérez et al. 2011; Robles 1919). In Sudan and Yemen, a skin pathology known as *sowda* is common. The clinical presentation of *sowda* is similar to the other forms of skin depigmentation associated with onchocerciasis and initially presents as localized hyperpigmented, hyperkeratotic plaques. These plaques are often found afflicting the legs, although in many cases only one leg is affected. The pathology is also known as lichenified onchodermatitis because in the later stages, the afflicted skin becomes lichenified and is associated with a strong patient immune response to infection (Murdoch et al. 1993; Hoerauf et al. 2002; Simonsen et al. 2014). Skin microfilariae in individuals presenting with *sowda* are difficult or impossible to detect. The clinical presentation of *sowda* is also strongly associated with a single nucleotide polymorphism in the interleukin-13 gene, a human gene associated with human immune response to parasitic infection. The development of the *sowda* presentation of onchocerciasis is, therefore, believed to be driven by an individual's immunological status determined by their heredity (Hoerauf et al. 2002; Simonsen et al. 2014).

Like the severity of the skin conditions, the extent of visual impairment caused by onchocerciasis varies from mild loss of visual field through to complete irreversible blindness. Prior to the WHO's control program, up to 46 % of men in some hyperendemic West African villages would be expected to be blinded by onchocerciasis before they died, following a long progressive loss of vision. The level of an individual's visual impairment correlates with their microfilarial load and the length of time they have been infected and whether or not they have head nodules. Anterior eye damage to the cornea is caused by sclerosing keratitis, which results from the host immune response to dying microfilariae that enter the cornea from the skin, via the conjunctiva. Prior to the development of sclerosing keratitis, opaque snowflake-shaped lesions, around dying microfilariae, can be seen on the cornea. This form of punctate keratitis heals, but the sclerosing keratitis form that develops from it does not. Turning the transparent cornea opaque from the margins, this form of ocular damage is accompanied by a growth in supplying blood vessels and causes a gradual and progressive reduction in visual field. The processes involved in posterior chamber damage are less well understood than those that cause damage to the cornea. Chorioretinal lesions at the outer side of the macula and inflammatory damage to the optic nerve are both correlated with the onchocerciasis and can often be found together with corneal damage (Muller 2002; Hoerauf et al. 2003; Bradley et al. 2005; Taylor et al. 2010).

While most of the well-recognized clinical manifestations of onchocerciasis are ocular or dermatological, many other forms of clinical manifestations have been attributed to onchocerciasis. Weight loss, dizziness, musculoskeletal pains, physiological arrested development, and neurological conditions like epilepsy and nodding syndrome have all been attributed to onchocerciasis, with varying degrees of research community acceptance (Taylor et al. 2010; Coffeng et al. 2014; Simonsen et al. 2014). Recently, *O. volvulus* adult parasites have been recorded emerging from the skin in a similar way to which *Dracunculus medinensis* "guinea worms" emerge from ankle blisters. Notably, however, unlike *D. medinensis*, these *O. volvulus* parasites have been found emerging from variety of sites around the body that lack any clear signs of blistering. Whether or not this is an unnatural *O. volvulus* presentation elicited by the WHO's mass drug treatment program or a pathology historically misidentified as being caused by guinea worms, owing to the formally widespread incidence of guinea worm disease, is presently unclear (Eberhard et al. 2010).

25.5 Diagnosis

The clinical presentation of heavy onchocercal infections, especially infections presenting with skin nodules, is often, in practice, sufficient to diagnose many cases of *O. volvulus* infections. The WHO, for example, used a simple nodule survey, in combination with predicted vector breeding sites, to produce the rapid epidemiological maps it used for APOC (Bradley et al. 2005; Molyneux 2009). The existence and locomotive activity of adult parasites in nodules can be verified by ultrasound, and *O. volvulus* larvae can be detected in simuliid vectors by PCR, a process increasing being used in onchocerciasis epidemiological monitoring and often referred to as xenomonitoring (Simonsen et al. 2014; Vlamink et al. 2015; Lamberton et al. 2015; Eisenbarth et al. 2016). Most clinical diagnoses are, however, confirmed with the identification of *O. volvulus* microfilariae obtained from an ~2 mm diameter (~1.5 mg) "skin-snip" biopsy, which are ideally obtained using a sterile Walser or Holth corneoscleral punch (WHO 1997; Simonsen et al. 2014).

Onchocerca volvulus microfilariae can be positively identified by a range of methods, but are most commonly identified by their morphology from skin snips, which have been incubated in water or saline solution, using light microscopy and Giemsa or hematoxylin staining (WHO 1997).

25.5.1 Differential Diagnosis[1]

In Africa, the skin-dwelling microfilariae of *M. streptocerca* can be found sympatrically with *O. volvulus* and pose the greatest risk to accurate

[1]See Chaps. 24, 26, and 27 for comparison to lymphatic filariasis, mansonelliasis, and loiasis, also for treatment (CBM).

microfilarial identifications. In Latin America, the normally blood-dwelling *M. ozzardi* parasites (which are reported to sometimes also occur in the skin) are the only parasites that pose a difficulty for this form of identification and only in the Amazonia onchocerciasis focus where the two parasites are found sympatrically (WHO 1997; Post et al. 2003; Bradley et al. 2005). Microfilariae-based diagnosis of onchocerciasis can also be performed or verified using DNA detection methods. PCR and sequencing of targeted *O. volvulus* DNA sequences from skin-snip DNA extracts are generally regarded as the gold standard of onchocerciasis diagnosis and have been shown (with a number of gene targets) to be both a very specific and sensitive approach for diagnosis (Alhassan et al. 2014; Ricciardi and Ndao 2015; Vlamink et al. 2015). A field-friendly LAMP-based DNA detection method has also been reported, but its utility to onchocerciasis clinical epidemiological monitoring has still to be rigorously tested (Alhassan et al. 2014; Vlamink et al. 2015).

Whether detecting filarial parasites in human blood or tissue samples or in their vectors, multi-copy genes are usually targeted for most filarial PCR parasite detection assays (Post et al. 2009; Tang et al. 2010; Lamberton et al. 2015; Eisenbarth et al. 2016). PCR-based assays that detect the ov-150 repeat *O. volvulus*-specific repeat sequence have historically been the most widely used PCR assay for onchocerciasis epidemiological monitoring (Alhassan et al. 2014; Vlamink et al. 2015; Thiele et al. 2016). However, the Tang et al. (2010) ITS-1 ribosomal DNA-targeting assay has been used for a range filarial diagnostics requirements and is increasingly being used and recommended for onchocerciasis diagnostics (Hagiya et al. 2014; Medeiros et al. 2015; Ricciardi and Ndao 2015; Suwa et al. 2015). Users of the Tang et al. (2010) assay avoid the complicated and time-consuming processing steps that are required for most ov-150-based assays and benefit from the assay's ability to detect, and potentially identify, a range of different filariae that could be interfering with onchocerciasis epidemiological monitoring (Tang et al. 2010; Suwa et al. 2015; Vlamink et al. 2015; Thiele et al. 2016). Recently, the Tang et al. (2010) assay was developed into a single-step real-time PCR assay (Thiele et al. 2016). While this new real-time PCR assay's infrastructural equipment requirements will restrict its roll-out and its utility to epidemiological monitoring has not yet been rigorously tested, this assay, or ITS-1-targeting real-time PCR assays like it, could prove an important tool for onchocerciasis elimination efforts as xenomonitoring and light parasitic infection detection become more important (Thiele et al. 2016). Mitochondrial DNA-targeting assays are also increasingly being used for filarial diagnostics and, like rDNA-targeting assays, can, potentially, be used to detect and identify several different species of filarial parasites in a single assay. However, the recent discovery of NUMTs in several filarial parasite genomes is likely to restrict their usefulness (Lamberton et al. 2015; Eisenbarth et al. 2016; Luz et al. unpublished data).

Numerous types of immunological diagnostic assays, which usually detect an individual's exposure to *O. volvulus* parasite using venous blood or skin snips, have been reported and used in onchocerciasis epidemiological monitoring and diagnostics (Rodríguez-Pérez et al. 2011; Alhassan et al. 2014; Vlamink et al. 2015). In most cases these assays work on the basis of detecting human antibodies against *O. volvulus* proteins or protein extracts and therefore by implying present or past exposure to the *O. volvulus* parasites and thus cannot easily be used to discriminate between present and past infections (Alhassan et al. 2014; Simonsen et al. 2014; Vlamink et al. 2015). Observed cross-reactivity between these assays and control sera (obtained from people thought to only have exposed to related filarial parasites) has, however, suggested that many lack specificity and has therefore compromised the utility of most (Rodríguez-Pérez et al. 2011; Alhassan et al. 2014; Luz et al. 2014; Simonsen et al. 2014).

Only the ov-16 protein-based onchocerciasis diagnostic assay is widely used and commercially available in a rapid format card test (Rodríguez-Pérez et al. 2011; Alhassan et al. 2014; Vlamink et al. 2015); however, even the reliability of this test has been questioned in *M. ozzardi* endemic areas (Luz et al. 2014; Vlamink et al. 2015; Botto et al. 2016). The reality is that the ov-16 assay is typical of filarial immunodiagnostic assays in that it has not been rigorously tested for sera cross-reactivity in all the areas in which it has been deployed. The importance of this sort of testing was recently highlighted by the discovery that ICT cards (a widely used test for *W. bancrofti* detection) cross-react with sera from *W. bancrofti*-free areas in several parts of Africa (Alhassan et al. 2014; Bakajika et al. 2014; Vlamink et al. 2015; Wanji et al. 2015). Whether ov-16 false positives emerge from recrudescence monitoring in Bioko (where all known onchocerciasis vectors have be eliminated) and a broad spectrum of filarial parasites occur will prove an interesting test of the reliability of ov-16 assays for recrudescence monitoring in Africa in general (Traoré et al. 2009).

A Mazzotti reaction to DEC (diethylcarbamazine) treatment was historically regarded as the gold standard for an *O. volvulus* infection; however, safety concerns related to this approach mean it is no longer widely used (Bradley et al. 2005; Molyneux 2009; Taylor et al. 2010).[2] A safer DEC-based patch test has, however, now been developed for onchocerciasis diagnosis. This test works by applying DEC (in Nivea® cream) to a small portion of the skin and then waiting 24 h for a localized immune response (a papular skin rash). Although the test was regarded by some authoritative sources to be a rapid reliable and sensitive method of diagnosing onchocerciasis, others have reported that it lacks specificity because it cross-reacts with *M. streptocerca* (Molyneux 2009; Simonsen et al. 2014).

The recent reports of *O. volvulus* adult parasites emerging from the skin (see above) present a new problem for the clinical diagnosis of onchocerciasis. Thus far, parasites emerging in this way have principally been identified by their morphological features, which have been facilitated by the usually large size of adult females and the close-set annular ridges running the length of their cuticle. The existence of just two underlying transverse striae between cuticular ridges (which can be seen only in tissue section) allows *O. volvulus* worms to be distinguished from many other *Onchocerca* parasite species that can cause zoonotic filariasis. As with skin-snip analysis, definitive identifications in such circumstances can be obtained by DNA extraction, PCR, and sequencing (Crosskey 1990; Orihel and Eberhard 1998; Muller 2002; Bradley et al. 2005; Molyneux 2009; Eberhard et al. 2010; Tang et al. 2010; Taylor et al. 2010).

[2]See Chaps. 26 and 27 for the risks of Mazzotti reaction (CBM).

25.6 Treatment

Although in Latin America onchocerciasis was initially treated by the surgical removal of skin nodules, nodulectomies (where skin nodules are removed under local anesthetic) are, now, not often performed anywhere, except when nodules occur on the head (because of the increased risk of blindness associated with head nodules). Onchocerciasis infections today are almost exclusively treated with ivermectin. Ivermectin is a macrocyclic lactone that kills microfilariae and temporarily sterilizes adult females, preventing them from releasing more of the symptom-inducing microfilariae. As part of the WHO's mass drug treatment program, ivermectin is typically given orally annually or biannually in 6 mg tablets at a dose of 150 μg/kg of body weight. Ivermectin is a well-tolerated onchocerciasis treatment and can be safely used by most onchocerciasis sufferers that are not also heavily infected with *Loa loa* parasites (i.e., those with >30 microfilariae per microliter in their blood). It is not usually recommended that ivermectin be given to pregnant women, although some reports of where this has occurred accidently suggested that pregnant women can be safely treated with ivermectin too (Muller 2002; Bradley et al. 2005; Taylor et al. 2010; Simonsen et al. 2014).

The severity of the side effects that do sometimes occur following ivermectin treatment seems to be correlated with an individual's microfilarial load and to be caused by the drug's microfilaricidal activity. In hyperendemic settings around

20 % of those treated for the first time can be expected to show some form of adverse reaction to the treatment; around 5 % can be expected to have severe adverse reactions. Subsequent doses of ivermectin (which kill fewer parasites) rarely have adverse consequences. The most commonly reported adverse consequences are increased itching and skin rashes. These symptoms usually manifest 2–3 days following treatment and then subside quickly thereafter. Musculoskeletal pains, fevers, lymph gland pain, and swelling are also reported as mild side effects of ivermectin treatment. More serious, but rare, adverse consequences reportedly include severe forms of postural hypotension, bronchospasm (in asthmatics), and bullae (Greene et al. 1985; Kläger et al. 1996; Taylor et al. 2010). In individuals infected with *L. loa* parasites, ivermectin treatment can induce very serious adverse events (SAEs) including encephalopathy and may even cause death, and thus the WHO has therefore historically left many areas where loiasis and onchocerciasis are found to co-occur untreated (Boussinesq 2006; Simonsen et al. 2014). As the WHO moves its onchocerciasis program focus from control to elimination in Africa, it will need to treat such individuals. In the WHO's (2013) PENDA report, it was tentatively proposed that a test-and-treat proposal may be adopted for onchocerciasis control in some loiasis co-endemic areas. The idea is being that individuals that test negative for *L. loa* infections or test positive for light *L. loa* infections would receive a standard ivermectin treatment and those testing positive for heavy *L. loa* infections (i.e., those with parasitemia counts of >30 per microliter of blood) would receive a nonstandard treatment. Heavily infected individuals might be, for example, given a course of doxycycline (see below) and/or a nonstandard dose and/or treatment course of ivermectin.

For about 40 years prior to the discovery of ivermectin as an effective treatment for onchocerciasis, diethylcarbamazine (or DEC) was a routinely used treatment for *O. volvulus* infections. Like ivermectin, DEC works very effectively as a microfilaricide; it can, however, cause very severe adverse consequences (often referred to as Mazzotti reactions), which are thought to be a consequence of its fast-acting microfilaricidal activity. The adverse effects caused by DEC are similar in nature to those caused by ivermectin and are similarly correlated with patient parasitic load, but DEC side effects can be life-threatening and are often much more severe when they do occur; the side effects caused by DEC also occur more often than the side effects caused by ivermectin. As well as the short-term increase in the amount of itching and other dermatological symptoms seen with ivermectin, DEC can also exacerbate visual impairment and evoke headaches, musculoskeletal pains, fever, postural hypotension, and bronchospasm. To minimize the impact of these side effects, DEC was often administered in gradually increasing doses. Before the widespread use of ivermectin, DEC was administered once daily at 0.5–1 mg/kg for the first 2 days and then twice daily at 2 mg/kg until a total dose of 30 mg/kg had been delivered. Because of the adverse inflammatory immune response triggered by DEC, it was typically given with betamethasone and aspirin to dampen the side effects (Greene et al. 1985; Muller 2002; Bradley et al. 2005; Taylor et al. 2010).

Doxycycline given at a dose of 200 mg per day for 6 weeks, with or without a follow-up dose of ivermectin (given 4–6 months post-doxycycline treatment), has been shown, like DEC and ivermectin, to drastically reduce levels of microfilariae in the skin. The treatment, which targets the filarial worms' intracellular bacterial *Wolbachia* endosymbionts, sterilizes adult females (possibly permanently) before eventually killing most of them. In one study >60 % of the treated adult female worms were dead 2 years post-doxycycline treatment, suggesting that it can be a curative treatment. Although doxycycline treatment can cause minor side "effects" like diarrhea, as a treatment choice, it benefits from the fact that it does not provoke the inflammatory reactions of the fast-acting filaricides, like ivermectin and DEC, and is not thought to cause SAEs. With ivermectin resistance and *L. loa* parasite coinfection both becoming increasingly important issues for the WHO's nascent onchocerciasis global elimination "program," doxycycline is likely to prove an increasing important tool for treatment (see above). Despite its clinical benefits, however, the long and intensive course of drug administration required for doxycycline treatment makes it difficult and expensive to mass distribute for international control and elimination programs. It is, therefore, unlikely to replace ivermectin treatment where ivermectin treatment is an effective therapeutic option. Other antibiotic treatments and/or any other *Wolbachia*-targeting treatment (a number of which are presently in development) may one day be used in a similar way to doxycycline treatment, but none have, as yet, been as rigorously trialed for this purpose as doxycycline has been (Hoerauf et al. 2003, 2008; Taylor et al. 2010, 2014).

Moxidectin is a macrocyclic lactone filaricide that is closely related to ivermectin and which has historically only been used in veterinary medicine. Recently, however, in a small-scale clinical trial in Ghana, the drug's utility and safety for onchocerciasis treatment were assessed. In this trial moxidectin was given in a single 8 mg dose and was seen to have a greater efficacy than the WHO's recommended dose of ivermectin, which was trialed alongside it. Moxidectin was also seen to evoke similar adverse consequences to those evoked by ivermectin; the proportion of people affected by these adverse consequences and the severity of them was also similar to what was observed for the ivermectin treatments. Following on from this pilot study, moxidectin is now being trialed on a larger scale to improve the picture of its safety profile. If ongoing trials show the drug to be safe for large-scale distribution, moxidectin may be used as a substitute for ivermectin. Under such circumstances and assuming the larger study supports the pilot study's findings on efficacy, moxidectin may possibly make an appealing option for onchocerciasis control in places where suboptimal ivermectin treatment responses exist or emerge (Awadzi et al. 2014; Turner et al. 2015b).

Many other drugs have been trialed for onchocerciasis treatment, often in combination with the abovementioned drugs; however, no other drugs have proven important chemotherapeutic treatments. Albendazole,[3] for example, is a broad-spectrum anthelmintic, which is sometimes used in combination with DEC and ivermectin to treat other filariae, but in trials was proven an ineffective treatment for onchocerciasis. In a recent Cochrane systematic review, the value of ivermectin and doxycycline used in combination for onchocerciasis was compared with using ivermectin alone. The study concluded that there was too little good-quality published data to draw any firm conclusions about whether this combination approach was better than the standard approach of using ivermectin alone (Bradley et al. 2005; Simonsen et al. 2014; Abegunde et al. 2016).

Although prior to the discovery of the macrofilaricidal properties of doxycycline, suramin (sold under the trade names Antrypol and Bayer 205) was also known to have an unambiguous macrofilaricidal effect on *O. volvulus* parasites; it was never widely used as a treatment. This is probably because of the toxicity of this drug, which meant it could only be administrated safely (as intravenous injection) in a hospital setting, which in turn made it expensive and impractical, especially for disease control programs. Suramin is also known to sometimes induce severe side effects which can result in death. Suramin is, thus, now largely an obsolete treatment for onchocerciasis (Muller 2002; Bradley et al. 2005; Simonsen et al. 2014) (see Chap. 26).

Although a research program to develop a vaccine for onchocerciasis was recently launched and the public health benefits an effective vaccine could provide have been shown, there is, presently, no effective vaccine for onchocerciasis and no rigorously tested clinically proven prophylactic medicines (Bradley et al. 2005; Simonsen et al. 2014; Turner et al. 2015b). Experiments on chimpanzees have shown that when ivermectin is co-injected with infectious stage *O. volvulus* (L_3), the animals are, subsequently, less likely to develop microfilariae-patent infections than those that do not receive ivermectin (Soboslay et al. 1992). It seems likely, thus, that ivermectin treatments provide some prophylactic protection and

[3]Albendazole is useful for the treatment of loiasis (see Chap. 27) (CBM).

may protect against new infections while it remains in the blood of those treated with it. Recent studies on onchocerciasis transmission in Cameroon have also implied that exposure to the cattle parasite may provide zooprophylactic protection against *O. volvulus* infections although how this could be manipulated to help with public health targets has not yet been investigated (Eisenbarth et al. 2016).

As doxycycline can kill both the adults and microfilariae of *O. volvulus* by killing their *Wolbachia*, it seems likely that it (or in fact any antibiotic or anti-*Wolbachia* treatment that can kill the *Wolbachia* of *O. volvulus*) may also provide prophylactic protection against new infections. Doxycycline is commonly used as a prophylactic against malaria and may thus be providing those who take it with protection against *O. volvulus* infections or indeed any other filarial parasites that can be effectively treated by targeting their *Wolbachia* (Bradley et al. 2005; Hoerauf et al. 2008; Taylor et al. 2010; Simonsen et al. 2014).

25.7 Epidemiology and Control

Although some species of great apes can be hosts to *Onchocerca volvulus* parasites and natural infections of them have been recorded, they are not generally regarded as important reservoirs of the parasite. This view is strongly supported by recent vector monitoring studies that have shown that in multiple foci in Africa and Latin America, vector transmission of the parasite can be halted with treatment of the human host alone. Thus all evidence suggests that the historical view that humans are the main or exclusive reservoirs of *O. volvulus* parasites is probably correct (Bradley et al. 2005; Adler et al. 2010; Traoré et al. 2012).

The level of severity of onchocercal disease is generally correlated with the level of microfilariae found in the skin, which in turn seems to be correlated with an individual's exposure to the bites from *O. volvulus*-infected simuliid vectors (which is usually measured for an entire onchocerciasis focus as an annual transmission potential—ATP). Onchocerciasis foci with the highest ATPs tend to be foci with both the highest microfilariae incidence (among the population) and infection intensities (number of parasites per microgram of the skin in an infected individual), which are also positively correlated with one another. Because of this, foci with high ATPs and community microfilarial loads (CMFLs), a measure that combines incidence and average microfilarial infection intensities within a focus, tend also to have high incidences of severe skin diseases and blindness. Blindness and skin nodules are therefore more commonly found in hyperendemic onchocerciasis foci than in mesoendemic and hypoendemic foci, and more cases of the most serious skin conditions and blindness are found more commonly among the old of hyperendemic villages. Within foci, people that work outdoors and close to the simuliid vector breeding sites are at greater risk of infectious bites, having high microfilarial loads and thus with developing more severe skin and ocular pathologies than those who do not (Bradley et al. 2005; Basáñez et al. 2009; Molyneux 2009; Adler et al. 2010).

The precise parasite strain that an individual is infected with may have an important impact on the pathology of onchocerciasis. Although the genetic basis of these differences has not yet been well characterized, people who live in the African savannah and have comparable levels of infections with those living in the forest can be expected to have more severe ocular symptoms and less severe dermatological symptoms than those living in the forest (Bradley et al. 2005; Taylor et al. 2010). While there are other explanations for these observations, the notion that differences between savannah and forest strains of the parasite are at least in part responsible for differences in clinical pathology has been supported by some (though not all) vectorial studies. These studies suggest that the two parasite strains develop differently in a number of different simuliid vectors and by studies on the parasites' intrastrain differences in *Wolbachia* bacteria levels (Adler et al. 2010).

Onchocerca volvulus parasites from the African savannahs (where more severe ocular pathologies occur) have more intracellular

Wolbachia than forest forms do. As *Wolbachia* has been shown in animal models to elicit the inflammatory host immune responses responsible for most of the onchocerciasis ocular pathologies, this is perhaps not surprising. What role genetic variation within these bacteria or within their host filariae has in the disease pathology is still, however, not fully understood (Basáñez et al. 2009; Taylor et al. 2010). Although it is thought that general onchocerciasis genetic immunity may exist as individuals that live in hyperendemic areas and seem to have similar parasite exposure profiles can have widely varying clinical presentations, only the genetic basis of *sowda* could be regarded as well characterized (Hoerauf et al. 2002; Bradley et al. 2005; Taylor et al. 2010; Simonsen et al. 2014).

Regional differences in vectorial fauna are also known to greatly impact on *O. volvulus* transmission dynamics and to be changing as a result of WHO-led disease control programs. *Simulium guianense s.l.* is, for example, presently regarded as the most important vector of *O. volvulus* in Latin America because of its role in transmission in the hyperendemic highland areas of the Amazonia onchocerciasis focus, which is the only Latin American focus where *O. volvulus* transmission is still ongoing (Basáñez et al. 2009; Rodríguez-Pérez et al. 2011; Shelley et al. 2010; Crainey et al. 2014). Before WHO onchocerciasis control efforts began, however, *S. guianense s.l.* was one of the world's least important vectors of *O. volvulus* because the Amazonia focus, though geographically large, contains only a small *O. volvulus*-infected human population (in comparison with the other highly populous hyperendemic foci in Latin America and Africa).

Sometimes vectorial transmission dynamics differ between foci because of the differing vector densities; sometimes they vary because the vector populations are made up of different species, and sometimes they can vary because some vector species have substructuring and thus contain different forms with different behavioral traits at different geographic localities. Even within the Amazonia focus, *S. guianense s.l.* populations appear to have different transmission roles in different areas because of differences in behavior. *Simulium guianense s.l.*, found breeding in the mesoendemic lowland areas of the Amazonia onchocerciasis focus, for example, have never been recorded biting humans, suggesting that they are zoophilic like other populations (found outside the focus) that appear to be and thus are not involved in lowland area *O. volvulus* transmission (Basáñez et al. 2009; Shelley et al. 2010; Rodríguez-Pérez et al. 2011; Crainey et al. 2014). Similarly, *S. exiguum* has been reported to occur in all 13 WHO-recognized onchocerciasis geopolitical foci in Latin America, but is thought to have only ever been an important vector of *O. volvulus* in Columbia, Ecuador, and Mexico. Some of the other members of complex that were reported elsewhere were not thought to be important vectors because they were regarded as zoophilic, while other populations that were recorded as anthropophilic were thought to be too small to be epidemiologically important, and others were thought to be unimportant because they were observed to be refractory to parasite development. Vector population size, refractoriness, and the host-feeding preference are not, however, the only aspects of its behavior that are thought to impact on onchocerciasis epidemiology. The high proportion of head nodules (and corresponding high levels of ocular pathologies) that Robles reported *O. volvulus* causing have been, for example, explained by the upper-body biting preference of the Guatemalan vector that was responsible for its transmission—*S. ochraceum s.l.* (Basáñez et al. 2009; Adler et al. 2010; Shelley et al. 2010; Rodríguez-Pérez et al. 2011; Crainey et al. 2014).

Vector biology, ecology, and diversity can affect *O. volvulus* epidemiology in less direct ways too. Numerous *O. volvulus* vector species with "forest" and "savannah" forms and cytospecies have been recorded. In some cases, these two variant behavioral forms are clearly of different epidemiological relevance, but in other cases epidemiological differences may exist but are more difficult to discern. For example, *S. limbatum* inside the forested region of the Amazonia focus are molecularly distinct from the *S. limbatum* found in the savannah areas outside the focus. As the Yanomami people (see Fig. 25.6), who are presently serving as the principal reservoir for the parasites at this focus, rarely leave the forest, the savannah forms of *S. limbatum* (though physiologically capable of developing the *O. volvulus* parasites) in practice probably have no role in their transmission (Shelley et al. 2010; Conceição et al. 2013). In Africa most *O. volvulus* transmission is carried out by black flies from the *S. damnosum* species complex (see Table 25.1), which is composed of over 50 cytospecies. Although these cytospecies are defined by their polytene chromosome banding patterns, they are usually also divided into forest and savannah groups (Adler et al. 2010; Post et al. 2011, 2013). While there is little doubt that the different cytoforms have different epidemiological roles in different geographical localities, a full-resolved picture of this has still not emerged (Adler et al. 2010; Lamberton et al. 2014, 2015). For example, while it has been proposed that savannah parasite strains and savannah cytospecies (and conversely forest parasite strains) have coevolved together and that this has shaped African onchocerciasis epidemiology, this perspective remains controversial. And similarly, while *Wolbachia* have been detected from onchocerciasis vectors, how these bacteria impact on vector biology in not yet full known (Adler et al. 2010; Crainey et al. 2010).

Physiological onchocerciasis vector competencies have been shown to vary substantially in a number of ways too. In studies where different vector species were fed on the same volunteer, it has been seen repeatedly that the amount of parasites ingested by a vector and the amount of infectious L_3 parasites that then develop from a blood meal depend on the exact vector species concerned. The bases of these variations are complex. For example, vector species that have well-developed cibarial armatures, which have "teeth" that can shear microfilariae, are typically less efficient vectors than those that lack them; however, when they feed on heavily infected people, they are often more likely to develop infectious L_3 parasites (than if they feed on a lightly infected individual) and may also suffer a low incidence of microfilariae-induced mortalities than an unarmed species fed on the same individual. This means that the epidemiological relevance of unarmed and armed vector species is likely to vary with foci CMFL and thus that the vectors that were the most important in disease transmission at a foci prior to control may not be the vectors that pose the greatest risk of causing onchocerciasis recrudescence (Basáñez et al. 2009; Adler et al. 2010; Shelley et al. 2010).

Onchocerca volvulus parasites have also been seen to develop more efficiently and quickly in some vector species than others even when they have been ingested in similar numbers and by vectors with morphologically similar cibarial armatures. In some cases these kinds of vector competency differences have been attributed to differences in the vectors' immune response to parasitic infections and sometimes to the nature of or speed of the development of the vectors' peritrophic matrix, which forms a hard barrier around blood meals which protects vectors from parasitic invasions, although there may be hitherto unknown forces at work too (Basáñez et al. 2009; Adler et al. 2010; Shelley et al. 2010).

Climatic variation is thought to strongly affect onchocerciasis epidemiology in a plethora of complex and difficult to predict ways (Cheke et al. 2015). Both African and Latin American *O. volvulus* transmission has been reported to follow seasonal patterns. These patterns are thought to emerge principally as a response to changes in parous vector biting rates, which are themselves thought to respond to complex climate-influenced ecosystem changes (Shelley et al. 2010; Cheke et al. 2015). Seasonal changes in rainfall, evaporation, and mountain snowmelt, for example, are thought to affect vector densities by dramatically altering the availability and favorability of vector river breeding sites (Shelley et al. 2010; Rodriguez-Perez et al. 2011; Post et al. 2013; Cheke et al. 2015; Lamberton et al. 2015, 2016). As *O. volvulus* development is also temperature sensitive (with ingested *O. volvulus* microfilariae not developing into the infectious L_3 stage larvae below 18 °C), onchocerciasis transmission and geographical distribution are likely to also be directly influenced by temperature. It has, for this reason, been suggested that the observed absence of onchocerciasis transmission in certain mountainous climates in Africa can be explained by their altitude-related cooler climates (Crosskey 1990). Climatic and seasonal variations are also thought to affect vector densities in less direct ways, for example, by altering vector predation and/or parasitism levels (Laird 1981; Walsh and Ocran 1985).

References

Abegunde AT, Ahuja RM, Okafor NJ (2016) Doxycycline plus ivermectin versus ivermectin alone for treatment of patients with onchocerciasis. Cochrane Database Syst Rev 1:1–14

Adler PH, Cheke RA, Post RJ (2010) Evolution, epidemiology, and population genetics of blackflies (Diptera: Simuliidae). Infect Genet Evol 10:846–865

Alhassan A, Li Z, Poole CB et al (2014) Expanding the MDx toolbox for filarial diagnosis and surveillance. Trends Parasitol 31:391–400

Awadzi K, Opoku N, Attah S et al (2014) A randomized, single-ascending-dose, ivermectin-controlled, double-blind study of moxidectin in *Onchocerca volvulus* infection. PLoS Negl Trop Dis 8:1–18

Bakajika DK, Nigo MM, Lotsima JP et al (2014) Filarial antigenemia and *Loa loa* night blood microfilaremia in an area without bancroftian filariasis in the Democratic Republic of Congo. Am J Trop Med Hyg 91:1142–1148

Botto C, Basañez MG, Escalona M et al (2016) Evidence of suppression of onchocerciasis transmission in the Venezuelan Amazonian focus. Parasit Vectors 9:1–18

Basáñez MG, Churcher TS, Grillet ME (2009) Onchocerca-*Simulium* interactions and the population and evolutionary biology of *Onchocerca volvulus*. Adv Parasitol 68:263–313

Blacklock D (1927) The insect transmission of *Onchocerca volvulus* (Leuckart, 1893): the cause of worm nodules in man in Africa. Br Med J 1:129–133

Boussinesq M (2006) Loiasis. Ann Trop Med Parasitol 100:715–731

Bradley JE, Witworth J, Basáñez MG (2005) Onchocerciasis. In: Collier L, Balows A, Sussman H (eds) Topley and Wilson's microbiology and microbial infection. Arnold Hodder, London, pp 781–801

Cheke RA, Basáñez MG, Perry M et al (2015) Potential effects of warmer worms and vectors on onchocerciasis transmission in West Africa. Philos Trans R Soc Lond B Biol Sci 370:1–14

Coffeng LE, Stolk WA, Zouré HG et al (2014) African programme for onchocerciasis control 1995-2015: updated health impact estimates based on new disability weights. PLoS Negl Trop Dis 8:1–8

Conceição PA, Crainey JL, Almeida TP et al (2013) New molecular identifiers for *Simulium limbatum* and *Simulium incrustatum s.l.* and the detection of genetic substructure with potential implications for onchocerciasis epidemiology in the Amazonia focus of Brazil. Acta Trop 127:118–125

Cook CG, Zumla AI (2009) Manson's tropical diseases, 23rd edn. Elsevier Saunders, Philadelphia, p 1830

Crainey JL, Wilson MD, Post RJ (2010) Phylogenetically distinct *Wolbachia* gene and pseudogene sequences obtained from the African onchocerciasis vector *Simulium squamosum*. Int J Parasitol 40:569–578

Crainey JL, Mattos-Gloria A, Hamada N et al (2014) New tools and insights to assist with the molecular identification of *S. guianense s.l.*, main *Onchocerca volvulus* vector within the Amazonia Onchocerciasis focus. Acta Trop 131:47–55

Crainey JL, Silva TR, Encinas F et al (2016) The mitogenome of *Onchocerca volvulus* from the Brazilian Amazonia focus. Mem Inst Oswaldo Cruz 111:79–81

Crosskey RW (1990) The natural history of blackflies. Wiley, Chichester

Crump A, Morel CM, Omura S (2012) The onchocerciasis chronicle: from the beginning to the end? Trends Parasitol 28:280–288

Davies JB (1994) Sixty years of onchocerciasis vector control: a chronological summary with comments on eradication, reinvasion, and insecticide resistance. Annu Rev Entomol 39:23–45

Drame PM, Meng Z, Bennuru S et al (2016) Identification and validation of *Loa loa* microfilaria-specific biomarkers: a rational design approach using proteomics and novel immunoassays. MBio 16:1–8

Eberhard M, Ruiz-Tiben E, Korkor A et al (2010) Emergence of *Onchocerca volvulus* from Skin Mimicking Dracunculiasis medinensis. Am J Trop Med Hyg 83:1348–1351

Eisenbarth A, Achukwi MD, Renz A (2016) Ongoing transmission of *Onchocerca volvulus* after 25 years of Annual ivermectin mass treatments in the Vina du Nord River Valley, in Northern Cameroon. PLoS Negl Trop Dis 6:1–16

Franz M, Schulz-Key H (1981) Scanning electron microscope studies on the anterior region of the larvae of *Onchocerca volvulus* in the vector. Trans R Soc Trop Med Hyg 75:141–142

Frempong KK, Walker M, Cheke RA et al (2016) Does increasing treatment frequency address suboptimal responses to ivermectin for the control and elimination of river blindness? Clin Infect Dis 62:1338–47

Fülleborn F (1908) Über filaria volvulus (Leuckart). Archiv fuer schiffs-und Tropen-hygiene xii, Beihefte

Golden A, Stevens EJ, Yokobe L et al (2016) A recombinant positive control for serology diagnostic tests supporting elimination of *Onchocerca volvulus*. PLoS Negl Trop Dis 10:1–16

Greene B, Taylor H, Cupp E et al (1985) Comparison of ivermectin and diethylcarbamazine in the treatment of onchocerciasis. N Engl J Med 33:133–138

Hagiya H, Terasaka T, Kimura K et al (2014) Filarial chyluria as a rare cause of urinary retention. Intern Med 53:2001–2005

Hildebrandt J, Eisenbarth A, Renz A et al (2012) Single worm genotyping demonstrates that O. ochengi females simultaneously produce progeny sired by different males. Parasitol Res 111:2217–2221

Hoerauf A, Kruse S, Brattig NW et al (2002) The variant Arg110Gln of human IL-13 is associated with an immunologically hyperreactive form of onchocerciasis (sowda). Microbes Infect 4:37–42

Hoerauf A, Büttner DW, Adjei O et al (2003) Onchocerciasis. Br Med J 326:207–210

Hoerauf A, Specht S, Büttner M et al (2008) *Wolbachia* endobacteria depletion by doxycycline as antifilarial therapy has macrofilaricidal activity in onchocerciasis: a randomized placebo-controlled study. Med Microbiol Immunol 197:295–311

Kelly-Hope LA, Cano J, Stanton MC et al (2014) Innovative tools for assessing risks for severe adverse events in areas of overlapping Loa loa and other filarial distributions: the application of microstratification mapping. Parasit Vectors 7:307

Kelly-Hope LA, Unnasch TR, Stanton MC et al (2015) Hypo-endemic onchocerciasis hotspots: defining areas of high risk through micro-mapping and environmental delineation. Infect Dis Poverty 4:1–7

Kläger S, Whitworth J, Downham M (1996) Viability and fertility of adult *Onchocerca volvulus* after 6 years of treatment with ivermectin. Trop Med Int Health 1:581–589

Laird M (1981) Blackflies: the future for biological methods in integrated control. Academic, London

Lamberton P, Cheke RA, Walker M et al (2014) Onchocerciasis transmission in Ghana: biting and parous rates of host-seeking sibling species of the *Simulium damnosum* complex. Parasit Vectors 7:1–23

Lamberton P, Cheke RA, Winskill P et al (2015) Onchocerciasis transmission in Ghana: persistence under different control strategies and the role of the simuliid vectors. PLoS Negl Trop Dis 9:1–27

Lamberton P, Cheke RA, Walker M et al (2016) Onchocerciasis transmission in Ghana: the human blood index of sibling species of the *Simulium damnosum* complex. Patasit Vectors 9:1–19

Luz S, Crainey JL, Shelley AJ et al (2014) Outstanding insecurities concerning the use of an Ov16-based ELISA in the Amazonia onchocerciasis focus. Mem Inst Oswaldo Cruz 109:506–508

Manson P (1893) Disease of the skin in tropical climates. In: Davidson A (ed) Hygiene and diseases of warm climates. Young J. Pentland, Edinburgh and London

Medeiros JF, Almeida TA, Silva LB et al (2015) A field trial of a PCR-based *Mansonella ozzardi* diagnosis assay detects high-levels of submicroscopic *M. ozzardi* infections in both venous blood samples and FTA card dried blood spots. Parasit Vectors 8:1–8

Molyneux DH (2009) Filaria control and elimination: diagnostic, monitoring and surveillance needs. Trans R Soc Trop Med Hyg 103:338–341

Molyneux DH (2015) Eradication and elimination: facing the challenges, tempering expectations. Int Health 5:299–301

Molyneux DH, Ward SA (2015) Reflections on the Nobel Prize for Medicine 2015—the public health legacy and impact of avermectin and artemisinin. Trends Parasitol 31:605–607

Molyneux DH, Hopkins A, Bradley MH et al (2014) Multidimensional complexities of filariasis control in an era of large-scale mass drug administration programmes: a can of worms. Paras Vector 7:1–8

Morales-Hojas R (2009) Molecular systematics of filarial parasites, with an emphasis on groups of medical and veterinary importance, and its relevance for epidemiology. Infect Genet Evol 9:748–759

Morales-Hojas R, Cheke RA, Post RJ (2006) Molecular systematics of five *Onchocerca* species (Nematoda: Filarioidea) including the human parasite, *O. volvulus*, suggest sympatric speciation. J Helminthol 80:281–290

Muller R (2002) Worms and human disease, 2nd edn. CABI, London and Wallingford

Murdoch M, Hay R, Mackenzie C et al (1993) A clinical classification and grading system of the cutaneous changes in onchocerciasis. Br J Dermatol 129:260–269

Orihel T, Eberhard M (1998) Zoonotic filariasis. Clin Microbiol Rev 11:366–381

PENDA (2013) http://www.who.int/apoc/en_apoc_strategic_plan_2013_ok.pdf

Post RJ, Adams Z, Shelley AJ et al (2003) The morphological discrimination of microfilariae of *Onchocerca volvulus* from *Mansonella ozzardi*. Parasitology 127:21–27

Post RJ, Crainey JL, Bivand A et al (2009) Laser-assisted microdissection for the study of the ecology of parasites in their hosts. Mol Ecol Resour 9:480–486

Post RJ, Onyenwe E, Somiari SAE (2011) A guide to the *Simulium damnosum* complex (Diptera: Simuliidae) in Nigeria, with a cytotaxonomic key for the identification of the sibling species. Ann Trop Med Parasitol 105:277–297

Post RJ, Cheke RA, Boakye DA et al (2013) Stability and change in the distribution of cytospecies of the *Simulium damnosum* complex (Diptera: Simuliidae) in southern Ghana from 1971 to 2011. Paras Vector 205:1–11

Prout W (1901) A Filaria found in Sierra Leone? Filaria Volvulus (Leuckhart). Br Med J 1:209–211

Ricciardi A, Ndao M (2015) Diagnosis of parasitic infections what's going on? J Biomol Screen 20:6–21

Robles R (1919) Onchocercose humaine au Guatemala produisant la cécité et "l'erysipele du littoral" (Erisipela de la costa). Bull Soc Pathol Exot 12:442–460

Rodríguez-Pérez MA, Unnasch TR, Real-Najarro O (2011) Assessment and monitoring of Onchocerciasis in Latin America. Adv Parasitol 77:175–226

Simonsen PE, Fischer PU, Hoerauf A, Weil GJ (2014) The filariases. In: Farrar J, Hotez P, Junghanss T, Kang G, Lalloo D, White NJ (eds) Manson's tropical diseases, 23rd edn. Elsevier Saunders, Philadelphia, pp 737–765

Sharp NAD (1926) A contribution to the study of *Onchocerca volvulus* Leuckart, with some observations on its prevalence in Nigeria. Trans R Soc Trop Med Hyg 19:382–388

Shelley AJ, Hernandez LM, Maia-Herzog M et al (2010) The Blackflies (Diptera: Simuliidae) of Brazil. Pensoft, Sofia

Soboslay PT, Weiss N, Dreweck CM et al (1992) Experimental onchocerciasis in chimpanzees. Antibody response and antigen recognition after primary infection with Onchocerca volvulus. Exp Parasitol 74:367–380

Suwa UF, Calero RB, Rubio M et al (2015) Continuing *Onchocerca volvulus* transmission in Homoxi region of the Amazonia onchocerciasis focus (2012-2013). Poster presentation at the 51° Congresso da sociedade Brasileira de Medicina Tropical, Centro de Eventos do Ceará, Fortaleza, Brazil, 16 June 2015

Tang THT, Lopez-Velez R, Lanza M et al (2010) Nested PCR to detect and distinguish the sympatric filarial species *Onchocerca volvulus, Mansonella ozzardi and Mansonella perstans* in the Amazon Region. Mem Inst Oswaldo Cruz 105:823–828

Taylor MJ, Hoerauf A, Bockarie M (2010) Lymphatic filariasis and onchocerciasis. Lancet 376:1175–1185

Taylor MJ, Hoerauf A, Townson S et al (2014) Anti-*Wolbachia* drug discovery and development: safe macrofilaricides for onchocerciasis and lymphatic filariasis. Parasitology 141:119–127

Thiele EA, Cama VA, Lakwo T et al (2016) Detection of Onchocerca volvulus in skin snip by microscopy and real-time polymerase chain reaction: implications for monitoring and evaluation activities. Am J Trop Med Hyg 94:906–911

Traoré S, Wilson MD, Sima A et al (2009) The elimination of the onchocerciasis vector from the island of Bioko as a result of larviciding by the WHO African program for Onchocerciasis control. Acta Trop 111:211–218

Traoré M, Sarr M, Badji A et al (2012) Proof-of-principle of onchocerciasis elimination with ivermectin treatment in endemic foci in Africa: final results of a study in Mali and Senegal. PLoS Negl Trop Dis 6:1–14

Trpis M, Wergin W, Murphy C (2001) Development of *Onchocerca volvulus* (Filarioidea: Onchocercidae) in the West African blackfly *Simulium yahense* (Diptera: Simuliidae) in Liberia. J Parasitol 87:1265–1272

Turner HC, Walker M, Lustigman S et al (2015a) Human onchocerciasis: modelling the potential long-term consequences of a vaccination programme. PLoS Negl Trop Dis 9:1–19

Turner HC, Walker M, Attah SK et al (2015b) The potential impact of moxidectin on onchocerciasis elimination in Africa: an economic evaluation based on the phase II clinical trial data. Parasit Vectors 8:1–12

Vlaminck J, Fischer PU, Weil GJ (2015) Diagnostic tools for onchocerciasis elimination programs. Trends Parasitol 31:571–582

Walsh JF, Ocran MH (1985) Mermithid parasitism of *Simulium damnosum* s.l. and the transmission of human onchocerciasis. Trans R Soc Trop Med Hyg 6:843–844

Wanji S, Amvongo-Adjia N, Koudou B et al (2015) Cross-reactivity of filariasis ICT cards in areas of contrasting endemicity of *Loa loa* and *Mansonella perstans* in Cameroon: implications for shrinking of the lymphatic filariasis map in the Central African region. PLoS Negl Trop Dis 9:1–24

WHO (1997) Bench Aids for the diagnosis of filarial infections. WHO, Geneva

Mansonelliasis 26

Jansen Fernandes Medeiros, James Lee Crainey, Felipe Arley Costa Pessoa, and Sérgio Luis Bessa Luz

Abstract

Mansonelliasis is caused by one of three species of filarial parasite from the genus *Mansonella*: *Mansonella ozzardi*, *M. perstans*, or *M. streptocerca*. Mansonelliasis caused by *M. streptocerca* is thought to be limited to the equatorial rainforest areas of West and Central Africa. The more broadly distributed *Mansonella perstans* parasite is thought to be infecting around 114 million people in continental Africa alone and to also occur in certain parts of South America. *Mansonella ozzardi* is, however, the principal cause of mansonelliasis in the New World and the only known cause in Central America and the Caribbean. Mansonelliasis infections are typically diagnosed by the identification of microfilariae, which can be found in the blood (at all times of day) of those infected with *M. perstans* and *M. ozzardi* and in the skin of those infected with *M. streptocerca*. The three parasites can be distinguished from each other and the microfilariae of other filarial parasites by their morphological features or by PCR. There is, presently, no universally agreed symptom-set for the clinical diagnosis of mansonelliasis or guidelines for its treatment. Infections are often regarded as either mostly or completely asymptomatic and there have thus been few studies investigating its treatment. Presently, there are no large-scale internationally coordinated programs for mansonelliasis control. DEC, ivermectin, and doxycycline have all been shown to clear *Mansonella* microfilariae from some (though usually not all) infected individuals.

Due to the intimate relationship among lymphatic filariasis, onchocercosis, mansonelliasis, and loiasis on diagnosis, distribution, and treatment, all four chapters need to be checked when solving problems related to any of them (CBM).

J.F. Medeiros
Fundação Oswaldo Cruz - Fiocruz Rondônia,
Rua da Beira, 7671, Lagoa, Porto Velho, Rondônia,
Brazil, 76812-245
e-mail: jmedeiro@gmail.com

J.L. Crainey • F.A.C. Pessoa • S.L.B. Luz (✉)
Leônidas e Maria Deane Institute (Fiocruz Amazônia), Rua Teresina, 476, Adrianópolis,
Manaus, Amazonas, Brazil, 69027-070
e-mail: james.lee@fiocruz.br;
felipe.pessoa@fiocruz.br; sergio.luz@fiocruz.br

C.B. Marcondes (ed.), *Arthropod Borne Diseases*, DOI 10.1007/978-3-319-13884-8_26

Doxycycline treatment may be curative and seems to have the fewest side effects; it also, however, requires the longest treatment course and has only been tested on *M. perstans*.

Keywords
Mansonella ozzardi • *Mansonella perstans* • *Mansonella streptocerca* • Blackflies • Biting midge vectors

26.1 Introduction

Mansonelliasis can be caused by one of three species of parasite from the genus *Mansonella* (Filarioidea, Spirurida, Onchocercidae): *M. ozzardi*, *M. perstans*, and *M. streptocerca* (Fig. 26.1). The distribution of these parasites has been difficult to establish as the symptoms they cause are not universally agreed upon, and parasite incidence surveys are invasive and more difficult to justify than similar surveys for more serious diseases (Shelley and Coscarón 2001; Muller 2002; Simonsen et al. 2011).

In a recent multi-locus sequence typing (MLST) analysis of the Onchocercidae filarial parasites, *Mansonella* parasites were shown to be more closely related to parasites (traditionally) classified as belonging to the Splendidofilariinae subgenus than was previously realized. This MLST analysis also cast doubt over the validity of the Onchocerinae subgenus, which the *Mansonella* genus has traditionally been placed in, but did not cast doubt over the validity of the *Mansonella* genus itself (Lefoulon et al. 2015). While this analysis only included two *Mansonella* species, its findings were in agreement with other recent morphological and molecular analyses of

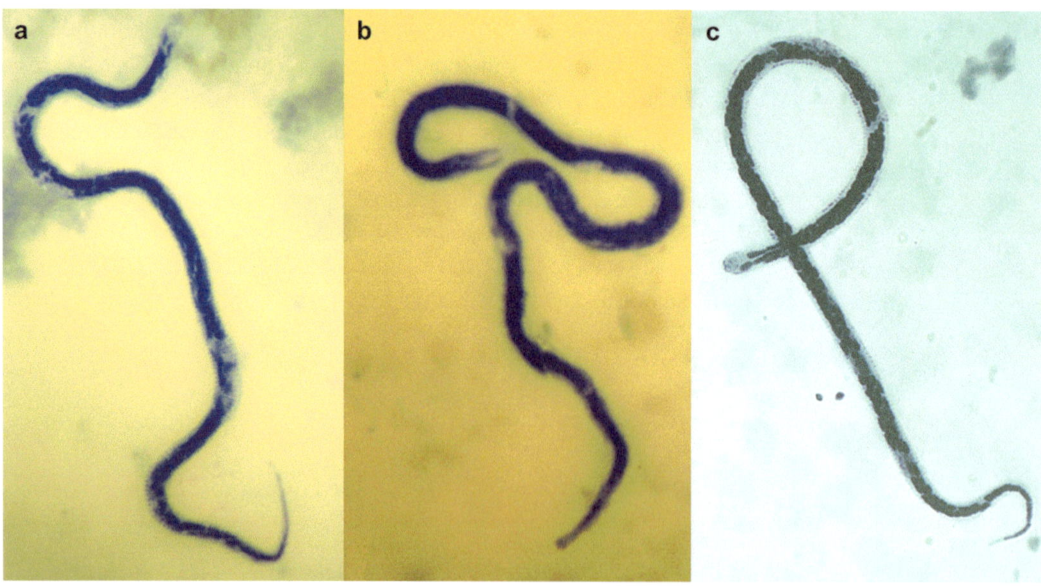

Fig. 26.1 Microfilariae of *Mansonella ozzardi*. (**a**) *Mansonella perstans* (**b**) and *Mansonella streptocerca* (**c**). *Source*: Fig. 26.1a: Medeiros et al. (2009) Acta Amaz 40:779–780. Fig. 26.1b: Medeiros and Py-Daniel (2009) Mansonelose. In: Marcondes CB (ed) Doenças transmitidas e causadas por artrópodes. Editora Atheneu, Rio de Janeiro. Fig. 26.1c: CDC: Center for Disease Control and Prevention

filarial parasites from the superfamily Filarioidea, which have provided strong support for the existence of *Mansonella* as a valid genus within the Onchocercidae family of filarial worms and also confirmed the close evolutionary relationship between the three mansonelliasis-causing parasites (Casiraghi et al. 2004; Ferri et al. 2009, 2011; Bain et al. 2015).

The origin of how the three mansonelliasis-causing *Mansonella* parasites came to infect humans is, however, far less clear. Based on the limited data available, *M. perstans* and *M. streptocerca* are both generally thought to have coevolved with their hosts from more primitive primate-infecting ancestors, whereas these same primate ancestors of humans are not thought to have been infected by the contemporary ancestors of *M. ozzardi* parasites. It is presently thought, instead, that present-day human *M. ozzardi* parasites have evolved from ancestors that were originally infecting mammals from either the Carnivora order or Sciuromorpha suborders and that present-day human parasitism started after a host switch. This has been postulated based on the morphological similarity of *M. ozzardi* and the *Mansonella* parasites *M. llewellyi* and *M. interstitium*, which infect raccoons and squirrels, respectively (Bain 2002; Morales-Hojas 2009; Bain et al. 2015).

The localization of adult *Mansonella* parasites appears to vary among the species, although it should be noted that only a small number of adult parasites (of any of the three species) have been recovered from humans, and it is not clear exactly where all those recovered from monkeys were living. The adults of *M. streptocerca* have been recovered from humans and been recorded in the dermis of the upper thorax and shoulders, whereas *M. perstans* and *M. ozzardi* adults have both been reportedly recovered from the peritoneal cavity. Adult female *Mansonella* parasites are viviparous and can release large numbers of microfilariae until they die. In the case of *M. perstans* and *M. ozzardi*, these microfilariae are released into and found in the circulating blood; *M. streptocerca* microfilariae are found subcutaneously. The larvae of human *Mansonella* parasites develop from microfilariae through three stages inside an insect vector. The exact insect vector species they infect depends on the geography and on the species of *Mansonella* parasite concerned; so far only simuliid blackflies and ceratopogonid midges (both from the order Diptera) have been successfully incriminated in *Mansonella* parasite transmission (Meyers et al. 1978; Orihel and Eberhard 1998; Shelley and Coscarón 2001; Muller 2002; Simonsen et al. 2011).

The clinical symptoms of *Mansonella* parasite infections are ill defined, but some, if not most, infections appear to be asymptomatic. Pruritus is commonly reported to be associated with infections from all three parasites; fevers, articular pains, headaches, and various ocular pathologies are commonly attributed to *M. perstans* and *M. ozzardi* infections. *Mansonella ozzardi* infections are associated with ocular lesions and *M. perstans* have been reported to be associated with conjunctiva nodules. Dermatitis of the shoulders and thorax seems to be a common symptom of *M. streptocerca*, but does not appear to be associated with the other *Mansonella* parasite infections. Numerous other species-specific symptoms have also been described (and are discussed in more depth below). In addition to these, there is also the distinct possibility that mansonelliasis may cause other (even severe) pathologies that are yet to be characterized. For example, *Mansonella* parasite infections in East African are known to cause "bung-eye" [sic]—a severe ocular pathology characterized by yellowish (worm containing) nodules in the bulbar conjunctiva (and which can often cause edema of the eye lids and sometimes proptosis). Whether *M. perstans*, *M. streptocerca*, a hitherto uncharacterized *Mansonella* human parasite, or a zoonotic *Mansonella* parasite infection is responsible for this pathology is still to be determined. Definitively established knowledge of the pathology of *Mansonella* parasitism is, in comparison with other filarial diseases, scarce and resolving this problem, arguably, represents the most pressing issue for *Mansonella* research today. At the moment, existing knowledge of the pathology of human *Mansonella* parasitism is too poor to feasibly permit an accurate assessment of the disease's global burden (GB), and this may well be resulting in disproportionate neglect of the disease. For example, in a recent, GB of disease study commissioned by the major tropical disease funding body the "Bill and Melinda Gates Foundation," mansonelliasis was omitted entirely from the 291 individually named diseases that were assessed for their relative importance for funding prioritization (Muller 2002; Murray et al. 2012; Simonsen et al. 2011; Vianna et al. 2013).

On a practical level, the medical importance of *Mansonella* parasites is very often regarded as limited to how they can adversely affect disease control programs that target other filarial parasites. The misidentification of *M. streptocerca* as *O. volvulus* led to hundreds of Ugandans receiving ivermectin without a clinical justification (Fischer et al. 1997). The misidentification of *M. ozzardi* infections as *O. volvulus* has resulted in the false reporting of "new" onchocerciasis foci, and, more recently, possible cross-reactivity between *M. perstans* sera and the ICT cards used in *W. bancrofti* detection has raised questions about the accuracy of modern lymphatic filarial distribution maps (Shelley 2002; Wanji et al. 2015). As all three human *Mansonella* parasites have broad distributions that overlap extensively with other filarial parasites, this medical importance is not trivial. However, the view that *Mansonella* parasite infections themselves are largely benign is clearly the most prevalent view among public health policy makers as *Mansonella* parasites are not presently the targets of any major intentional control programs. Mansonelliasis is not included in the World Health Organization's (WHO's) list of the 17 most neglected tropical diseases, despite the fact that 114 million Africans have been estimated to be infected with *M. perstans* alone (Shelley and Coscarón 2001; Muller 2002; Post et al. 2003; Tang et al. 2010; Simonsen et al. 2011; Luz et al. 2014).

Most diagnoses of *Mansonella* parasite infections are performed using light microscopy and parasite morphology. At present, there are no commonly used immunology-based assays (like ELISAs) for the detection of *Mansonella* parasites or for the detection of *Mansonella* parasite exposure. PCR assays for the reliable and accurate diagnosis of *Mansonella* infections have been developed, but are not yet routinely used for *Mansonella* incidence assessments. Data from the few PCR surveys that have been conducted, however, suggest that traditional microscopy-based surveys of *Mansonella* incidence estimates could be dramatically underestimating the true levels of population parasitism and thus that PCR should be more widely used for this purpose (Tang et al. 2010; Simonsen et al. 2011; Marcos et al. 2012; Medeiros et al. 2015).

26.2 *Mansonella ozzardi*

The first published report of *M. ozzardi* is probably that of Fredrick P. Henry in the 1896 Proceeding of the Academy of Natural Science of Philadelphia. In this report Henry records a new species of filaria (*Filaria demarquayi*) discovered by the British parasitologist Sir Patrick Manson, in a similar way in which Manson himself first reported Leuckart's discovery of *Onchocerca volvulus* in 1892 (Manson 1892; Henry 1897). Manson's subsequent description of *M. ozzardi*, which was published in 1897, reported *M. ozzardi* from microfilariae observed in the peripheral blood of indigenous people living in British Guiana and the Caribbean island of Saint Vincent (Manson 1897). Manson initially thought the parasites he observed from these two localities were different species and recorded them as such: *F. demarquayi* and *F. ozzardi*. As early as 1899, however, it became apparent that Caribbean *F. demarquayi* (which Manson was then referring to as *F. demarquaii*) were in fact the same species as the South American *F. ozzardi*. Despite the fact (even by Manson's own reckoning) the name *F. demarquayi* had precedence over the name *F. ozzardi*, the species became known as "*Filaria ozzardi*" (ostensibly because of a misidentified adult specimen) and then as *Mansonella ozzardi*, following several genus reassignments (Manson 1917; Orihel and Eberhard 1982; Bain 2002; Morales-Hojas 2009; Bain et al. 2015).

26.2.1 Parasite Morphology

26.2.1.1 Adult Stage

Very few adult specimens of *M. ozzardi* have been recovered and described. To remove the possibility of our adult *M. ozzardi* description being corrupted by misidentified *M. perstans* specimens, our account of *M. ozzardi* adults, given here, is based solely on adult *M. ozzardi* parasites obtained from experimental monkey infections established with *M. ozzardi* microfilariae obtained from the blood of humans living in the Caribbean (where no other sheathless filarial blood parasites have ever been recorded). The anterior ends of adults are smoothly rounded and have annular swellings, and the bodies of adult worms are otherwise uniform in diameter until their terminal tail ends, which are tapered and have four papillae. The adults lack precloacal papillae and have regressed buccal capsules leading to simple thin tubular esophagi. The reproductive systems of both males and females are recorded to overlap with the esophageal region (Orihel and Eberhard 1982; Bain et al. 2015).

Adult females are described with bodies that are approximately 32.2–61.5 mm in length and up to 0.13–0.16 mm in diameter. Nerve rings are recorded to be 0.22–0.28 mm proximal of the anterior end of the adult worms; the vulvas are recorded to be 0.52–0.66 mm. The first four most distal of female annular swellings are recorded as being the most prominent. Approximately 5 % of the female body is filled by highly coiled ovaries; large ovoid ovejectors and long muscular vaginas are also described for the species. Paired uterine branches and tails between 0.17 and 0.25 mm are reported features of the females; the terminal papillae of females are described as lobed (Orihel and Eberhard 1982; Bain et al. 2015).

Adult male worms are recorded to have body lengths between 24 and 28.4 mm and to be 0.07–0.08 mm in diameter. Males are also recorded to have 11 pericloacal papillae (with three more papillae immediately distal to them), esophagi between 0.77 and 1.02 mm in length, and nerve rings 0.17–0.2 mm proximal to their anterior ends. The tail regions of the male worms are recorded to be between 0.12 and 0.15 mm in length and to be flattened dorsoventrally (Orihel and Eberhard 1982; Bain et al. 2015).

26.2.1.2 Microfilariae Stage

The microfilariae of *M. ozzardi* (Fig. 26.1a) lack the sheath which envelopes the microfilariae of many other species belonging to other genera of filariae. They have been recorded as measuring between 149 and 254 μm in length and between 0.5 and 5 μm in width and as having long, slender, anucleate tails. The cephalic head space has been measured to range from between 3.8 and 6.7 μm and has been recorded to contain a thread running through it that can be made visible with alcian blue staining. *Mansonella ozzardi* microfilariae tails have been measured to range from around 3.1–8.4 μm in length and up to 2.4–3.4 μm in width. Tails have been measured as thin as 0.5 μm at a point close to their tips. The nerve rings of *M. ozzardi* microfilariae are narrow (between 2.4 and 5.3 μm in width) and often do not have the distinct anucleate region that is typical of filarial parasite microfilariae (Laurence and Simpson 1968; WHO 1997; Post et al. 2003; Bain et al. 2015).

26.2.1.3 Larval Stage

The larval stages of *M. ozzardi* develop inside their insect vector host in three distinct stages separated by molts (Fig. 26.2). The larvae of *M. ozzardi* are capable of developing in a diverse range of ceratopogonid midge and simuliid blackfly hosts, but seem to show similar patterns of development in both. The first sausage-stage (L_1) (Fig. 26.2a) larvae have pointed tails. Larvae recovered 3 days postinfection: measure between 91.3 and 202.9 μm in length and between 11.8 and 22 μm in width; have tails between 6 and 15 μm in length; have nerve rings that are around 26 μm proximal to their anterior ends and have excretory cells that are about 34 μm proximal to the same point. Late developing L_1 larvae (recovered after 6 days postinfection) are recorded as measuring between 229.7 and 349.3 μm in length and between 13.1 and 21.3 μm in width. The second larval stage (L_2) (Fig. 26.2b) have a more conventional worm shape, with a slightly bulbous

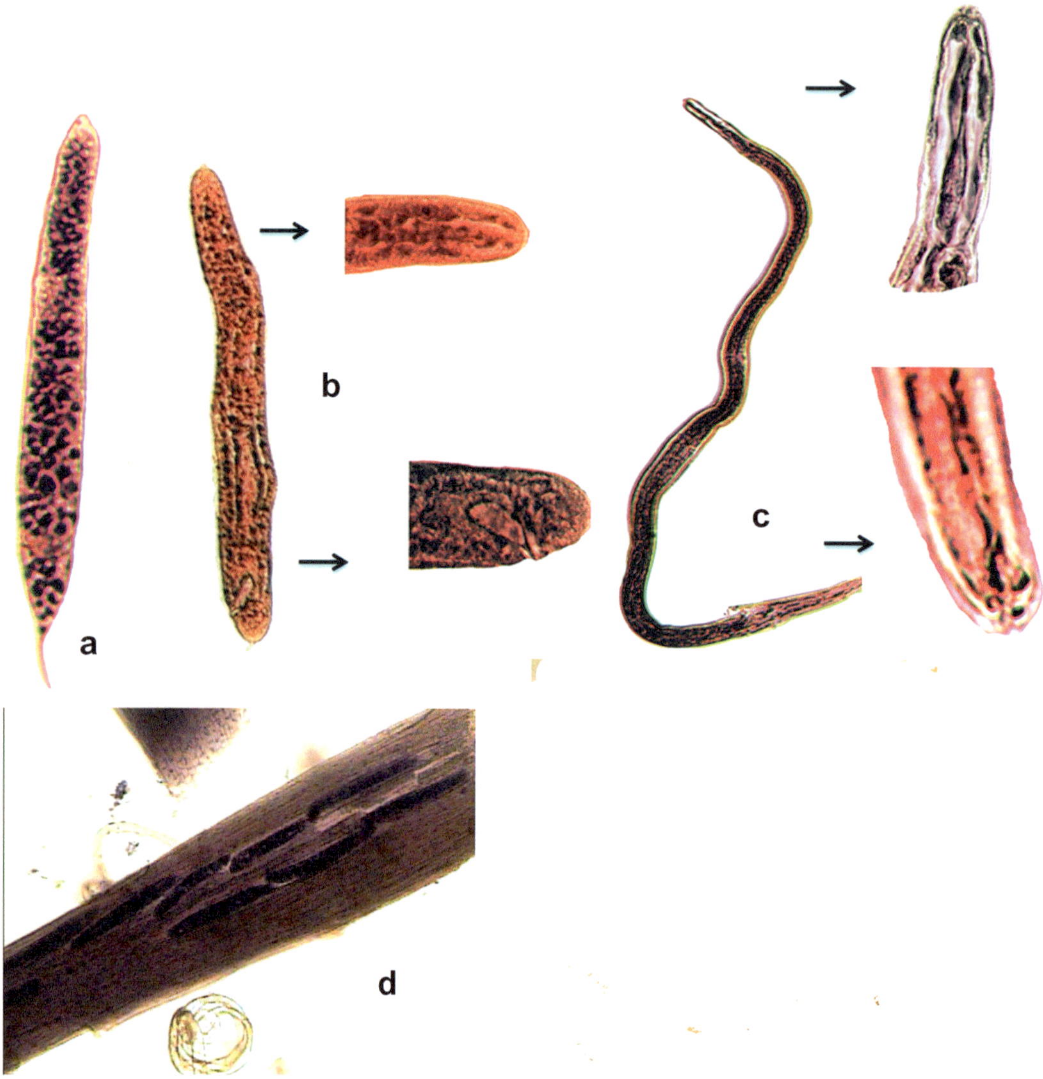

Fig. 26.2 Larval stages of *Mansonella ozzardi*. (**a**) L_1 larva. (**b**) L_2 larva showing some internal structure developing. (**c**) L_3 with some internal structures developed. (**d**) Dissected thoracic muscle of black flies with some L_2 larvae. *Source*: Medeiros and Pessoa

posterior end. L_2 larvae are recorded as measuring between 235 and 564 μm in length and 16.4–28 μm in diameter. Nerve rings and excretory cells measure approximately 38–59 μm and are 50–81 μm proximal to the anterior tips of the larvae; esophagi have been reported to be between 161 and 301 μm in length and their thin-walled intestines between 77 and 138 μm. L_2 rectums are recorded to measure between 17 and 40 μm; L_2 larval tails have been recorded to be between 33 and 43 μm long. A part of the L_1 cuticle can often be seen still attached to posterior region of the anal plug of L2 larvae, following the first larval molt; following the second molt the anal plug and the cuticle are shed completely. Infectious stage L_3 (Fig. 26.2c) larvae have been recorded to have long bodies measuring between 470 and 846 μm with fairly uniform diameters measuring between 9.6 and 23 μm in width, with tapered ends. Nerve rings are reported to be located 60–75 μm proximal to the anterior end of larvae. Esophagi have been recorded to be between 338 and 563 μm; the rectum of L_3 larvae is slightly wider than the intestine (which measures between 90 and 123 μm) they are recorded measuring 18–32 μm in length. L_3 tails are between 35 and 59 μm long and 8.9–13 μm in diameter and have bifid terminal ends. The male genital primordium is in the posterior end of the worm 22–58 μm proximal to the esophageal-intestinal junction. The female genital primordium is believed to be located 92–161 μm proximal to the anterior end of larvae. The posterior end of L_3 larvae also has four caudal papillae. To the best of our knowledge, no significant difference in the development of larvae within different vector species has yet been recorded; although the size that L_3 larvae can grow to seems to vary among species (Shelley et al. 1980; Lowrie et al. 1982; Tidwell et al. 1980; Tidwell and Tidwell 1982; Yarzabal et al. 1985).

26.2.2 Transmission Cycle

The development of *M. ozzardi*, through three larval stages (L_1–L_3), inside their insect hosts is synchronous (Fig. 26.3); both ceratopogonid biting midges and simuliid blackflies (Fig. 26.4) are known to support the development of infectious stage L_3 larvae. Larval stages of the *M. ozzardi* have also been observed to develop in some mosquito (Culicidae) species, although not in all the species that have been tested and larvae have never been observed to develop through to the infectious L_3 stage in mosquitoes. In several feeding trials, where black flies and biting midges were fed on the same volunteer, black flies (simuliids) were seen to ingest more microfilariae than the tested *Culicoides* species. In simuliid hosts, microfilariae migrate from the gut and have been detected in the thorax of simuliids as soon as 4 h

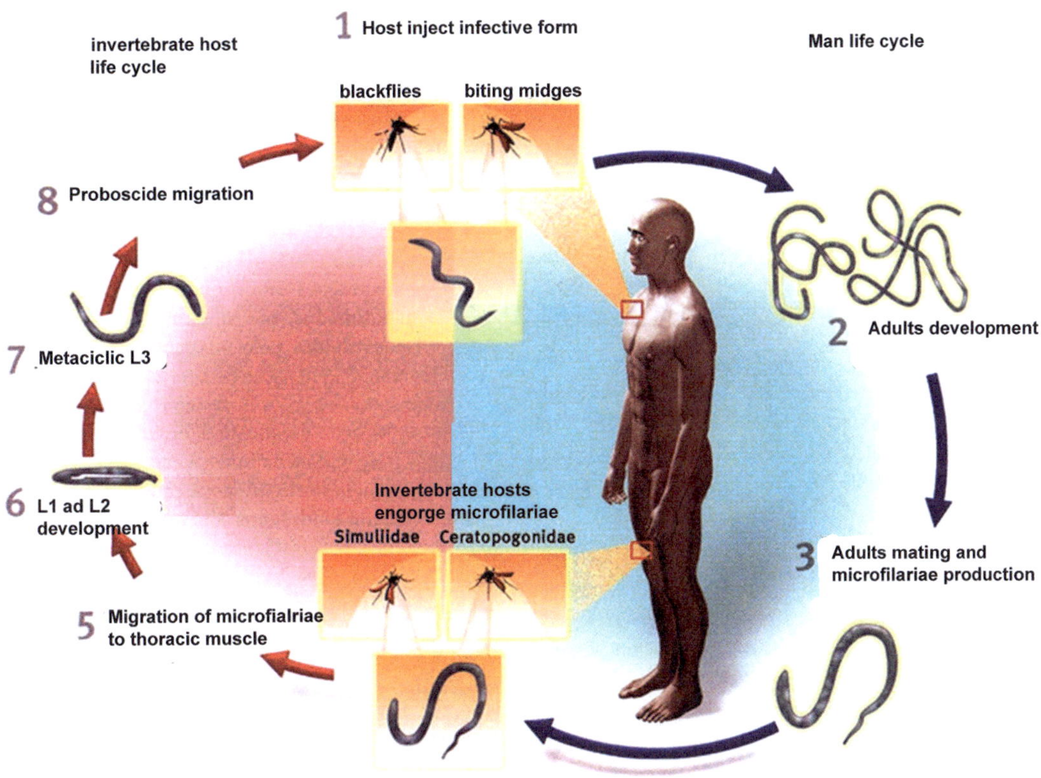

Fig. 26.3 Life cycle of *Mansonella* sp. *Source*: Modified from Medeiros et al. (2006) Ciência Hoje 38:68–70

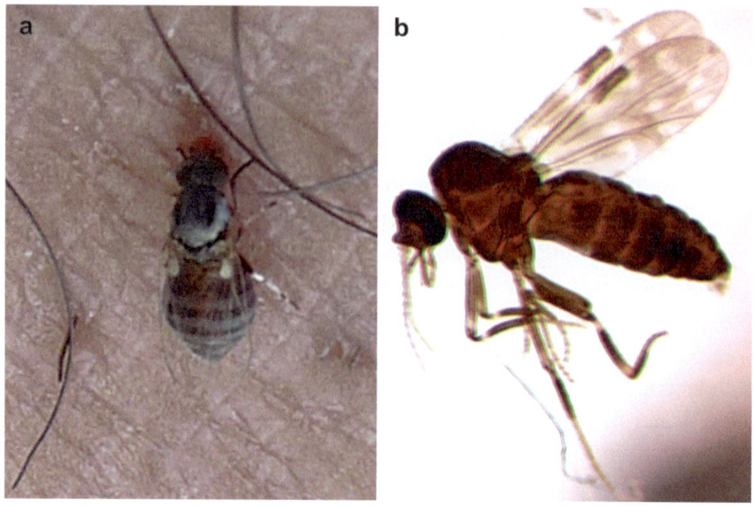

Fig. 26.4 Simuliid black flies (Diptera: Simuliidae) (**a**) and ceratopogonid midges (Diptera, Ceratopogonidae) (**b**), vectors of *Mansonella ozzardi*. *Source*: Fig. 26.4a: Medeiros and Pessoa. Fig. 26.4b: Medeiros et al. (2006) Ciência Hoje 38:68–70

after a blood meal. After the microfilariae have entered their development site, they lose their motility and develop into the sausage-stage (L_1) larvae. These L_1 larvae have been recorded between 2 and 5 days after blood meals in multiple species of both *Simulium* and *Culicoides* and are typically found developing aligned to the thoracic muscle fibers used for flight (See Fig. 26.2d). L_2 phase larvae are normally found 4–5 days after a blood meal (in both *Culicoides* and *Simulium* vectors) and usually in the thorax. The large L_3 infectious stage larvae often begin to appear 5 days after blood-feeding in both *Culicoides* and *Simulium* vectors and after 9 days L_3 larvae can be found in the heads and mouth parts of *Simulium*, *Culicoides* and *Leptoconops* vector species (Shelley et al. 1980; Tidwell et al. 1980; Lowrie et al. 1982, 1983; Tidwell and Tidwell 1982; Yarzabal et al. 1985; Shelley and Coscarón 2001).

Very little is known of how *M. ozzardi* L_3 larvae entering human blood develop into adults and mate, it is not even clear at this stage exactly where in the human body the adult parasites persist. Both adult male and female *M. ozzardi* have been recovered from African patas monkeys (*Erythrocebus patas*) experimentally injected with L_3 *M. ozzardi* microfilariae. Female adult forms develop from infectious L_3 larvae and begin to produce and release microfilariae into circulating blood (at levels detectable in 1 ml samples of blood) after 149 days. In the patas monkey the levels of circulating microfilariae show variable patterns of progress before reaching a peak of about 15 microfilariae per 20 μl of blood. Typically, microfilariae levels rise rapidly (by over 50-fold) in the 3 weeks immediately after patency. In some artificially infected patas monkeys, this peak can be maintained for 16 weeks, but in others it declines. The *M. ozzardi* life cycle is completed when a competent insect vector blood feeds on an infected human and the microfilariae enter its guts (Orihel et al.1981; Muller 2002).

26.2.3 Clinical Presentation

The pathology of *M. ozzardi* is ill-defined. Mansonelliasis caused by *M. ozzardi* parasites is often regarded as mostly asymptomatic with studies reporting between ~30 and 60 % of those infected as such. Although it is generally agreed that some individuals infected with *M. ozzardi* are adversely affected by *M. ozzardi* infections, there is no universally agreed set of symptoms that define the clinical presentation of *M. ozzardi* infections. Articular pains, headaches, corneal lesions, "coldness in legs," pruritus, dizziness, fevers and eosinophilia are all frequently attributed to *M. ozzardi* infections; however, many of these symptoms are common to many other tropical diseases, like malaria, which are often also found where *M. ozzardi* is endemic. Some ocular pathologies, which appear to occur more rarely, like, for example, corneal lesions, are now quite well established as pathologies that are associated with *M. ozzardi* infections; however, these pathologies do not seem to occur everywhere *M. ozzardi* infections do. Similarly, while epidemiological studies have shown that many of the above-mentioned symptoms are positively correlated with *M. ozzardi* parasitism, different epidemiological studies have reported apparently conflicting results. For example, one study in the Brazilian Amazon reported that articular pains, "coldness in legs" and fever are all significantly positively correlated with *M. ozzardi* parasitism, but that headaches and lymphadenopathy were not. However, a study conducted nearby in another part of the Brazilian Amazon found that articular pains, headaches and lymphadenopathy were all significantly positively correlated with *M. ozzardi* parasitism and "coldness in legs" was not. One possible explanation for these types of inconsistencies in findings is that there are differences in sensitivity of parasitism misdiagnosis caused by differences in the abundance of sub-patent infections. While both studies used the same light-microscopy method of diagnosis, a recent PCR survey has shown the existence of high-levels of sub-patent (sub-microscopic) *M. ozzardi* infections in the Brazilian Amazon region, which have hitherto not been considered in *M. ozzardi* epidemiological studies (Martins et al. 2010; Vianna et al. 2013; Adami et al. 2014; Medeiros et al. 2015).

26.2.4 Diagnosis

Clinical diagnosis of *M. ozzardi* parasitism is focused on microfilariae in the circulating blood and mostly commonly performed using light-microscopy-based methods. Blood slide preparations and light-microscopy can be used, in combination with species-specific morphological characters and staining to definitively identify *M. ozzardi* parasitemias. Morphologically, the unsheathed microfilariae of *M. ozzardi* can be distinguished from other microfilariae by their anucleate tails and/or alcian blue-staining of their cephalic armature. Microfilarial levels in the circulating blood can, however, sometimes be too low to be easily detected with a simple blood film and so parasite concentrating methods are sometimes used to improve the sensitivity of light-microscope based detection. The Knott and polycarbonate membrane filter methods are widely used to resolve this problem; however, these methods can distort some morphological structures of *M. ozzardi* microfilariae and compromise identifications (WHO 1997; Medeiros et al. 2010). Although a reliable and highly sensitive PCR method of *M. ozzardi* identification has been developed and can be used for the reliable identification of *M. ozzardi* parasites in vector and blood samples, this technique is still not widely used (Tang et al. 2010; Medeiros et al. 2015). Immunology-based molecular diagnostic tools, like ELISAs, which are routinely used for the study of other filarial parasites like *O. volvulus*, are still to be developed for *M. ozzardi* diagnostics (Post et al. 2003; Alhassan et al. 2014; Luz et al. 2014).

26.2.5 Treatment

There is presently no known cure for *M. ozzardi* infections, no recommended prophylactic treatment for preventing infections and only limited treatment options.

Chemotherapeutic mass-treatment programs for the control of *M. ozzardi* have been designed and conducted in a diverse range of settings including the Caribbean, the Amazon and Argentina and have typically used ivermectin. From these studies it is clear that: a single dose of ivermectin delivered orally (in tablets) at a dose of between 0.15 and 0.2 mg per kilogram of body weight is highly effective at removing circulating *M. ozzardi* microfilariae from the blood system; that the drug takes affect within 24 h, and that the drug's effect last for at least 12 months. Similar results have been obtained with onchocerciasis treatment regimes. In the case of the *O. volvulus* parasites, ivermectin seems to work by killing microfilariae and temporarily sterilizing the adult females that are responsible for producing them. Consistent with the notion that the drug operates similarly on *M. ozzardi* parasites, circulating parasites are usually detectable in around 18–36 % of *M. ozzardi* parasitized patients tested 3–4 years post ivermectin treatment, although it is important to note that circulating parasites detected more than 5 months after treatment could derive from re-infection events. The efficacy of the filaricide diethylcarbamazine (DEC) has also been investigated for its utility for *M. ozzardi* parasitemia treatment. In a single study conducted in Trinidad, a dose of 6 mg per kg of body weight of DEC (delivered monthly for a year) as part of an effective *Wuchereria bancrofti* control program, was shown to have little impact on *M. ozzardi* parasitism (Chadee et al. 1995; Shelley and Coscaron 2001; Muller 2002; Basano et al. 2014).

Although at present there is no known curative treatment for *M. ozzardi* infections, *M. ozzardi* parasites have been shown to harbor the bacterial endosymbionts *Wolbachia* that are also present in most other filarial worms and which have been successfully targeted for filarial worm treatments. In a randomized control trial using the antibiotic doxycycline (given daily for 6 weeks), similar microfilaricide results were obtained as have been obtained with ivermectin. This study, however, also showed that in ~3 % of *M. perstans* parasitism cases, *Wolbachia*-free *M. perstans* microfilariae persisted in circulating blood following treatment showing the endosymbiont is not necessary for *M. perstans* survival and thus potentially not as good a *Mansonella* drug target for treatment as it is for other filarial worm infections. At present even the prevalence of *Wolbachia* within *M. ozzardi* is not established and thus the utility of antibiotics for curing or treating *M. ozzardi* infections is still completely unknown. As doxycycline is already used as a malarial prophylactic in areas where *M. ozzardi* is present, it is entirely possible that it is already acting as a *M. ozzardi* prophylactic too, although to the best of our knowledge its effectiveness as such, or indeed the effectiveness of other antibiotics that could be working similarly, has never been investigated (Casiraghi et al. 2001; Coulibaly et al. 2009; Phillips et al. 2014).

26.2.6 Epidemiology and Control

Mansonella ozzardi is usually only recorded to occur in the New Word, where it is known to have a broad distribution spanning from Mexico to Argentina. *Mansonella ozzardi* is found in many parts of the Amazon rainforest and several Caribbean islands including Desirade Island, Dominica, Guadeloupe, Hispaniola, Martinique, St Lucia, St Vincent, Trinidad, and probably St Kitts. There are also credible, albeit long-forgotten, reports of *M. ozzardi* occurring in Papua New Guinea, and thus, the geographical range of this parasite, and indeed mansonelliasis, may extend well into the Eastern Hemisphere (Shelley and Coscarón 2001; Muller 2002; Crainey et al. 2016).

From a global perspective, the *M. ozzardi* parasites found in the arc of Caribbean islands running from Trinidad to Hispaniola appear to be spread exclusively by ceratopogonid biting midges, whereas in continental South America *Simulium* blackflies and *Culicoides* biting midges both seem to play a role. Multiple vector species are often involved in transmission at any given foci and extrapolating dynamics from one focus to the next is often unsafe. For example, while simuliids appear to have a greater importance in the transmission of *M. ozzardi* in the Amazon region than *Culicoides* species do, in Argentina this situation appears to be reversed. Parasite development within both *Culicoides* and *Simulium* vector species appears to be temperature sensitive and also transmission rates appear to be affected by vector population densities, which fluctuate with climatic conditions, suggesting that in some foci there may be a seasonal aspect to *M. ozzardi* epidemiology (Shelley et al. 1980; Tidwell and Tidwell 1982; Shelley and Coscarón 2001).

In comparison with other filarial diseases, *M. ozzardi* epidemiology is both very complex and very poorly understood, although its epidemiology is perhaps better understood than the epidemiology of the other mansonelliasis causing parasites. In Haiti, it has been noted that *M. ozzardi* parasitism maps closely to the coastal breeding sites of its known vectors. Similarly, but in contrast, it has also been noted that the distribution of mansonelliasis in the Amazon rainforest seems to follow the riverine breeding sites of the Amazon region's simuliid vectors of *M. ozzardi*, in a similar way that onchocerciasis maps to riverine breeding sites of the simuliid vectors of *O. volvulus* in both Africa and Latin America.[1] In both cases, however, these simple relationships betray the more complex vector transmission dynamics that lie beneath them. In Haiti, one coastal vector *Leptoconops becquaerti* breeds in the white calcareous sands of tourist beaches, whereas the two other known vectors breed nearby in black (*C. furens*) and red (*C. barbosai*) tidal mangroves. All three vector species appear to also have varying biting habits and vector competencies. In the Brazilian Amazon although all the known vectors share similar larval breeding sites, the behavioral ecology of the known Amazonian simuliid vectors of *M. ozzardi* is so diverse that the region's mansonelliasis epidemiology seems to be no less complex than that of the Caribbean (Lowrie et al. 1982, 1983; Lowrie and Raccurt 1984; Shelley 1988).

The WHO describes *M. ozzardi* as aperiodic; however, the evidence for aperiodicity being a universal feature of *M. ozzardi* parasitism is weak and disputed. After monitoring multiple circulating blood parasite levels over time Nathan et al. (1978) concluded that blood parasitism levels are aperiodic; however, studying the same data set, Pichon (1983) found evidence for cryptic periodicity in some of those parasitized. How this fascinating observation may be associated with *M. ozzardi* epidemiology has been left largely unexplored. For other filarial parasites, blood periodicity has been recorded and shown to be positively correlated with vector biting habits. One possible explanation is thus that this asynchronous periodicity is a reflection of the diverse and varying biting habits of *M. ozzardi* vectors. For example, in contrast to the day-biting habits of most simuliid vectors of *M. ozzardi*, most of the *Culicoides* species that have been studied seem to bite throughout the night (with dawn and dusk biting peaks). The levels of parasites circulating in the blood vary during the course of a *M. ozzardi* infection and vary between infected individuals as does the course of infection; how human immunology and other factors may be effecting this is also not well known nor is how long a *M. ozzardi* infection last. Several demographic risk factors have been associated with *M. ozzardi* infections. It is known, for example, that those who have outdoor working professions (like fisherman and farmers) in the Amazon region (where simuliids seem to be the principal vectors) are at greater risk of being infected than those who do not. Age is also strongly positively correlated with infection status in a diverse range of settings. The cause of this is unknown, but is often attributed to the oldest residence having the greatest risk of exposure to infectious vector bites. Parasite prevalence levels are also known to fluctuate dramatically across time and space. Parasite incidence estimates in continental South America are usually much higher than estimates for the Caribbean islands (Nathan et al. 1978; Pichon 1983; WHO 1997; Medeiros et al. 2008, 2009; Martins et al. 2010; Adami et al. 2014).

26.3 *Mansonella perstans*

Mansonella perstans was the first of the mansonelliasis causative parasites to be discovered, and, like *M. ozzardi*, was discovered by Sir Patrick Manson. Like the other two mansonelliasis causing *Mansonella* parasites species, it was first described using the morphological features of its microfilariae; in the case of *M. perstans*, however, these descriptions came from microfilariae found in the blood of indigenous West Africans. The parasite was initially named *Filaria sanguinis hominis minor*. Later it was renamed as *Filaria sanguinis hominis perstans* and then *Filaria perstans*. In the following years, the placement of the species in the genus was

[1]Onchocerciasis is more commonly known as "river blindness" because of the most severe ocular pathology it can cause and its epidemiological link to rivers.

changed a number of times, which also resulted in a series of renaming: *Acanthocheilonema perstans*, *Tetrapetalonema perstans* followed by *Dipetalonema perstans*. After a review of the genus *Mansonella* in the 1980s, the presently used name, *M. perstans* (Fig. 26.1b), became adopted (Eberhard and Orihel 1984; Muller 2002; Bain et al. 2015).

26.3.1 Morphology

Generally, the morphological features and morphogenesis of *M. perstans* are similar to what has been described for *M. ozzardi*. Adult worms recovered from indigenous Africans from various parts of Africa (where *M. ozzardi* does not occur) are recorded as white, cylindrical, thread-like and without a striated cuticle. Females measure between 50 and 80 mm in length and between 80 and 120 μm in diameter; males appear to be smaller with lengths of between 35 and 45 mm and diameters ranging from 50 to 60 μm; adult females have been reported to have bifurcated tails. The microfilariae lack the sheath typical (Fig. 26.1b) of many other human filariae and have been recorded to range between 155 and 225 μm in length and 4 and 5 μm in diameter. The cephalic space is small and features a double row of nuclei that reach the end of the tail, which is blunt. The sausage-stage L_1 larvae have been recorded as about 80 μm in length and about 15 μm in diameter. The L_2 larvae are ill defined and appear to have a similar form to those of *M. ozzardi*. Infectious stage L_3 larvae are recorded as measuring between about 600 and 900 μm in length; with blunt cleft miter shaped tails and four posterior papillae (Sharp 1928; Eberhard and Orihel 1984; WHO 1997; Muller 2002; Simonsen et al. 2011; Bain et al. 2015).

26.3.2 Transmission Cycle

Mansonella perstans is often reported infecting great apes like chimpanzees and gorillas. However, direct experimental evidence and molecular genetic evidence to show that the same *M. perstans* that infect humans are also able to infect apes is still lacking. Certainly, *M. perstans* is endemic in parts of Africa where infected apes are not known or likely to exist and so even if under some circumstances apes are involved in *M. perstans* transmission they do not appear to be a necessary part of the parasite's life cycle. Like all other human *Mansonella* parasites, little is known about the morphogenesis process that turns infectious L_3 stage larvae (that enter the blood through the bite of an infected *Culicoides* vector) into adults (Muller 2002; Simonsen et al. 2011).

Adult *M. perstans* are rarely detected, but have been recovered during autopsies. Adult *M. perstans* parasites are reportedly found within the abdominal cavity, pericardial, pleural and peritoneal, as well as in retroperitoneal, perirenal tissues and the subcutaneous tissues. In contrast to *M. ozzardi*, there have been no animal model experiments and so how long it may take for an adult female to mate and begin releasing microfilariae after entering the blood as an infectious L_3 larvae is unknown. Blood-circulating *M. perstans* microfilariae that are ingested by *C. austeni* become very active for the first few minutes; after 6 h they are much less so and can be found in the stomach wall; by about 24 h, ingested microfilariae have typically lost 1/3 of their length, have taken the shape of a porcupine quill and the visibility of distinct internal nuclei is lost. On the third day postinfection, the sausage-stage L_1 larvae are typically visible. L_2 stage larvae are typically observable 4–6 days postinfection. From day 8 onwards, mobile L_3 infectious stage larvae are detectable in the insect vectors' heads and mouthparts. Infectious larvae appear to be able to exit their *Culicoides* vectors through the proboscis or by puncturing the labrum membrane (See Fig. 26.5). Sharp (1928) provided detailed accounts of this process in which a growing mass of wriggling worms can be seen to form a bulge from the labrum membrane before the L_3 larvae escape from their *Culicoides* vectors, and also reported punctured labrum from previously infected flies, which were not offered further blood meals suggesting there may be only a limited window of time during which vectors remain infective (Sharp 1928; Muller 2002; Simonsen et al. 2011).

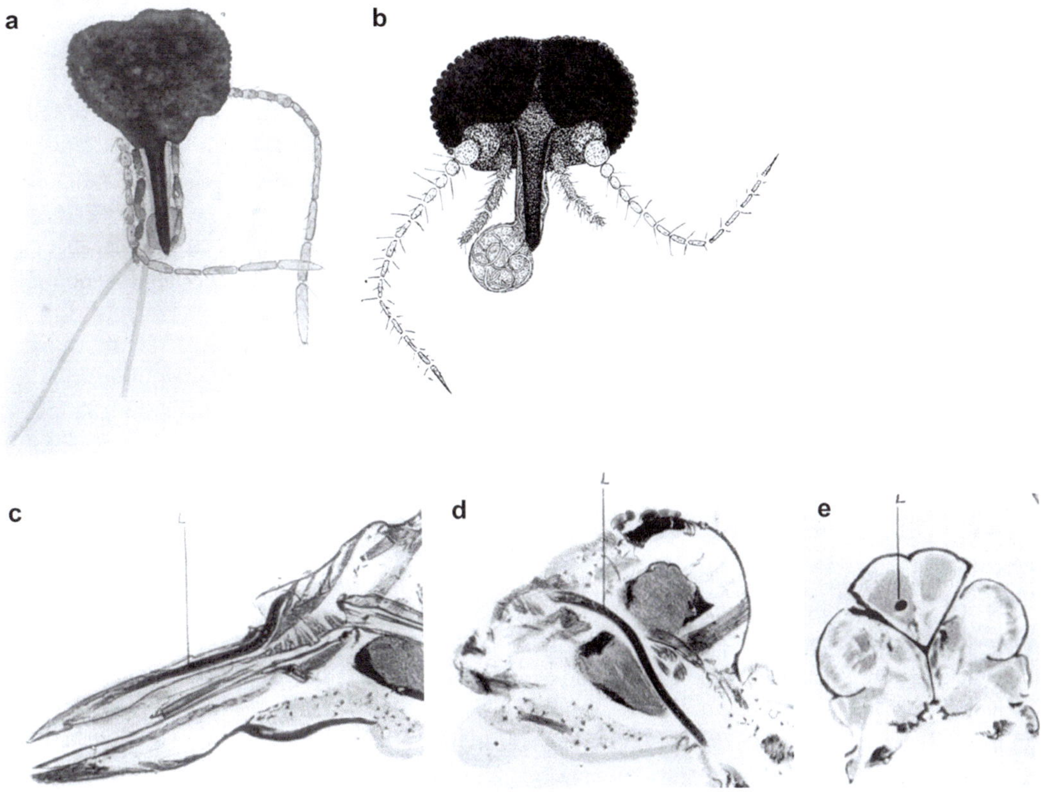

Fig. 26.5 Larvae in heads of *Culicoides* biting midges. (**a**) Larvae of *Mansonella perstans* emerging from the proboscis of *Culicoides austeni*. (**b**) Larvae of *M. perstans* about to emerge from expanded labrum. (**c**) Section through the proboscis showing L_3 stage *M. perstans* larva (L) in the labrum-epipharynx. (**d**) Section through the head showing L_3 stage *M. perstans* larva (L) passing between the fore and hindbrain. (**e**) Section through the thorax showing a larva (L) developing in the thoracic muscle. All taken from Sharp (1928)

26.3.3 Clinical Presentation

Like the other *Mansonella* parasite infections, *M. perstans* infections are very frequently asymptomatic, and like the other human *Mansonella* parasite infections, there is no universally agreed set of symptoms to define *M. perstans* infections. The prevailing view among clinicians and public health policy makers seems to be that *M. perstans* infections are largely benign; there are certainly no large-scale control efforts to limit their spread, anyway. This view seems to have formed because of the present paucity of definitive data linking *M. perstans* infections to disease pathology. There has, however, been a diverse range of symptoms attributed to *M. perstans* parasitism including "bung-eye" [sic], pruritus, edema, fevers, headaches, arthralgia, pericarditis, severe abdominal pain in the liver area, meningoencephalitis, and neuropsychiatric disorders. In addition to the possibility that *M. perstans* causes symptoms that are difficult to characterize because they manifest so rarely, there are also a number of other alternative explanations that could explain why a pathogenic *M. perstans* has not yet been clearly demonstrated as such. For example, like all the human *Mansonella* parasites, *M. perstans* infections occur in areas where most of these symptoms can be caused by other kinds of disease and are usually diagnosed by light microscopy (which misses submicroscopic infections); similarly, there is also some evidence that coinfections might be required to elicit some symptoms; in a study in Buruli, 20 %

(3 of 15) of the patients that were coinfected with *Mycobacterium ulcerans*[2] reported pruritus, whereas the condition was not reported by any of the other 51 mono-infected or uninfected individuals (Orihel and Eberhard 1998; Simonsen et al. 2011; Phillips et al. 2014).

26.3.4 Diagnosis

Recent studies have shown that a key *W. bancrofti* detection assay (the ICT card test), which has been used extensively for WHO lymphatic filariasis mapping, can cross-react and give false-positive diagnoses for people from *W. bancrofti*-free areas. If *M. perstans* is found to have an important role in this cross-reactivity, *M. perstans* diagnostics that can assist with the parasites' accurate mapping may prove an important tool for the WHO's nascent lymphatic filariasis and onchocerciasis elimination program (Alhassan et al. 2014; Wanji et al. 2015). *Mansonella perstans* parasitism is usually diagnosed by detecting and identifying their microfilariae. These microfilariae are normally recovered from the circulating blood (from where they can be collected 24 h a day), but have also been recovered and identified from the ascetic fluid, pleural effusions, bone marrow, and skin. Although PCR assays have been developed and shown to be affective for this purpose, most diagnosis of *M. perstans* has historically been done with parasite morphology (see above), using light microscopy and thick blood smears. *Mansonella perstans* can be distinguished from other blood-circulating parasites found in Africa by their small size and their tails (which are nucleated) and the fact that they are not enveloped with a sheath (WHO 1997; Muller 2002; Simonsen et al. 2011). To meet the emergent needs of the WHO's lymphatic filariasis and onchocerciasis elimination program, however, *M. perstans* diagnostics will likely need to be able to detect parasites in areas where long-term anti-helminthic treatment programs are ongoing and, therefore, in areas where those affected are likely to have *M. perstans* parasitemias with very low parasite counts (Alhassan et al. 2014). Although various parasite concentrating methods such as membrane filtration and the centrifugation-based Knott method can be used in combination with light microscopy to increase the sensitivity of *M. perstans* detection, there are reports that such methods can damage microfilariae and make their identifying features difficult to see (WHO 1997; Medeiros et al. 2010). PCR-based diagnostics (which are both highly sensitive and specific) could be deployed to perform this function. The ribosomal ITS-1 targeting protocol of Tang et al. (2010) makes an appealing option for this as it can detect all other common human filarial parasites and is highly sensitive and specific (WHO 1997; Muller 2002; Medeiros et al. 2010; Tang et al. 2010; Simonsen et al. 2011).

26.3.5 Treatment

At present, there is no known universally effective treatment or cure for *M. perstans* nor is there a proven prophylactic against the acquisition of *M. perstans* infections. A recent preliminary trial in Mali treating *M. perstans*-infected individuals with 200 mg of doxycycline (daily for 6 weeks) suggested that (like for other filarial worms that harbor the intracellular bacteria *Wolbachia*) some infections can be treated and possibly even completely cured with the antibiotic. This same trial, however, also showed the existence of *M. perstans* parasites that lack *Wolbachia* and which are unaffected by the treatment. In a similar trial using the same treatment regime in Ghana, however, doxycycline was found to be almost completely ineffective at clearing *M. perstans* microfilariae from the blood of those treated. The apparent absence of *Wolbachia* in *M. perstans* from Uganda and Gabon also suggests the treatment would be ineffective there too. Thus while anti-*Wolbachia* treatments (like doxycycline) represent an appealing option for the treatment of some *M. perstans* infections—because they may be able to kill both the adult and microfilarial stages of *M. perstans* parasites with few side

[2]A bacterium causing the disease called Buruli ulcer (https://en.wikipedia.org/wiki/Buruli_ulcer) (CBM).

effects for those being treated—they will never be able to treat all *M. perstans* infections and thus traditional antifilarial drugs seem set to have a future in *M. perstans* treatment for some time to come (Coulibaly et al. 2009; Simonsen et al. 2011; Phillips et al. 2014).

Trials with the traditional standard antifilarial drugs have also produced variable results for *M. perstans* infections, and most of those tested are very unlikely to be curative. These antifilarial drugs can also have undesirable side effects, especially when those treated have *Loa loa* infections (see Chap. 27 for more details), which is a common form of parasitic infection in *M. perstans* endemic areas. *Mansonella perstans* infections, which are most often asymptomatic and generally regarded as benign (see above), are therefore very often left untreated in endemic areas. Individuals who are nonnative to endemic areas and have acquired their infections while living, working, or traveling through a *M. perstans* endemic area are, however, more commonly treated, and thus a lot of the existing knowledge about drug treatment regimes derives from small studies performed on such infections which may give a rosier picture of the drug's true efficacy.[3] The treatment of nonnative tourism-acquired type infections, as the doxycycline treatment mentioned above, has usually involved prolonged drug treatment regimes. Such case study type tests have evaluated the use of albendazole, DEC, mebendazole, and thiabendazole for *M. perstans* treatment. Trials with all of these drugs have reported microfilariae blood clearance in at least some of the studies that have tested them. Mebendazole (given at 100 mg twice daily for 28 days) appears to be the most commonly chosen treatment for tourism-acquired type infections, with the adverse side effects associated with alternative therapies (like DEC and thiabendazole) and the limited effectiveness of albendazole probably part of the reason why. Small trials testing the use of combination therapies, using DEC in combination with thiabendazole and with mebendazole and mebendazole in combination with levamisole, have all suggested combination therapies can be even more effective than any single drug used alone. Trials of all high single-dose drug treatments including treatments using ivermectin and albendazole and the two drugs in combination have proven largely ineffective at clearing blood-circulating *M. perstans* microfilariae (Muller 2002; Bregani et al. 2006; Akue et al. 2011; Simonsen et al. 2011).

In 1893, Manson suggested that "a pure and properly conserved water supply is all that is necessary to secure immunity from *F. perstans* and any disease it may produce." Manson was suggesting this as a prophylactic measure rather than a prophylactic medicine as he believed, at the time, that *M. perstans* infections were acquired through drinking water.

Sadly, knowledge of the parasite's life cycle has improved since the nineteenth century (in part because of the work of Manson) and prophylactic medicines have not developed all that much. Thus while it is now known that using a DEET-based insect repellent is more likely to be an effective prophylactic measure against acquiring a *Mansonella* infection than acquiring a "properly conserved water supply," there is presently no clinically proven prophylactic drug available for this purpose.

To the best of our knowledge, there have been no trials of *M. perstans* prophylactics, since doxycycline is routinely used as a prophylactic against malaria in many mansonelliasis endemic areas and may therefore already be providing this function against *Wolbachia*-infected human *Mansonella* parasites for individuals that use it in areas where these parasites occur (like Mali). However, it is unlikely to function as such in areas where they appear not to occur, like Gabon,

[3] Such studies most likely contain a disproportionate number of light *M. perstans* infections, as the length of time an individual lives in an endemic area seems to be correlated with the level of an infection they have. Data from these tourism-acquired type studies cannot therefore be safely extrapolated from for the treatment of individuals living in endemic areas. Should public health policy makers decide that large-scale *M. perstans* control is desirable, planning of a large-scale *M. perstans* control program should take account of this, as data from other filarial treatment trials suggest that individuals with heavy levels of parasitism are the most likely to present with severe adverse consequences to drug treatments such as DEC and ivermectin.

Ghana, and Uganda (Muller 2002; Bregani et al. 2006; Akue et al. 2011).

26.3.5.1 Epidemiology

In a recent expert review, *M. perstans* was proposed to be endemic in at least 33 sub-Saharan African countries with a broad distribution spanning as far north as Sudan and Mauritania and as far south as Mozambique and Angola and that the parasite was affecting around 114 million people. It has also been reported to occur in a number of South America countries including Guyana, Suriname, Colombia, and Venezuela, and there is also at least one first-hand account of the parasite occurring in Papua New Guinea (Sharp 1928; Muller 2002; Simonsen et al. 2011; Crainey et al. 2016).

The epidemiological link between *M. perstans* and banana cultivation was first recognized by Dr Cuthbert Christy at the beginning of the twentieth century before the parasites' mode of transmission had been elucidated. In his reports for the Royal Society, Christy provided maps of the distribution of *M. perstans* in Uganda and recorded that the distribution of *M. perstans* parasites in Uganda almost exactly matched the areas of banana cultivation. Christy, similarly, noted that the parasite was usually absent from areas that did not have banana cultivation and that the intensity of *M. perstans* parasitemia also matched the intensity of banana cultivation, with areas of intense banana cultivation not only having the highest incidences of parasitemias but also having the most heavy-infected individuals too (Christy 1903). The epidemiological link that Christy presented was viewed, by contemporaries, as so strong that some argued (in the Lancet) about whether the parasite had an insect vector and whether it might be acquired by direct contact with and/or consumption of bananas or banana plant products (Bastian 1904a, b; Low 1904). This link was further reinforced when the man-biting adult-stage vectors of *M. perstans* were finally incriminated in transmission and then found to use banana tree stumps for the development of their juvenile stages (Sharp 1928).

Thus far, only three species of vector (all *Culicoides*) have been shown to be capable of developing infectious stage L_3 *M. perstans*, *C. grahamii*, *C. inornatipennis*, and a species named as *C. austeni*. All three of these vectors are known to use banana tree stumps and litter for their larval breeding sites and to travel only short distances for blood meals. As the global distribution of *M. perstans* (although patchy) is largely continuous across most of West and sub-Saharan Africa, just as banana cultivation is, it is tempting to assume that these larval breeding sites (like the river larval breeding site of *O. volvulus* vectors) can be used to predict parasite distribution across the parasites' entire range. While this is possibly true, it is worth noting that the three known vector species are not the only species suspected in *M. perstans* transmission. *Culicoides milnei*, *C. hortensis*, *C. krameri*, *C. vitshumbiensis*, *C. pycnostictus*, and *C. ravus* have all, with varying degrees of confidence, been proposed as possible vectors.

Undoubtedly, of course, there will be variations in the vectorial dynamics across the range *M. perstans* and, from what little information there is, the evidence suggests that these variations could be quite dramatic. For example, detailed studies of *C. austeni* in Gambia have shown it to be part of a species complex with diverse physiological and behavioral variations, with some populations even being autogenesis (Linley et al. 1983; Simonsen et al. 2011). Thus while banana plantations could very well play a central role in *M. perstans* vectorial dynamics, it should not be forgotten that *Culicoides* vectors are known to breed in a diverse variety of sites ranging from the brackish tidal mangroves and salt marshes to the riverbanks and swamps, tree holes, rubbish pits, rotting plants, and mud and also not to forget that there are many reports of *M. perstans* occurring where banana cultivation is not thought to occur (especially historic reports from the rainforest areas of South America) (Low 1904; Sharp 1928; Linley et al. 1983). Recently, Stensgaard et al. (2016) produced a nationwide geostatistical risk map for the parasites' distribution in Uganda based on school survey parasitological data and remotely sensed geographic and several bioclimatic factors. Maps like these have been used to delineate *L. loa*, *O. volvulus*, and co-endemic areas to help identify areas that might be unsuitable for simple ivermectin drug treatment programs (see Chap. 25 and Kelly-Hope et al. 2015). If *M. perstans* infections are found to have a role in ICT card cross-reactivity, similar maps that could delineate *M. perstans and W. bancrofti* distribution could help to identify areas where ICT cards could not be safely used for lymphatic filariasis diagnostics and therefore ultimately help in the parasite's elimination (Stensgaard et al. 2016).

Blood parasitemia levels can vary dramatically from one to over 12,000 microfilariae per milliliter. In a *M. perstans* incidence survey in Gabon, the prevalence and the intensity of infections was found to be significantly higher in villages from forested areas than in lake land or Savannah area villages, and within the forested region, blood parasitemia levels in the mountainous areas were found to be significantly higher than other parts. Consistent with the notion that vectors are an important driver in *M. perstans* prevalence and infection intensities, *M. perstans* have been reported as having many of the same risk factors of infection that are recorded for *M. ozzardi* and which are often linked with increased risk of exposure to vector biting, for example, age, sex, and outdoor work. People infected with *M. perstans* in Ghana have also recently been reported to be more likely to be infected with *Mycobacterium ulcerans*, which is also purported to be vector-borne. Although the study size was small and the correlation not significant, the results of this study are consistent with exposure to vector biting being a risk factor for contracting *M. perstans* infections (Linley et al. 1983; Akue et al. 2011; Simonsen et al. 2011; Dolo et al. 2012; Phillips et al. 2014).

While reports of *M. perstans* in apes (like chimpanzees) are common, apes are not generally thought to be epidemiologically important reservoirs of *M. perstans* parasites. This notion is supported by the observation that very high and intense *M. perstans* infections can be found in places where apes cannot be. Most investigated *Culicoides* species seem to travel only short distances from their breeding sites to take their blood meals. *Onchocerca volvulus*-infected simuliid vectors seem to travel much shorter distances from their breeding sites than their uninfected counterparts[4] (Dalmat 1955). Since *M. perstans* also develops in the flight muscles of its vectors, infected flies very likely have reduced mobility and thus probably also travel less far from their breeding sites than uninfected vectors do. Assuming that *Culicoides* species are indeed the principal or sole vectors of *M. perstans*, the short distances that they are predicted to travel from their breeding sites would therefore probably limit their range of transmission, supporting the notion that great apes (like chimpanzees) do not play an important role in *M. perstans* epidemiology, as such great apes like chimpanzees are not normally found living wild close to sites of human habitation (Linley et al. 1983; Simonsen et al. 2011).

In a guide to human filarial diagnosis, the WHO describes the levels of circulating *M. perstans* microfilariae as aperiodic; however, there are reports dating from as early as 1912 suggesting that circulating blood levels of *M. perstans* have mixed periodicity, with some patients showing signs of diurnal and others showing signs of nocturnal periodicity. The more recent accounts of cryptic periodicity in *M. ozzardi* parasitism may therefore be regarded as reminiscent of these older (1912) accounts of mixed periodicity first recorded for *M. perstans*. As with the reports of cryptic periodicity in *M. ozzardi*, the cryptic periodicity of *M. perstans* and how it may be related to vector ecology and the parasite's epidemiology have hitherto not been well investigated. Similarly, interesting reports that *M. perstans* may be a species complex are yet to be thoroughly investigated, despite the fact that these reports appear to be supported by the recent discovery of *Wolbachia*-positive and *Wolbachia*-negative *M. perstans* populations (Sharp 1928; Linley et al. 1983; WHO 1997; Coulibaly et al. 2009; Simonsen et al. 2011; Phillips et al. 2014).

[4]Forty thousand black flies fed on *Onchocerca volvulus*-infected patients were liberated, to check the distance they could fly, an experiment of questionable ethics (see https://catalog.hathitrust.org/Record/006151533 [page 332]) (CBM).

26.4 *Mansonella streptocerca*

Mansonella streptocerca was the last of the mansonelliasis parasites to be discovered. As with the other human parasites from the genus, it was first described on the basis of its microfilariae, which were recovered from skin snips obtained from native Ghanaians in the early 1920s (Macfie and Corson 1922).

26.4.1 Parasite Morphology

26.4.1.1 Adults

Adult *M. streptocerca* worms have been recovered from humans (coiled) in tissues of the dermis; female worms recovered from bonobo chimpanzees have been recorded measuring 23–27.6 mm in length. Females recovered from humans are reportedly up to 0.085 mm in diameter and have nerve rings between 0.01 mm and 0.015 mm from their anterior ends and esophagi 0.4–0.5 mm in length; they reportedly have four cephalic papillae arranged dorsoventrally in an elongated rectangle and four external labial papillae arranged in a square.

The hitherto described *M. streptocerca* adult females can be distinguished from the adult females of *M. perstans* by the position of their vulvas, which is recorded to occur between 0.4 and 0.54 mm from the anterior end of both primate and human-derived adult *M. streptocerca* worms; *M. perstans* specimens obtained from Africans have vulva openings between 0.825 and 1.22 mm from their anterior ends. The paired uteri of *M. streptocerca* females are reported to be straight and to run the length of the body from about 2 mm from either end; the ovaries and ovary ducts (which are 0.02–0.025 mm in diameter) are confined to the last 2 mm of the posterior end of the worms and join to form a vagina. The female anus is recorded to be between 0.1 and 0.15 mm from the posterior end of the worm; the tails of *M. streptocerca* females obtained from bonobos are 0.116–0.22 mm. Male worms obtained from humans are reported to have smooth cuticles and to have morphological features "consistent with" the adult males obtained from chimpanzees and thus to be between 13 and 18.1 mm in length and to be up to 0.05 mm in width. The tails of male worms obtained from chimpanzees are recorded to be between 0.067 and 0.074 mm; the left spicules are recorded to be between 0.333 and 0.349 mm and the right to be between 0.117 and 0.120 mm; the esophagi are recorded to range from 0.383 to 0.390 mm in length; and nerve ring is reported to be 0.2–0.202 mm from the posterior end of worms (Neafie et al. 1975; Meyers et al. 1977; Muller 2002; Bain et al. 2015).

26.4.1.2 Microfilariae and Larvae

The microfilariae of *M. streptocerca* lack the sheaths that many other microfilariae have and are slender with narrowing tails, measuring between about 180 and 240 μm long and between about 5 and 6 μm in width. The cephalic space ranges from 3 to 5 μm in length, and the first four oval nuclei are arranged in a single row, followed by a row of seven to ten rounded nuclei. The tails of the microfilariae are reported to be bluntly rounded with nuclei until the end. The caudal space is reported to be 1 μm in length and the caudal end is often described to take the shape of a "shepherd's crook." Infectious stage L_3 larvae have been measured as 574 μm in length and to be obtainable 8 days postinfection (Macfie and Corson 1922; Neafie et al. 1975; Meyers et al. 1977; Eberhard and Orihel 1984; WHO 1997; Muller 2002; Bain et al. 2015).

26.4.2 Transmission Cycle

Mansonella streptocerca is certainly the least well studied of the mansonelliasis-causing *Mansonella* parasites and thus comparatively little is known of its life cycle. The biting midge *C. grahamii* (which is a known vector of *M. perstans*) can also support the development of *M. streptocerca*. Infectious stage larvae can develop from microfilariae in *C. grahamii* within 8 days and, like the other human parasites of the genus, do so in the thorax muscles. As with *M. perstans*, nothing is known about how infectious stage larvae become adults and how long it takes for a fertilized female to begin producing microfilariae (Fig. 26.1c). In line with the other human parasites from the genus, *M. streptocerca* parasites are reported to develop to infectious stage larvae inside *Culicoides* vector species within 8 days. Like *M. perstans* infections, *M. streptocerca* have also been recorded in apes including gorillas (*Gorilla gorilla*) and chimpanzees (*Pan* spp.). Because, however, *M. streptocerca* can be found in areas where primate parasite reservoirs cannot be, primates are not thought to play a necessary or indeed an important part *of M. streptocerca* epidemiology (Macfie and Corson 1922; Muller 2002).

26.4.3 Clinical Presentation

Mansonella streptocerca infections were not initially thought to be associated with skin disease, and even though there are now clearly established clinical manifestations for infections, these are usually mild and are not always detectable in every person infected and thus infections are not regarded as having much clinical importance. The clinical manifestations of infections when they do occur are most difficult to distinguish from leprosy and onchocerciasis. Dermatitis including papular dermatitis around the thorax and shoulders is the most common symptom of *M. streptocerca* infections. Pruritus is also a common symptom and it has also been suggested that *M. streptocerca* may sometimes cause lymphedema and elephantiasis (Macfie and Corson 1922; Meyers et al. 1972; Fischer et al. 1997; Muller 2002).

26.4.4 Diagnosis

Mansonella streptocerca is presently diagnosed by the identification of microfilariae obtained from skin biopsies. Such microfilariae can be identified by PCR or by their morphology and movements observed under light microscopic examination (WHO 1997; Fischer et al. 1998; Muller 2002).

26.4.5 Treatment

There are presently no large-scale international control programs targeting *M. streptocerca* for mansonelliasis control. Similarly, there are no universally agreed best practice clinical guidelines for the treatment of *M. streptocerca* infections. Diethylcarbamazine (DEC) and ivermectin have both been tested as possible drugs for the treatment of *M. streptocerca* infections and both have been shown to be effective at eliminating *M. streptocerca* microfilariae from the skin snips of those treated. However, while ivermectin (given once orally at a dose of 150 g per kilogram of body weight) has been seen to clear microfilariae from the skin of about 55 % of infected people treated with it, it probably does not kill adults and therefore cannot be regarded as curative. In contrast, a daily dose of 3–6 mg per kilogram of body weight of DEC, given for 28 days, seems to clear microfilariae from the skin and to kill adult parasites also (Meyers et al. 1978; Fischer et al. 1997; Muller 2002).

Despite the potential clinical benefits, historically, DEC treatment has not been used much in the treatment *of M. streptocerca* infections as many people who are infected with *M. streptocerca* are also often infected with *O. volvulus*. As *O. volvulus* infections cause a more seriously debilitating disease (onchocerciasis) than *M. streptocerca* infections do the treatment of *O. volvulus* infections takes precedence over the treatment of *M. streptocerca* infections. The WHO recommends that onchocerciasis should be treated with a single dose of ivermectin given orally at 150 μg per kilogram of body weight (twice a year), and thus, this is how most *M. streptocerca* infections are "treated." As the WHO's African program for onchocerciasis elimination draws to a close and fewer and fewer onchocerciasis foci exist, the appeal of DEC treatment for *M. streptocerca* infections will increase. The fact, however, that DEC and ivermectin treatments can cause serious adverse effects similar to the Mazzotti reaction described for the ivermectin treatment of *M. streptocerca*-infected patients and because both ivermectin and DEC can also cause severe adverse consequences (including death) for those infected with *Loa loa* (which like *O. volvulus* has a distribution that overlaps extensively with *M. streptocerca*), the best practice treatment of *M. streptocerca* in *O. volvulus* free zones is not presently obvious (Kelly-Hope et al. 2015).

To date *Wolbachia* has not been recorded in *M. streptocerca* and thus whether it can be targeted for *M. streptocerca* treatment with doxycycline or other anti-*Wolbachia* therapies remains an open question. It is also not entirely certain that such treatments could avoid the adverse side effects reported for ivermectin and DEC or if such a treatment might be safely adapted for use as a prophylactic. On this basis of present knowledge, like for *O. volvulus* treatment, it seems likely the best practice treatment of *M. streptocerca* will involve first establishing whether a patient is infected with *L. loa* (and if so, infected to what extent) and then treating those uninfected with *L. loa* (and possibly those light *L. loa* infections) with a specially formulated course of DEC. On the basis of the present knowledge, it is probably best to leave those heavily infected with *L. loa* (and *M. streptocerca*) untreated; however, if *M. streptocerca* is shown to be infected with *Wolbachia* and treatable with antibiotics, new best practice treatment guidelines for such patients may very well emerge (Fischer et al. 1997; Coulibaly et al. 2009).

26.4.6 Epidemiology

The distribution of the skin-dwelling *M. streptocerca* (which requires painful skin biopsies to establish) is even less well known than that of the other blood-dwelling parasites, but it is thought to be restricted to Africa. *Mansonella streptocerca* infections have been reported to occur in tropical rainforest areas of Angola, Cameroon, the Central African Republic, the Democratic Republic of Congo, Equatorial Guinea, Nigeria, and Uganda (Muller 2002; Simonsen et al. 2011).

From the few epidemiological studies that have been performed, it appears that prevalence rates vary widely. For example, in some geographically close parts of the Democratic Republic of Congo, incidence estimates ranging from 2 to 100 % have been made. Similar variations have also been recorded in Uganda, Nigeria, Ghana, and the Central African Republic. As with the other *Mansonella* infections within foci incidence measures, adults are usually more likely to be found infected than children are. As so little is known of the parasite's biology and that of the vectors involved in its transmission, it is difficult to know what factors may be driving these fluctuations and restricting the range of *M. streptocerca* distribution (which is not as geographically well dispersed or as common as *M. perstans*). Although *M. streptocerca* infections have been recorded in apes, they are not thought to be important reservoirs or to play much of a role in human transmission dynamics. In blood-feeding experiments *C. grahamii* was observed to take up more *M. streptocerca* than *M. perstans* from a coinfected individual (Macfie and Corson 1922; Fischer et al. 1997; Muller 2002).

References

Adami Y, Rodrigues G, Alves M et al (2014) New records of *Mansonella ozzardi*: a parasite that is spreading from the state of Amazonas to previously uninfected areas of the state of Acre in the Purus River region. Mem Inst Oswaldo Cruz 109:87–92

Akue J, Nkoghe D, Padilla C et al (2011) Epidemiology of concomitant infection due to *Loa loa* and *Mansonella perstans* in Gabon. PLoS Negl Trop Dis 5, e1329

Alhassan A, Li Z, Poole CB et al (2014) Expanding the MDx toolbox for filarial diagnosis and surveillance. Trends Parasitol 31:391–400

Bain O (2002) Evolutionary relationships among filarial nematodes. In: Parasites WC (ed) The filaria. Springer, New York and UK

Bain O, Mutafchiev Y, Junker K et al (2015) Review of the genus *Mansonella* Faust, 1929 sensu lato (Nematoda: Onchocercidae), with descriptions of a new subgenus and a new subspecies. Zootaxa 3918:151–193

Basano SD, Fontes G, Medeiros JF et al (2014) Sustained clearance of *Mansonella ozzardi* infection after treatment with ivermectin in the Brazilian Amazon. Am J Trop Med Hyg 90:1170–1175

Bastian HC (1904a) Note on the probable mode of infection by the so-called *Filaria perstans*, and on the probability that this organism really belongs to the genus *Tylenchus* (Bastian). Lancet 163:286–287

Bastian HC (1904b) The anatomical characters of the so-called *Filaria perstans* and on the mode of infection thereby. Lancet 163:643–645

Bregani ER, Rovellini A, Mbaidoum N et al (2006) Comparison of different anthelminthic drug regimens against *Mansonella perstans* filariasis. Trans R Soc Trop Med Hyg 100:458–463

Casiraghi M, Favia G, Cancrini G et al (2001) Molecular identification of *Wolbachia* from the filarial nematode *Mansonella ozzardi*. Parasitol Res 87:417–420

Casiraghi M, Bain O, Guerrero R et al (2004) Mapping the presence of *Wolbachia pipientis* on the phylogeny of filarial nematodes: evidence for symbiont loss during evolution. Int J Parasitol 34:191–203

Chadee D, Tilluckdharry C, Rawlins S et al (1995) Mass chemotherapy with diethylcarbamazine for the control of Bancroftian filariasis: a twelve-year follow-up in northern Trinidad, including observations on *Mansonella ozzardi*. Am J Trop Med Hyg 52:174–176

Christy C (1903) The distribution of sleeping sickness, *F. perstans* etc... in Equatorial Africa. Royal Society reports of the Sleeping Sickness Commission, no. 2, p 3–8

Coulibaly YI, Dembele B, Diallo AA et al (2009) A randomized trial of doxycycline for *Mansonella perstans* infection. New Engl J Med 361:1448–1458

Crainey JL, Ribeiro da Silva TR, Luz SL (2016) Historic accounts of *Mansonella* parasitaemias in the South Pacific and their relevance to lymphatic filariasis elimination efforts today. Asian Pac J Trop Med 9:205–210

Dalmat H (1955) The black flies (Diptera, Simuliidae) of Guatemala and their role as vectors of onchocerciasis. Smithsonian Institution, Washington, DC

Dolo H, Coulibaly YI, Dembele B et al (2012) Filariasis attenuates anemia and proinflammatory responses associated with clinical malaria: a matched prospective study in children and young adults. PLoS Negl Trop Dis 6, e1890

Eberhard ML, Orihel TC (1984) The genus *Mansonella* (syn. *Tetrapetalonema*): a new classification. Ann Parasitol Hum Comp 59:483–496

Ferri E, Barbuto M, Bain O et al (2009) Integrated taxonomy: traditional approach and DNA barcoding for the identification of filarioid worms and related parasites (Nematoda). Front Zool 6:1–12

Ferri E, Bain O, Barbuto M et al (2011) New insights into the evolution of *Wolbachia* infections in filarial nematodes inferred from a large range of screened species. PLoS One 6:1–17

Fischer P, Büttner DW, Bamuhiiga J et al (1998) Detection of the filarial parasite *Mansonella streptocerca* in skin biopsies by a nested polymerase chain reaction-based assay. Am J Trop Med Hyg 58:816–820

Fischer P, Bamuhiiga J, Buttner D (1997) Treatment of human *Mansonella streptocerca* infection with ivermectin. Trop Med Int Health 2:191–199

Henry FP (1897) Remarks on filaria. Proc Acad Natl Sci Phila 48(1896):271–275

Kelly-Hope LA, Unnasch TR, Stanton MC et al (2015) Hypo-endemic onchocerciasis hotspots: defining areas of high risk through micro-mapping and environmental delineation. Infect Dis Poverty 4:1–7

Laurence B, Simpson M (1968) Cephalic and pharyngeal structures in microfilariae revealed by staining. J Helminthol 42:309–330

Lefoulon E, Bain O, Bourret J et al (2015) Shaking the tree: multi-locus sequence typing usurps current Onchocercid (Filarial Nematode) phylogeny. PLoS Negl Trop Dis 9:1–19

Linley JR, Hoch AL, Pinheiro FP (1983) Biting midges (Diptera: Ceratopogonidae) and human health. J Med Entomol 20:347–364

Low GC (1904) *Filaria perstans* and the suggestion that it belongs to the genus *Tylenchus* (Bastian). Lancet 163:420–421

Lowrie R, Raccurt C (1984) Assessment of *Culicoides barbosai* as a vector of *Mansonella ozzardi* in Haiti. Am J Trop Med Hyg 33:1275–1277

Lowrie R, Orihel T, Eberhard M (1982) *Culicoides variipennis*, a laboratory vector for the Amazon form of *Mansonella ozzardi*. Am J Trop Med Hyg 31:166–167

Lowrie R, Raccurt C, Eberhard M et al (1983) Assessment of *Leptoconops bequaerti* as a potential vector of *Mansonella ozzardi* in Haiti. Am J Trop Med Hyg 32:1013–1015

Luz SL, Crainey JL, Shelley AJ et al (2014) Outstanding insecurities concerning the use of an Ov16-based ELISA in the Amazonia onchocerciasis focus. Mem Inst Oswaldo Cruz 109:506–508

Macfie J, Corson J (1922) A new species of filarial larva found in the skin of natives in the Gold Coast. Ann Trop Med Parasitol 16:465–471

Manson P (1892) The geographical distribution, pathological relations, and life history of *Filaria sanguinis hominis diurna* and of *Filaria sanguinis hominis perstans*, in connexion with preventive medicine. The Trans. 1th Internatl. Cong. Hyg. and Demog., London, 1 Aug 1891 (Sect. 1), p 88

Manson P (1897) On certain new species of nematode haematozoa occurring in America. Br Med J 2:1837–1838

Manson P (1917) Tropical disease: a manual of diseases of warm climates, 6th edn. Cassell, London

Marcos LA, Arrospide N, Recuenco S et al (2012) Genetic characterization of atypical *Mansonella (Mansonella) ozzardi* microfilariae in human blood samples from northeastern Peru. Am J Trop Med Hyg 87:491–494

Martins M, Pessoa FAC, Medeiros MB et al (2010) *Mansonella ozzardi* in Amazonas, Brazil: prevalence and distribution in the municipality of Coari, in the middle Solimões River. Mem Inst Oswaldo Cruz 105:246–253

Medeiros JF, Py-Daniel V, Barbosa UC et al (2008) Current profile of *Mansonella ozzardi* (Nematoda: Onchocercidae) in communities along the Ituxi river, Lábrea municipality, Amazonas, Brazil. Mem Inst Oswaldo Cruz 103:409–411

Medeiros JF, Py-Daniel V, Barbosa UC et al (2009) *Mansonella ozzardi* in Brazil: prevalence of infection in riverine communities in the Purus region, in the state of Amazonas. Mem Inst Oswaldo Cruz 104:74–80

Medeiros JF, Pessoa FAC, Martins M (2010) Importância do método de gota espessa de sangue no diagnóstico de filárias simpátricas no Amazonas, Brasil. Acta Amaz 40:779–780

Medeiros JF, Almeida TA, Silva LB et al (2015) A field trial of a PCR-based *Mansonella ozzardi* diagnosis assay detects high-levels of submicroscopic *M. ozzardi* infections in both venous blood samples and FTA card dried blood spots. Parasit Vectors 8:1–8

Meyers WM, Connor DH, Harman LE et al (1972) Human streptocerciasis. A clinico-pathologic study of 40 Africans (Zairians) including identification of the adult filaria. Am J Trop Med Hyg 21:528–45

Meyers WM, Neafie RC, Moris R et al (1977) Streptocerciasis: observation of adult male *Dipetalonema streptocerca* in man. Am J Trop Med Hyg 26:1153–1155

Meyers WM, Moris R, Neafie RC et al (1978) Streptocerciasis: degeneration of adult *Dipetalonema streptocerca* in man following diethylcarbamazine therapy. Am J Trop Med Hyg 27:1137–1147

Morales-Hojas R (2009) Molecular systematics of filarial parasites, with an emphasis on groups of medical and veterinary importance, and its relevance for epidemiology. Infect Genet Evol 9:748–759

Muller R (2002) Worms and human disease. CABI, London/Wallingford

Murray CJL, Vos T, Lozano R (2012) Disability-adjusted life years (DALYs) for 291 diseases and injuries in 21 regions, 1990–2010: a systematic analysis for the Global Burden of Disease Study 2010. Lancet 380:2197–2223

Nathan MB, Bartholomew C, Tikasingh E (1978) The detection of *Mansonella ozzardi* microfilariae in the skin and blood with a note on the absence of periodicity. Trans R Soc Trop Med Hyg 72:420–422

Neafie RC, Connor DH, Meyers WM (1975) *Dipetalonema streptocerca* (Macfie and Corson, 1922): description of the adult female. Am J Trop Med Hyg 24:264–267

Orihel TC, Eberhard ML (1982) *Mansonella ozzardi*: a redescription with comments on its taxonomic relationships. Am J Trop Med Hyg 31:1142–1147

Orihel TC, Eberhard ML (1998) Zoonotic filariasis. Clin Microbiol Rev 11:366–381

Orihel T, Lowrie RJ, Eberhard M et al (1981) Susceptibility of laboratory primates to infection with *Mansonella ozzardi* from man. Am J Trop Med Hyg 30:790–794

Phillips R, Frimpong M, Sarfo F et al (2014) Infection with *Mansonella perstans* nematodes in Buruli ulcer patients, Ghana. Emerg Infect Dis 20:1000–1003

Pichon G (1983) Crypto-periodicity in *Mansonella ozzardi*. Trans R Soc Trop Med Hyg 77:331–333

Post RJ, Adams Z, Shelley AJ et al (2003) The morphological discrimination of microfilariae of *Onchocerca volvulus* from *Mansonella ozzardi*. Parasitology 127:21–27

Sharp N (1928) *Filaria perstans*: its development in *Culicoides austeni*. Trans R Soc Trop Med Hyg 21:371–396

Shelley AJ (1988) Biosystematics and medical importance of the *Simulium amazonicum* and *S. exiguum* complex in Latin America. In: Service M (ed) Biosystematics of haematophagous insects. Systematics Association, Oxford

Shelley AJ (2002) Human onchocerciasis in Brazil: an overview. Cad Saude Publica 18:1167–1177

Shelley AJ, Coscarón S (2001) Simuliid blackflies (Diptera: Simuliidae) and ceratopogonid midges (Diptera: Ceratopogonidae) as vectors of *Mansonella ozzardi* (Nematoda: Onchocercidae) in northern Argentina. Mem Inst Oswaldo Cruz 96:451–458

Shelley AJ, Dias A, Moraes MAP (1980) *Simulium* species of the Amazonicum group as vectors of *Mansonella ozzardi* in the Brazilian Amazon. Trans R Soc Trop Med Hyg 74:784–788

Simonsen P, Onapa A, Asio S (2011) *Mansonella perstans* filariasis in Africa. Acta Trop 120: S109–S120

Stensgaard AS, Vounatsou P, Onap AW (2016) Ecological drivers of *Mansonella perstans* infection in Uganda and patterns of co-endemicity with lymphatic filariasis and malaria. PLoS Negl Trop Dis 10:1–16

Tang TH, López-Vélez R, Lanza M et al (2010) Nested PCR to detect and distinguish the sympatric filarial species *Onchocerca volvulus*, *Mansonella ozzardi* and *Mansonella perstans* in the Amazon Region. Mem Inst Oswaldo Cruz 105:823–828

Tidwell MA, Tidwell MA (1982) Development of *Mansonella ozzardi* in *Simulium amazonicum*, *S. argentiscutum* and *Culicoides insinuatus* from Amazonas, Colombia. Am J Trop Med Hyg 3:1137–1141

Tidwell MA, Tidwell MA, Muñoz de Hoyos P (1980) Development of *Mansonella ozzardi* in blackflies of the *Simulium sanguineum* group from eastern Vaupés, Colombia. Am J Trop Med Hyg 29:1209–1214

Vianna LMM, Martins M, Cohen MJ et al (2013) *Mansonella ozzardi* corneal lesions in the Amazon: a cross-sectional study. BMJ Open 2, e001266

Wanji S, Amvongo-Adjia N, Koudou B et al (2015) Cross-reactivity of filariais ICT cards in areas of contrasting endemicity of *Loa loa* and *Mansonella perstans* in Cameroon: implications for shrinking of the lymphatic filariasis map in the Central African region. PLoS Negl Trop Dis 9:1–24

WHO (1997) Bench Aids for the diagnosis of filarial infections. WHO, Geneva

Yarzabal L, Basanez MG, Ramirez Perez J et al (1985) Experimental and natural infection of *Simulium sanchezi* by *Mansonella ozzardi* in the Middle Orinoco region of Venezuela. Trans R Soc Trop Med Hyg 79:29–33

Loiasis

27

Sébastien Pion and Cédric Chesnais

Abstract

Loa loa is a long-lived parasite (up to 20 years of age) which is transmitted through the bites of deer flies (mainly *Chrysops dimidiata* and *C. silacea*). *L. loa* is endemic to Africa, from southeastern Benin in the west to southern Sudan in the east, and down south to Angola and possibly Zambia. Loiasis was first identified during the slave trade period. At that time, the disease was depicted by the spectacular passage of the adult worm under the bulbar conjunctiva of the eye ("eye worm") and the transient episodes of subcutaneous migratory edema referred to as "Calabar swelling." Since then, loiasis has also been associated with more severe symptoms such as cardiopathy (involving hypereosinophilia), nephropathy, and hormonal disturbance. Despite its endemic presence in Central Africa, with possibly tens of millions individuals concerned by the disease, loiasis has drawn little attention until the implementation of mass drug administration for lymphatic filariasis and onchocerciasis in the region. Indeed, post-therapeutic serious adverse events (SAEs) have occurred in individuals harboring high density of *L. loa* microfilariae. Solutions to prevent those SAEs are currently under evaluation in the field, but specific operational research on loiasis should nonetheless be encouraged.

Keywords

Occult loiasis • Eye worm • Angioedema • Post-therapeutic serious adverse events • *Chrysops* • Diethylcarbamazine • Ivermectin

Due to the intimate relationship among lymphatic filariasis, onchocercosis, mansonelliasis, and loiasis on diagnosis, distribution, and treatment, all four chapters need to be checked when solving problems related to any of them (CBM).

S. Pion (✉) • C. Chesnais
Joint Research Unit 233 "TransVIHMI", Institut de Recherche pour le Développement, INSERM U1175, Montpellier University, Agropolis Avenue 911, PO Box 64501, 34394 Montpellier, France
e-mail: sebastien.pion@ird.fr; cedric.chesnais@ird.fr

27.1 History

The filarial worm *Loa loa* is exclusively found in Africa. Nonetheless, the first scientific description is a case report from Santo Domingo (current Dominican Republic). In 1770, André Mongin, a French surgeon, documented the subconjunctival

C.B. Marcondes (ed.), *Arthropod Borne Diseases*, DOI 10.1007/978-3-319-13884-8_27

migration of the adult worm in a black slave. In 1778, Dr. François Guyot, surgeon in the French navy, underlined the harmful nature of *L. loa* infection when reporting that a number of African slaves, gathered together before they were sent to America, suffered from recurrent ocular symptoms. He successfully extracted a worm from one of the slaves' eye. The first *L. loa* microfilariae were observed more than a century later in 1890, by ophthalmologist Dr. Stephen Mackenzie. The latter sent the specimen to Sir Patrick Manson for identification, and the parasite was named *Filaria* diurnal, in accordance with its diurnal periodicity in the peripheral bloodstream. In 1894, ophthalmologist Dr. Argyll Robertson from Edinburgh Royal Infirmary extracted two worms, a male and a female, from the conjunctiva of a patient who had spent the previous eight years in Calabar, Nigeria. Dr. Argyll Robertson gave the first exhaustive description of both the parasite and clinical pathology. Eventually, in 1904, Dr. Emile Brumpt depicted endemic cases in Africa, and, in 1905, Dr. Charles Wardell Stiles gave the definitive name to the parasite: *Loa loa*.

27.2 Morphology and Life Cycle

Adult stages of *L. loa* are thin and round worms. Males are about 3 cm long and 0.4 mm in diameter and females, 5–7 cm long and 0.5 mm in diameter. They live within the upper layer of the epidermis and in the conjunctive tissues. Females have long been thought to live up to 15 years, but a recent case report of a patient from whom a female *L. loa* was removed from the right eyelid during an acute episode of periorbital edema 21 years after visiting an endemic area indicates that they may live much longer (Richardson et al. 2012). The life expectancy of males has been more difficult to assess and is still a matter of speculation. After mating, embryos – or microfilariae – develop in the females' uterus. Those embryos are sheathed in a vitelline membrane covered with a thin chitinous layer. Microfilariae are 205–300 μm long and 6–8 μm wide. They are produced following a periodical fashion (Fain and Maertens 1973) and circulate in the peripheral bloodstream following a diurnal cycle: at nighttime, they actively accumulate in the capillaries and small vessels of the lung, whereas during the day, they are passively carried away in the bloodstream. This periodicity has a genetic basis, but would be modulated by oxygen rate in the blood vessels, body temperature, as well as some chemical processes (Boussinesq 2006).

During a blood meal taken on an infected host, microfilariae are ingested by the female tabanid vector (see below). The sheathed embryos migrate in the whole vector's body, and their development takes preferably place inside adipose cells of the abdomen, head, and thorax. Within 10–12 days, two successive episodes of molting turn the microfilariae into infective L_3 larvae. L_3 larvae are about 2 mm long and 26 μm diameter and are very motile. Straight after the second molting, the larvae leave the adipose cells, cross the hemocele barrier, migrate toward the head, and aggregate in the labium.

During the next blood meal, the vector lacerates the epidermis leading to a localized hemorrhage. The human blood is mixed to an anticoagulant salivary secretion produced by the vector. It is then sucked up by the labrum, and the resulting internal pressure leads to the bursting of the thin labio-pharyngeal membrane. The infective larvae are thus dropped off on the skin and will actively penetrate the skin through the wound created by the bite. They will then migrate in the subcutaneous layers. The maturation from infective larvae to the sexed adult stages takes about 3 months, period during which the larvae will molt twice. Experimental infections in nonhuman primates have shown that the prepatent period, i.e., the time between inoculation of infective larvae and appearance of microfilariae, is around 140 days (Orihel and Eberhard 1985).

27.2.1 Vectors

The two most important vectors of *L. loa* in humans are *Chrysops silacea* and *C. dimidiata*, two species typically found in forested areas. Those two species show a very well-marked diurnal activity, with a peak in the morning and another one during the afternoon; the light intensity would be the main factor determining the vectors' activity. Other *Chrysops* species are also natural vector of *L. loa*: *C. zahrai* in mountain-forest areas and at the savannah-forest fringe and *C. longicornis* and *C. distinctipenis* in savannah areas. However, these three species have a low degree of anthropophily and mainly feed on animals; as a consequence, they are very rarely infective for humans. As such, their role in the transmission of *L. loa* is considered as incidental (Duke 1958).

Chrysops langi and *C. centurionis* are potential vectors of *L. loa*, but, unlike the pre-cited species, the latter species are canopy-dwelling and mainly active at dusk and during the night when they are presumed to feed mainly on sleeping monkeys. These specific habits in space and time of activity constitute a virtual ecological barrier between the transmission of human and simian *L. loa*: in nature, the diurnally periodic human *L. loa* is transmitted among humans, at the ground level, by the day-biting species *C. silacea* and *C. dimidiata*, whereas the nocturnally periodic parasite is being transmitted at night, in the forest canopy, by *C. langi* and *C. centurionis* (Duke 1958).

27.3 Clinical Manifestations

27.3.1 Occult Loiasis

A peculiar aspect of loiasis is that a significant proportion of infected subjects, whose infection has been ascertained by previous eye worm episodes, do not present with microfilaremia (Fain 1978). Even in villages where the intensity of transmission is very high, the prevalence of microfilaremia recorded in the oldest and most infected age groups rarely exceeds 60 %.

Besides this, in high-endemicity settings, it has been observed that, in young children, *L. loa*-specific antibodies are very frequent, whereas microfilaremia is almost absent (Goussard et al. 1984). This so-called "occult loiasis" pattern is associated with immunological mechanisms (Akue et al. 1997). Complex segregation analysis has shown that only a fraction of the population (around 40 %) would be genetically predisposed to become microfilaremic when infected with *L. loa* (Garcia et al. 1999).

27.3.2 Frequent Symptoms and Clinical Profiles

The most specific manifestations of loiasis are the passage of the adult worm under the bulbar conjunctiva of the eye ("eye worm") and the transient episodes of subcutaneous migratory edema referred to as "Calabar swelling," named after its first description, in 1895, in the Nigerian town of Calabar.

The subcutaneous migration of an adult *Loa* manifests itself as a palpable serpiginous lace moving by about 1 cm per minute. The migration under the bulbar conjunctiva may take from several minutes to more than a day. It is generally associated with local irritation of the eye, with itching and photophobia. Passage under the palpebral conjunctiva, which is less frequent, is often associated with palpebral or periorbital edema. In some unusual instances, the adult worm and microfilariae can also penetrate the eye where they provoke retinal hemorrhage and uveitis.

Calabar swellings mostly occur on the limbs, but can also affect the face and other body parts. Calabar swellings are acquired angioedemas taking place in subcutaneous and hypodermic tissues. They would occur following the release of microfilariae by a mature female worm (Fain and Maertens 1973) and are suspected to result from the discharge of allergens produced during the lysis of microfilariae (Eveland et al. 1975). According to other investigators, Calabar swellings would rather be caused by antigenic products emitted by adult worms and can occur at distance from the "source" worms (Negesse et al. 1985).

In loiasis endemic areas, about two-thirds of the population would experience at least one eye worm episode in a lifetime and 75 % at least one episode of Calabar swelling. To our knowledge, only one study has analyzed in depth their respective frequency in an endemic population, as well as the possible association between those two clinical signs and the presence of microfilaremia (Chesnais 2014). The frequency of eye worm seems positively correlated with that of Calabar swelling; however, they would be independent of the presence of *Loa* microfilaremia. The main result of this analysis was that four different clinical profiles could be defined:

1. *Amicrofilaremic individuals presenting with clinical signs*: they compose the so-called "occult loiasis" group. Among them, two subgroups can be defined: (a) those presenting both eye worm and Calabar swelling history (about 30 % of the total population, constant across all age groups) and (b) those presenting with Calabar swelling only (about 10 % but this proportion tends to decrease with age, indicating that some of these individuals are likely to become microfilaremic at some point).
2. *Microfilaremic individuals presenting with no clinical signs*: this profile is the less common and represents only 2–5 % of the population.
3. *Microfilaremic individuals presenting with one or the two clinical signs* (*most frequently both of them*): they represent about 40 % of the total population; this proportion tends to increase from childhood to adulthood but is quite constant afterwards.
4. *Individuals with no sign of loiasis* (*including no microfilaria in the blood*): this group represents about 10 % of the total population. Since this proportion is relatively constant across age groups and does not correspond to a particularly nonexposed group of individuals, it suggests that they could be refractory to the infection, likely resulting from predisposition factors.

During the same study, a special focus was given to the description of Calabar swelling episodes. First of all, in the study population, Calabar swelling was almost systematically (95 %) associated with pruritus. Then, using multifactorial statistics, three types of episodes were identified:

1. Short (<2 days) and painless episodes of edema, with no specific frequency or localization on the body.
2. Long (>7 days) and painful edema: they are usually generalized edema, progressively affecting several parts of the body, but are rarely recurrent (<2 episodes during the previous year).
3. Edema lasting 3–5 days, moderately painful, essentially localized on the limbs and recurrent (≥4 episodes during the previous year).

This analysis shows the spectrum of Calabar swelling episodes and suggests the existence of immuno-tolerance profiles in relation with genetic or immunologic factors.

Loiasis is also associated with pruritus (related or not to edema), urticaria-like cutaneous eruption, arthralgia, and headaches. These symptoms are generally considered non-severe but would be a major cause of medical consultation in Central Africa. Yet because they are not specific of loiasis and could be due to a number of infectious or noninfectious causes, the fraction attributable to loiasis is currently a matter of speculation.

27.3.3 Infrequent Severe Symptoms

On rare occasions, loiasis could also cause severe condition such as nephropathy. A number of cases of pericarditis have been attributed to loiasis. This would result from chronic and intense hypereosinophilia. However, as for mild symptoms, it is unclear whether pericarditis is ascribable to idiopathic eosinophilia or to loiasis-induced hypereosinophilia. Lastly, *Loa* microfilaremia has also been incriminated in hormonal disturbance delaying puberty and disrupting menstruation and potentially leading to hypogonadism.

27.3.4 Perspectives

To date, the clinical picture of loiasis has essentially been characterized in case reports, either presenting imported cases of natives or expatriates returning to their country of origin, and, for the most part, issued by medical teams from non-endemic countries. A comprehensive study of the clinical manifestations of loiasis, including the kidney, heart, and other systemic symptoms, in endemic populations has long been awaited (Metzger and Mordmüller 2014). Such a study, considered together with the distribution of loiasis prevalence throughout its endemic area, would allow the estimation of morbidity metrics such as disability-adjusted life years (DALYs). This critical work could possibly lead to the recognition of loiasis as a proper neglected tropical disease, worthwhile of specific control measures.

27.4 Diagnosis

27.4.1 Clinical Evidence

In endemic areas, hypereosinophilia together with transient episodes of subcutaneous swelling of the upper limbs is highly indicative of loiasis. However, the only pathognomonic sign of loiasis is the "eye worm." Yet the eye worm episodes do not occur very often, nor in every infected individual, so the diagnosis should consider clinical evidence in relation to the patient's history of exposure to loiasis (refer to Boxes 27.1 and 27.2, Fig. 27.1).

Box 27.1

Diagnostics in Practice: When to Consider Loiasis?

Loiasis should be considered for any patient with a history of exposure to *L. loa*. Exposure to loiasis includes short-term stays that may have taken place within the previous 20 years.

Patients with Clinical Signs Evocative of Loiasis: Eye Worm and Angioedema

Any patient presenting an episode of *subconjunctival passage of a worm* should be considered infected with *L. loa*. As suggested by a case from India with no history of travel in an endemic area, loiasis transmission may exceptionally occur outside West and Central Africa. When the worm can be extracted from the eye of a patient with no past exposure to the parasite, the specimen should nonetheless be genetically identified using specific probes.

In any other case, a thorough interview should be conducted to determine precisely the history of possible exposure to *L. loa*. We recommend the use of the detailed maps provided by the WHO (2010) to evaluate the degree of exposure to loiasis. Indeed, in endemic areas, the prevalence of infection may virtually reach 90 % (see above). Thus, the occurrence of a clinical sign evocative of loiasis (angioedema, pruritus) in a person who has stayed or lived in a hyperendemic area will be a strong argument to support the diagnosis of loiasis. In a complementary manner, if the patient has no history of strong exposure to *L. loa*, neurotic angioedema, which often appears between 10 and 20 years old, should be considered. Moreover, differential diagnosis for pruritus can be sought.

Systematically, the next step should consist of a blood exam composed by a complete blood count to measure eosinophilia and for the research of *L. loa* microfilaremia during daytime (possibly around midday). The research of *L. loa* microfilariae should be preferably done by qPCR. If qPCR is not available, a thick blood smear can be done. In the latter case, a differential diagnostic should be made with *Wuchereria bancrofti* infection (microfilariae are of similar size and morphology). Despite inverted periodicities in the peripheral blood (diurnal for *L. loa* and nocturnal for *W. bancrofti*), their concomitance in a blood sample has occasionally been observed. The presence of *L. loa* mf in the blood definitely confirms the diagnosis. Nonetheless, a negative PCR or blood smear exam does not allow discarding loiasis.

In individuals with no blood mf, differential diagnosis has to be made. Results of the complete blood counts will guide the diagnosis, especially the presence of hypereosinophilia. The next step should be a thorough differential diagnostic including, and this is not exhaustive, tropical parasitosis (e.g., strongyloidiasis) and allergic reactions. If no differential diagnostic can be made, a trial treatment can be considered (see Box 27.2 for modalities).

(continued)

Patients with Clinical Signs Non-evocative of Loiasis (Ocular Impairment, Cardiopathy, Nephropathy, Neuropathy, Functional Asplenia)

In patients presenting with organ function impairment, the diagnosis of loiasis can be made by elimination. Usual causes of impairment should be sought first. If no evidence of a specific cause can be found, loiasis has to be considered. In addition to the blood exam to detect *L. loa* mf (see above), the presence of *L. loa* (either as mf or adult worms) may also be searched in biopsies of impaired organs. Should all tests be negative, no treatment for loiasis shall be recommended.

Box 27.2

Treatment in Practice

Treating loiasis is a sensitive question. The vast majority of affected people live in the most deprived health settings. For the time being, there is no available drug suitable for large-scale treatment. Yet, in 2012, a sensible individual treatment strategy has been proposed (Boussinesq 2012). The recommended protocols depend entirely on the microfilarial density and do not consider the symptoms. Some protocols may require medical surveillance, either as ambulatory visits or hospitalization for several days. The aim of this surveillance is the early detection and medical case management of possible adverse events. As a matter of fact, this treatment protocol is hardly applicable in low-resource settings. Four protocols have been defined, according to the *L. loa* microfilarial density in the peripheral blood and associated risk of adverse events. The aim of the protocol is to bring the patient into the safest group, in which he can safely receive the curative drug (DEC).

See treatment tables for details.

27.4.2 Parasitological Evidence

Evidencing the presence of microfilariae in the blood ascertains the diagnosis. The research of microfilariae is done by performing a blood smear, preferably between 10:00 a.m. and 4:00 p.m., to fit with the diurnal periodicity. The fresh blood sample can be examined on the spot and, where microfilariae are present, they can be seen wiggling. However, the blood smears are usually dried, dehemoglobinized, stained with Giemsa stain, and then examined using optical microscopy. The volume of blood sampled can be calibrated using a capillary (typically 70 or 50 μL), and by counting all microfilariae present on the slides, one can assess the microfilarial density. Performing a calibrated blood smear is quite easy, does not require sophisticated material, and has thus been extensively used during field surveys. The main shortcoming of blood smears is its lack of sensitivity when the microfilarial density is low. In addition, because of the proportion of occult loiasis (infection without microfilaremia) in endemic areas, it is critical to keep in mind that, more than ever, the absence of evidence (microfilariae in the blood) is not an evidence of absence of infection. As seen above, in areas where loiasis is highly endemic, up to 90 % of the population is infected with *L. loa*.

A number of blood-processing techniques have been proposed to improve the sensitivity of the blood smear, essentially by concentrating the microfilariae: blood filtration through a membrane (Millipore or Nucleopore), sedimentation of blood mixed with formalin, centrifugation of blood after hemolysis with saponin or formalin, and density gradient centrifugation. However, these techniques requiring some specific non-portable material, they are not practical for large-scale epidemiological surveys.

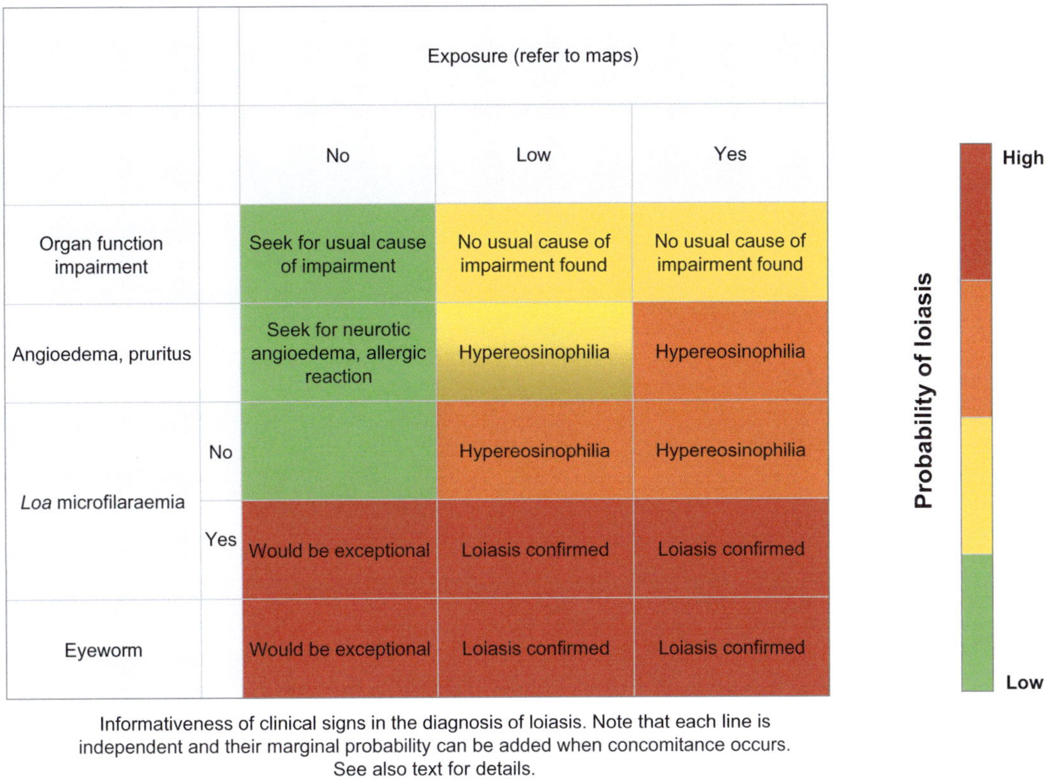

Fig. 27.1 Informativeness of clinical signs in the diagnosis of loiasis. Note that each line is independent and their marginal probability can be added when concomitance occurs. See also text for details

27.4.3 Serologic and Biomolecular Assays

Serological assays have been developed. Unlike the blood smear, they can be used to detect an occult infection but overall have shown low sensitivity. In addition, they cannot discern active and past infection, nor they can quantify the density of microfilariae.

In recent years, the research for a specific and sensitive *L. loa* diagnostic has been based on DNA detection. Real-time PCR (rtPCR)-based methods have shown very good sensitivity and specificity and, in addition, may provide accurate quantification of microfilariae (Fink et al. 2011). This being said, these methods require a centralized, well-equipped laboratory and relatively expensive reagents and may not be easily practicable in resource-limited countries for the time being. To encompass this barrier, colorimetric loop-mediated isothermal amplification (LAMP) assays have been developed. LAMP assays do not require sophisticated equipment, such as a thermocycler or an automated fluorescence reader, and can be done with reduced cost and time compared to rtPCR methods. They have recently shown very encouraging results (Drame et al. 2014), and since they can provide a colorimetric readout within minutes of processing, they are potentially adaptable to move out of the laboratory and to be performed in field conditions.

27.4.4 Novel Portable Technologies for Rapid Testing

Very recently, attempts have been made to develop a rapid, point-of-care, and field-friendly diagnostic method for *L. loa* microfilaremia. This research has been specifically motivated by the need to expand the large-scale treatments pro-

- Patients with >30,000 mf/mL

Treatment	Drug dosage	Where	Duration	Control	Objective	Parasite-specific adverse events	Adjunctive therapy
Apheresis		Hospital	Repeat session if necessary	Mf density right after the session	Microfilarial density <30,000 mf/mL	None	None
Or							
Albendazole	200 mg twice daily	Ambulatory visits	21 days, repeat the course if necessary and absence of drug-specific AEs	– Mf density at 6 months – Complete blood count + liver function at day 10	Microfilarial density <30,000 mf/mL	Calabar swelling	Anti-histaminic or corticoides

- Patients with 8000–30,000 mf/mL

Treatment	Drug dosage	Where	Duration	Control	Objective	Parasite-specific adverse events	Adjunctive therapy
Albendazole	200 mg twice daily	Ambulatory visits	21 days, repeat the course if necessary and absence of drug-specific AEs	– Mf density at 6 months – Complete blood count + liver function at day 10	Microfilarial density <8000 mf/mL	Calabar swelling	Anti-histaminic or corticoides
Or							
Ivermectin	150–200 μg/kg	Hospital 3–4 days	Single oral dose, repeat if necessary	Day 30	Microfilarial density <8000 mf/mL	– Pruritus – Angioedema – Arthralgia – Headache – Fever	Anti-histaminic or corticoides plus symptomatic treatment if needed

- Patients with 2000–8000 mf/mL

Treatment	Drug dosage	Where	Duration	Control	Objective	Parasite-specific adverse events	Adjunctive therapy
Ivermectin	150–200 μg/kg	Ambulatory visits	Single oral dose, repeat if necessary	Day 30	Microfilarial density <2000 mf/mL	– Pruritus – Angioedema – Arthralgia – Headache – Fever	Anti-histaminic plus symptomatic treatment if needed

- Patients with <2000 mf/mL

Treatment	Drug dosage	Where	Duration	Control	Objective	Parasite-specific adverse events	Adjunctive therapy
DECa (in amicrofilaremic patients)	Start with 50 mg/day in 2–3 doses, then double the dose every day up to 400 mg/day (in 2–3 doses)	First week in hospital then ambulatory visits	21–28 days, repeat the course twice if necessary	None	Absence of symptoms	None expected	
Or							
DEC (in microfilaremic patients)	Start with 3–6 mg/day in 2–3 doses, then double the dose every day up to 400 mg/day (in 2–3 doses)	First week in hospital then ambulatory visits	21–28 days, repeat the course twice if necessary	Day 30	Mf clearance, absence of symptoms	– Pruritus – Angioedema – Arthralgia – Headache – Fever	Anti-histaminic or corticoides plus symptomatic treatment if needed
Or							
Ivermectin if patient refractory to DEC	150–200 μg/kg	Ambulatory visits	Single oral dose every 3 months during 2 years	Every 6 months	Mf clearance, absence of symptoms		

aDEC: Diethylcarbamazine

grams for onchocerciasis and lymphatic filariasis (LF) to areas where loiasis is also endemic (see Sect. 27.8). In those areas, individuals with heavy *L. loa* microfilarial densities (>30,000 microfilariae per mL of blood) are at risk of developing serious adverse events (SAEs) when receiving a standard dose of ivermectin. Those individuals represent only a small fraction of the population, but, because of the life-threatening nature of SAEs, it is currently preferable to exclude the whole population, i.e., several million people, from mass drug administration (MDA) with ivermectin. Being able to detect the individuals at risk of SAEs, and to exclude them from ivermectin treatments according to a test and (not) treat strategy, would open up large areas for onchocerciasis and lymphatic filariasis elimination. In addition, implementing MDA with ivermectin in *L. loa* endemic areas would also be highly beneficial against loiasis.

Over the past year, two methodologies emerged as being potentially useful as point-of-care rapid tests. The first is a handheld automated cell counter (Scepter 2.0, Millipore) that was developed as a rapid and convenient method for enumerating purified population of cells using microfluidic-based pipette tip. The Scepter has been repurposed for the quantification of blood microfilariae (Bennuru et al. 2014). A small quantity of blood (20 μL) is collected by finger prick, mixed in 80 μL of 10 % saponin solution to lyse the blood cells, and analyzed using 60 μm sensors. In field setting conditions, results were obtained in less than 2 min from the time of blood draw (finger prick). Microfilaria counts correlated well with counts obtained with a calibrated blood smear [correlation coefficient = 0.89, CI (0.78–0.94), $p < 0.0001$], but the Scepter showed only 66 % sensitivity (versus calibrated blood smear) to detect microfilarial densities above the critical threshold of 30,000 mf/mL. Although the Scepter appears as a potentially functional point-of-care diagnostic tool, a comprehensive cost analysis would be necessary to help establish whether the introduction of this tool would be worth the investment of resources for a better clinical outcome.

The second methodology, which has also recently been under development, is based on automated counting of *L. loa* microfilariae in peripheral blood using a smartphone embedded in a portable microscopic device, the *CellScope Loa*. The principle of this test is to record five 3 s movies of a 20 μL rectangular capillary filled with blood collected by finger prick. The movies are then analyzed by a dedicated mobile phone software application, and all microfilariae present in the field (each field corresponds to 2.73 μL of whole blood) are numbered (D'Ambrosio et al. 2015).

The CellScope Loa performance (versus the gold standard calibrated blood smear method) was assessed, in 2015, in about 15,000 individual residents of the central region of Cameroon, where *L. loa* is highly endemic. About 2 min was necessary to screen a patient for SAE risk with the CellScope Loa, including time to fill the capillary with blood, load the capillary into the microscope, trigger the software application to collect a series of videos, and receive the results from automated image processing. The CellScope Loa showed 95 % sensitivity and 99 % specificity to identify individuals with >30,000 mf/mL (unpublished results). Cost-effectiveness analysis of a test and (not) treat strategy based on the CellScope Loa is currently in progress.

27.5 Chemotherapy

A number of anthelmintic molecules have been tested on *L. loa*. For the time being, none of them meet with the ideal properties of 100 % efficacy and 100 % safety. We present below the molecules of interest and possible safe strategies to treat *L. loa* infection.

27.5.1 Diethylcarbamazine

Since the early 1950s, diethylcarbamazine (DEC) has been the treatment of choice for loiasis. DEC has a major advantage over other drugs active on *L. loa*: it is macrofilaricidal, i.e., induces the death of adult worms, whereas other drugs are only microfilaricidal and/or embryotoxic. However, DEC should be used with much caution, since it may induce severe adverse events, such as encephalitis, in some patients.

DEC ingestion brings about a concentration of *L. loa* microfilariae in the liver, where they undergo antibody opsonization and are then phagocytized by Küpffer cells. Microfilaria clearance from the human bloodstream is often attained within several days following the intake of a relatively small dose of DEC (see below). Different treatment protocols have resulted in total disappearance of clinical signs (e.g., Klion et al. 1994). Those protocols consist in 3–4 weeks courses with an increasing dose of DEC, and the death of adult worms requires one or more courses of treatment. However, likewise ivermectin, DEC can induce mild to severe adverse events, including fatal encephalopathies (Fain 1978; Carme et al. 1991a). The probability of occurrence of those adverse events, as well as their severity, is positively correlated with the microfilarial density. As a consequence, no treatment should be given without prior assessment of the microfilarial density. The treatment strategy will thereafter be defined according to the level of microfilarial infection (see Box 27.2).

27.5.2 Ivermectin

27.5.2.1 Ivermectin at Standard Dose (150–200 μg/kg)

The first trials of ivermectin (200 μg/kg, unique oral dose) on *L. loa* concluded to its safety and efficacy (e.g., Carme et al. 1991b). On average, the microfilarial densities are reduced by 85 %, 95 %, and 90 %, respectively 1 week, 3 months, and 1 year after a single oral dose of ivermectin (unpublished observations). In addition, ivermectin treatment often brings about an improvement in the clinical signs, especially in the itching and the frequency of Calabar swellings.

However, with the large-scale treatments with ivermectin implemented in the mid-1990s to control onchocerciasis, a number of cases of encephalopathy, usually associated with coma, occurring 3–4 days after a treatment with ivermectin, were soon reported from different Central African countries (Twum-Danso and Meredith 2003). In 1997, it was demonstrated that, likewise post-DEC SAEs, high *L. loa* microfilarial density was the main risk factor for post-ivermectin SAEs (Gardon et al. 1997). After a standard dose of ivermectin, the risk of occurrence of non-neurological SAEs (i.e., developing a functional impairment requiring at least 1 week of full-time assistance) would be increased above 8000 mf/mL, and the risk of neurological SAEs would be significantly increased above 30,000 mf/mL. Based on retrospective observations, the incidence of SAEs with functional impairment is 2.5–4.0/1000 (M. Boussinesq personal communication). The occurrence of those SAEs has seriously hampered the programs for onchocerciasis and lymphatic filariasis control/elimination in Africa (see Sect. 27.8).

27.5.2.2 Ivermectin at Lower Dosage to Prevent the Occurrence of SAEs

Assuming a dose-effect relationship of ivermectin on *L. loa*, two trials were conducted to assess whether a lower dosage of ivermectin would bring about a more progressive decrease than a standard dose. In the first trial, a dose of 3 mg (i.e., 50 μg/kg for a 60 kg person) tended to decrease slightly less rapidly the *L. loa* microfilaremia than the standard dose, but the difference between the two groups was not significant (Kamgno et al. 2000). In the second trial, lower doses were tested: a single dose of 1.5 mg (i.e., 25 μg/kg for a 60 kg person) and two doses of 1.5 mg given at a 2-week interval. Those two protocols brought about a lower mean reduction rate in the *L. loa* microfilaremia than a standard dose of 150 μg/kg. However, there was great between-patients variability in the decrease, ranging from an absence of response to a decrease similar to what is expected after a standard dose. In addition, the side effects (none of them was serious) were not less frequent in the low-dose groups.

27.5.3 Benzimidazoles: Mebendazole and Albendazole

27.5.3.1 Mebendazole

Mebendazole has been shown to be a safe and effective microfilaricidal on a number of filarial species (*Brugia pahangi*, *W. bancrofti*, *O. volvu-*

lus, and *M. perstans*). Mebendazole, likewise other benzimidazoles, would inhibit the binding of tubulin into microtubules. As a result, the production of neuromuscular mediator, as well as glucose absorption, would be irreversibly reduced leading to the paralysis of the parasite.

The different trials of mebendazole on *L. loa* have shown uneven efficacy between trials or even between patients. The first trial showed that 2 g of mebendazole given daily for 3–4 weeks brings about a slight decrease in microfilaremia (Richard-Lenoble et al. 1985). Following the same treatment dosage, no decrease in microfilaremia during the 21-day treatment course nor after this period was obtained (Burchard and Kern 1987). The most interesting effect was obtained with either three daily doses of 100 mg, or one or three daily dose of 500 mg, during 21 days (Van Hoegaerden et al. 1987). A progressive but significant reduction in microfilaremia was obtained in 16 of 19 treated patients. However, no change in microfilaremia was observed in the three remaining treated individuals.

The limited absorption and rapid metabolism of mebendazole has meant that the drug has to be administered at high and prolonged doses to obtain an effect on *L. loa*. The frequency of adverse effects that occur because of the high doses necessary to achieve reasonable blood levels has limited its utility (Klion and Nutman 2011). In addition, such a dosage is unpractical for use in routine but may constitute a preliminary treatment to progressively decrease the microfilarial density below the threshold which ivermectin or DEC can be safely administered.

27.5.3.2 Albendazole

Antifilarial activity of albendazole on *L.* loa has also been assessed in several trials. They show that, likewise mebendazole, a prolonged dosage (200 mg twice a day during 3 weeks) can result in a progressive and significant decrease in microfilarial density over a 6-month follow-up (Klion et al. 1993). Shorter protocols (400 mg daily during 3 days, 600 mg single dose, 800 mg daily during 3 days) or repeated treatments (six 800 mg doses given every 2 months) brought out mixed results, with significant decrease in some individuals, but a lack of effect in a significant proportion of others.

No side effects were recorded during these trials, but the occurrence of a neurologic complication (delirium) following the treatment of a symptomatic *L. loa*-infected patient with a short course of albendazole (2 × 200 mg) was reported (Blum et al. 2001). The latest observation should warrant a thorough monitoring of patients during any course of treatment including albendazole.

27.5.4 Antimalarial Drugs

Properties of antimalarial drugs on filariae have been evaluated. Results of in vivo trials of chloroquine and amodiaquine have been of particular interest. Chloroquine resulted in a marked decrease in *O. volvulus* microfilarial loads and had a macrofilaricidal effect when injected into the onchocercal nodules, but not when administered orally. Amodiaquine showed a macrofilaricidal effect on *W. bancrofti* in one patient treated with a high dose (2400 mg over 4 days). Those observations inspired a controlled trial to assess the effect of standard doses of quinine (600 mg twice a day for 5 days), chloroquine (600 mg once a day for 3 days), amodiaquine (800 mg once a day for 3 days), and artesunate (200 mg once a day for 3 days) on *L. loa* microfilaremia. Unfortunately, no significant change in the loads was recorded in any of the treatment groups during a 90-day follow-up (Kamgno et al. 2010). For supplemental information, in a comprehensive literature review, see Kamgno et al. (2010).

27.5.5 Post-therapeutic Serious Adverse Treatments

27.5.5.1 Physiopathogenesis of Serious Adverse Events

The pathogenesis leading to SAEs is not clearly known. Different mechanisms may explain the pathogenesis of SAEs, and it is likely that they act concurrently (Boussinesq et al. 2003).

The first mechanism would be an obstruction of the cerebral microcirculation, due to the paralysis of microfilariae present in blood vessels. Retinal lesions and retinal hemorrhages occurring during post-ivermectin SAEs (Fobi et al.

2000) suggest that an embolism process could take place directly in the brain capillaries.

The second mechanism would be that the microfilariae, fleeing from the blood vessels after the treatment was absorbed, cross the hematoencephalic barrier and penetrate the cerebral tissue. Although this mechanism of crossing of blood-brain barrier has not been demonstrated, *L. loa* microfilariae have been found in the brain of a person deceased after a post-DEC encephalitis (van Bogaert et al. 1955). Once in the brain, inflammatory processes resulting from the presence of microfilariae or from microfilariae-secreted antigens may gradually lead to encephalitis.

27.5.6 Treatment of SAEs

The first-line strategy to handle post-therapeutic *L. loa*-related SAEs is to prevent their occurrence. As indicated above, the quantitative assessment of microfilarial density should be systematic prior to treatment, and the result should guide the treatment strategy.

However, the current strategy of control programs for lymphatic filariasis and onchocerciasis elimination relies on large-scale treatment (without individual diagnostic) of the population with ivermectin (plus albendazole for lymphatic filariasis). In loiasis endemic areas where onchocerciasis is meso- to hyperendemic, it is considered that the benefits of MDA with ivermectin for the populations justify the risk of SAE occurrence in a small fraction of the population. Thus, guidelines for the management of SAEs have been issued (The Mectizan® Expert Committee, The Technical Consultative Committee 2004). Basically, the treatment of these SAE cases relies on the provision of adequate hydration and nutrition and the prevention of decubitus complications by nursing care and antibiotics. Although no strictly controlled trial has been conducted, corticosteroids do not seem to improve the condition in the short term and may even induce iatrogenic complications (Boussinesq et al. 2003). Apart from steroids and antibiotics, no chemotherapeutic agent has been used regularly in the management of *L. loa*-related SAEs, and no specific treatment is currently recommended.

27.6 Epidemiology

27.6.1 Geographical Distribution

Literature reviews of loiasis surveys have been published in the 1970s and then comprehensively updated in 1997 (Boussinesq and Gardon 1997). *Loa loa* is restricted to Africa, with a distribution that goes from southeastern Benin in the west to southern Sudan in the east, and down south to Angola and possibly Zambia. In the early 2000s, the critical need for evaluation of loiasis endemicity in areas co-endemic with onchocerciasis (see Sect. 27.8) has led to the development of the Rapid Assessment Procedure for Loiasis (RAPLOA) (WHO 2010). This method is based on the prevalence of history of "eye worm" in a community, assessed through interviews of a sample of adult inhabitants. RAPLOA has been implemented in nearly 5000 communities covering an area of 2500 × 3000 km centered on the Congo Basin forest. Statistical kriging procedure was used to produce spatially smoothed contour maps of the interpolated prevalence of eye worm and provide a global map of loiasis based on actual survey data (Zouré et al. 2011; see Fig. 27.2).

There is no precise estimate of the population actually infected by *L. loa*. The relationship between the prevalence of eye worm in a community and the prevalence of actual infection (including microfilaremic and amicrofilaremic individuals) has not been assessed, so the latter cannot be derived from the former. This being said, it seems reasonable to assume that tens of millions of people are currently infected by *L. loa*. Evaluation of accurate figures should be on the agenda of health officers.

The highest loiasis endemicity occurs in rainforest, where the vegetal cover is dense and is therefore highly favorable to the contact between humans and vectors. Yet loiasis is not exclusively distributed in forest settings and has also been found in numerous savannah settings. Loiasis can be considered as a typically rural parasitosis. However, a case report of an infected person with a 3-day stay in the town of Lagos, Nigeria, as unique history of exposure, suggests that *Chrysops* bites may exceptionally occur in urban areas (Richardson et al. 2012).

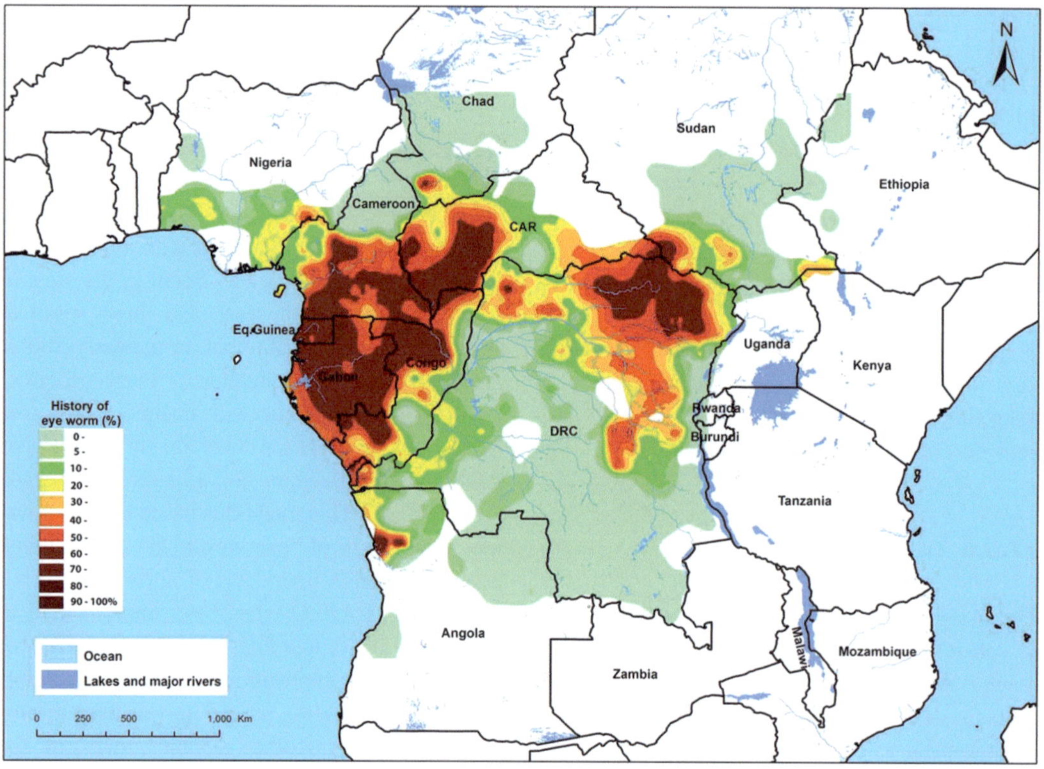

Fig. 27.2 Map of the estimated prevalence of eye worm history in Africa. (Zouré et al., doi:10.1371/journal.pntd.0001210.g004)

On a larger scale, regular *L. loa* transmission does not occur outside its endemic area, but, again, a case report from India with no history of travel in an endemic country (Bhedasgaonkar et al. 2011) suggests that potential vectors able to support the development and transmission of *L. loa* from an infected returning traveler are present across the world.

27.6.2 Animal Reservoir

A number of Cercopithecidae monkeys are naturally infected with *Loa* spp. (Orihel and Moore 1975). The simian parasites are morphologically slightly different from the ones found in humans: for example, the mean length of adult females in naturally infected *Mandrillus leucophaeus* is 8–10 cm, whereas it is 3.8–6 cm for mature worms collected from humans (Fain and Maertens 1973). Whether simian parasites can infect humans, and vice versa, is a question of relevance because of possible epidemiological implications.

Experimental attempts to infect a human with a nocturnally periodic *Loa* strain have proven unsuccessful (Duke 2004). However, monkeys have successfully been inoculated with diurnal *Loa* provided by a human (Duke and Lewis 1964). The same author showed that hybridization of simian and human parasites was also possible and produces fertile hybrids, though not highly fecund. However, for ecological reasons mainly related to the biology of vectors (see above), hybridization is unlikely to occur under natural conditions, and it is believed that human and simian parasites are in a process of divergent evolution (Orihel and Moore 1975).

27.7 Prophylaxis

DEC has prophylactic effect against *L. loa*. Two treatment schedules have been proposed for previously unexposed individuals: a weekly oral dose of 300 mg (Nutman et al. 1988) or two daily doses of 50 mg (about 5 mg/kg) during three consecutive days once each month (Duke 1963). However, because of the risk of adverse events, DEC should only be given to persons assuredly free of *O. volvulus* or *L. loa* infection.

27.8 Challenges Associated with Loiasis

To date, *L. loa* seats at the table of the neglected tropical diseases as the uninvited guest. Indeed, loiasis has no dedicated control program but, quite ironically, hampers the existing programs for the two other major filariases, lymphatic filariasis and onchocerciasis.

First of all, implementation of elimination programs for lymphatic filariasis and onchocerciasis in areas where loiasis is co-endemic is problematic because individuals harboring high density of *L. loa* microfilariae may develop SAEs when treated with ivermectin or DEC. Not only those SAEs had a fatal outcome in hundreds of individuals, and probably left irreversible sequelae in many more, but it led officials to suspend MDA for community in needs. Indeed, because the geographical distributions of loiasis, onchocerciasis, and lymphatic filariasis overlap in Central Africa, millions of people concerned with lymphatic filariasis or onchocerciasis have been excluded from MDA. Different solutions have since been considered to implement safe MDA in loiasis endemic areas. In 2012, WHO proposed a provisional strategy for controlling lymphatic filariasis in areas with co-endemic loiasis but where onchocerciasis is absent (WHO 2012). This includes MDA with albendazole alone (preferably semiannually) together with integrated vector management. Though sensitive, the evidence base for this policy was thin at the time when it was issued, and more information was needed on the efficacy of MDA with albendazole alone for LF control or elimination in community settings. The first parasitological evaluation was conducted in October 2014, 24 months after its outset, i.e., 6 months after four rounds of MDA with albendazole alone. This evaluation survey showed dramatic reductions of *W. bancrofti* infection levels, with microfilarial densities in microfilariae (mf) carriers reduced by 86.3 % from baseline values and total clearance of microfilaremia achieved in 71 % of those individuals who were initially microfilaremic (Pion et al. 2015). These intermediate results suggest that the strategy of semiannual MDA with albendazole could be highly effective and opens up Central Africa for lymphatic filariasis elimination in areas where loiasis is co-endemic.

Areas where loiasis is co-endemic with onchocerciasis constitute a different situation. As indicated above, to date, in loiasis endemic areas where onchocerciasis is meso- to hyperendemic, the benefits of MDA with ivermectin for the populations justify the risk of SAEs. In those areas, specific monitoring of SAEs has been set up and is conducted during MDA campaigns. Now, with the paradigm shift from control to elimination of onchocerciasis, hypo-endemic areas would have to also receive MDA with ivermectin. In those areas, the balance between the benefits for the population (from which only small fraction is concerned by onchocerciasis) and the risks of SAEs is arguable. A "test and treat" strategy may be of interest in such a context. For example, individuals for which *O. volvulus* infection is confirmed (using a point-of-care test) (PATH 2014) and who do not present a risk of SAE (checked using the CellScope Loa) could receive ivermectin in total safety. Individuals with onchocerciasis but who should not receive ivermectin because of a high *L. loa* microfilarial density could receive a course of doxycycline. The ways and means of such a strategy are currently being discussed by expert committees.

A second issue caused by the presence of *L. loa* has been recently pointed out. As seen above, in Africa, the main intervention strategy to control/eliminate lymphatic filariasis is MDA with albendazole in combination with DEC (or ivermectin in countries where onchocerciasis is endemic). The method of choice to map lymphatic filariasis is to examine samples of the population using a field-friendly immunochromatographic card test (ICT) which detects *W. bancrofti* circulating filarial antigens (CFA) in a blood drop. During mapping surveys, when the prevalence of lymphatic filariasis is found to be 1 % or above in a community, decision is made to implement MDA in the whole health district for about 7 years as a way to achieve lymphatic filariasis elimination. In fact, given the small size of most rural communities in Central Africa, the presence of one individual with a positive ICT in a community triggers MDA for tens of thousands. This effective strategy has led to the implementation of MDA in 22 countries, treating more than 100 million individuals in 2012. In 2012, concerns were raised regarding the specificity of the ICT test. In two independent surveys, the presence of *W. bancrofti* could not be confirmed, neither by PCR (Bakajika et al. 2014) nor by the highly specific antigen Wb123 (Pion et al. 2014), despite significant proportions of positive ICTs. Further analysis indicated that ICT positivity was highly associated with *L. loa* microfilarial density. As a consequence, the mapping of LF in areas where loiasis is also present may have to be revised, by remapping using more specific tools, such as PCR or Wb123 assays on blood collected at nighttime. In addition, some surveys may have misled some communities to receive MDA for lymphatic filariasis, where this disease is absent. Decision should be made to possibly refocus the interventions in places of higher priority.

To conclude, another way to tackle *L. loa*-related issues may be to set up a specific elimination program for loiasis. Efficient drugs exist, procedures to safely use these drugs have been proposed, and vector control could be part of a global strategy. Not only, this would open up large areas for lymphatic filariasis and onchocerciasis elimination but would also bring a substantial relief for the populations affected with *L. loa*, a definitely deleterious parasite for humankind.

References

Akue JP, Hommel M, Devaney E (1997) High levels of parasite-specific IgG1 correlate with the amicrofilaremic state in *Loa loa* infection. J Infect Dis 175: 158–163

Bakajika DK, Nigo MM, Lotsima JP et al (2014) Filarial antigenemia and *Loa loa* night blood microfilaremia in an area without bancroftian filariasis in the Democratic Republic of Congo. Am J Trop Med Hyg. doi:10.4269/ajtmh.14-0358

Bennuru S, Pion SD, Kamgno J et al (2014) Repurposed automated handheld counter as a point-of-care tool to identify individuals 'at risk' of serious post-ivermectin encephalopathy. PLoS Negl Trop Dis 8, e3180. doi:10.1371/journal.pntd.0003180

Bhedasgaonkar S, Baile RB, Nadkarni S et al (2011) *Loa loa* macrofilariasis in the eyelid: case report of the first periocular subcutaneous manifestation in India. J Parasit Dis 35:230–231. doi:10.1007/s12639-011-0043-6

Blum J, Wiestner A, Fuhr P, Hatz C (2001) Encephalopathy following *Loa loa* treatment with albendazole. Acta Trop 78:63–65

Boussinesq M (2006) Loiasis. Ann Trop Med Parasitol 100:715–731. doi:10.1179/136485906X112194

Boussinesq M (2012) Loiasis: new epidemiologic insights and proposed treatment strategy. J Travel Med 19:140–143

Boussinesq M, Gardon J (1997) Prevalences of *Loa loa* microfilaraemia throughout the area endemic for the infection. Ann Trop Med Parasitol 91:573–589

Boussinesq M, Gardon J, Gardon-Wendel N et al (2003) Clinical picture, epidemiology and outcome of Loa-associated serious adverse events related to mass ivermectin treatment of onchocerciasis in Cameroon. Filaria J 2(Suppl 1):S4. doi:10.1186/1475-2883-2-S1-S4

Burchard GD, Kern P (1987) Failure of high dose mebendazole as a microfilaricide in patients with loiasis. Trans R Soc Trop Med Hyg 81:420

Carme B, Boulesteix J, Boutes H et al (1991a) Five cases of encephalitis during treatment of loiasis with diethylcarbamazine. Am J Trop Med Hyg 44:684–690

Carme B, Ebkili B, Mbitsi A et al (1991b) Essai thérapeutique de l'ivermectine au cours de la loase à moyenne et forte microfilarémie. Ann Soc Belg Med Trop 71:47–50

Chesnais CB (2014) Recherches épidémiologiques appliquées à la lutte contre les filarioses en Afrique centrale. PhD thesis, Université de Montpellier, France

D'Ambrosio MV, Bakalar M, Bennuru S et al (2015) Rapid, point-of-care quantification of *Loa loa* microfilariae in human whole blood with a mobile phone microscope. Sci Transl Med 7:286. doi:10.1126/scitranslmed.aaa3480

Drame PM, Fink DL, Kamgno J et al (2014) Loop-mediated isothermal amplification for rapid and semiquantitative detection of *Loa loa* infection. J Clin Microbiol 52:2071–2077. doi:10.1128/JCM.00525-14

Duke BOL (1958) Studies on the biting habits of *Chrysops*. V. The biting-cycles and infection rates of *C. silacea*, *C. dimidiata*, *C. langi* and *C. centurionis* at canopy level in the rain-forest at Bombe, British Cameroons. Ann Trop Med Parasitol 52:24–35

Duke BOL (1963) Studies on the chemoprophylaxis of loiasis II. Observations on diethylcarbamazine citrate (Banocide) as a prophylactic in man. Trop Med Parasitol 57:82–96

Duke BOL (2004) Failed attempts at experimental transplantation and transmission of nocturnally-periodic simian *Loa* from monkey to man. Filaria J 3:5. doi:10.1186/1475-2883-3-5

Duke BO, Lewis DJ (1964) Studies on factors influencing the transmission of onchocerciasis. III. Observations on the effect of the peritrophic membrane in limiting the development of *Onchocerca volvulus* microfilariae in *Simulium damnosum*. Ann Trop Med Parasitol 58:83–88

Eveland LK, Yermakov V, Kenney M (1975) *Loa loa* infection without microfilaraemia. Trans R Soc Trop Med Hyg 69:354–355

Fain A (1978) Les problèmes actuels de la loase. Bull World Health Organ 56:155–167

Fain A, Maertens K (1973) Notes sur la ponte des microfilaires chez *Loa loa* et sur le degré de maturité des vers en migration. Bull Soc Pathol Exot 66:737–742

Fink DL, Kamgno J, Nutman TB (2011) Rapid molecular assays for specific detection and quantitation of *Loa loa* microfilaremia. PLoS Negl Trop Dis 5, e1299. doi:10.1371/journal.pntd.0001299.t003

Fobi G, Gardon J, Santiago M et al (2000) Ocular findings after ivermectin treatment of patients with high *Loa loa* microfilaremia. Ophthalmic Epidemiol 7:27–39

Garcia A, Abel L, Cot M et al (1999) Genetic epidemiology of host predisposition microfilaraemia in human loiasis. Trop Med Int Health 4:565–574

Gardon J, Gardon-Wendel N, Demanga-Ngangue et al (1997) Serious reactions after mass treatment of onchocerciasis with ivermectin in an area endemic for *Loa loa* infection. Lancet 350:18–22. doi:10.1016/S0140-6736(96)11094-1

Goussard B, Ivanoff B, Frost E et al (1984) Age of appearance of IgG, IgM, and IgE antibodies specific for *Loa loa* in Gabonese children. Microbiol Immunol 28:787–792

Kamgno J, Gardon J, Boussinesq M (2000) Essai de prévention des encéphalopathies à *Loa loa* post-ivermectine par l'administration d'une faible dose initiale. Med Trop 60:275–277

Kamgno J, Djomo PN, Pion SD et al (2010) A controlled trial to assess the effect of quinine, chloroquine, amodiaquine, and artesunate on *Loa loa* microfilaremia. Am J Trop Med Hyg 82:379–385. doi:10.4269/ajtmh.2010.09-0573

Klion AD, Nutman TB (2011) Loiasis and Mansonella infections. In: Richard L (ed) Tropical infectious diseases: principles, pathogens, and practice, 2nd ed. Saunders and Elsevier Edinburgh, pp 1163–1175. Source: http://www.sciencedirect.com/science/article/pii/B9780702039355001427

Klion AD, Massougbodji A, Horton J et al (1993) Albendazole in human loiasis: results of a double-blind, placebo-controlled trial. J Infect Dis 168:202–206

Klion AD, Ottesen EA, Nutman TB (1994) Effectiveness of diethylcarbamazine in treating loiasis acquired by expatriate visitors to endemic regions: long-term follow-up. J Infect Dis 169:604–610

The Mectizan® Expert Committee, The Technical Consultative Committee (2004) Recommendations for the treatment of Onchocerciasis with Mectizan® in areas co-endemic for Onchocerciasis and Loiasis. http://www.who.int/apoc/publications/englishmectccloarecs-june04.pdf

Metzger WG, Mordmüller B (2014) *Loa loa*—does it deserve to be neglected? Lancet Infect Dis 14:353–357. doi:10.1016/S1473-3099(13)70263-9

Negesse YY, Lanoie LOL, Neafie RCR, Connor DHD (1985) Loiasis: "Calabar" swellings and involvement of deep organs. Am J Trop Med Hyg 34:537–546

Nutman TB, Miller KD, Mulligan M et al (1988) Diethylcarbamazine prophylaxis for human loiasis. Results of a double-blind study. N Engl J Med 319:752–756. doi:10.1056/NEJM198809223191204

Orihel TC, Eberhard ML (1985) *Loa loa*: development and course of patency in experimentally-infected primates. Trop Med Parasitol 36:215–224

Orihel TC, Moore PJ (1975) *Loa loa*: experimental infection in two species of African primates. Am J Trop Med Hyg 24:606–609

PATH (2014) New test will combat major cause of preventable blindness in Africa. http://www.path.org/news/press-room/703/. Accessed 26 Feb 2015

Pion SD, Montavon C, Chesnais CB et al (2014) Correlation between high *Loa loa* microfilaremia and levels of circulating filarial antigens used to detect *Wuchereria bancrofti* infection. ASTMH 63rd annual meeting, New Orleans, 2–6 Nov 2014

Pion SD, Chesnais CB, Bopda J et al (2015) The impact of mass drug administration with albendazole alone on lymphatic filariasis and soil-transmitted helminth infections: a community-based study in the Republic of Congo. Am J Trop Med Hyg 92:959–966

Richard-Lenoble D, Kombila M, Burnier I, Maganga ML (1985) Filarioses au Gabon: traitement par le mebendazole des filarioses à *M. perstans* et *Loa loa*. Bull Soc Pathol Exot 78:485–491

Richardson ET, Luo R, Fink DL et al (2012) Transient facial swellings in a patient with a remote African travel history. J Travel Med 19:183–185. doi:10.1111/j.1708-8305.2012.00612.x

Twum-Danso NAY, Meredith SEO (2003) Variation in incidence of serious adverse events after onchocerciasis treatment with ivermectin in areas of Cameroon co-endemic for loiasis. Trop Med Int Health 8:820–831. doi:10.1046/j.1365-3156.2003.01091.x

Van Bogaert L, Dubois A, Janssens PG et al (1955) Encephalitis in *Loa loa* filariasis. J Neurol Neurosurg Psychiatry 18:103–119

Van Hoegaerden M, Ivanoff B, Flocard F et al (1987) The use of mebendazole in the treatment of filariases due to *Loa loa* and *Mansonella perstans*. Ann Trop Med Parasitol 81:275–282

WHO (2010) Map of the estimated prevalence of eye worm history in Africa. WHO. http://www.who.int/apoc/raploa/en/

WHO (2012) Provisional strategies for interrupting lymphatic filariasis transmission in loiasis-endemic countries. Report on the meeting on lymphatic filariasis, malaria and integrated vector management. http://apps.who.int/iris/bitstream/10665/75139/3/WHO_HTM_NTD_PCT_2012.6_eng.pdf

Zouré HGM, Wanji S, Noma M et al (2011) The geographic distribution of *Loa loa* in Africa: results of large-scale implementation of the Rapid Assessment Procedure for Loiasis (RAPLOA). PLoS Negl Trop Dis 5, e1210. doi:10.1371/journal.pntd.0001210.t002

Dirofilariosis

28

Filipe Dantas-Torres, Emanuele Brianti,
and Domenico Otranto

Abstract

Dirofilarioses are diseases caused by spirurid nematodes of the genus *Dirofilaria*, which are transmitted by several mosquito species. The most important species causing dirofilariosis in dogs are *Dirofilaria immitis* and *Dirofilaria repens*, which may also affect other mammals, including cats and humans. *Dirofilaria immitis* causes heartworm disease in dogs and may eventually lead to the formation of pulmonary nodules in humans, which may be often confounded with malignant tumors. *Dirofilaria repens* in dogs is usually asymptomatic, but in humans the parasite may localize in different tissues and organs (e.g., the eyes, lungs, and testicles) and cause disease. The diagnosis of cardiopulmonary dirofilariosis is usually based on clinical signs and laboratorial tests, including the modified Knott test for the detection of circulating microfilariae and rapid serological tests, targeting antigens of or antibodies to the parasite. Molecular tools have also been recently developed but are still mostly restricted to research centers and universities. This chapter reviews general aspects related to the etiology, epidemiology, clinical signs, diagnosis, treatment, and control of dirofilariosis.

Keywords

Dirofilaria • Dogs • Cats • Humans • Mosquitoes • Heart • Lung • Eye • Nodules

F. Dantas-Torres (✉)
Department of Immunology, Aggeu Magalhães Research Center, Oswaldo Cruz Foundation, Recife, PE, Brazil

Department of Veterinary Medicine, University of Bari, Valenzano, Italy
e-mail: filipe.dantas@cpqam.fiocruz.br

E. Brianti
Department of Veterinary Sciences, University of Messina, Messina, Italy
e-mail: ebrianti@unime.it

D. Otranto
Department of Veterinary Medicine, University of Bari, Valenzano, Italy
e-mail: domenico.otranto@uniba.it

C.B. Marcondes (ed.), *Arthropod Borne Diseases*, DOI 10.1007/978-3-319-13884-8_28

28.1 Early History

Francesco Birago was the first to observe the etiological agent of heartworm disease in 1629. In 1856, Leidy described the agent as *Filaria immitis* and later on the species was transferred to the genus *Dirofilaria*, which was described by Railliet and Henry in 1911. The earliest report of human infection by *Dirofilaria* dates back to 1885, when Addario removed a worm from the eyelid of a woman in Milan, Italy, which he named as *Filaria conjunctivae*. Dirofilarioses have long been known as diseases of dogs. In Brazil, for instance, the first report of heartworm infection in a dog dates back to 1878, whereas the first human and feline cases were described in 1887 and 1921, respectively. In 1900, Grassi and Noè demonstrated the development of *Dirofilaria immitis* in mosquitoes. Four years later, Bancroft reported the experimental transmission of *D. immitis* to susceptible dogs by the bite of an infected *Culex fatigans* (syn. *Culex quinquefasciatus*), definitely proving the role of mosquitoes as biological vectors of the parasite.

Since these pioneer works, much information has been generated regarding the life cycle, pathogenesis, diagnosis, treatment, and prevention of *D. immitis* infection, as it will be summarized in the next sections of this chapter.

28.2 Etiology

Dirofilariosis is caused by nematodes of the genus *Dirofilaria* (Spirurida: Onchocercidae). The genus includes ~27 valid species (Table 28.1) and ~15 species of questionable validity. Interestingly, a case of human dirofilariosis reported in northern Brazil was caused by a nematode morphologically similar but genetically different from *D. immitis* from dogs living in the same region (Otranto et al. 2011b). Similarly, a *Dirofilaria* species related to but genetically different from *D. repens* was described in rural dogs in Chile (López et al. 2012). Moreover, a new candidate species named "*Candidatus Dirofilaria hongkongensis*" was reported from Hong Kong, China (To et al. 2012). These findings indicate that the number of *Dirofilaria* species may be larger, particularly among wild animals.

While different species of *Dirofilaria* have been reported in dogs and humans, *D. immitis* and *D. repens* are by far the most common and relevant species from both veterinary and medical perspectives. For this reason, the following sections of this chapter will focus primarily on these two species.

Microfilariae (first-stage larvae—L_1) of *D. immitis* present a straight tail and a spindle-shaped cephalic extremity, being 290–330 μm in length and 5–7 μm in diameter (Fig. 28.1). Those of *D. repens* are longer (350–385 μm in length and 7–8 μm in diameter), presenting a curved tail and a rounded cephalic extremity (Fig. 28.1). Adults of *D. immitis* have a filiform appearance, with females measuring 250–300 mm in length and 1–1.3 mm in diameter and males measuring 120–200 mm in length and 0.7–0.9 mm in diameter. Females of *D. repens* are 100–170 mm in length and 4.6–6.3 mm in diameter, whereas males are 50–70 mm in length and 3.7–4.5 mm in diameter. Adults of *D. immitis* live in the pulmonary artery and right ventricle of dogs (Fig. 28.2), whereas adults of *D. repens* reside in subcutaneous tissues of vertebrate hosts, being also found in the abdominal cavity and within connective muscular fasciae. Nonetheless, aberrant nodular locations have been reported in both dogs (e.g., ocular) and humans (e.g., ocular, pulmonary, testicular).

28.3 Life Cycle

In the definitive host, the life cycle of *D. immitis* begins when an infected female mosquito feeds on a susceptible animal, like a dog. During blood feeding, infective third-stage larvae (L_3) migrate actively from the mosquito proboscis to the skin of the host. Between 3 and 12 days post-infection, L_3 molt into fourth-stage larvae (L_4). Molting into preadult nematodes (fifth-stage larvae—L_5) occurs 50–70 days post-infection. The preadults reach the pulmonary artery and right ventricle of dogs from 70 to 85 days post-infection and develop into sexually mature adults at 120–180

Table 28.1 Species of the genus *Dirofilaria*, their hosts, and geographical distribution (adapted from Dantas-Torres and Otranto 2013)

Species	Host families	Distribution
D. ailure	Procyonidae	China
D. freitasi	Bradypodidae	Brazil
D. immitis	Canidae, Felidae, Hominidae, many others	Cosmopolitan
D. lutrae	Mustelidae	Florida, Louisiana
D. spectans	Hominidae (1 case), Mustelidae	Brazil
D. acutiuscula	Canidae, Caviidae, Felidae, Tayassuidae	South America, USA
D. bonnie	Muridae	Java
D. cancrivori	Procyonidae	Guyana
D. corynodes	Cercopithecidae	Africa, Thailand
D. genettae	Felidae, Viverridae	Nigeria
D. granulosa	Felidae	Africa, Asia
D. incrassata	Bradypodidae, Procyonidae	Brazil, Central America
D. linstowi	Cercopithecidae	Sri Lanka
D. macacae	Cercopithecidae	Indochina
D. macrodemos	Bradypodidae	Guyana, Panama
D. magnilarvata	Cercopithecidae, Hominidae, Hylobatidae	Malaya
D. minor	Felidae	Vietnam
D. pagumae	Viverridae	Indochina
D. panamensis	Bradypodidae	Panama
D. repens	Canidae, Felidae, Hominidae, Viverridae	Europe, Asia, Africa
D. sachsi	Bovidae	East Africa
D. striata	Canidae, Felidae, Hominidae (1 case), Tayassuidae	Brazil, Venezuela, USA
D. subdermata	Erethizontidae	North America, South Africa
D. sudanensis	Felidae, Hyaenidae	Sudan
D. tawila	Cercopithecidae	Senegal, Liberia
D. tenuis	Hominidae, Procyonidae	North America
D. ursi	Felidae, Hominidae, Ursidae	Asia, North America

The taxonomic status of some species may be arguable/doubtful

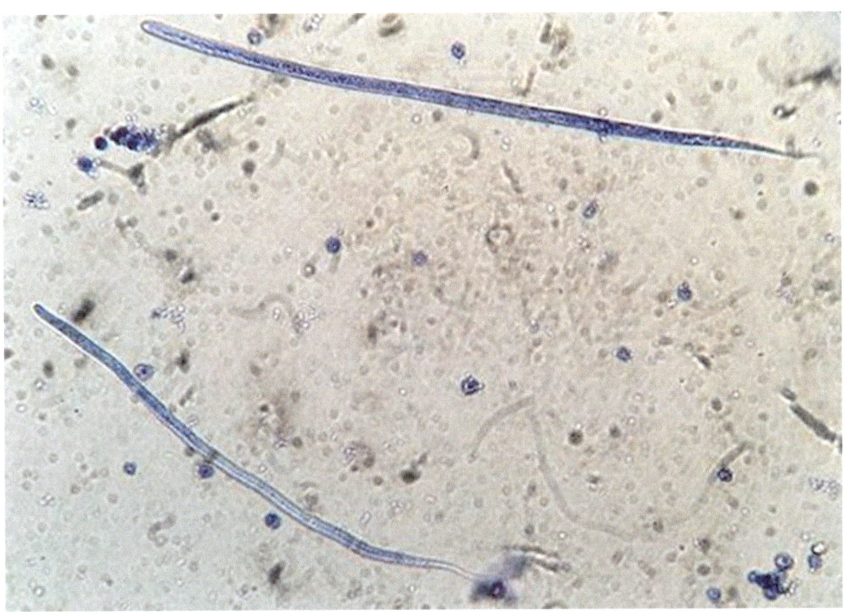

Fig. 28.1 *Dirofilaria repens* (on the *top*) and *Dirofilaria immitis* (on the *bottom*) microfilariae (courtesy of Dr. Luigi Venco)

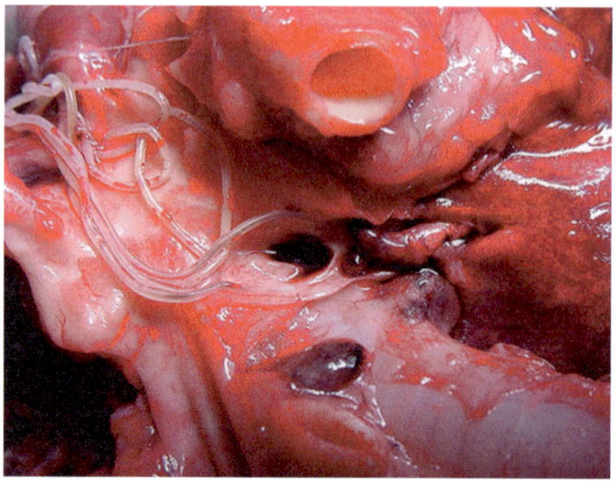

Fig. 28.2 *Dirofilaria immitis* adults in the main pulmonary artery (courtesy of Dr. Luigi Venco)

days post-infection (McCall et al. 2008). After mating, females start producing first-stage larvae, which usually occur 6–9 months post-infection. Microfilariae have an estimated life-span of 2 years, whereas adults may live for up to 7 years.

Microfilariae are ingested when a mosquito feeds on an infected host and, within approximately 24 h, reach the Malpighian tubules where they molt into second-stage larvae (L_2) between 8 and 10 days post-infection. About 3 days later, L_2 molt into L_3, which migrate to the mosquito mouthparts where they remain until the next blood meal. The L_3 development time in mosquitoes is strongly dependent on temperature. These parameters have been successfully applied to mathematical models for the prediction of *D. immitis* distribution in different continents (Cuervo et al. 2015).

28.4 Clinical and Laboratorial Findings

The great majority of the dogs infected by *D. immitis* are asymptomatic. Indeed, in hyperendemic areas, a high percentage of dogs will present circulating microfilariae, but no apparent clinical sign. The appearance of clinical signs depends on several factors, including adult worm burden and localization.

Cardiopulmonary or heartworm disease is pretty much a result of the direct damage caused by adult worms and their antigenic products, including *Wolbachia* symbiotic bacteria (Kramer et al. 2005). When present, clinical signs include coughing, hemoptysis, dyspnea, weakness, and reduced tolerance to exercise. Syncope associated with excitement or increased physical activity may occur. If congestive heart failure is present, dogs may develop ascites, peripheral edema, hydrothorax, and hydropericardium. Cardiac murmurs over the right side of the thorax and abnormal cardiac rhythm, due to tricuspid valve insufficiency and atrial fibrillation, respectively, may be present. Eventually, dogs may suddenly die following respiratory distress. Thromboembolism caused by natural or drug-induced death of adult worms may lead to life-threatening dyspnea and hemoptysis. Anemia, thrombocytopenia, proteinuria with renal insufficiency, and azotemia may be present in some cases.

Adult worms live in the pulmonary arteries and right ventricle, but in massive infections, adult worms may enter the right atrium and eventually the caudal vena cava. Obstruction of the venous return to the right ventricle may cause an acute life-threatening condition known as vena cava syndrome. This syndrome is more frequent in small-sized dogs, being characterized by volumetric and pressure overloads in the right atrium and caudal vena cava, resulting in a significant elevation in venous pressure and difficulty in return circulation (Simón et al. 2012). Dogs with vena cava syndrome may present with severe dyspnea, tricuspid valve murmurs, acute intravascular hemolysis, hemoglobinuria, and disseminated intravascular coagulation. The risk of death is eminent and immediate surgery to remove as much adult worms as possible is recommended.

Dirofilaria immitis-infected cats may eventually display clinical sings, even though subclinical infections are more common. Cats usually have a relatively low adult worm burden (2–4 worms) and low microfilaremia, being typically amicrofilaremic (McCall et al. 2008). Cats with heartworm disease may present with cough, dyspnea, and wheezing (Dillon 1998). Clinical signs are usually more evident when immature worms arrive and die in the pulmonary arteries, inducing arterial and parenchymal inflammatory responses. Neurological signs such as ataxia, syncope, blindness, or vestibular alterations can be seen depending on the localization of the migrating worms (Dillon 1998). Sudden death with no apparent clinical sign may also occur in acute cases in cats. Ferrets usually remain asymptomatic, but when present, clinical signs include dyspnea and coughing. Human patients may seldom present with cough (with pleural or non-pleural thoracic pain), hemoptysis, dyspnea, fever, and other nonspecific signs such as malaise and myalgia (Simón et al. 2012).

In contrast to *D. immitis*, *D. repens* does not usually cause disease in dogs and cats. Some dogs may present multifocal nodular dermatitis and prurigo papularis dermatitis (Simón et al. 2012). Indeed, although the disease often runs asymptomatically, a plethora of nonspecific dermic alterations has been reported (Albanese et al. 2013). In a literature review of 100 subcutaneous dirofilarioses in dogs and 31 in cats, several dermatological signs (e.g., pruritus, erythema, papules, focal or multifocal alopecia, hyperkeratosis, crusting, and nodules) were reported (Tarello 2011). Non-dermatological signs (e.g., conjunctivitis, anorexia, vomiting, fever, lethargy, lymphadenomegaly, and pale mucous membranes) have also been reported in some cases (Tarello 2011), but the involvement of *D. repens* as a cause for these specific alterations remains poorly understood. Cases of intravitreous infection in dogs may be caused by *D. repens* (Hermosilla et al. 2006) and *D. immitis* (Dantas-Torres et al. 2009). In human patient, subcutaneous dirofilariosis may be clinically characterized by the presence of a subcutaneous nodule. Ocular infections by *D. repens* in humans have been increasingly reported in the Old World, being associated with symptoms like low visual acuity, floaters, loss of sight, blepharedema, and palpebral ptosis, and moderate ocular discomfort occurs (Stringfellow et al. 2002).

28.5 Diagnosis

The diagnosis of *Dirofilaria* spp. infections is usually performed through the classical detection of circulating microfilariae. The thick blood film method may be used to detect circulating microfilariae, but it may lack in sensitivity in dogs with low parasite load. Concentration methods may increase the sensitivity of microscopic detection of microfilariae. These include the modified Knott and other tests to concentrate microfilariae. Once detected, microfilariae can be differentiated by morphological analysis (Liotta et al. 2013; Magnis et al. 2013), by the histochemical staining of anatomical regions with phosphatase activity (Peribáñez et al. 2001), or even by molecular methods (Latrofa et al. 2012a, b). Despite using concentration methods, false-negative results may occur in single-sex infections (i.e., adult male worms only), low numbers of circulating microfilariae, and past administration of microfilaricidal drugs.

Molecular methods based on the polymerase chain reaction (PCR) have been standardized for the detection and identification of *Dirofilaria* species. Different ribosomal and mitochondrial genes have been proven useful for species identification, including the mitochondrial cytochrome c oxidase subunit 1 (*cox*1). For instance, a single-step multiplex PCR has been developed for the simultaneous detection and discrimination of *D. immitis*, *D. repens*, and other filarial nematodes infecting dogs (Latrofa et al. 2012b). In the same way, a high-sensitive real-time PCR was developed to detect, discriminate, and quantify DNA of *D. immitis* and *D. repens* in the canine host and mosquito vectors (Latrofa et al. 2012a). Nonetheless, these molecular methods are still more restricted to research centers and universities.

Commercial serological tests are available for the detection of heartworm antigens and have been proven feasible for in-clinic use. They present good sensitivity and specificity and are available as rapid enzyme-linked immunosorbent assay (ELISA) and immunochromatographic test systems. These tests can also detect most of the occult infections consisting of at least one mature female worm. Still, false-negative results may occur in cases of low parasite load and low antigenemia and in dogs infected by only adult male worms. Also, dogs with circulating microfilariae of *D. repens* may be positive in the microscopic examination but serologically negative, as these tests are highly specific for *D. immitis* antigens.

Imaging diagnostic tools (e.g., radiography and echocardiography) may provide insights into the clinical status of individual patients. Typically, radiographic findings of heartworm vascular disease include enlarged, tortuous, and often truncated peripheral intralobar and interlobar branches of the pulmonary arteries, particularly in the diaphragmatic lobes. In the presence of suggestive clinical signs or radiographic findings, echocardiography should be considered. The presence and severity of pulmonary hypertension can be precisely determined by Doppler echocardiography. Moreover, echocardiography may allow the veterinary practitioner to visualize adult heartworms as two parallel hyperechoic lines in the main pulmonary artery (Fig. 28.3), left and right interlobar branches, or right heart atrium and ventricle (Simón et al. 2012).

The diagnosis of *D. immitis* infection in other hosts, such as cats, may be troublesome, due to low worm burdens, male-only infections, and nonspecific radiographic findings (Lee and Atkins 2010). Serological tests targeting antigens of or antibodies to *D. immitis* are available for cats and present good sensitivity and specificity. While antigenic tests are possible to have false-negative results, antibody assays may lead to false-positive results (Berdoulay et al. 2004). Radiography and echocardiography may be the only tools for diagnosis of the infection in cats, ferrets, and humans. Alterations in the radiographic profiles (e.g., interstitial pulmonary patterning) are transient in cats (Selcer et al. 1996), which rarely present alterations in the cardiac silhouette (Simón et al. 2012). On the other hand, cardiomegaly and severe pleural effusion are considered the main radiographic findings in ferrets (Supakorndej et al. 1995).

In most human cases, solitary peripheral subpleural pulmonary nodules with well-defined borders ("coin lesion") may be visualized on chest radiographies. Cases of ocular dirofilariosis in humans and dogs are typically diagnosed upon standard ophthalmic examination (Dantas-Torres et al. 2009; Otranto et al. 2011a, b).

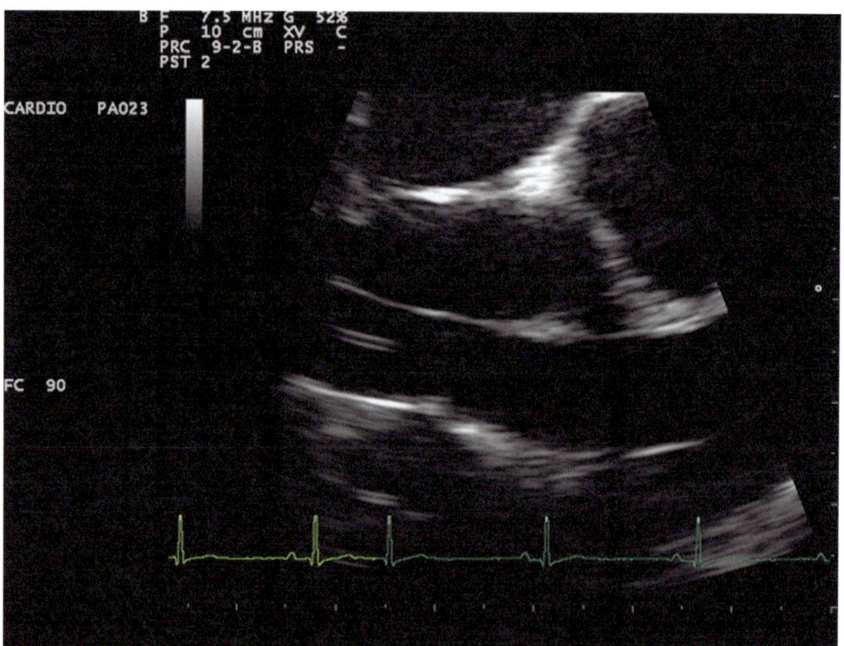

Fig. 28.3 Echocardiography: the parallel hyperechoic lines in the right pulmonary artery represent an adult heartworm (courtesy of Dr. Luigi Venco)

28.6 Treatment

The organic arsenical compound melarsomine dihydrochloride is the only effective adulticide drug available for treating dogs suffering from dirofilariosis. The treatment requires a series of precautionary measures to avoid severe complications associated with adulticide therapy, including fatal pulmonary thromboembolism. The standard protocol consists of the administration of two intramuscular injections every 24 h in doses of 2.5 mg/kg. Nonetheless, this protocol kills only 90% of the adult worms (Simón et al. 2012). Currently, the two-step treatment is the most widely accepted regimen, with can reduce the risk of thromboembolism. In practice, the treatment consists of an initial injection of 2.5 mg/kg and followed by two further injections of 2.5 mg/kg (at a 24 h interval), which are given 50–60 days after the initial one.

The risk of thromboembolism is eminent and may be clinically unapparent in some cases. However, life-threatening respiratory distress can occur in severely affected dogs. It is recommended to restrict physical activity following adulticide therapy, for 30–40 days (Simón et al. 2012). The administration of heparin and glucocorticosteroids (prednisolone at 2 mg/kg daily for 4–5 days) is also recommended. These measures can reduce the risk of thromboembolism.

In dogs with high worm burden, especially for those with vena cava syndrome, surgical therapy is recommended instead of adulticide therapy. Under general anesthesia and with echocardiographic guidance, heartworm removal can be carried out using flexible alligator forceps, which is introduced via the jugular vein (Arita et al. 2003); other methods have also been successfully used (Yoon et al. 2013). In contrast with adulticide therapy, the surgical extraction of *D. immitis* adults can reduce the risk of thromboembolism. Surgery is the recommended procedure in cases of canine ocular dirofilariosis (Dantas-Torres et al. 2009). Similarly, surgical removal is the only treatment option for cases of ocular and pulmonary dirofilariosis in human patients.

Neither adulticide therapy with melarsomine dihydrochloride nor surgical therapies are recommended in cats, due to the eminent risk of pulmonary thromboembolism and anaphylactic shock (Simón et al. 2012). Corticosteroid-based supportive therapy is recommended for symptomatic cats as well as for asymptomatic cats that exhibit radiographic evidence of disease when antigen and/or antibody tests are positive. Reduced doses of prednisolone may relieve respiratory distress (Dillon 1998). If a cat presents with severe clinical signs, high doses of prednisolone (1–2 mg/kg, three times a day) are advised. Nonetheless, the best recommendation for cats (e.g., also ferrets) is the administration of prophylaxis with macrocyclic lactones.

Several alternative protocols have been proposed for the treatment of heartworm infection in dogs, including long-term administration of macrocyclic lactones alone or in combination with doxycycline (targeting *Wolbachia* endosymbiotic bacteria) (Kramer et al. 2008; Bazzocchi et al. 2008; Grandi et al. 2010). The European Scientific Counsel Companion Animal Parasites (ESCCAP) and the American Heartworm Society do not recommend slow killing of *D. immitis* adults with long-term macrocyclic lactone therapy. Indeed, long-term therapy cannot prevent the onset of cardiopulmonary damage in dogs and potentially favor the selection of resistant subpopulations of *D. immitis*.

While several protocols have been proposed (Traversa et al. 2011; Hellmann et al. 2011), there is no consensus regarding the treatment for subcutaneous dirofilariosis caused by *D. repens*, as no adulticide therapy is known for dogs. Considering the zoonotic potential, infected dogs should be treated monthly for 12 months with microfilaricide drugs. In human cases, surgery may also be necessary to remove *D. repens* adults from subcutaneous nodules or from other locations, including from the eye.

28.7 Epidemiology

Dirofilaria immitis is a cosmopolitan filarial nematode, whereas *D. repens* is endemic in the Old World (Dantas-Torres and Otranto 2013). Studies on the epidemiology of *D. immitis* infections in dogs have been carried out in different parts of the world, and the epidemiological scenarios may vary widely. In the same way, the level of exposure of dogs and other hosts to mosquito vectors may vary according to socioeconomic and landscape characteristics of each population. In Brazil, for example, *D. immitis* is more prevalent in dogs living in coastal regions (Labarthe et al. 2014), but a recent study pointed out that riverine dogs in the Amazon region are also exposed to *D. immitis* (Soares et al. 2014). This indicates that the risk of heartworm infection may be currently underestimated in some Brazilian regions.

The prevalence of *D. immitis* may vary widely from country to country and even within a country. A 1.3 % nationwide prevalence has been recently estimated in the United States, ranging from 0.4 to 2.9 % according to region (Little et al. 2014). The prevalence of *D. immitis* in some cities in Southeastern United States may be much higher than the national or regional prevalence (Bowman et al. 2009). In other parts of the Americas, the prevalence of *D. immitis* is much higher than in the United States, reaching over 50 % in some areas. In the same way, the prevalence of *D. immitis* and *D. repens* in dogs in Europe may be very high in some endemic foci. In Serbia, for instance, researchers examined 193 blood samples from privately owned dogs resident in four provinces in Vojvodina and found *D. repens* microfilariae in 49.2 % of 193 dogs (Tasić et al. 2008). As another example, a recent study found *D. immitis* antigens in 27.3 % of 304 dogs from central Portugal (Vieira et al. 2014).The prevalence and geographical distribution of *D. immitis* are changing (either reducing or increasing), and this process has been associated to factors such as climate change, movements of infected dogs, and introduction of new mosquito vectors (Otranto et al. 2009; Morchón et al. 2012). However, it is important to highlight that the relevance of each of these factors may vary, as climate change will probably have a minimal impact on *D. immitis* transmission in some areas (Cuervo et al. 2015).

Besides dogs, cats, and ferrets, several other animals may serve as hosts of *D. immitis*, including wild canids (Tolnai et al. 2014). Wild felids have been occasionally found infected by *D. immitis* in Europe, but circulating microfilariae were detected in a few cases only (reviewed in Otranto et al. 2015). From public health perspective, dogs are the prime hosts of *D. immitis*, owing to their high rates of infection and their close relationship with humans. For instance, a study conducted in the island of Gran Canaria, Spain, an overall prevalence of 19 % was found in dogs, reaching 43 % in the autochthonous breed of Canarian Warren hound (Montoya-Alonso et al. 2011). Moreover, cats were also highly exposed to *D. immitis* (33 %), and a seropositivity of 12 % was reported in humans. Positivity in both cats and humans was correlated to the prevalence in dogs. The correlation between canine and human risk of exposure has also been suggested for *D. repens* in Sicily, Italy (Otranto et al. 2011a).

The risk factors for *D. immitis* infection in dogs may vary widely from region to region, but in a recent study in central Portugal, dogs aged between 4 and 9 years, dogs living in a rural environment, large breed dogs, and dogs living outdoors were significantly associated with *D. immitis* positivity (Vieira et al. 2014). Indeed, living outdoor, being frequently bitten by mosquitoes, may increase the risk of *D. immitis* in dogs. Guard dogs, for example, were at a higher risk in Southern Italy (Giangaspero et al. 2013). A recent study identified median household income, annual temperature, county elevation, and presence of the mosquitoes *Ochlerotatus trivittatus*[1] (=*Aedes trivittatus*), *Jarnellius sierrensis* (=*Aedes sierrensis*), and *Cx. quinquefasciatus* as significant predictors of heartworm prevalence across the United States (Wang et al. 2014).

The constant presence of efficient mosquito vectors is an important risk factor for dogs living in *D. immitis*-endemic areas. The introduction of the invasive mosquito species *Stegomyia albopicta* (=*Aedes albopictus*) has probably contributed to the spreading of *D. immitis* in Southern Italy (Otranto et al. 2009; Giangaspero et al. 2013). *Hulecoeteomyia koreica* (=*Aedes koreicus*) is another invasive species recently introduced in Northern Italy and a potential vector of

[1] See footnote 1 in Chap. 2.

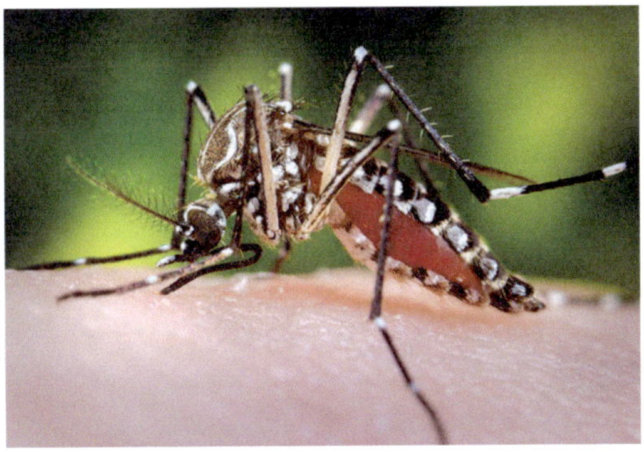

Fig. 28.4 *Aedes aegypti* female (courtesy of CDC/Prof. Frank Hadley Collins, Center for Global Health and Infectious Diseases, University of Notre Dame)

D. immitis (Montarsi et al. 2015). The number of potential vectors for both *D. immitis* and *D. repens* is impressive, as several mosquito species of different genera have been implicated in their transmission. Since *D. immitis* and *D. repens* grow under laboratory conditions in the same mosquito species with similar developmental times, they often occur in sympatry in the Old World.

Considering the worldwide distribution of *D. immitis*, the primary and secondary vector species may vary from country to country and even from region to region within a given country. Different populations of a same mosquito species may also have varying vector efficiency. In Brazil, for instance, the most important vectors are *Stegomyia aegypti* (=*Aedes aegypti*) (Fig. 28.4), *S. albopicta*, *Ochlerotatus scapularis* (=*Aedes scapularis*), *Ochlerotatus taeniorhynchus* (=*Aedes taeniorhynchus*), and *Cx. quinquefasciatus* (Fig. 28.5), but the vector efficiency of some populations of these mosquito species seems to vary.

28.8 Control and Prophylaxis

The control of dirofilariosis is based on mosquito control and treatment of infected dogs. Effective mosquito control at population level may be unrealistic in most situations, particularly in tropical and subtropical zones, where mosquitoes are naturally abundant. Nonetheless, dogs and cats living in areas where mosquitoes and other arthropod vectors (e.g., phlebotomine sand flies

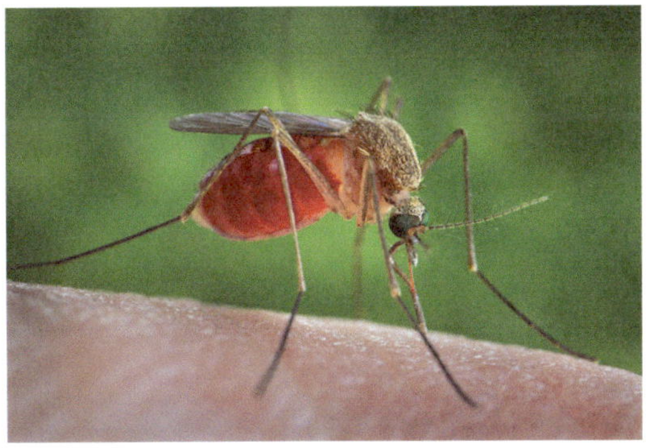

Fig. 28.5 *Culex quinquefasciatus* female (courtesy of CDC/James Gathany)

and ticks) are abundant should be protected with repellents and insecticides.

Treatment of infected dogs with an adulticide drug is also important to reduce the availability of circulating microfilariae for mosquito vectors in a given area. However, in addition to represent a high risk for the animal, adulticide treatment is not always possible for many reasons, including the unavailability of melarsomine dihydrochloride in some countries (e.g., Brazil). Therefore, the best evidence-based practice currently recommended for the prevention of dirofilariosis in dogs and cats is the monthly chemoprophylaxis with macrocyclic lactones.

The American Heartworm Society recommends the year-round chemoprophylaxis in the United States. In Europe, the ESCCAP recommends year-round treatment in some areas but not in others. Nonetheless, the increasing geographical range of *S. albopicta* in several Mediterranean countries indicates that year-round chemoprophylaxis should be used (Otranto et al. 2013). Drugs may be available as tablets/chewables, injectable, or topical presentations and may be applied monthly or at a 6-month interval. Macrocyclic lactones available for use in dogs include ivermectin, milbemycin oxime, moxidectin, and selamectin (Table 28.2).

It is important to emphasize that adult *D. immitis* infection must be ruled out by testing for circulating antigen or microfilariae before starting any chemoprophylaxis regimen. If a dog is already infected, the veterinary practitioner should recommend adulticide therapy. Chemoprophylaxis should start around 4 weeks

Table 28.2 Prevention of dirofilariosis in dogs and cats (adapted from ESCCAP—http://www.esccap.org/uploads/docs/ih38c2d6_ESCCAP_Guidelines_GL5_01Oct2012.pdf)

Compound	Presentation	Dog	Cat
Ivermectin	Tablets/chewables	6–12 μg/kg	24–71 μg/kg
Milbemycin oxime	Flavor tablets	0.5–1 mg/kg	2–4 mg/kg
Moxidectin	Tablets	3–6 μg/kg	
	Injectable	0.17 mg/kg	
	Topical	2.5–6.25 mg/kg	1–2 mg/kg
Selamectin	Topical	6–12 mg/kg	6–12 mg/kg

after adulticide therapy, and dogs should be annually retested for circulating antigens and blood microfilariae.

The resistance to macrocyclic lactones has been now confirmed in the United States and has raised a lot of questions regarding current schemes for heartworm chemoprophylaxis. Nevertheless, the American Heartworm Society has recently emphasized that the year-round chemoprophylaxis scheme should be maintained in the whole country.

References

Albanese F, Abramo F, Braglia C et al (2013) Nodular lesions due to infestation by *Dirofilaria repens* in dogs from Italy. Vet Dermatol 24:255–256

Arita N, Yamane I, Takemura N (2003) Comparison of canine heartworm removal rates using flexible alligator forceps guided by transesophageal echocardiography and fluoroscopy. J Vet Med Sci 65:259–261

Bazzocchi C, Mortarino M, Grandi G et al (2008) Combined ivermectin and doxycycline treatment has microfilaricidal and adulticidal activity against *Dirofilaria immitis* in experimentally infected dogs. Int J Parasitol 38:1401–1410

Berdoulay P, Levy JK, Snyder PS et al (2004) Comparison of serological tests for the detection of natural heartworm infection in cats. J Am Anim Hosp Assoc 40:376–384

Bowman D, Little SE, Lorentzen L et al (2009) Prevalence and geographic distribution of *Dirofilaria immitis*, *Borrelia burgdorferi*, *Ehrlichia canis*, and *Anaplasma phagocytophilum* in dogs in the United States: results of a national clinic-based serologic survey. Vet Parasitol 160:138–148

Cuervo PF, Rinaldi L, Cringoli G (2015) Modeling the extrinsic incubation of *Dirofilaria immitis* in South America based on monthly and continuous climatic data. Vet Parasitol 209:70–75

Dantas-Torres F, Otranto D (2013) Dirofilariosis in the Americas: a more virulent *Dirofilaria immitis*? Parasit Vectors 6:288

Dantas-Torres F, Lia RP, Barbuto M et al (2009) Ocular dirofilariosis by *Dirofilaria immitis* in a dog: first case report from Europe. J Small Anim Pract 50:667–669

Dillon R (1998) Clinical significance of feline heartworm disease. Vet Clin North Am Small Anim Pract 28:1547–1565

Giangaspero A, Marangi M, Latrofa MS et al (2013) Evidences of increasing risk of dirofilarioses in southern Italy. Parasitol Res 112:1357–1361

Grandi G, Quintavalla C, Mavropoulou A et al (2010) A combination of doxycycline and ivermectin is adulticidal in dogs with naturally acquired heartworm disease (*Dirofilaria immitis*). Vet Parasitol 169:347–351

Hellmann K, Heine J, Braun G et al (2011) Evaluation of the therapeutic and preventive efficacy of 2.5 % moxidectin/10 % imidacloprid (Advocate(®), Bayer animal health) in dogs naturally infected or at risk of natural infection by *Dirofilaria repens*. Parasitol Res 109:S77–S86

Hermosilla C, Pantchev N, Dyachenko V et al (2006) First autochthonous case of canine ocular *Dirofilaria repens* infection in Germany. Vet Rec 158:134–135

Kramer L, Simón F, Tamarozzi F et al (2005) Is *Wolbachia* complicating the pathological effects of *Dirofilaria immitis* infections? Vet Parasitol 133:133–136

Kramer L, Grandi G, Leoni M et al (2008) *Wolbachia* and its influence on the pathology and immunology of *Dirofilaria immitis* infection. Vet Parasitol 158:191–195

Labarthe NV, Pereira Paiva J et al (2014) Updated canine infection rates for *Dirofilaria immitis* in areas of Brazil previously identified as having a high incidence of heartworm-infected dogs. Parasit Vectors 7:493

Latrofa MS, Dantas-Torres F, Annoscia G et al (2012a) A duplex real-time polymerase chain reaction assay for the detection of and differentiation between *Dirofilaria immitis* and *Dirofilaria repens* in dogs and mosquitoes. Vet Parasitol 185:181–185

Latrofa MS, Weigl S, Dantas-Torres F et al (2012b) A multiplex PCR for the simultaneous detection of species of filarioids infesting dogs. Acta Trop 122:150–154

Lee AC, Atkins CE (2010) Understanding feline heartworm infection: disease, diagnosis, and treatment. Top Companion Anim Med 25:224–230

Liotta JL, Sandhu GK, Rishniw M et al (2013) Differentiation of the microfilariae of *Dirofilaria immitis* and *Dirofilaria repens* in stained blood films. J Parasitol 99:421–425

Little SE, Beall MJ, Bowman DD et al (2014) Canine infection with *Dirofilaria immitis*, *Borrelia burgdorferi*, *Anaplasma* spp., and *Ehrlichia* spp. in the United States, 2010–2012. Parasit Vectors 7:257

López J, Valiente-Echeverría F, Carrasco M et al (2012) Morphological and molecular identification of canine filariae in a semi-rural district of the Metropolitan Region in Chile. Rev Chil Infectol 29:248–289

Magnis J, Lorentz S, Guardone L et al (2013) Morphometric analyses of canine blood microfilariae isolated by the Knott's test enables *Dirofilaria immitis* and *D. repens* species-specific and *Acanthocheilonema* (syn. *Dipetalonema*) genus-specific diagnosis. Parasit Vectors 6:48

McCall JW, Genchi C, Kramer LH et al (2008) Heartworm disease in animals and humans. Adv Parasitol 66:193–285

Montarsi F, Ciocchetta S, Devine G et al (2015) Development of *Dirofilaria immitis* within the mosquito *Aedes* (*Finlaya*) *koreicus*, a new invasive species for Europe. Parasit Vectors 8:177

Montoya-Alonso JA, Carretón E, Corbera JA et al (2011) Current prevalence of *Dirofilaria immitis* in dogs, cats and humans from the island of Gran Canaria, Spain. Vet Parasitol 176:291–294

Morchón R, Carretón E, González-Miguel J et al (2012) Heartworm disease (*Dirofilaria immitis*) and their vectors in Europe—new distribution trends. Front Physiol 3:196

Otranto D, Capelli G, Genchi C (2009) Changing distribution patterns of canine vector borne diseases in Italy: leishmaniosis vs. dirofilariosis. Parasit Vectors 2(Suppl 1):S2

Otranto D, Brianti E, Gaglio G et al (2011a) Human ocular infection with *Dirofilaria repens* (Railliet and Henry, 1911) in an area endemic for canine dirofilariasis. Am J Trop Med Hyg 84:1002–1004

Otranto D, Diniz DG, Dantas-Torres F et al (2011b) Human intraocular filariasis caused by *Dirofilaria* sp. nematode, Brazil. Emerg Infect Dis 17:863–866

Otranto D, Dantas-Torres F, Brianti E et al. (2013) Vector-borne helminths of dogs and humans in Europe. Parasit Vectors 6:16

Otranto D, Cantacessi C, Dantas-Torres F et al (2015) The role of wild canids and felids in spreading parasites to dogs and cats in Europe. Part II: Helminths and arthropods. Vet Parasitol 213:24–37

Peribáñez MA, Lucientes J, Arce S et al (2001) Histochemical differentiation of *Dirofilaria immitis*, *Dirofilaria repens* and *Acanthocheilonema dracunculoides* microfilariae by staining with a commercial kit, Leucognost-SP. Vet Parasitol 102:173–175

Selcer BA, Newell SM, Mansour AE et al (1996) Radiographic and 2-D echocardiographic findings in eighteen cats experimentally exposed to *D. immitis* via mosquito bites. Vet Radiol Ultrasound 37:37–44

Simón F, Siles-Lucas M, Morchón R et al (2012) Human and animal dirofilariasis: the emergence of a zoonotic mosaic. Clin Microbiol Rev 25:507–544

Soares HS, Camargo LM, Gennari SM et al (2014) Survey of canine tick-borne diseases in Lábrea, Brazilian Amazon: 'accidental' findings of *Dirofilaria immitis* infection. Rev Bras Parasitol Vet 23:473–480

Stringfellow GJ, Francis IC, Coroneo MT et al (2002) Orbital dirofilariasis. Clin Exp Ophthalmol 30:378–380

Supakorndej P, Lewis RE, McCall JW et al (1995) Radiographic and angiographic evaluations of ferrets experimentally infected with *Dirofilaria immitis*. Vet Radiol Ultrasound 36:23–29

Tarello W (2011) Clinical aspects of dermatitis associated with *Dirofilaria repens* in pets: a review of 100 canine and 31 feline cases (1990–2010) and a report of a new clinic case imported from Italy to Dubai. J Parasitol Res 2011:578385

Tasić A, Rossi L, Tasić S et al (2008) Survey of canine dirofilariasis in Vojvodina, Serbia. Parasitol Res 103:1297–1302

To KK, Wong SS, Poon RW et al (2012) A novel *Dirofilaria* species causing human and canine infections in Hong Kong. J Clin Microbiol 50: 3534–3541

Tolnai Z, Széll Z, Sproch Á et al (2014) *Dirofilaria immitis*: an emerging parasite in dogs, red foxes and golden jackals in Hungary. Vet Parasitol 203:339–342

Traversa D, Aste G, Di Cesare A et al (2011) Efficacy of a single administration of a spot-on solution containing imidacloprid 10%/moxidectin 2.5% in eliminating *Dirofilaria repens* microfilariae in naturally infected dogs. Vet Parasitol 179:107–112

Vieira AL, Vieira MJ, Oliveira JM et al (2014) Prevalence of canine heartworm (*Dirofilaria immitis*) disease in dogs of central Portugal. Parasite 21:5

Wang D, Bowman DD, Brown HE et al (2014) Factors influencing U.S. canine heartworm (*Dirofilaria immitis*) prevalence. Parasit Vectors 7:264

Yoon WK, Choi R, Lee SG et al (2013) Comparison of 2 retrieval devices for heartworm removal in 52 dogs with heavy worm burden. J Vet Intern Med 27: 469–473

Thelaziosis

29

Domenico Otranto and Filipe Dantas-Torres

Abstract

Thelaziosis (or eye worm infection) is a disease caused by spirurid nematodes of the genus *Thelazia*, which are transmitted by several species of secretophagous, nonbiting flies (e.g., *Musca autumnalis*, *Musca larvipara*, and *Phortica variegata*) which feed on ocular secretions, tears, and conjunctiva of many animal species, including humans. When present under the lids and the nictitating membrane, in nasolachrymal ducts, conjunctival sacs, and excretory ducts of lachrymal glands, these worms may cause mild to severe eye disease in affected animals. Hosts of *Thelazia* spp. comprise a wide range of wild and domestic animals and some may also infest humans. Several studies have been carried out in the United States, Europe, and Asia, but there is still little information regarding the biology and ecology of many species of *Thelazia*. Due to their zoonotic relevance, *Thelazia callipaeda* has been more studied. This chapter reviews aspects related to the etiology, epidemiology, clinical signs, diagnosis, treatment, and control of thelaziosis. Some aspects of the biology and ecology of *Thelazia* nematodes are also discussed.

Keywords

Thelaziosis • *Thelazia* • *Musca* • Muscidae • Drosophilidae • *Phortica* • Eye disease

D. Otranto
Department of Veterinary Medicine, University of Bari, Valenzano, Italy
e-mail: domenico.otranto@uniba.it

F. Dantas-Torres (✉)
Department of Veterinary Medicine, University of Bari, Valenzano, Italy

Department of Immunology, Aggeu Magalhães Research Center, Recife, Brazil
e-mail: filipe.dantas@cpqam.fiocruz.br

29.1 Background

Thelaziosis is a disease caused by parasitic nematodes of the genus *Thelazia* (Spirurida: Thelaziidae), which are also commonly referred to as eye worms, due to their localization in the orbital cavities and associated tissues (e.g., under the eyelid, lachrymal ducts, and nictitating membrane) of several animal species, including

C.B. Marcondes (ed.), *Arthropod Borne Diseases*, DOI 10.1007/978-3-319-13884-8_29

humans. All *Thelazia* species are transmitted by secretophagous, nonbiting flies which feed on ocular secretions, tears, and conjunctiva of animals (Otranto and Traversa 2005). In the fly body, the parasite goes through larval development, and the third-stage larvae (L_3) are transmitted when flies feed on the ocular secretions. The presence of larvae and adults of *Thelazia* spp. under the lids and the nictitating membrane, in nasolachrymal ducts, conjunctival sacs, and excretory ducts of lachrymal glands (accordingly to the different species), may cause mild (e.g., lachrymation, ocular discharge, epiphora, and conjunctivitis) and, sometimes, severe eye disease (e.g., keratitis and ulcers). Although several studies have been carried out in the United States, Europe, and Asia, little is known about the biology and epidemiology of many species of *Thelazia*. For their zoonotic role, *Thelazia callipaeda* has been widely studied in many aspects of their biology, ecology, pathogenesis in hosts, and treatment, and information in this chapter are mainly focused on this species.

29.2 Etiology

Among the 16 species of *Thelazia* described so far (Table 29.1), five species (i.e., *Thelazia gulosa*, *Thelazia skrjabini*, *Thelazia rhodesi* of cattle, *Thelazia lacrymalis* of equids, *T. callipaeda* (Figs. 29.1 and 29.2), and *Thelazia californiensis* of wild and domestic carnivores, lagomorphs and humans) are the most commonly retrieved, thus causing veterinary and/or medical concern in many geographical areas of the world. Conversely, reports regarding the other *Thelazia* species are sporadic, and information about their morphology, biology, and epidemiology is scarce.

Table 29.1 Species of *Thelazia*, hosts parasitized, vector(s), and geographic distribution (adapted from Otranto and Traversa 2005)

Species	Hosts parasitized	Vector(s)	Geographic distribution
Thelazia brevispiculata	Cattle	Unknown	China
Thelazia bubalis	Buffalos	Unknown	India
Thelazia californiensis	Canids, felids, domestic and wild ruminants, humans	*Fannia benjamini*, *F. canicularis*[a]	North America
Thelazia callipaeda	Canids, felids, lagomorphs, humans	*Phortica okadai*[a], *P. variegata*	Italy, ex-USSR, Far Eastern countries
Thelazia depressa	Mustelids	Unknown	Africa
Thelazia erschowi	Pigs	Unknown	Ex-USSR
Thelazia ferulata	Cattle	Unknown	China
Thelazia gulosa	Domestic and wild ruminants	*Musca amica*, *M. autumnalis*, *M. domestica*, *M. larvipara*, *M. osiris*, *M. vitripennis*	Europe, Asia, North America, Australia
Thelazia hsüi	Cattle	Unknown	China
Thelazia iheringi	Agouties	Unknown	South America
Thelazia kansuensis	Cattle	Unknown	China
Thelazia lacrymalis	Equids, domestic and wild ruminants	*M. autumnalis*[a]	Europe, North and South America
Thelazia leesei	Camels	*M. lucidula*	Ex-USSR, India
Thelazia petrowi	Cattle	Unknown	Ex-USSR
Thelazia rhodesi	Domestic and wild ruminants, horses	*M. autumnalis*, *M. convexifrons*, *M. crassirostris*, *M. domestica* ?, *M. hervei*[a], *M. larvipara*, *M. sorbens*	Europe, Asia, Africa, North and South America
Thelazia skrjabini	Domestic and wild ruminants	*M. amica*, *M. autumnalis*[a], *M. hervei*[a], *M. osiris*	Europe, Asia, North America, Australia

[a]Under experimental conditions;
? = reported, but confirmation is needed

Thelazia callipaeda males and females (7.5–13 mm and 12–18.5 mm in length and 340–430 μm and 370–510 μm in width, respectively) present striate cuticle on the entire body surface and a buccal capsule with a typically hexagonal mouth opening. Males have a ventrally curved caudal end, 15 pairs of papillae on the ventral surface, and two spicules (Fig. 29.3), which greatly differ one from each other for shape and size (Otranto et al. 2003b). In the female, the vagina opens anteriorly to the esophagus-intestinal junction (Fig. 29.4) (this is a distinct character with respect to *T. californiensis*) and in the posterior half of the body. Immature eggs or germ cells are frequently observed. The first-stage larvae (L_1, 382–400 μm long) present a shell membrane (i.e., embryonated egg) and are arranged in a row in the distal uterus of the adult females (Otranto et al. 2003b).

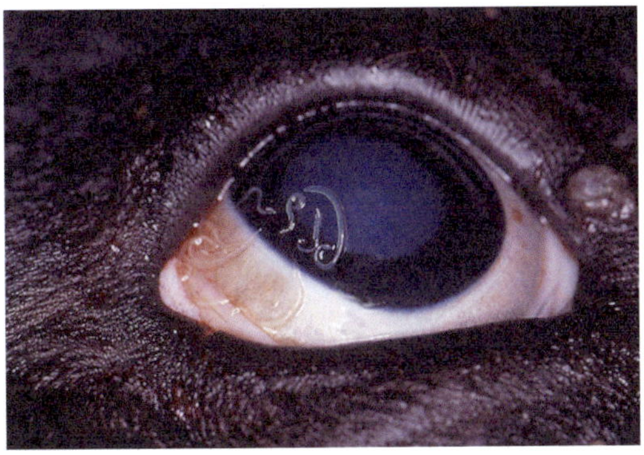

Fig. 29.1 *Thelazia lacrymalis* in a horse

Fig. 29.2 *Thelazia callipaeda* in a cat

29.3 Life Cycle

Depending on the species, adults of *Thelazia* nematodes live under the nictitating membrane or in the lachrymal ducts of the definitive hosts, where the mature females release the L_1s into the lachrymal secretions. The biological cycle requires an arthropod intermediate host, represented by a nonbiting dipteran, which feeds on the lachrymal secretions of animals (Fig. 29.5) (reviewed in Otranto and Traversa 2005). Thirteen species of *Musca* have been incriminated in the transmission of eye worms of cattle and

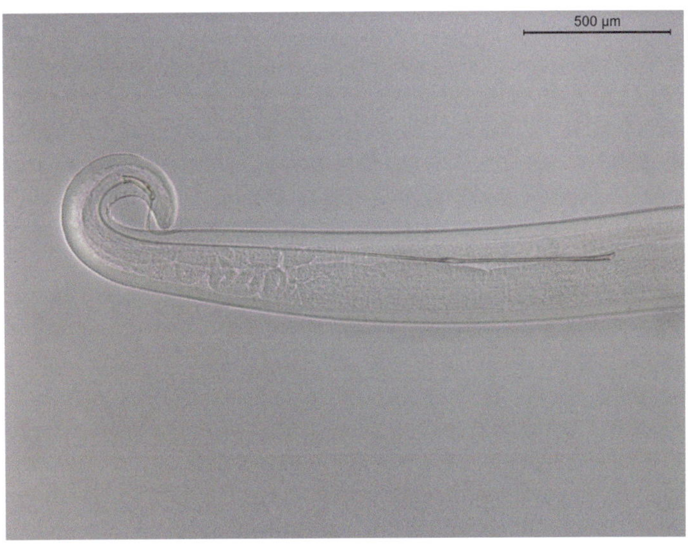

Fig. 29.3 *Thelazia callipaeda* male with long and short spicules

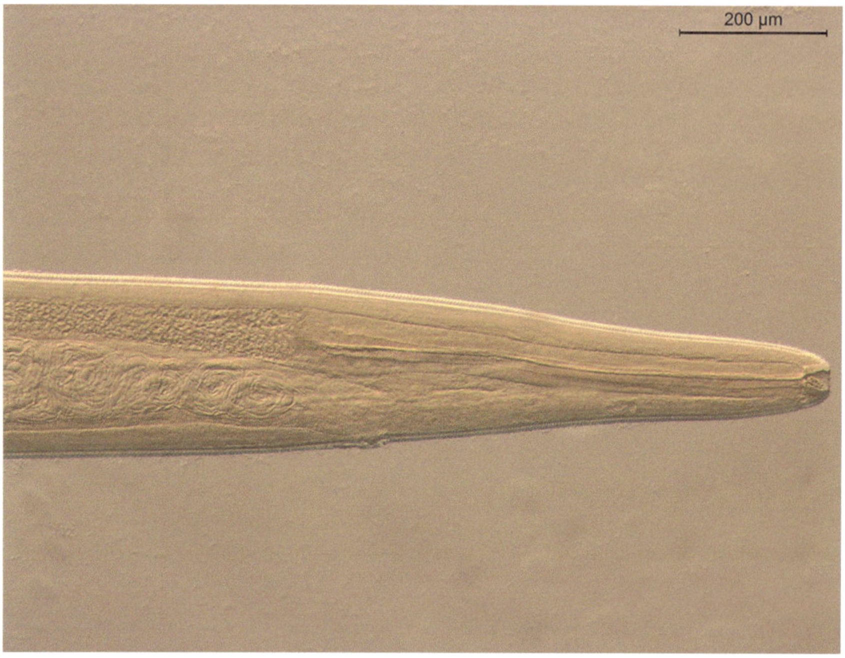

Fig. 29.4 *Thelazia callipaeda* female with the vulva located anterior to the esophagus-intestinal junction

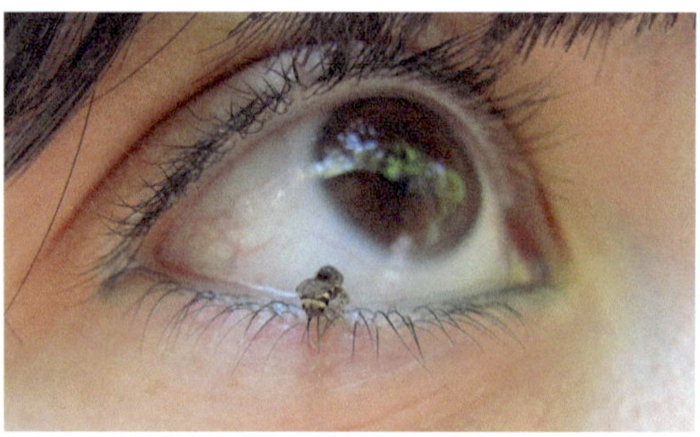

Fig. 29.5 *Phortica variegata* feeding of lachrymal secretions on a human eye

horses, but only face flies (i.e., *Musca autumnalis* and *M. larvipara*) have been demonstrated, both under experimental and natural conditions, to act as vectors in the United States, Canada, and Italy (Table 29.1). For example, *M. autumnalis* is proven vector of *T. gulosa*, *T. skrjabini*, and *T. rhodesi*, whereas *M. larvipara* was demonstrated to be the vector of *T. rhodesi*.

In the intermediate host, the L_1 develops to L_2 and finally to L_3, the infective stage. After being ingested by the intermediate host, the larvae remain encapsulated in the testes or egg follicles (*T. rhodesi*, *T. skrjabini*, and *T. callipaeda*) of the vector and then migrate through the arthropod coelom to the labella. The infective stages emerge from the labella of infected flies while they feed on the lachrymal secretions of a receptive animal. Animal infestation is mostly related to the presence of face flies during summer and fall. Over the last 10 years, knowledge on the biology of *T. callipaeda* in the definitive host and in its vector has been greatly enhanced (Otranto et al. 2004, 2006a). *Phortica variegata* and *P. okadai* (Diptera, Drosophilidae, Steganinae) have been shown to act as vectors of *T. callipaeda* in Europe and China, respectively (Otranto et al. 2005a). This finding is of relevance considering that, unlike the vectors of other *Thelazia* species (i.e., *Musca* spp.), *Phortica* flies are drosophilids (commonly known as "fruit flies"), which feed on vegetable substrate and also on lachrymal secretions of humans and carnivores. The mature females release L_1 into the lachrymal secretions, which are subsequently ingested by *P. variegata* (Otranto et al. 2004).

In the intermediate host L_1 undergoes development until L_3 in about 14–21 days from the infestation, and they may survive in overwintering flies for 6 months (Otranto et al. 2005b). After transmission to a definitive host, *T. callipaeda* develop into the adult stage within ~35 days (Otranto et al. 2004). In the definitive host, the existence of seasonality in the reproductive cycle of females is associated with the presence/absence of the vector (Otranto et al. 2004). Indeed, *T. callipaeda* L_1 are present in the lachrymal secretions of naturally infected dogs during summer season (they are ready to be ingested by flies feeding about the eyes) whereas fourth-stage larvae (L_4) in spring and early fall (Otranto et al. 2004).

29.4 Clinical Aspects

The clinical signs of thelaziosis are induced by the presence of both immature and adult stages in the eyes of the definitive hosts. In particular, asymptomatic infestations occur mainly when only male nematodes parasitize animals, whereas evident clinical alterations are more frequently observed in the presence of gravid females, thus suggesting the importance of larvae in the pathogenesis of the conjunctivitis (Otranto and Traversa 2005). The lateral serration of *Thelazia* spp. cuticle causes mechanical damages to the conjunctival and corneal epithelium resulting in ocular discharge, which is of importance for the transmission of worms through flies feeding on lachrymal secretions containing L_1s. In cattle, clinical signs of bovine keratitis by *Moraxella bovis* (pinkeye) and of thelaziosis are overlapping, and both occur when secretophagous flies are present. In particular, pinkeye most likely develop on traumatic lesions caused by *Thelazia* spp., and its incidence increases proportionally with the fly population density (flies may carry *M. bovis* and *Thelazia* spp.) (Otranto and Traversa 2005).

Adults and larvae of *T. callipaeda* cause eye disease with symptoms in dogs, cats, and humans ranging from mild (e.g., lachrymation, ocular discharge, and epiphora) to severe (e.g., conjunctivitis, keratitis, and corneal opacity or ulcers) according to the mechanical damage induced by lateral serration of their cuticle on the conjunctiva and cornea and to the worm burden. Human patients present mild conjunctivitis, follicular hypertrophy of the conjunctiva, foreign body sensation, excessive lachrymation, and ocular secretions, itchiness, congestion, swelling, hypersensitivity to light, and keratitis, depending on the number of nematodes present in the eye, their location, and the host immune response. They usually complain for floating "filaments" on the eye surface. In addition, patients often scratch their eyes, and bacterial secondary infestation by *Pasteurella*, *Chlamydia*, and/or *Staphylococcus* may complicate the primary diagnosis of thelaziosis with swelling of the eyelids, congestion of the conjunctiva, and yellow secretions from affected eyes with a purulent inflammation (Shen et al. 2006).

29.5 Diagnosis

The diagnosis of thelaziosis relies on the detection of the nematodes in the eye and associated tissues of the infested hosts and on their morphological and/or molecular identification. A thorough clinical and ophthalmological examination of patients with suspected thelaziosis reveals the presence of adult and/or larval nematodes located mostly in the conjunctival sac or in the medial or lateral canthus of the eye. Adult nematodes appear as a thin, creamy-white thread in the conjunctival sac, whereas small-sized larvae are difficult to be detected and identified by the naked eye. In general, nematodes can be collected by cotton swabs or by flushing saline from the medial canthus. The latter methodology is the only way for recovering L_3 or L_4 at the early stage of infestation. A local anesthesia with 1 % dicaine may be recommended for detection and recovery of *T. callipaeda* in dogs, cats, and humans (Shen et al. 2006). Knowing the seasonality of human thelaziosis may help the correct etiological diagnosis of the disease, since it allows the differential diagnosis with allergic conjunctivitis (often caused by pollens), which occurs frequently in spring and summer. For epidemiological studies, a postmortem examination at the slaughterhouses may be useful for diagnosing thelaziosis in cattle and horses.

Because of the ease with which *Thelazia* species can be detected and identified morphologically, molecular studies are more frequently used for research purposes, mainly in eco-epidemiological and phylogenetic studies, rather than for the diagnosis of species. Many target genes have been studied. For example, the interspecific and intraspecific differences in the sequence of the first internal transcribed spacer (ITS1) of the ribosomal DNA (rDNA) have been evaluated for *T. callipaeda*, *T. gulosa*, *T. rhodesi*, *T. skrjabini*, and *T. lacrymalis*, which incidentally are the most common eye worm species (Otranto and Traversa 2004). Since the intraspecific differences showed to be markedly lower than the interspecific variations, the ITS1 was demonstrated to be a powerful genetic marker for the molecular identification of eye worms and a useful tool for developing molecular epidemiological techniques to study *Thelazia* transmission patterns and prevalence in definitive and intermediate hosts (Otranto and Traversa 2004). Indeed, this ribosomal target has been used in a molecular marker in epidemiological surveys on the vectors of *T. gulosa*, *T. rhodesi*, and *T. skrjabini* (Otranto et al. 2003c) and to investigate the vector of *T. callipaeda* under natural conditions (Otranto et al. 2006b). The analysis of mitochondrial gene encoding for the cytochrome c oxidase subunit 1 (*cox*1) by both PCR-coupled sequencing and by PCR-based single-strand conformation polymorphism (SSCP) of the most variable portion (v-*cox*1) of the *cox*1 gene demonstrated the absence of genetic variation among individual nematodes from Europe, and the existence of seven different haplotypes within *T. callipaeda* isolates from Asia, with distinct genetic differences between the European and Asian specimens (Otranto et al. 2005b).

29.6 Treatment

Ivermectin and levamisole are efficacious for treating horse and cattle thelaziosis. In addition, antibiotic ointments are recommended when concomitant bacterial infections are present. Different treatment strategies have been tested in dogs and cats. For example, topic instillation of organophosphates or moxidectin 1 % showed to be effective against *T. callipaeda* (Lia et al. 2004). A spot-on formulation of imidacloprid 10 % and moxidectin 2.5 % showed a high efficacy (95.23 %) to control of canine thelaziosis within 9 days after treatment (Bianciardi and Otranto 2005) and, later on, as high as 100 % from day 7 until day 28 post treatment (Otranto et al. 2016). The administration of an injectable sustained-release formulation of moxidectin provided one season protection against *T. callipaeda* infestation in dogs from an endemic area of Northern Italy (Rossi et al. 2007). Similarly, the efficacy of oral formulations of milbemycin oxime in dogs (Ferroglio et al. 2008) and of milbemycin oxime/praziquantel (Motta et al. 2012) was demonstrated in dogs and cats.

In human patients, adults and larvae *T. callipaeda* can be removed mechanically (following local anesthesia) either by rinsing of the conjunctival sac with sterile physiological saline or with fine forceps or cotton swabs. Clinical signs, excluding secondary infections with other pathogens, generally resolve rapidly after the removal of the worms.

29.7 Epidemiology

Bovine thelaziosis is caused by *T. gulosa*, *T. rhodesi*, and *T. skrjabini*, which are present worldwide; *T. gulosa* and *T. skrjabini* have been reported mainly in the New World and in Southern Italy (Geden and Stoffolano 1982; O'Hara and Kennedy 1989) and *T. rhodesi* in Europe and in Eastern countries. Equine thelaziosis caused by *T. lacrymalis* is common in the United States, Canada, and Europe (Otranto and Traversa 2005).

Bovine and equine thelaziosis are commonly reported in North America and Europe not only due to the susceptibility of these animals to *Thelazia* but mainly because of the exposure of grazing animals to face flies transmitting eye worms. Although for long time *T. callipaeda* has been known as "oriental eye worm" due to its distribution in Far Eastern countries (including India, Thailand, China, and Japan), it was demonstrated that this nematode is endemic in Italy (Otranto et al. 2003a), Southwestern France (Dorchies et al. 2007), Switzerland (Malacrida et al. 2008), Spain (Miró et al. 2011), Portugal (Vieira et al. 2012), and recently in countries of the Balkans (Hodzic et al. 2014). In particular, *T. callipaeda* is highly endemic in Southern Italy with prevalence of infestation in some municipalities up to 60.14 % of dogs (Otranto et al. 2003a). In the same areas, cats, foxes, wolves, and beech martens have been found infested by the eye worm (Otranto et al. 2003a, 2009).

The high prevalence and intensity of infection in adult cattle and dogs, rather than in young, has been explained by the long survival of *Thelazia* (more than 1 year), the likely absence of a protective immunological response and the reinfections occurring throughout the life of animals (Otranto et al. 2003a). However, although adult parasites are present throughout the year, two peaks of *Thelazia* spp. infection have been recorded in cattle: in early summer (adults that had overwintered) and in late summer in relationship to the presence of the vectors (Otranto and Traversa 2005). In a study aiming to monitor under natural conditions *P. variegata* in an area highly endemic for dog thelaziosis, the highest number of flies was retrieved in July–August (Otranto et al. 2006b). In this study, an overall prevalence of infected flies of 1.34 % was detected (Otranto et al. 2006b). In addition, suitable environments for the geographic distribution and development of *P. variegata* across Italy and Europe (including the Balkan area) were predicted using a desktop implementation of the Genetic Algorithm for Rule-Set Prediction, and all recent reports of *T. callipaeda* fall within the suitable area indicated by the model (Dorchies et al. 2007; Malacrida et al. 2008; Miró et al. 2011; Vieira et al. 2012).

The investigation on the vectors of *Thelazia* has been carried out on species affecting cattle in North America, Slovakia, and ex-USSR by dissection of experimentally and naturally infested flies. Among several species of face flies studied, *M. autumnalis* and *M. larvipara* have been incriminated in the transmission of eye worms (Otranto and Traversa 2005) with a prevalence of infestation in North America ranging from 0 to 5.7 % (mean value 2.62 %), with an average larval burden of 2.3–4.2 larvae per fly (Otranto and Traversa 2005). In a molecular epidemiological survey carried out for the identification of *T. gulosa*, *T. skrjabini*, and *T. rhodesi* in their fly vectors in field conditions, a mean infestation rate of 2.86 % was recorded in 5190 flies collected around the eyes of free-ranging cattle, allowing the assessment of new vector associations for different species of *Musca* and *Thelazia* (Otranto et al. 2003c).

Since the first report of human thelaziosis in a man from Beijing, China, human thelaziosis has been described in several areas of the former Soviet Union, Asia (e.g., China, Korea, Japan, Indonesia, Thailand, Taiwan, and India) and, more recently, Europe (France, Italy, and Spain). Human cases are predominantly associated to poor, rural communities with low health and socioeconomic standards, where domestic dogs and other animals (e.g., cats and foxes) living in close vicinity with humans are heavily affected (Shen et al. 2006; Otranto and Dutto 2008; Fuentes et al. 2012). Nonetheless, the prevalence and relevance of human thelaziosis has been increasing in China over the last two decades with 370 new cases reported in elders and children from 17 provinces in the southeastern part of the country in 2005 (Shen et al. 2006).

29.8 Control and Prophylaxis

Environmental treatment against flies and hygiene of stables may be useful to prevent thelaziosis in both cattle and horses, mostly during the face fly season. For controlling human infestation, the contact with the fly intermediate host should be avoided by the use of bed nets even though strategies aiming to control the fly populations should also be considered (Shen et al. 2006). Unfortunately, the approaches above are often ineffective, because the biology and feeding habits of *Phortica* flies are poorly known. For this reason prevention through the use of bed nets to protect children while sleeping, keeping their faces and eyes clean, and treating all infested domestic animals (which may act as reservoirs for human infection) with effective drugs is also recommended (Shen et al. 2006). No information is available on the usefulness of any drug as repellents on animals against *P. variegata* for preventing *T. callipaeda* infestation. An oral formulation of milbemycin oxime in infested dogs was suggested to be effective as chemoprophylaxis approach to reduce the prevalence of dog thelaziosis and therefore the risk for human infestations in endemic areas (Ferroglio et al. 2008).

References

Bianciardi P, Otranto D (2005) Treatment of dog thelaziosis caused by *Thelazia callipaeda* (Spirurida, Thelaziidae) using a topical formulation of imidacloprid 10% and moxidectin 2.5%. Vet Parasitol 129:89–93

Dorchies P, Chaudieu G, Siméon LA et al (2007) First reports of autochthonous eye worm infection by *Thelazia callipaeda* (Spirurida, Thelaziidae) in dogs and cat from France. Vet Parasitol 149:294–297

Ferroglio E, Rossi L, Tomio E et al (2008) Therapeutic and prophylactic efficacy of milbemycin oxime (Interceptor) against *Thelazia callipaeda* in naturally exposed dogs. Vet Parasitol 154:351–353

Fuentes I, Montes I, Saugar JM et al (2012) Thelaziosis in humans, a zoonotic infection, Spain, 2011. Emerg Infect Dis 18:2073–2075

Geden CJ, Stoffolano JG (1982) Development of the bovine eyeworm, *Thelazia gulosa* Railliet and Henry 1910 in experimentally infected female *Musca autumnalis* De Geer. J Parasitol 68:287–292

Hodzic A, Latrofa MS, Annoscia G et al (2014) The spread of zoonotic *Thelazia callipaeda* in the Balkan area. Parasit Vectors 7:328

Lia RP, Traversa D, Agostini A et al (2004) Field efficacy of moxidectin 1 per cent against *Thelazia callipaeda* in naturally infected dogs. Vet Rec 154:143–145

Malacrida F, Hegglin D, Bacciarini L et al (2008) Emergence of canine ocular thelaziosis caused by *Thelazia callipaeda* in southern Switzerland. Vet Parasitol 157:321–327

Miró G, Montoya A, Hernández L et al (2011) *Thelazia callipaeda*: infection in dogs: a new parasite for Spain. Parasit Vectors 4:148

Motta B, Schnyder M, Basano FS et al (2012) Therapeutic efficacy of milbemycin oxime/praziquantel oral formulation (Milbemax®) against *Thelazia callipaeda* in naturally infested dogs and cats. Parasit Vectors 5:85

O'Hara JE, Kennedy MJ (1989) Prevalence and intensity of *Thelazia* spp. (Nematoda: Thelazioidea) in a *Musca autumnalis* (Diptera: Muscidae) population from central Alberta. J Parasitol 75:803–806

Otranto D, Dutto M (2008) Human thelaziasis, Europe. Emerg Infect Dis 14:647–649

Otranto D, Traversa D (2004) Molecular characterization of the first internal transcribed spacer of ribosomal DNA of the most common species of eye worms (Thelazioidea: Thelazia). J Parasitol 90:185–188

Otranto D, Traversa D (2005) *Thelazia* eyeworm: an original endo- and ectoparasitic nematode. Trends Parasitol 1:1–4

Otranto D, Ferroglio E, Lia RP et al (2003a) Current status and epidemiological observations of *Thelazia callipaeda* (Spirurida, Thelaziidae) in dogs, cats and foxes in Italy: a "coincidence" or a parasitic disease of the Old Continent? Vet Parasitol 116:315–325

Otranto D, Lia RP, Traversa D, Giannetto S (2003b) *Thelazia callipaeda* (Spirurida, Thelaziidae) of carnivores and humans: morphological study by light and scanning electron microscopy. Parassitologia 45:125–133

Otranto D, Tarsitano E, Traversa D et al (2003c) Molecular epidemiological survey on the vectors of *Thelazia gulosa*, *Thelazia rhodesi* and *Thelazia skrjabini* (Spirurida: Thelaziidae). Parasitology 127:365–373

Otranto D, Lia RP, Buono V et al (2004) Biology of *Thelazia callipaeda* (Spirurida, Thelaziidae) eyeworms in naturally infected definitive hosts. Parasitology 129:627–633

Otranto D, Lia RP, Cantacessi C et al (2005a) Nematode biology and larval development of *Thelazia callipaeda* (Spirurida, Thelaziidae) in the drosophilid intermediate host in Europe and China. Parasitology 131:847–855

Otranto D, Testini G, De Luca F et al (2005b) Analysis of genetic variability within *Thelazia callipaeda* (Nematoda: Thelazioidea) from Europe and Asia by sequencing and mutation scanning of mitochondrial cytochrome *c* oxidase subunit 1 gene. Mol Cell Probes 19:306–313

Otranto D, Brianti E, Cantacessi C et al (2006a) The zoophilic fruitfly *Phortica variegata*: morphology, ecology and biological niche. Med Vet Entomol 20:358–364

Otranto D, Cantacessi C, Testini G et al (2006b) *Phortica variegata* as an intermediate host of *Thelazia callipaeda* under natural conditions: evidence for pathogen transmission by a male arthropod vector. Int J Parasitol 36:1167–1173

Otranto D, Dantas-Torres F, Mallia E et al (2009) *Thelazia callipaeda* (Spirurida, Thelaziidae) in wild animals: report of new host species and ecological implications. Vet Parasitol 166:262–267

Otranto D, Colella V, Crescenzo G et al (2016) Efficacy of moxidectin 2.5% and imidacloprid 10% in the treatment of ocular thelaziosis by Thelazia callipaeda in naturally infected dogs. Vet Parasitol 227:118–121

Rossi L, Rigano C, Tomio E et al (2007) Use of sustained-release moxidectin to prevent eyeworm (*Thelazia callipaeda*) infection in dogs. Vet Rec 161:820–821

Shen JL, Gasser RB, Chu D et al (2006) Human thelaziosis—a neglected parasitic disease of the eye. J Parasitol 92:872–875

Vieira L, Rodrigues FT, Costa A et al (2012) First report of canine ocular thelaziosis by *Thelazia callipaeda* in Portugal. Parasit Vectors 5:124

Habronemosis

30

Annunziata Giangaspero and Donato Traversa

Abstract

Equid habronemosis is a widespread parasitic disease caused by the dioecious nematodes *Habronema microstoma*, *Habronema muscae*, and *Draschia megastoma* (Spirurida, Habronematidae). These parasites live as both larvae and adults in the stomach of domestic and wild equids (i.e., horses, donkeys, mules, and zebras), while as larvae they live on the skin and occasionally on other sites, such as ocular and genital mucosa and lung parenchyma. Their life cycle requires an intermediate host, represented by dung-inhabiting secretophagous or hematophagous muscid flies. Habronemosis causes mild to severe clinical symptoms in affected animals, depending on the parasite's stage of development stage and on localization. Larvae and adults living in the stomach cause gastric habronemosis, while larvae on the skin cause cutaneous habronemosis (known as "summer sores"). When larvae infect other sites, such as the ocular and genital mucosa of horses, they cause mucocutaneous habronemosis. Habronemosis causes severe economic losses, which mainly regard sportive horses, because their performance is impaired and the infection is unaesthetic. Clinical diagnosis of habronemosis is quite difficult, because the symptoms may overlap those of other diseases. Molecular diagnosis is a useful tool for both clinicians and epidemiologists. Macrocyclic lactones are effective against clinical habronemosis, while disease prevention relies on control of flies.

Keywords

Habronema microstoma • *Habronema muscae* • *Draschia megastoma* • Epidemiology • Clinical aspects • Diagnosis • Control

A. Giangaspero (✉)
Department of Science of Agriculture, Food and Environment, University of Foggia,
Via Napoli 25, 71100 Foggia, Italy
e-mail: annunziata.giangaspero@unifg.it

D. Traversa
Faculty of Veterinary Medicine, Teaching Veterinary Hospital, University of Teramo,
Località Piano d'Accio snc, 64100 Teramo, Italy

C.B. Marcondes (ed.), *Arthropod Borne Diseases*, DOI 10.1007/978-3-319-13884-8_30

30.1 Etiology

Habronemosis is caused by *Habronema microstoma* Creplin, 1849 (syn. *Habronema majus* Schneider 1866); *Habronema muscae* Carter, 1865; and *Draschia megastoma* (previously named *Habronema megastoma*)[1] (Spirurida: Habronematidae).

H. microstoma and *H. muscae* adults differ in color (*H. microstoma* is white and *H. muscae* is yellowish to orange) but are similar in size (8–30 mm long), with a lateral ala, a pair of lateral lips, and a buccal capsule divided into an anterior, domed part and a posterior cylindrical part. *H. microstoma* has trilobate pharynx teeth. *D. megastoma* adults (7–13 mm long) are white, and the head is separated from the rest of the body by a constriction. The stoma is funnel shaped, with a wall formed by two separated lateral vales without teeth (Naem 2007).

The eggs (40–50 μm × 10–16 μm) are elongate (*Habronema*) or cylindrical (*Draschia*) and contain larvae, which may hatch during their intestinal transit (Euzeby 1961; Erschow 1956).

30.2 Life Cycle

H. muscae, *H. microstoma*, and *D. megastoma* adults live in the stomach wall or freely on the mucosal surface of *margo plicatus* of equids. After mating, females produce embryonated eggs, which may hatch either during intestinal transit or in the environment after release via feces. The developing first-stage larvae (L_1) are motile, show a marked hygrotropism and thermotropism, and can survive up to a week in suitable environmental conditions. Eggs and/or larvae are then ingested by dung-inhabiting muscid larvae. *Musca domestica* and *Stomoxys calcitrans* are considered to be the main vectors of *H. microstoma* and *H. muscae*, respectively. *H. microstoma* and *H. muscae* develop further in the flies until they reach the infective third stage (L_3), which occurs at about the same time as the fly *imago* emerges from the *puparium*, so that there is synchronous development with the insect (Ransom 1911; Roubaud and Descauzeaux 1922a; Anderson 2000). When infected flies settle on the animals, the L_3s are stimulated by the warm surface and migrate into the fly head and proboscis. The feeding fly deposits them onto moist skin areas (e.g., around nostrils, mouth, skin lesions, eyes, and/or genitalia) of the definitive host (Roubaud and Descauzeaux 1922b). When deposited around the nostrils and/or mouth, the L_3s are swallowed by the animal host and develop into adults in its stomach. The parasites remain as larvae and do not achieve sexual maturity when they are deposited at other cutaneous sites or at mucocutaneous transition sites (eyes, nostrils, genital mucosa) or when, as rarely occurs, they reach internal organs (i.e., lungs) as erratic forms (Hiepe et al. 1985).

[1]Although improperly, the term "habronemosis" is commonly used also for the disease caused by *Draschia megastoma*.

30.3 Clinical Signs and Histopathological Aspects

Clinical symptoms of infection by *H. microstoma*, *H. muscae*, and *D. megastoma* vary according to the parasite's stage of development and its localization. Adults exert a dual effect on the host; due to their size, they have a mechanical-irritative effect, and they produce toxic metabolites. Adults are found only in the stomach of infected animals and cause *gastric habronemosis*, whereas larvae can affect the skin (*cutaneous habronemosis*), mucosa (*mucosal habronemosis*), or lungs (*pulmonary habronemosis*).

30.3.1 Gastric Habronemosis

Habronema spp. are localized at the level of gastric mucosa glands (Fig. 30.1) causing atrophy, mechanical irritation of the stomach, and secretory and functional disorders with varying degrees of severity, such as anorexia/dysorexia, digestive disorders, diarrhea, progressive weight loss, ulcers, and postprandial colic (Erschow

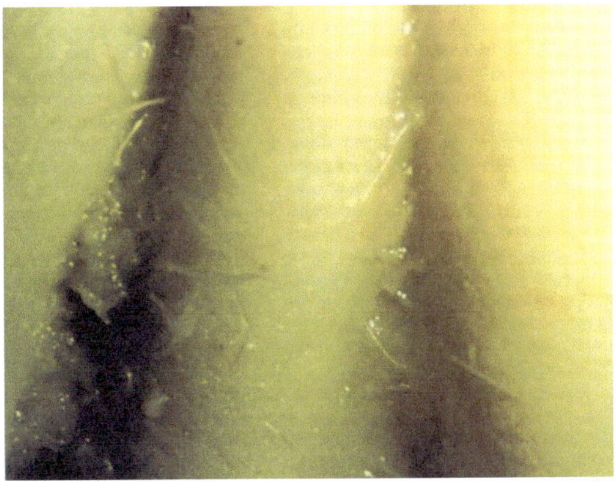

Fig. 30.1 Gastric habronemosis in a horse at slaughterhouse (original: A. Giangaspero)

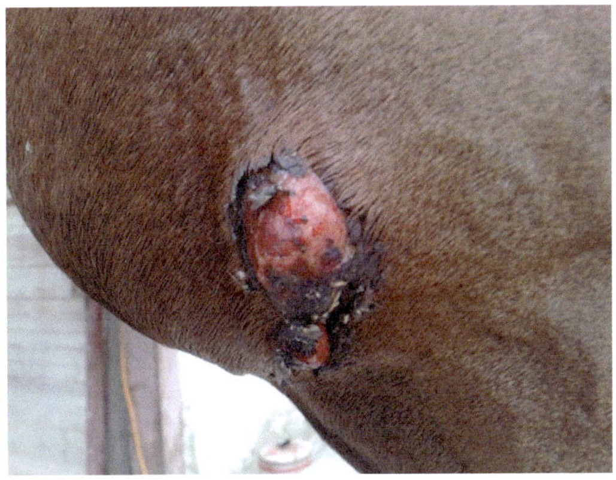

Fig. 30.2 Cutaneous habronemosis in a 15-year-old sorrel Anglo-Arab male with a 4-year history of recurrent skin disease (original: A. Giangaspero)

1956; Reddy et al. 1976). Adult *D. megastoma* produces large tumors, which may impede the motility of the stomach or obstruct the pyloric opening of the stomach. The lesions produced by this nematode occasionally cause acute hemorrhages into the stomach or perforation of the stomach wall leading to acute peritonitis and death (Dunn 1969; Soulsby 1982).

30.3.2 Cutaneous Habronemosis

This is the most severe form, also known as "summer sores". It is characterized by the appearance of lesions that tend to disappear spontaneously in the cold months but reappear in the following warm season. L_3 is laid on the wounds by flies and lesion is caused by the spine at the posterior end of larvae and by a local allergic reaction. During winter, larvae can remain in lesions in a dormant state and then reactivate later when environmental temperatures rise (Herd and Donham 1981). The body regions most affected are the chest, fetlocks, and the inner side of the legs. Clinical manifestations vary depending on the origin/time of the lesion.

Skin lesions can be single or multiple (Fig. 30.2) and are severe, proliferative, exuberant and granulomatous, often bleeding, itching, and ulcerated and may contain necrotic, caseous, or calcified granules (Murray et al. 1978; Scott and Miller 2003).

If the lesion originates from a preexisting lesion (as is usually the case), the wound develops into the following forms: dry, wet, and edematous. A dry lesion appears as a generally circular alopecic area covered by grayish scales. A wet lesion is characterized by a slight discharge with hair agglutination, while an edematous lesion is irregular and hairy, 3–5 cm in diameter, with edema and small nodules (Fig. 30.3).

The lesions usually appear during periods of intense fly activity (i.e., spring and summer) and often regress in winter. Nevertheless, certain equids present yearly recurrence and the lesions evolve into non-healing granulomatous tumor-like masses; these may attract more flies, leading to a super-infection (Vasey 1981; Anderson 2000; Scott and Miller 2003).

Histologically, the wounds are filled with eosinophils, macrophages, lymphocytes, and a few plasmacells. In peripheral areas, vascular and fibro-connective proliferation can be detected, with aggregates of eosinophils in coagulation necrosis (Petrella et al. 2010) (Fig. 30.4).

30.3.3 Muco-cutaneous Habronemosis

Muco-cutaneous habronemosis is caused by the presence of larvae on ocular and periocular areas (conjunctiva, medial canthus, nasolacrimal ducts)

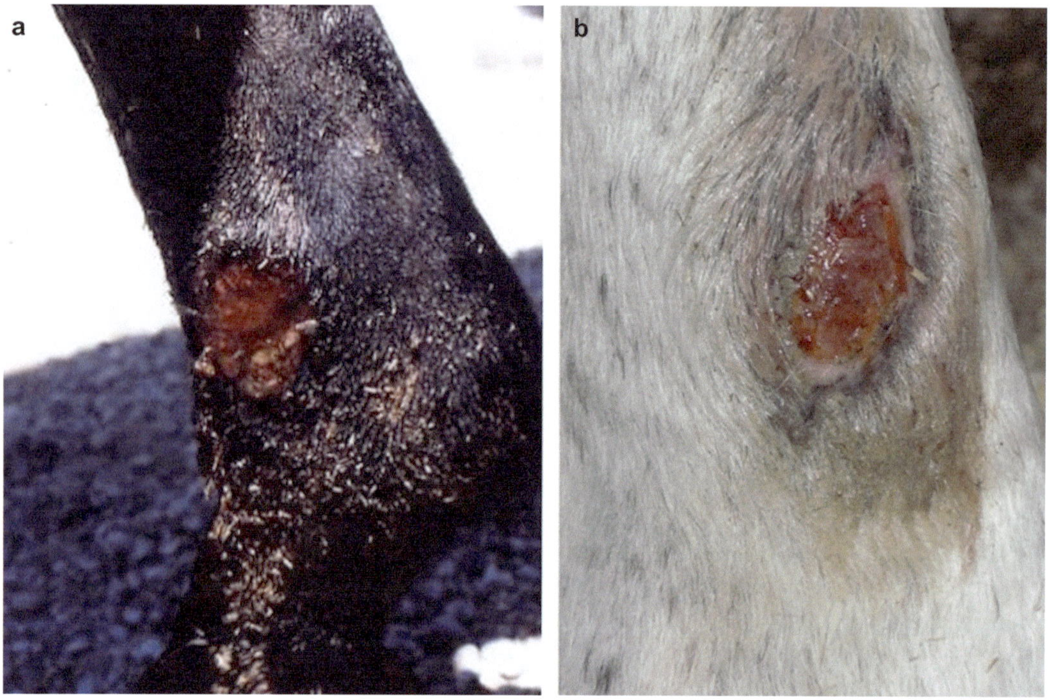

Fig. 30.3 Cutaneous habronemosis in a horse (originals: A. Giangaspero (**a**), D. Traversa (**b**))

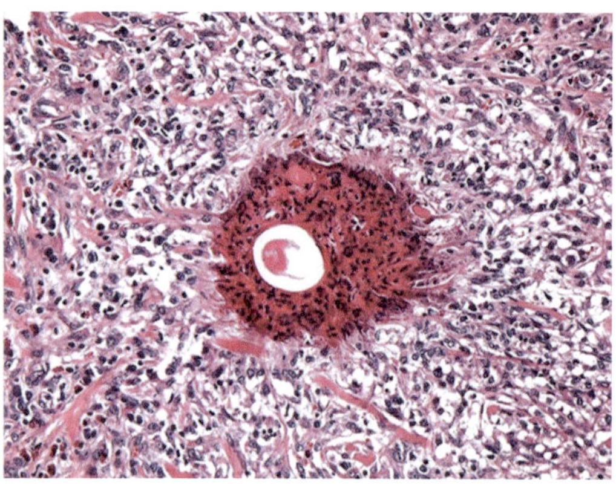

Fig. 30.4 Histological examination of a summer sore: numerous infiltrates of eosinophils and parasitic forms related to *Habronema* (original: A. Petrella)

and commissure of the lips or genitalia (urethral process, glans, prepuce, vaginal fornix). When larvae are released inside the eye corner, infected animals may present marked photophobia and lacrimation (Rebhun et al. 1981). Chronic *Habronema* lesions cause inflammation of the urethral process and present increased size with different degrees of fibrosis (Stick 1979). Histological examination of the mucous membranes reveals yellowish areas of granulation tissue, infiltrated by eosinophils and affected by collagenolytic phenomena (Rebhun et al. 1981).

30.3.4 Pulmonary Habronemosis

Pathogenesis is unclear, because it is not fully understood how larvae reach the lungs (Erschow 1956; Soulsby 1982; Lyons et al. 1985). Larvae on skin wounds may travel to the lungs via the bloodstream, whether larvae deposited in the nostrils or mouth mucosa can migrate to lung via the trachea (Bain et al. 1969). In any case, the parasites damage the peribronchial tissue because they create large nodules (0.2–2 cm) containing larvae or their residues (Schuster et al. 2010). *D. megastoma* larvae may also be found in small nodules in the brain (Mayhew et al. 1982).

30.4 Diagnosis

Clinical diagnosis of gastric habronemosis is challenging, since signs may overlap those of other conditions. Copromicroscopic diagnosis of gastric habronemosis is based on the detection of larvated eggs by flotation or of larvae using the Baermann technique or on coproculture; however, all these techniques are unreliable for their very low sensitivity, even for horses with a high intensity of infection. An alternative diagnostic method is gastric lavage followed by microscopic examination of the sedimented washings, but this approach is invasive, risky, and time-consuming, and it requires the infected animal to be sedated and restrained (Euzeby 1961; Dunn 1969; Soulsby 1982).

Clinical diagnosis of cutaneous and mucocutaneous forms is also problematic, as granulomatous lesions are similar to those of other diseases, such as botryomycosis, pythiosis, onchocercosis, equine sarcoid, and squamous cell carcinoma (Inzana and Carter 1990; Pascoe 1990; Chaffin et al. 1995; Pusterla et al. 2003). The surface of the lesion must be scraped at several points and the tissue digested for 12–18 h at 37 °C using an acid pepsin solution to detect larvae in the skin (Murray et al. 1978).

A molecular approach (i.e., a semi-nested PCR) to detect *Habronema* DNA has been described (Traversa et al. 2004a, b, Giangaspero et al. 2005) which allows detection and identification of *H. microstoma* and *H. muscae* irrespective of their life cycle stages, with important implications for clinicians. From a practical clinical point of view, this tool is useful for the diagnosis of both gastric and cutaneous habronemosis, which are sometimes difficult to differentiate from other gastric and skin diseases of equids with similar clinical signs (Traversa et al. 2004b, 2007). This genetic test can also be used for ecological and epidemiological purposes, e.g., to investigate the prevalence of habronemosis in live equids and to study infection patterns of both *H. microstoma* and *H. muscae* in their intermediate hosts via the specific detection of their DNA in muscid flies (Traversa et al. 2006, 2008).

30.5 Epidemiology

Habronemosis is distributed worldwide; incidence is high in tropical and subtropical areas, but it is also prevalent in temperate regions, including Mediterranean countries. Gastric habronemosis has been reported to affect as many as 70 % of native horses in Central and North America (Foster and Ortiz 1937; Lyons et al. 1983; Reinemeyer et al. 1984); Australia (Waddell 1969; Mfitilodze and Hutchinson 1989). Africa has the highest registered rates of infection, since 85–100 % of examined donkeys and horses have been found to be infected by *Habronema* spp. (Pandey et al. 1981, 1992; Vercruysse et al. 1986; Pandey and Eysker 1988). In European countries, prevalence ranges from 1 to 62 % (Bauer 1986; Poland Gawor 1995; Hoglund et al. 1997; Tieri et al. 1998; Borgsteede and van Beek 1998; Molini et al. 2004; Traversa et al. 2006).

As many as 500 individual parasites may be detected in a single animal, although up to 4000 immature forms have occasionally been detected in geographical areas where parasite pressure is very high (Pandey et al. 1992).

Several fly genera have been considered as potential vectors of habronemosis (*Musca, Fannia*, *Sarcophaga*, *Haematobia*, *Stomoxys*). Importantly, the role of *S. calcitrans* and *M. domestica* in transmitting *H. microstoma* and *H. muscae* has been confirmed in field conditions (Traversa et al. 2008). *Musca domestica* and *S. calcitrans* (Diptera: Muscidae) are actually the species most closely associated with the host animals' environment. Both larval and adult stages of *M. domestica* are polyphagous; adults feed on secretions from the eyes, nose, and mouth. Larval stages of *S. calcitrans* infect every type of substrate, but bedding straw is the preferred medium. Both males and females are hematophagous, mainly feeding on animals' legs and flanks.

Habronemosis presents a seasonal trend, which reflects the seasonality of its intermediate hosts. Summer is the peak infection season in temperate climates, whereas in tropical areas, it occurs in spring-early summer and autumn (Pandey et al. 1981, 1992).

30.6 Prophylaxis and Treatment

Essential measures to reduce the incidence and prevent the recurrence of habronemosis (Gasthuys et al. 2004; Pusterla et al. 2003; Rebhun et al. 1981) are integrated fly control and proper removal and disposal of manure. Horses can also be protected from flies by using licensed repellents (Riha et al. 1986).

Different drugs have been described for treating habronemosis, including corticosteroids used to reduce inflammatory hypersensitivity reactions. Ivermectin (200 μg/kg) (Xiao et al. 1994; Costa et al. 1998) and moxidectin (400 μg/kg) (Xiao et al. 1994; Costa et al. 1998; Bauer et al. 1998) have been used to kill larvae and adult worms in the stomach (Pusterla et al. 2003; Herd and Donham 1981). Surgical intervention is indicated in cases of summer sores refractory to medical treatment.

References

Anderson RC (2000) The superfamily Habronematoidea. In: Nematode parasites of vertebrates: their development and transmission, 2nd edn. CABI Publishing, Guildford, pp 428–430

Bain AM, Rofe JC, Hotson IK et al (1969) *Habronema megastoma* larvae associated with pulmonary abscesses in a foal. Aust Vet J 45:101–102

Bauer C (1986) Infestation with stomach parasites in horses in North Germany. Dtsch Tierarztl Wochenschr 93:383–389

Bauer C, Cirak VY, Hermosilla C et al (1998) Efficacy of a 2 per cent moxidectin gel against gastrointestinal parasites of ponies. Vet Rec 143:558–561

Borgsteede FH, van Beek G (1998) Parasites of stomach and small intestine of 70 horses slaughtered in The Netherlands. Vet Q 20:31–34

Chaffin MK, Schumacher J, McMullan WC (1995) Cutaneous pythiosis in the horse. Vet Clin North Am Equine Pract 11:91–103

Costa AJ, Barbosa OF, Moraes FR et al (1998) Comparative efficacy evaluation of moxidectin gel and ivermectin paste against internal parasites of equines in Brazil. Vet Parasitol 80:29–36

Dunn AM (1969) Habronemiasis and Drascheiasis (sic. Draschiasis). In: Veterinary helminthology. W. Heinemann Medical Books Ltd, London, pp 199–200

Erschow VS (1956) Parasitology and parasitic diseases of livestock. State Publishing House for Agricultural Literature, Moscow

Euzeby J (1961) Les maladies vermineuses des animaux domestiques et leurs incidences sur la pathologie humaine. Maladies dues aux nemathelminthes, Fascicule Premier. Vigot Freres, Paris, pp 252–292

Foster AO, Ortiz OP (1937) A further report on the parasites of a selected group of equines in Panama. J Parasitol 23:360–364

Gasthuys FMR, Van Heerden M, Vercruysse J (2004) Conjunctival habronemiasis in a horse in Belgium. Vet Rec 154:757–758

Gawor JJ (1995) The prevalence and abundance of internal parasites in working horses autopsied in Poland. Vet Parasitol 58:99–108

Giangaspero A, Traversa D, Otranto D (2005) A new tool for the diagnosis *in vivo* of habronemosis in horses. Equine Vet J 37:263–264

Herd RP, Donham JC (1981) Efficacy of ivermectin against cutaneous *Draschia* and *Habronema* infection (summer sores) in horses. Am J Vet Res 42:1953–1955

Hiepe T, Buchwalder R, Nickel S (1985) Veterinämedizinische Helminthologie. In: Hiepe T (ed) Lehrbuch der Parasitologie. Gustav Fischer Verlag, Jena, p 419

Hoglund J, Ljungstrom BL, Nilsson O et al (1997) Occurrence of *Gasterophilus intestinalis* and some parasitic nematodes of horses in Sweden. Acta Vet Scand 38:157–165

Inzana TJ, Carter GR (1990) Fungi causing systemic or deep infections. In: Carter GR, Cole JR (eds) Diagnostic procedures in veterinary bacteriology and mycology. Academic Press Inc, San Diego, CA, pp 433–455

Lyons ET, Tolliver SC, Drudge JH et al (1983) Parasites in Kentucky thoroughbreds at necropsy: emphasis on stomach worms and tapeworms. Am J Vet Res 44:839–844

Lyons ET, Tolliver SC, Drudge JH et al (1985) Parasites in lungs of dead equids in Kentucky: emphasis on *Dictyocaulus arnfieldi*. Am J Vet Res 46:924–927

Mayhew IG, Lichtenfels JR, Greiner EC et al (1982) Migration of a spiruroid nematode through the brain of a horse. J Am Vet Med Ass 180:1306–1311

Mfitilodze MW, Hutchinson GW (1989) Prevalence and intensity of non-strongyle intestinal parasites of horses in northern Queensland. Aust Vet J 66:23–26

Molini U, Traversa D, Splendiani F et al (2004) *Thelazia* spp. and *Habronema* spp. in horses of Abruzzo region: an epidemiological update. Parassitologia 46 (Suppl 1):52

Murray DR, Ladds PW, Campbell RSF (1978) Granulomatous and neoplastic diseases of the skin of horses. Aust Vet J 54:338–341

Naem S (2007) The comparative morphology of three equine habronematid nematodes: SEM observations. Parasitol Res 101:1303–1310

Pandey VS, Eysker M (1988) Parasites of the stomach in donkeys of the highveld of Zimbabwe. Vet Q 10:246–248

Pandey VS, Ouhelli H, Elkhalfane A (1981) Epidemiological observation on stomach worms of horses in Morocco. J Helminthol 55:155–160

Pandey VS, Ouhelli H, Verhulst A (1992) Epidemiological observations on stomach worms of donkeys in Morocco. Vet Res Commun 16:273–279

Pascoe RR (1990) A colour atlas of equine dermatology. Wolfe Publishing Ltd, Ipswich

Petrella A, Traversa D, Maggi A et al (2010) Usefulness of a routine molecular approach to solve a skin lesion enigma: the case of "summer sores". Parassitologia 52:329

Pusterla N, Watson JL, Wilson WD et al (2003) Cutaneous and ocular habronemiasis in horses: 63 cases (1988–2002). J Am Vet Med Assoc 222:978–982

Ransom BH (1911) The life history of a parasitic nematode *Habronema muscae*. Science 34:690–692

Rebhun WC, Mirro EJ, Georgi ME et al (1981) Habronemic blepharoconjunctivitis in horses. J Am Vet Med Assoc 179:469–472

Reddy AB, Gaur SNS, Sharma UK (1976) Pathological changes due to *Habronema muscae* and *H. megastoma* (*Draschia megastoma*) infection in equines. Indian J Anim Sci 46:207–210

Reinemeyer CR, Smith SA, Gabel AA et al (1984) The prevalence and intensity of internal parasites of horses in the U.S.A. Vet Parasitol 15:75–83

Riha J, Minar J, Skaloud J et al (1986) The effect of selected preparations for the protection of draft horses against dipterous blood-sucking insects. Vet Med (Praha) 31:173–179

Roubaud E, Descauzeaux J (1922a) Deuxième contribution à l'étude des mouches dans leurs rapports avec l'évolution des habronèmes d'équides. Bull Soc Pathol Exot 15:978–1001

Roubaud E, Descauzeaux J (1922b) Evolution de l'*Habronema muscae* Carter chez la mouche domestique et de l'*Habronema microstoma* Schneider chez le stomoxe (Note préliminaire). Bull Soc Pathol Exot 15:572–574

Schuster RK, Sivakumar S, Kinne J et al (2010) Cutaneous and pulmonal habronemosis transmitted by *Musca domestica* in a stable in the United Arab Emirates. Vet Parasitol 174:170–174

Scott DW, Miller WH Jr (2003) Parasitic diseases. In: Scott DW, Miller WH Jr (eds) Equine dermatology. Saunders, St. Louis, MO, pp 357–360

Soulsby EJL (1982) Helminths, arthropods and protozoa of domesticated animals, 7th edn. Baillière, Tindall and Cassell, Eastbourne, pp 285–287

Stick JA (1979) Amputation of the equine urethral process affected with habronemiasis. Vet Med Small Anim Clin 74:1453–1457

Tieri E, Giangaspero A, Gatti A et al (1998) L'abronemosi degli equini. ODV 5:7–17

Traversa D, Giangaspero A, Galli P et al (2004a) Specific identification of *Habronema microstoma* and *Habronema muscae* (Spirurida, Habronematidae) by PCR using markers in ribosomal DNA. Mol Cell Probes 18:215–221

Traversa D, Giangaspero A, Iorio R et al (2004b) Semi-nested PCR for the specific detection of *Habronema microstoma* or *Habronema muscae* DNA in horse faeces. Parasitology 129:733–739

Traversa D, Iorio R, Capelli G et al (2006) Molecular cross-sectional survey of gastric habronemosis in horses. Vet Parasitol 141:285–290

Traversa D, Iorio R, Petrizzi L et al (2007) Molecular diagnosis of equid summer sores. Vet Parasitol 150:116–121

Traversa D, Otranto D, Iorio R et al (2008) Identification of the intermediate hosts of *Habronema microstoma* and *Habronema muscae* under field conditions. Med Vet Entomol 22:283–287

Vasey JR (1981) Equine cutaneous habronemiasis. Compend Cont Educ Pract Vet 3:290–298

Vercruysse J, Harris EA, Kaboret YY et al (1986) Gastrointestinal helminths of donkeys in Burkina Faso. Z Parasitenkd 72:821–825

Waddell AH (1969) A survey of *Habronema* spp. and the identification of third-stage larvae of *Habronema megastoma* and *Habronema muscae* in section. Aust Vet J 45:20–21

Xiao L, Herd RP, Majewski GA (1994) Comparative efficacy of moxidectin and ivermectin against hypobiotic and encysted and other equine parasites. Vet Parasitol 53:83–90

Part III

Diseases Caused by Arthropods

Flies

31

Carlos Brisola Marcondes
and Patricia Jacqueline Thyssen

Abstract

Here we discuss the role of several dipterans (flies, horseflies, mosquitoes, and others) in causing disease in human beings and other animals, through various mechanisms such as sucking of blood, contamination of foods and other materials, and infestation of the body. The use of maggots for cleaning and enhancing wound healing and in forensic entomology is also reviewed in this chapter. Other noxious Diptera (mosquitoes, black flies, sand flies, biting midges, and horse flies) are briefly discussed.

Keywords

Diptera • Maggot therapy • Forensic entomology • Biting flies • Myiasis

Diptera is a group containing two-winged flies, which have only one pair of functional wings (i.e., the fore wings) and modified hind wings that form club-like halteres. Diptera includes about 153,000 species, grouped into 160 families (Pape and Thompson 2013). Several of these insects are of medical importance due to transmission of parasites and due to direct harm caused by blood sucking and damage to animal tissues. Flies included in superfamilies *Muscoidea* and *Oestroidea* can cause damage by blood sucking, contamination of food, and infestation of animal tissues, also called myiasis. Other dipteran families will be briefly discussed below.

31.1 Bloodsucking Flies

As discussed in Chap. 4, several insects developed the habit of blood sucking. Some flies are well adapted for obtaining blood from vertebrates, while others take advantage by rasping the skin to cause bleeding and obtaining blood flowing from the wound. Adult males and females can both suck blood.

C.B. Marcondes (✉)
Department of Microbiology, Immunology and Parasitology, Center of Biological Sciences, Federal University of Santa Catarina,
88040-900 Florianópolis, SC, Brazil
e-mail: cbrisolamarcondes@gmail.com

P.J. Thyssen
Department of Animal Biology, Institute of Biology, State University of Campinas, UNICAMP,
13083-862 Campinas, SP, Brazil
e-mail: thyssenpj@yahoo.com.br

C.B. Marcondes (ed.), *Arthropod Borne Diseases*, DOI 10.1007/978-3-319-13884-8_31

Bloodsucking flies are included in the tribe *Stomoxyini* of Muscidae and in Glossinidae. Species of veterinary importance include some species in family Hippoboscidae, which can parasitize sheep, pigeons, and other birds, while species from the families Streblidae and Nycteribiidae parasitize bats. Some species in other families (e.g., Athericidae) feed on vertebrates, confirming a polyphyletic origin of the blood-feeding habit (see Chap. 2). Tribe *Stomoxyini* includes 10 genera and 40 species, with *Haematobia*, *Haematobosca*, and *Stomoxys* being the genera with the greatest medical importance. Both sexes of biting flies suck blood (see Chap. 4).

31.1.1 *Stomoxys calcitrans* (Stable Flies)

Among the 18 species in *Stomoxys*, only *S. calcitrans* is considered cosmopolitan. The others occur mostly in tropical regions, including 12 in the African continent, 4 in the Asian continent, and 1 in Asia and Africa (Zumpt 1973). These flies bite horses, cattle, pigs, sheep, dogs, and sometimes humans. They are similar to house flies but lighter in coloration, with the dorsal surface of abdomen slightly checkered, the M_{1+2} wing vein softly curved, and the proboscis evident in dorsal or lateral view, with short palps (Fig. 31.1a). They usually land on animal hosts with the posterior

Fig. 31.1 Several adult flies of medical, veterinary, and forensic importance. (**a**) *Stomoxys calcitrans* (Muscidae), with diagnostic biting mouthpart and bean-shaped eye (*arrow*); (**b**) *Haematobia irritans* (Muscidae), with biting mouthpart; (**c**) *Musca domestica* (Muscidae); (**d**) *Dermatobia hominis* (Oestridae); (**e**) *Cochliomyia macellaria* (Calliphoridae); (**f**) *Lucilia eximia* (Calliphoridae); (**g**) *Chrysomya albiceps* (Calliphoridae); (**h**) *Peckia* (*Pattonella*) *intermutans* (Sarcophagidae); (**i**) *Tabanus* sp. (Tabanidae), with long biting mouthpart

part of the abdomen nearer to the surface than the head and with the head downturned. *Neivamyia*, a poorly known Neotropical genus with four species, should probably be included in *Stomoxys*.

31.1.1.1 Biology

Stomoxys calcitrans larvae breed on wet to dry substrates, primarily on cattle and chicken dung, but also on deteriorating cut grass, onion leaves, and residues from sugarcane processing. Adults of both sexes viciously bite several animals. Adults can disperse several kilometers from a breeding ground. Mostly in Europe and the Eastern USA, they also can bite humans; they are important pests in Gulf coast regions from Alabama to Florida. The biology, economic importance, and control of *S. calcitrans* were recently reviewed (Salem 2012).

31.1.1.2 Veterinary and Medical Importance

Multiple bites are a nuisance to animals and can reduce weight gain and milk production. They are more concentrated in and near stables, rarely invading households, but can also bite in pastures. Dogs can be severely bitten, particularly on the ears, especially if weakened by babesiosis or other disease. Lions and other sylvatic felids may even be killed by multiple bites, as occurred in an outbreak in 1962 affecting lions in Ngorongoro Crater (Tanzania), in which the population dwindled from 75–100 to 12 individuals (Packer et al. 1991). Zoo managers at São Paulo noticed the preference of these flies for felids and have utilized traps for their control.

Stomoxys flies caused mortality in bushbucks (*Tragelaphus scriptus*) in Kenya in 1993, and *S. omega* greatly contributed to mortality of ungulates in the Northern Congo in 1997. *Stomoxys calcitrans* can also viciously bite humans in certain regions, which along with horse flies (Tabanidae) contributed to an anthrax outbreak in Chad that resulted in 33 human deaths, as well as the deaths of many donkeys and horses (Lamarque et al. 1989).

Stable flies have been incriminated as mechanical vectors of several trypanosomes, bacteria, and viruses of cattle and wild beasts. Baldacchino et al. (2013) reviewed the transmission of pathogens by *S. calcitrans*. Several trypanosomes (see Chaps. 18 and 19), *Bacillus anthracis* (anthrax), *Escherichia coli* (bovine mastitis), and several viruses (bovine leukosis, infectious equine anemia, West Nile, vesicular stomatitis, Bluetongue, etc.) may be transmitted by *S. calcitrans*. Eigen et al. (2002) hypothesized the transmission of HIV from great apes to humans by this species. The parasitic nematodes *Thelazia* and *Habronema* (see Chaps. 29 and 30) are also transmitted between equines by *S. calcitrans*.

In the USA and Brazil, the annual loss for the cattle industry due to *S. calcitrans* is at least U$2.221 million and U$335.5 million, respectively. The frequency and severity of outbreaks are increasing in Brazil due to the rapid expansion of sugarcane production and the use of vinasse for fertilization of sugarcane fields, which stimulates the development of immature stages and the production of adults.

31.1.1.3 Control

Although insecticides may be used on animals employing methods similar to those recommended for control of *Haematobia irritans* (see below), there is a strong tendency for development of resistance. Since the flies are only on the animals for short periods of time, the insecticides act mostly as repellents. Traps with light in the UV range (360 nm) and the blue range (450–550 nm) or suction (see *Haematobia* control below) may also reduce the number of *Stomoxys* on dairy cows. They are also attracted to traps with CO_2, ammonia, and phenylpropanoid compounds and to Nzi traps with a blue-colored canvas plus attractants, also useful for *Glossina* and horse flies. A sterile insect release program (SIR) combined with an integrated management program was moderately successful in St. Croix, Virgin Islands. However, large populations and immigration of wild flies limit the efficacy of SIR programs.

Residues adequate for the development of immature forms of the fly, such as waste from henhouses, sugarcane processing by-products, and plant tissues (e.g., onion, grass), must be protected from colonization by flies; the material should be covered by black plastic, buried or processed. Piles of manure should be turned over or covered to expose immature forms to predators and parasites and to disturb development.

The use of avermectin for control of biting flies (and helminths and ticks) should be evaluated, as these drugs impair the development of immature forms in the dung. However unintended effects must be carefully evaluated, especially since the development of dung-processing and dung-recycling beetles may also be negatively affected. This disturbance on dung degradation seems to be more evident in Europe and other temperate regions, where the beetle fauna is less diversified, than in Argentina and Brazil (Suárez 2002).

31.1.2 *Haematobia irritans* and *Haematobia exigua* (Horn Flies, Buffalo Flies)

These flies are brownish in color and smaller (4–5 mm) than *S. calcitrans*, with distally dilated maxillary palps measuring 2/3 of the proboscis (Fig. 31.1b). *Haematobia exigua* is smaller than *H. irritans*. In Asia, morphological and molecular characters associated with differences between populations may enable separation into distinct species.

31.1.2.1 Biology

Although they can bite dogs and donkeys, as well as horses when pastured with or near cattle, *H. irritans* prefer to bite bovines. They tend to concentrate mostly around the head and in zebuines, on the hump, and the shoulder. These flies can bite from 24 to 38 times per day (Oyarzún et al. 2008) and can inhabit animal hosts in huge numbers (e.g., sometimes 10,000). These flies have been reported on buffalos on Marajó Island (Brazil) (Fig. 31.3a). After the bovine defecates, females lay eggs on the fresh dung and return to the animal, where they stay during the daylight hours. Feces of other ruminants are not convenient, due to a greater tendency to lose water.

In Brazil, adults emerge in 8–9 days in the summer and 20–28 days in the winter. In northern higher latitudes, pupae enter diapause and adults emerge from March to May. In some temperate areas of Chile, adults are active between November and May. In tropical areas such as Queensland (Australia) and central Brazil, quantity of rain is more important than temperature to regulate the population size.

Adult flies, mostly young females, can migrate up to 12 km between herds; physical barriers such as trees do not prevent migration. The quantity of *H. irritans* per animal seems to be greater in taurine than in zebuine cows in Brazil, and size and color influence colonization. Dark-colored coats are associated with thinner skin and a higher concentration of hemoglobin and attract more flies (Breijo et al. 2013). The distribution of flies on the host's body can be influenced by rain, sunlight, and temperature. Male and adult hosts are more heavily attacked than females and young animals, and some animals tend to have higher infestations (low carriers vs. high carriers), for reasons unknown.

Haematobia irritans expanded its range from the original Eurasian distribution to include North America between 1884 and 1886 and, in the last 30 years, has also entered South America. *Haematobia exigua* tends to bite buffalo and cattle and is distributed in Australasia.

31.1.2.2 Veterinary Importance

Multiple bites are a nuisance to cattle, causing unrest and reducing weight gain and milk production, as well as causing hardening of leather and devaluation due to multiple inflammatory reactions and scars. Males, which usually attract more flies, may have impaired reproductive capability. Although densities below 100–200 flies/animal are considered tolerable, results of studies comparing cattle treated and untreated have been very variable (Oyarzún et al. 2008). Studies show the weaning weight of control groups varied between 10 and 12.5 kg, and the average daily gain varied between 0.04 and 0.12 kg (Oyarzún et al. 2008).

Annual damage is estimated at 1 billion USD in the USA, and despite more than a century of struggle against *H. irritans*, the damage caused by this fly has not been significantly reduced.

Damage is estimated at 150 million USD in Brazil. In central Brazil, with a mean infestation of 500 flies per animal, the annual loss of weight per animal was calculated as 40 kg. If all animals were parasitized, the total loss would amount to 1.4 million tons. Damage to cattle hide is considerable, mostly in infestations higher than 100 per animal.

Haematobia irritans can transmit the nematode *Stephanofilaria stilesi* in Central Queensland (Australia), but the transmission of other parasites is jeopardized by its low tendency to change hosts. This species has been found carrying *D. hominis* eggs and may also carry *Mycobacterium bovis* and can (in addition to *S. calcitrans*) facilitate mastitis caused by *Staphylococcus* spp. and teat atresia due to multiple bites.

31.1.2.3 Control

Haematobia irritans tends to develop resistance to insecticides in few years of use, preventing the indiscriminate utilization of these substances. Adequate recycling of dung by insects (e.g., using coprophagous beetles) along with utilization of insecticides via ear tags and pulverization in adequate seasons is considered the best control method.

In central Brazil, where a cooler and drier season in the middle of the year alternates with a warmer one with more rain, EMBRAPA (Brazilian Agricultural Research Corporation) proposed a scheme for control:(a) introduction of dung-removing beetles (preferably the African species *Onthophagus gazella*); (b) if the population attains 200 flies/animal, organophosphorus insecticide application at the end of the dry season and pyrethroid pour-on or ear tags. Ear tags may be utilized for 3 months at most, and insecticides should never be applied during the dry season except during exceptionally large infestations.

One study in Australia showed 96 % mortality of *H. irritans* larvae in the experimental group with dung beetles present, indicating the importance of beetles for control. No ruminants were present in Australia until the end of the eighteenth century, and their presence after that time required the introduction of dung-processing beetles for recycling feces.

Several other insects are also used in horn flies' biological control: parasitoid microhymenopterans such as *Spalangia* spp. and *Muscidifurax* sp. (Hymenoptera: Pteromalidae), fire ants, and predatory immatures of *Orthelia* and *Gymnodia* (Diptera: Muscidae).

Ear tags with chlorfenapyr (AC303630), a pyrrole compound, were efficient for control of the fly (Guglielmone et al. 2000), and pyrethroid-resistant flies are much more susceptible to this substance. The use of diflubenzuron in pastures negatively affected the fly (Tomberlin et al. 2007), and insect growth regulators such as methoprene and fenoxycarb can also be utilized for the control (Oyarzún et al. 2008). The combination of chlorfenapyr and alphacypermethrin is useful for the control of resistant *Anopheles* and *Culex quinquefasciatus* and should be tested for *Haematobia*. Cypermethrin and chlorpiriphos were also highly effective for the control of *H. irritans*. Several avermectins, primarily abamectin for *H. exigua*, are efficient for the control of larval stages, but the excretion of most of them into milk is a disadvantage.

Since the flies may move from animals at night, a tent trap was designed with two plastic layers (one black colored with holes and a transparent external, one) where cattle pass (Fig. 31.3a) (Tozer and Sutherst 1996); the flies pass through the holes in the black plastic layer and are killed by heat from the sun. Another tent trap was tested utilizing a panel made from fabric straps to brush off the flies from the animals and a black light to attract them to an electrocution device (Watson et al. 2002). A third trap that aspirates flies from dairy cattle was highly efficient for reduction in quantities of *S. calcitrans*, *Musca autumnalis*, and *H. irritans* in North Carolina (USA) (Denning et al. 2014). These traps[1] were effective for reducing the number of flies per animal, but utilization is more adequate for calm dairy cows.

[1]A tent trap developed in Mauritius had already been described in 1930 (Loughnan 1930 *apud* Bruce 1938).

31.1.3 *Haematobosca*

These flies occur mostly in the Afrotropical region; however, some species occur in the Oriental and Holarctic regions. There is some controversy as to whether they initiate wounds or whether they enlarge wounds caused by horse fly bites. They congregate around wounds on the body of several ungulates, including moose and cattle. *Haematobosca alcis* attacks moose in the Nearctic region, biting mostly the back legs (Fig.31.3b) and congregating around sores "much like petals on a daisy" (Fig.31.3c). These open sores seldom become infected with bacteria, which suggest that the flies may produce antibiotic substances. Some species have also been suspected as vectors of trypanosomes.

31.1.4 *Glossina* spp. (Glossinidae) (Tsetse Flies)

The 22 (or 23) known species of tsetse flies are brownish in color and measure from 6 to 14 mm. They are distributed from the Indian to the Atlantic Ocean and from the Sahara desert to the Kalahari, absent only in the extreme north and south of the continent. Adults are long-lived (half-life >28 days) (Jordan 1993; Krafsur 2009). Females feed a single larva in their "uterus," with "milk" rich in lipids, proteins, and tyrosine, giving birth to a third-instar larva each 9 days which then burrows into the soil and transforms into a black puparium. Tsetse flies are typical *k* strategists, compensating for low reproductive rate with high survival rate (97 % per day).

Tsetse flies are mostly associated with bush vegetation, and adults usually feed on mammals; however, some species feed on reptiles and birds. They can transmit several trypanosomes, mostly among wild large mammals of Africa and can transmit *Trypanosoma brucei gambiense* and *Trypanosoma brucei rhodesiense* to humans, *Trypanosoma brucei* to cattle, and other *Trypanosoma* species to pigs and wild animals (see Chaps. 18 and 19).

31.2 Secretion-Feeding Flies

31.2.1 *Musca autumnalis* and *Musca larvipara* (Face Flies)

These flies are widely distributed in the Palearctic (from India and Kazakhstan to East of Europe and North of Africa) and Nearctic regions and feed on mammalian secretions (Krafsur and Moon 1997); having well-developed prestomal teeth, they can reopen healing wounds. The flies probe and consume secretions or discharge from the eyes, nostrils, mouths, teats, and other body parts of cattle, horses, and bison, causing considerable nuisance if numerous (>15 per animal). Bacteria causing pinkeye (*Moraxella bovis*) can be transmitted by these flies, and possibly *Brucella abortus*, *Thelazia*, *Habronema*, and *Parafilaria bovicola* (see Chaps. 29 and 30). Insecticides and traps have not been effective for their control due to short time of residence on the host and short generation time. Sterile insect technique (SIT) similar to that successfully utilized for *Cochliomyia hominivorax* has been proposed.

31.2.2 *Musca sorbens*

Feeding mostly on excrement and secretions, this fly also can rasp wounds. It is very common around garbage dumps, pigpens, and carrion, swarming on foodstuffs. Immature forms prefer human feces for oviposition. It is widely distributed on North Africa and Asia and attacks humans, mostly children and in the brighter hours of the day, transmitting bacteria.

31.2.3 *Musca crassirostris* (Indian Cattle Flies)

With strong prestomal teeth, these flies can scrape the skin to obtain blood. They are widely distributed in the Afrotropical region, North Africa, the Middle East, and the Oriental region in Taiwan and Lombok. It has been incriminated as vector of *Thelazia rhodesii* in India.

31.2.4 Drosophilidae (Fruit Flies)

Some of these flies have been incriminated as vectors of *Thelazia callipaeda* (*Phortica variegata* and *P. okadai* in Asia and *P. variegata* in Italy) (see Chap. 29). Certain species of drosophilids can be involved with pseudo- or accidental myiasis (see myiasis classification below). Females, attracted by the smell of fruit and other fermented vegetables, lay eggs which when ingested may cause intestinal problems.

31.3 Contaminating and Irritating Flies

Several flies can carry bacteria, protozoa, helminths, and viruses, which can cause high rates of morbidity and mortality; thus, control is of great importance.

31.3.1 *Musca domestica* (House Fly) and Other Synanthropic Flies

Musca domestica is globally distributed and has caused great problems in rural and urban areas. Adults are 7–8 mm long with gray to black coloration, with four dark stripes on the mesonotum and wings with M_{1+2} angulated (Fig. 31.1c). Other smaller (*Fannia* spp.) or larger (*Sarcophaga* spp.) flies have similar biology, but a smaller role in human and animal health. Some blowflies in the family Calliphoridae (*Chrysomya*[2] spp.), called sometimes as "latrine flies," play an important role in mechanical transmission of pathogens.

31.3.1.1 Biology

Immature flies develop in several types of deteriorating organic matter (manure, garbage, meat, etc.), in drier substrate than is adequate for *S. calcitrans*. To continue their natural development cycle, third-instar larvae spontaneously leave the resource and often burrow into the soil to pupate. They are well adapted to farms and backyards and easily invade human occupations. Adults can disperse for several kilometers, but usually concentrate around breeding places where the food level is satisfactory for them.

31.3.1.2 Damage

House flies can transmit many pathogens through regurgitation through the proboscis, defecation, or when walking on food, on the skin, and on hospital and kitchen equipment. Hundreds of bacteria, helminths, protozoa, and viruses have been associated with *M. domestica* and other synanthropic flies (Greenberg 1973). Analysis of bacteria in house flies in hospitals indicated high level of multiple resistance to antibiotics, mostly in *Klebsiela* spp., indicating contamination of flies in these establishments.

Their role is greater for the transmission of shigellosis than of salmonellosis due to the smaller number of bacteria needed for transmission of *Shigella* spp. *Chlamydia trachomatis*, a Gram-negative bacterium causing trachoma (a potentially blinding disease), is primarily transmitted by direct contact, but eye-frequenting flies (*M. domestica* and *M. sorbens*) may be important for transmission in some areas. *Musca domestica* is rarely reported causing accidental myiasis and for this reason has been tested for use in maggot therapy.

31.3.1.3 Control

After the dream of a "world without flies," after the discovery of chlorate insecticides in the 1940s, house flies have developed resistance to most insecticides, which have rapidly become useless. Measures for control of *M. domestica* and other synanthropic flies are similar to those recommended for *S. calcitrans*; however, they are less selective in the choice of material for immature forms than *S. calcitrans*, which is restricted to manure and vegetable material. Garbage and other organic materials must be made inaccessible to house flies and larvae destroyed by solar heat or iso-

[2]The use of "mya" and not "myia" in this genus, as would be correct, is due to a mistake in the original name.

lated from the soil to prevent the burying and molting to pupae. Manure and garbage piles must be covered by heavy-gauge (6 mm) black plastic kept in place by stones.[3] Since immature forms of *M. domestica* can be useful as food for some fish (e.g., *Oligusarcus argenteus*), garbage can be accumulated in containers with a screen on the bottom, over water for fish to feed on dropping larvae.

Some hymenopteran parasitoids of house fly pupae (see above in *H. irritans*) that prey on mites and fungi (e.g., *Beauveria bassiana* and others) have been tested with promising results.[4] In addition to controlling flies, food, hospital instruments, and other materials must be protected by screening of doors and windows. The adoption of extra screen doors and screening of windows resulted in great success for control of intestinal infections in the USA, mainly in southern areas. The eradication or control of synanthropic flies and the improvement of toilet facilities and personal hygiene in underdeveloped countries have caused a drastic reduction in the incidence of gastrointestinal diseases in children.

31.3.2 *Muscina stabulans*

These flies are found in and near houses and in precarious conditions. Predominantly present in rural areas and scarce in towns, adults feed mostly on fruits and sap flowing from trees. Immature forms develop mainly in human feces and rarely in carrion. Larvae may compete with other larvae, and third instars also can prey on other larvae, including *M. domestica* and *S. calcitrans*. *Muscina stabulans* has also been incriminated as vector of many bacteria and viruses (Greenberg 1973).

[3]Stafford III (2008) recommended the utilization of old tires, but they could constitute good breeding places for *Aedes aegypti*.

[4]These hymenopteran parasitoids also have some usefulness for other flies, like *Chrysomya* spp. and fruit flies (*Anastrepha* spp. and *Ceratitis capitata*).

31.4 Myiasis-Causing Flies

Myiasis [from myia (=fly); in Modern Greek the spelling of the word is μύγα] was proposed by Hope (1840) as a term to identify a particular type of association between flies and live vertebrates[5] which is apparently little beneficial to the host, considering that the most severe cases culminate in large lesions, weakness, or even death. A certain number of definitions of myiasis are available in the literature; however, the most widely accepted currently is that proposed by Zumpt (1965), which is "the infestation of live human and vertebrate animals with dipterous larvae, which, at least for a certain period, feed on the host's dead or living tissue, liquid body substances, or ingested food."

There seems to be no universal consensus regarding the classification of myiasis. Several publications, especially those within the medical scope, prefer to distinguish the different forms by taking into account the location of the infestation on the host's body. Thus, in a more simplified way, and probably under the influence of Bishopp (apud Patton 1922) and James (1947), a wide range of records and reports of myiasis can be found, including ocular, oral, nasal, anal, urogenital, enteric, traumatic, furuncular, and bloodsucking, among others; not uncommonly the causal agent of myiasis is misidentified or is not diagnosed to the species level. "Primary" and "secondary" myiasis are terms often used to define clinical conditions in which the infestation begins without or with a preexisting lesion, respectively.

In parasitological terms, myiasis type can be grouped according to the relationship between the dipteran and its host (Zumpt 1965). In (1) *obligatory* myiasis, flies feed on body tissues or other substances from their living hosts; (2) a *facultative* relationship describes when the flies usually consume dead organic matter and opportunistically feed on dead tissues of live vertebrates; (3) *accidental*, also called pseudo-

[5]Before that, it was common to refer to animal diseases caused by insect larvae, in general, as scholechiasis, according to Kirby and Spence (1815).

myiasis, occurs via the accidental swallowing of immature flies (eggs or larvae) or unusual larval penetration of natural orifices of animals.

The emergence of the habits causing myiasis and the transition to parasitism may have arisen from a broad adaptability process (e.g., the high degree of specificity of Oestridae to their hosts) or by complete lack of specialization of a group to exploit the resources available in the nature (e.g., Calliphoridae and Sarcophagidae) (James 1969). Zumpt (1965) proposed routes for the evolution of this behavior derived from ectoparasitism or saprophagy. In this way, flies that feed on feces or decomposing corpses, since they are attracted by these odors, adapted to lay eggs in wounds of living animals, and their larvae have come to feed on the necrotic tissues. Some of these species have also developed the ability to invade living tissue, with the probable emergence of appropriate enzymes for this purpose.

The registry of parasitized vertebrates, both wild and domestic, is numerous and includes amphibians, reptiles, birds, and mammals, including humans. Likewise, the number of fly families containing species that cause myiasis is also extensive, although most are associated with facultative or accidental cases. The list below provides an idea of the degree of importance and/or involvement of some taxa (family, genus, or species) with their respective types of parasitism:

- *Obligatory*: Calliphoridae (*Auchmeromyia*, *Chrysomya bezziana*, *Cochliomyia hominivorax*, *Cordylobia*, and *Lucilia*), Muscidae (*Passeromyia* and *Philornis*, both in which species affect wild birds and are found in the Old World and the New World, respectively. For additional details, see Guimarães and Papavero (1999)), Oestridae (Oestrinae – *Oestrus*, *Rhinoestrus*, and *Cephalopina*; Gasterophilinae – *Gasterophilus*; Hypodermatinae – *Hypoderma*; Cuterebrinae – *Cuterebra*; and *Dermatobia*), and Sarcophagidae (*Wohlfahrtia*).
- *Facultative*: Calliphoridae (*Calliphora*, *Cochliomyia*, *Chrysomya*, *Lucilia*, *Phormia*, and *Protophormia*), Fanniidae (*Fannia*), Muscidae (*M. domestica*) and Sarcophagidae (spp.).
- *Accidental*: some families which contain species involved Anisopodidae, Calliphoridae, Drosophilidae, Ephydridae, Fanniidae, Muscidae, Phoridae, Piophilidae, Psychodidae, Sarcophagidae, Scenopinidae, Sepsidae, Sphaeroceridae, Stratiomyidae, Syrphidae, Tachinidae, Tephritidae and Tipulidae; in most cases, myiasis is intestinal, urogenital, and/or nasopharyngeal.

The last extensive revision on the geographical distribution, economic importance, and status of control techniques for myiasis of humans and domestic animals was brilliantly conducted and published by Hall and Wall (1995). Thus, we will address this work to provide information herein on the biology and importance of the species most commonly involved with myiasis globally and update information on possible control mechanisms.

31.4.1 *Cochliomyia* spp.

Blow flies (=Calliphoridae) are common and conspicuous due to their size, noise, and metallic coloration. The colors vary widely in the various species; thus, it is difficult to identify genera or species by color alone within the family Calliphoridae. For the *Cochliomyia* genus, it is useful to observe the dark vertical stripes on the mesonotum (Fig. 31.1e). Four species (average size of 8–10 mm) are valid within the genus: *Cochliomyia aldrichi*, *C. hominivorax* (=*Cochliomyia americana*), *C. macellaria* (=*Cochliomyia fontanai*), and *Cochliomyia minima*.

Cochliomyia hominivorax, also referred to as the New World screwworm, is among the most important species causing obligatory myiasis and may be associated with large economic damage when it affects domestic animals, particularly with regard to livestock (Hall and Wall 1995). Larvae feeding on the skin of the host cause a condition known as traumatic myiasis (Fig. 31.3g), which can be fatal in some cases.

In the forensic field, cases of myiasis in humans may be useful to confirm mistreatment (Thyssen et al. 2012). There are more than 7000 documented cases in the literature of myiasis exclusively caused by this species.

Cochliomyia macellaria are known as facultative or secondary agents of myiasis because their larvae feed only on necrotic tissue, being commonly found in animal carcasses or exposed meat. Furthermore, *C. macellaria* is usually among the first colonizers of carrion, more abundant on corpses exposed to warm and directly sunlit areas. Thus, the importance of this species has been highlighted for use in the forensic field for determination of the time of death in some cases. Like any other flies, it can vector the bacteria that cause dysentery; however, their behavior as a causal agent of secondary myiasis can be explored for use in surgical maggot therapy.

Cochliomyia hominivorax is distributed almost throughout South America, in Central America, the Caribbean Islands, and in the southern states of the USA. It is confined to the Nearctic and Neotropical regions, but the potential for these flies to spread is considered a real problem. For example, in the year 1989 an infestation of sheep with *C. hominivorax* outside of the American continent (i.e., in Northern Africa) was documented.

Adults preferentially live in forested areas, but they travel to open pastures looking for hosts and are capable of ovipositing up to 400 eggs per cycle (Guimarães and Papavero 1999). They can disperse long distances and are only active in summer or in warm climates (temperatures above 16 °C). The life cycle is similar to any other Diptera. Females are attracted to the smell of wounds and lay eggs grouped in masses on the edge of an open wound (of whatever nature, e.g., umbilicus of young animals) or in natural body openings (e.g., the nose, ears, mouth, anus, and eyes). The infestation in natural openings occurs when there are injuries, poor hygiene, or in association with bacterial infections (Fig. 31.3d). The larvae hatch within 24 h and begin to feed on living tissue, reaching maturity within 5 days. Mature larvae leave the wound and burrow into loose soil to pupate. Adults emerge in 8 days and live on average 3–4 weeks, depending on the temperature and humidity. Females obtain protein to support egg production from wounds, excrement, or decaying animal matter.

It was the first insect to be controlled and eradicated using sterile insect release, in a program known as sterile insect technique (SIT), originally proposed by Knipling (1955). This technique consists of releasing sterilized males on the field in numbers sufficient to overwhelm the wild population, with the idea that sterilized individuals will compete with wild "normal" males. Notable success has been obtained in the some areas subjected to SIT, for example, in Curacao. It is estimated that the cost of the program was about 100 million US dollars, and a few months for eradication be reported as effective.

Control costs have been estimated at US$2 million per year. Control of myiasis in animals is achieved through the use of chemical compounds, mainly organophosphorus compounds, macrocyclic lactones and, more recently, a foamy spray based on spinosad. Concerns about insecticide resistance to organophosphorus compounds should be taken seriously. Studies are currently underway to develop a synthetic odor bait to replace the use of wounded animals (i.e., sentinels) to attract female *C. hominivorax*. The synthetic bait is formulated to mimic natural wound fluid from animals and could be used at research stations that monitor *C. hominivorax* in eradicated regions in addition to helping decrease *C. hominivorax* populations in infested regions (Cork and Hall 2007).

Larvae of *C. hominivorax* and *C. macellaria* are extremely similar (Fig. 31.2i). The most effective way to differentiate the two is to note the absence of the distinctive pigmented tracheal tubes on *C. macellaria* (Fig. 31.2u).

31.4.2 *Chrysomya* spp.

There are many species of *Chrysomya* (Calliphoridae), with about 15 occasionally involved in myiasis. They are most commonly

Fig. 31.2 Larvae of several flies of medical, veterinary, and forensic importance. From (**a**) to (**n**), (**u**), (**y**), and (**z**) lateral or dorsal view. From (**o**) to (**t**) and from (**v**) to (**x**) posterior view showing respiratory spiracles with diagnostic value. In (**a**) and (**o**) *Cuterebra* sp. (Cuterebridae); (**b**) *Hypoderma* (*Hypoderma*) *bovis* (Oestridae); (**c**) and (**p**) *Hypoderma* (*Hypoderma*) *lineatum* (Oestridae); (**d**) and (**q**) *Oestrus ovis* (Oestridae); (**e**) and (**r**) *Gasterophilus intestinalis* (Oestridae); (**f**) *Gasterophilus nasalis* (Oestridae); (**g**) and (**s**) *Dermatobia hominis* (Oestridae); (**h**) and (**t**) *Sarcophaga* sp. (Sarcophagidae); (**i**) and (**u**) *Cochliomyia hominivorax* (Calliphoridae), showing the pigmented tracheas; (**j**) *Chrysomya megacephala* (Calliphoridae); (**k**) and (**v**) *Chrysomya albiceps* (Calliphoridae); (**l**) and (**w**) *Lucilia eximia* (Calliphoridae); (**m**) and (**x**) muscid fly; (**n**) *Fannia* sp. (Fanniidae); (**y**) *Auchmeromyia* sp. (Calliphoridae); (**z**) *Cordylobia anthropophaga* (Calliphoridae)

found feeding and breeding on meat, fish, dairy products, animal carcasses, garbage, and excrement and therefore are vectors of numerous pathogens of humans and domestic animals. Some species prefer to lay their eggs on dead organisms and have a particularly important role in forensic entomology because they are usually the first to colonize and are the most abundant flies found feeding on carrion where they occur.

Adult flies (average size 10–12 mm) usually have a blue or green color, with horizontal dark bands delimiting the thoracic segments, plumose arista, and wings that are usually clear (Fig. 31.1g). The larvae of the most species are called screwworms because of the rings of spines on the body, making them similar in appearance to woodworking screws (Fig. 31.2k). Other blowfly larvae are also called screwworms even if the spines along the body are less evident to the naked eye (e.g., see the difference between larvae in i and k in Fig. 31.2). *Chrysomya rufifacies* larvae have the common name of "hairy maggot blowfly," and the same could be applied to *C. albiceps*. The appearance of *C. albiceps* and *C. rufifacies* larvae makes it easy to distinguish them from other more closely related species, but it is not easy to make a correct morphologic diagnosis between the two species or between other *Chrysomya* species in first and second instars that do not resemble these two.

Chrysomya was originally confined to the Afrotropical, Oriental, Palaeartic, and Australasia regions. Within these regions they are most common in tropical humid climates. *Chrysomya albiceps*, *C. megacephala*, and *C. putoria* arrived in South America about 35 years ago and became established in the Neotropical region and the first in the USA (Wells 1991). The establishment of these species in the New World has affected the native fauna by displacing several native species, among them *Cochliomyia* and *Lucilia* (Wells and Greenberg 1992). *Chrysomya albiceps* and *C. rufifacies* are known for being voracious predators of other blowfly species during the larval stage. In South America, *C. albiceps* is frequently confused with *C. rufifacies*, but the records for the latter must certainly be revised due to problems with misidentification (Grella et al. 2015).

While other *Chrysomya* species are facultative parasites of minor importance to human or animal health, *C. bezziana*, also referred as the Old World screwworm, is among the most important species causing obligatory myiasis in a wide range of mammals, including livestock (cattle, sheep, dogs, and pigs) and humans (more frequently observed in the oral cavity or ears), particularly in the Oriental region. When the infestation is in the head region, starting at a wound, or in discharge from the nose, it is likely to be fatal. Infestations in living or dead tissues take the form of a batch of larvae forming a suppurating superficial lesion in muscle or the body wall. Usually, several females laying their eggs in the same place on the animal body are attracted to pheromones expelled from the first female, so many larvae can be found in the same wound; the results are serious.

The *Chrysomya* life cycle is basically of the muscid type. The females, which are attracted to the odor, obtain the protein to support egg production from wounds or decaying animal matter. About 100–200 eggs are laid at the adult feeding site. The obligate parasitic species burrow into the wound area with their posterior spiracles (see this anatomical structure in Fig. 31.2v) on facing the surface. When mature, the larvae leave the wound and burrow in loose soil to pupate. Adults can disperse long distances and are only active in summer or in warm climates. In the forensic field, this species may not be useful to determine the time of death since it colonizes live hosts.

Organophosphorus insecticides (e.g., coumaphos, dichlofenthion, and fenchlorphos) have been applied on wounds with larvae. These products act on the larvae, which leave the wound and fall to the ground. Without a host to feed, the larvae die. Dicyclanil is used as insect growth regulator on castration wounds in cattle and has been quite effective in preventing infestations by *C. bezziana*. Ivermectin has been administered for the treatment of infestations.

A few insecticides are available for use in control of adult *C. bezziana*. Proper spraying and dipping the animal in insecticides can help in management or to prevent infestation by the insect. Unfortunately, *C. bezziana* has not been eradicated using the SIT, perhaps due to the extensive use of pesticides. Vaccination of livestock was explored both in vitro and in vivo, but this issue is still unresolved.

31.4.3 *Lucilia* spp.

This genus belonging to Calliphoridae is also known in the Neotropics as *Phaenicia*. Although the largest number of species is found in the Palearctic region, they are found in all regions. They are typical blowflies of cool humid climates; however, it is interesting that they have dispersed with humans throughout the world. Adult flies (average size 10–14 mm), known as greenbottles, are metallic dark green and are often shiny, sometimes with coppery reflections and numerous bristles on the thorax (Fig. 31.1f). The larva is similar in shape to other blowflies (Fig. 31.2l), except by the anatomical structure of the posterior spiracles (Fig. 31.2w).

The most common species are *Lucilia cuprina*, *L. eximia*, and *L. sericata*. Adults are noted for strong flight with a buzzing sound and for gathering in places with dead animals or similar odors. These blow flies are household pests and, because of their feeding habits, have the same potential for disease transmission as other fly species. However, they do not appear to be as important as *M. domestica*. Contamination of dirty bandages and wounds may occur (by *L. sericata*), causing facultative myiasis in humans. As in other species of blowflies, larvae feed and develop in a wide variety of carrion, including human corpses, making them important in a forensic context. They have been reported as the first colonizers of carcasses, although the potential of their presence to provide useful information regarding date and time of death is remote when their niche overlaps with predatory *Chrysomya* species. Sanford et al. (2014) demonstrated the need for precaution when evaluating *L. eximia* in forensic cases (a more detailed explanation is below in the Forensic Entomology section). *Lucilia sericata* has been the most promising species for use in maggot therapy or MDT. There is a section below exposing what it is and what are the purposes of the MDT.

The life cycle of *Lucilia* is basically the same as other types of blow flies. In general, female adults are attracted to tissues with previously existing necrosis to lay their eggs (50–100). Larvae usually develop without causing harm to live hosts, but may cause facultative myiasis and sometimes obligatory myiasis. These species occur particularly in woolly animal breeds where the fleece is easily soiled, thus attracting adult blowflies. The infestations may be restricted to the skin surface; however, small wounds and soft skin may be invaded and larvae may spread under the skin. Larvae feed on necrotic tissue for approximately 7 days depending on the temperature. Infestations include many larvae, and secondary bacterial infections may occur.

Lucilia sericata is involved in obligatory parasitism of sheep (for this reason the infestation is known as a “sheep strike”). This species is not a major cause of blow fly strike in most regions, but causes serious economic problems for sheep farmers in Africa, Australia, Northern Europe, and New Zealand. In the UK, infestation can reach 80 % of the flock. *Lucilia cuprina* was also confirmed as an agent of primary myiasis in Australia in the 1930s and has been registered since then as dominant species in the region.

Simple measures such as shearing sheep regularly may help to reduce the degree of infestation in an area, as well as removing feces and dead animal carcasses which can serve as an alternative food substrate for adult flies. Furthermore, moving the sheep from warm, humid, and sheltered areas to open areas can help to eliminate favorable conditions for fly development.

The insecticides employed do not differ much from those recommended for *C. bezziana*, though they can be costly. It is estimated that several million dollars are spent annually in loss of animals and attempt to control flies. The insecticides last from 6 to 16 weeks depending on the exact type of product (pyrethroid, cyromazine, or dicyclanil) used. Trapping systems such as sticky paper may also be used to control fly numbers.

Research has been conducted to test the use of essential oils and natural plant extracts to kill larvae. There has been no significant advancement in vaccine production or the use of SIT in the control of *Lucilia*; however, one type of genetic control has been tested in *L. cuprina*, which is radiation exposure to induce mutations that can confer sterility in adults.

31.4.4 *Cordylobia* and *Auchmeromyia*

These genera are calliphorid flies, with several species being parasites of humans and other animals. *Cordylobia* (=*tumbu fly*) larvae are similar to other species causing myiasis and are also known as mango fly, perhaps because adult color is predominantly yellow brown. *Auchmeromyia* (=*floor maggot*) larvae feed on blood, but do not remain in the host, yet according to above definitions this type of parasitism is classified as myiasis.

There are two most important and best-known species of *Cordylobia*, *C. anthropophaga* (which infest humans, dogs, cats, and some wild mammals) and *C. rodhaini* (which infest antelopes, rodents, and rarely humans). In the genus *Auchmeromyia*, *A. senegalensis* (*A. luteola* is one possible synonym) is important because it feeds on humans; pigs and warthogs are also parasitized, but whether it is the same species that infests humans remains unknown. Adults of both genera are difficult to distinguish from many other species of calliphorid flies. The larvae of *Cordylobia* are stout and covered with small sharp spines, while *Auchmeromyia* larvae are dark colored and have no spines and have rough tubercles on the central segments of the ventral surface (Fig. 31.2y, z).

Both genera are found only in the Afrotropical region, where adults feed on carcasses and excrement. Adult females lay eggs on soil near or within the housing or burrows of their hosts. *Cordylobia* can even be seen laying eggs on clothing that is hanging to dry (Zumpt 1965). The life cycle of both genera is basically the same as other blowfly species, except for females laying the eggs on the soil, i.e., out of contact with the host. Newly hatched *Cordylobia* larvae seek a host and penetrate into their skin, causing a large abscess (=furuncular myiasis), while *Auchmeromyia* larvae remain free-living and repetitively seek blood from hosts, especially at night, causing only irritation.

There are no control measures or eradication program proposals for these flies. For the treatment of myiasis by *Cordylobia*, one proposed method is asphyxiation of the larva for subsequent removal of the insect with the aid of a forceps. Caution is recommended when drying clothes in the environment for avoidance of *Cordylobia*, and it is recommended that people sleep on beds (or hammocks) to reduce parasitism by *Auchmeromyia*.

31.4.5 *Oestrus, Rhinoestrus,* and *Cephalopina* (Nasal Botflies)

Oestrid flies include many genera and species. The larvae of all species are obligatory parasites, usually of ungulate mammals but occasionally of humans. Some species (in the subfamily Oestrinae) are referred to as "nasal botflies" because of their habit of burying into the nasal cavity or throat of hosts.

Oestrus, *Rhinoestrus*, and *Cephalopina* together have greater veterinary importance than medical. Other genera include *Gedoelstia*, *Przhevalskiana*, *Cephenemyia*, and *Pharyngobolus*, affecting mostly wild animals. Following is a brief list of animals most often infested by these species and their distribution:

- *Oestrus* (widely distributed in most regions and climates): sheep, goats, and antelopes.
- *Rhinoestrus* (in the Palearctic, Oriental, and Afrotropical regions): horses, donkeys, and antelopes.
- *Cephalopina* (in areas where camels occur in the Palearctic and Oriental regions): camels, and infestations have been recorded with over 100 larvae per animal.
- *Gedoelstia* (in the Afrotropical region): antelopes and sheep.
- *Przhevalskiana* (in Palaeartic region): antelopes and goats.

- *Cephenemyia* (in the Nearctic region): deer.
- *Pharyngobolus* (in Africa): African elephants.

Adults of Oestridae are peculiar because they have a large, broad head with small, widely separated eyes (Fig. 31.1d) and mouthparts that are reduced to small knobs. In addition, adult flies have the habit of flying close to their hosts (with their distinctive buzzing), and this causes irritation. Larvae are very robust and have distinct rows of spines on each segment (from Fig. 31.2a–g); the morphology of posterior spiracles has been used for diagnostic purposes (from Fig. 31.2o to Fig. 31.2s).

Oestrus ovis females are larviparous. The advantage of this lifestyle is that the females deposit small batches of larvae directly into the nose of their hosts, increasing the chance of offspring survival. Larvae can remain within the nasal sinus of their hosts and sometimes in the throat tissues. They grow, and when the larvae are fully developed, they move down the nasal passage and drop into the soil to pupate. Larval development takes 30–50 days, and each larva accumulates sufficient food reserves to maintain the adult stage, and once adults do not feed. Typically 1–15 larvae are found in each infested host. Some cases of high infestations (up to 80 individuals) can lead to irritation of the mucosa, malnutrition, and death. The same can occur when mature larvae are unable to escape from the nasal sinus and die.

Cases involving humans are rare and occur when people live near flocks of sheep and goats. Adult females deposit their larvae in humans, preferentially in the conjunctival sacs.

The pesticides employed do not differ much from those recommended for other myiasis-causing flies. Preventative medication has been used in developed countries in sheep and other domestic animals, but without success.

31.4.6 *Gasterophilus* (Bot flies)

Gasterophilus larvae feed in the alimentary canal of horses, but there are species that infest donkeys, mules, and zebras. Larvae do not complete their development in humans. Sometimes the species of the subfamily Gasterophilinae (Oestridae) are mentioned as belonging to the family Gasterophilidae. Probably originating in the Palearctic and Afrotropical regions, they have been distributed to various parts of the world in association with their hosts.

The adult bot fly is very short-lived, and does not feed. They resemble honey bees in size and color, with a dense covering of setae. Larvae are very large and have an armature of very large spines encircling the segments (Fig. 31.2e, f). Other stomach bots of the Gasterophilinae are *Gyrostigma* sp., which infest rhinoceros; *Platycobboldia* sp., which infest African elephants; and *Cobboldia* sp., which infest Indian elephants.

Gasterophilus nasalis and *G. intestinalis* are the most important species from a veterinary point of view. Adult females are known for their persistence in flight around the hosts, causing a fear response. The infestation starts in the mouth when the host licks the eggs that were deposited onto hairs by females. Larvae migrate to the stomach where they attach to the inner lining, but their bodies remain exposed to the stomach cavity. When ready to pupate they release their hold on the wall of the gut and are passed during defecation. Very soon after deposition, they pupariate, and very distinctive puparia are sometimes found in the dung.

Infestations rarely cause problems, but heavy infestation of the stomach may cause secondary complications. Organophosphates and ivermectin are employed to control and treat infestations, but due to the type of parasitism, they are not always easily, correctly, or appropriately administered.

31.4.7 *Hypoderma* (Warble Flies)

Adults are large, stout flies and resemble bees like *Gasterophilus*. Several species of Hypodermatinae (Oestridae) occur in the Palearctic, Nearctic, and Neotropical regions, in warm humid and cool humid climates. *Hypoderma* (*Hypoderma*) *bovis* and *H.* (*H.*) *lineatum* infest cattle and sometimes horses or sheep, but the life cycle is not completed in the latter. Deer are also parasitized by *Hypoderma diana* in the Scottish Highlands. The adults have vestigial mouthparts and therefore do not feed. They have flight behavior similar to *Gasterophilus* regarding persistence on the host. Larvae are large and very stout and can be easily separated by posterior spiracle morphology on the last body segment (Fig. 31.2r).

Adult females lay eggs on hairs. The newly hatched larvae burrow through intact skin and migrate through connective tissue along the esophagus and the spine, until they reach a point under the skin where they stop to feed along the back; this infestation is known as furuncular myiasis. When fully grown, larvae burrow out of the skin and drop into the ground to pupariate.

Both *Hypoderma* species can cause a great amount of damage to the health of the cattle and to their hides. The flies have been eradicated in many countries in Europe, and in other world regions, preventive treatment includes subcutaneous use of ivermectin.

31.4.8 *Cuterebra* and *Dermatobia*

In both genera, all species are highly specialized obligate parasites, some with medical and veterinary importance. *Cuterebra* infests rodents or rabbits and occasionally humans and small livestock. *Dermatobia hominis*, which can also be called human bot fly, “berne” in Brazil, and “tórsalo” or “nuche” in Spanish-speaking countries (also “tupe” in Ecuador), cause serious problems for cattle via frequent infestations. Infestation in humans, pigs, monkeys, birds, cats, and dogs can also occur.

Cuterebra is distributed in the Nearctic and Neotropical regions, while *Dermatobia* occurs in the Neotropical region. As in Gasterophilinae, sometimes species of this subfamily, Cuterebrinae, are placed in the family Cuterebridae.

Adult *Cuterebra* are large and dark colored. They do not have bristles, and unlike *Hypoderma* the thorax and abdomen are densely covered in short setae. The larvae are very distinct, being robust with a dense covering of spines in the form of scales with serrated edges (Fig. 31.2a).

Adult *Dermatobia* are large with a metallic, blue abdomen and a yellowish-brown thorax and head (Fig. 31.1d). The head projects forward and has long sides, and the eyes are widely separated. The wings are the same as in blowflies. Larval shape varies depending on the stage, but in most cases the anterior segments are clearly defined by the large and sparse spines. The posterior segments are narrow, which gives the larvae an almost pear-shaped appearance (Fig. 31.2g, s). Although larvae are recognizable, misidentification sometimes occurs (Marcondes 2014); thus, caution should be taken in morphological examination and when shipping material to experts.

In both genera, adults do not feed and have life cycles similar to other dipterans such as blow flies. *Dermatobia hominis* has a particular behavior related to their degree of specialization: females lay eggs on other insects, usually in the abdomen, where they remain firmly fixed such that the insect will carry them. This is called phoresy, and though not exclusive to *Dermatobia*, it is not commonly seen among myiasis-causing flies. Many insect species can participate in the transport of *D. hominis* eggs, as female oviposition choice seems to be nonspecific. However among the most common and relevant are *Sarcopromusca*, *Musca*, *Stomoxys*, and mosquitoes,[6] probably due to the abundance and frequency with which they are observed living near humans and domestic animals. Larvae hatch immediately in response to heat stimulation, CO_2, and probably vibratory movements of the host, after which they penetrate the skin and develop therein. The adult stage is relatively brief and consists primarily of mating, host seeking, and oviposition (Catts 1982). *Dermatobia* and *Cuterebra* larvae develop individually in subcutaneous tissue (Fig. 31.3e), remaining at one site until development is complete. In both cases infestations are also called furuncular myiasis (Fig. 31.3f), though *Cuterebra* form larger abscesses in the subcutaneous tissue.

[6]Although larger mosquitoes (*Psorophora* sp.) have been considered as the only ones with these eggs, some small sabethine mosquitoes were recently found with eggs in the Brazilian state of São Paulo (CBM).

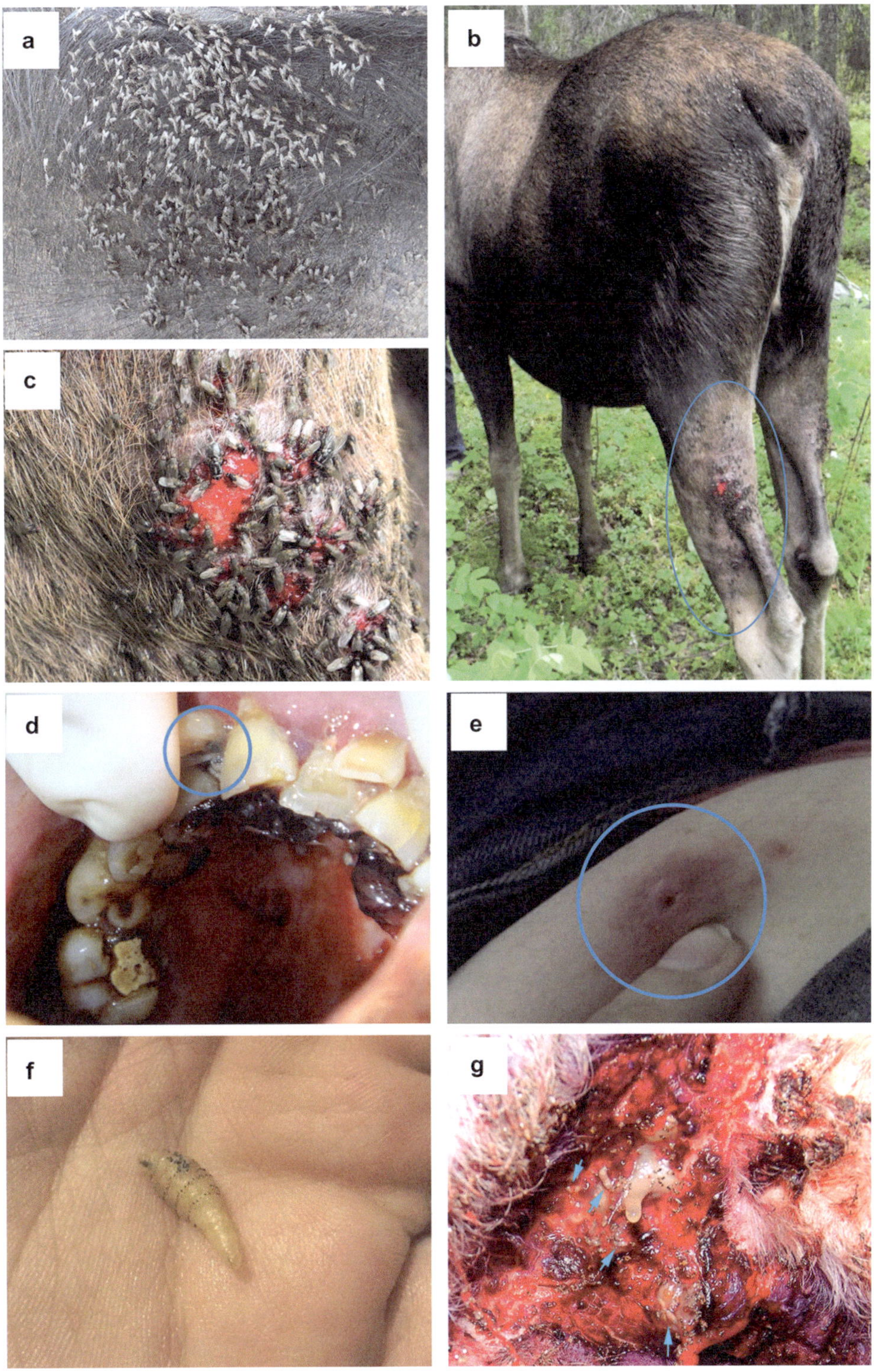

Fig. 31.3 Behavior and types of interactions between flies and their hosts. (**a**) *Haematobia irritans* biting buffalo in Marajó Island (Brazil) (Photo courtesy of JD Barbosa-Neto, UFPA); (**b**) and (**c**) *Haematobosca alcis* biting moose in the Artic (Photo courtesy of K White, Wildlife Research Biologist, Alaska Department of Dept. Fish and Game Division of Wildlife Conservation). Aspects of myiasis in the hosts. (**d**) and (**g**) primary myiasis by *C. hominivorax* in an elderly patient during hospitalization and in a preexisting wound in a dog, respectively; (**e**) and (**f**) primary myiasis by *D. hominis*

A single animal (e.g., bovine) may become infested with numerous *D. hominis* larvae, resulting in stress, loss of production, hide damage, or even death. The parasitism in cattle can reach up to 120 days. A geoprocessing group in the Rio de Janeiro, Brazil, analyzed spatiotemporal associations between factors involved in the seasonal distribution of *D. hominis* larvae in cattle and found that most infestations occur in warmer seasons (50 % in spring and 35 % in summer).

Efforts to control *D. hominis* have not been effective because no technique has been developed for mass production for this fly. In addition, dispersion, sexual behavior, and adult aggregation sites in the field are poorly known. Thus, parasitism by this fly continues to be treated primarily with chemicals.

31.4.9 *Wohlfahrtia* spp.

In this genus three species are important: *Wohlfahrtia magnifica*, distributed in the Palearctic and Oriental regions and living in tropical humid climates; *W. nuba*, also distributed in the Palearctic and Oriental regions and living in dry climates; and *W. vigil*, distributed in the Nearctic region. Adults are 6–10 mm on average and have three longitudinal stripes along the thorax, a typical characteristic of Sarcophagidae (Fig. 31.1h). Larvae are similar to blow flies (Fig. 31.2h), though the posterior spiracles are within a deep depression on the last segment (Fig. 31.2t).

Wohlfahrtia species infest humans, particularly in the nose or eyes, and many species of livestock (e.g., cattle, horses, sheep, goat, pigs) and other animals (e.g., donkeys, camels, dogs, cats, rodents, and birds). Females do not seem to be attracted by the smell of wounds, but only lay their eggs on living hosts. The life cycle is similar to that of other blow flies; however, the larvae are produced directly by the female, i.e., hatch from an egg inside the oviduct. *Wohlfahrtia* larvae are then deposited onto small wounds or mucous membranes of the host after which they penetrate intact skin.

Both *W. magnifica* and *W. nuba* are responsible for superficial myiasis in domestic animals, and although few larvae can be found in each infestation, they typically form a large wound. The results can be serious for the hosts. *Wohlfahrtia vigil* produces furuncular myiasis, in which an individual larva is found under the skin in the abscess.

There are no control measures or eradication program proposals for these flies. Neither a mechanical device nor a repellent mixture (butoxypolypropylene glycol + 0.6 % pyrethrins) prevented *W. vigil* myiasis in the laboratory, and moxidectin and ivermectin also were not effective for the treatment of infestation by this fly in sheep.

31.4.10 Other Species of Myiasis-Causing Flies

Some genera are worth highlighting here, due to the frequency with which they are reported causing facultative myiasis by occasional infestation of humans or other domestic animals. There is no parasite-host specificity, and most species feed and develop in carcasses or feces; these species thus have more forensic importance than medical or veterinary. These include:

- *Calliphora* spp. (Calliphoridae): found in all types of habitats. Human myiasis cases are rare. There is one case of fatal myiasis by *C. vicina* in a tortoise.
- *Phormia* spp. (Calliphoridae): a considerable number of auricular, anal, umbilical, traumatic dermal, opthalmic, and scalp myiasis by *Phormia regina* (known as black blowfly) have been reported in humans.
- *Protophormia* spp. (Calliphoridae): *Protophormia terraenovae* (known as bluebottle fly) has strong economic importance due to causing facultative myiasis in cattle and occasionally infests humans and wild animals.
- *Fannia* spp. (Fanniidae): few species infest humans, very rarely causing urinary or vaginal myiasis (see larval morphology in Fig. 31.2n).

- Sarcophagid flies (Sarcophagidae): widely (globally) distributed with complex family taxonomy. Human myiasis has been caused by species of *Sarcophaga*, and infestation in general is superficial (in previous wounds) and accidental (by ingestion) (see larval morphology in Fig. 31.2h, t).

31.5 Forensic Entomology

Within the forensic context, the study of insects and other arthropods associated with decomposing animal organic matter (i.e., necrophagous insects) is of great value to investigative processes. A growing number of studies in recent years have indicated this importance. After death animal tissues, including that of humans, are attractive for a wide variety of insect species seeking this substrate as a primary food source or for the development of their offspring, thus constituting a regular component of the cadaveric fauna when no barriers prevent the establishment of this natural process.

The interest in expanding the search for ways to promote greater accuracy of investigative work, particularly involving insects, may be relatively new; however, forensic entomology as a science is well established. Benecke (2001) documented the brief history of forensic entomology describing its development beginning in thirteenth century, leading to the present state of the field today.

Although many other organisms (bacteria, fungi, and even other arthropods) can be found in decomposing corpses, flies (in particular Calliphoridae, Sarcophagidae, and Muscidae) are the most abundant and frequent decomposers of carcasses. They are among the first organisms to find and colonize a corpse, succeeded by beetles, and can be seen up to the finality of the remains.

31.5.1 Flies and Time of Death

Eggs, larvae, pupae, and adult specimens of flies collected from cadavers have been used as clues not only to estimate the postmortem interval (PMI), i.e., the time between the establishment of the death and the discovery of the body, but also for a number of other parameters related to the circumstances of death. However, PMI is one of the most important data points within this scientific field, receiving noteworthy attention from the first of studies, as evidenced by the number of published articles exposing and discussing methods to achieve a secure and reliable estimate of this interval. Estimation of this time interval makes it possible to associate suspects with cases involving violent death.

The PMI can usually be estimated using methods linked to cadaveric phenomena which begin immediately after death, such as cutaneous evaporation, rigor mortis, algor mortis, and livor mortis, either together or separate. When compared to other methods, forensic entomology has been a very effective instrument of criminal investigation (Kashyap and Pillai 1989).

There are two ways to estimate PMI using insects. One involves determination of the immature lifetime that is rearing in the corpse after death and on the time it is collected, called PMI minimum (PMI_{min}); another way is through the arrival sequence of different species of insects colonizing the cadaver (process known as ecological succession), called PMI maximum (PMI_{max}) (Catts 1992).

The PMI_{min} should be calculated while a corpse is in the initial stage of decomposition, specifically up to approximately 120 h, because at this stage only a few species of female flies lay eggs. Once larvae are present, researchers should prioritize the collection of larger specimens because in theory larger larvae should be older and specimen age yields important information. On the other hand, the PMI_{max} is calculated when a corpse is in an advanced stage of decomposition (i.e., having been postmortem at least 120 h) since this is the minimum time in which the primary colonizer insects start abandoning the substrate for pupation.

Several studies have demonstrated that temperature, relative humidity, latitude, and sometimes altitude have a major influence on the corpse decomposition processes, as well as the insect physiological processes leading to changes in development rate, fertility, survival, and mortality. Larval density, competition between species, types of animal tissues, and the presence of toxic substances must also be observed and recorded due to occasional direct interference with the development, size, and weight of necrophagous species. This is especially important because some mathematical models (formulas) used to estimate insect age are based on information related to weight and length. For a more detailed understanding of these issues, see Goff et al. (1989) and Greenberg (2002).

Adult females classified as necrophagous or saprophagous do not place their offspring on a live host (Nuorteva 1977). An understanding of the life cycle and the biological, ecological, and behavioral characteristics inherent to each species is thus crucial to avoid mistakes in the forensic context, as well as to ensure correct species identification. Any misidentification can cause irreparable problems in the chain of legal action, since biological or behavioral data from misidentified species can be used to generate legal conclusions. Sanford et al. (2014) illustrated this by referencing a case in which a man had a necrotic lesion while alive (=myiasis caused by *L. eximia*) causing difficulty in PMI_{min} determination, which can easily be overestimated in such an atypical case.

Some insect species are easily visually identified, while others appear similar yet have quite different growth rates, behavior, and habitat preferences; species may not be morphologically distinct or may have highly variable morphological characteristics. Cases of misidentification are not rare (Grella et al. 2015); therefore, an expert in taxonomy should always be consulted. Most larvae of fly species with forensic importance are white or cream-colored and regardless of species are legless and lack a distinctly visible head (from Fig. 31.2h to Fig. 31.2m—note that in Fig. 31.2n, the larvae deviate from the pattern mentioned above). The adults and immature insects, especially eggs and first instars, are much more difficult to identify due to inadequate morphological differences between the various species, a paucity of available characters useful for diagnostic purposes, the existence of undescribed species, and the small number of available taxonomic keys (Thyssen 2010). Faced with these difficulties, forensic entomologists can utilize molecular tools, which allow faster identification of species regardless of life stage and sample integrity (Wells and Stevens 2008).

31.5.2 Corpse Displacement

Insect specimens or body fragments along with the knowledge of their geographical distribution can be useful for investigating whether a corpse was displaced after death, since the movement of the corpse from its original location can also transport colonizing fauna. Thus, the discrepancy between the composition of insects present on the corpse and the composition of species in the geographical area where the corpse was found can provide information about victim displacement (Hall (1990).

31.5.3 Entomotoxicology and Cause of Death

It is not always easy to determine the actual cause of death, and when this involves abusive or illegal use of certain substances, errors may occur in the PMI estimation. For example, the development time of some species of forensic importance can be affected by the action of drugs or drug metabolites present in the body. Some substances, when ingested by the larvae, can accelerate or decrease the time required to reach the pupae stage.

Larvae of cadaver-feeding species as well as the puparia have been used to discover toxic substances, especially when conventional methods do not allow detection (Beyer et al. 1980). Toxicological analyses have been performed using methods such as chromatography (Carvalho et al. 2001) and immunohistochemistry (Souza et al. 2013).

31.5.4 Abuse and Neglect

Insects can be associated with cases of inappropriate treatment of seniors in clinics and nursing homes, as well as mistreatment of children and disabled people (Thyssen et al. 2012). Cases of negligence deserve special attention because they may produce an unreliable PMI estimation. A *C. megacephala* (Calliphoridae) larva was found, commonly seen in Hawaii and associated with feces, in the diaper of a dead infant; the fly must have laid eggs in the child's feces while he/she was living (as evidenced by the act of defecation); thus, experts concluded that the PMI calculations only represented a time interval of death, i.e., there were likely some hours between larval feeding and actual time of death. Knowing that this fly lays its eggs in feces and the child remained a period life (evidenced by the act of defecation), the experts reviewed the PMI calculations estimated so that this only represents the interval of death, i.e., discounting some hours in which the larvae fed on a child's feces that took a short time to die.

31.5.5 Food Contamination and Material Damage

This topic involves cases in which financial reparation may be sought after encounter of whole insects or their fragments in food and other products stored, handled, or manufactured. Flies, beetles, cockroaches, and moths are among the insects most commonly found in biscuits, cereal or breakfast bars, flours, and grains, which indicate hygienic problems in the preparation or preservation of food. Live fly larvae have been found countless times in our research lab[7] in snacks that contained processed meat, fish, roasted meats, and sweets; these infestations are partially due to consumer carelessness in home food product storage.

The responsibility of the consumer versus trade or supplier stems from insect identification and associated eating habits and, when possible, an estimate of the lifetime of immatures in the food, in conjunction with information (date and time) provided by tax assessments emitted by the establishment where the product was purchased. This type of analysis is important for prevention of fraud by the consumer committed against bakeries, restaurants, markets, and companies.

Certain insects can cause structural or economic damage and are classified as pests because they often tend to attack furniture, door frames, carpets, libraries, and electronic devices and generally (initially) go unnoticed. This type of contamination can result from negligence or inefficiency of control methods, in which case services for this purpose are requested by individual clients, institutions, municipalities, or companies. Termites are responsible for such damage in most cases, followed by cockroaches and ants in fewer cases.

31.5.6 Artifacts

Cockroaches and ants may be responsible for creating or dispersing blood stains, thereby generating artifacts in the environment that may cause confusion in the search for evidence useful for solving a suspicious death (Brown et al. 2000). Some species of insects can also feed on fresh blood and regurgitate it later elsewhere, negatively impacting the interpretation of casework.

Another common error linked to PMI estimation is derived from the misuse of liquid fixatives or the fixing process for larvae collected in a corpse, because immature length is a parameter that can be used to estimate insect development time. The probability of 70 % ethanol causing shrinking in preserved larvae is very low when compared to the use of formalin, although both types of fixatives can be used in routine. However, in either case investigators must first immerse the larvae for a few seconds in boiling water (approximately 70 °C) to prevent retraction of the larval body segments, which could cause errors in the estimated time of death of over 24 h.

31.5.7 Main Species of Forensic Importance

Several families of Diptera, beetles, and other insects and arthropods have been documented in animal carcasses or human corpses; however, the species that should be indicated as the most

[7]Unpublished data from the Laboratory of Entomology, State University of Campinas, São Paulo, Brazil

important are those classified as necrophagous or saprophagous. This is because they are often linked to this type of food resource in search of nutrition and for the development of their offspring. Other species classified as omnivores, parasites, or predators may eventually be found in a corpse.

Table 31.1 provides a list of the most important necrophagous dipteran species in human remains and carcasses throughout the world, as well as the

Table 31.1 Genera and/or species of dipterans most often registered in human remains and carcasses around the world and their respective biogeographical regions

Family	Genera/species	Biogeographical region
Calliphoridae	*Calliphora coloradensis*	Neotropical, Nearctic
	C. latifrons	Neotropical
	C. loewi	Palearctic
	Calliphora vicina	Neotropical, Afrotropical
	C. vomitoria	Neotropical, Palearctic
	Chrysomya albiceps	Afrotropical, Oriental, Nearctic
	C. megacephala	Afrotropical, Oriental, Australian, Neotropical, Nearctic
	C. putoria	Afrotropical, Nearctic
	C. rufifacies	Australian, Oriental, Nearctic
	Cochliomyia macellaria	Nearctic, Neotropical
	Hemilucilia segmentaria	Neotropical
	Somomya callipes (=*Compsomyiops callipes*)	Nearctic
	Cynomya cadaverina	Neotropical
	C. mortuorum	Neotropical, Palearctic
	H. segmentaria	Nearctic
	Lucilia cluvia	Nearctic, Neotropical
	L. coeruleiviridis	Neotropical
	L. cuprina	Oriental, Nearctic, Afrotropical
	L. eximia	Nearctic, Neotropical
	Lucilia illustris	Nearctic, Palearctic, Neotropical
	L. sericata	Palearctic, Neotropical, Nearctic
	L. silvarum	Palearctic, Neotropical
	Paralucilia fulvinota	Neotropical
	P. regina	Palearctic, Neotropical, Nearctic, Australian
	P. terraenovae	Neotropical, Palearctic, Nearctic
Fanniidae	*Fannia canicularis*	Palearctic, Neotropical, Nearctic, Afrotropical, Oriental, Australian
	F. scalaris	Palearctic, Neotropical, Nearctic, Afrotropical, Australian
Muscidae	*Hydrotaea aenescens*	Neotropical, Australian, Nearctic, Palearctic
	H. chalcogaster	Oriental, Neotropical, Palearctic, Afrotropical, Australian
	H. ignava (=*Ophyra leucostoma*)	Palearctic, Neotropical, Nearctic
	M. domestica	Oriental, Neotropical, Palearctic, Afrotropical, Australian, Nearctic
	M. stabulans	Palearctic, Neotropical, Nearctic
	S. calcitrans	Nearctic, Australian, Palearctic
	Synthesiomyia nudiseta	Palearctic, Nearctic, Australian, Neotropical
Phoridae	*Conicera tibialis*	Palearctic, Australian, Nearctic
	Megaselia (*Megaselia*) *scalaris*	Nearctic, Australian, Neotropical, Palearctic

(continued)

Table 31.1 (continued)

Family	Genera/species	Biogeographical region
Piophilidae	*Piophila (Piophila) casei*	Oriental, Neotropical, Palearctic, Afrotropical, Australian, Nearctic
Sarcophagidae	*Peckia (Pattonella) intermutans*	Nearctic
	Sarcophaga (Bercaea) africa (=*Sarcophaga haemorrhoidalis*)	Afrotropical, Australian
	Microcerella spp.	Nearctic, Neotropical
	Oxysarcodexia spp.	Nearctic, Neotropical, Australian
	Peckia spp.	Nearctic, Neotropical, Australian
	Ravinia spp.	Oriental, Neotropical, Palearctic, Afrotropical, Australian, Nearctic
	Sarcophaga spp.	Oriental, Neotropical, Palearctic, Afrotropical, Australian, Nearctic
	Tricharaea (*Sarcophagula*) spp.	Nearctic, Neotropical
Sepsidae	*Sepsis* spp.	Oriental, Neotropical, Palearctic, Afrotropical, Australian, Nearctic
Stratiomyidae	*Hermetia illucens*	Oriental, Neotropical, Palearctic, Afrotropical, Australian, Nearctic
Syrphidae	*Ornidia obesa*	Oriental, Neotropical, Palearctic, Afrotropical, Australian, Nearctic

biogeographical regions in which they can be found. The reader will note that several of these species have been reported in previous sections of this chapter, such as flies causing myiasis or acting as pathogen vectors. Thus, in these cases the biological cycle of each species has been presented. For more details and information on species not reported here, see the published studies and books on the topic forensic entomology recommended in *Further reading*.

31.6 Maggot Debridement Therapy

The observation by some physicians, primarily in the military, that wounds improved when infested by maggots led to the development of maggot debridement therapy (MDT) after the First World War.

Maggots feeding exclusively on dead tissues can efficiently clean and promote the healing of wounds of several etiologies. The beneficial effect is produced by (1) removing dead tissues by mechanical activity of mouth hooks and excretion of enzymes; (2) killing ingested bacteria, including methicillin-resistant *Staphylococcus aureus* (MRSA)[8, 9]; (3) excreting allantoin and urea, which are useful to promote healing, and ammonia, which increases pH from the acidic condition more convenient to bacteria; and (4) crawling on the wound, stimulating the healing.

Potential problems for MDT include escape of larvae, non-utilization on wounds near blood vessels and cavities and, theoretically, secondary effects of ammonia when utilizing more than a few thousand larvae; in sheep, high larval loads have been associated with production of ammonia, alteration of consciousness, and death, and transient hyperammonemia was observed in a patient after applying 10,000 larvae on his leg. Pain has been observed in some patients and is manageable by analgesics. Since *Proteus* and *Pseudomonas* bacteria are not always killed by MDT, they can proliferate (Fleischmann et al. 2004). Used larvae must be discarded as infectious waste. Fear and anxiety can be alleviated by

[8] In a test conducted in Petri dishes, *Candida albicans*, *Pseudomonas aeruginosa*, Vancomycin-resistant *Enterococcus*, and MRSA were all lysed by larvae.

[9] In classes, a common observation is that a bacterium resistant to digestion by a larva is as absurd and frightening as a cockroach resistant to be smashed by a shoe (CBM).

education, by demonstration of past results and by reminding patients that the maggots neither proliferate nor damage living tissues. Easy and rapid access to larvae is necessary, because they cannot be stored on shelves or in refrigerators like other medications.

Although MDT does not cure the condition causing diabetic ulcers (e.g., careful control of glucose level and adequate orthopedic shoes are fundamental for the control of diabetes), it promotes the rapid healing of wounds, improves the condition of patients, and prevents amputation and deaths. Several studies have shown the efficiency of this therapy, mostly in diabetic wounds, but also in chronic leg ulcers, pressure ulcers, acute and chronic ulcers, and leishmaniasis, among others (Sherman et al. 1995, Sherman and Shimoda 2004, Dumville et al. 2009).

The success of MDT in comparison with antibiotics and healing gels induced its development in many countries, including the UK, Germany, the USA, Israel, and Belgium, as well as Australia, Brazil, Egypt, China, Colombia, Japan, Malaysia, Mexico, New Zealand, Poland, South Africa, Turkey, and others.

Although some people have utilized maggots from nature, which are certainly contaminated by bacteria, for modern and efficient MDT, flies of appropriate species are reared in laboratories, and eggs are sterilized by the immersion in hypochloride solutions, after which first-instar larvae are applied on the wound contained by adequate covering. *Lucilia sericata* has been most frequently utilized around the world; however, other species such as *L. cuprina* (accidentally, in Egypt), *L. eximia* (purposefully, in Colombia), *C. megacephala*, and *C. macellaria* (in Brazil) also have produced good results (Echeverri et al. 2010, Masiero et al. 2015, Pinheiro et al. 2015). Other flies can be tested if it is known that the larvae eat only dead tissues, if they are easy to rear, and if the females produce eggs (and not larvae), which can be sterilized.

Facilities for the production of maggots can be relatively simple, and maggots can be sent by mail to several parts of the country with lower cost than other treatments. This therapy is very useful, and its international spread will certainly be a great contribution to human health. MDT can also be utilized for wounds in domestic animals if adequate containing procedures are utilized (Jones and Wall 2008).

31.7 Other Diptera

31.7.1 Culicidae (Mosquitoes)

Mosquitoes are present in all continents except Antarctica and practically all islands. Approximately 3600 species have been described, and Brazil,[10] Indonesia, Malaysia, and Thailand have the greatest diversity (Foley et al. 2007). Eggs are laid in water or on surfaces vulnerable to submergence, in extremely diverse water containers and bodies (lagoons, temporary ponds, tree holes, bromeliads, carnivorous plants, artificial containers, etc.).

Mosquito bites can be very annoying, mostly in warm seasons in temperate or cold regions like Canada and Siberia, attacking in great numbers[11] and even killing large ungulates. Several species are vectors of arboviruses (yellow fever, dengue, encephalitis, etc.), filarial worms, and protozoa.

The control of each species depends on the biology, which must be previously understood. Insecticides may be useful in certain situations, but the type and form of utilization should be carefully chosen. Environmental modification such as draining or alteration of edges of lakes and elimination of water-filled containers may be useful for the control of several species.

Several taxonomic modifications have been proposed including splitting the previous genus *Aedes*, and such changes have gradually been accepted by experts.

[10] For example, 75 species have already been reported in the poorly studied and small Brazilian state of Santa Catarina (37 of them in a forest in the capital), a quantity similar to that of the UK (36 spp.).

[11] See Chap. 3 for their utilization in torture.

31.7.2 Simuliidae (Black Flies)

Black flies are stout lower Diptera ("Nematocera"[12]), humpbacked, and frequently black. Adults fly long distances from their breeding places. Immature forms live in lotic environments, usually rich in oxygen. Besides transmitting some worm species (*Onchocerca* and *Mansonella*, see Chaps. 26 and 27), black flies may be very annoying and cause great financial loss, mostly in terms of tourism.

They can be controlled by the application of insecticides and (preferably) toxins from some bacteria (*Bacillus thuringiensis var. israelensis*—Bti) to water. These toxins must be applied just below the ponds formed in rivers and creeks because they become useless after precipitation in still water. Brushing of surfaces containing immature forms, removal of branches and leaves on the water surface, and recuperation of riparian forest to prevent dispersion of adult flies are also useful. Careful identification of species and determination of importance is also fundamental for localizing the immature forms and for evaluation of the effects of control measures.

31.7.3 Biting Midges (Ceratopogonidae)

Biting midges are small (1–4 mm) nematoceran Diptera that, in addition to transmitting some viruses and worms, can be very annoying due to allergenic saliva and huge numbers[13] and which can cause financial loss, mostly involving tourism and devaluation of properties. They can transmit some arboviruses (see Chap. 10) and helminths (see Chap. 26). Up to 15,000 females were collected in one light trap night, and a 2-min collection produced 300 individuals in Adirondack Mountains of New York state. They can cause *sweet itch* in horses, an allergic reaction to multiple bites.

Although they are popularly associated with swamps, several species breed in rotten banana stumps, cocoa fruits, tree holes, and damp manure. Careful planning of building in areas subject to biting midges must be adopted, and environmental management[14] needs to be considered. If no other control measure can be utilized, the problem may be reduced by UBV application of organophosphate insecticides in adequate hours, usually around dusk, window screening impregnated with pyrethroid insecticides, and utilization of repellents.

31.7.4 Sand Flies (Phlebotominae)

Besides transmitting several species of *Leishmania* (see Chap. 16) and arboviruses (see Chap. 10), sand flies can cause a skin irritation ("urticaria multiformis endemica" called "harara" in Arabic) in Northern Africa. They may cause measles-like lesions in children due to multiple bites of *Lutzomyia longipalpis*, as observed by CBM in Paraiba state, northeastern Brazil.

Other insects of this family ("moth flies") may occur in sewage, latrines, and bathrooms. Their presence in vast numbers can be an annoyance for sewage work personnel. Intestinal and urinary myiasis have been reported to be caused by larvae of *Telmatoscopus* and *Psychoda,* respectively. Additionally, some species have been found contaminated by *Nocardia* sp. in a hospital in Brazil. Studies of German hospitals were recently reviewed, with bacteria of 45 species found on *Clogmia albipunctata.*

[12]Nematocera was shown to be a paraphyletic group, in which Culicomorpha includes Culicoidea, with Culicidae and others, and Chironomoidea includes Simuliidae, Ceratopogonidae, Chironomidae, and Thaumaleidae. Nematocera (*nematos* = thread, filamentous, *cera* = antenna) indicates a "multisegmented" antennae, but, in reality, all insect antennae are divided in three segments, and secondary subdivisions of third segment (flagellum) should better be called subsegments.

[13]As an example, in Florianópolis in 2014, a fugitive from the law decided to submit himself to the police rather than endure the relentless attacks of the biting midges in his mangrove hiding place.

[14]The filling of coastal swamps with sand in Jamaica, in the 1950s, caused the replacement of a disturbing species (*Culicoides furens*) by a greater pest (*Leptoconops becquaerti*).

31.7.5 Horse Flies/Deer Flies (Tabanidae)

These large Diptera (Fig. 31.1i) transmit the *Loa loa* worm (see Chap. 27) and several bacteria, protozoa, and viruses and may be a serious nuisance for cattle and humans. Attack by horse flies (and blood-feeding Muscidae and other insects) can induce herd grouping for protection (i.e., of those situated in the center) and reduction of exposed skin area. Since they breed in water containers in natural environment, their control is difficult, and traps have been utilized. Topic application of pyrethroids on the skin of domestic animals can limit biting.

31.7.6 Chironomidae and Other Diptera

Chironomidae a large and ecologically important group, usually in the bottom of feeding pyramid, with some species having medical importance. Chironomidae and Chaoboridae may occur in massive quantities and can disrupt human activities. In some East African lakes, swarms of emerging *Chaoborus edulis* may be so dense that can cause suffocation, allergy, and traffic accidents. Local people capture the swarming midges and use them to make cakes, which are dried in the sun and consumed. These midges enter in the eyes, ears, nose, and mouth, fall into food and beverages, and stain walls and windows, and the dead bodies often accumulate in large piles that can become putrid and odoriferous, sometimes serving as breeding media for stable flies.

Some Chloropidae flies, mostly *Hyppelates* spp. ("eye gnats"), can transmit *Chlamydia* and other bacteria. Some Drosophilidae have been incriminated as vectors of *Thelazia* helminths (see Chap. 29).

References

Baldacchino F, Muenworn V, Desquesnes M et al (2013) Transmission of pathogens by *Stomoxys* flies (Diptera: Muscidae): a review. Parasite 20:26

Benecke M (2001) A brief history of forensic entomology. Forensic Sci Int 120:2–14

Beyer JC, Enos WF, Stajic M (1980) Drug identification through analysis of maggots. J Forensic Sci 25:411–412

Breijo M, Rocha S, Ures X et al. (2013) Accessibility of blood affects the attractiveness of cattle to horn flies. Med Vet Entomol. doi:10.1111/mve.12008

Brown RE, Hawkes RI, Parker MA et al (2000) Entomological alteration of bloodstain evidence. In: Byrd JH, Castner JL (eds) Forensic entomology: the utility of arthropods in legal investigations. CRC, Washington, DC

Bruce WG (1938) A practical trap for the control of horn flies on cattle. J Kansas Entomol Soc 11:88–93

Carvalho LML, Linhares AX, Trigo JR (2001) Determination of drug levels and the effect of diazepam on the growth of necrophagous flies of forensic importance in southeastern Brazil. Forensic Sci Int 120:140–144

Catts EP (1982) Biology of New World bot flies: Cuterebridae. Annu Rev Entomol 27:313–338

Catts EP (1992) Problems in estimating the PMI in death investigations. J Agric Entomol 9:245–255

Cork A, Hall MJR (2007) Development of an odour-baited target for female New World screwworm, *Cochliomyia hominivorax*: studies with host baits and synthetic wound fluids. Med Vet Entomol 21:85–92

Denning SS, Washburn SP, Watson DW (2014) Development of a novel walk-through fly trap for the control of horn flies and other pests on pastured dairy cows. J Dairy Sci 97:4624–4631

Dumville JC, Worthy G, Bland JM et al. (2009) Larval therapy for leg ulcers (VenUs II): ramdomised controlled trial. Br Med J. doi:10.1136/bmj.b773

Echeverri MIW, Álvarez CR, Higuita SEH et al (2010) *Lucilia eximia* (Diptera: Calliphoridae), una nueva alternativa para la terapia larval y reporte de casos en Colombia. Iatreia 23:107–116

Eigen M, Kloft WJ, Brandner G (2002) Transferability of HIV by arthropods supports the hypothesis about transmission of the virus from apes to man. Naturwissenschaften 89:185–186

Fleischmann W, Grassberger M, Sherman R (2004) Maggot therapy: a handbook of maggot-assisted wound healing. Thieme, Stuttgart and New York

Foley DH, Rueda LM, Wilkerson RC (2007) Insight into global mosquito biogeography from country species records. J Med Entomol 44:554–567

Goff ML, Omori AI, Goodbrod JR (1989) Effect of cocaine in tissue on the development rate of Boettcherisca peregrina (Diptera: Sarcophagidae). J Med Entomol 26: 91–93

Greenberg, B (1973) Flies and disease, Vol. II Biology and disease transmission. Princeton Univ. Press, Princeton

Greenberg B (2002) Problems in estimating the time of death. In: Greenberg B, Kunich JC (eds) Entomology and the law: flies as forensic indicators. Cambridge University Press, Cambridge

Grella MD, Savino AG, Paulo DF et al (2015) Phenotypic polymorphism of *Chrysomya albiceps* (Wiedemann)

(Diptera: Calliphoridae) may lead to species misidentification. Acta Trop 141:60–72

Guglielmone AA, Volpogni MM, Scherling M (2000) Chlorfenapyr ear tags to control Haematobia irritans (L.) (Diptera: Muscidae) on cattle. Vet Parasitol 93:77–82

Guimarães JH, Papavero N (1999) Myiasis in man and animals in the Neotropical region. Ed Plêiade, São Paulo

Hall RD. (1990) Medicocriminal entomology. In: Catts EP, Haskell NH (eds.) Entomology & death: a procedural guide. Joyce's Print Shop, Clemson

Hall M, Wall R (1995) Myiasis of humans and domestic animals. Adv Parasitol 35:257–334

Hope FW (1840) On insects and their larvae occasionally found in the human body. Philos Trans R Soc Lond B Biol Sci 2:256–271

James MT (1947) The flies that cause myiasis in man. Miscellaneous publications, vol 631. Department of Agriculture, Washington DC

James MT (1969) A study in the origin of parasitism. Bull Entomol Soc Am 15:251–253

Jones G, Wall R (2008) Maggot-therapy in veterinary medicine. Res Vet Sci 85:394–398

Jordan AM (1993) Tsetse flies (Glossinidae). In: Lane RP, Crosskey RW (eds) Medical insects and arachnids. Springer, Dordrecht, pp 333–388

Kashyap VK, Pillai VV (1989) Efficacy of entomological method in estimation of postmortem interval: a comparative analysis. Forensic Sci Int 40:245–250

Kirby W, Spence W (1815) An introduction to entomology, vol I, 3rd edn. Longman, Brown, Green and Longmans, London

Knipling EF (1955) Possibilities of insect control or eradication through the use of sexually sterile males. J Econ Entomol 48:459–462

Krafsur ES (2009) Tsetse flies: genetics, evolution, and role as vectors. Infect Genet Evol 9:124–141

Krafsur ES, Moon RD (1997) Bionomics of the face fly, *Musca autumnalis*. Annu Rev Entomol 42:503–523

Lamarque D, Haessler C, Champion R (1989) Anthrax in Chad, a continuing zoonosis. Med Trop 49:245–251

Marcondes CB (2014) *Dermatobia hominis* (Diptera: Cuterebridae) in Africa and the need for caution in its taxonomy. J Infect Public Health 7:73–74

Masiero FS, Nassu MP, Soares MP et al (2015) Histological patterns in healing chronic wounds using *Cochliomyia macellaria* (Diptera: Calliphoridae) larvae and other therapeutic measures. Parasitol Res 114:2865–2872

Nuorteva P (1977) Sarcosaprophagous insects as forensic indicators. In: Tedeshi CG, Eckert WG, Tedeshi LG (eds) Forensic medicine: a study in trauma and environmental hazards, vol 2. Saunders, Philadelphia, pp 1072–1095

Oyarzún MP, Quiroz A, Birkett MA (2008) Insecticide resistance in the horn fly: alternative control strategies. Med Vet Entomol 22:188–202

Packer C, Puser AE, Rowley H et al (1991) Case study of a population bottleneck: lions of the Ngorongoro crater. Conserv Biol 5:219–231

Pape T, Thompson FC (2013) Systema Dipterorum, Version 1.5. http://www.diptera.org/. Accessed 30 Jan 2015

Patton WS (1922) Notes on the myiasis-producing Diptera of man and animals. Bull Entomol Res 12:239–261

Pinheiro MARQ, Ferraz JB, Junior MAA et al (2015) Use of maggot therapy for treating a diabetic foot ulcer colonized by multidrug resistant bacteria in Brazil. Indian J Med Res 141:340–342

Salem A (2012) *Stomoxys calcitrans* (L. 1758): morphologie, biologie, rôle vecteur et moyens de lute. Dissertation, Université de Toulouse. http://ethesis.inp-toulouse.fr/archive/00002044/01/salem.pdf

Sanford MR, Whitworth TL, Phatak DR (2014) Human wound colonization by *Lucilia eximia* and *Chrysomya rufifacies* (Diptera: Calliphoridae): myiasis, perimortem, or postmortem colonization? J Med Entomol 51:716–719

Sherman RA, Shimoda KJ (2004) Presurgical maggot debridement of soft tissue wounds is associated with decreased rates of postoperative infection. Clin Infect Dis 39:1067–1070

Sherman RA, Wyle F, Vulpe M (1995) Maggot therapy for treating pressure ulcers in spinal cord injury patients. J Spinal Cord Med 18:71–74

Souza CM, Lima CGP, Alves MJ Jr et al (2013) Standardization of histological procedures for the detection of toxic substances by immunohistochemistry in Dipteran larvae of forensic importance. J Forensic Sci 58:1015–1021

Stafford III KC (2008) A guide to biology, dispersal, and management of the house fly and related flies for farmers, municipalities, and public health officials. Conn Agric Exp Stn, New Haven (Bul 1013)

Suárez VH (2002) Helminthic control on grazing ruminants and environmental risks in South America. Vet Res 33:563–573

Thyssen PJ (2010) Keys for identification of immature insects. In: Amendt J, Goff ML, Campobasso CP et al (eds) Forensic entomology: novel arthropods, environments and approaches. Springer, Dordrecht

Thyssen PJ, Nassu MP, Costella AMU et al (2012) Record of oral myiasis by *Cochliomyia hominivorax* (Diptera: Calliphoridae): case evidencing negligence in the treatment of incapable. Parasitol Res 111:957–959

Tomberlin JK, Lohmeyer KH, Kattes D (2007) Treatment of pastures with diflubenzuron suppresses horn fly, *Haematobia irritans* (Diptera: Muscidae) development. J Agric Urban Entomol 24:95–101

Tozer RS, Sutherst RW (1996) Control of horn fly (Diptera: Muscidae) in Florida with an Australian trap. J Econ Entomol 89:415–420

Watson DW, Stringham SM, Denning SS et al (2002) Managing the horn fly (Diptera: Muscidae) using an electric walk-through fly trap. J Econ Entomol 95:1113–1118

Wells JD (1991) *Chrysomya megacephala* (Diptera: Calliphoridae) has reached the continental United States: review of its biology, pest status, and spread around the world. J Med Entomol 28: 471–473

Wells JD, Greenberg B (1992) Interaction between *Chrysomya rufifacies* and *Cochliomyia macellaria* (Diptera: Calliphoridae): the possible consequences of an invasion. Bull Entomol Res 82:133–137

Wells JD, Stevens JR (2008) Application of DNA-based methods in forensic entomology. Annu Rev Entomol 53:103–120

Zumpt F (1965) Myiasis in man and animals in the Old World. Butherworths, London

Zumpt F (1973) The Stomoxyine biting flies of the world. Diptera: Muscidae: taxonomy, biology, economic importance and control measures. Gustav Fischer Verlag, Stuttgart

Further Reading

Amendt J, Goff ML, Campobasso CP et al (2010) Forensic entomology: novel arthropods, environments and approaches. Springer, Dordrecht

Byrd JH, Castner JL (2010) Forensic entomology: the utility of arthropods in legal investigations, 2nd edn. CRC, Boca Raton

Catts EP, Haskell NH (1990) Entomology and death: a procedural guide. Joyce's Print Shop, Clemson, SC

Gennard DE (2007) Forensic entomology: an introduction. Wiley, Chichester

Greenberg B (1971) Flies and disease, vol I, Ecology, classification and biotic associations. Princeton University Press, Princeton

Greenberg B (1973) Flies and disease, vol II, Biology and disease transmission. Princeton University Press, Princeton

Hanski I (1987) Carrion fly community dynamics: patchiness, seasonality and coexistence. Ecol Entomol 12:257–266

Papavero N, Guimarães JH (1999) Myiasis in man and animals in the neotropical region: bibliographic database. Plêiade, São Paulo

Smith KGV (1986) A manual of forensic entomology. Cornell University Press, Ithaca, p 205

Sukarsih PS, Satria E et al (2000) Vaccination against the Old World screwworm fly (*Chrysomya bezziana*). Parasite Immunol 22:545–552

Thyssen PJ, Souza CM, Shimamoto PM et al (2014) Rates of development of immatures of three species of *Chrysomya* (Diptera: Calliphoridae) reared in different types of animal tissues: implications for estimating the postmortem interval. Parasitol Res 113:3373–3380

Tomberlin JK, Benbow ME (2015) Forensic entomology: international dimensions and frontiers. CRC, Boca Raton

Wells JD, LaMotte LR (1995) Estimating maggot age from weight using inverse prediction. J Forensic Sci 40:585–590

Sucking and Chewing Lice

32

Carlos Brisola Marcondes
and Pedro Marcos Linardi

Abstract

Lice are obligatory, permanent parasitic insects belonging to order Phthiraptera, which have a developed proboscis to suck blood from capillaries (suborder Anoplura) of mammals, or chewing mouth pieces, adapted to eat hairs and feathers, and sometimes also the skin and blood of birds and mammals (suborders Amblycera, Ischnocera, and Rhynchophthirina, formerly known as Mallophaga). Three of ca. 500 known species of Anoplura (*Pediculus humanus*, *Pediculus capitis*, and *Pthirus pubis*) parasitize humans, causing primary and secondary skin lesions, and *P. humanus* can transmit exanthematic typhus, relapsing fever, and trench fever. Several of the ca. 4000 species of chewing lice can cause cutaneous lesions in domestic animals. Infestation by sucking body lice has currently been associated with bad conditions of hygiene, usually in wars and prisons, but head lice (*P. capitis*) has had a resurgence, associated with people crowding and precarious head cleaning. Control of body sucking lice is obtained by cleaning of clothes and bathing, but that of head lice needs careful collective surveying, associated with the use of louse combs, and eventually insecticides may be very carefully utilized. "Mallophaga" can be controlled by the application of insecticides to the bodies of infested animals. Due to the strong association of sucking lice with humans, mostly of eggs, adhered to hair or clothes, they can be utilized for forensic indications.

Keywords

Sucking lice • Chewing lice • *Pediculus* • Pediculosis • Phthiriasis

C.B. Marcondes (✉)
Department of Microbiology, Immunology and Parasitology, Center of Biological Sciences, Federal University of Santa Catarina,
88040-900 Florianópolis, SC, Brazil
e-mail: cbrisolamarcondes@gmail.com

P.M. Linardi
Department of Parasitology, Institute of Biological Sciences, Federal University of Minas Gerais,
P. O. Box - 486, Belo Horizonte, Minas Gerais, Brazil
e-mail: linardi@icb.ufmg.br

C.B. Marcondes (ed.), *Arthropod Borne Diseases*, DOI 10.1007/978-3-319-13884-8_32

32.1 Introduction

Sucking lice have parasitized humans for all their history, and infestation by head lice probably preceded that by body louse, developed when humans started to utilize clothes. Transmitting relapsing fever, exanthematic typhus, and trench fever, the last firstly characterized in the WW1, have caused great mortality and suffering. *Pediculus* has two species, one inhabiting the hairs and the other the rest of the body; the body louse, with improving hygiene after WW2, has become restricted to precarious environments, but low body care, associated with long hairs, and crowding has caused and recrudescence of occurrence of head lice. *Pthirus pubis*, mostly associated with pubic region and usually sexually transmitted, has also been frequently found in eyelashes of children, causing severe consequences.

Chewing lice are still adapted to chew the hairs, feathers, and eventually skin, and some species already have adapted to ingest blood. Some species have mouthparts in the tip of a beak, an interesting demonstration of the steps for the evolution of blood-sucking habits.

Since lice are strongly associated with hosts, their phylogeny parallels that of hosts, sometimes with different speeds, but this has also been contested (Barker 1994). Extinction of hosts frequently means extinction of their lice. Strong association of lice with host bodies and low (or null, in the case of eggs) vagility make them useful for forensic objectives.

Fig. 32.1 *Pediculus* sp. female (Anoplura)

Fig. 32.2 *Pthirus pubis* male (Anoplura)

32.2 Morphology and Biology

Lice are small wingless insects (1–6 mm in length for sucking lice and 1–11 for chewing lice) in which their bodies are flattened dorsoventrally and their legs have one or two tarsal claws to grasp on hairs, well adapted to their shapes, contributing for the specificity of the majority of the species. Heads of Anoplura are narrower than the thorax (Figs. 32.1 and 32.2), and mouthparts can be seen, after clarification, in the head, while heads of "Mallophaga" are wider than the thorax (Fig. 32.3), and chewing mouthparts can be seen under the head.

The chewing lice are divided into three suborders: Rhynchophthirina, parasitizing African elephants and boars, with a long beak, in whose tips are inserted the mandibles; Amblycera, with antennae usually hidden in grooves and with

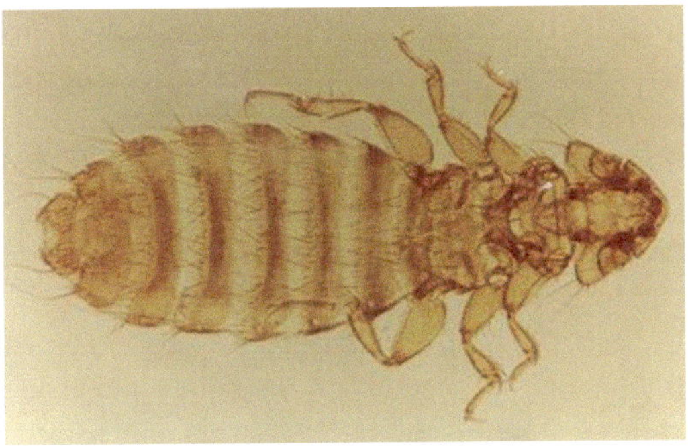

Fig. 32.3 *Menacanthus stramineus* female (Mallophaga). From Marcelo de Campos Pereira (http://www.icb.usp.br/~marcelcp/Imagens/f-piolho1.jpg), in which the wide head can be noticed

Fig. 32.4 *Pediculus humanus*: adults and nits glued to cloth

maxillary palps; and Ischnocera, with exposed antennae and without maxillary palps, both groups including parasites of mammals or birds. Because cladistic studies proved Mallophaga be a paraphyletic group among the Phthiraptera, with Anoplura and Ischnocera being more related themselves than Ischnocera and Amblycera, the name Mallophaga as order or suborder must be rejected. However, Mallophaga as a common name continues to be employed, in virtue of its consecrate use.

Both sucking and chewing lice spend their entire lives on a host. Lice eggs (*nits*, *lendeas*, etc.) are strongly glued on hairs at the preferred regions of the body of hosts. *Pediculus humanus* is an exception, whose eggs are glued to clothes of humans ("naked apes") (Fig. 32.4). The paurometabolous cycle has three nymphal stages, all of them developing on the host and with feeding habits similar to those of adults, in about 1 month.

Body heat and food are essential for survival and sucking lice can support cold and fast for a few hours. "Mallophaga" are more mobile and resistant than Anoplura, enduring a few days out of hosts, but both kinds of lice are very difficult to rear in the laboratory. In the same manner, Amblycera are also generally more mobile than Ischnocera. Birds in humid regions have more chewing lice than those in arid ones, even comparing the same species in different environments (Johnson and Clayton 2003).

In birds, Ischnocera lice feed only on feathers, while several Amblycera feed on blood and even on other lice, mite, exuviae, egg cases, and fungi. Some chewing lice are transported between hosts by Hippoboscidae flies and even fleas. In mammals, some Amblycera are adapted to feed on blood, while some Ischnocera may accidentally ingest it.

Endosymbiotic bacteria concentrated in structures called mycetomes are present in Anoplura, Rhynchophthirina, and most Amblycera and Ischnocera, in which they play an important role in the nutritional physiology of lice, since certain experiments showed that human lice deprived of these bacteria did not feed, survive, or reproduce (Johnson and Clayton 2003).

32.3 Hosts

Anoplura are hematophagous that feed directly from the capillaries, infesting exclusively mammals and being very specific; they choose sites of infestation on the hosts' body less subjected to grooming. "Mallophaga" parasitize birds and mammals, showing varied diet that includes cells of skin peeling, withdrawn products of feathers, and blood.

Among the Anoplura, 532 species are currently known, included in 49 genera and 15 families. Several families contain only one or a few genera, due to the extinction of several groups of prehistoric mammals taken as hosts, and that coevolved with the anoplurans (Durden and Musser 1994). Sucking lice are found in the orders Artiodactyla, Carnivora, Dermoptera, Hyracoidea, Insectivora, Lagomorpha, Macroscelidea, Perissodactyla, Primates, Rodentia, Scandentia, and Tubulidentata. Approximately 65 % of the species of mammals are suspicious of hidden sucking lice.

Currently, among the three suborders of chewing lice, almost 4500 valid species and subspecies in nine families are known, of which 554 species (12.4 %) in five families infest mammals and other 3910 species (87.6 %) in four families parasitize birds (Price et al. 2003). Chewing lice infest thirteen orders: Artiodactyla, Carnivora, Dasyuromorphia, Didelphimorphia, Diprotodontia, Hyracoidea, Paucituberculata, Peramelemorphia, Perissodactyla, Primates, Proboscidea, Rodentia, and Xenarthra. Regarding the birds, the chewing lice are found in all 27 known orders (Price et al. 2003) and 157 (91 %) of the 173 families listed (Howard and Moore 1991).

Among the hosts, the transmission of lice usually occurs by direct contact.

Three species of Anoplura parasitize humans: *Pediculus humanus*, *Pediculus capitis*, and *Pthirus pubis*. Although the first had been also divided into two subspecies (*P. humanus humanus* and *P. humanus capitis*), morphometric differences and molecular evidence of separation between the head and body populations are strong arguments to consider them as distinct species (Leo et al. 2005), but due to small and badly defined differences (Light et al. 2008; Olds et al. 2012), they have been frequently considered as ecotypes of the same species. They are here considered as separate species mostly due to reproductive isolation.

Pediculus humanus is distributed in several areas of the human body and glues their eggs on clothes. A molecular analysis of head and body lice suggested that the last diverged from the first at least by 83,000 years and possibly as early as 170,000 years, and they were transported to Europe with modern humans, already using clothes (Toups et al. 2010). Currently, *P. humanus* is mostly found in adults, while *P. capitis* is more frequent on children.

Pthirus pubis is mostly distributed in the genital area and transmitted by sexual contact. It diverged from *P. gorillae* probably at three million years ago, much more recently than the divergence between hosts, probably due to habitat sharing, predating, or other forms of contact (Light and Reed 2009).

Some species of other genera have potential importance in human and veterinary medicine. *Haematopinus* includes 21 species and parasitizes ungulates (Artiodactyla and Perissodactyla). *Linognathus* and *Solenopotes* (Linognathidae) include, respectively, 53 and 12 species, parasitizing Artiodactyla and Carnivora. Three species of *Linognathus* parasitize ovines; one of them also found in bovines; one is found in dogs, mostly long haired; and *S. capillatus* parasitizes bovines. Species of *Hopopleura* infest rodents, and some species can have a role in the transmission of rickettsiae among these animals. Sylvatic Suidae, Tayassuidae, Camelidae (camels, lamas, vicuñas, and guanacos), and several deers (*Cervus* sp.), with economic potential, can host several species of *Haematopinus*, *Solenopotes*, and other genera (Durden and Musser 1994).

Although lice are usually highly host specific, some species from domestic animals can be found in feral populations of these animals or their relatives. For example, one species of Anoplura (*Linognathus setosus*) and two of Mallophaga (*Trichodectes canis* and *Heterodoxus spiniger*) found on dogs also can parasitize coyotes, wolves, feral dogs, and some other wild canids (Durden 2001).

Durden (2001) lists several references of taxonomic publications, and Linardi (2011) has details on morphology and identification of Anoplura and Mallophaga and keys for identification of species from Brazil.

Chewing lice have also been observed from other economically significant mammals: alpaca, llamas, and guanacos (*Bovicola breviceps*); being more resistant and mobile and less specific than Anoplura, chewing lice can survive on soil and even be transported by biting flies, like *Haematobia irritans*.

Tables 32.1 and 32.2 list the principal species of lice that infest domestic mammals and birds,

Table 32.1 Principal species of sucking lice (Anoplura) that infest humans and domestic mammals

Species of mammal lice	Common name	Family	Host(s)	Sites of infestation
Pediculus capitis	Head louse	Pediculidae	Human	Head
P. humanus	Body louse	Pediculidae	Human	Body, clothes
Pthirus pubis	Crab louse, pubic louse	Pthiridae	Human	Pubic region, eyelashes
Haematopinus asini	Horse sucking louse	Haematopinidae	Horse, donkey	Crine, tail, trunk
H. eurysternus	Short-nosed cattle louse	Haematopinidae	Cattle	Neck, dewlap, back, base of horns
H. quadripertusus	Cattle tail louse	Haematopinidae	Cattle	Tail, ears, eyelids
H. suis	Hog louse	Haematopinidae	Swine	Head, base of ears, among legs
H. tuberculatus	Buffalo louse	Haematopinidae	Asiatic buffalo, cattle	Tail
Linognathus africanus	African blue louse	Linognathidae	Goat, sheep, deer	Sides of the body
Linognathus ovillus	Sheep face louse	Linognathidae	Sheep	Face
L. pedalis	Sheep foot louse	Linognathidae	Sheep	Foot, belly
L. setosus	Dog sucking louse	Linognathidae	Dog, other canids	On the neck, shoulders, under the collar
L. stenopsis	Goat sucking louse	Linognathidae	Goat	Head, tail
L. vituli	Long-nosed cattle louse	Linognathidae	Cattle	Dewlap, shoulders, over the body
Solenopotes capillatus	Little blue cattle louse	Linognathidae	Cattle	Head, dewlap, topline, and under the tail

respectively. Durden (2001) contains a complete list of sucking and chewing lice parasitizing wild animals.

32.4 Epidemiology

32.4.1 Lice as Infesting Agents

The direct effect of human lice bites due to the reaction of the host's tissues to the foreign proteins in the injected saliva causes more or less intense irritation depending on the sensitivity of each individual (Busvine 1966; Weidhaas and Gratz 1982). Insensitive individuals might act as possible sources for the dissemination of lice.

Infestation by sucking lice usually causes intense pruritus, skin abrasions, and lesions, which combined with unhygienic conditions, may lead to dermatitis and impetigo, especially infected by staphylococci and other bacteria. The skin of heavily and chronically infected people by body lice, mostly homeless, may exhibit a thickening and yellowing coloration, known as "vagabond's disease" or melanoderm. Scratching due to pruritus causes lesions and hair loss. Psychological consequences of head lice infestation, mostly in children, including bullying and segregation, can occur. Violent scratching or irritation might almost be taken as an indication of non-habitual or accidental infestation. An accidental infestation can cause intolerable irritation which makes the day a nightmare and the night a *nuit blanche*. Some desensitization can occur, and people with severe infestation may be undisturbed. When serious infestations of head lice are associated with poor social conditions and inadequate diets, children may exhibit anemic iron deficiency aggravated by hematophagy. In high infestations, the lice population can exceed 1000 individuals.

The bite of the crab lice gives also rise to a similar irritation, causing a characteristic blue spot from 0.2 to 3 cm across (Busvine 1966), not found in the bite of the head or body lice. *Pthirus*

Table 32.2 Principal species of chewing lice (Mallophaga) that infest domestic mammals and birds

Mammals				
Amblycera	Common name	Family	Host(s)	Sites of infestation
Heterodoxus spiniger	Australian dog louse	Boopiidae	Dog	Over the body
Trimenopon hispidus	Guinea pig louse	Trimenoponidae	Guinea pig	Over the body
Gliricola porcelli	Slender guinea pig louse	Gyropidae	Guinea pig	Over the body
Gyropus ovalis	Oval guinea pig louse	Gyropidae	Guinea pig	Over the body
Ischnocera				
Bovicola bovis	Cattle biting louse	Trichodectidae	Cattle	Tail, shoulders, sides of the body
B. caprae	Goat biting louse	Trichodectidae	Goat	Back
B. equi	Horse biting louse	Trichodectidae	Horse	Head, crine, tail
B. ovis	Sheep biting louse	Trichodectidae	Sheep	Over the body
Felicola subrostrata	Cat biting louse	Trichodectidae	Cat	Head, neck
Trichodectes canis	Dog biting louse	Trichodectidae	Dog	Head, neck
Birds				
Amblycera				
Colpocephalum turbinatum	Pigeon louse	Menoponidae	Pigeon	Tail wings
Hohorstiella lata	–	Menoponidae	Pigeon	Shafts of head feathers
Menopon gallinae	Shaft louse	Menoponidae	Chicken	Wings' feathers, around the anus
Menacanthus cornutus	–	Menoponidae	Chicken	Skin of chest and back
M. pallidulus	–	Menoponidae	Chicken	Skin of head
M. stramineus	Chicken body louse	Menoponidae	Chicken, turkey	Skin surface of back and belly, tail wings
Trinoton querquedulae	Large duck louse	Menoponidae	Duck	Barbs and barbules from down feathers
Ischnocera				
Anaticola anseris	Slender goose louse	Philopteridae	Goose	Skin, wings
Campanulotes bidentatus	Small pigeon louse	Philopteridae	Pigeon	Fluffy parts of body feathers
Chelopistis meleagridis	Large turkey louse	Philopteridae	Turkey	Feathers of neck and chest
Coloceras piageti	–	Philopteridae	Pigeon	Feathers of neck and back
Columbicola columbae	Slender pigeon louse	Philopteridae	Pigeon	Feathers of wings and anterior region of the body
Cuclotogaster heterographus	Chicken head louse	Philopteridae	Chicken, turkey, partridge	Feathers of head and neck
Goniocotes gallinae	Pluff louse	Philopteridae	Chicken, pigeon	Feathers of the back
Goniodes dissimilis	Brown chicken louse	Philopteridae	Chicken	Feathers of the back
G. gigas	Large chicken louse	Philopteridae	Chicken	Feathers of the back
Lipeurus caponis	Wing louse	Philopteridae	Chicken	Feathers of the anterior wings
Oxylipeurus polytrapezius	Slender turkey louse	Philopteridae	Turkey	Wings

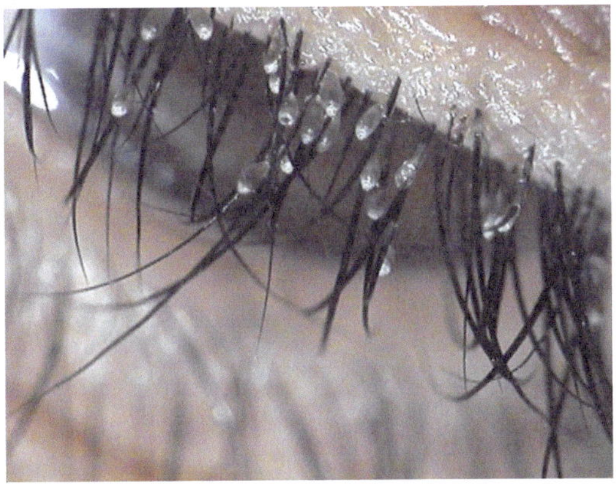

Fig. 32.5 Infestation by *Pthirus pubis* (eggs) on a children's eye (from Anane et al. 2013, authorized by Soc. Franç. Ophtalmol)

pubis is found mostly in people with promiscuous sexual activity. A previous study indicated 30 % of crab louse-infested patients with other sexually transmitted diseases. This louse has also been observed on eyelashes (phthiriasis palpebrarum) (Fig. 32.5), causing blepharoconjunctivitis; the association of this infestation to sexual abuse of children needs to be investigated, mostly if associated with suspicion on contacting adults. The semitransparent louse, deeply burrowing in the eyelid margin, is sometimes difficult to see.

Although the three species of *lice* are cosmopolitan, the infestation by *P. humanus* is less observed in industrialized countries, where washing and changing clothes are frequent habit. Currently, this infestation is practically limited to adult populations with gregarious life form, in character of promiscuity (prisons, asylums, charities, soldiers in campaign, and others), or between beggars and vagrants. This is not the case of *P. capitis*, whose resurgence in recent years has become a serious public health problem throughout the world, being more prevalent on children and young people. In England, *P. pubis* has been found in the proportion of 1 % in heads of children, and there should therefore be other means of transmission, apart from sexual contact (Mumcuoglu 1996).

The resurgence of head lice, in the last four decades, has allowed this pediculosis to become one of the most important parasitic infections in schools, around the world. Its account, with regard to prevalence, service, and/or distribution by continents, can be found in more than one hundred publications, only in the period 1975/1985, many of them found in Linardi et al. (1995).

Being an exclusively anthropophilic parasite, the pediculosis capitis has been linked with periods of greater social upheaval, war periods, and circumstances of high promiscuity, in consequence of the reduction in hygiene, increased contact between people, and the negligence on the part of the parasitized. The decrease of hygiene and increasing infestation could be related, but not as cause and effect (Maunder 1977). Both would be resulting from a single cause, the negligence, affecting mainly least favored social classes.

More recently, other factors, acting associated with the already known, determined the resurgence of this parasitosis in pandemic level: development of resistance (natural selection) of lice to insecticides and drugs marketed (Maunder 1977; Sholdt et al. 1979), authorities not considering it as a serious problem, and the lack of inspection in certain groups (e.g., the elderly) that would hand out the infestation. Relatively to host sex, females are most affected, in all age groups. In the USA, this infestation is rare in blacks (Slonka et al. 1976). In Malaysia, Indians are more parasitized than Malaysians and these more than Europeans residing there (Sinniah et al. 1981). The infestation is also low in Chinese (Busvine 1966). In Canada, lice are more frequent among the American Indians than among Caucasian origin (Ewasechko 1981). Racial differences could be related to genetic aspects of hair, in its oval shaped or rounded in cross section, facilitating or hindering the presence of lice (Ashcroft 1969), or even related to cultural habits and socioeconomic status of the communities, such as the use of oils and creams in hair after the bath, creating an environment unfavorable to the insect. In some indigenous communities, e.g., of Brazil and Central Africa, the ingestion of lice, after manual grooming, usually tribal, may cause the decrease of ectoparasites in those communities.

As regards the color of the hair, it has been noticed that the lice of blond individuals are paler than those on individuals of black or brown hair, as a result of insect adaptation to the environment, even on the occasion of the nymphal stadia. The pediculosis capitis comes finding satisfactory conditions for their persistence in Brazil, thanks to the high concentration of young population in major urban centers, in precarious conditions of housing and basic sanitation (Linardi et al. 1995). Once established, the infestation has a seasonal fluctuation, being more frequent in the period April/September, concordant with the cold-dry season.

According to recent authors, the pediculosis capitis appears not to be related to the size of the hair, but to the diameter and spacing between the hair strands, preferring the thicker and more densely deployed.

Due to the infestation, the animals become too agitated, biting the skin. The act of scratching can cause injuries aggravated by the presence of opportunistic organisms. Production decline might be due to disturbance of feeding and resting of the infested animals. Pigs and cattle, when parasitized by large lice of the genus *Haematopinus*, may become anemic due to the spoliative action, caused by intense blood feeding. The infestation by *H. eurysternus* may cause edema and hyperemia, although such effects are not always noticed due to normal skin coloration (Nelson et al. 1977).

Extreme infestations of 50,000 lice per host have been observed in coyotes and European red squirrels, and laboratory rats and mice can be heavily infested by *Hopopleura* spp. and *Polyplax* spp., becoming unthrifty and restless. Behavior modifications like kicking, licking, and reduced feeding times have been observed in several domesticated and wild mammals (see Durden 2001).

The most common mode of transmission is through direct contact between people with passage of adults and/or nymphs. This condition is encountered during the childish games, school transportation, overcrowded schools, sharing same bed by several members of the same family, etc.

Although the indirect transmission of adults and/or nymphs is through contact with people using objects belonging to other infested people (combs, brushes, hats, caps, shower caps, pillowcases, etc.), such a method would be limited by the short survival of the insect out of the human head. The transmission of pediculosis capitis through eggs of ectoparasite would be an unlikely event, given the small likelihood of occurrence of successive events: accommodation of nits in another individual and successful development to the adult stage.

Relatively to Mallophaga, no chewing lice parasitize humans. In other animals, they can cause skin lesions, aggravated by scratching, rubbing on trees and fence posts, and grooming. Infestation is particularly serious on young animals, e.g., chicks, attacked by the head louse (*Cuclotogaster heterographus*). Damage to abdominal contour feathers of rock doves (*Columba livia*) reduces the thermal isolation, inducing an increase in heat conductance, which causes an increase in metabolic rate and spend of fat; this causes a reduction in winter survival and in attractiveness of rock dove males (Clayton 1990). This effect on metabolic rate and their consequences are stronger in European bee-eaters (*Merops apiaster*) (Hoi et al. 2012).

Poultry lice can reduce food consumption, body mass, and egg production. They can be so annoying to cattle that patches of skin can become raw and encrusted, as a result of host grooming (Johnson and Clayton 2003), and fences, barns, and other structures can be destroyed by rubbing against them. Infestation of ca. 10 % of the cattle in the northern USA causes losses of many millions of dollars.

Since Ischnocera lice of birds are usually restricted to feathers, they usually cause less damage and preening activity, and being less mobile than Amblycera, they have not been associated with transmission of potential pathogens, while the last has been incriminated as vectors of several helminths (see below), by ingestion during preening activity (Clayton 1990; Clayton and Tomkins 1994).

Trichodectes canis can ingest eggs of the tapeworm *Dipylidium caninum* from dog feces or perianal region, from which cysticercoid larvae are produced, and these can be ingested by other animals or even by children. The menoponid chewing louse *Trinotum anserinum* transmits a heartworm of swans and geese.

32.4.2 Lice as Vectors of Diseases

According to Traub et al. (1978), species of lice of the genus *Polyplax* and *Hoplopleura* are vectors of murine typhus (endemic typhus) among rodents. Of the three species of human lice, only *Pediculus humanus* is involved in the transmission of diseases, although the head lice can still be infected experimentally. Once the man is the only reservoir of diseases carried by them, the importance of *Pediculus humanus* will depend on the existence of an outbreak of the disease in a given area and the prevalence of this louse in the local community. *Pediculus humanus* is the natural vector of exanthematic typhus, trench fever, and relapsing fever (Weidhaas and Gratz 1982).

The exanthematic or epidemic typhus or also louse-borne typhus is caused by *Rickettsia prowazekii*, which has caused epidemics, mostly in conditions of war and in prisoners (see Chaps. 3 and 12) (Robinson et al. 2003). Lice acquire the pathogen from the blood of infected individuals. About half of the lice that feed on the blood of infected individuals become infective. The rickettsias, multiplying in the insect gut, invade the stomach wall and destroy the cells, allowing the blood ingested vent the louse's body, making him pink. The rickettsias appear in the feces of lice, 2–6 days after the blood meal. These feces can contaminate humans through minor injuries (bruises, scrapes on the skin) or infect mucous membranes, especially the conjunctiva. Generally, the feces are deposited in the human skin when the insect feeds, with the rickettsias penetrating the skin through small cuts or abrasions, caused by injuries resulting from the bite of lice. The disease can also be gained by crushing of lice, between fingers or teeth, as is customary in some underdeveloped countries and regions. As a rule, the lice abandon the feverish patient, passing to other individuals, so spreading the disease. Lice die within two weeks after the infection. The rickettsias may remain viable in the feces and dead lice for at least 60 days. In this way, it is dangerous to handle the clothing or bedding of patients with typhus.

This disease is currently restricted to high-altitude regions in Africa (Burundi, Ethiopia, and Rwanda), Asia (Himalayas and other highest and cold regions), and Central and South America (Andes). The Brill-Zinsser disease, a milder form of ET, transmitted by aspiration of lice feces, can be found in Europe and North America. Individuals with Brill-Zinsser disease can be sources of infection for lice or to new outbreaks. Although most cases occur during wars and social upheavals, the prevalence is higher in winter (see Chap. 12).

Trench fever, also known as "5-day fever," "quintan fever," "urban trench fever," and "Wolhynia fever," is a moderately serious disease transmitted by body lice. It was very frequent during the two world wars, causing over a million cases, only during the WW1 (see Chap. 3). The etiologic agent is *Bartonella quintana* (formerly known as *Rickettsia quintana* or *Rochalimaea quintana*) (see Chap. 12).

The disease is characterized by intermittent fever (initially for 5 days), generalized pains, multiple relapses, and mortality. The incubation period is ca. 2 weeks. The onset of symptoms is usually sudden, with high fever, severe headache, pain on moving the eyeballs, soreness of the muscles of the legs and back, and frequently hyperesthesia of the shins. Pain in the legs is the most constant symptom. The rickettsias, after the multiplication in the intestine of lice, are eliminated in feces. The transmission occurs through contamination of a skin abrasion with infectious feces of lice. Some authors admit transmission through the louse-bite wound with the feces of an infected body louse. Infected individuals typically undergo a long riquetsemia; the rickettsias have been detected in the blood of infected individuals until 8 years after the acute phase of the disease.

It has been currently found in all continents, except in Australia and Antarctica, and in several cities (Seattle, Marseille, Tokyo, etc.), mostly associated to homelessness, alcoholism, and poverty (Ohl and Spach 2000). Difficulty to change clothes for sleeping, as observed in trenches and prisons, and in difficult social conditions, as observed in Lipetsky (Russia) in 1997, can enhance infestation by body lice and cause outbreaks of diseases they transmit. Although head lice have been found infected by *B. quintana* in Nepalese children, their role in the transmission still needs to be proved.

Relapsing fever is caused by *Borrelia recurrentis*. The lice ingest the spirochetes from the blood of infected individuals, which are often found during the febrile accesses. The spirochetes can be found in the stomach of infected lice for a few hours, being subsequently digested. In certain cases, some of them may persist, reappearing after a week on general body cavity, where it will remain throughout the life of the insect. In this way, the transmission happens only when the lice are crushed on the skin or mucous membranes of an individual, thus freeing the spirochetes. Generally, this occurs in highly infested and individuals who have a habit of crushing their lice between their fingers or teeth (Weidhaas and Gratz 1982) (see Chap. 13).

Currently, relapsing fever occurs primarily in Ethiopia and Sudan. Some minor or occasional outbreaks have been found in East Africa and in Central and South America.

Sucking lice have been found infected by *Yersinia pestis* in the Democratic Republic of Congo and can transmit experimentally the bacteria between rodents. Due to restriction of transmission by fleas during European winters, lice could be suspected for such occurrence. A demonstration of infection of lice associated with corpses from medieval epidemics by these bacteria, similar to the observation of infection by *R. prowazekii* in lice associated with Napoleonic soldiers buried in Vilnius, would be desirable (see Chap. 3).

The dog sucking louse *Linognathus setosus* has been shown immatures of the filarial nematode *Dipetalonema reconditum*. Another louse, *Haematopinus suis*, is a vector of the virus of the swinepox, a potentially fatal disease, characterized by lesions on the belly of infected animals. This louse has also been incriminated in the transmission of *Mycoplasma suis* and *M. parvum* that cause swine mycoplasma infection and African swine fever virus, respectively (Mullen and Durden 2009).

Several fungi, bacteria, protozoa, and viruses have been found in sucking and chewing lice, but their role in transmission of some of them still needs to be proved (Durden 2001).

32.5 Diagnosis

Diagnosis of infestation is easily done by collection of mobile lice or eggs on fur or clothes, with the fingers, but preferably utilizing combs or jeweler's forceps and throwing the lice on ethanol 70 °GL, avoiding crushing them between the nails, to prevent contamination by bacteria. Since eggs are glued near the skin, their presence only far from it can indicate an extinct infestation. Particular attention must be paid to areas adjacent to alopecia and skin discoloration and to body parts inaccessible to grooming (Durden 2001). Dandruffs are easily differentiated from nits, which are firmly adhered to hairs (Fig. 32.3).

Diagnosis of infestation by chewing lice is obtained by the same technique utilized for sucking lice. Excessive emphasis on host specificity, more than on the lice themselves, is responsible for many synonymies, mostly for Mallophaga (Johnson and Clayton 2003), and care is needed when identifying these lice.

32.6 Control

Since the species of sucking lice that infest humans have different habitats, they require different methods of control. Several methods have been utilized for the control of head lice. The most efficient, rarely viable, is the total elimination of hairs. Table 32.3 analyzes methods for the control of head lice.

Table 32.3 Methods for the control of head lice (modified from Linardi 2011)

Methods	Stages	Mode of action	Efficiency/indication	Contraindications/critics
Manual collection	Adults and nymphs	Destruction in fire. Immersion in ethanol[a]	Very efficient if collectively developed	Do not kill lice between nails, to prevent contamination by bacteria
	Eggs	Smashing	+	–
Washing	Adults and nymphs	Destruction or dislodging from hairs	Low or inefficient	When submerged, lice close their spiracles. Time of exposition too low
Frequent combing or brushing	Adults and nymphs	Removal or maiming of lice, reducing their survival	Highly efficient, mostly if associated with oil or cream	The comb may damage the hair
	Eggs	Removal and damage	The comb must be moved in two directions, to also break the eggs	
Hot air	Eggs	Desiccation of hairs	Efficient, when used daily for a few minutes	Low efficiency for adults and nymphs, which can move on the hair
Total shave of hairs	Adults, nymphs, and eggs	Total removal of habitat of lice	Very efficient due to immediate result. Utilized in some conditions, like intern schools, reformatories, and prisons	May cause bullying and shame to parasitized individuals
Short cut of hairs	Adults, nymphs, and eggs	Partial removal of habitat of eggs and of shelter for mobile lice	Partial efficiency. Only hairs cut until 0.7 cm from the skin	Eggs at more than 0.7 cm from the skin are empty or contain dead nymphs
Oils, creams, and Vaseline	Adults and nymphs	Immobilization of insects and difficulty to fix the eggs on hairs	Very efficient if associated with utilization of combs. The hair must be carefully massaged	–
	Eggs		Idem	
Saline solution	Adults, nymphs, and eggs	Destruction by exosmosis (?)	Relative	How can the presence of lice on some marine mammals (Pinnipedia) be explained?
Electric device[b]	Adults	Removal, mutilation, and death	–	–

[a]Useful for utilization in classes and research

[b]In Brazil this electric device was released a few years ago by the Laboratory Catarinense and named Robi Comb

The principal insecticides used for lice control are pyrethroids (permethrin, deltamethrin, bioallethrin), malathion, lindane, and benzyl benzoate. Although presented in several formulations (soap, shampoo, and lotion), the insecticide in lotion will always be more advantageous, due to the effect of the concentration and the time of action on the host.

Although good results have been obtained for the control of epidemic typhus by the utilization of DDT on the body of soldiers and citizens of Naples, in 1943, this insecticide is not being used anymore, and other less toxic and persistent insecticides must be used. Any application of insecticides, in lotions or shampoos, must be carefully done, due to the risk of intoxication. Several children have been killed by the utilization of insecticides of agricultural indication, sometimes organophosphorus, in the world.[1] Ivermectin, either orally (200–400 μg/kg) or topically (1, 0.5, and 0.25 %), was shown to be effi-

[1]In 2008, a 3-year-old girl, in the southern Brazilian state of Santa Catarina, whose mother had applied Diazitop® (Diazinon) on her head, died, and more than ten cases of intoxicated children were registered in that year.

cient for the control of lice, but its repeated utilization is not ethically acceptable. Three 26 mg doses, at 7-day intervals, almost eliminated body lice from homeless men from a shelter in Marseilles, reducing the total quantity from 1866 to 6 lice, but after 45 days, sharing the shelter with infested people, index of infestation went from 14.5 to 60.7 % (Foucault et al. 2006).

For body lice, cleaning of clothes in hot water or vapor,[2] with a bath, is usually very efficient.

For crab lice, control is shaving or cutting the infested hair to remove adults, immature, and eggs glued to hairs. For more information concerning control of human lice, see Burgess (2004).

Sucking lice from small domestic animals may be controlled mostly by the utilization of brushes, associated with oil, ivermectin, or shampoos/lotions containing insecticides, preferably pyrethroids. Hair cutting may also be utilized, but this procedure is contraindicated in the winter, in cold regions, besides exposing animals to insect biting.

Since chewing lice are more resistant to conditions in the soil than Anoplura, treatment of animal's bodies, similar to those used for sucking lice, must be associated with careful cleaning of installations, removing debris (straw, residues of food, feces, etc.), and applying insecticides.

In a study on the treatment of *cattle* infested by at least two species of sucking lice and one species of chewing lice by ten insecticides applied pour-on, only permethrin was not efficient (Campbell et al. 2001). The performance of moxidectin (1 mg/kg injected or a long-acting oil-based formulation) was good for *L. vituli* and *S. capillatus* (Anoplura) but not for *B. bovis* (Mallophaga) (Clealee et al. 2004). For *chickens*, the application of cypermethrin (by immersion) and permethrin at 2 % and propoxur at 1 % (powder dusting of feathers) was efficient for the control of three species of Mallophaga (Prelezov 2008). For dogs, the application of 1 or 2 ml of 65 % permethrin pour-on was efficient for *T. canis* infestation, and these lice were eliminated from dogs treated in days 0 and 28 by fipronil (FRONTLINE® spray and spot-on and FRONTLINE® Plus for dogs) (Pollmeier et al. 2002). The same drug, alone or associated with 12 % methoprene, also eliminated the biting louse *Felicola subrostratus* from *cats* (Pollmeier et al. 2004) and in a new formulation (Eliminall®) also was efficient to eliminate fleas, ticks, mites, and chewing lice from dogs (Kuzner et al. 2013).

[2]See Chap. 3 and Remarque's "All Quiet on the Western Front" for references to lice control in the WW1.

References

Anane S, Malek I, Kamoun R et al (2013) Phthiriasis palpebrarum: diagnosis and treatment. J Fr Ophtalmol 36:815–819

Ashcroft MT (1969) Racial differences in *Pediculus h. capitis* infestation in Guyana. Trans R Soc Trop Med Hyg 63:547

Barker SC (1994) Phylogeny and classification, origins, and evolution of host associations of lice. Int J Parasitol 24:1285–1291

Burgess IF (2004) Human lice and their control. Annu Rev Entomol 49:457–481

Busvine JR (1966) Insects and hygiene. Methuen & Co, London

Campbell JB, Boxler DJ, Davis RL et al (2001) Comparative efficacy of several insecticides for control of cattle lice (Mallophaga: Trichodectidae and Anoplura: Haematopinidae). Vet Parasitol 96:155–164

Clayton DH (1990) Mate choice in experimentally parasitized rock doves: lousy males lose. Am Zool 30:251–262

Clayton DH, Tomkins DM (1994) Comparative effects of mites and lice on the reproductive success of rock doves (*Columba livia*). Parasitology 110:195–206

Clealee RM, Lloyd JE, Smith JE et al (2004) Persistent activity of moxidectin long-acting injectable formulations against natural and experimentally enhanced populations of lice infesting cattle. Vet Parasitol 120:215–227

Durden LA (2001) Lice (Phthiraptera). In: Samuel WM, Pybus MJ, Kocan AA (eds) Diseases of wild mammals, 2nd edn. Iowa State University Press, Ames, pp 3–17

Durden LA, Musser GG (1994) The mammalian hosts of the sucking lice (Anoplura) of the world: a host-parasite list. Bull Soc Vector Ecol 19:130–168

Ewasechko CA (1981) Prevalence of head lice (*Pediculus capitis* 'De Geer') among children in a rural, central Alberta School. Can J Publ Health 72:249–252

Foucault C, Ranque S, Badiaga S et al (2006) Oral ivermectin in the treatment of body lice. J Inf Dis 193:474–476

Hoi H, Krištofík J, Darolová A et al (2012) Experimental evidence for costs due to chewing lice in the European bee-eater (*Merops apiaster*). Parasitology 139:53–59

Howard R, Moore A (1991) A complete checklist of the birds of the world. Academic, London, 622 p

Johnson KP, Clayton DH (2003) The biology, ecology, and evolution of chewing lice. In: Price RD, Hellenthal RA, Palma RL et al (eds) The chewing lice: world checklist and biological overview, vol 24. Illinois Natural History Survey Special Publications, Champaign, IL, pp 449–476

Kuzner J, Turk S, Grace S et al (2013) Confirmation of the efficacy of a novel fipronil spot-on for the treatment and control of fleas, ticks and chewing lice on dogs. Vet Parasitol 193:245–251

Leo NP, Hughes JM, Yang X et al (2005) The head and body lice of humans are genetically distinct (Insecta: Phthiraptera, Pediculidae): evidence from double infestations. Heredity 95:34–40

Light JE, Allen JM, Long LM et al (2008) Geographic distribution and origins of human head lice (*Pediculus humanus capitis*) based on mitochondrial data. J Parasitol 94:1275–1281

Light JE, Reed DL (2009) Multigene analysis of phylogenetic relationships and divergence times of primate sucking lice. Mol Phylogenet Evol 50:376–390

Linardi PM (2011) Piolhos (sugadores e mastigadores). In: Marcondes CB Entomologia médica e veterinária. Editora Atheneu, São Paulo, pp 279–334

Linardi PM, De Maria M, Botelho JR et al (1995) Alguns fatores epidemiológicos relativos à infestação humana por *Pediculus capitis* (Anoplura: Pediculidae) em Belo Horizonte, Minas Gerais, Brasil. Rev Bras Entomol 39:921–929

Maunder JW (1977) Parasites and man. Human lice—biology and control. J R Soc Health 97:29–32

Mullen G, Durden L (2009) Medical and veterinary entomology, 2nd edn. Academic/Elsevier, San Diego

Mumcuoglu KY (1996) Control of human lice (Anoplura: Pediculidae) infestations: past and present. Am Entomol 42:175–178

Nelson WA, Bell JF, Clifford CM et al (1977) Interaction of ectoparasites and their hosts. J Med Entomol 13:389–428

Ohl ME, Spach DH (2000) *Bartonella quintana* and urban trench fever. Clin Inf Dis 31:131–135

Olds BP, Coates BS, Steele LD et al (2012) Comparison of the transcriptional profiles of head and body lice. Ins Mol Biol 21:257–268

Pollmeier M, Pengo G, Jeannin P et al (2002) Evaluation of the efficacy of fipronil formulations in the treatment and control of biting lice, *Trichodectes canis* (De Geer, 1778) on dogs. Vet Parasitol 107:127–136

Pollmeier M, Pengo G, Longo M et al (2004) Effective treatment and control of biting lice, *Felicola subrostratus* (Nitzsch in Burmeister, 1838), on cats using fipronil formulations. Vet Parasitol 121:157–165

Prelezov P (2008) Comparative testing of some insecticides for control of mallophagosis in chickens. Trakia J Sci 6(Suppl 1):78–81

Price RD, Hellenthal RA, Palma RL et al (eds) (2003) The chewing lice: world checklist and biological overview. Illinois Natural History Survey Publications, Champaign, 501 p

Robinson D, Leo N, Prociv P et al (2003) Potential role of head lice, *Pediculus humanus capitis*, as vectors of *Rickettsia prowazekii*. Parasitol Res 90:209–211

Sholdt LL, Holloway ML, Fronk WD (1979) The epidemiology of human pediculosis in Ethiopia. Navy Disease Vector Ecology and Control Center, Jacksonville

Sinniah SB, Sinniah D, Rajeswari B (1981) Epidemiology of *Pediculus humanus capitis* infestation in Malaysian school children. Am J Trop Med Hyg 30:734–738

Slonka GF, McKinley TW, McCroan JE et al (1976) Epidemiology of an outbreak of head lice in Georgia. Am J Trop Med Hyg 25:739–743

Toups MA, Kitchen A, Light JE et al (2010) Origin of clothing lice indicates early clothing use by anatomically modern humans in Africa. Mol Biol Evol 28:29–32

Traub RT, Wisseman MD Jr, Farhang-Azad A (1978) The ecology of murine typhus—a critical review. Trop Dis Bull 75:237–317

Weidhaas DE, Gratz NG (1982) Lice. WHO, Geneva (WHO/VBC/82.858)

Fleas and Diseases

33

Pedro Marcos Linardi

Abstract

Adult fleas are blood-feeding insects that feed on birds and largely on mammals, acting as parasites and vectors of infectious diseases. Tungiasis is the most well-known direct effect caused by flea parasitism, with the sand flea *Tunga penetrans* occurring both on domestic and wild animals and also on humans, causing pain, pruritus, and secondary bacterial infection. Bites of some fleas may cause allergic dermatitis on humans and animals. In heavy infestations, the cat flea (*Ctenocephalides felis felis*) can also cause iron deficiency anemia in young animals. As biological vectors, fleas transmit viral, rickettsial, and bacterial diseases to humans, including bubonic plague, murine typhus, flea-borne spotted fever, bartonellosis, tularemia, and others. They are intermediate hosts for some trypanosomatids (*Trypanosoma lewisi*), cestodes (*Hymenolepis* spp., *Dipylidium caninum*), and filarial worms (*Dipetalonema reconditum*) between animals or between animals and humans, as well as lodge other monoxenic organisms, such as gregarines, microsporidians, and other flagellates (*Leptomonas*). Flea control depends on whether the purpose is to control fleas as household pest of animals or humans or to control the transmission of flea-borne diseases. Mechanical (natural) and chemical methods, as well as insect growth regulators (IGRs) or insect development inhibitors (IDI), can be used, applying insecticides to the hosts (pets) or to the indoor and outdoor environments. In case of *X. cheopis*, rodent control must be preceded or accompanied by the flea control. Concerning *T. penetrans*, the treatment is the surgical extraction of the embedded fleas; in humans, tetanus immunization is also recommended.

Keywords

Fleas • Sand fleas • Siphonaptera • Pulicidae • Flea-borne diseases

P.M. Linardi (✉)
Department of Parasitology, Institute of Biological Sicences, Federal University of Minas Gerais, P. O. Box - 486, Belo Horizonte, Minas Gerais, Brazil
e-mail: linardi@icb.ufmg.br

C.B. Marcondes (ed.), *Arthropod Borne Diseases*, DOI 10.1007/978-3-319-13884-8_33

33.1 Morphology, Classification, and Distribution

Fleas are wingless insects, usually dark colored, with a laterally compressed body of about 1.5–4 mm length and fitted with bristles angled backward that permit easy movement through the hairs or feathers on the host's body (Fig. 33.1a–d). Most of the known species present combs that are more robust and sclerotized bristles aimed to fixing and locomoting between the hairs of their hosts (Fig. 33.1c). The legs are adaptated for jumping, with the hind longer than the two prior legs. The tough body is able to withstand great pressure, likely an adaptation to survive attempts to eliminate them by smashing or scratching. Sexual dimorphism is pronounced, with females larger than males; the posterior part of females is rounded, while that of males is upturned, to accommodate the copulatory apparatus in last segments.

Currently almost 3000 species and/or subspecies are known, included in 238 genera and 15 families of the order Siphonaptera (Lewis 1998), distributing from Arctic to Antarctica. Also included in this order are the tungids or sand fleas that are penetrating fleas.

33.2 Hosts

Hosts of fleas are endothermic animals, of which approximately 94 % are mammals. According to Marshall (1981), the following orders of mammals have been found infested by species of fleas: Rodentia (74 %), Insectivora (8 %),

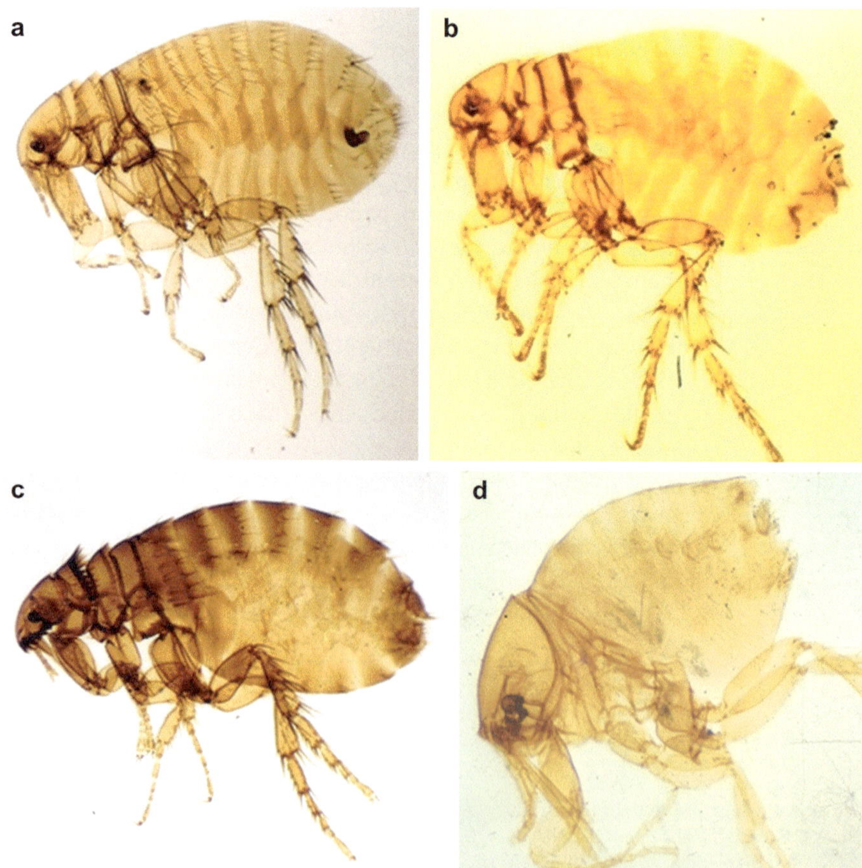

Fig. 33.1 Adult fleas: **a**—*Xenopsylla cheopis* (female), **b**—*Pulex irritans* (female), **c**—*Ctenocephalides felis felis* (female), **d**—*Tunga penetrans* (female)

Marsupialia (5 %), Chiroptera (5 %), Lagomorpha (3 %), and Carnivora (3 %). Less than 1 % of records are found in Monotremata, Cingulata, Pilosa, Pholidota, Hyracoidea, Artiodactyla, Perissodactyla, and Proboscidea. In Primates, only humans are regarded as a usual host. Parasitism on birds is secondary and must have originated from fleas parasitic on mammals.

About 30 % of the total number of flea species are known from only a single host and a single record. Occurrences of fleas on reptiles are accidental, although they are able to digest blood of these hosts. Likewise, fleas are able to feed even on hemolymph of ticks (Krasnov 2008).

Host specificity is the rule rather than the exception (Traub 1980). It is doubly enforced because larvae are essentially free-living insects with nutritional and environmental requirements different from the adults. Specific associations, characterized by the uniqueness of hosts, may support the identification of flea species as well as to subsidize mastozoologists because some fleas and their hosts may have coevolved, with primitive hosts tending to have primitive fleas and the most advanced mammals being infested by the most evolutionarily recent fleas.

However, from the epidemiological viewpoint, the eclecticism is fundamental, because the exchange of fleas between hosts can reveal the proximity of faunas and the airing of diseases.

Among the mammals, the order Rodentia is the most important because: (1) it taxonomically contains the largest number of species parasitized; (2) it geographically presents a large distribution; (3) ecologically, rodent species occupy various niches in different ecotopes; (4) epidemiologically, some of these species function as reservoirs of infections transmitted by fleas (e.g., plague, murine typhus, tularemia).

33.3 Biology

Adults of both sexes feed directly from capillaries (solenophages), but the females requiring blood for maturation of their eggs feed more than males. The repast is prolonged after the overflowing expansion so that the blood extravasated serves as larval food for a great number of flea species.

Some species feed at frequent intervals, usually once a day or more often, like *X. cheopis*, though most can withstand long periods of starvation. The feeding may occur both during the day as at night.

Each meal lasts about ten minutes, with two to three meals a day. In *Ctenocephalides felis felis*, after a period of 48 h, females and males increase in body weight, respectively, 140 % and 19 % (Dryden and Rust 1994), with the females being able to ingest daily an average of almost 14 μl of blood, which corresponds to 15 times its own body weight. The blood of normal host is digested faster than the unusual hosts, with the temperature rise accelerating the digestion (Bibikova 1977). Thus, the number of meals would be correlated with the speed of digestion, which in turn would be governed by the temperature conditions.

As a holometabolous insect, the life cycle from egg to adult may be completed in approximately 25–30 days, depending on conditions of temperature, humidity, and food obtained by the larvae (Fig. 33.2). Hatching occurs within 1–2 days, 24–36 h after the first blood repast, with the hatching proportion ranging from 6.2 % in *Polygenis tripus* (Cerqueira and Linardi 1979) to 74.2 % in *C. felis felis* (Linardi et al. 1997).

The eggs are ovoid or ellipsoidal, whitish, measuring between 300 and 700 μm, and usually deposited among the hairs or in the nests of the hosts or also in areas where the pets most frequently rest or sleep. The number of eggs laid varies with the species and nutrition of the females: *X. cheopis* lays about 400 throughout its existence and *C. felis felis* almost 1800 over a period of 50 days (Dryden 1989).

Each of the three larval instars molts each 3 days; an exception is *Tunga penetrans*, with only two larval instars. Larvae are wormlike, whitish, and legless, with chewing mouthparts, and live freely in nests of their hosts, feeding of excrement of adult fleas embedded in organic debris and ejecta of their hosts (Fig. 33.3). In each larval instar, the length is increased reaching more than double at the end of the third instar. The first instar carries a dorsal structure on the

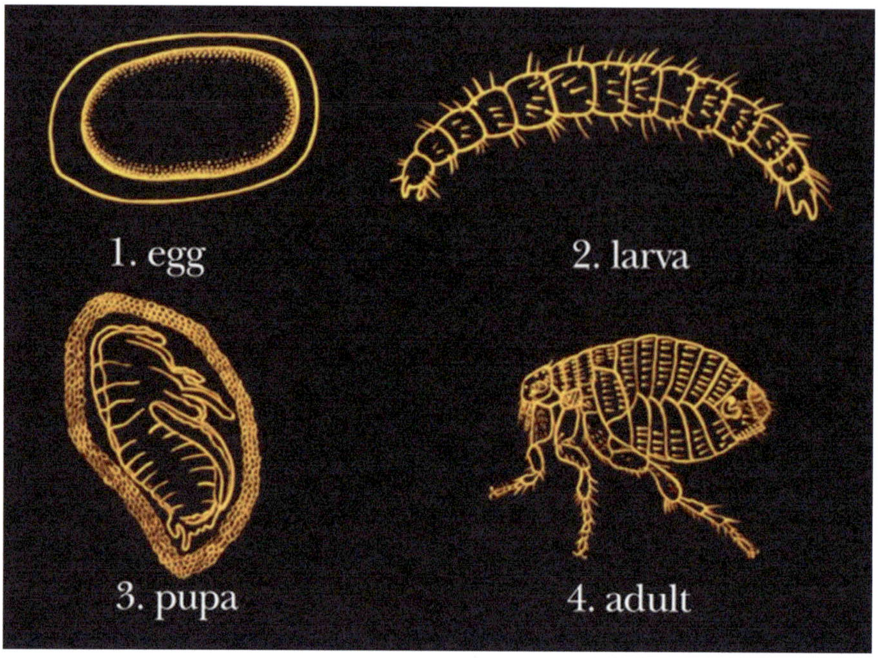

Fig. 33.2 Life cycle of a flea

Fig. 33.3 Third-instar larva of *Ctenocephalides felis felis* (×22)

head, intended to break the eggs during the eclosion. In *C. felis felis*, larvae present positive geotaxis (follow gravitation) and thigmotaxis (react to mechanical contact) and negative phototaxis (move in the opposite direction to light), as well as some type of hygrotactic response. This behavior enables the larvae to find secure locations and hide to protect against desiccation. As a result of such actions, more than 80 % of these fleas develop at the base of the carpets of the households, where they move for more than 46 cm. When confined to sand, the larvae penetrate up to 2.3 mm, in order to avoid the light. After a short prepupa phase in which the immature presents itself in the form of U, it follows the pupa, which is the stage of resistance, and, later, the emergent adult who usually stays within the cocoon. The emergence of adults from pupae is stimulated by mechanical pressure, e.g., the offset of the host in the vicinity or its trampling on cocoons, and heating, provided by the body temperature of a potential host seated on them. The opening of doors or windows in closed environments also favors the emergence. In *C. felis felis*, adult emergence from pupae is 93.3 % (Linardi et al. 1997), with the females emerging before the males in approximately a week (Linardi and Nagem 1972). Flea populations are evenly distributed in environments, with about 50 % eggs, 35 % larvae, 10 % pupae, and 5 % adults.

Longevity varies by species and depends on climatic conditions and food situation. They might live a year and a half under ideal conditions, including the temperature, food supply, and humidity, but generally, an adult flea only lives for 2 or 3 months. According to Bacot (1914) under similar air temperatures and relative humidities in laboratory, the duration in days of life span for four cosmopolitan flea species were, respectively, for starving and fed adults: *P. irritans* (125 and 513), *X. cheopis* (100 and 38), *C. canis* (234 and 58), and *Nosopsyllus fasciatus* (106 and 95). It also has been demonstrated that starving "body" fleas usually survive less time than starving "nest" fleas.

Cat fleas find their hosts by visual and thermal stimuli. Light, carbon dioxide, and air currents only stimulate locomotion. The positive phototaxis observed in adults of *C. felis felis* constitutes the basis for the use of light traps on controlling this species, once adults guide and move up to 8.4 m in the direction of the light source. Moreover, the finding of a host is also influenced by the action of certain products of the olfactory excretion on fleas. Currently, the sex of the hosts is a factor that has been increasingly considered in relation to the infestation, due to the ovarian dependence observed in a species specific of an American rabbit (Mead-Briggs 1964).

33.4 Flea Species of Medical and Veterinary Importance

Most fleas have a preferred host, but in the absence of their normal host, many of them will take a blood meal from a wide variety of animals (Pratt and Stark 1973). Some species are cosmopolitans or present a very broad distribution, due to association with humans and their pets and livestock or because they have been dispersed by commensal rodents, among them are *Xenopsylla cheopis* (Fig. 33.1a), *Pulex irritans* (Fig. 33.1b), and *Ctenocephalides felis felis* (Fig. 33.1c). Others, as *Tunga penetrans* (Fig. 33.1d), are well known in Latin America and Africa, because the females penetrate the skin of their hosts, undergo hypertrophy, and remain attached for long period. Sporadic occurrence of tungiasis has been reported in individuals from the USA, Europe, New Zealand, and Asia, where sand flea infestations were diagnosed after travel to infested regions. However, because of their unsuitable climatic and environmental conditions, the tungid did not establish in these regions.

Table 33.1 lists the principal flea species that attack the humans and that can also infest other animals, as well as their distribution on the world. Some of these fleas are also found on floor of houses and bedclothes.

33.5 Epidemiology

Among the various groups of ectoparasites that infest mammals, members of the order Siphonaptera are distinguished by their dual action as infesting agents and diseases' vectors. Adult fleas, being obligate blood-sucking parasites, provoke irritations on the skin of the host and can produce allergic reactions and secondary infections. Furthermore, fleas are reservoirs of various zoonoses and are known to be involved in the transmission of the agents of murine typhus, plague, trypanosomiasis, and helminthosis, both between synanthropic mammals and from them to humans.

The time that adults spend on parasitic action must be considered, with the following modalities being observed: (1) fleas that live on the body of the host, with the females penetrating under the skin of the same and leading to the growth of the body and the formation of neosomes (hypertrophied females + host inflammatory tissues), in virtue of permanent feeding (*Tunga* spp.); (2) fleas that live on the skin and coat intermittently (most of the known species, such as *Ctenocephalides* spp., *Xenopsylla* spp., *Polygenis* spp.); and (3) fleas that do not live on the body of the host, only looking it to exercise the hematophagy (inhabitants of nests (bird fleas) and others found in the environment of the host, such as *Pulex irritans*).

33.6 Fleas as Infesting Agents

The direct effects on humans and animals have been noticed mainly for the species *C. felis felis*, *P. irritans*, and *T. penetrans*. Among the arthropods, the cat flea is the most important

Table 33.1 Principal flea species of public health importance

Flea species	Common names	Family	Principal host (s)	Distribution
Xenopsylla cheopis	Oriental rat flea	Pulicidae	Commensal rodents	Cosmopolitan
Pulex irritans	Human flea	Pulicidae	Domestic and wild mammals	Cosmopolitan
Ctenocephalides felis spp.	Cat flea	Pulicidae	Dogs, cats, livestock	Cosmopolitan
C. canis	Dog flea	Pulicidae	Dogs	Cosmopolitan
Nosopsyllus fasciatus	Northern rat flea	Ceratophyllidae	Rats	Cosmopolitan
Leptopsylla segnis	Mouse flea	Leptopsyllidae	Rats and mice	Cosmopolitan
Echidnophaga gallinacea	Sticktight flea	Pulicidae	Poultry	Cosmopolitan
Ceratophyllus gallinae	European hen flea	Ceratophyllidae	Chicken, poultry	Cosmopolitan
Xenopsylla brasiliensis	–	Pulicidae	Commensal rodents	Europe, India, Brazil, Africa
X. astia	–	Pulicidae	Gerbils and rats	Asia, Africa
Nosopsyllus londinensis	–	Ceratophyllidae	Rats and mice	Egypt, USA, UK
Monopsyllus anisus	–	Ceratophyllidae	Rats	Russia, China, Canada, USA, UK
Orchopeas howardi	Squirrel flea	Ceratophyllidae	Rodents	USA, England
Dasypsyllus gallinulae	Moorhen flea	Ceratophyllidae	Wild birds	USA, Europe, Malayan subregion, Nepal
Synosternus pallidus	Hedgehog flea	Pulicidae	Wild carnivores, rats	West Africa, Asia, India, Europe
Hoplopsyllus anomalus	Rodent flea	Pulicidae	Ground squirrels	USA, Mexico
Diamanus montanus	Rock squirrel flea	Ceratophyllidae	Rodents	USA, Mexico
Ceratophyllus niger	Western hen flea	Ceratophyllidae	Chickens	USA
Cediopsylla simplex	Rabbit flea	Pulicidae	Rabbits	USA
Stivalius cognatus	–	Stivaliidae	Rats	Java
Tiamastus cavicola	–	Rhopalopsyllidae	Caviids	Bolivia, Chile, Peru, Argentina
Polygenis bohlsi jordani	Wild rodent flea	Rhopalopsyllidae	Wild rodents	Brazil
Tunga penetrans	Sand flea, chigger, chigoe flea	Tungidae	Pigs, dogs, rats, edentates	Latin America, Africa
T. trimamillata	Sand flea	Tungidae	Cows	Ecuador, Peru, Brazil

ectoparasite of dogs and cats worldwide (Rust and Dryden 1997). In the USA flea-related diseases constitute a high proportion of the dermatological cases reported to veterinarians. The direct impact of the infestations on the hosts provokes the following actions:

33.6.1 Irritative Action

The effect of the flea bite and inoculation of saliva can cause allergic reactions of varying intensity to their hosts. In humans, the hypersensitivity caused is known as "prurigo[1]" of Hebra. Bites of *P. irritans* can cause severe allergic dermatitis, which is manifested as erythematous edematous papules with hemorrhagic puncta on the lower extremities, especially on the ankles. In pets, such as cats and dogs, it is estimated that over 50 % of cases seen in veterinary dermatological clinics are related to the irritation caused by flea bite and known as allergic dermatitis (Bevier-Tournay 1989). Infested animals become restless and, in an attempt to get rid of fleas, bite and scratch the skin, pulling the hairs and scarifying the cutaneous tissues. Krämer and Mencke (2001) separate the flea bite dermatitis (FBD) from flea allergy dermatitis (FAD). While FBD is characterized as a reaction in the form of skin irritation, accompanied by itching and invasion of lymphocytes in the area next to the bite, FAD is a disease in which

[1]Prurido is the itching feeling, while in prurigo there are prurido and papules (CBM).

a hypersensitivity is produced in response to the introduction of antigen of the salivary glands of fleas. In FBD, the lesions are seen along the dorsum, particularly on the base of the tail. On the other hand, allergic dermatitis is manifested in lumbar-sacral dorsal area, forming a typical triangular image or an inverted V, which may be subsequently extended to the thighs, belly, and flanks. The neck, legs, and base of the ears can also be compromised. On other hosts, the reactions may present greater severity, depending on the presence of certain species of fleas not specific to them. Although some vaccines to kill fleas (*C. felis felis*) have been developed, the results were contradictory among researchers (Opdebeeck and Slacek 1993; Heath et al. 1994).

33.6.2 Spoliative Action

Considering that several species of fleas continue to practice hematophagy, even after being filled with blood, high infestation on small animals can lead to anemia, as observed in rats and other small experimental animals when they are introduced into a vigorous colony of fleas. Fleas bite their hosts several times a day, eating amounts of blood that increase significantly their body weight. Consequently, in small rodents the spoliation may be more intense, since they can host a large number of fleas in a small body surface (Linardi et al. 1984). Dogs and cats try to eliminate their fleas through skin bites. In humans, such action is not well evidenced, because they pluck their own ectoparasites. Attack by fleas (*C. canis*) in very large numbers has directly caused a fatal exsanguination in North American fox pups (Jellison 1959). When in heavy infestations, *C. felis felis* can also cause iron deficiency anemia in young animals (Harvey et al. 1982), in the same way *Ceratophyllus gallinae* in poultry.

33.6.3 Inflammatory Action

This is mainly evidenced by penetrating or semipenetrating flea species, whose holes left in the body of their hosts become prone to infection by opportunistic agents.

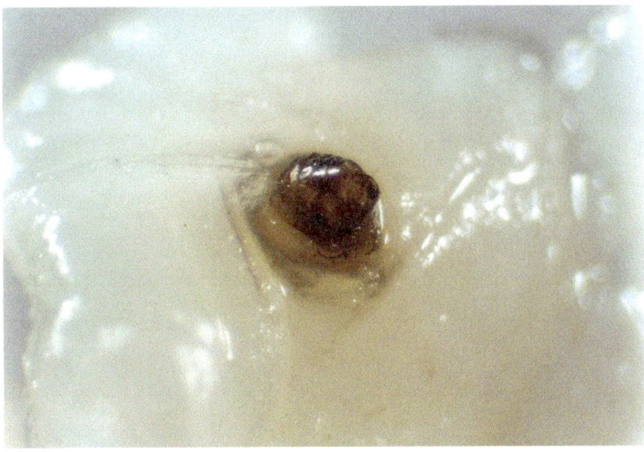

Fig. 33.4 Female of *Tunga* penetrating the skin (×40)

Tunga penetrans is the smallest known flea species, measuring about 1 mm in length, whose females penetrate the skin of their hosts (Fig. 33.4), and after mating, the gravid females undergo hypertrophy, becoming neosomes measuring 1 cm of diameter. This species may remain attached to the host for a period of more than 5 weeks, and later it is absorbed or sloughed from the host epidermis by tissue repair mechanisms (Eisele et al. 2003). *Tunga penetrans* has been referred in the literature as *sand flea*, *sandflöh*, *puce de sable*, *jigger*, *chigger*, *chigoe*, *chique*, *nigua*, *bicho-de-pé*, and *jatecuba*, among other names, having been found on hosts belonging to eight different orders of mammals, including Cingulata, Pilosa, Artiodactyla, Perissodactyla, Carnivora, Rodentia, Primates, and Proboscidea. However, in spite of the broad spectrum and category of hosts, the pig has been considered the most important animal reservoir for *T. penetrans* (Ugbomoiko et al. 2008). Humans and dogs might be considered secondary hosts or even true or essential hosts undergoing the process of adaptation. A review on the neosomes of tungid fleas that infest wild and domestic animals and concentrated mainly on hosts, infestation, sites of attachment, and impact on the hosts was recently presented by Linardi and Avelar (2014).

The infestation, known as tungiasis, is endemic or potentially endemic to 89 countries with varying degrees of incidence and prevalence, which vary in relation to the area and population studied. This species occurs commonly between the toes, sole plant, and periungual regions when parasitizing humans but also can be found on other sites, such as the hands, soles, elbows,

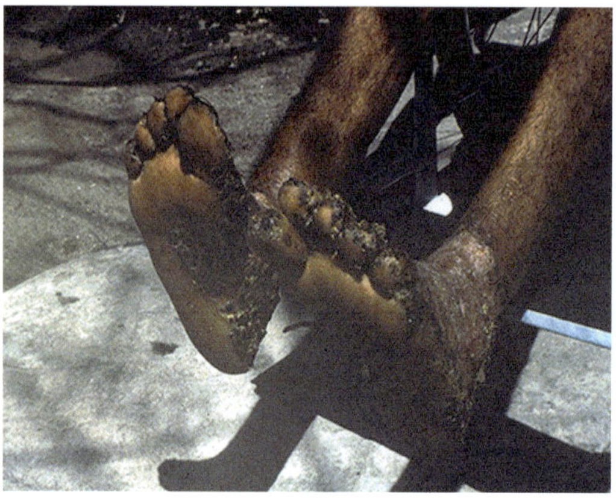

Fig. 33.5 Tungiasis: a cluster of lesions at the feet of a man

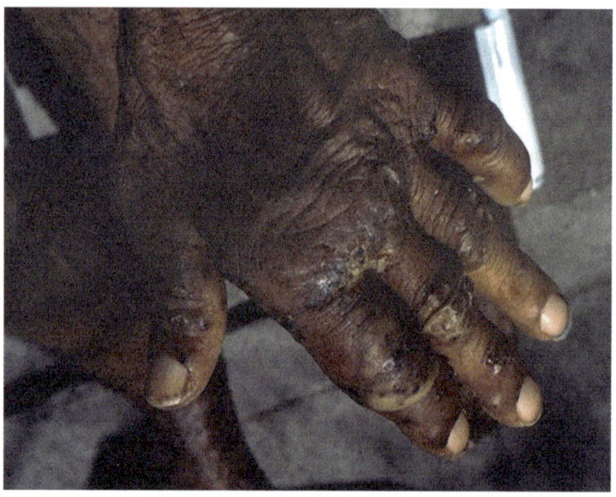

Fig. 33.6 Tungiasis: a cluster of lesions with secondary infection and suppurations at the hand of the same man

scrotum, neck, anus, gluteal area, genital region, buttocks, heels, groin, eyelids, face, etc. (Figs. 33.5 and 33.6). Severe cases characterized by multiple sites throughout the body on a single person have also been observed. In Brazil, when cutaneous lesions are numerous and close together, located in clusters along the heel, they receive the title of "honeycomb." Tungiasis may cause difficulties of posture and locomotion of hosts, necrosis of bone and tendon, and even loss of the toes. In humans, the intense pain and itching are perceived as the most irritating symptoms. Because most infestations occur on the feet, severe inflammation with deep skin fissures cause difficulties of posture and prevent individuals from walking normally. In some case, they may cause deformation of the digits and the loss of toenails.

Secondary infections primarily caused by *Staphylococcus aureus* and Gram-negative bacteria lead to the formation of pustules, suppuration, and ulcers that may also be a port of entry for *Clostridium tetani* (tetanus), *Clostridium perfringens* (gas gangrene), and *Paracoccidioides brasiliensis* (paracoccidioidomycosis). Some pathogenic microorganisms such as *Streptococcus pyogenes*, *Bacillus* spp., *Enterococcus faecalis*, *Pseudomonas* spp., Enterobacteriaceae, and other anaerobic bacteria also have been found in lesions (Feldmeier et al. 2002). Other complications include cases of lymphangitis, lymphadenitis, and sepsis. From molecular techniques, *Wolbachia* has also identified in neosomes of swines or humans in localities of Ghana, Brazil, Ecuador, Burundi, Kenya, and Zaire.

On pigs, *T. penetrans* occurs particularly on the feet, snout, and scrotum but also on the mammary glands and in the calcaneal region. When the lesions are located in the hoofs, the animals present walking difficulties; in the same manner, when present in the teats, they obstructed the galactophorous channel, causing agalactia and mastitis in lactating sows, blocking milk production and subsequent involution of the mammary glands, and consequently provoking starvation of the piglets.

In cattle, as in other ectoparasitosis, the lesions may affect weight gain and milk production. Whether the infestation occurs on the paws, the breeders may be prevented from fecundating the cows because of the lack of support for the coverage.

On both dogs and cats, neosomes can be seen around the claws, on the pads. Dogs can also be parasitized on the muzzle and in the nipples. According to Heukelbach et al. (2004), infested dogs change their position constantly, lick their pads, and are reluctant to stand up. In severe cases, dogs are unable to walk and may even die, presumably due to the secondary infection of lesions, leading to septicemia. Severe inflammation with loss of digits can also be observed in rat tungiasis, with the lesions occurring both on the feet and tails.

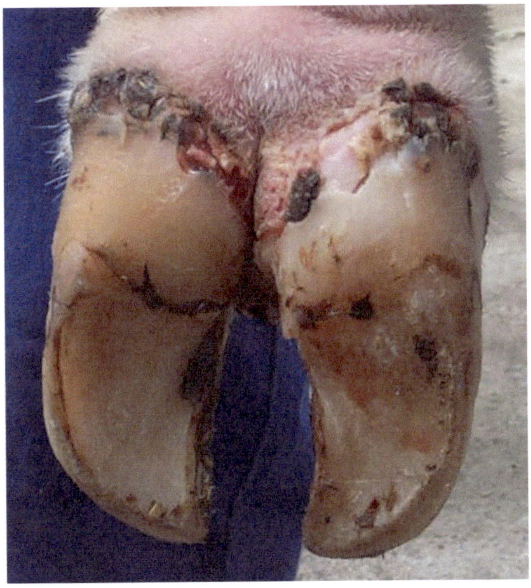

Fig. 33.7 Female *Tunga trimamillata* on the hoof of Holstein-Zebu cow

Neosomes of *T. trimamillata* are found along the coronary band, in the coronary and digital cushions, and sometimes on the sole of the hoof on bovines, goats, and sheeps (Fig. 33.7). In cattle, they can also be observed on the perianal area, on the udder of cows, and prepuce of bulls. However, the lesions are often complicated by secondary infection with ulceration on the upper edge of the nails, as a consequence of the rupture of the neosomes caused by the excoriation of the animals and colonization of flies with the formation of myiasis by screwworms. On pigs, the neosomes tend to be localized on the calcaneal region and scrotum in males and the udder in females. According to Pampiglione et al. (2009), the local inhabitants of the Andean regions may differentiate the two species of *Tunga* that infest humans: when the infestation is less annoying and caused by small sand fleas from swines, it is *T. penetrans*; otherwise, the tungiasis from *T. trimamillata* is more painful and caused by larger sand fleas.

Tungiasis can be considered a useful indicator of social conditions and hygiene practices. According to Karunamoorthi (2013), the prevalence of tungiasis may reach more than 50 % of the population in some of the hyperendemic zones, with often recurrent and sometimes massive parasitic infestations that are responsible for superinfections.

In Brazil alone, more than 10^6 individuals are at risk for severe tungiasis (Heukelbach et al. 2001). In some villages of Tanzania and Kenya, the shared use of pins and other unsterilized equipment for jigger removal has increased the risk for possible spread of AIDS and other diseases in the local populations.

The natural habitat is primarily the sandy and warm soil of deserts and beaches. On tropical beaches, the dispersion of *T. penetrans* is ensured by infested dogs. Additionally, horse manure used to fertilize soils aids in sand flea dissemination because *T. penetrans* and *T. trimamillata* also live in stables and stock farms as well as in the soil and dust close to farms. In some municipalities of Rio Grande do Sul State, Brazil, trucks carrying sand and sheaves of grass used for construction of houses can also act in the dissemination of these species.

33.6.4 Fleas as Disease Vectors

The role of fleas in the transmission of human diseases or in the maintenance of epizootic diseases among animals can be interpreted in two different levels: as a biological vector, in which only the multiplication of the etiological agent occurs inside the insect (viruses and bacterial diseases), or as intermediate host, in which the participation of the insect is required to close the life cycle of the endoparasite (protozoosis and helminthiasis) (Linardi and Guimarães 2000).

Fleas act in the transmission of virus diseases (myxomatosis), bacterial diseases (rickettsiosis, salmonellosis, tularemia, plague), and protozoosis (murine trypanosomiasis) and as intermediate hosts of tapeworms and filarial worms, as well as infection by tylenchid nematodes or infestation by other arthropods (Jellison 1959; Jenkins 1964; Bibikova 1977; Gratz and Brown 1983; Linardi and Guimarães 2000; Traversa 2013).

Table 33.2 presents the principal flea-borne diseases and certain particularities of their pathogens.

Table 33.2 Flea-borne diseases to humans and other animals

Disease	Pathogen	Localization of pathogen in the flea	Reproduction of pathogen in flea	Reservoirs	Flea species vectors	Method of transmission	Pathogenic effect on the flea
Viruses							
Myxomatosis	Myxoma virus	Digestive tract	–	Rabbits: *Oryctolagus cuniculus*, *Sylvilagus*	*Spilopsyllus cuniculi*	Mechanical inoculation	+
Bacteria							
Murine typhus	*Rickettsia typhi*	Digestive tract	+	Commensal rats	*Xenopsylla cheopis*, *Ctenocephalides f. felis*	Feces and crushing	–
Flea-borne spotted fever	*Rickettsia felis*	Digestive tract	+	Cats	*Ctenocephalides f. felis*	Crushing, bite	–
Cat scratch disease	*Bartonella henselae*	Digestive tract	+	Cats and wild rodents	*Ctenocephalides f. felis*, *C. canis*, *Polygenis gwyni*	Feces	–
Salmonellosis	*Salmonella enteritidis*, *S. typhimurium*	Digestive tract	+	Rodents and man	*X. cheopis*, *Nosopsyllus fasciatus*	Mechanical inoculation and contamination	+
Tularemia	*Francisella tularensis*	Digestive tract	–	Rodents and lagomorphs	*Diamanus montanus*, *Cediopsylla simplex*, *Neopsylla setosa*, *Ctenophthalmus assimilis*	Mechanical inoculation and contamination	–
Bubonic plague	*Yersinia pestis*	Digestive tract	+	Wild and commensal rodents	*X. cheopis*, *X. brasiliensis*, *X. astia*, *N. fasciatus*, *Polygenis* spp.	Bite	+
Protozoa							
Murine trypanosomatids	*Trypanosoma lewisi*	Digestive tract	+	Synanthropic rodents	*X. cheopis*, *N. fasciatus*	Feces	–
Helminths							
Dipylidiasis	*Acanthocheilonema* (=*Dipylidium caninum*)	Body cavity	–	Dogs	*C. felis felis*, *C. canis*, *P. irritans*	Ingestion, crushing	+
Hymenolepiasis	*Hymenolepis diminuta*, *Rodentolepis* (=*Hymenolepis*) *nana*	Body cavity	–	Synanthropic rodents	*X. cheopis*, *N. fasciatus*, *Leptosylla segnis*, *P. irritans*, *Ctenocephalides* spp.	Ingestion, crushing	+
Canine filariasis	*Dipetalonema reconditum*	Body cavity	–	Dogs	*C. felis felis*, *C. canis*	Active penetration after biting	+

The following are the main data concerning these etiologies and other relevant comments:

33.6.4.1 Myxomatosis

Myxoma virus (MV) is a poxvirus of the *Leporipoxvirus* genus that evolved in *Sylvilagus* lagomorphs and is the causative agent of myxomatosis in European rabbits (*Oryctolagus cuniculus*). It can also occur in *Sylvilagus* (cottontail rabbits), only causing localized skin tumors, contrary to the European rabbit (*Oryctolagus cuniculus*), which are more severely affected. The pathogen infects the digestive tract of the insect, being mechanically transmitted by fleas of these rabbits, *Spilopsyllus cuniculi*, or mosquitoes. The disease has been recorded in Australia, New Zealand, Europe, the USA, and South America. Although the virus does not reproduce inside the flea, it causes some pathogenic effect on the same.

33.6.4.2 Rickettsiosis[2]

Concerning murine typhus or flea-borne typhus, also known as flea-borne rat, urban, and endemic typhus, the etiologic agent is *Rickettsia typhi* (formerly *Rickettsia mooseri*), and the reservoirs are commensal rodents. Pathogen transmission takes place through the feces of *X. cheopis*, especially when they are crushed by the fingers. Many recent reports stress this as a reemerging disease, which spreads through travelers returning from endemic regions. In the USA, *R. typhi* was also found to be maintained in the cat flea, *Ctenocephalides felis*, collected from opossums. Another contamination type is possibly by the aerosol route. This rickettsia infects endothelial cells in mammalian hosts and midgut epithelial cells in the flea host. The most common clinical manifestations are high fever, severe headache, chills, myalgia, weakness, and nausea. Because it infects and reproduces into the digestive tract of flea, they are excreted along with its feces for the rest of its life but have no pathogenic action on the fleas.

Murine typhus occurs in epidemics or is prevalent throughout the world and accounts for widespread illness in areas infested with many rats and fleas. This rickettsiosis can cause severe and even fatal cases because it is often unrecognized and substantially underreported (Azad and Beard 1998). In the last years, murine typhus outbreaks have been reported in several countries, including Japan, Algeria, Tunisia, Greece, Australia, Indonesia, China, Kuwait, and Thailand, and thousands of human cases occurring annually in the USA. Some human cases have been reported in Brazil, in the States of Minas Gerais and São Paulo[3] (see Chap. 9).

Another species, *Rickettsia felis*, is the etiologic agent of an emerging rickettsiosis, recognized as flea-borne spotted fever, also called cat-flea typhus, with few human cases reported from the USA, Mexico, Brazil, France, Egypt, Germany, Laos, South Korea, Spain, Taiwan, Thailand, and Tunisia. However, the geographic distribution of *R. felis* in wild-caught arthropods has been reported in almost 30 countries. The cat flea, *Ctenocephalides felis*, is currently the only known biological vector of *R. felis*, although also evidenced by molecular techniques in other species of fleas, as in *Polygenis (Neopolygenis) atopus* collected from opossums by Horta et al. (2007) in the State of São Paulo, Brazil, and other ectoparasites, such as ticks and mites. The maintenance of *R. felis* in the environment is most likely a function of stable vertical transmission, through transstadial and transovarial transmission within cat flea populations. Clinical signs commonly involve fever, headache, and rash, but this rickettsiosis is not lethal for fleas. A review concerning distribution and arthropods infected, transmission mechanisms, current understanding of vertebrate infection, and human disease was presented by Reif and Macaluso (2009).

Bartonella are Gram-negative bacteria that cause various human diseases and have various arthropods, such as lice, ticks, and fleas, as vec-

[2] For more details, see Chap. 12 (CBM).

[3] As observed for many diseases, this one has been found where experts work and are alert. "If a tree falls in a forest and no one is around to hear it, does it make a sound?" (Berkeley 1710) (CBM).

tors, and the transmission to humans may also occur by scratches or bites from reservoir hosts, especially cats. The role of fleas as competent vectors of *Bartonella* spp. has been poorly studied. *Bartonella henselae*, the etiologic agent of cat scratch disease, is transmitted among cats and humans by the feces of *C. felis felis,* being *C. canis* also considered a suspected vector. The cat flea is also considered as a potential vector for *B. clarridgeiae* and *B. koehlerae*. In man, the initial lesion consists of a papule, pustule, or vesicle that develops at one week and enlarged lymph nodes at two to three weeks, after a cat bite or scratch, usually on the arm (see Chap. 9). Current known records for other species of *Bartonella* detected in fleas worldwide were presented by Bitam et al. (2010) including France, Japan, Algeria, Peru, Portugal, Thailand, England, Egypt, New Zealand, the USA, and China, with the following species of bacteria and fleas involved: *B. clarridgeiae* in *C. felis felis* and *Ctenophthalmus lushniensis*; *B. quintana* in *C. felis felis*, *P. irritans*, and *P. simulans*; *B. koehlerae* in *C. felis felis*; *B. elizabethae* in *X. cheopis*, *L. segnis*, and *Archeopsylla erinacei*; *B. tribocorum* in *X. cheopis*; *B. rochalimae* in *Pulex* sp.; *B. taylorii* and *B. grahamii* in *Ctenophthalmus nobilis*; and *Bartonella* spp. in *N. fasciatus*, *Oropsylla hirsuta*, and *Stenoponia tripectinata*.

By using of molecular techniques, another group of bacteria of the genus *Wolbachia* has also been found in fleas: *Wolbachia* sp. in *Tunga penetrans* and *W. pipientis* in cat fleas from France. Any effects that they might have on fleas, however, are unknown.

33.6.4.3 Salmonellosis

Both *Salmonella enteritidis* and *S. typhimurium* reproduce in the digestive tract of *X. cheopis* and *Nosopsyllus fasciatus* fleas, causing an enteric disease in rats that during the defecation contaminate the environment. The salmonellae are accumulated in the proventriculus of the fleas, thus interrupting its normal function and suppressing their vitality. Rodents and man are considered as reservoirs, and the method of transmission is guaranteed both by inoculation and contamination.

33.6.4.4 Tularemia[4]

Francisella tularensis is located in the digestive tract of several species of fleas, although not reproducing nor being harmful to them, since they are eliminated in the flea's feces after two months. It has as reservoirs rodents and lagomorphs. Although ticks are much more efficient vectors, the role of fleas, such as *Diamanus montanus* and *Cediopsylla simplex* in the USA, as well as *Ctenophthalmus assimilis* and *Neopsylla setosa* in ex-USSR, in the human transmission or in the spread of disease among rodents and rabbits must not be underestimated. The transmission is made by mechanical inoculation and contamination (see Chap. 8).

33.6.4.5 Plague

This is undoubtedly the main disease transmitted by fleas, given its morbidity, mortality, geographical distribution, and historical record, resulting in the deaths of an estimated 75–200 million people and peaking in Europe in the years 1346–1353. It is primarily an infection of wild rodents by *Yersinia pestis*, and although commensal rats are the principal reservoirs, both wild and domestic carnivores, by carrying fleas infective to those rodents or even for man, can also participate as occasional sources (see Chap. 11). The transmission takes place by the bite of infective fleas after bacteria reproduce and block the proventriculus of the insect, preventing it to feed but increasing significantly the number of bites (Fig. 33.8).

More than 200 flea species, included in 82 genera, can potentially transmit the plague bacterium, and more than 20 flea species bite easily humans. However, *X. cheopis* is considered to be the most important vector of human plague, not only because of the ease to become infected, partially blocked, and transmit the disease but also because of its close association with humans, its tendency to bite them when starved; its successful adaptation to urban conditions, especially port cities; and its worldwide distribution (Bahmanyar and Cavanaugh 1976). Other important vectors of bubonic plague are *X. brasiliensis*, in Africa, South America, and India and *Nosopsyllus fasci-*

[4]For more details on tularemia and plague, see Chap. 11 (CBM).

atus, with almost worldwide distribution. *Xenopsylla astia*, *C. felis*, *C. canis*, and *P. irritans*, although found in ports and on ships, are considered a relatively weak vectors. Figure 33.9 depicts the different modalities of transfer between various hosts of *Yersinia pestis* in the USA, involving a wide variety of mammal species, such as rodents, rabbits, wild and domestic carnivores, and humans. In South America, especially in Brazil, species of *Polygenis* are vectors of sylvatic plague, and five basic types of plague cycles can be epidemiologically characterized:

(1) Synanthropic rats → rat fleas (*X. cheopis*) → man. This is the basic cycle on bubonic plague epidemic.
(2) Wild rodents → wild rodent fleas (*Polygenis* spp.) → human. Less often, *Polygenis* species bite the humans directly. This occurs in individuals whose occupations put them in close contact with wild rodents or circumstances arising from the penetration into the wild environment, such as on the occasion of the cereals harvested in the Brazilian Northeast.
(3) Wild rodents → wild rodent fleas (*Polygenis* spp.) → synanthropic rats → rat fleas (*X. cheopis*) → human. More often, fleas from wild rodents (*Polygenis* spp.), interchanging of hosts, can bring the plague of the wild for synanthropic rodents. These, with their own fleas (*X. cheopis*), keep the plague in the urban area and/or peri-urban areas. In this interchange it is more common to find *Polygenis* on *Rattus* than *X. cheopis* on wild rodents.
(4) Human → human fleas (*Pulex irritans*) → man. This mode can be found in certain areas where an initial case establishes an infective pathway maintained by fleas of this species between humans.
(5) Human → human (pneumonic plague). In this type, a primary case of bubonic plague developing secondary pneumonic plague infects the contacts via the respiratory route.

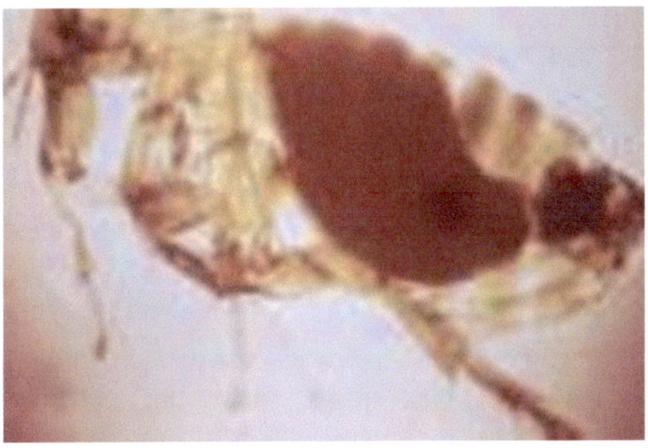

Fig. 33.8 Female *Xenopsylla cheopis* with the proventriculus and midgut partially blocked (reproduced of CDC)

Inside the insects, the persistence of the bacillus varies according to the flea species, as observed in Chapada do Araripe, Pernambuco State, located at endemic area of plague in Brazil, where 0.79 % of fleas from 29,512 flea pools examined were infected, with the following infection rates by flea species: *Polygenis bohlsi jordani* + *Polygenis tri-*

Fig. 33.9 Promastigotes of *Leptomonas ctenocephali* in gut's macerate of *C. felis felis* stained by Giemsa (×1000)

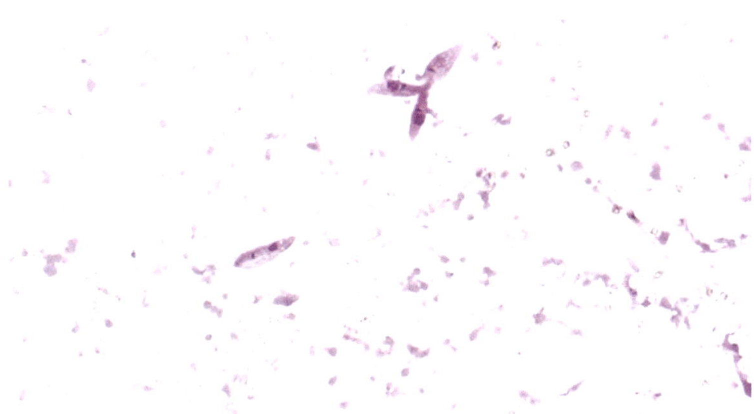

pus (1.2 %), *X. cheopis* (0.40 %), *Pulex irritans* (0.11 %), and *Ctenocephalides* spp. (0.07 %) (Brasil et al. 1989). In North America, *Diamanus montanus*, *Opisocrostis hirsutus*, *Thrassis bacchi*, *Hystrichopsylla linsdalei*, *Orchopeas sexdentatus*, *Hoplopsyllus anomalus*, *Monopsyllus wagneri*, and *Echidnophaga gallinacea* presented some value for vector efficiency among laboratory mice and guinea pigs. *Xenopsylla conformis*, *X. skrjabini*, and *Oropsylla silantiewi* found in the former USSR are also effective vectors because the blood meals frequently cause blockage of their proventriculi. Infected fleas live on average three days, eventually 11.

The estimation of flea population density has an important meaning concerning the epidemiological surveillance of plague. Although the fluctuations in the flea population may be due to patchy distribution of fleas on rodents, seasonal variations due to climate changes, size of the hosts, and fleas free-living at the time of the trapping, the abnormal concentrations on individual hosts may reflect a decrease in host population due to epizootic plague. The flea index is calculated dividing the total number of fleas recovered from all the animals trapped by the number of rats. To demonstrate seasonal trends, the flea indices must be calculated monthly or weekly in the same area in which the rodents are trapped. Due to the paramount importance of *X. cheopis* in the transmission of plague, particular attention is paid to the *X. cheopis* index. A specific index of over one for *X. cheopis* must be regarded as indicative of a dangerous situation.

33.6.4.6 Trypanosomatids

Until now, about 30 species of monoxenous and heteroxenous trypanosomatids have been recorded in fleas, with predominance to the genus *Trypanosoma* (18 species). The most common species is *Trypanosoma lewisi*, which infects commensal rats, having the rat fleas *X. cheopis* and *N. fasciatus* as their principal vectors. The flagellates multiply inside the insect but have no pathogenic action on it. The transmission is essentially performed via fecal contamination by fleas.

Blastocrithidia and *Leptomonas* are genera of monoxenous trypanosomatids found in fleas. Among the 15 considered valid species of *Leptomonas*, six have been found in fleas, with *L. pulexsimulantis* developing in *Pulex simulans* and *L. ctenocephali* in *C. felis felis*. Recently, *L. ctenocephali* (Fig. 33.9) was reported in 1.7 % of cat fleas collected in Belo Horizonte City, Brazil (Avelar et al. 2007, 2011). As noted by Pacheco et al. (1998) monoxenous trypanosomatids of fleas can cause opportunistic infections in immunodeficient individuals, with the infections in humans possibly occurring per os.

The PCR technique has also allowed recognizing DNA of *Leishmania chagasi* in *Ctenocephalides felis felis* removed from naturally infected dogs, thus opening the possibility of mechanical transmission of canine visceral leishmaniasis by ingestion of infected fleas (Coutinho and Linardi 2007; Ferreira et al. 2009).

33.6.4.7 Tapeworms

Dipylidium caninum is a usual cestode in domestic carnivores that eventually parasites man. The infection of fleas (*Ctenocephalides* spp., *Pulex irritans*) occurs in their larval stages, by the ingestion of eggs of the cestode from the feces of dogs or direct intake of proglottis. Inside the fleas (larvae or adults), the eggs develop for cysticercoids. When ingested by dogs or cats during grooming, the infected fleas release the cysticercoids in the digestive tract of vertebrates, thus restoring the biological cycle. The infection is pathogenic for fleas. Once the life cycle of *D. caninum* can also be complemented by ingestion of crushed fleas or by the licking of their contents, this parasite can infect children when handling dogs. Therefore, the control of *Ctenocephalides* spp. in domiciliary and peridomiciliary environments will act to reduce the potential risk of this helminthiasis in the infant population.

Fleas also act as intermediate hosts of two rat tapeworms, respectively, *Hymenolepis diminuta* and *Rodentolepis nana* (=*H. nana*) that are common in flea species parasites of synanthropic rodent, such as *X. cheopis*, *N. fasciatus*, *L. segnis*, or else *Ctenocephalides* and *Pulex*. The cysticer-

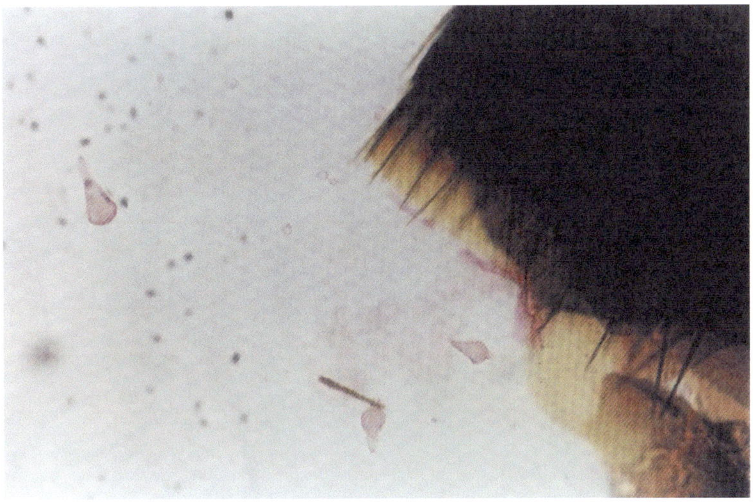

Fig. 33.10 Hymenolepids in body cavity of *Polygenis tripus* stained by Giemsa (×100)

coids lodge in the intestinal tract and body cavity of fleas, being harmful to them (Fig. 33.10). The transmission is by ingestion or crushing of infective adult fleas. Three other species of *Hymenolepis*, respectively, *H. murina*, *H. microstoma*, and *H. citelli* have also been found in flea larvae.

33.6.4.8 Canine Filariasis

The filarial worm *Acanthocheilonema* (= *Dipetalonema) reconditum* is a pathogenic and exclusive parasite of dogs that develops in the body cavity of fleas *Ctenocephalides* spp. The transmission between dogs takes place through the active penetration of microfilariae in the tissues, at the time of the flea hematophagy.

33.6.4.9 Tylenchids Infections

Worms of Allantonematidae (Nematoda: Tylenchida) have been observed in the body cavity of fleas. In the Neotropics, Linardi et al. (1981) reported such records for the first time in *P. tripus*, being also the first record of the family Rhopalopsyllidae (Fig. 33.11). About 11.2% of fleas examined were infected, and because the occurrence of allantonematids may cause damage to the parasitized host, such as sterilization, atrophy of genitalia, and lethality, the authors raised the possibility of its use for biological control and, consequently, prophylaxis of the plague.

33.6.4.10 Arthropod Infection and/or Infestation

Some parasitoids such as larvae of Chalcidoidea (Hymenoptera) have been observed in certain species of fleas. Some species of mites, e.g., *Rhizoglyphus echinopus*, have also been found in association with certain species of fleas, both on phoresy and in body cavity.

33.6.4.11 Other Parasites of Fleas

The literature contains multiple records of other parasites in fleas, such as *Bacillus thuringiensis*, *Beauveria bassiana*, *Escherichia coli*, *Hepatozoon erhardovae*, *Listeria* sp., *Nosema* sp., and gregarines. Beard et al. (1990) listed various organisms as endosymbionts of 60 species of fleas, developing mutualistic, commensalistic, or parasitic actions, and, except a few exceptions, little is known about their biology, prevalence, and degree of pathogenicity to natural populations of fleas. *Listeria* can persist during relatively long periods in alimentary canal of experimentally infected *Nosopsyllus consimilis*,

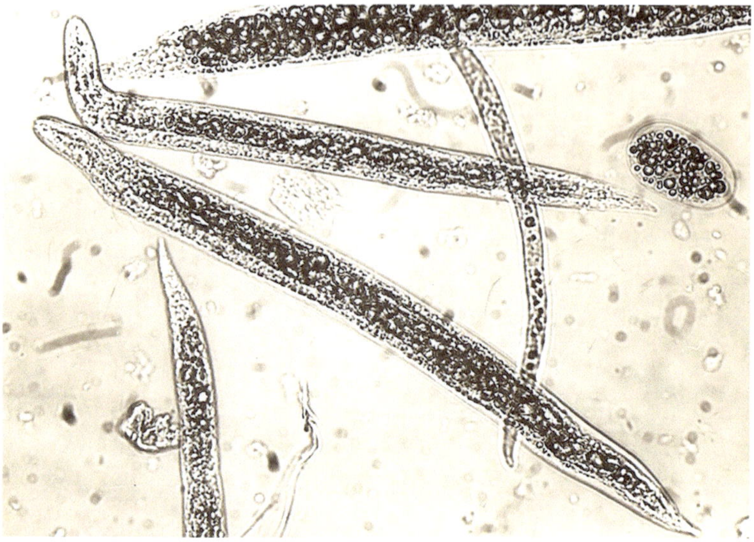

Fig. 33.11 Allantonematid worms in body cavity of desiccated *Polygenis tripus* (×276)

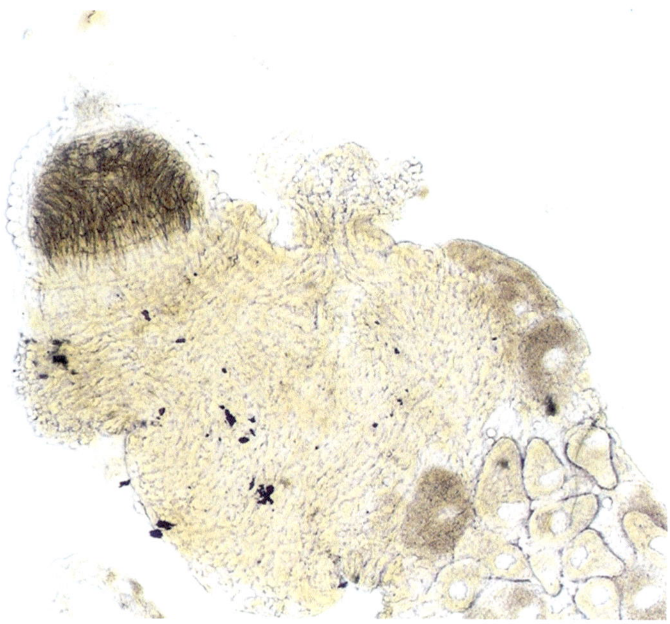

Fig. 33.12 Trophozoites of gregarine *Steinina* sp. (Actinocephalidae) found in the midgut of *C. felis felis*

and since this bacillus penetrates in the flea's muscular tissue and body cavity, it could possibly be transmitted via flea bites. *Nolleria pulicis* (Chitridiopsidae) and a gregarine, *Steinina* sp. (Actinocephalidae) (Fig. 33.12), were found by Avelar and Linardi (2008) in C. *felis felis* collected from dogs in Belo Horizonte, MG. Consequently, the dissection of fleas, associated with other techniques, such as PCR and the inoculation of flea macerated in laboratory animals, is a fundamental step for the elucidation of such issues. To what extent some endosymbionts might be pathogenic for fleas and thus be used in biological control? On the other hand, some monoxenous trypanosomatids found in fleas should deserve special attention, in view of the possibility to act as opportunistic agents causing infections in mammals (Pacheco et al. 1998; Garin et al. 2001).

33.7 Control

Considering that no species of flea is specific to humans and that they can suck both domestic and wild animals, a good control program should include indoors, outdoors, and, necessarily, the parasitized animals. It is also of paramount importance to know which species of flea is the object of control, as well as whether the purpose is to control or prevent transmission of flea-borne diseases or to control fleas as household pests of animals or man.

Concerning *T. penetrans* and tungiasis, the control consists of local excision or sterile curettage of the neosomes. Parasitized dogs can be treated with a subcutaneous application of ivermectin and healing ointments and repellents. In humans, prophylactic measures include the wearing of shoes, improving hygiene, use of gloves for handling manure and sweeping floors, and the restriction of free movement of people and infested animals among households.

Anti-tetanus prophylaxis is recommended to prevent secondary infection. Both within and around the residences, the pyrethroids cypermethrin and deltamethrin can be used as a measure of environmental control. Given the diversity of habitats, integrated actions are required for the control of tungiasis, involving clinicians, veterinarians, biologists, public health specialists, technicians, and companies responsible for vector control. An overview of human tungiasis, including risk factors, treatment, repellents, prevention, control, and healthcare stakeholders, has been presented by Heukelbach (2006) and Karunamoorthi (2013).

In the case of infestation by *X. cheopis* and other vectors of plague or murine typhus, since most fleas from dead rodents seek other animals or humans for their blood meals, some surveillance measures that include rodent control should be preceded or, at least, accompanied by a flea control program. In this case, the insecticide formulated as a dusting powder should be applied to rodent burrows and runways to ensure that rodents will come in contact with them by picking up the dust on their fur, and consequently fleas will be killed by spread of the dust on the fur when the rodent grooms itself. Insecticides used in rodent flea control include the organochlorines, DDT (10 %) and lindane (3 %); carbamates, bendiocarb (1 %), carbaryl (5 %), and propoxur (1 %); organophosphates, diazinon (2 %), fenitrothion (2 %), jodfenphos (5 %), malathion (5 %), and pirimiphos-methyl (2 %); and synthetic pyrethroids, deltamethrin (0.005 %) and permethrin (0.5 %). Where resistance to the organochlorines is known to occur, dusts of one of the organophosphates, carbamates, or pyrethroids should be used as an alternative (Gratz and Brown 1983).

To control *Pulex* spp., emulsions, solutions, or suspensions of bendiocarb (0.24 %), deltamethrin (0.005 %), jodfenphos (1 %), lindane (1 %), malathion (2 %), natural pyrethrins (0.2 %), permethrin (0.125 %), and pirimiphos-methyl (1 %) can be used as sprays on mattress, cracks, and crevices in the floor and bed.

The control of *Ctenocephalides* spp. includes alternative methods of mechanical and chemical control. Biological control using predators (ants, tenebrionids, and staphylinids beetles) or parasites (nematode *Steinernema carpocapsae*) against the larval stages of fleas can also be employed, however, in peridomestic environment.

The following are methods that do not use insecticides, which can be employed for controlling *Ctenocephalides*:

(1) Pet grooming: to remove adult fleas on the pets everyday
(2) Flea comb: to trap fleas in its teeth as the comb is brushed through the pet's hair
(3) Vacuuming: to collect adult and immature fleas inside the home, every other day
(4) Washing floors: to stop the fleas' life cycle, once a week

(5) Lighted trap: to hold and trap fleas using an electric light near pet's resting and sleeping areas. The positive phototaxis has been the basis for the use of light traps, because fleas orient, jump, and are attracted toward a light source, such as in the ancient traps using a candle inside a dish with water. A light trap utilizing castor oil was described and patented (https://www.google.com/patents/USRE34402)
(6) Desiccant powders as diatomaceous earth: to kill larvae on carpets by dehydration
(7) Oral repellents: to repel fleas after ingesting by an animal, including brewer's yeast, thiamine, sulfur, garlic, cod liver oil, vitamin C and vinegar, apple cider, and vitamin B1
(8) Herbal and citrus extract repellents: to repel fleas from animal's body and as shampoos during bathing or soaking, including citrus extracts, cedarwood oil, eucalyptus oil, walnut oil, coconut oil, peppermint oil, citronella oil, *d*-limonene, and linalool.

Ultrasonic flea collars and other similar artifacts did not prove to be effective for controlling fleas.

The insecticides used to control fleas on pets parasitized can be found under various formulations: shampoos, sprays, soaps, collars, powders, and oral systemics. In Brazil, recently were released: (a) lufenuron, an insect development inhibitor (IDI) that prevents the formation of chitin in larva L_1, especially in the cephalic appendage egg breaker, preventing them of hatching (Program®—Novartis); (b) fipronil, a phenylpyrazole applied as spray (Frontline®—Rhodia-Mérieux); (c) imidacloprid + chloronicotinyl + synthetic nitroguanidine, to be topically applied on the neck of the animal (Advantage®); and (d) selamectin, an avermectin as active product (Revolution®).

Some insecticides used to control fleas inside the home can also be extended to the peridomestic habitats, due to the residual action, photostability, and preparation in aqueous medium, which facilitates the absorption by soils and environmental protection. Thus, the microencapsulation would be the most recommended, in view of such purposes and perspectives.

The integrated control against fleas represents currently one of the best measures, by associating two different types of chemicals (an insecticide + development inhibitor or an insecticide + a growth regulator) with a mechanical method in residences (aspiration and steam cleaning on carpets). Therefore, while the insecticide acts killing fleas immediately, the inhibitor of development or growth regulator will prevent fleas for reproducing; in turn, the mechanical intervention will reduce the number of immature stages present in the environment.

Table 33.3 summarizes the main mechanical and chemical methods that can be used for the control of pet fleas and tungids when infesting animals and environments.

A good strategy would be to employ, alternatively, natural and chemical methods to reduce the repetitive chemical treatment, lowering the likelihood of development of resistance to insecticides. Otherwise, modifications in climatic conditions, movements of hosts and goods, changes in animal phenology and human behavior, and increase of wildlife are presently concurring in the geographic spread of parasites, including pet fleas (Traversa 2013). In recent years, borates or boric acid applied to carpets have been used with some success for the control of larvae in the intradomiciliar environment.

It should be noted that the continued use of the vacuum cleaner inside the housing is very important, since it collects not only eggs, larvae, pupae, and adults newly emerged but, above all, the feces of fleas and other nutrients essential for larval feeding. It is estimated that for every flea present in a given housing, there is an occurrence of approximately 100 other juveniles, including eggs, larvae, and pupae.

Table 33.3 Some natural and chemical methods for controlling pulicid and tungid fleas

Environments	Natural methods	Chemical methods
On infested animals	**Pulicidae** Pet grooming wash the coat Frequent brushing or combing (fine-toothed comb) **Tungidae** Surgical extraction of the parasite using a sterile needle after disinfection. Permanent use of footwear. Use of gloves for handling manure. Anti-tetanic prophylaxis	OP: coumaphos, metriphonate, fenthion, chlorpyrifos, trichlorfon IGR: methoprene IDI: lufenuron (Program) PI: deltamethrin, permethrin, FI: "Frontline®" IM+CL+NI: "Advantage" C: propoxur, carbaryl, bendiocarb AV: revolution
Inside of dwellings	**Pulicidae** House sweeping and subsequent incineration of sweep. Use of vacuum cleaner. Washing of the floor of the households Frequent washing of the "bed" of the animal	C: bendiocarb OP: dichlorvos, chlorpyrifos, diazinon, propetamphos PI: K-Othrine® IGR: methoprene, fenoxycarb, pyriproxyfen, alsystin, cyromazine PI+OP or PI+IGR PI (permethrin)+IGR (methoprene)+pyrethrins synergistic Borates
In the peridomestic environment	**Pulicidae** Frequent sweeping kennel and other shelters, with subsequent incineration of sweep. Prevent the airing of manure and organic matter. Vegetation management. Soil management. Prevent contact or exchange of the animal with others outside the home **Tungidae** Use of fire, directed to land piled up, after sweeping and tickler. Cleaning the area. Capture of stray dogs. Prohibitions or restrictions on the transit of persons, vehicles loaded with sand, grass and manure in not infested areas	OP: propetamphos (considerable residual activity and photostability), diazinon, chlorpyrifos, malathion OP: microencapsulated (chlorpyrifos, diazinon) OP: malathion PI: K-Othrine® (spray the manure to be used in gardening) OP: microencapsulated

AV avermectin, *C* carbamate, *CL* chloronicotinyl, *FI* fipronil (=phenylpyrazole), *IDI* insect development inhibitor, *IGR* insect growth regulator, *IM* imidacloprid, *NI* synthetic nitroguanidine, *OP* organophosphate, *PI* synthetic pyrethroid

References

Avelar DM, Linardi PM (2008) Seasonality and prevalence rates of *Steinina* sp. (Eugregarinorida: Actinocephalidae) in *Ctenocephalides felis felis* (Siphonaptera: Pulicidae) from dogs captured in Belo Horizonte, Minas Gerais, Brazil. J Med Entomol 45:1139–1142

Avelar DM, Bussolotti AS, Ramos MCA et al (2007) Endosymbionts of *Ctenocephalides felis felis* (Siphonaptera: Pulicidae) obtained from dogs captured in Belo Horizonte, Minas Gerais, Brazil. J Invertebr Pathol 94:149–152

Avelar DM, Melo MN, Linardi PM (2011) Morphology and growth characteristics of cultured *Leptomonas ctenocephali* from *Ctenocephalides felis felis* (Siphonaptera: Pulicidae) of dogs in Brazil. Vet Parasitol 180:394–398

Azad AF, Beard CB (1998) Rickettsial pathogens and their arthropod vectors. Emerg Infect Dis 4:179–186

Bacot AW (1914) A study of the bionomics of the common rat fleas and other species associated with human habitations. J Hyg 13(Suppl):447–654

Bahmanyar MA, Cavanaugh DC (1976) Plague manual. WHO, Geneva

Beard CB, Butler JF, Hall DW (1990) Prevalence and biology of endosymbionts of fleas (Siphonaptera: Pulicidae) from dogs and cats in Alachua County, Florida. J Med Entomol 27:1050–1061

Berkeley G (1710) A treatise concerning the principles of human knowledge. Akron Rhames, Dublin

Bevier-Tournay DE (1989) Fleas and flea control. Curr Vet Ther 10:586–592

Bibikova VA (1977) Contemporary views on the interrelationships between fleas and pathogens of human and animal diseases. Annu Rev Entomol 22:23–32

Bitam I, Dittmar K, Parola P et al (2010) Fleas and flea-borne diseases. Int J Infect Dis 14:667–676

Brasil DP, Carvalho FG, Almeida CR et al (1989) Studies on natural infection with *Yersinia pestis*. Rev Soc Bras Med Trop 22:177–181

Cerqueira EJL, Linardi PM (1979) Ciclo evolutivo de *Polygenis tripus* (Jordan, 1933) em laboratório, com descrição da larva (Siphonaptera, Rhopalopsyllidae). Rev Bras Biol 39:419–423

Coutinho MTZ, Linardi PM (2007) Can fleas from dogs infected with canine visceral leishmaniasis transfer the infection to other mammals? Vet Parasitol 147:320–325

Dryden MW (1989) Host association, on-host longevity and egg production of *Ctenocephalides felis felis*. Vet Parasitol 34:117–122

Dryden MW, Rust MK (1994) The cat flea: biology, ecology and control. Vet Parasitol 52:1–19

Eisele MJ, Heukelbach J, Van Mardk E et al (2003) Investigations on the biology, epidemiology, pathology and control of *Tunga penetrans* in Brazil. I. Natural history of tungiasis in man. Parasitol Res 90:87–99

Feldmeier H, Heukelbach J, Eisele M et al (2002) Bacterial superinfection in human tungiasis. Trop Med Int Health 7:559–564

Ferreira MGPA, Fattori KR, Souza F et al (2009) Potential role for dog fleas in the cycle of *Leishmania* spp. Vet Parasitol 165:150–154

Garin YF, Sulahia A, Méneceur P et al (2001) Experimental pathogenicity of a presumed monoxenous trypanosomatid isolated from humans in a murine model. J Eukaryot Microbiol 48:170–176

Gratz NG, Brown WA (1983) XII. Fleas—biology and control. WHO, Geneva (WHO/VBC/83.874)

Harvey JW, French TW, Meyer DJ (1982) Chronic iron deficiency anemia in dogs. J Am Anim Hosp Assoc 18:946–960

Heath A, Arfsten A, Yamanaka M et al (1994) Vaccination against the cat flea *Ctenocephalides felis felis*. Parasite Immunol 16:187–191

Heukelbach J (2006) Revision on tungiasis: treatment options and prevention. Expert Rev Anti Infect Ther 4:151–157

Heukelbach J, Oliveira FAS, Hesse G et al (2001) Tungiasis: a neglected health problem of poor communities. Trop Med Int Health 6:267–272

Heukelbach J, Costa AML, Wilcke T et al (2004) The animal reservoir of *Tunga penetrans* in severely affected communities of north-east Brazil. Med Vet Entomol 18:329–335

Horta MC, Labruna MB, Pinter A et al (2007) *Rickettsia* infection in five areas of the state of São Paulo, Brazil. Mem Inst Oswaldo Cruz 102:793–801

Jellison WL (1959) Fleas and disease. Annu Rev Entomol 4:389–414

Jenkins DW (1964) Pathogens, parasites and predators of medically important arthropods. Anotated list and bibliography. Bull World Health Organ 30(Suppl):1–30

Karunamoorthi K (2013) Tungiasis: a neglected epidermal parasitic skin disease of marginalized populations—a call for global science and policy. Parasitol Res 112:3635–3643

Krämer F, Mencke N (2001) Flea biology and control. Springer, Berlin

Krasnov BR (2008) Functional and evolutionary ecology of fleas. A model for ecological parasitology. Cambridge University Press, Cambridge

Lewis RE (1998) Resumé of the Siphonaptera (Insecta) of the world. J Med Entomol 35:377–389

Linardi PM, Avelar DM (2014) Neosomes of tungid fleas on wild and domestic animals. Parasitol Res 113:3517–3533

Linardi PM, Guimarães LR (2000) Sifonápteros do Brasil. Museu de Zoologia USP/FAPESP, São Paulo

Linardi PM, Nagem RL (1972) Observações sobre o ciclo evolutivo de *Ctenocephalides felis* (Bouché, 1835) (Siphonaptera, Pulicidae) e sua sobrevida fora do hospedeiro. Bol Mus Hist Nat UFMG Zool 13:1–22

Linardi PM, Cerqueira EJL, Williams P (1981) *Polygenis tripus* (Siphonaptera: Rhopalopsyllidae) naturally infected by Allantonematidae (Nematoda: Tylenchida). J Med Entomol 18:41–43

Linardi PM, Botelho JR, Cunha HC et al (1984) Ectoparasitos de roedores da região urbana de Belo Horizonte, MG. I. Interação entre ectoparasitos e hospedeiros. Mem Inst Oswaldo Cruz 79:239–247

Linardi PM, De Maria M, Botelho JR (1997) Effects of larval nutrition on the postembryonic development of *Ctenocephalides felis felis* (Siphonaptera: Pulicidae). J Med Entomol 34:494–497

Marshall AG (1981) The ecology of ectoparasitic insects. Academic, London

Mead-Briggs AR (1964) The reproductive biology of the rabbit flea *Spilopsyllus cuniculi.* J Exp Biol 41:371–402

Opdebeeck JP, Slacek B (1993) An attempt to protect cats against infestation with *Ctenocephalides felis felis* using gut membrane antigens as vaccine. Int J Parasitol 23:1063–1067

Pacheco RS, Marzochi MCA, Pires MQ et al (1998) Parasite genotypically related to a monoxenous trypanosomatid of dog's flea causing opportunistic infection in an HIV positive patient. Mem Inst Oswaldo Cruz 93:531–537

Pampiglione S, Fioravanti ML, Gustinelli A et al (2009) Sand flea (*Tunga* spp.) infections in humans and domestic animals: state of the art. Med Vet Entomol 23:172–176

Pratt HD, Stark HE (1973) Fleas of public health importance and their control. U.S. Public Health Service Home Atlanta (CDC) 74–8267

Reif RE, Macaluso KR (2009) Ecology of *Rickettsia felis*: a review. J Med Entomol 46:723–736

Rust MK, Dryden MW (1997) The biology, ecology, and management of the cat flea. Annu Rev Entomol 42:451–473

Traub R (1980) The zoogeography and evolution of some fleas lice and mammals. In: Starcke H, Traub R (eds) Fleas. Balkema, Rotterdam, pp 93–172

Traversa D (2013) Fleas infesting pets in the era of emerging extra-intestinal nematodes. Parasit Vectors 6:59–74. doi:10.1186/1756-3305-6-59

Ugbomoiko S, Ariza L, Heukelbach J (2008) Pigs are the most important animal reservoir for *Tunga penetrans* (jigger flea) in rural Nigeria. Trop Doct 38:226–227

Diseases Caused by Acari (Ticks and Mites)

34

Carlos Brisola Marcondes
and Filipe Dantas-Torres

Abstract

The subclass Acari includes ticks (order Ixodida) and mites (Mesostigmata, Trombidiformes, Sarcoptiformes, several other orders). Ticks can transmit several pathogens and cause direct damage (e.g., blood depletion, skin lesions, and paralysis) to their host. Mites can cause lesions in the skin (e.g., mange-causing mites) of their hosts and affect their respiratory tract. Some free-living mites and their discarded cuticles can be aspired by and cause allergic reactions in humans. Systematic control of ticks on domestic animals and, eventually, in the environment is required. Control of mange-causing mites is usually done by treating affected patients and by avoiding sharing of clothes. Cleaning of houses, carpets, and bedclothes and reduction on humidity are efficient for the control of dust mites.

Keywords

Ticks • Mites • Mange • Tick paralysis • Eczema • Respiratory allergy

In addition to transmitting several disease agents (refer to Chaps. 9, 10, 12, 13, 14, 15, 21, 22, 23), ticks (order Ixodida) and mites (Mesostigmata, Trombidiformes, Sarcoptiformes, several other orders) can cause severe damage due to their direct action on their hosts. For example, tick bites may be painful and can result in skin lesions that may favor secondary bacterial infections. Ticks can also cause severe blood loss, resulting in anemia. Some ticks may also induce allergies, toxicosis, and paralysis, with potential fatal outcomes for the host (Hall-Mendelin et al. 2011). Some mites can also feed on blood, cause skin lesions (mange), and induce allergies. This chapter is focused on direct damages, diseases, and syndromes caused by ticks and mites.

C.B. Marcondes (✉)
Department of Microbiology, Immunology and Parasitology, Center of Biological Sciences, Federal University of Santa Catarina,
88040-900 Florianópolis, SC, Brazil
e-mail: cbrisolamarcondes@gmail.com

F. Dantas-Torres
Department of Immunology, Aggeu Magalhães Research Centre, Recife, PE, Brazil
e-mail: filipe.dantas@cpqam.fiocruz.br

C.B. Marcondes (ed.), *Arthropod Borne Diseases*, DOI 10.1007/978-3-319-13884-8_34

34.1 Ticks

The order Ixodida is subdivided into three families: Argasidae, Ixodidae, and Nuttalliellidae. Members of the family Ixodidae are usually referred to as hard ticks because of their dorsal scutum, which is present on all developmental stages (Fig. 34.1). This family includes about 715 species, including several vectors of medico-veterinary significance (Dantas-Torres et al. 2012). The family Argasidae includes the so-called soft ticks that lack a dorsal scutum in the nymphal and adult stages (Fig. 34.2); larvae may have a dorsal shield. The generic classification of Argasidae is currently under dispute, but this family is represented by about 200 species. The third family, Nuttalliellidae, is a more primitive one and includes only one living representative, *Nuttalliella namaqua*, whose medical significance is probably minor.

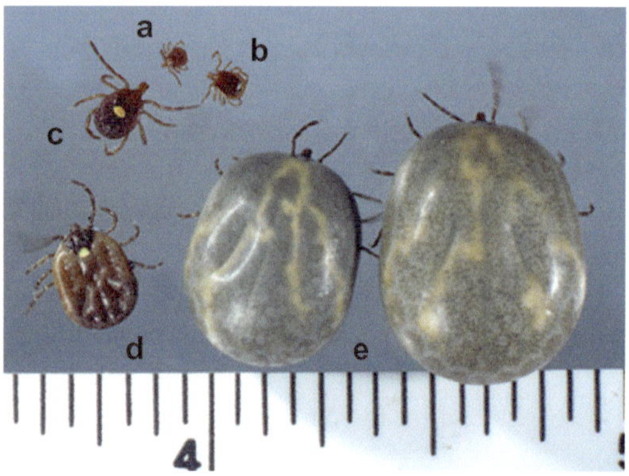

Fig. 34.1 Nymph (**a**), male (**b**), female (**c**), partially engorged female (**d**), and two fully engorged adults (**e**) of *Amblyomma americanum*. Image courtesy of James Occi, from CDC

Fig. 34.2 Ventral (*left*) and dorsal (*right*) views of *Ornithodoros turicata*, a known vector of TBRF spirochetes in the Midwestern and Southwestern USA. Image courtesy of James Occi, from CDC

Tick larvae are six legged, whereas nymphs and adults are eight legged. They feed on blood from various animal species. Their morphology and life cycles have been detailed in several textbooks (e.g., Krantz and Walter 2009). They can remain on the same host for all their cycle or move between hosts, depending on tick species and their biological strategies. All stages of hard ticks typically remain attached to host, feeding for long periods, whereas adults of soft ticks search for their host, take a short meal (usually for some minutes), and then drop off and hide again. In Argasidae, the feeding pattern is quite variable, with immature stages of some species spending long periods of time sucking blood on their hosts (Jongejan and Uilenberg 2004), whereas others do not feed at all. Even some adults of soft ticks (species of *Antricola*) have nonfunctional mouthparts and apparently do not feed on blood (Ribeiro et al. 2012).

34.1.1 Anemia

Tick infestation can cause significant blood loss, mainly when a great number of individuals infest a young and/or undernourished animal host. *Dermacentor albipictus*, called the winter tick in North America, can cause anemia, loss of winter coat due to excessive grooming behavior and even death of elks and moose. It has been estimated that females of the giant sloth tick *Amblyomma varium* can swallow approximately 20 ml and weigh up to 5 g when engorged. Death of hosts of several species caused by tick blood sucking has been reported (Balashov 2007).

34.1.2 Skin Lesions

Tick bites may cause severe skin lesions on companion animals (dogs and cats) and livestock. Multiple lesions caused by tick bites can cause a

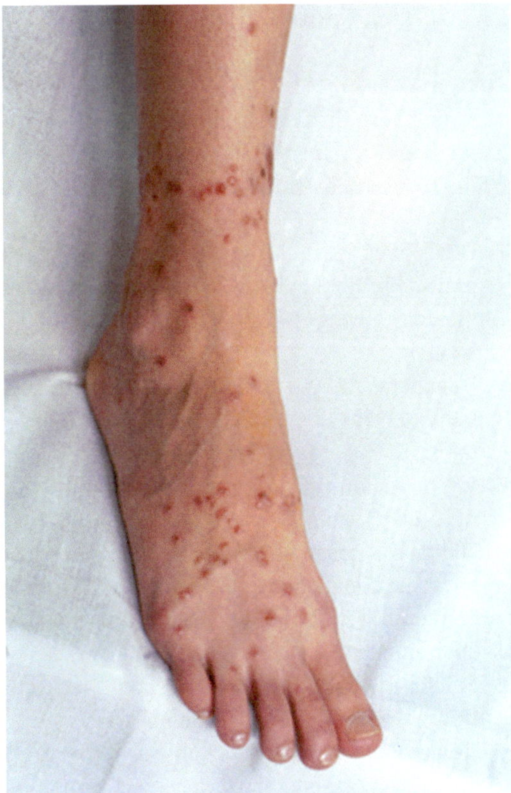

Fig. 34.3 Reaction to *Amblyomma* tick bite (courtesy of João Luiz Costa Cardoso)

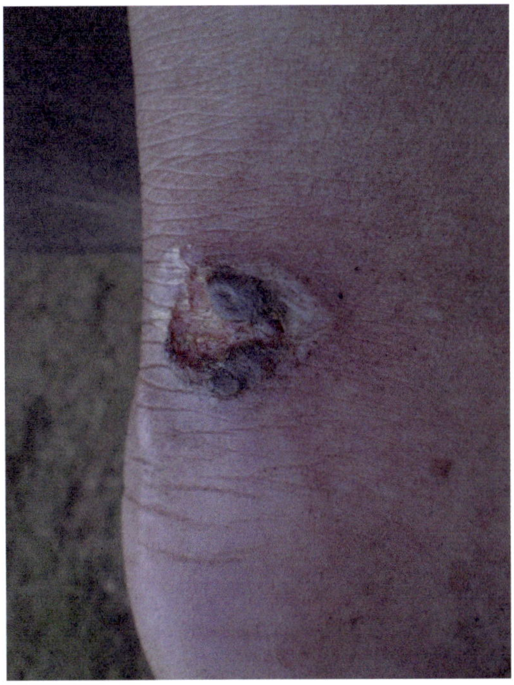

Fig. 34.4 Reaction to tick bite (courtesy of João Ricardo Martins)

thickening of the skin and devaluate the leather, causing a great economic impact on the livestock industry in some countries. They can also facilitate the occurrence of myiasis (maggot infestations).

Tick bites may also lead to local inflammatory reaction (Figs. 34.3 and 34.4) and favor secondary bacterial infection. Prompt and proper tick removal is important to reduce the risk of pathogen transmission and secondary bacterial infections (Dantas-Torres 2007).

34.1.3 Paralysis and Other Forms of Toxicoses

Tick paralysis (TP) is a disease syndrome caused by some tick species, during blood feeding, when they inoculate toxins present in their saliva, inducing an ascendant flaccid paralysis, sometimes evolving to respiratory paralysis and even death, if the tick is not timely removed. Before the onset of the actual paralysis, there may be a period of vague complaints, irritability, tiredness, pain, or paresthesia in the lower extremities for 12–24 h (Rose 1954).

Nearly 70 tick species are reputed as capable of inducing paralysis (reviewed by Hall-Mendelin et al. 2011; Otranto et al. 2012), the most important being *Ixodes holocyclus* in Australia; *Dermacentor andersoni*, *Dermacentor variabilis*, and *Argas radiatus* in North America; *Ixodes rubicundus* in South Africa; *Rhipicephalus evertsi evertsi* and *Argas walkerae* in Ethiopia; and *A. radiatus* in the Nearctic region of North America. Adult females are usually involved, and the toxicity in the saliva increases with time of feeding.

Ixodes holocyclus can cause paralysis in humans, dogs, cats, sheep, cattle, goats, pigs, and horses in Australia (Hall-Mendelin et al. 2011). It has been estimated that 10,000 companion animals are affected annually in this country. A glycoprotein called *holocyclotoxin* has been isolated and characterized from the saliva of *I. holocyclus*. This toxin inhibits transmitter release at neuromuscular junctions and the release of acetylcholine (Cooper and Spence 1976).

Clinically, tick paralysis is characterized by lethargy, weakness and unsteadiness of gait, ataxia, loss of appetite, dilated pupils, ascending symmetrical paralysis, slurred speech, and depressed deep tendon and gag reflexes (Hall-Mendelin et al. 2011 and references cited therein). Several other signs such as labored breathing and bradycardia have also been reported (Grattan-Smith et al. 1997; Hall-Mendelin et al. 2011). The diagnosis is usually made based on clinical signs and the presence of ticks. Differential diagnoses may include the Guillain–Barré syndrome and Lyme disease. The patient usually recovers within a day after tick removal. Eventually, the patient's clinical condition may further deteriorate after tick removal and respiratory support may be required in an intensive care unit. While tick antiserum is available, it may cause acute allergic reactions.

Tick toxicoses have been reported in livestock around the world. For instance, infestation by *Ornithodoros savignyi* (the sand tampan) was reported to kill cattle in Africa, only 6 h of exposition in an infested pen (Mans et al. 2004). Sweating sickness, with two similar but milder forms (*Mhlosinga* and *Magudu*), has been reported in cattle and sometimes in other artiodactyls from several regions of Africa and from Sri Lanka and India. These forms of toxicoses are induced by *Hyalomma* ticks and characterized by a profuse moist eczema, normally observed after 5–7 days after tick infestation. *Rhipicephalus appendiculatus* (the brown ear tick) causes a leucocytotropic disease in cattle, followed by relapses of tick-borne diseases like babesiosis, anaplasmosis, and spirochetosis.

Egg toxins (ixovotoxins) are present in larvae of several species of ixodid ticks, being injected by tick saliva and causing, by the inhibition of proteases, hyperesthesia and anorexia, serous nasal and eye discharge, apparent paresis (different from TP), and loss of voice in experimentally infested guinea pigs.

Although trials for the production and utilization of antitoxins have been developed, prompt tick removal is essential. Protection against bites and control of ticks, to be detailed below, are also important.

34.1.4 Tick Control

Ticks are very resistant to environmental conditions and very difficult to control. Some recommendations for personal protection (Stafford 2004, 2005) are listed below:

- Boots or at least closed shoes and long pants, preferably light colored, to make easier to see the ticks, always tucked into the socks, should be used.
- Repellents (e.g., DEET, permethrin, and picaridin) may be applied on clothes and/or on the skin, as per label recommendations.
- When walking in a trail, keep to the center of the trail, to minimize contact with adjacent vegetation, where ticks are usually present. Examine companion animals and clothes when coming back home after walking in wooded environments. Since common laundry is not efficient to kill ticks, clothes should be carefully inspected and washed in hot water.
- Remove the tick using a thin-tipped tweezers or forceps, grasping the tick as close to the skin surface as possible. Pull the tick upward with steady, even pressure. This should remove the tick with the mouthparts intact. If the tick hypostome remains in the skin, an inflammatory reaction may develop. The tick bite area must be disinfected. The precocious removal of the tick is particularly important, also for reducing the risk of pathogen transmission. Preserve the tick alive or in 70–80 % ethyl alcohol for identification.
- Environmental management includes reduction of tall vegetation around houses, playgrounds, and other areas used by humans, trying to promote exposition to sunlight and reduce humidity. This may be effective to reduce the population of some, but not all, tick species in the peri-domestic environment.

Tick control strategies may vary according to tick and animal species. It may also vary according to particular situations, such as the availability of some products in local veterinary markets. Even though alternative methods such as anti-tick vaccines and biological control have been extensively studied, chemical control remains the most important strategy. Several drugs are available, formulated as collars, stop-on, pour-on, sprays, dips, and shampoos. For companion animals, pyrethroid-impregnated collar may protect dogs from tick bites for up to 8 months.

34.2 Mites

34.2.1 Blood-Feeding Mites

Some species of mites have adapted to blood feeding, causing not only disturbance and blood depletion but also being able to transmit pathogens to animals and humans. Dermanyssidae (e.g., *Dermanyssus*) and Macronyssidae (e.g., *Ornithonyssus*) (Fig. 34.5) have been implicated as vectors of viruses, bacteria, and helminths (reviewed by Moro et al. 2005).

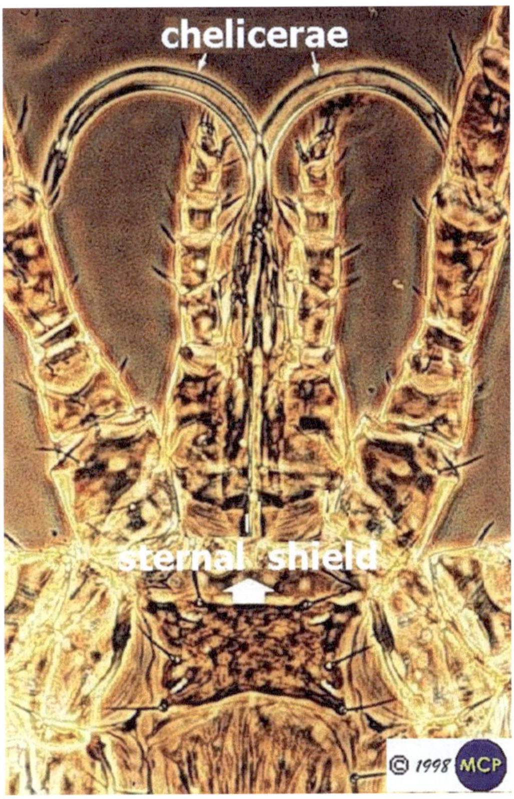

Fig. 34.5 Chelicerae of genus *Ornithonyssus*, much stouter than those of *Dermanyssus*, and the chelae are easily visible under ordinary magnification (courtesy of Marcelo Campos Pereira)

Dermanyssus gallinae (the red mite) infests poultry and is the most important ectoparasite affecting egg-laying chickens in several countries, causing disturbance and anemia. With a very short 7-day cycle, large populations can rapidly be established in poultry houses. Humans[1] and several other animals can occasionally be bitten. The mites remain most of the times hidden in the cracks and crevices of the poultry house and the control still mostly based on the use of synthetic acaricides, although resistance and treatment failure are widely reported (Sparagano 2009). Other methods such as heating poultry houses between flocks, lighting (short-cycle intermittent light/dark periods), and use of inert substances (diatomaceous earth, kaolin, and silicas) may be useful as well. Hygiene practices including the cleaning of houses with water can greatly reduce the number of mites and mite eggs, but this should be done in association with other measures, including application of acaricides.

Ornithonyssus sylviarum (the northern fowl mite) and *O. bursa* (the tropical fowl mite) are bird-associated mites. They are very important for egg-laying hens, with a reduction in profit of U$0.07–0.10 per hen in a 10-week period and a calculated reduction, for 28,000 hens, of U$1960 to U$2800 (Mullens et al. 2009). Spending more time on the hosts, their control is mostly directed to them, with higher risk for contamination of meat and eggs. Since feathers are repellent for water solutions and powder, the use of lambda-cyhalothrin in oil solution is better to prevent contamination of cages and workers, but care should be taken to avoid the permanence of residues in animals and eggs.

Some bloodsucking mites from rodents such as *Liponyssoides sanguineus* (formerly *Allodermanyssus sanguineus*) and *Ornithonyssus bacoti* can bite humans and transmit, respectively, *Rickettsia akari* and *R. typhi* (see Chap. 12).

[1] A *nuit blanche* in a hammock, after trying to collect sand flies in a small henhouse, in 1985, was never forgotten (CBM).

Ornithonyssus bacoti can cause serious disturbance for rats in animal facilities, and permethrin (sprayed in the installations and in cotton balls weekly furnished in mouse cages) associated to careful cleaning of all materials may be effective in controlling *O. bacoti* (Hill et al. 2005). The maintenance of rodents as pets may result in human exposure to this mite.

34.2.2 Mange-Causing Mites

Infestation by some mites can result in an inflammatory skin disease generally referred to as mange, characterized by skin lesions, hair loss, and/or itching. Although the clinical presentation may be indicative of the mite species involved, skin scrapings and other methods (e.g., adhesive tape) should be used for collecting the mites and confirming the specific diagnosis.

34.2.2.1 Sarcoptic and Notoedric Mange

Sarcoptic mange or scabies is caused by *Sarcoptes scabiei*, a worldwide-distributed burrowing mite that may affect several animal species. Indeed, about 100 species of wild and domestic mammals have been found infested by *S. scabiei*, including dogs, cats, swine, donkeys, and rodents. However, different varieties may infest distinct hosts, such as *S. scabiei* var. *canis* in dogs and *S. scabiei* var. *hominis* in humans. The transmission among different species is not common and frequently results in a self-limiting infestation. Canids are highly susceptible, including dogs (Fig. 34.6) and wild canids. The disease can be fatal for coyotes and foxes, mostly in outbreak situations, as described in red foxes (*Vulpes vulpes*), which literally led to extinction on a Danish island, and can be transmitted among different carnivore species and pose a threat for some endangered species, like wombats and mountain gorillas (reviewed by Kołodziej-Sobocińska et al. 2013).

In humans, the most common signs and symptoms of scabies are itching, especially at night, and papular itchy rash. Scratching the rash can cause skin sores and favor secondary bacterial infections. Lesions may affect much of the body

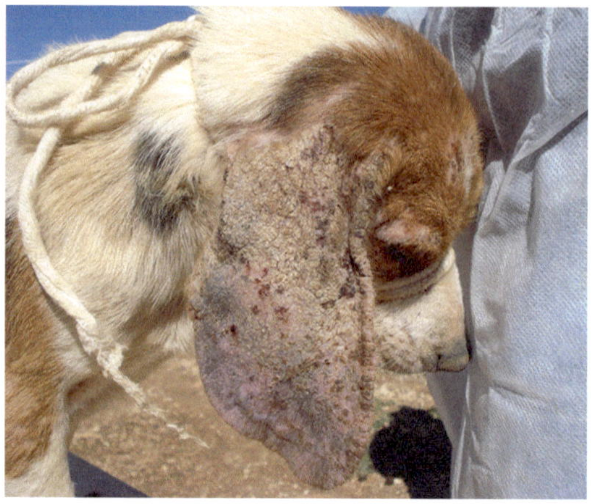

Fig. 34.6 Sarcoptic mange in a dog. Original

or be limited to some sites (e.g., the wrist, elbow, armpit, webbing between the fingers, nipple, penis, waist, beltline, and buttocks). Lesions in the head may be seen in young children but usually not in old children or in adults. Some patients may present thick crusts of skin that contain large numbers of mites and mite eggs. This condition is referred to as crusted (Norwegian) scabies, which is a severe form of scabies usually seen in patients who are immunocompromised, elderly, disabled, and/or debilitated. They are very contagious to other persons and should be treated timely to avoid the risk of outbreaks. Transmission is through direct skin-to-skin contact or less frequently by fomites (e.g., clothes). Sexual transmission may also occur, especially in sexually active young people.

Treatment of scabies in humans is commonly performed with topical acaricides (e.g., permethrin, lindane, and benzyl benzoate) or with oral ivermectin (not approved by FDA for the treatment of scabies). Permethrin is the drug of choice in the USA. Two (or more) applications of a permethrin cream 5 %, each about a week apart, may be needed, particularly in patients with crusted scabies. Oral ivermectin (200 μg/kg) is safe and effective, also for treating crusted scabies. Two or more doses (at least 7 days apart) may be needed. Topical 12.5 % benzyl benzoate may present a higher cure rate than oral ivermectin (single dose). Clothing, towels, and bedclothes used by the infested person in the 48 h prior to treatment should be laundered using the hot cycle or dry-cleaned (Mumcuoglu and Gilead 2008).

Treatment of sarcoptic mange in dogs is usually performed with selamectin and topical moxidectin/imidacloprid, which are approved for this use. High-dose ivermectin has been widely used but are not usually recommended due to toxicity in ivermectin-sensitive dogs and due to treatment failures.

Notoedric mange is caused by *Notoedres cati*, a close relative of *S. scabies*. This mite is a parasite of cats, rats, rabbits, and occasionally humans. Cats may present with skin lesions in the pinna of the ears, face, paws, and hindquarters. Typically, lesions include alopecia, hyperkeratosis, gray crusts of the skin, and pruritus, often accompanied by secondary pyoderma. Several drugs are reputed effective, including ivermectin (200–300 μg/kg), fipronil, and selamectin.

34.2.2.2 Psoroptic Mange

Psoroptic mange is caused by the scab mite *Psoroptes ovis*, which can affect cattle, sheep, goats, rabbits, and horses. Traditionally, the genus *Psoroptes* used to include five species (i.e., *P. equi*, *P. ovis*, *P. cuniculi*, *P. cervinus*, and *P. natalensis*), but some authors have argued for their conspecificity and grouped them into a single taxon, i.e., *P. equi* (Zahler et al. 2000). A recent taxonomic revision of the genus argued that, based on the rules of priority, the correct name should be *P. ovis* (O'Connor and Klimov 2015).

The disease is usually characterized by the presence of papules, scales, and crusts (excoriation), which sometimes can cover several parts of the body. Weight loss and decreased milk production and even death may occur in cattle. In sheep, the mites occur almost exclusively in woolly parts of the body, sometimes invading ears; animals can become emaciated and anemic. In goats, it starts on the ear and spreads on adjacent regions, sometimes on all the body. The mites can cause *otitis externa* in horses, usually accompanied by head shaking.

Treatment of psoroptic mange can be done with topical, oral, or injectable drugs. Spray dipping (useful for small herds) and vat dipping are also used, though time consuming and expensive. Drugs used include doramectin, ivermectin, moxidectin, and many others. The treatment of choice will depend on drugs available in each region and also to farmers' preference to particular modes of administration.

34.2.2.3 Chorioptic Mange

Chorioptic mites can infest cattle, small ruminants, and horses worldwide. For a long time, the taxonomy of the genus *Chorioptes* has been an issue of long debate (Zahler et al. 2001; Lusat et al. 2011; Bochkov et al. 2014). Currently, the genus is thought to include six valid species (*C. bovis*, *C. panda*, *C. texanus*, *C. crewei*, *C. mydaus*, and *C. sweatmani*), four (*C. bovis*, *C. panda*, *C. texanus*, and *C. sweatmani*) of which have recently been differentiated also by robust genetic data (Bochkov et al. 2014).

Infested cattle and small ruminants are usually symptomless, but when present, clinical signs include rubbing and scratching of the affected area (base of tail, perineum and the udder, as well as legs) and damage to the hide, which is often not observed until after slaughter. Chorioptic mange is more evident in affected horses, which usually shake, scratch, rub, stamp, and bite the affected areas regularly. Common signs include crusty skin lesions on the pastern, fetlock, and cannon as well as thickening of the skin and hair loss in chronic cases. The disease is more common during colder months, particularly autumn and winter. Presumptive diagnosis can be confirmed by skin scrapings from the edge of a fresh lesion and observation of mites under a microscope. For the treatment, the affected area should always be clipped and treated topically or systemically. In cattle, pour-on doramectin, eprinomectin (both pour-on and injectable formulations), and pour-on moxidectin are effective and approved in the USA. For horses, topical fipronil is effective in killing the mites, though not specifically approved for this use. Lime sulfur dip is also effective against chorioptic mites and should be applied according to label directions for species-specific dilution. Topical antibiotic and anti-inflammatory ointment may also be applied, and horses with deep secondary bacterial infection may require oral antibiotics.

34.2.2.4 Otodectic Mange

Otodectes cynotis (the ear mite) can be found in dogs, cats, rabbits, and ferrets worldwide. Clinical signs of otodectic mange are shaking of the head and scratching at the ears, associated with inflammation of the ear canals. Accumulation of cerumen and serous to purulent exudates may be frequent, especially when associated with secondary bacterial infections. Diagnosis is based on clinical signs and observation of mites on an otoscope or on swabs of the ear canal. Topical selamectin and moxidectin/imidacloprid are efficacious for ear mites in both dogs and cats. Cleaning of the ear canal is also recommended.

34.2.2.5 Demodectic Mange

Demodex mites are normal inhabitants of the skin of mammals but may eventually overgrow and cause disease. These mites have short legs and an elongated form, being easily distinguished from other mange-causing mites. Humans may be infested by *Demodex folliculorum* and *D. brevis*, cats by *D. cati* and *D. gatoi*, and dogs by *D. canis*, *D. injai*, and another candidate species designated as *D.* sp. "cornei." A probable new species of *Demodex* from cats, differentiated by molecular biology, was recently reported (Ferreira et al. 2015).

In humans, the prevalence of *Demodex* mite infestation may reach 100 % in some age groups, but most individuals are asymptomatic. Some patients may develop skin lesions that are usually caused by an overgrowth of mites associated with immune suppressive conditions (corticosteroid, cytostatic therapy, malignant neoplasia, hepatopathies, lymphosarcoma, and HIV infection) (Ken et al. 2014). Overgrowth of *Demodex* mites has also been associated with suppurative or granulomatous inflammation as well as with other skin lesions such as such as rosacea, pityriasis, and blepharitis. The treatment of human demodicosis is usually performed with topical products (e.g., crotamiton cream, permethrin cream, and metronidazole). In some cases (e.g., severe skin lesions, HIV coinfection), oral ivermectin may be recommended.

Canine demodicosis may be localized or generalized. Localized demodicosis usually occurs in puppies less than 6 months of age and is usually characterized by a self-limiting disease that resolves spontaneously without treatment. Normally dogs with localized demodicosis present a mild, patchy alopecia on the head or limbs, with no pruritus. Generalized demodicosis, however, is a moderate to severe disease, particularly in adult dogs and is usually associated to an overgrowth of mites attributable to occur an underlying systemic disease or immunodeficiency (Singh and Dimri 2014). Clinically, generalized demodicosis is characterized by alopecia (Fig. 34.7) and erythema, which are often accompanied by secondary superficial or deep pyoderma, which may lead to pruritus (Fig. 34.8). Lymph nodes may be enlarged.

Localized demodicosis in cats is characterized by alopecia and crusts, scaling around the face, neck, and eyelids, with varying degree of

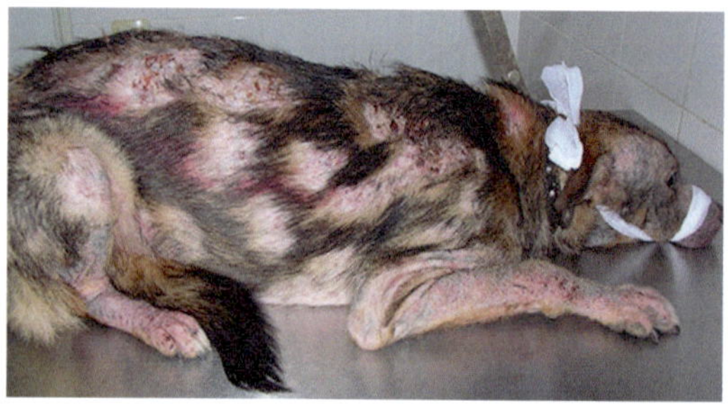

Fig. 34.7 Mongrel dog with a generalized demodectic mange (with permission from Kusnetsova et al., Elsevier, ©2012)

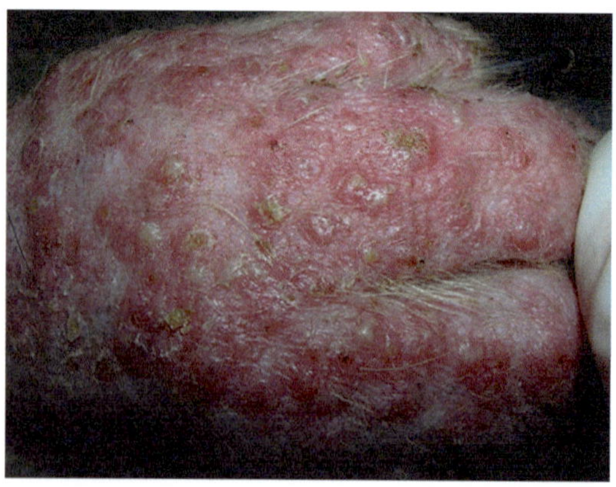

Fig. 34.8 Pad of the dog of previous figure (with permission from Kusnetsova et al., Elsevier, ©2012)

hyperpigmentation. Generalized demodicosis by *D. cati* is usually associated with underlying systemic disease (Foley 1995), differently from what occurs with *D. gatoi* (not associated with underlying disease). Treatment for cats may be performed with lime sulfur dips (weekly for 6 weeks with 3.1 % solution; skin irritation may occur), oral ivermectin (once weekly at 0.3 mg/kg for four consecutive weeks; side effects may occur), and topical amitraz solution (at 0.0125–0.025 %, every 5–7 days for 4–6 weeks; side effects may occur).

Dogs with localized demodicosis usually recover spontaneously without treatment. On the other hand, cases of generalized demodicosis may require extended (sometimes for several months) and aggressive therapy. Any underlying disorders should also be investigated and treated accordingly. Amitraz dip at 250 ppm (every 2 weeks) is approved for the treatment of generalized demodicosis. Other treatment options include high-dose oral ivermectin, oral milbemycin oxime, topical moxidectin, and injectable doramectin. Care should be taken to avoid the use of ivermectin in sensitive breeds (e.g., Collies, Shetland Sheepdogs, and other herding breeds).

34.2.2.6 Furfuraceous Mange (Hairclasping Mites or "Walking Dandruff" Mites)

Some non-burrowing mites can cause skin desquamation, producing a mobile dandruff, in several animals (rats, cats, and dogs), sometimes also affecting humans. Species of *Cheyletiella* have been incriminated as causing this mange in dogs (*C. yasguri*) and cats (*C. blakei*). Treatment by single dose of topical moxidectin is usually efficient.

The cat fur mite *Lynxacarus radovskyi* causes a similar disease in cats, which damages hairs and causes desquamation, sometimes affecting their entire body. Transmission may occur easily between co-housed cats; all cats may be affected in a house or cat shelter. The severity of pruritus is variable and apparently not related to the incidence of mites. Mites and eggs may be numerous and give "peppered" appearance to the cat's coat. Treatment with ivermectin at 300 μg/kg (IM), administered subcutaneously, is effective. Pyrethrin products (e.g., pyrethrin-based miticidal shampoo) and lime sulfur dips are also effective.

34.2.2.7 Knemidocoptic Mange

Knemidocoptes mites cause mange in several Passeriformes (e.g., canaries, sparrows, robins, etc.), Galliformes (e.g., chickens, turkeys), and Psittaciformes (e.g., parrots, parakeets, etc.). Deforming lesions occur in legs and head, affecting circulation and sometimes causing death. The mites burrow into the feather follicles and stratum corneum, mostly in face, feet, and cere, feeding on keratin. Unfeathered regions of the body (beak, eyelids, legs, and vent) are most commonly affected. In canaries and other passerines, lesions are located mostly in the legs, and in psittacines the disease may break out several months after contagion. Lesions can facilitate secondary bacterial infection and cause lameness. In pigeons, *Knemidocoptes laevis* causes mange known as "depluming scabies"; the birds pluck the feathers until becoming deplumed.

Unfeathered areas of the body infested by *Knemidocoptes* can have a honeycombed proliferation of tissue, with characteristic holes (Worell 2013), but skin scrapings must be performed to confirm the diagnosis. Although transmission occurs by direct and indirect contact among birds, susceptibility varies, possibly influenced by genetics. Treatment with spot-on solutions of ivermectin is usually effective.

34.2.3 The Air-Sac Mite and the Cyst Mite of Chickens

Many mites parasitize birds, some of them causing disease (Proctor and Owens 2000; Shivaprasad 2002). *Cytodites nudus* (the air-sac mite) parasitizes air sacs and lungs of chickens, turkeys, and several other birds (e.g., pheasants, pigeons, canaries, and mallards). The mites may be noticed as white spots on the bronchi, lungs, air sacs, and abdominal organs. When in large numbers, they can cause weakness, weight loss, pneumonia, peritonitis, pulmonary edema, and eventually death of affected birds. Secure diagnosis can only be done by necropsy and histopathological examination of the lung, observing the 0.5 mm mites and immunological reaction. Treatments reported in the literature included ivermectin, a nearby dichlorvos pest strip, topical moxidectin, and a pyrethrin/piperonyl butoxide spray. Other treatment options such as administration of malathion by aerosol have been reported as effective.

Laminosioptes cysticola (the fowl cyst mite, the subcutaneous mite, or the flesh mite) is a parasite found in subcutaneous tissues of chickens and turkey, where they cause the formation of calcified nodules. Generally, these mites are nonpathogenic but may reduce the value of carcass when in large numbers. Therapy and prophylaxis are unknown, but destroying affected birds has been indicated to control the prevalence of infestation within the flocks.

34.2.4 Respiratory Allergy Caused by Dust Mites

Dust mites feed mostly on skin scales and other detritus in the floor and in blankets and pillows, and their bodies, fecal matter, exuviae, and possibly saliva can be aspirated from dust, causing respiratory allergy and atopic dermatitis (Arlian and Platts-Mills 2001). Although they are more frequently found in carpets, fabric-covered furniture (sofas, chairs), and mattresses in homes, they can live and breed in any place with adequate food and water conditions. Clothes probably serve as a vehicle for transporting mites, but not for reproduction, and they have also found in automobile seats of persons whose homes contained live mites. Stuffed toys seem to have small quantities of mites. Mites and allergens have been reported in buses, cinemas, churches, banks, libraries, hospital, museums, and hotels (Arlian et al. 2002).

Humidity, high temperatures, low ventilation, and quantity of dust are favorable for proliferation of mites in houses. The quantity of mites and the access of patients to them are important for the development of allergy. Evaluation of mite occurrence and density, by the collection of mites or allergens, must be done by the collection of several samples in each house, and comparison among houses must consider seasonal fluctuation.

Ideal conditions for both mites are 22 °C and 75 % relative humidity (RH), and populations are mostly regulated by the seasonal fluctuation on the last. Although tolerance to low RH depends on temperature, if RH is maintained fewer than 51 % in a house, dust mites will die from dehydration. In dry climates, they cannot survive, but the utilization of evaporative coolers provides adequate conditions for their survival. The quantity of hours (if possible >8 h) of low RH is important, and *Dermatophagoides farinae* is more tolerant to low RH than *D. pteronyssinus*.

Tolerance to high temperatures is important for the control. The exposition of mites to water at 50 °C kills all *D. farinae* in some minutes. The addition of laundry products at this temperature, but not at 35 °C, is efficient to kill all mites. Weekly washing of bed linens at warm water will reduce mite populations, and, since allergens are water soluble, they will be removed.

Most dust mites belong to the family Pyroglyphidae but also to Acaridae, Glyciphagidae, and Chortoglyphidae. Although occurrence of dust mites has been mostly studied in Europe and the USA, they are widely distributed in the world. The most frequent species in temperate regions are *D. farinae*, *D. pteronyssinus*, and *Euroglyphus maynei*. In tropical and subtropical areas, *Blomia tropicalis* and *D. pteronyssinus* are the most frequent species. *Lepidoglyphus destructor* and *Tyrophagus putreus* are frequent in farms in Austria, and a great diversity of species may be found in houses and annexes. Some storage mites, like the above cited and *T. putrescentiae*, can also cause allergy.

Dust mites can cause rhinitis, asthma, and atopic allergy in humans and several domestic animals. Allergy, manifested as rhinitis, asthma, and/or atopic/eczema dermatitis syndrome (AEDS), causing sleeplessness, can impair the quality of life. The prevalence of asthma in the USA and several other countries has been increasing, and several studies have reported a strong association between allergy to dust mites and these diseases (Millián and Díaz 2004). Allergic rhinitis and asthma are frequently associated to atopy, characterized by a tendency to produce IgE antibodies against normally harmless substances.

Besides humans, dogs have been shown to develop allergy to dust mites, and an allergen (Der F 15) was shown to be important for these animals. In the presence of *D. farinae* and *D. pteronyssinus*, dogs present more positive skin tests for the last, while humans present for both. In a study of 150 dogs in France, 80 % presented reaction to at least one mite, with predominance of *D. farinae*, followed by *D. pteronyssinus*.

Treatment by antihistamines and other usual medicines can improve the health conditions of allergic patients, but the long-term maintenance is not convenient. Sublingual immunotherapy (SLIT) for desensitization against dust mites and other allergens has been used with good results (Tonnel et al. 2004; Wilson et al. 2005).

Reducing the exposure by controlling mites and their detritus is essential for the solution of the problems caused by dust mites (Arlian and Platts-Mills 2001). Dust mites may be controlled by reduction of RH, obtained by air conditioning, dehumidifier or ventilation, and exposition of the house to sunlight. Mattresses, carpets, and other house materials must be exposed for several hours to direct sunlight, as frequently as possible. Foam mattresses tend to have more mites than spring ones. Vacuum aspiration, ensuring that vacuum has good quality (preferably double) bags, reduces the mite population and the quantity of allergens. Reduction of dust in houses is highly desirable to prevent asthma crises and to reduce need of treatment. Weekly cleaning of houses is important, and utilization of bleach was shown to be useful for the reduction of asthma, including sensitization by dust mites (Nickmilder et al. 2006). Casing pillows and covering beds with specially manufactured protective casing, made from plastic, vapor-permeable materials, finely woven fabrics, or nonwoven materials, with pore size less than 10 μm, to block both allergens and mites after careful cleaning and reduction of RH is useful. Replacement of blankets has a limited effect, because they become infested in a short time. Weekly laundry of bedding at temperatures ≥55 °C is efficient, and the utilization of detergents and products containing chlorine bleach increases mortality of mites. Laundering at lower temperatures does not kill all mites and transfers mites for clean clothes but reduces the quantity of allergens. Dry cleaning is efficient in killing mites but does not destroy all allergens. If possible, the quantity of carpets and similar materials, which can host mites and their detritus, in the house, mostly in humid climates must be reduced. Application of insecticides in houses has a poor effect on mites and is dangerous.

References

Arlian LG, Morgan MS, Neal JS (2002) Dust mite allergens: ecology and distribution. Curr Allergy Asthma Rep 2:401–411

Arlian LG, Platts-Mills TAE (2001) The biology of dust mites and the remediation of mite allergens in allergic disease. J Allergy Clin Immunol 107(3 Suppl): S406–S413

Balashov YS (2007) Harmfulness of parasitic insects and acarines to mammals and birds. Entomol Rev 87:1300–1316

Bochkov AV, Klimov PB, Hestvik G et al (2014) Integrated Bayesian species delimitation and morphological diagnostics of chorioptic mange mites (Acariformes: Psoroptidae: *Chorioptes*). Parasitol Res 113:2603–2627

Cooper BJ, Spence I (1976) Temperature-dependent inhibition of evoked acetylcholine release in tick paralysis. Nature 263(5579):693–695

Dantas-Torres F (2007) Rocky Mountain spotted fever. Lancet Infect Dis 7:724–732

Dantas-Torres F, Bruno B, Chomel BB, Otranto D (2012) Ticks and tick-borne diseases: a One Health perspective. Trends Parasitol 29:516

Ferreira D, Sastre N, Ravera I et al (2015) Identification of a third feline *Demodex* species through partial sequencing of the 16S rDNA and frequency of *Demodex* species in 74 cats using a PCR assay. Vet Dermatol 26:239–253

Foley RH (1995) Feline demodicosis. Small Anim Parasitol 17:481–486

Grattan-Smith PJ, Morris JG, Johnston HM et al (1997) Clinical and neurophysiological features of tick paralysis. Brain 120:1975–1987

Hall-Mendelin S, Craig SB, Hall RA et al (2011) Tick paralysis in Australia caused by *Ixodes holocyclus* Neumann. Ann Trop Med Parasitol 105:95–106

Hill WA, Randolph MM, Boyd KL et al (2005) Use of permethrin eradicated the tropical rat mite (*Ornithonyssus bacoti*) from a colony of mutagenized and transgenic mice. Contemp Top Lab Anim Sci 44:31–34

Jongejan F, Uilenberg G (2004) The global importance of ticks. Parasitology 129:S3–S14

Ken KM, Shockman SC, Sirichotirana M et al (2014) Dermatoses associated with mites other than *Sarcoptes*. Semin Cutan Med Surg 33:110–115

Krantz GW, Walter DE (2009) A manual of acarology, 3rd edn. Texas Tech Univ Press, Lubbock, 807 p

Kusnetsova E, Bettenay S, Nikolaeva L et al (2012) Influence of systemic antibiotics on the treatment of dogs with generalized demodicosis. Vet Parasitol 188:148–155

Lusat J, Bornstein S, Wall R (2011) *Chorioptes* mites: re-evaluation of species integrity. Med Vet Entomol 25:370–376

Mans BJ, Gothe R, Neitz AWH (2004) Biochemical perspectives on paralysis and other forms of toxicoses caused by ticks. Parasitology 129:S95–S111

Millián E, Díaz AM (2004) Allergy to house dust mites and asthma. P R Health Sci 23:47–57

Moro V, Chauve C, Zenner L (2005) Vectorial role of some Dermanyssoid mites (Acari, Mesostigmata, Dermanyssoidea). Parasite 12:99–109

Mullens BA, Owen JP, Kuney DR et al (2009) Temporal changes in distribution, prevalence and intensity of northern fowl mite (*Ornithonyssus sylviarum*) parasitism in commercial caged laying hens, with a comprehensive economic analysis of parasite impact. Vet Parasitol 160:116–133

Mumcuoglu KY, Gilead L (2008) Treatment of scabies infestation. Parasite 15:248–251

Nickmilder M, Carbonelle S, Bernard A (2006) House cleaning with chlorine bleach and the risks of allergic and respiratory diseases in children. Pediatr Allergy Immunol 18:27–35

O'Connor BM, Klimov PB (2015) Review and resolution of some nomenclatural issues regarding the genus *Psoroptes* (Acari: Psoroptidae), scab-mites of domestic and wild animals. Exp Appl Acarol 66:337–345

Otranto D, Dantas-Torres F, Taralloa VD et al (2012) Apparent tick paralysis by *Rhipicephalus sanguineus* (Acari: Ixodidae) in dogs. Vet Parasitol 185: 325–329

Proctor H, Owens I (2000) Mites and birds: diversity, parasitism and coevolution. Trends Ecol Evol 15:358–364

Ribeiro JMC, Labruna MB, Mans BJ et al (2012) The sialotranscriptome of *Antricola delacruzi* female ticks is compatible with non-hematophagous behavior and an alternative source of food. Insect Biochem Mol Biol 42:332–342

Rose I (1954) A review of tick paralysis. Can Med Assoc J 70:175–176

Shivaprasad HL (2002) Pathology of birds—an overview. In: Davis CL (ed) Foundation conference on gross morbid anatomy of animals, 48 p

Singh SK, Dimri U (2014) The immune-pathological conversions of canine demodicosis. Vet Parasitol 203:1–5

Kołodziej-Sobocińska M, Zalewski A, Kowalczyk R (2013) Sarcoptic mange vulnerability in carnivores of the Białowieża Primeval Forest, Poland: underlying determinant factors. Ecol Res 29:1118. doi:10.1007/s11284-013-1118-x

Sparagano O (2009) Control of poultry mites: where do we stand? Exp Appl Acarol 48:1–2

Stafford KC III (2004) A integrated guide for homeowners, pest control operators, and public health officials for the prevention of tick-associated disease. Connecticut Dept Public Health, New Haven, Available at: www.caes.state.ct.us

Stafford KC III (2005) Tick bite prevention and the use of insect repellents. Connecticut Dept Public Health, New Haven

Tonnel AB, Scherpereel A, Douay B et al (2004) Allergic rhinitis due to house dust mites: evaluation of the efficacy of specific sublingual immunotherapy. Allergy 59:491–497

Wilson DR, Torres Lima M, Durham SR (2005) Sublingual immunotherapy for allergic rhinitis: systematic review and meta-analysis. Allergy 60:4–12

Worell AR (2013) Dermatological conditions affecting the beak, claws, and feet of captive avian species. Vet Clin North Am Exot Anim Pract 16:777–799

Zahler M, Hendrikx WML, Essig A et al (2000) Species of the genus *Psoroptes* (Acari: Psoroptidae). Exp Appl Acarol 24:213–225

Zahler M, Hendrikx WM, Essig A et al (2001) Taxonomic reconsideration of the genus *Chorioptes* Gervais and van Beneden, 1859 (Acari: Psoroptidae). Exp Appl Acarol 25:517–523

Other Irritating Arthropods (Beetles, Bugs, Centipedes, Etc.)

35

Carla de Lima Bicho and Gerson Azulim Müller

Abstract

Environmental changes, mainly caused by increasing urbanization, generate enabling conditions, sometimes not desirable, which favor the proximity between the humans and the population of other animals, including arthropods. Many of these species produce irritant substances responsible for causing important clinical manifestations in their victims. Those manifestations can range from discomfort and pain caused by contact with the toxin to various cutaneous symptoms, including blistering, contact dermatitis and allergic reaction, among others. In many instances, death occurs. Arthropods annoying of public health importance, such as species of insects (class Insecta), centipedes (class Chilopoda) and millipedes (class Diplopoda), which have not been previously discussed, are dealt here. Among the insects, species of the orders Coleoptera (beetles), Hymenoptera (ants, bees and wasps), Lepidoptera (moths) and Hemiptera (bugs haematophagous and non-haematophagous) are detailed. This section aims to provide basic allowances for the identification of these species and emphasises clinical aspects, diagnosis, treatment and prophylactic measures of diseases caused by them.

Keywords

Insect • Chilopoda • Diplopoda • Public health importance • Poison

As referred in the introduction, spiders and scorpions are not included in the book, due to the availability of recent good reviews on problems caused by these arthropods (CBM).

C. de Lima Bicho (✉)
Department of Biology, Center of Biological Sciences and Health, State University of Paraiba, 58429-500 Campina Grande, PB, Brazil
e-mail: clbicho@gmail.com

G.A. Müller
Federal Institute of Education, Science and Technology Farroupilha, Panambi Campus, 98280-000 Panambi, RS, Brazil
e-mail: gecoazul@hotmail.com

Certain day-to-day situations may favour proximity of the human population to other organisms such as arthropods, which can result in somewhat undesirable consequences. This is particularly so when the species involved are arthropods that defend themselves by means of irritating substances which can trigger severe clinical manifestations in their victims. Such manifestations may include discomfort and pain caused by contact with a toxin or even death. The principal aim of

C.B. Marcondes (ed.), *Arthropod Borne Diseases*, DOI 10.1007/978-3-319-13884-8_35

this chapter is thus to provide basic information on the diagnosis and treatment of health problems caused by species of irritating arthropods, as well as making available other details which may be relevant and important in this context.

35.1 Class Insecta

35.1.1 Order Coleoptera

The insects in this order are popularly known as beetles. The name Coleoptera (from the Greek, *koleos* = sheath and *pteron* = wing) refers to the first pair of sclerotized wings or 'elytra', which protect the membranous hindwings when at rest. There are approximately 350,000 described species worldwide, distributed among 175 families, of which approximately 15 are reported in the literature to contain species of medical importance. Here we will consider the health problems triggered by blister-forming species of the most important families (Staphylinidae, Meloidae and Oedemeridae) as a function of the toxins they produce (pederin and cantharidin).[1]

35.1.1.1 Staphylinidae: Pederin

The family Staphylinidae comprises species commonly known as 'rove beetles'. These are easily recognised by their short elytra, which expose more than half of the abdominal segments. There are several genera of medical importance, with species of the genus *Paederus* responsible for most of the reports available in the literature. At least 20 species trigger profiles of dermatitis, of which *Paederus columbinus* and *P. brasiliensis* (South America), *P. sabaeus* (Africa) and *P. fuscipes* (Europe and Asia) (Krinsky 2009a) are among the best known.

The individuals belonging to this genus have a slender, elongated body (about 1 cm in length) and are normally bicoloured (red or orange and black or metallic blue). They are very agile and prefer to crawl rather than to fly. While moving or when attacked, they raise the posterior part of the abdomen in a characteristic defensive posture. They are found in humid localities, with abundant organic material of plant or animal origin and are frequently associated with plantations of maize, beans, rice, potatoes, tobacco or sugar cane. They are predatory, feeding on small insects and other arthropods.

Dermatitis caused by *Paederus* spp. occurs due to the presence of a potent toxin present in the haemolymph, known as pederin. This substance is a crystalline amide, soluble in water and alcohol. It has caustic, blister-forming properties which inhibit to protein synthesis and karyokinesis. Only certain adult females have pederin since they have the necessary endosymbiotic bacteria (*Pseudomonas* spp.) to produce it. Larvae and males can however acquire the toxin through transovarian transmission or ingestion (Ghoneim 2013c).

Clinical Aspects

Release of pederin occurs when the insect is compressed accidentally or touched by the victim when it crawls on the skin. Contact with humans is favoured by the fact that these beetles are attracted to sources of light at night. Being the most exposed parts of the body, the face, neck and arms are the localities most affected. 'Kissing' (or 'mirror image') lesions may also be seen. These occur when there is juxtaposition of an area of the body affected by the toxin with another previously intact surface, such as the inner thighs.

The toxin causes burning, erythema and blistering. The lesion frequently presents a linear trajectory which extends for up to 15 cm in length and corresponds to the area of the skin touched by which the insect from the moment of contact. The zoodermatitis caused by *Paederus* spp. is thus described as 'linear dermatitis'. An erythematous lesion appears within few hours of initial contact with the toxin, and during the next 48 h, the patient may experience a burning sensation and pruritus. Within 48 h of appearance of the erythema, blisters and vesicles develop which coalesce in 2 days. Nearly a week after contact, the blisters and vesicles exfoliate leaving

[1]Chapter 3 cites the utilization of toxins from Chrysomelidae beetles on arrows by San bushmen, that can kill antelopes and giraffe (CBM).

pigmented scars, which may persist for up to 2 months. Several weeks later, a dark scar forms. In the most severe cases, the patient may report localised pain, fever, vomiting, joint pain and adenopathy. The fingers may spread pederin to other areas of the body, such as the genitals and eyes, producing conjunctivitis, irritation, blepharitis and exfoliative keratitis in the latter. In Africa, the conjunctivitis caused by this toxin is known as 'Nairobi eye' (Cressey et al. 2013; Ghoneim 2013c).

Diagnosis and Treatment

This dermatitis requires differential, clinical and histological diagnosis to distinguish it from generalised pustular psoriasis and subcorneal pustular dermatosis. Its clinical appearance may also be confused with other health problems, such as herpes simplex, herpes zoster, contact dermatitis and phytophotodermatitis (Borroni et al. 1991; Cressey et al. 2013).

There is a wide spectrum of histopathology changes, such as epidermal necrosis with intraepidermal blistering during the early stage and acanthosis with mitotic figures in the advanced stage. The presence of acantholytic foci, distant from the site of the lesion, suggests that the toxin may indirectly induce this process (Borroni et al. 1991).

Recognition of indicators such as the occurrence of exposed lesions on areas of the body, linear dermatitis, the presence of 'kissing lesions' and results of histopathological analysis would ensure that the correct diagnosis was made.

If contact has been made with the toxin, it should be cleaned immediately by washing the affected area with soap and running water. Damp compresses of physiological serum or potassium permanganate as well as antimicrobials should be applied to the lesions. The use of topical corticosteroids is also recommended, and systemic antibiotics may be used to control secondary infection. According to certain authors, coriander (*Coriandrum sativum*; Apiaceae) ground up and applied to the cutaneous areas affected at the onset of symptoms, eases the burning sensation and reduces the severity of the profile, as well as evolution of dermatitis. If the eyes are affected, they should be washed with clean water before antibiotics and corticoids are administered.

35.1.1.2 Meloidae and Oedemeridae: Cantharidin

The Meloidae are known as the 'true blister beetle' or 'oil beetles' since when perturbed, they release a toxin which is able to produce vesicles on the skin of the victim. This toxin (cantharidin) causes a condition similar to that produced by species of *Paederus* but with noninflammatory blisters and vesicles.

Cantharidin is a terpene soluble in several organic solvents but practically insoluble in water. It is present in the haemolymph and gonad and synthesised only by the males, before being stored and transferred to the female at the moment of copulation. The female then covers her eggs with the compound to protect them from predators. When disturbed, the adults release a substance from the joints of the legs and/or antennal pores, while the larvae release it from the oral cavity. The quantity of toxin in the body of the insect generally corresponds to 1–5 % of its dry weight (Ghoneim 2013b; Russell et al. 2013).

The beetles of the family Meloidae are widely distributed across the world, although with no species records from New Zealand and most of the islands of Polynesia. They have an elongated body (measuring up to 40 mm in length), with varied colours and patterns. They possess a set of distinctive characters, such as a head that is rectangular when viewed from above, a cylindrical prothorax that is narrower than the head and a base of the elytra, which have a soft texture. The legs are long and slender.

The adults are phytophagous, and many species are potential pests of various crops; the larvae are predatory and feed on eggs of other insects. The dispersal capacity of these insects is reduced, and most species are diurnal.

Nearly 20 species belonging to the genera *Epicauta* (America, Africa and India), *Lytta* (Europe and China), *Mylabris* (India and Nigeria), *Cylindrothorax* (Africa, India and Asia) and *Psalydolytta* (Africa) are known to produce lesions on the human skin if handled (Krinsky 2009a). A study carried out at the beginning of the nineteenth century on the southern European species *Lytta vesicatoria* (Spanish fly) was the precursor for the study of cantharidin, leading to its artificial purification and crystallization (Ghoneim 2013b).

The Oedemeridae have a cosmopolitan distribution and are known as 'false blister beetles' or 'pollen-feeding beetles'. As in the meloids, their body fluids contain cantharidin. They are soft-bodied beetles that generally present brilliant and showy colouration (blue, green, golden) and measure 5–25 mm in length. The characteristics that differentiate them from the Meloidae are that the pronotum is wider anteriorly, tapering towards the elytra. These are elongated and sometimes divergent apically, with rounded extremities. The legs are long and slender.

The larvae feed on dead wood and dry stalks of herbaceous plants. The adults eat pollen and nectar from flowers of various families. With the exception of some species with crepuscular or nocturnal habits, they are frequently observed on flowers during the day. The adults act as pollinators, pollen grains remaining adhered to pubescent regions of the body. The flowering periods of certain plants and emergence of the adults are directly synchronized.

Health problems triggered by species of the genera *Oxycopis*, *Oxacis*, *Alloxacis*, *Sessinia*, *Eobia* and *Thelyphassa* are reported from the USA, America Central, Australia, New Zealand and regions of the Pacific (Krinsky 2009a). Attracted in great number to lights at night, the adults may become a nuisance to those who frequent open-air recreational facilities, such as restaurants and tennis courts. An example of this is the North American species *Oxycopis mcdonaldi* (Arnett 2000).

Clinical Aspects

The cutaneous eruptions caused by cantharidin appear more rapidly (2–3 h after contact) but are less severe than those triggered by pederin. Exposed regions of the body are the most affected.

When ingested or absorbed through the skin, cantharidin is highly toxic to mammals, including man. On ingesting this substance, a victim feels a rapid burning sensation in the mouth and throat, followed by dysphagia, vomiting and haematemesis. Absorption of the toxin is directly related to the quantity of lipids in the digestive tract. Thus, the greater the quantity of fatty food present at the moment of ingestion of the substance, the greater the damage to the gastrointestinal mucosae. Lumbar pains, dysuria, proteinuria and haematuria may all be manifested, as well as coagulopathies and cases of flaccid paralysis, similar to that seen in Guillain-Barré syndrome. Cantharidin may produce renal insufficiency and even death[2]; the lethal dose has been estimated at 1 mg/kg body weight. Gas chromatography and mass spectrometry may be used to confirm poisoning (Ghoneim 2013a).

Treatment

For cantharidin to be dissolved, it is necessary to sterilise the area affected with acetone, ether or alcohol and then wash it with soap and water. Topical steroids may also be applied to the unbroken skin. In case of ingestion of cantharidin, it is recommended that the patient drinks a lot of water. Milk should not be given since it enhances absorption of this substance. Vomiting should not be induced. Chemical adsorption using activated charcoal may be prescribed, although there is still no evidence that cantharidin has the capacity to bind to this material and reduce absorption by the digestive system (Moed et al. 2001).

Blister-forming species of beetle should be recognised and avoided whenever possible. If the insect lands on the skin, it should be carefully flicked off with any object to hand. If contact is made with cantharadin, the affected as well as the hands should be washed thoroughly with soap and water, to prevent other parts of the body from being affected. The use of a mosquito net on a bed, as well as protecting lights with a net, avoids the danger of insect falling on sleepers. It is also

[2]Cantarella, produced from cantharidin (or arsenic), was allegedly utilised by the Borgia family to kill their enemies (CBM).

advisable to carry out minute inspection of the walls and ceilings close to light sources before going to bed.

35.1.2 Order Hymenoptera

35.1.2.1 Formicidae

The members of this family are the ants, eusocial insects whose members are organised into castes (groups of workers, queens and males), with clear division of labour among them. Approximately 13,000 species have been described worldwide, some of which present medical importance.

Triggering of Allergenic Profiles

The genus *Solenopsis* (subfamily Myrmicinae), distributed widely from South and Central America to the southern USA, contains two important species, the 'black imported fire ant' (*S. richteri*) and the 'red imported fire ant' (*S. invicta*). Morphological, biological and behavioural aspects of these species are very similar.

The worker caste is composed of polymorphic wingless individuals, whose antennae have ten segments, the compact clave being formed by the two basal segments. Two distinct nodules are also visible between the propodaeum and gaster (petiole and postpetiole). The colouration of the body varies from pale yellow through reddish brown to shiny black, and its length varies from 3 to 6 mm. The nests of these ants are most often constructed in open-air localities but also around buildings. They occasionally may infest installations and electrical equipment. Their members are omnivorous and opportunistic, that is, they feed on almost all types of plants and animals, as well as a variety of household foodstuffs.

Originally from South America (Brazil and Uruguay), this species invaded North America at the beginning of the twentieth century, probably via the Port of Mobile (Alabama). They now constitute a serious problem in various areas where they have been introduced. Annual costs due to 'imported fire ants' in the USA exceed US 2 billion dollars. Approximately 50 % of residents of the regions infested by these ants are affected at least once per year, with 6–10 % requiring hospital treatment (Della-Lucia 2003).

'Imported fire ants' turn aggressive and attack in large numbers if their nests are disturbed. Defense of the nest is the responsibility of the workers, which are sterile females incapable of producing eggs. The ovipositor has thus been modified in a structure known as a sting, which has an associated venom gland. The substance produced in this gland consists mainly of oleaginous alkaloids, the most important fraction being Solenopsin A which has a cytotoxic, protein-inhibitory action. Although it has little local effect, it is able to trigger allergic reactions in sensitive individuals. The alkaloids destroy the epidermal cells and attract polymorphonuclear neutrophils to the site, resulting in the development of pustules (Conceição et al. 2006).

When defending itself, the ant attaches itself to the skin of victim with the aid of its mandibles, and it, if these are not removed by mean of a rotational movement, is able to inflict painful stings (approximately ten). As a consequence of this, an urticariform papule forms with associated pruritus. Within 24 h, the papule is replaced by a sterile pustule, which heals within a week. However, if the pustule bursts, there may commonly be secondary infections, resulting in abscesses, cellulitis, erysipelas and necrosis of extremities. The allergic process is manifested in the form of urticaria, angioedema, a sensation of thoracic oppression, nausea, vomiting, anaphylaxia and anaphylactic shock and may even end in death of the person affected (Conceição et al. 2006; Haddad and Larsson 2015).

Diagnosis is basically clinical, and histopathological examination may reveal subcorneal pustules similar to those seen in pustulose psoriasis.

Initial treatment involves the application of local cold compresses. Topical corticoids should be prescribed or, in the cases of severe accidents, systemic corticoids that ameliorate pruritus and pain but do not affect resolution of the lesions. The pain may be alleviated with the use of analgesic drugs, and antihistamines taken orally are always indicated (Haddad and Larsson 2015).

As a prophylactic measure, one should avoid areas where there are high concentrations of imported fire ants. Care should be taken to avoid nests when working in areas where they may occur and suitable protective clothing worn. Any nests that are discovered should not be disturbed.

In the USA, the stings of other species of 'fire ants' such as *S. geminata* and *S. xyloni*, as well as the 'harvester ants' (*Pogonomyrmex* spp.), may trigger allergic reactions in humans. The harvester ant is considered to be one of the most venomous of all insects, and its sting generally understood to be the most painful of any North American ant (Hoffman 2010; Russell et al. 2013).

The 'giant tropical bullet' ant, *Paraponera clavata* (subfamily Paraponerinae), is a large species (1.8–2.5 cm in length) that nests in soil, generally at the bases of trees which possess extrafloral nectaries. Nectar and insects (termites) constitute important items of its diet. It presents a wide distribution throughout South and Central America; in Brazil, it is commonly found in the Amazon forest. It may be recognized by the pair of blunt horns located on the first thoracic segment and by the triangular petiole, inclined towards the propodeum. Other popular names for *P. clavata* are 'true tocandira', 'Cabo Verde ant', 'lesser giant hunting ant', 'conga ant' and '24-hour ant'.

Its stings are responsible for causing intense pain which tends to increase over time, persisting for 24 h. The venom contains enzymes (nucleotidase, alkaline phosphatase and phospholipase), haemolytic and procoagulant factors, as well as a neuropeptide known as poneratoxin (PoTX). As a consequence of the action of the venom, the victim develops erythema and local edema that may be accompanied by systemic phenomena (chills, tremors, sweating, vomiting, diaphoresis, lymphadenopathy and cardiac arrhythmias), insensitivity of the affected member and paralysis. They may produce glottal oedema or even death, if not treated in sufficient time (Haddad et al. 2005).

The genus *Dinoponera* (subfamily Ponerinae) is a strictly South American group which contains ants known as 'tocandiras'. In contrast to most species, the nest of this ant is constructed in holes excavated in the soil, to a depth of nearly 2 m, and there are no queens. Its diet is composed of fruit, seeds and small arthropods. Originally from the Amazon region, the 'giant Amazonian ant' (*D. gigantea)* measures approximately 3.0 cm in length and has an extremely painful sting. The victim on being stung perceives the formation of a whitish patch at the site, followed by a small oedema. The pain extends slowly along the entire member, reaching its maximum intensity after about 12 h. After between 24 and 48 h, this pain diminishes and disappears. Edema and erythema are visible at 48 h after contact with the venom. Pallor, chills, nausea, vomiting, lymphadenopathy and tachycardia are other clinical manifestations that may also be observed (Haddad et al. 2005).

Species of the genus *Pachycondyla* (subfamily Ponerinae) are often associated with cases of anaphylaxia, particularly the species *P. chinensis* ('Asian needle ant') and *P. sennaarensis* ('samsum ant'). There are records of allergic and anaphylactic reactions caused by *P. chinensis* in several countries of Asia, in the USA and New Zealand. The species *P. sennaarensis* occurs in Iran, Arabia and the African Tropics, producing dermal lesions and systemic reactions in its victims (Lee et al. 2009; Nikbakhtzadeh et al. 2009).

In Australia, species of the genus *Myrmecia* (subfamily Myrmeciinae) contain large black ants (sometimes over 36 mm long) which are popularly known as 'bulldog ants' due to their aggressive behaviour. These ants possess large eyes and long, slender mandibles. They are found in native forests. The workers tend to be solitary and feed principally on nectar and honeydew. The larvae feed on small insects captured by the workers.

The 'jack jumper ant' (*M. pilosula*, known also as the 'hopper ant' or 'jumping jack') and the 'bull ant' (*M. pyriformis*) are noted for their painful stings and allergenic venom. The symptoms triggered by the stings of *M. pilosula* are very similar to those of 'fire ants'. Patients with severe allergic reaction may have difficulty in breathing and speaking due to swelling of the tongue and throat, tightness in the chest, cough, abdominal pain, nausea, vomiting and loss of consciousness. The 'jack jumper ants' constitute a significant cause of anaphylactic reactions in

Tasmania, sometimes with fatal results. In Queensland, the 'green-head ant', *Rhytidoponera metallica* (subfamily Ectatomminae), is also reported to be responsible for triggering allergic profiles (Klotz et al. 2005).

In Europe, allergic reactions to ants are uncommon, and there are few recorded cases, most of these involving *Formica rufa* (subfamily Formicinae). This species is known as the 'red wood ant' and bites rather than stings. Ants belonging to the genus *Formica* also squirt perceived assailants with formic acid and other nitrogenous compounds that are produced in the venom gland (Klotz et al. 2005).

Transport of Microorganisms

The ants found in hospital environments may transmit various species of bacteria, both gram positive and gram negative. The bacteria isolated from these insects are those related to classic hospital infections, such as *Escherichia coli*, *Pseudomonas* spp. and *Klebsiella* spp. (gram-negative), as well as *Staphylococcus aureus* and *Streptococcus* spp. (gram-positive) (Garcia and Lise 2013).

Nosocomial infections acquired in hospitals constitute a serious public health problem and are among the principal causes of mortality and morbidity worldwide. These infections demand a greater period of hospitalisation of the patient, generating increased expenses due to the extra costs of their treatment. In the USA, nearly two million people contract hospital infections each year, a fact which generates additional costs in the order of five billion dollars each year. Approximately 25 % of hospital infections occur in intensive care units (ICUs), increasing the internment period and total costs of ICU by more than 20 % (Sydnor and Perl 2011). Hospital infections which involve vectors are classified as *exogenous*, since they are caused by microorganisms that do not belong to the patient's own microbiota.

Hospitals are highly propitious localities for the establishment and propagation of ants, due to a number of factors. These include failures in the building structure, the close proximity of human residences favoring the migration of the insects, the packaging of medicines which may serve as hiding places, the circulation of people with belongings which also may harbour ants and the abundance of different types of foodstuffs that provide additional attractions (Zarzuela et al. 2002). However the finding of ants in these localities is not necessarily a sign of dirt due to neglect, given that some species are equally attracted to sterile materials.

Certain characteristics found on the bodies of ants favor the transport of pathogens, such as the presence or absence of setae, the structure of the cuticle and distribution of exocrine glands which facilitate the adherence and the survival of microorganisms (Fontana et al. 2010). The fact that they move so rapidly and normally range over extensive areas suggests that as well as acting as vectors of pathogenic species, foraging ants may also act as important vehicles for the dispersion of drug resistance in these environments.

Interest in the ant x hospital x health human triad was stimulated by studies published in the 1970s on the 'pharaoh ant' (*Monomorium pharaonis*) (subfamily Myrmicinae), in hospitals throughout Europe, when various bacterial species of medical importance were isolated (Beatson 1972).

The 'pharaoh ant', also known as the 'sugar ant', measures 1.5–2 mm in length and is an almost-transparent pale yellow in colour, with a dark brown gaster. Among all the ants considered to be urban pests, it is probably the most widely distributed species in the world (Americas, Europe, SE Asia and Oceania).

The species is polygamous, i.e. it contains many queens, a fact which allows the colony to fragment rapidly into other colonies. It has cryptic habits and its nests may be very small, making its detection even more difficult. It is attracted to fatty or to sugary foods, which may be used as baits to determine its presence. Although *M. pharaonis* is considered to be a pest of the interior of buildings, the search for food (foraging) may also occur outside. The problem of infestations by *M. pharaonis* is related both to variety of pathogens to which it exposed in hospitals and its diminutive size, which allows it to infiltrate the surgical dressings of patients, intravenous tubes and medical instruments (Beatson 1973).

Studies have shown that the diversity of ants in hospitals is higher in tropical than in temperate countries, where *M. pharaonis* predominates. In Brazil and other regions of South America, *Tapinoma melanocephalum* (subfamily Dolichoderinae) and *Paratrechina longicornis* (subfamily Formicinae) are the dominant species in urban and hospital environments (Garcia and Lise 2013; Castro et al. 2015).

Commonly known as the 'ghost ant', *T. melanocephalum* is widely distributed in tropical and subtropical latitudes throughout the world, in the Old and New Worlds. The name 'ghost' is due to the fact that the gaster and appendages (antennae and legs) are almost transparent to the naked eye; the head and thorax are dark brown. Its extremely small size (1.3–1.5 mm long), together with its pale coloration, makes it very difficult to see.

The workers move rapidly and with a jerky motion. When squashed, they release a smell similar to that of rotten coconuts. Because of this disagreeable odour, the species is known as the 'corpse ant' in Malaysia.

Colonies of these ants containing numerous queens (polygamy) occupy spaces that are too small or unstable to support huge numbers, both inside buildings and in the open air. The nesting habits of the 'ghost ant' are similar to those of the 'pharaoh ant'. In houses, 'ghost ant' infestations are confined to environments that provide the high degree of humidity necessary for the survival of this species. The workers may be encountered in kitchens and bathrooms, sinks, counters and floors. Although they have a very diverse diet, they appear to show a preference for carbohydrates.

Morphologically *T. melanocephalum* and *M. pharaonis* are very similar. However, they may be easily differentiated by the petiole and/or postpetiole and by the characteristic coloration of each species. The former species has only a petiole, while 'pharaoh' ant bears both a petiole and postpetiole. Furthermore, in the 'ghost ant', the petiole is not visible dorsally, a detail that distinguishes it from other bicoloured species, which are also larger.

The 'longhorn crazy ant' or 'crazy ant' *Paratrechina longicornis* is so-called due to its rapid, erratic movements. 'Hairy ant', 'black crazy ant' and 'slender crazy ant' are other common names for this species. This ant is found predominantly in the hottest regions of the tropics and subtropics but may also occur in temperate areas, where it is considered to be a household pest. The head, thorax, petiole and gaster are dark brown to black, the body generally measuring 2.3–3 mm in length. Its most distinctive characteristics include the presence of strong, erect whitish or greyish bristles distributed sparsely over the body, as well as 12-segmented antennae lacking a clave but with an extraordinarily long scape (basal segment). The legs of this species are long and slender.

Paratrechina longicornis is omnivorous and apparently presents seasonal preferences in its diet. During the summer months, it exploits foods rich in proteins, while in spring and autumn, it feeds on honeydew (a sweet liquid eliminated by leafhoppers).

This species may nest in several types of environments, both dry and humid. Inside buildings, it may be found under floors or close to sources of heat. The nests may be transported to other localities when conditions become critical. One peculiarity of *P. longicornis* is that the workers do exploit established trails as frequently as other ant species and, commonly forage at great distances from their nests, making their control particularly difficult.

Quali-quantitative microbiological studies have demonstrated the risk that these ants represent to the health of hospital communities, given their capacity to transport pathogenic microorganisms. Species of *Staphylococcus*, *Escherichia*, *Pseudomonas*, *Enterobacter*, *Bacillus*, *Streptococcus* and *Klebsiella* are the bacteria most frequently recorded in association to these ants in hospitals. Other ants may also be encountered, such as the 'Argentine ant' *Linepithema humile* and species of the genera *Solenopsis*, *Pheidole* and *Camponotus* (Garcia and Lise 2013; Castro et al. 2015).

Contamination in hospital environments may involve not only bacteria but also other microorganisms such as fungi. In recent years, there has been an increase in the number of infections caused by fungal species which were not previously considered to be pathogenic. These infections are frequently associated with immunocompromised patients, such as those with neoplastic conditions undergoing treatment and acquired or primary immunodeficiencies as well as recipients of solid organ or bone marrow transplants. Species of *Candida*, *Cryptococcus* and *Aspergillus* constitute the principal causes of systemic infections among this group of patients (Wanke et al. 2000). These genera are frequently found in the bodies of ants which coexist in such environments, such as in *T. melanocephalum*, in *P. longicornis* and in species of *Pheidole* and *Solenopsis* (Garcia and Lise 2013).

The great capacity of ants to maintain associations with bacteria including gram-positive cocci and both gram-negative and gram-positive bacilli, as well as yeast-like and filamentous fungi, is notorious. However, studies that investigate the transportation of fungi by these vectors are much less advanced than research on bacteria and more needs to be done.

35.1.2.2 Apidae (Subfamily Apinae)

In the Americas, for example, hybridisation of European 'honey bees' *Apis mellifera* and *Apis mellifera ligustica* introduced in the Americas in the nineteenth century by European immigrants, with the 'African bee' *Apis mellifera scutellata*, that is much more aggressive and that was introduced on the continent in the mid-twentieth century, produced a lineage of insects known as 'Africanized bees', which retained much of the morphological and behavioral characteristics of *Apis mellifera scutellata*. These bees, because of the range of their distribution (the southern part of South America to the southern region of the USA) and their aggressiveness, are responsible for hundreds of attacks on humans and domestic animals every year, often resulting in death.

Another example is that of *Apis dorsalis*, known as the 'giant honeybee' of Asia, which is considered by experts the most dangerous stinging insect in the world. These bees are found from Vietnam to the forests of India and usually build a single hive to house the colonies, which can be up to three metres in diameter, and, when they see an approaching intruder, they can pursue it for a distance of up to 100 m. In the Vietnam War, this species was used as a biological weapon by the Viet cong from the north, who set small explosive charges near the hives built in the trees and detonated them when enemy soldiers approached, thus generating small fragments containing several angry workers, attacking all who were in their range of action (see Chap. 3).

The worker bees of these species have a modified ovipositor that allows the venom to be injected, causing pain and physical discomfort to aggressors taken as a threat to the colony. The toxicity of these poisons and their allergenic activities are associated with the presence of enzymes such as hyaluronidase and phospholipase, small molecules such as peptides (e.g. secarpin) and biogenic amines (e.g. histamine) and large peptides such as apamin, melittin and mast cell degranulating peptide.

The treatment of people stung by bees consists of careful removal of the stinger and poison sac that get stuck in the victim's skin after the sting to hinder the injection of more venom. In the case of local toxic reactions in which there is only edema, erythema and pain at the sting site, it is recommended to apply cold compresses on site and to use painkillers. In the case of more extensive local reactions, it is suggested to make use of non-steroidal anti-inflammatory drugs and antihistamines. In systemic toxic reactions due to multiple stings, the prognosis is often severe, especially when the victim has cardio-pulmonary disease or has been stung more than 500 times. Empirically, the use of high doses of antihistamines and corticosteroids has been shown to be effective in combating inflammatory effects and histamine intoxication in such cases. Finally, in the occurrence of systemic allergic reactions, therapy should take into account the severity of the case, where epinephrine, corticosteroids and antihistamines can be used, along with cardiorespiratory support measures. It is estimated that approximately 3 % of victims of bee stings develop systemic allergic reactions (McKenna 1992).

35.1.2.3 Vespidae (Wasps)

Wasps differ from bees by having little body hair and the ability to sting multiple times a potential intruder near the colony. Wasp species of medical importance have variable sizes and degrees of aggressiveness. In Brazil, the dreaded wasps of the species *Agelaia vicina*, which are small, about an inch long, and dark in colour with yellowish stripes on the head, thorax and abdomen, are extremely aggressive and attack any simple approach. In Australia and New Zealand, there is the *Vespula germanica*; this species was introduced from Europe and spread through these countries with ease, where in recent years, attacks have been recorded with fatalities.

In East Asia, *Vespa mandarinia* and subspecies *Vespa mandarinia japonica*, commonly known as 'giant wasps' of Asia, can be up to five inches long and have a wingspan of up to 8 cm. These wasps can, in their stings, inject large amount of venom into their victims. In China, hundreds of people were hospitalized, and dozens of them have died due to intense attacks by these insects, which motivated the leaders of some cities to enact a state of emergency and to arrange for some technical teams to destroy the nests of these insects.

With regard to composition and activity, wasp venoms contain acid phosphatase, hyaluronidase, phospholipase and hydrophobic peptides such as mastoparans, which cause degranulation of mast cells for release of histamines, besides stimulating the release of adenylic acids and catecholamines by adrenal cells. Some mastoparans cause haemolysis and feature myotoxic activity. The pain associated with wasp stings is caused by the presence of kinins, derivatives of bradykinin. Except for the removal of the sting with poison sac, all treatment measures described above for bee stings can be used on the victims of wasp stings (Yanagawa et al. 2007).

35.1.3 Order Lepidoptera

The insects of the order Lepidoptera are popularly known as butterflies and moths. Their main feature, in the winged life stage, is the presence of scales covering their wings. From a medical point of view, the moths of the genus *Hylesia* are important because the bristles present on the abdomen of females of various species on contact with the human skin can cause dermatitis. This genus is commonly found in the Neotropics. In addition, there are records of the occurrence of respiratory symptoms in forest workers who come in contact with the species *Orgyia pseudotsugata* in the USA.

The larvae (caterpillars) are mainly responsible for causing harm to humans by Lepidoptera. This occurs by contact of the victim with structures located in the insect integument that releases toxins that can cause intense pain, erythematous papules and/or pruritic lesions. In Europe, the species *Thaumetopoea processionea* can be highlighted, which is especially dangerous to people's eyes. In the USA, the most important species are *Automerisio*, *Hemileucamaia*, *Lagoa crispata*, *Megalopyge crispata*, *Megalopyge opercularis*, *Nygmiaphoeorrhoea* and *Norapecretata*.

In Brazil, the most important species of Lepidoptera is *Lonomia obliqua*, whose venom may cause the condition known as haemorrhagic diathesis. After skin contact with the bristles of the caterpillars of this species, the victim develops a local inflammatory process followed by symptoms such as nausea, headache and fever. After a period of 24–48 h, the victim may develop a severe haemorrhagic disease, which in some cases, leads to acute renal failure. Treatment of patients who have come in contact with this insect varies according to the severity of manifestations, ranging from symptomatic treatment with the use of painkillers and cold compresses on the site, up to the administration of antilonomic serum.

Contact with caterpillars can be considered an occupational hazard for workers from rural areas, with the highest frequency of these events occurring in the warmer and wetter periods. Regarding the treatment of skin manifestations arising from contact with these insects, it is suggested that the site be washed with cold running water. Pain relief may be achieved with systemic administration of analgesics. Usually the patient's symptoms subside after 24 h without complications (Battisti et al. 2010).

35.1.4 Order Hemiptera

The order Hemiptera contains the well-known bugs (Heteroptera) and the cicadas, leafhoppers, whiteflies, mealy bugs and aphids (Homoptera). There are about 90,000 described species worldwide, distributed among approximately 140 families.

The medically important species in this order belong to the suborder Heteroptera. Although the approximately 40,000 described species possess a wide diversity of sizes, colours and forms, they also present certain characteristics which allow them to be recognized easily. One of the most distinctive is the structure of the fore wings, which are divided into a tough, thickened basal part and a membranous apical region. These wings are known as 'hemelytra'. The mouthparts are modified into a biting-sucking rostrum. Phytophagous heteropterans use the rostrum to suck the sap of plants, while others ingest the haemolymph of insects (entomophagous) and some take vertebrate blood (haematophagous). Most species are terrestrial although some are aquatic.

The bloodsucking heteropterans belong to the families Reduviidae (kissing bugs) and Cimicidae (bedbugs), both including species that parasitize man. Since the kissing bugs and the disease they transmit (Chagas' disease) were covered in Chap. 17, the focus here will be directed towards the bed bugs.

The Cimicidae have oval bodies flattened dorsoventrally and covered in short bristles. The hemelytra have the form of scales and are rudimentary, the second pair of wings being absent. The rostrum is short and robust, the eyes prominent, the ocelli absent and the antennae short, with few segments. The adults are small, varying from 4–7 mm in length.

Two species are closely associated with man, the 'common bed bug' (*Cimex lectularius*) and the 'tropical bed bug' (*C. hemipterus*). These can be distinguished morphologically by the prothorax, in which *C. lectularius* is broader laterally and has much more flattened margins than in *C. hemipterus*. There are also species that are normally ectoparasites of bats and birds (the 'bat bug' *C. adjunctus* and 'swallow bug' *Oeciacus vicarius*) but may bite man when their traditional hosts are absent. In West Africa, *Leptocimex boueti*, which occurs in human dwellings in rural areas, takes blood from both humans and bats (Criado et al. 2011; Russell et al. 2013). This species is recognized by the thorax which is not quite as long as the head, by the elongated third antennomere and by its long slender legs.

Temperature is the preponderant factor influencing the biological cycle, the behaviour and even the geographical distribution of these species. In temperate regions, *C. lectularius* is present by virtue of its greater resistance to low temperatures, whereas *C. hemipterus* is widely distributed in tropical and subtropical areas (Cannet et al. 2015).

In the domestic environment, bed bugs may remain hidden for many hours in cracks and holes in the walls, behind wallpaper, in curtains, carpets, furniture, mattresses and pillows. Feeding activity is carried out preferentially at night and at dawn, when they emerge from their hiding places to bite. The flattened form of the body facilitates penetration and locomotion in these hiding places. The stains of blood and faeces left on furniture, as well as the characteristic sickly sweet smell that they produce (similar to that of corander), are all signs of household infestation by bedbugs.

These insects may also be found in other environments, such as schools, airports, theaters, ships, hotels and hospitals. The peridomicile, principally chicken houses, may also be a focus of infestation. Bed bugs may disperse from one infested site to another actively by means of ventilation ducts and doorways or passively, in baggage and furniture originating from infested localities.

Although hepatitis B virus and HIV and many other pathogens have been isolated from bed bugs, there is no epidemiological evidence that they are involved in the transmission of causative agents of human disease. Some types of alphavirus (WEE, Buggy Creek and Fort Morgan virus) have been isolated from *O. vicarius*, indicating a possible role in the maintenance of FMV among migrating birds (see Chap. 9). However, even without being confirmed vectors of human pathogens, bed bugs currently constitute a re-emerging public health problem in several parts of the world, their infestations having both psychological and economic implications. The resurgence of these species appears to be related to insecticide resistance as a result of changes in control measures, as well as increased international commerce and travel (Krinsky 2009b; Criado et al. 2011). They can be found in several kinds of houses, including hotels[3] and, due to multiple bites, can disturb sleeping and cause entomophobia.

35.1.4.1 Clinical Aspects[4]

Blood feeding may be carried out on any region of the human body. However, the face, neck, shoulders, arms and hands are the principal areas affected. The saliva of bed bugs has anticoagulant and anaesthetic properties, which allows them to feed undisturbed for long periods. The bite is painless but causes swelling, irritation and small localized hemorrhages. Depending on the degree of immunological response by the individual, the result of feeding can be manifested in several ways, including papules, pustules and small blisters with erythema and circular indurations, as well as kissing lesions which may resemble multiform erythema. Scratching may produce excoriations liable to secondary infection. The cutaneous reactions are autolimiting and disappear without treatment (Criado et al. 2011; Doggett et al. 2012).

As well as cutaneous problems, chronic infestations may cause nervous disorders, lethargy, diarrhea and anaemia in children (Doggett et al. 2012).

35.1.4.2 Diagnosis and Treatment

The histopathological manifestations of bed bug bites are similar to those triggered by the bites of other insects. There is a perivascular infiltrate with lymphocytes, histiocytes and eosinophils, as well as mastocytes on the surface of the dermis. Depending on the type of cutaneous lesion caused by the bite of the bed bug, differential clinical diagnosis can be based on comparison with herpetiform dermatitis, eczema, scabies, skin reactions related to drug use and reactions to the bites of other arthropods (Criado et al. 2011).

The affected area should be washed with soap and water to reduce the risk of infection. Ice packs should be applied to diminish the swelling. If the symptoms are severe, pruritus may be controlled with antihistamines and topical steroids. Antibiotics may be prescribed if there are secondary bacterial infections (Doggett et al. 2012).

In general, infestations are seen in human populations of low socio-economic levels, despite not always being due to low standards of hygiene and house conservation. Nevertheless good hygiene and housekeeping remain the best ways of preventing bedbug infestations.

35.1.4.3 Other Non-Haematophagous Heteropterans of Importance

Predatory bugs may also bite man if they feel threatened. The bite is very painful, and allergic reactions to the saliva of the insect may occur. Glandular secretions liberated by phytophagous bugs may also trigger cutaneous lesions in humans.

Predators

The members of the family Belostomatidae are voracious carnivores and have a diversified diet which include aquatic insects, snails and small vertebrates such as frogs, fish and birds. They are frequently found in lakes and quiet watercourses, localities which are often visited by humans. Finally, they may also occur in swimming pools. In some places they are known as 'electric light bugs' due to the fact that they are attracted in great numbers to bright artificial lighting.

The chestnut-coloured body is oval and flattened dorsoventrally, the size being very variable (10–140 mm in length). The antennae are shorter than the head and remain retracted under the prominent eyes. The biting-sucking

[3]Even president F. D. Roosevelt and other participants of Yalta conference, in February 1945, were disturbed by bedbugs (CBM).

[4]For further details on reactions to the bites of bloodsucking insects, see Chap. 36 (CBM).

mouthparts are well developed, enclosed within a sheath (labium) which surrounds the four stylets (two mandibular and two maxillary). The anterior legs are adapted to capture prey and the posterior ones are flattened and fringed to help propel the insect through the water, although belostomatids are poor swimmers.

When handled without due care or trod on, these bugs react with bites that are generally considered to be the most painful of those of all insects. The bite involves penetration of the stylets, with the inoculation of proteolytic saliva. Bacilli of *Mycobacterium ulcerans* ('Buruli ulcer') may colonize the salivary glands of the Belostomatidae and inoculated into the human skin at the moment of the bite. A type of mycobacteriosis known as 'Buruli ulcer' and related to water bug bites has been reported in several tropical and subtropical countries. Its clinical profile is characterised by nodules, areas of edema and plaques and the most obvious symptom being the presence of a large ulcer that normally occurs on the arms or legs (Boleira et al. 2010; Marion et al. 2014).

The 'giant water bug' or 'toe biter' as *Lethocerus americanus*, known in North America, is a species associated with painful accidents, whose severity depends on the sensitivity of the individual bitten.

One of the first symptoms to appear after the bite is a throbbing pain, similar to that triggered by the stings of bees or wasps. The pain increases over a matter of hours, progressing to oedema and light reddening and accompanied by fever. During the next 24 h, the affected area tends to become more reddish and extends progressively along the affected member. Between the third and fourth day, there is a reduction in the oedema, its immediate area assuming a reddish-blue tone. The linear rash trajectory disappears and is replaced by pruritus. The pain remains intense. During the second week, papules and pruritus can be seen although the oedema tends to diminish. During the third and fourth week, the pruritus and the purplish tone of the skin disappear, but light local inflammation persists, ceasing after a few days. Clinical treatment should be symptomatic, based on the use of antihistamines, analgesics and antipruritic pomades (Diaz and Pefaur 2006).

Since they are widely distributed insects present in aquatic environments that are often shared with bathers, people should be instructed on how to recognize them, as well as be alert to the type of accidents that they may cause.

Other Hemiptera associated with aquatic ecosystems also may be responsible for painful bites when handled, as is the case for the genera *Pelocoris* and *Ambrysus* (family Naucoridae) (Polhemus 1979) and *Laccotrephes* (family Nepidae).

Phytophagous

Phytophagous species of the family Pentatomidae may be associated with dermatitis that manifest themselves in a similar way to those caused by the beetles that release the substances pederin and cantharidin, although the effects in humans are milder.

The insects of this family are exclusively terrestrial and numerous in tropical and subtropical regions. The body is shield shaped and the antennae have five antennomeres, biting-sucking mouthparts (forming a rostrum) with four subsegments and a triangular scutellum. Their length varies between 2 and 20 mm, those of medium size predominating.

The pentatomids are well known for their production of secretions with disagreeable odours, liberated by glands situated on the thorax (adults) or abdomen (nymphs) and are thus responsible for their common name of 'stink bugs'. Such secretions may cause cutaneous lesions in humans.

In Brazil, accidents have been recorded with the species *Antiteuchus mixtus* (subfamily Discocephalinae) and *Edeste maculata* (subfamily Edessinae). A burning sensation is usually the first manifestation to be reported with subsequent development of edematous and erythematous plaques (1–3 cm in diameter) with microblisters on its surface. The lesions are pruritic. Histopathological analyses have shown moderate inflammatory reactions, with lymphocytes on the surface of the dermis, dilated capillaries and erythrocytes. The use of topical corticosteroids is recommended for treatment (Haddad et al. 2002).

In the rainy season in India, *Chilocoris assmuthi* (Cydnidae), popularly known as 'burrower bugs', are responsible for producing small brown asymptomatic stains on the feet (uppers and soles), necks and chests of people who have been in contact with these insects. If one of these bugs is pressed between the fingers for a few seconds, reddish-brown spots appear almost immediately at the point of contact. Similar lesions can be observed if one of these insects is rubbed against the forearm. A more uniform dark color develops as the lesions grow. This pigment is difficult to remove with soap and water, and the use of acetone is therefore recommended. If no measures are taken to remove the pigment, it fades gradually over a period of 5–6 days (Malhotra et al. 2015). The species in question is distributed throughout Asia (India, Nepal, Pakistan, Sri Lanka and Thailand) but is rare in urban areas.

35.2 Class Chilopoda

Chilopoda, popularly known as 'centipedes', have a flat body divided into the head and articulated trunk. The head has sensory organs, including antennae and eye spot (not all species), two pairs of maxillae and one pair of mandibles. In the trunk, there is one pair of legs per segment. The first segment of the trunk has one pair of forcipules, also known as poisonous claws, which are curved towards the ventral midline and have a sharp stinger at the end, which is connected to a poison gland. In the last segment, there are two anal legs, which have defensive and aggressive functions. Centipedes are usually brown but may have aposematic colouration, depending on the species. They can be up to 30 cm in length and display 15–191 pairs of legs (Ruppert et al. 2004) (Fig. 35.1).

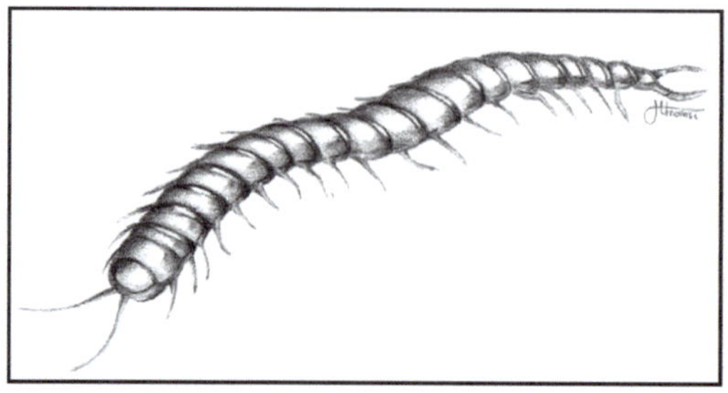

Fig. 35.1 A chilopod: presence of one pair of legs per segment

This group is composed of approximately 3500 identified species worldwide. Centipedes are predatory animals with preferentially nocturnal habits and can be found in the soil, under rocks and under leaves, bark and logs of trees. In urban areas, they occur in and around the home and can be found in gardens, vegetable gardens and under bricks and rubble. Their movements are fast, and unlike the millipedes, they cannot coil up in a spiral. Higher species diversity occurs in tropical and subtropical regions, where the majority of accidents are recorded in the warm and rainy seasons. In some areas, centipede bites may represent up to 17 % of all accidents registered with venomous animals (Haddad et al. 2012; Kania and Klapec 2012).

The poison of centipedes consists of proteins with enzymatic action, such as esterases, hyaluronidases and proteases (e.g. metalloproteases), non-enzymatic proteins such as myotoxins, cardiotoxins, neurotoxins and cytotoxins and non-peptide compounds, such as serotonin and histamine. Thus, symptoms are characterized by itching and intense pain, of burning type, at the bite site, which generally ceases within 24 h. The bite site shows mild edema and hyperthermia. In some cases, there may be at the site paraesthesia, severe oedema, erythema, slight bleeding, local necrosis, lymphangitis and lymphadenitis, and systemic symptoms may include fever, chills, tremors, sweating, dyspnea, headache, vomiting, dizziness and anxiety. Serious toxic responses, as rhabdomyolysis, myocardial ischemia, proteinuria and acute renal failure, have been reported after the bite of *Scolopendra heros*, known as the 'giant desert centipede'. In the literature there are records of fatal cases, both children and adults (Undheim and King 2011).

Centipede bites can also result in Wells' syndrome, requiring the administration of corticosteroids topically or by intralesional injections. However, in most cases, the treatment should be only symptomatic for pain relief and reduction of swelling, such as rest, application of cold packs, and if necessary administration of anaesthetics, analgesics, antipyretics and antihistamines. Vaccination against tetanus should be considered if the patient was not immunised in the last 10 years. To reduce the risk of infection, it is suggested that the affected area is washed with soap and water (James et al. 2011; Haddad et al. 2012).

The diagnosis based on the patient's condition is very difficult, unless the animal is taken together with the patient to be identified. However, centipede bites are characterized by bleeding marks in the form of a 'v' due to the large forcipules. This feature should be taken into account, especially when children are bitten by these animals, since some species of *Scolopendra* can be as large as 20 cm long, where their bites can be mistaken for snake bites. Thus, recognition of the characteristic V shaped of bites of centipedes is important to avoid inappropriate treatment of these patients with antivenom. Recently, various kinds of centipedes began being sold in pet shops as exotic animals, which can increase the number of accidents with these animals involving children (James et al. 2011).

The genus *Scolopendra* (family Scolopendridae) includes the most important group of centipedes, since it has a cosmopolitan distribution and comprises the species that are responsible for most human accidents, besides having a more toxic poison than most other centipedes. Among the more than 90 species of the genus, we can highlight the following: *S. viridicornis*, occurring in Brazil; *S. heros*, in the USA; *S. subspinipes*, in Italy; *S. gigantea*, in Turkey; *S. validates*, in Iran; *S. morsitans*, in Australia; and *S. multidens*, in Japan.

The following preventive measures are advised to avoid bites of these animals: keep backyards, gardens and land clean; prevent the accumulation of rubble, dry leaves, household garbage and construction material in the vicinity of houses; avoid heavy foliage along the walls of houses; avoid inserting bare hands in holes, under stones and rotting logs; inspect clothes and shoes before putting them on; always wear shoes and gloves to work in gardens and with building material; seal door-sills as well as cracks and holes in the walls and floor; place screens on windows; keep beds away from walls; and avoid contact of bedding with floor (James et al. 2011; Haddad et al. 2012).

35.3 Class Diplopoda

Diplopoda, popularly known as 'millipedes', have an elongated body and generally cylindrical. The body is divided into the head, with the sense organs and mouthparts, and trunk, which consists of a series of rings formed by two segments (diplosegments). In the majority of each of these rings are seen two pairs of legs and two pairs of spiracles. Millipedes vary between 2 mm and 30 cm, and most have a dark colour (Fig. 35.2). Some species can be red or orange. To compensate for the lack of speed to escape predators, millipedes have developed a number of protective mechanisms, such as the glands that secrete repellent substances. These structures are present in several species of the group and communicate with the external environment through small pores which are located in pairs on the sides of almost all diplosegments (Ruppert et al. 2004).

There are approximately 12,000 millipede species described worldwide. These animals are scavengers, feeding on decaying plant matter. They have nocturnal habits and live, preferably under leaves, rocks, tree bark and fallen logs on the ground. In general, most species move slowly over the ground. Millipedes have higher species diversity in tropical areas, where they can occur at densities of up to 40 individuals/m^2. They have marked seasonality, and their appearance usually occurs in warmer and wetter periods.

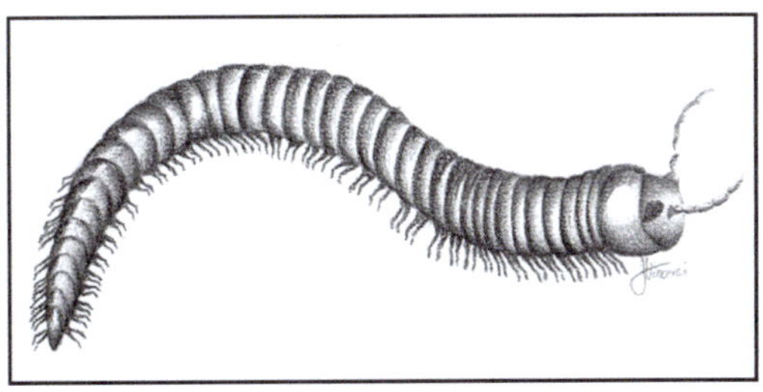

Fig. 35.2 A diplopod: presence of two pairs of legs per ring

The composition of the repellent discharge varies between species, where aldehydes, quinones (e.g. benzoquinone), phenol, iodine, chlorine and hydrogen cyanide have been identified. The last substance, when released, has the odor of bitter almonds and is toxic to invertebrates and small vertebrates. Hence, when disturbed, millipedes coil up into a tight spiral, keeping the head in the centre, and then slowly release the secretion, which is generally yellowish. Some species of millipedes, however, can emit jets or spray as far as 30 cm (Arab et al. 2003; James et al. 2011).

Millipedes may even be found on hanging clothes to dry or inside shoes, which can cause accidental contact of these animals with humans and release their defense secretion. Once in contact with the human skin, the secretion can cause a burning sensation, resembling a mild sunburn. The site becomes instantly hyperchromatic, with color ranging from light yellow to dark brown, due to the action of quinones. When contact with the secretion is prolonged, more severe lesions are observed, with the formation of intense erythema, vesicles, local pain and paraesthesia. Accidents involving these animals can also manifest as an acute digital ischemia evidenced by the sudden change in temperature and toe colour (blue toe syndrome), which initially becomes cold and pale and later appears cyanotic. On contact with the ocular mucosa or the periorbital tissues, secretions of millipedes may cause serious eye damage. The clinical manifestations in these cases are described as a burning sensation in the eye region, oedema, conjunctivitis and keratitis. The occurrence of blindness is a very possible sequel of this type of accident (Haddad et al. 2012).

The recommended treatment in such cases is limited to extensive washing of the skin and/or mucosa to limit the toxic effects of secretion. The use of ether or alcohol in recent accidents should be encouraged, given the solvent properties of these substances. Furthermore, if necessary, the topical corticosteroids and administration of analgesics and anti-inflammatories are recommended.

If accidentally ingested, the millipedes may cause infection since they carry on their integument a wide variety of bacteria potentially pathogenic to man, such as *Pantoea agglomerans*, *Salmonella arizonae* and *Serratia marcescens*. In addition, in the literature there are few reports that some species of millipedes (*Brachyiulus lusitanus* and *Nopoiulus kochii*) can cause pseudoparasitism in the alimentary tract of children who had accidentally ingested them. Patients, at that time, had stomach ache and reported a burning sensation in the throat. Laboratory tests of stool specimens of patients showed a small amount of blood in the faeces, and there was no evidence of other potentially pathogenic organisms. Patients were treated with niclosamide, and after a few days, millipedes were expelled dead or dying in the faeces (Mowlavi et al. 2009).

The diagnosis of an accident with millipedes is usually performed when the animal is brought by the victim to medical care unit. In children, burns caused by millipedes have been misinterpreted as signs of child abuse, due to the characteristics of the lesions caused by their secretions (James et al. 2011).

Among the millipede taxa responsible for most accidents with humans, species of the genus *Polyconoceras* occur in Oceania, among other places, and can cause accidents with eye injuries because they have the ability to squirt their defensive secretion. In North America, a species that stands out is *Narceus americanus*; it is a very abundant species. *Rhinocricus padbergi* occurs from Central America to South America, and most accidents are reported in Brazil. Preventive measures are the same as recommended for centipedes.

Acknowledgements We thank Leonardo Vincensi for the drawings of the Chilopoda and Diplopoda. We are also grateful to Grzegorz Kania, Department of Biology and Parasitology, Medical University, Lublin, Poland, for sending us their scientific articles.

References

Arab A, Zacarian GG, Fontanetti S et al (2003) Composition of the defensive secretion of the Neotropical millipede *Rhinocricus padbergi* Verhoeff 1938 (Diplopoda: Spirobolida: Rhinocricidae). Entomotropica 18:79–82

Arnett RH Jr (2000) False blister beetles (Insecta: Coleoptera: Oedemeridae). http://edis.ifas.ufl.edu/pdffiles/in/in31100.pdf. Accessed 24 Mar 2015

Battisti A, Holm G, Fagrell B et al (2010) Urticating hairs in arthropods: their nature and medical significance. Annu Rev Entomol 56:203–220. doi:10.1146/annurev-ento-120709-144844

Beatson SH (1972) Pharaoh's ants as pathogens vectors in hospitals. Lancet 1:425–427

Beatson SH (1973) Pharaoh's ants enter giving-sets. Lancet 1:606

Boleira M, Lupi O, Lehman L, Asiedu KB et al (2010) Buruli ulcer. An Bras Dermatol 85:281–301. doi:10.1590/s0365-05962010000300002

Borroni G, Brazzelli V, Rosso R, Pavan M (1991) Paederus fuscipes dermatitis: a histopathological study. Am J Dermatopathol 13:467–474. doi:10.1097/00000372-199110000-00007

Cannet A, Akhoundi M, Berenger JM et al (2015) A review of data on laboratory colonies of bed bugs (Cimicidae), an insect of emerging medical relevance. Parasite 22:21. doi:10.1051/parasite/2015021

Conceição LG, Haddad V Jr, Loures FH (2006) Pustular dermatosis caused by fire ant (*Solenopsis invicta*) stings in a dog. Vet Dermatol 17:453–455

Cressey BD, Paniz-Mondolfi AE, Rodriquez-Morales AJ et al (2013) Dermatitis linearis: vesicating dermatosis caused by *Paederus* species (Coleoptera: Staphylinidae). Case series and review. Wilderness Environ Med 24:124–131. doi:10.1016/j.wem.2012.11.005

Criado PR, Belda W Jr, Criado RFJ et al (2011) Bedbugs (Cimicidae infestation): the world-wide renaissance of an old partner of human kind. Braz J Infect Dis 15:74–80. doi:10.1590/s1413-86702011000100014

Castro MM, Prezoto HHS, Fernandes EF et al (2015) The ant fauna of hospitals: advancements in public health and research priorities in Brazil. Rev Bras Entomol 59:77–83. doi:10.1016/j.rbe.2015.02.011

Della-Lucia TMC (2003) Hormigas de importancia econômica em la región Neotropical. In: Fernandéz F (ed) Introducción a las hormigas de la región Neotropical, vol XXVI. Instituto de Investigación de Recursos Biológicos Alexander von Humboldt, Bogotá, pp 337–349

Diaz A, Pefaur JE (2006) Envenenamiento por un insecto de la familia Belostomatidae (Insecta: Hemiptera). Rev Fac Med 29:125–128

Doggett SL, Dwyer DE, Penas PF et al (2012) Bed bugs: clinical relevance and control options. Clin Microbiol Rev 25:164–192

Fontana R, Wetler RMC, Aquino RSS et al (2010) Disseminação das bactérias patogênicas por formigas em dois hospitais do nordeste do Brasil. Neotrop Entomol 39:655–663. doi:10.1590/S1519-566X2010000400029

Garcia FRM, Lise F (2013) Ants associated with pathogenic microorganisms in Brazilian hospitals: attention to a silent vector. Acta Sci Health Sci 35:9–14. doi:10.4025/actascihealthsci.v35i1.10471

Ghoneim KS (2013a) Cantharidin toxicosis to animal and human in the world, a review. Standard Res J Toxicol Environ Health Sci 1:1–16

Ghoneim KS (2013b) Enhancement of research interests in physiology and biochemistry of blister beetles (Coleoptera: Meloidae): a review. Int Res J Biochem Bioinform 3:75–92

Ghoneim KS (2013c) Human dermatosis caused by vesicating beetle products (Insecta), cantharidin and paederin: an overview. World J Med Med Sci 1:1–26

Haddad V Jr, Larsson CE (2015) Anaphylaxis caused by stings from the *Solenopsis invicta*, lava-pés ant or red imported fire ant. An Bras Dermatol 90:22–25. doi:10.1590/abd1806-4841.20153420

Haddad V Jr, Cardoso JLC, Moraes RHP (2002) Skin lesions caused by stink bugs (Insecta: Heteroptera: Pentatomidae): first report of dermatological injuries in humans. Wilderness Environ Med 13:48–50

Haddad V Jr, Cardoso JLC, Moraes RHP (2005) Description of an injury in a human caused by a false tocandira (*Dinoponera gigantea*, Perty, 1833) with a revision on folkloric, pharmacological and clinical aspects of the giant ants of the genera *Paraponera* and *Dinoponera* (subfamily Ponerinae). Rev Inst Med Trop Sao Paulo 47:235–238. doi:10.1590/S0036-46652005000400012

Haddad V Jr, Cardoso JL, Lupi O et al (2012) Tropical dermatology: venomous arthropods and human skin, Part II. Diplopoda, Chilopoda, and Arachnida. J Am Acad Dermatol. doi:10.1051/parasite/2009162161

Hoffman DR (2010) Ant venoms. Curr Opin Allergy Clin Immunol 10:342–346. doi:10.1097/ACI.0b013e328339f325

James DW, Berger GT, Elston MD (2011) Andrews' diseases of the skin: clinical dermatology. Saunders Elsevier, Philadelphia

Kania G, Klapec T (2012) Seasonal activity of millipedes (Diplopoda) – their economic and medical significance. Ann Agric Environ Med 19:646–650

Klotz JH, de Shazo RD, Pinnas JL et al (2005) Adverse reactions to ants other than imported fire ants. Ann Allergy Asthma Immunol 95:418–425

Krinsky WL (2009a) Beetles (Coleoptera). In: Mullen G, Durden L (eds) Medical and veterinary entomology, 2nd edn. Academic Press/Elsevier, San Diego, pp 97–109

Krinsky WL (2009b) True bugs (Hemiptera). In: Mullen G, Durden L (eds) Medical and veterinary entomology, 2nd edn. Academic Press/Elsevier, San Diego, pp 80–96

Lee EK, Jeong KY, Lyu D-P et al (2009) Characterization of the major allergens of *Pachycondyla chinensis* in ant sting anaphylaxis patients. Clin Exp Allergy 39:602–607. doi:10.1111/j.1365-2222.2008.03181.x

Malhotra AK, Lis JA, Ramam MM (2015) Cydnidae (Burrowing Bug) pigmentation: a novel arthropod dermatosis. JAMA Dermatol 151:232–233. doi:10.1001/jamadermatol.2014.2715

Marion E, Chauty A, Yeramian E et al (2014) A case of guilt by association: water bug bite incriminated in *M. ulcerans* infection. Int J Mycobacteriol 3:158–161. doi:10.1016/j.ijmyco.2014.01.004

McKenna WR (1992) Killer bees: what the allergist should know. Pediatr Asthma Allergy Immunol 6:275–285. doi:10.1089/pai.1992.6.275

Moed L, Shwayder TA, Chang MW (2001) Cantharidin revisited: a blistering defense of an ancient medicine. Arch Dermatol 137:1357–1360

Mowlavi G, Naddaf SR, Rezaeian M et al (2009) Apparent pseudoparasitism of the alimentary canal of a 5-year-old child by the millipede *Brachyiulus lusitanus* (Diplopoda: Julidae). Parasite 16:161–163. doi:10.1051/parasite/2009162161

Nikbakhtzadeh MR, Akbarzadeh K, Tirgari S (2009) Bioecology and chemical diversity of abdominal glands in the iranian samsum ant *Pachycondyla sennaarensis* (Formicidae: Ponerinae). J Venom Anim Toxins Incl Trop Dis 15:509–526

Polhemus JT (1979) Family Naucoridae - creeping water bugs, saucer bugs. In: Menke AS (ed) The semiaquatic and aquatic Hemiptera of California (Heteroptera: Hemiptera). Bulletin of the California Insect Survey. University of California Press, Berkeley, pp 131–138

Ruppert EE, Fox RS, Barnes RD (2004) Invertebrate zoology: a functional evolutionary approach. Brooks Cole Thomson, Belmont

Russell RC, Otranto D, Wall RL (eds) (2013) The encyclopedia of medical and veterinary entomology. CAB International, Oxfordshire

Sydnor ERM, Perl TM (2011) Hospital epidemiology and infection control in acute-care settings. Clin Microbiol Rev 24:141–173. doi:10.1128/CMR.00027-10

Undheim EAB, King GF (2011) On the venom system of centipedes (Chilopoda), a neglected group of venomous animals. Toxicon 57:512–524. doi:10.1016/j.toxicon.2011.01.004

Wanke B, Lazera MS, Nucci M (2000) Fungal infections in the immunocompromised host. Mem Inst Oswaldo Cruz 95:153–158. doi:10.1590/S0074-02762000000700025

Yanagawa Y, Morita K, Sugiura T et al (2007) Cutaneous hemorrhage or necrosis findings after *Vespa mandarinia* (wasp) stings may predict the occurrence of multiple organ injury: a case report and review of literature. Clin Toxicol (Phila) 45:803–807. doi:10.1080/15563650701664871

Zarzuela MFM, Ribeiro MCC, Campos-Farinha AEC (2002) Distribuição de formigas urbanas em um hospital da região sudeste do Brasil. Arq Inst Biol 69:85–87

Allergy to Hematophagous Arthropod Bites

36

Hélio Amante Miot and Hermenio C. Lima

Abstract

Allergies to hematophagous arthropod bites are inflammatory reactivity to arthropod salivary components. They vary in intensity and quality dependent upon the arthropod specie and the individual immune response to specific proteins of the insect's saliva. Individuals who were not previously exposed show mild localized reactions not beyond those expected by pharmacological substances present in arthropods' saliva. Allergic reactions are immunological in their nature and the diversity derived from hypersensitivity reactions with different levels of participation of the immune system components. Some are mainly derived from a humoral immune response, and others are based predominantly on T-lymphocyte-mediated hypersensitivities. The majority of these undesired biological answers are self-limited, and few may cause a systemic reaction. This article intends to discuss the immunological ingredients of this evolutionary interaction.

Keywords

Insect bites and stings • Insect proteins • Hypersensitivity • Salivary proteins and peptides

36.1 Introduction

Arthropods constitute a large phylum of living creatures that are of major medical importance because of infectious disease transmission as well as hypersensitivity reactions due to their bites. The interaction of arthropods with other animals is associated to defense, predation, feeding, and species perpetuation needing (Pollack and Marcus 2005; Singh and Mann 2013).

H.A. Miot (✉)
Department of Dermatology, UNESP Medical School, Campus Universitário de Rubião Jr, Botucatu, SP, Brazil
e-mail: heliomiot@fmb.unesp.br

H.C. Lima
Division of Dermatology, Department of Medicine, Michael G. DeGroote School of Medicine, McMaster University, 1280 Main Street West, Hamilton, ON, Canada, L8S4L8
e-mail: hlima@mcmaster.ca

C.B. Marcondes (ed.), *Arthropod Borne Diseases*, DOI 10.1007/978-3-319-13884-8_36

Hematophagous arthropods developed and evolved independently several times in all parts of the planet. The specific need to obtain blood from hosts led arthropods to develop sophisticated adaptations in their mouthparts and elaborate salivary substances to ease this task. Their interactions with humans lead to immunologic reactions and transmission of several infectious diseases.

Allergic reactions to saliva substances are immunological in nature. Each salivary constituent may start different immune reactions. They involve immunoglobulin and T-lymphocyte-mediated hypersensitivities (Tan and Campbell 2013). Allergies to arthropod products are individual biological inflammatory reactivity that exceeds the homeostatic response to insect bite and salivary components in intensity and quality.

Arthropod hematophagy has resulted in a large diversity of human medical conditions. Most of the encounters with biting arthropods result in more or less pronounced localized reactions. While most are self-limited, some of these hypersensitivity reactions may cause systemic allergic reactions such as anaphylaxis, high fever, lymphadenopathy, and hepatosplenomegaly (Karppinen et al. 2012; Singh and Mann 2013).

It is intended here to discuss various aspects of allergic reactions to blood-sucking arthropod bites. Basic concepts of these eventual excessive reactions as common to all will be addressed.

36.2 Interface Between Hematophagous Arthropods and Their Blood Supply

Most female hematophagous arthropods seek for mammals' blood to obtain albumin as a protein source for egg formation. Others attack as a defense mechanism or for egg deposition. These processes led to skin damage and inflammatory reactions.

The skin consists of cells, extracellular fluids, and secretions. These elements participate and react simultaneously or sequentially when an insect tries to obtain blood from the dermis. Components involved in this interaction are the following. Keratinocytes and Langerhans cells are the principal components in the epidermis. Macrophages and dendritic cells, lymphocytes, mast cells, fibroblasts, neural terminals (axons), and endothelial cells from blood vessels and lymphatic vessels are part of the human skin dermis setting. Complement, coagulation, and kinins are proteins of interstitial fluid in addition to natural antibodies. We call this structure built from millions of years of evolution of the skin immune unit.

Hematophagous arthropods have developed several mechanisms and apparatuses for obtaining blood. They have to break the dynamic stability of the skin to obtain its blood supply. The rupture of the skin structure triggers early inflammatory processes such as coagulation cascade, which leads to structure repair. Their bites will likely cause more itch than pain and start other inflammatory reactions after the blood sucking. It is from this process that the immune system sensitization will start triggering an allergic reaction in future bites.

Arthropods can also affect skin structure when they seek their prey or defend against a possible aggressor. In these two situations, arthropods may or may not possess body structures for inoculation of substances, which eventually will cause a reaction in affected individuals. These reactions result from the injection of acidic substances or alkaloids, such as formic acid, or other pharmacologically active substances, such as venoms and toxins, which are usually proteins. Humans are occasionally affected by these circumstances. Usually, in addition to pain, there are four possible reactions after this fortuitous interaction with these arthropods: immediate local reaction, immediate regional and or systemic reaction, late reactions, and, less frequently, anaphylactic responses. The last one is not the focus of this material (Fitzgerald and Flood 2006; Tracy 2011).

36.3 Antigenic Sensitization

After finding an appropriate target for blood sucking, the majority of hematophagous insects puncture the skin repeatedly until a vessel is found (Choumet et al. 2012). This process breaks down tissues and cells. An injury of this order generates regulatory responses for tissue structure recovery. The pruritus or itch is part of this process. Itch is a reactive innate reflex, which can be interpreted as a phylogenetic gain. It may have been carried out during evolution to reduce tissue damage. Studies showed that the more reactive the target is to these tissue injury processes, the greater the likelihood that the bite will be interrupted (Kelly 2001). Pruritus becomes the main symptom associated from this interaction.

Insect eventually succeeds in obtaining blood. Antigens are exposed to and can sensitize the immune system from that point on. The variation of the antigen molecules to be sensitized can be as diverse as the number of hematophagous insect species that exist. The saliva of these insects is rich in substances necessary to obtain the blood or other types of food. Analysis of the *Anopheles gambiae* adult salivary gland showed more than 70 proteins and peptides possibly secreted in the saliva that facilitate presumably the blood and sugars meals (Calvo et al. 2006).

In addition to the species, the location, the number of bites, the substances present in the saliva, and the type of mouthparts used by multiple serotypes all may influence the antigen presence. Furthermore, the saliva of the same hematophagous arthropods contains a complex mixture of compounds. There are anticoagulants, vasodilators, histamine, putrescine, and esterase (Ribeiro and Garcia 1981).

These salivary antigens result from their molecular resistance to enzymatic destruction to the nonspecific inflammatory reactions at the insect bite site. The presence of these biochemical stable antigens results in their endocytosis by resident antigen-presenting cells (APC). This process can be amplified by the migration of APC to the bite site. Some arthropod bites can induce intense Langerhans cell migration mimicking a histiocytosis. The injection of these chemicals also modulates in the beginning of sensitization and allergic process development. The injection of bioactive immunomodulators can change the immune response in other directions. For example, sand fly saliva extracts can inhibit macrophage migration and APC activation (Lima and Titus 1996).

APC's play pivotal roles in the immune responses against insect saliva. Salivary gland antigens are captured in the periphery by APC's and presented to antigen-specific lymphocyte at draining sentinel lymph nodes. Expressing peptides derived from the saliva antigens in their surface, APC starts the specific immune response by stimulating lymphocytic cells. This process is studied more in dendritic and Langerhans cells. Macrophages are able to perform the same function but not extensively because the antigens are mainly located at the epidermis. In this location, lymphocyte population expansion occurs by clonal proliferative responses. Enlargement of regional draining lymph nodes has been observed in individuals after multiple insect bites.

Hypersensitivity reactions to hematophagous insect bites are immunological in nature. Individuals not previously exposed do not exhibit reactions beyond those expected by the pharmacological substances present in the insect saliva. Allergic manifestations are due to previous sensitization to specific antigens from insect saliva. Therefore, the variation of the type of sensitization induces a diverse reaction among human beings (Arlian 2002).

In summary, at the stage of sensitization, antigen molecular nature, antigen dose, the insect bite location, external factors, and individual aspects may influence the type and magnitude of the hypersensitivity response. Individuals can be clinically nonreactive, mild reactive, or hyperreactive to the insect bite depending on genetic and acquired factors. For example, individuals with higher risk of sensitization are those with the highest level of exposure, young children, and atopic subjects, those who had no previous exposure to indigenous mosquitoes, and those individuals with lymphoproliferative diseases caused by Epstein-Barr virus. Patients with mastocytosis are at greater risk of anaphylaxis by arthropod bites; on the other hand, endogenous populations (e.g., Amerindians) have minor skin reactions, beyond mild pruritus and papule formation, after native insect bites (Ellis and Day 2005; Demain et al. 2010).

36.4 Pathophysiological and Immunological Response Mechanisms of Clinical Reactivity After Insect Hematophagous Bite

The multiple variables affecting the quality of insect bite reactivity make it difficult to be predicted. However, it is well known that the reactions can be divided in immediate reactions and late regarding time. It also can be classified in cutaneous and or systemic depend upon regarding the involvement of other organ beyond the skin. Furthermore, humoral and cellular immune responses can be mounted against various antigens from contact with the hematophagous saliva antigen. Despite these classifications, the majority of reactions are dependent on antibodies induced by Th2 lymphocytes, and the levels of immunoglobulins can be correlated with intensity of reactions. Immunoglobulins IgE and IgG and delayed-type hypersensitivity (DTH) mediated by T lymphocytes have been described. Other reactions that mainly depend on cells are less frequently observed (Chen et al. 1998). These responses are excessive and characterize the hypersensitivity reaction against salivary gland components.

Patient's reactive is classically divided in *five stages* and are summarized in Table 36.1 (Mellanby 1946). These stages of clinical reactivity of the skin occur after multiple consecutive bites by the same insect species and were evidenced in animals and human beings in a typical chronological sequence. This sequence has been observed with many arthropods including mosquitoes, fleas, and other hematophagous bugs. The development of each stage is dependent on the exposure season and independent of age.

Table 36.1 Skin reaction stages after an insect bite

Stage	Description
1	Individuals without previous exposure. The first bites do not cause any visible immunological reaction but pharmacological reaction to insect saliva chemicals
2	The bites induce a reaction characterized by an erythematous papule after 24 h
3	Bites cause a biphasic reaction. Initially there is an immediate reaction 10–15 min after an insect bite. It is clinically characterized by an erythematous macular or a papular urticarial type lesion. Later there is another reaction 12–24 h after the bite
4	The latency between both reactions shortens due to an anticipation of the later reaction
5	A state of complete tolerance is reached with prolonged exposure to multiple bites

There is no association between the stages of insect bite reaction and the type of hypersensitivity. Therefore, hypersensitivity reaction to insect saliva remains to follow Gell and Coombs classification (Singh and Mann 2013).

The itch is the main symptom at *stage 1*, and there is no Gell and Coombs category. The pathophysiological model of the arthropod bite itch is the pruritoceptive at this stage. Therefore, the source of the symptom occurs in the skin and derives from inflammatory mediators triggered by local tissue damage and or reaction against peptides from saliva of arthropods. The subsequent release by various mediators at the damaged site is the main source of pruritus. These mediators stimulate the free terminations of neural cells, which lead to information to the brain. Kinins and substance P are among these products. As soon as they are released, they cause a direct neural stimulation and a rapid release of other inflammatory mediators (Tonussi and Ferreira 1999).

The itch stimulus is captured by free nerve endings in the dermis and carried out mainly by slow and amyelic axonal type C nerves and, to a lesser extent, by fast and myelinated type until the spinal cord posterior horn. This reflex leads to a not precise perception of the bitten area affected (alokinesis). Following the pathway, type C nerves intersect the marrow to the contra lateral position and are conducted directly to the thalamus by spinothalamic tract lateral cortex and then up to the postcentral, where one has the conscience of the pruritus. The pathway ascending to the limbic system by the same neural pathways that lead pain, pleasure, and tickling justifies the emergence of sensations associated with the act of scratching, as pleasure or irritability (Greaves 2007).

The detection process of tissue destruction by the nervous system leads to reflex act of removing the site from the point of application. This process is associated with the vasoconstriction reflex, followed by a vasodilatation. Therefore, the pruritus or a pain of low intensity may be caused by inoculation of some of the substances mentioned above, which cause direct or indirect neural stimulation (Durate et al. 1990).

The *stages 2 and 3* have signs of hypersensitivity type I, which is mediated by IgE molecules against substances present in the insects' saliva. Two types of immediate hypersensitivities based on IgE have been described. The IgE-type antibodies against the saliva of insects are associated with the immediate local reactions. In the presence of the saliva antigen, there is a skin vascular reactivity, erythema, after the release of histamine by connection to the antigen IgE on the surface of the mast cells. Follow reflex vasodilatation around the site of inoculation, and there is formation of edema by extravasation of fluid. This sequence is characterized by the reaction triple of Lewis resulting in an increase of temperature and local edema at the end of the process (Serafin and Austen 1987). Antihistamines prophylactically given are an effective treatment for the mosquito-bite weal and skin pruritus (Karppinen et al. 2012).

This immediate cutaneous reaction can be followed by generalized extensive urticaria reaction. Nonetheless, anaphylaxis was rarely described with hematophagous arthropods. The induction of asthma has been considered after the insect bite as a form of immediate systemic manifestation. This rare reaction tends to occur in individuals who have shown positive intradermal tests with extract of insect total body and who had a history of asthma and rhinitis that were positive to bronchial provocation test with these same antigens (Agarwal et al. 1991).

Another type of reaction dependent on IgE has been described. This is known as Prausnitz–Kustner reaction. This reaction occurs up to 6 h after the bite of the insect and can be reproduced transfer of antibodies. It results in a predominant eosinophilic infiltrate. The direct effect of saliva as inducer of this process has been removed, although neutrophil migration direct has been observed (Owhashi et al. 2001). This is more evident at *stage 3*.

Late hypersensitivities are mediated by type IgG antibodies and/or cellular response. Type IgG antibodies are associated with late reactions causing necrosis in Arthus-type cutaneous reactions, vesicles, bullae, and cellulitis. The IgG in these cases is mostly also dependent on Th2 lymphocytes such as IgG1 and IgG4 types. This type of humoral reactions manifests more often in the skin of individuals bitten repeatedly by the same insect. The reactions at the site of the bite are mainly plates topped by blisters, which become edematous and necrotic. Histology of the insect bite showed small vessel (leukocytoclastic or lymphocytic) vasculitis as the most important marker of this reaction. The systemic reactions are high fever, general malaise, and subsequently lymphadenopathy and hepatosplenomegaly and include rare anaphylaxis (Mizuki et al. 1998; Morsy 2012). This reflects the local deposition of immune complexes in endothelial cells beyond the insect bite. This reaction is classified type III or Arthus hypersensitivity reaction, which occurs when the antigen complex IgG or IgM in the capillaries activates a sequence of the events after complement cascade activation. Serum sickness, Churg-Strauss syndrome, and erythema multiforme are late manifestations of systemic humoral and cell reactions, respectively (Gaig et al. 1999; Deshazo et al. 2012).

Another reactive phenomenon can be induced by late cellular immunity. The evidence of cellular reactions to insect stings is limited. However, in experimental model, it showed the difficulty of transferring the reactivity by cellular immunity despite some success. In human beings, it has been shown sometimes signs of blast formation and lymphocyte transformation in vitro when stimulated with mosquitoes' antigens. Others demonstrate the durability and Th1 nature of DTH to sand fly bites in humans living in a cutaneous leishmaniasis-endemic area. A systemic Th2 response may explain why some individuals remain susceptible to disease (Oliveira et al. 2013).

Purely cellular reactions are difficult to observe in human beings, but there is a report of hypersensitivity to insect bites in a patient with agamaglobulinemias. Diseases with alteration of the T cell regulation have cellular immune response intensified. For example, cutaneous reactions may appear exaggerated in patients with atopic circumstances (Killby and Silverman 1967), infection by HIV/AIDS, deficiency of alpha-l-antitrypsin, congenital agammaglobulinemia, chronic lymphocytic leukemia, and viral infections with lymphocytosis of T lymphocytes or NK (Tokura et al. 2001). It is noteworthy that NK cells are not considered participants of hypersensitivity to hematophagous insects (Seon et al. 2013). However, in one study, NK cells CD56+ increased the T cell response against insect extract in vitro. Chronic prurigo is the main manifestation of a cellular T cell late response.

An intense cellular inflammatory reaction, known as Skeeter syndrome, was recently described (Simons and Peng 1999) and occurs within hours after the bite and is characterized by cardinal signs of inflammation. It is impossible to differentiate between an infection and an allergic response. The relationship of cause and effect with the bite is rarely apparent.

An important feature of hypersensitivity to hematophagous arthropods is clinical tolerance or *stage 5*. The natural desensitization usually occurs with the years of exposure as occurs in endogenous populations. Constant antigenic stimulation produces regulatory measures, which result in changes in the pattern of clinical responses cellular and humoral. In the serum of these individuals, there is IgG4 and not IgE against the salivary antigens (Peng et al. 2004). This process is part of the immunomodulation present in saliva of hematophagous arthropods. For example, maxadilan present in Phlebotomine sand fly saliva diminishes the production of IFN-gamma, IL-12, and TNF (Rogers and Titus 2003). The saliva of ticks can inhibit the differentiation and maturation of antigen-presenting cells. *Aedes aegypti* and *Anopheles gambiae* produce proteins that bind avidly to serotonin, histamine, and norepinephrine, decreasing the inflammation. Other studies suggested the participation of T regulators (Treg) inducing a tolerance in vitro and a potential therapeutic use of these cells in allergy (Hamza et al. 2013). Furthermore, cellular immune response against flea antigen suggests a participation of Th17 in the skin reaction and IL-10 for systemic regulatory control mechanism. This pattern of cytokine production also explains the clinical manifestation variability (Dominguez-Amorocho et al. 2013).

Therefore, the tolerance is beneficial for both the parties in this case. It is expected that the immunotherapy to reduce immune reactivity against insect bite will be possible with the use of recombinant antigens (Peng and Simons 2007).

Clinically, different arthropods led to different dermatologic lesions in different subjects. It is difficult to characterize the species of arthropods according to the skin lesion or symptoms. In general, mosquitoes cause pruritic erythematous and edematous papules that can ulcerate after scratch. Fleas and bedbugs led to weals, papulovesicles, and nodules, habitually in a linear pattern (breakfast, lunch, and dinner). Ticks usually interact with mammals for many hours and can produce infiltrated papules and skin nodules and evolve to pseudolymphomatous lesions (e.g., *lymphocytoma cutis*) (Singh and Mann 2013).

36.5 Prevention and Treatment of Allergy Caused by Bites

The best prevention strategy for arthropod allergy is the avoidance of biting. The knowledge of local fauna and arthropod habits and appropriate use of bed nets and repellents (e.g., DEET and picaridin) beyond adequate clothing reduces the probability of stings. Systemic antihistamines can prevent pruritus and skin reaction.

Treatment of skin reactions should be directed to the extension of the lesions. Small papule can be treated with mid potency topical steroids. The occlusion strengthens steroid potency protecting against scratch and secondary infection. Nodular or infiltrate lesions usually respond to intralesional steroids. Systemic antihistamines and steroids should be reserved for extensive cases. Anaphylaxis is a life-threatening condition and should be managed at emergency room (Bilo 2011).

36.6 Conclusion

The relationship of the immune system with allergenic challenges involves physiological and homeostatic aspects. The immune system does not react but interacts with the environment, and this interaction is initiated, most of the times, in the skin and mucous membranes. This interaction, causal or not, can have different results. Most of the times, it is a balance, invisible to other sensory systems, which is characterized by integrity of what we call health. Therefore, the immune system is capturing of sensory information immune environment and responding to this with necessary adjustments to maintain homeostasis.

The model of allergies to stings of hematophagous arthropods depicts this process. It can be silent but evolves to be clinically noticeable, finalizing by tolerance. It is almost a portrait of a coevolution between two species in a short period of time. In this phasic development, the reactivity to allergens present in the saliva of insects is, almost always, the natural progression for the reduction or absence of clinical symptoms. Understanding this phenomenon may result in accelerating this natural process and benefit many individuals exposed routinely or possibly to this natural interaction between two bodies. On the other hand, this awareness can be used to generate possible vaccines by blocking the process of obtaining blood from these insects.

Acknowledgments We would like to thank Tony and Blazenka Matic for their assistance and review.

References

Agarwal MK, Chaudhry S, Jhamb S et al (1991) Etiologic significance of mosquito (*Anopheles stephensi*) in respiratory allergy in India. Ann Allergy 67:598–602

Arlian LG (2002) Arthropod allergens and human health. Annu Rev Entomol 47:395–433

Bilo MB (2011) Anaphylaxis caused by Hymenoptera stings: from epidemiology to treatment. Allergy 66(Suppl 95):35–37

Calvo E, Pham VM, Lombardo F et al (2006) The sialotranscriptome of adult male *Anopheles gambiae* mosquitoes. Insect Biochem Mol Biol 36:570–575

Chen YL, Simons FE, Peng Z (1998) A mouse model of mosquito allergy for study of antigen-specific IgE and IgG subclass responses, lymphocyte proliferation, and IL-4 and IFN-gamma production. Int Arch Allergy Immunol 116:269–277

Choumet V, Attout T, Chartier L et al (2012) Visualizing non infectious and infectious *Anopheles gambiae* blood feedings in naive and saliva-immunized mice. PLoS One 7, e5046

Demain JG, Minaei AA, Tracy JM (2010) Anaphylaxis and insect allergy. Curr Opin Allergy Clin Immunol 10:318–322

Deshazo RD, Feldlaufer MF, Mihm MC et al (2012) Bullous reactions to bedbug bites reflect cutaneous vasculitis. Am J Med 125:688–694

Dominguez-Amorocho O, Duarte S, Gonzalez JM et al (2013) Differences in systemic and skin migrating-specific CD4 T cells in papular urticaria by flea bite. Int Arch Allergy Immunol 160:165–172

Durate ID, Lorenzetti BB, Ferreira SH (1990) Peripheral analgesia and activation of the nitric oxide-cyclic GMP pathway. Eur J Pharmacol 186:289–293

Ellis AK, Day JH (2005) Clinical reactivity to insect stings. Curr Opin Allergy Clin Immunol 5:349–354

Fitzgerald KT, Flood AA (2006) Hymenoptera stings. Clin Tech Small Anim Pract 21:194–204

Gaig P, Garcia-Ortega P, Enrique E et al (1999) Serum sickness-like syndrome due to mosquito bite. J Investig Allergol Clin Immunol 9:190–192

Greaves MW (2007) Recent advances in pathophysiology and current management of itch. Ann Acad Med Singapore 36:788–792

Hamza E, Akdis CA, Wagner B et al (2013) In vitro induction of functional allergen-specific CD4+ CD25high Treg cells in horses affected with insect bite hypersensitivity. Clin Exp Allergy 43:889–901

Karppinen A, Brummer-Korvenkontio H, Reunala T et al (2012) Rupatadine 10 mg in the treatment of immediate mosquito-bite allergy. J Eur Acad Dermatol Venereol 26:919–922

Kelly DW (2001) Why are some people bitten more than others? Trends Parasitol 17:578–581

Killby VA, Silverman PH (1967) Hypersensitive reactions in man to specific mosquito bites. Am J Trop Med Hyg 16:374–380

Lima HC, Titus RG (1996) Effects of sand fly vector saliva on development of cutaneous lesions and the immune response to *Leishmania braziliensis* in BALB/c mice. Infect Immun 64:5442–5445

Mellanby K (1946) Man's reaction to mosquito bites. Nature 158:554

Mizuki M, Ueda S, Tagawa S et al (1998) Natural killer cell-derived large granular lymphocyte lymphoma of lung developed in a patient with hypersensitivity to mosquito bites and reactivated Epstein-Barr virus infection. Am J Hematol 59:309–315

Morsy TA (2012) Insect bites and what is eating you? J Egypt Soc Parasitol 42:291–308

Oliveira F, Traore B, Gomes R et al (2013) Delayed-type hypersensitivity to sand fly saliva in humans from a leishmaniasis-endemic area of Mali is Th1-mediated and persists to midlife. J Invest Dermatol 133:452–459

Owhashi M, Harada M, Suguri S et al (2001) The role of saliva of *Anopheles stephensi* in inflammatory response: identification of a high molecular weight neutrophil chemotactic factor. Parasitol Res 87:376–382

Peng Z, Ho MK, Li C et al (2004) Evidence for natural desensitization to mosquito salivary allergens: mosquito saliva specific IgE and IgG levels in children. Ann Allergy Asthma Immunol 93:553–556

Peng Z, Simons FE (2007) Advances in mosquito allergy. Curr Opin Allergy Clin Immunol 7:350–354

Pollack RJ, Marcus LC (2005) A travel medicine guide to arthropods of medical importance. Infect Dis Clin North Am 19:169–183

Ribeiro JM, Garcia ES (1981) Platelet antiaggregating activity in the salivary secretion of the blood sucking bug *Rhodnius prolixus*. Experientia 37:384–386

Rogers KA, Titus RG (2003) Immunomodulatory effects of Maxadilan and *Phlebotomus papatasi* sand fly salivary gland lysates on human primary in vitro immune responses. Parasite Immunol 25:127–134

Seon HS, Roh JH, Lee SH et al (2013) A case of hypersensitivity to mosquito bites without peripheral natural killer cell lymphocytosis in a 6-year-old Korean boy. J Korean Med Sci 28:164–166

Serafin WE, Austen KF (1987) Mediators of immediate hypersensitivity reactions. N Engl J Med 317:30–34

Simons FE, Peng Z (1999) Skeeter syndrome. J Allergy Clin Immunol 104:705–707

Singh S, Mann BK (2013) Insect bite reactions. Indian J Dermatol Venereol Leprol 79:151–164

Tan JW, Campbell DE (2013) Insect allergy in children. J Paediatr Child Health 49:E381–E387

Tokura Y, Ishihara S, Tagawa S et al (2001) Hypersensitivity to mosquito bites as the primary clinical manifestation of a juvenile type of Epstein-Barr virus-associated natural killer cell leukemia/lymphoma. J Am Acad Dermatol 45:569–578

Tonussi CR, Ferreira SH (1999) Tumour necrosis factor-alpha mediates carrageenin-induced knee-joint incapacitation and also triggers overt nociception in previously inflamed rat knee-joints. Pain 82:81–87

Tracy JM (2011) Insect allergy. Mt Sinai J Med 78:773–783

Part IV

Additional Considerations

Geoprocessing and Expected Distribution of Diseases (Including Deforestation, Global Warming, and Other Changes)

37

Raquel M. Gleiser

Abstract

In epidemiology, in order to estimate risks and plan control measures, it is very important to predict when and where a disease may occur, even in areas not previously studied. It is also relevant for decision making to estimate or predict a priori the impacts resulting from changes in land use, land cover, and climate, among others, on the potential occurrence, distribution, or incidence of a disease. The meteorological and environmental products derived from satellite imagery can be used to monitor some conditions that favor or alternatively preclude the proliferation of vectors or affect the transmission of pathogens. In this chapter, we intend to make an introduction of the use of remote sensing and GIS technology for the study and surveillance of vector populations and for risk assessment of vector-borne diseases. In addition, some of the relations between anthropic and environmental changes and selected vector-borne diseases are revised.

Keywords

Geographic information systems • Remote sensing • Risk mapping • Risk models • Climate change • Biotic homogenization

In epidemiology, in order to estimate risks and plan control measures, it is very important to predict when and where a disease may occur, even in areas not previously studied. It is also relevant for decision making to estimate or predict a priori the impacts resulting from changes in land use, land cover, and climate, among others, on the potential occurrence, distribution, or incidence of a disease. Zoonosis risk models aim to assess the relative suitability of an environment for the establishment of a disease, that is, to

R.M. Gleiser (✉)
Centro de Relevamiento y Evaluación de Recursos Agrícolas y Naturales, Instituto Multidisciplinario de Biología Vegetal (IMBIV), Universidad Nacional de Córdoba (UNC), Consejo Nacional de Investigaciones Científicas y Técnicas (CONICET), Av. Valparaíso Sn (5016) and Facultad de Ciencias Exactas, Físicas y Naturales, UNC, Av. Vélez Sársfield 299, Córdoba, Argentina
e-mail: raquel.gleiser@unc.edu.ar

C.B. Marcondes (ed.), *Arthropod Borne Diseases*, DOI 10.1007/978-3-319-13884-8_37

identify areas where pathogen, vector, and host may coexist spatially and temporally. These models may be developed based on two broad alternative strategies: biology-based models and statistics-based models (Malone et al. 2006). Biology-based models depend on the knowledge (from previous laboratory and/or field studies) of the biological requirements of the species involved: maps are generated for those regions matching a set of conditions previously deemed suitable. Statistics-based models basically correlate data on the spatial (and/or temporal) distribution of a pathogen and/or vector (or a disease) with a set of environmental data (frequently derived from satellite imagery) to determine which variables are more (empirically) relevant to predict distribution and then build maps of areas matching those conditions. They are especially useful when little is known about the biological requirements of the organism of interest. Nevertheless, to avoid producing models that are statistically rather than biologically relevant, predictors should be used that are (potentially) ecologically relevant to the target species, describing the biological and ecological constraints of the species in the spatial range to be modeled (Estrada-Peña et al. 2014). Although either strategy has its limitations, there are several examples of useful applications as well that will be discussed in this chapter.

Among the statistics-based models, the ecological niche models allow approaching the limits of tolerance of the species of interest from the observed data distribution. These models in general relate spatial point data of a species occurrence with geographic information on the environmental characteristics (including climate) of these sites, compared to a random set of point data from the same study area. The aim is to define a set of decision rules that best summarize the factors associated with a species and to produce a hypothesis of the niche dimensions of the species (Peterson and Shaw 2003). This model can then be projected for a region to estimate the potential distribution of the species, based on the identification of zones that either have or lack the niche condition. Models associating climate variables to the current known distribution of a species are used for projecting the future distribution under different climate change scenarios. These models in general do not take into account interspecific interactions that may affect a species distribution in spite of otherwise adequate environmental conditions.

In any case, both strategies take advantage of the advances in geographic information systems (GIS), Global Positioning System (GPS), and satellite imagery availability. A GIS is a computer-based system designed to capture, store, manipulate, analyze, manage, and display digital geographical data. A GPS is a satellite navigation tool used to determine the position of an object or landscape element on the Earth's surface. Satellite imagery is captured with remote sensors and provides a digital record with excellent spatial and temporal references, covering larger land extensions compared to traditional land survey methods. Data provided by satellite imagery do not (generally) allow the identification of pathogens or vectors per se but may be used to characterize the environment where they develop. Over the last two to three decades, lower costs of electronic and computer equipment and progress in computing and technology of geographic information systems (GIS), Global Positioning Systems (GPS), and remote sensing have favored the development of models for surveillance of pests and vectors, as well as for predicting risk of diseases they transmit. With appropriate field studies they can be used to identify and map with good probability potential habitat of specific vectors and parasites. As historical databases, information on the vector or pathogen distribution in the past can be used to indicate possible future situations, pointing out areas or seasons of higher risk of emergence of problems, and the regions from which new risks could arise.

Currently, sports cameras and other similar economically accessible portable gadgets (such as cell phones) that record GPS-encoded videos can be used to collect fine-scale spatial data that could eventually be used for surveillance and risk mapping (Lee et al. 2016). For an example of these recent developments, see Curtis et al. (2013).

In this chapter, we intend to make an introduction of the use of remote sensing and GIS technology for the study and surveillance of vector populations and for risk assessment of vector-borne diseases. Also, we revise some of the relations between anthropic and environmental changes and selected vector-borne diseases.

37.1 Remote Sensing and Satellite Imagery

Remote sensors are instruments that can detect the nature of an object without being in direct contact with it. Familiar examples of remote sensors are our eyes and ears. Some instruments measure how surfaces reflect or absorb different bands of electromagnetic radiations, such as visible light, ultraviolet, and infrared rays. Figure 37.1 illustrates a wide spectrum of electromagnetic radiations that may be detected in nature. Each substance reflects or absorbs radiation in a characteristic pattern, resulting in unique "spectral signatures." Thus, we see plants green because chlorophyll absorbs most of the visible range radiations and partially reflects green wavelengths (Fig. 37.1). Through the analysis of the interactions between electromagnetic radiation and matter, the presence of a wide variety of phenomena, elements, components, and bodies both in the Earth's surface and in the oceans and the atmosphere can be detected and even accurately measured. A widely used index is the normalized difference vegetation index (NDVI) that is calculated as the difference between near-infrared and visible reflectance values (most reflected and absorbed, respectively, by plants) normalized over the sum of the two. Still, it should be noted that, although it would be expected that each land cover/Earth's surface material would have a unique distinctive spectral response pattern that would allow it to be reliably detected by visual or digital means, this is often not the case, due to variations in moisture content, soil type, and phenological state of the vegetation,

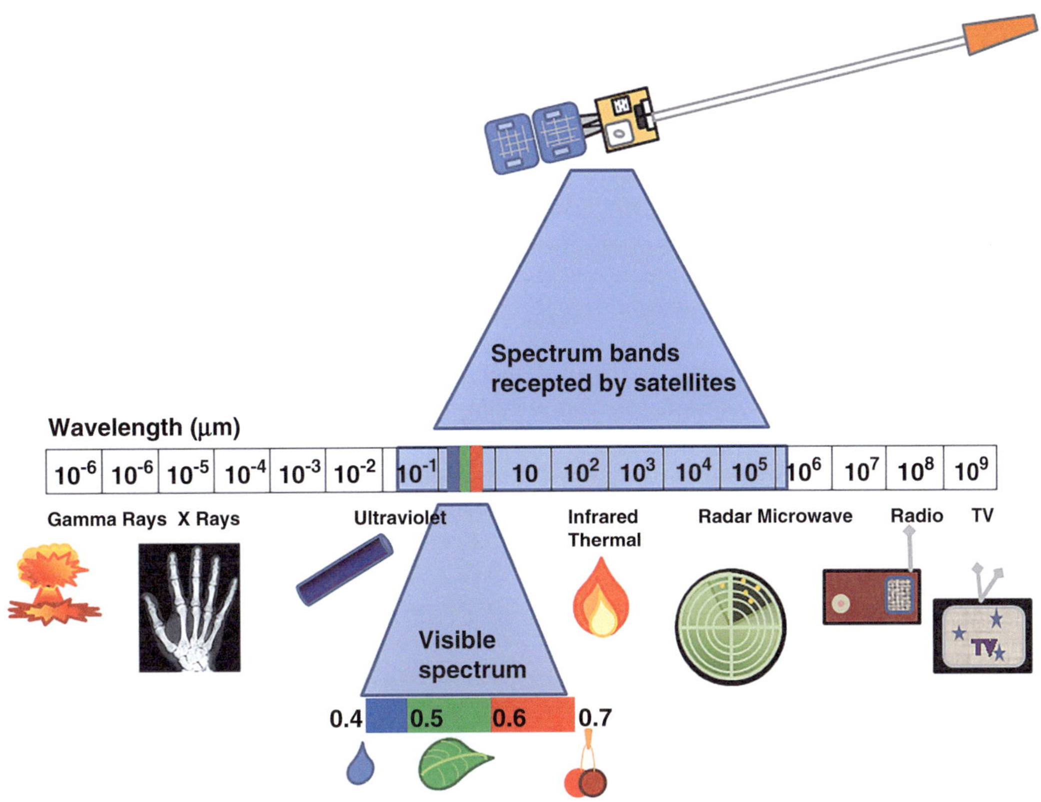

Fig. 37.1 Spectral electromagnetic radiations that can be detected in nature

among others. Rather, the combination of spectral bands and the time of year at which distinctive patterns can be found for each of the information classes of interest must be found.

The sensors on board satellites create the images as a sequence of small rectangular portions of a scene, known as picture elements or "pixels," arranged in a geometrically regular grid. The data sequence is readily reconstructed to form an image. The intensity of each pixel is represented by numbers, which indicate the different values of reflectance for a particular spectral band, facilitating the handling, processing, and data analysis using digital computers. Although the images are very seductive, actually the numerical data provide more information and allow a better analysis of the phenomena to be studied.

The process of developing interpreted maps from satellite images is known as classification and was traditionally achieved through visual interpretation of terrain features and manual delineation of their boundaries. Currently, human interpretation is added to the computer aided, where various algorithms are applied to assign pixel elements to groups of classes. The classification process can follow one of the two basic approaches, known generically as supervised and unsupervised classification. The *supervised classification* is a decision-making process based on a priori knowledge of the way in which certain ground covers or classes reflect the spectrum bands considered (spectral signatures). First, areas or sites are delimited whose characteristics are known in the field (called ground truth), and then the software generates a statistical description of how the different bands reflect the spectrum in these areas (land cover classes). Next, a statistical procedure is used to assess the likelihood for each pixel of the unknown areas of the image to belong to one of these classes. By contrast, with the *unsupervised classification*, classes are unknown at the beginning, and the process seeks to discover the common occurrence and distinct reflectance patterns in an image, with the assumption that these represent important classes of land cover. The identity of each class is then determined by combining experience and ground truth, namely, by visiting the area of study and observing the actual types of land cover.

Even considering that all electromagnetic radiation, one way or another, acts on the matter, there are restrictions on the type of radiation that can potentially be used to know remotely the Earth's surface materials, as the atmospheric gases absorb certain wave lengths. The electromagnetic registry tracks that can be used are known as "atmospheric windows." Additionally, only wavelengths in the microwave band pass through the clouds and other particles suspended in the atmosphere.

Since the 1960s several satellites bearing remote sensors were launched. Geostationary satellites orbit at about 41,000 km of altitude, and their movement is synchronized with the Earth at such a speed that they always face the same hemisphere. As they transmit very frequent images, they are particularly useful for climate observations, although their spatial resolution is low. Other satellites, known as polar orbiting or simply polar, cover an orbit that passes near the poles and progressively scans the surface of the Earth, sending images of the entire planet. They are in lower heights (between 600 and 950 km) and provide images of higher spatial resolution, but with lower temporal resolution. These satellites run synchronized orbits with the sun, that is, their orbits remain fixed relative to the sun, so that they routinely see the same point in the same local time.

Most of these sensors are optically passive because they only measure the energy reflected from the sun, or reemission of solar (or volcanic) radiation stored as heat in the Earth's surface, not allowing information of the Earth's surface through cloud cover. This can be a major problem in some regions or periods of the year, in which the most coverage coincides with the satellite passing time. An alternative is taking photographs from aircraft which may carry the same type of sensors as satellites. While the problem of clouds can be avoided, its use is limited to small areas, for practical problems and cost. Other options are the active sensors, such as synthetic aperture radar, which can provide high-resolution data uninterrupted by cloud cover. Table 37.1 summarizes the main characteristics of some satellites available to the public.

Table 37.1 Summary of the main characteristics of some satellites available to the public. Temporal resolution refers to the image acquisition rate for a particular site

Satellite[a]	Source	Orbit height (km)	Temporal resolution	Bands or channels[b]	Spatial resolution	Width imaging swath (km)	More information
Sun synchronous orbit							
SAR-Lupe (series)	Germany	500	<24 h	X (3.1 cm 9,65 GHz)	0.5 m; 1 m	5.5; 8	http://www.lsespace.com/
TerraSAR	Germany	514	1–3 days	X (31 mm, frequency 9.6 GHz)	1 m	10–100	http://ilrs.gsfc.nasa.gov/
IKONOS	USA (commercial)	681	3 days	Vis, NIV, Pan (5)	1 m, 4 m	11–14	http://www.digitalglobe.com
OrbView	USA (commercial)	470	<3 days	Vis, NIV, Pan (5)	1 m, 4 m	8	http://www.digitalglobe.com
QuickBird	USA (commercial)	450	1–3 days	Vis, NIV, Pan (5)	55 cm, 2.16 m	14.9	http://www.digitalglobe.com/
IRS (series)	India	817	5 and 24 days	Vis, NIR, MIR (4)	24 m; 2.5–360 m	70–7400	http://dos.gov.in/satellites.aspx
SPOT (series)	France	830	26 days			2250	https://earth.esa.int/
HRV-IR				Vis, NIR, MIR, Pan	5 m, 10 m, 20 m		
Vegetation	France (+other EU)		Daily	Vis, NIR, MIR, (4)	1000 m		http://www.vgt.vito.be/
Radarsat	Canada	798	24 days	Microwaves, C (5.3 GHz)	8–100 m	50–500	http://gs.mdacorporation.com/
	(+industry, NASA)						
Landsat	USA	705	16 days				http://www.nasa.gov/
				VNIR, SWIR, LWIR, pan (8)	15–100 m	170	
EOS/Terra	USA and multinational	705	16 days				
Terra ASTER	(Canada, Japan)			Vis, NIR, SWIR,TIR (14)	15–90 m	2330	http://terra.nasa.gov/
Terra-Água MODIS				Vis a SWIR (36)	250–1000 m		
CBERS (series)	Brazil-China	778					http://www.cbers.inpe.br/
WFI			5 days	Vis, IR	260 m	890	
CCD			3–26 days	Vis, NIR (4)	20 m	120	
IRMSS			26 days	pan + IR,TIR (4)	80 m	120	
EUMETSAT	Mainly Europe (intergovernmental organization)						http://www.eumetsat.int/
METOP/AVHRR		817	12 h	Vis/IR, TIR (6)	1000 m	1464	
HIRS				Vis (1), IR (19)	10,000 m	2160	
AMSU-A		837		Microwave (15ch; (23–90 GHz)	48 km	2074	

(continued)

Table 37.2 (continued)

Satellite[a]	Source	Orbit height (km)	Temporal resolution	Bands or channels[b]	Spatial resolution	Width imaging swath (km)	More information
NOAA	USA	870	12 h				http://www.ncdc.noaa.gov/ http://noaasis.noaa.gov/
AVHRR				Vis/NIR, TIR (6)	1100 m	2700	
Geostationary orbit							
GOES (series)	USA	35,800	≤26 min	Vis, IR, TIR, other atmospheric	1000–8000 m	Hemisphere	http://www.nsof.class.noaa.gov/ http://www.goes.noaa.gov/
EUMETSAT Meteosat (series)	(Intergovernmental organization)	36,000	15 min	Vis, IR (12)	1000–3000 m	Hemisphere	http://www.eumetsat.int/

Sun synchronous orbit = polar orbiting
Vis visible, *IR* infrared, *LWIV* long-wave infrared, *NIV/VNIV* near infrared, *SWIV* short-wave infrared, *TIV* thermal infrared, *WV* water vapor
[a]Only some sensors aboard the satellites are included on the Table
[b]The number in brackets indicates the approximate number of detected spectrum bands

Depending on the sensor characteristics, the satellite images may have different resolutions: spectral or radiometric, spatial or geometric, and temporal. The spectral resolution refers to the spectrum of the electromagnetic field represented in each band and is the basis for the development of spectral signatures identifying the various elements of landscape or Earth's surface features. The spatial resolution is the smallest area of land which can be individualized. For example, in an image with a spatial resolution of 1 km, each pixel represents the average reflectance spectrum within a 1 × 1 km parcel of land, and so elements measuring few meters in diameter cannot be discriminated. However, if they represent values sufficiently contrasting against their surrounding area, their presence may eventually be inferred by the effect they have on the average pixel reflectance. The temporal resolution indicates the time interval between two successive passes of the satellite over the same spot on the earth (see Table 37.1).

From the data collected by remote sensors, different products can be derived, including meteorological products, vegetation indices, digital elevation models (DEM), and groundcover. The meteorological products (precipitation, temperature, humidity), as exemplified in the following sections, can be used to monitor the conditions that favor or alternatively preclude the proliferation of vectors or affect the transmission of pathogens. Vegetation indices and land cover maps are used to identify or define suitable habitats for vectors or reservoirs, as well as suitable conditions (e.g., deforestation exposure). In turn, the vegetation can be an indicator of humidity and availability of larval habitats (Linthicum et al. 1991; Gleiser et al. 1997; Gleiser and Zalazar 2009). The DEM can be used to detect low and flood prone areas where larval habitats are more frequent or to analyze the distribution of the species, since the elevation/height can limit the distribution of vectors and their associated pathogens (McCann et al. 2014).

During the late 1990s, a series of very resourceful large-scale studies was developed to map the habitats of tsetse flies, based on the analysis of climate data and vegetation provided by remote sensors (Rogers et al. 1996). Their findings of correlations between NDVI, standard meteorological variables, and estimated mortality rate of *Glossina* allowed to predict regions where the flies could not occur. Further studies that integrated more information and more complex analyses improved the accuracy of the maps and showed habitat differences between subspecies, besides producing risk maps of disease transmission (Hendrickx et al. 1999).

NDVI may also be used as a surrogate ("proxy") to other variables that affect the problem populations. Thus, the abundance of *Culicoides imicola* (Diptera: Ceratopogonidae), vector of African horse sickness (AHS), is affected by soil moisture level, which in turn affects vegetation characteristics. Studies in Morocco showed a positive correlation between the average annual abundance of this gnat and the minimum annual average value of the vegetation index (Baylis et al. 1998). Similarly, the distribution and abundance of the tick *Rhipicephalus appendiculatus*, vector of *Theileria parva* group piroplasmas, are affected by the minimum deficit and the saturation temperature variables which were related to NDVI. Larval abundance is related positively with the NDVI values, due to the effect that the above climatic conditions have on variations in vegetation, detected by satellite. In South Texas, within the context of an eradication program of the Southern cattle tick, *Rhipicephalus (Boophilus) microplus* (Canestrini) (Ixodida: Ixodidae), sampling sites of the invasive tick were selected by first locating favorable white-tailed deer (tick host) habitat using NDVI and another vegetation index (modified soil-adjusted vegetation index) (Phillips et al. 2014).

Studies conducted in Kenya were pioneers in demonstrating that it is possible to combine passive and active remote sensing in surveillance and warning programs of animal diseases, because they provide ecological parameters associated with the earliest indicators of viral activity (Linthicum et al. 1991). Based on 3-week composed NDVI images from NOAA satellite (15 km resolution), a potential viral activity factor was developed that allowed prediction of time and areas of higher risk of mosquito-borne Rift Valley fever (RVF) epidemics. Outbreaks are associated with the amount and duration of rainfall flooding the mosquito habitats ("dambos"). By comparing dry and wet season data obtained from satellite with higher resolution, such as Landsat TM sensors, larval habitats of a size similar to that of the satellite resolution (50 m) can be identified. This is possible because the vegetation and the soil of dambos are characterized by higher moisture content than the surrounding areas, keeping the vegetation greener during the dry season, with less seasonal variation in the reflectance of vegetation. The NDVI values of SPOT allow not only to locate the potential larval habitat but also to differentiate those that are flooded and appear with higher values than adjacent areas.

A multi-criteria analysis (considering coverage variables, distance to rice paddies, human density, temperature, elevation, etc.) permitted the development of malaria risk maps for different geographical areas of Madagascar (Rakotomanana et al. 2007). Environmental variables were estimated from Landsat 7 ETM+images (resolution 30 m), SPOT 4 (resolution 10 m), Synthetic Aperture Radar Precision Images (SAR PRI) of ERS-2, and Advanced Synthetic Aperture Radar Image Mode Precision images (ASAR_IMP_1P) from Envisat. In Kenya's western plains, larval *Anopheles gambiae* s.l. habitats are generally small and murky water, and therefore it is difficult or impossible to detect them even in high-resolution images (ca. 1 m) as IKONOS. However, this type of image was useful for finding larval habitats indirectly, from the associations between the presence of habitats and specific types of land cover, which can be distinguished through image classifications (Mutuku et al. 2009). Landsat TM (30 m) and IKONOS (4 m) images were used to map coverage and changes in land use in Kenya and evaluate their relation with suitable habitats for *Cx. quinquefasciatus*. The images were classified into three classes: built, rice fields, and fallow. Most preponderance of positive aquatic habitats for *Cx. quinquefasciatus* was associated with areas in which temporal modifications were observed, mainly paddy to fallow and vice versa (Jacob et al. 2006).

In South Korea, Landsat 7 ETM+and a QuickBird images were used to identify by supervised classification classes of land use that were considered of interest to estimate malaria vector mosquito habitats. Significant correlations were found between land use classes and positive habitats for larvae and adult *Anopheles sinensis* abundance. The larvae were concentrated mainly in rice paddies, while adults were more common in bare soil and forest, although these relationships varied geographically. For this species and *An. pullus*, *An. sineroides*, and *An. lesteri*, negative correlations between the distribution of larvae and adults were detected, suggesting that adults leave the rice fields to search for human blood meals.

On the other hand, not always are images, even with high resolution, effective to locate larval habitats. In Belize, SPOT (10 and 20 m resolution) and IKONOS (1 and 4 m resolution) images were assessed to locate positive habitats for *An. darlingi*. In this region, the species is associated with bamboos and fallen trees constituents in watercourses. Although different habitat evaluation criteria (supervised and unsupervised classification) were explored, no image or test strategy was efficient to sort efficiently sets of bamboo, as these could be mixed with other plant species. Nor were significant correlations found between positive mosquito sites and land cover categories (Achee et al. 2006).

Another study in Connecticut (USA) on West Nile virus vectors also failed to identify predictors of abundance of *Cx. restuans* from Landsat TM, ETM +, MSS, and ETM, possibly because this species uses a large variety of larval habitats. However, significant correlations were found for *Ae. vexans*, *Cx. pipiens*, *Cx. salinarius*, and *Cs. melanura* and NDVI-derived parameters (Diuk-Wasser et al. 2006).

The examples described above generally classify images in land cover or land use categories or extract the reflectance values themselves. Images can also serve only as a reference to digitize (or locate) sites and landscape elements analyzed in the field, as shown in a study on the spatiotemporal pattern of reinfestation by *Triatoma guasayana* in northwest Argentina (Vazquez-Prokopec et al. 2005). In this paper, IKONOS images of

1 and 4 m resolution were used to digitize the sites visited in the field, vegetation, and other landscape elements such as houses, which soon served as reference to digitize premises and constructions visited on the field, but not visible in the images.

The epidemiological risk models developed for a particular region may serve as a reference or starting point for application in other areas or other similar diseases. However, direct extrapolation from one area to another is not recommended without validation studies in the field, as there may be subtle differences in the natural environment and/or the human activities that affect markedly the validity of the proposed model. For example, in Thailand the seropositivity risk factors for dengue differ between urban and rural sites. In rural areas, people living in the center of towns and further away from natural or agricultural coverage would have increased risk of seropositivity, while in periurban areas, risk factors would be linked to other factors, such as knowledge of the disease, use of preventive measures, and quality of homes, rather than the land cover features per se.

37.2 Changes in Land Use, Anthropogenic Effects, and Arthropod-Borne Diseases

Human activities such as agriculture, farming, and urbanization, among others, cause significant changes in the environment and generate mosaics of different types of coverage with similar patterns in different parts of the world. One of the most severe land uses is urbanization, an expanding process due to a change from rural to urban populations. There is great concern about the transformation and loss of natural environments on a large scale that is occurring as a result of human activities, which lead on one side to biodiversity loss and increased risk of species extinction, and secondly, an increase in the problems associated with pests and pathogens vectors.

Land use changes linked to human activity can influence the distribution and incidence of arthropod vector-borne diseases. For example, by modifying the availability and distribution of water collections, either intentionally (creating water reservoirs, prevention of flooding, wetland drainage) or unintentionally (changes in relief, accumulation of waste and artificial containers, etc.), the new conditions can benefit the production of diverse mosquitoes. In the Amazon basin, the reemergence of malaria has associated with deforestation, which has encouraged the proliferation of the predominant vector, *Anopheles darlingi*. This mosquito develops in sunny pools generated with the removal of forest cover, but not in intact shaded forests. In turn, the human advance on the forest facilitates contact with other species of mosquito vectors and pathogen reservoirs.

Cities are highly disturbed and fragmented complex systems characterized by the presence of patches of different types, sizes, and shapes, generally smaller in relation to natural environments, with extreme environmental conditions, particular microclimates, and high (or, alternatively, low) nutrient availability for different species. Strong impacts of urbanization on the environment include, among others, fragmentation and degradation of habitats, deposition of heavy metals and other pollutants, changes in flow and nutrient cycling, creation of heterogeneous microhabitats, and a change in temperature due to the presence of various greenhouse gases, forming new environments where the thermal variation is reduced, so that various species of animals and plants can establish and persist throughout the year in the same sites. These conditions have diverse impacts on the structure, function, and dynamics of ecosystems, affecting biodiversity, biogeochemical cycles, and climatic conditions. In turn, the complex nature of urban habitats can have different effects on biotic communities, fostering a strong dominance of those species best adapted to humans and their activities. Elevated levels of disturbance often favor small species that can quickly take advantage of unpredictable resources, while larger species and predators are often disadvantaged, but this pattern (Martinson and Raupp 2013) is not always verified. Studies in birds, mammals, and insects indicate that increasing the degree of urbanization increases the abundance of nonnative or cosmopolitan species, so it is possible to find increasing and decreasing gradients of species throughout the landscape. While some studies show that urbanization can promote increased levels of biodiversity due to the addition of exotic species that replace the native faster than these are lost, globally urbanization can produce a decrease in species diversity, since, in general, many exotic species manage to exploit these new heterogeneous habitats, replacing native species and producing a biological homogenization of these environments.

Biotic homogenization is a process by which native communities are gradually replaced by range-expanding, cosmopolitan, exotic communities. Among the many human activities, urbanization is the main promoter of biotic homogenization, one reason being that cities are physically very similar throughout the world. Synanthropic species are considered "global homogenizers" because they may be found in cities worldwide, reducing biodiversity at a global level and resulting in biotas from distant cities resembling more each other than their own native ecosystems. The mosquitoes *Ae. aegypti*, *Ae. albopictus*, and *Culex pipiens/quinquefasciatus*, the housefly *Musca domestica*, and the cockroaches *Periplaneta americana* and *Blattella germanica* are but a few examples of arthropods of medical relevance that can be found in many cities worldwide.

It is still unclear whether urbanization influences in the same way on the arthropod fauna in different parts of the world. While some groups of species respond consistently in different cities, others vary depending on local conditions (Martinson and Raupp 2013). For example, *Muscina stabulans*, a cosmopolitan fly of medical interest as a mechanic vector of pathogens, has shown a higher degree of synanthropism in temperate Buenos Aires, Argentina, compared to studies in more tropical areas in Brazil (Patitucci et al. 2010).

Very diverse communities of potential hosts have been associated with a lower risk of infection to various arthropod-borne diseases, probably reducing human exposure to pathogens by a "dilution effect." As a result, biodiversity losses associated with deterioration of natural ecosystems could have a negative effect on the risk of human infection. Keesing et al. (2006) discuss several mechanisms underlying the effects of species diversity on disease transmission. Among vector-borne diseases, some of them include encounter reduction between susceptible host and vector by redistributing vector meals, regulating the abundance of susceptible host species (by interspecific competition, predation etc.), and vector regulation. A connection between species diversity and transmission of vector-borne diseases of humans is the base for zooprophylaxis, which is the use of nonhuman animals to protect human health by diverting vectors searching for a meal away from humans. This approach has been proposed to reduce malaria transmission. It has also been suggested that under certain conditions, increased diversity could actually increase the risk of transmission, when the provision of additional meal sources increases vector numbers or activity, or when added species function as alternative sources of infection, especially when an added species in a species-poor community is actually the most competent reservoir, resulting in an amplification of disease risk.

Human activities can contribute to the expansion of the range of a species to regions where it had not occurred, a phenomenon known generally as biological invasion, by deliberate or accidental introduction. The introduction of vectors and/or pathogens can have direct impacts on human and domestic animals' health and on native fauna. It has been hypothesized that the spatial expansion of *Anopheles arabiensis* and *An. gambiae* s.l., mosquito vectors of malaria, could be linked to the development of agriculture in Africa, south of the Sahara, through the significant production of new larval habitats and availability of blood sources, both from humans and domestic animals. *Anopheles arabiensis* and *An. gambiae* s.s. are usually sympatric species in several African regions, although *An. arabiensis*, more adapted to arid areas, has a wider distribution. Actually, based on genetic-based studies, it seems the later species is expanding from the savannas to drought prone areas, where it could substitute *An. gambiae* s.s. as a main vector (Matthews et al. 2007).

Habitats modified as a result of urbanization are more susceptible to invasion or colonization by alien species than those undisturbed. One hypothesis to explain the success of invasive species is that they exploit "empty niches" not used by native species. Human activities generate empty niches or "niche opportunities" by providing resources, reducing or removing the threat of natural enemies or competing species, and/or altering the physical environment (e.g., temperature, creation of new habitats) improving habitability for the invader. For instance, *Aedes aegypti* (Diptera: Culicidae), the principal vector of dengue in the world, native to Africa, is a mosquito markedly adapted to anthropogenic environments that uses artificial water containers as larval habitats and females preferentially ingest human blood. Among the factors that favor their proliferation are rapid urban area growth with little planning, water storage containers in areas where the supply is low or unpredictable, and high production of disposable containers that are adequate larval habitats. Consistently, surveys of larval habitats used by *Ae. aegypti* in general concentrate in artificial containers (Vezzani and Carbajo 2008 and references therein). Although this mosquito may use natural cavities such as tree holes (as in its native range) that could allow it to disperse to natural environments, their presence is usually restricted to the urban environment (Mangudo et al. 2015).

Changes in the composition and structure of natural ecosystems by invasive species can alter pathogen transmission cycles by favoring (or alternatively reducing) the development conditions of the vectors and changing the contact of the vectors with reservoir and accidental hosts. For example, in an endemic area of Lyme disease in the United States, in areas dominated by exotic shrubs, the number of larvae and nymphs/ticks searching hosts was twice that in regions dominated by native species. In this region, feeding habits of white-tailed deer (tick hosts) favor the dominance of invader shrubs. Therefore, the invasion of exotic plants favored by deer grazing increases the risk of exposure to the vector of Lyme disease (Elias et al. 2006). The expansion of *Juniperus virginiana* tree from the Eastern United States into former grassland areas in the west may influence mosquito populations by positively or negatively altering host availability as a consequence of changes in the landscape. Moreover, populations of the invasive mosquito *Ae. albopictus* may benefit from high-quality *J. virginiana* litter components' input into container habitats. In experimental conditions, this plant material could support positive population growth rates as a function of initial larval density where the native leaf could not (Reiskind and Zarrabi 2011). *Lonicera maackii* is an Asian shrub that was introduced to the United States as an ornamental plant in the late nineteenth century. In the Eastern United States, the forest invasion by *L. maackii* alters forest understory structure and the oviposition behavior of container mosquito *Ochlerotatus triseriatus* (Say) (=*Aedes triseriatus*), the primary vector of La Crosse virus (LACV) in the United States and potential vector of other arboviruses (yellow fever, eastern equine encephalitis, and dengue). Greater oviposition rate is associated with plots surrounded by larger proportions of grassland or cropland in uninvaded forest, while in invaded forest, oviposition rates are not linked to the availability of open habitat. Since the presence and density of *L. maackii* may be difficult to assess or be undetectable in the most commonly used satellite images, invasion by this invasive plant may negatively affect the accuracy of remote sensor-based risk models (Conley et al. 2011).

The introduction of an alien species can, moreover, alter the abundance of other and even cause local extinction of native species, thus contributing to loss of biodiversity and landscape homogenization. It has been shown, for example, that an increase in larval competition between *Ae. aegypti* and *Ae. albopictus* mosquitoes increases the proportion of the second species that transmits dengue virus. The presence of *Ae. albopictus* decreases *Oc. triseriatus* larval survival, but in turn, from surviving larvae emerge larger adult females that are more likely to develop disseminated virus infections, thereby affecting the risk of pathogen transmission. The *Chrysomya albi-*

ceps fly along with two others of the same genus was introduced to the New World about 35 years ago and since then has spread throughout America as far south as Argentina. Since its introduction, it is showing a strong impact on the structure of local communities, replacing (removing?) native species such as *Cochliomyia macellaria* and *Lucilia sericata.* During the larval stage, *Chrysomya* exhibits aggressive behavior based on a combination of competition and predation, including killing and eating prey that uses the same resource (intraguild predator).

Other human activity factors besides land cover change that have been cited as main contributors to the reemergence of vector-borne diseases such as dengue and other flavivirus include human population density, poverty and associated poor sanitation conditions, inadequate vector (mosquito) control, and migrations and international travel. International commerce of used tires started and/or aided a rapid global dispersal of species such as *Ae. aegypti* and *Ae. albopictus*, by dispersal of their drought-resistant eggs. The latter species is native from Asia and has medical relevance as an aggressive and efficient vector of diverse flavivirus (including dengue, chikungunya, yellow fever, West Nile fever), alphavirus, and bunyaviruses. Risk maps of global *Ae. albopictus* dispersal based on ecological niche modeling combined with risk analysis due to tire importation and proximity to infested countries indicated that all South American countries had adequate areas for the establishment of the vector, though Suriname, Ecuador, and Chile are still free from this vector (and in Argentina the species has only been reported from a small area, northeast of the country). If the restrictions and controls on used tire commerce are maintained, Chile would be in an advantageous situation because the areas that are adequate for the mosquito are geographically isolated from the neighboring infested countries.

The patterns of daily activities and housing characteristics in a community may increase or decrease vulnerability. For example, air conditioning and TV reduce time outdoors and contact with mosquito vectors (Reiter et al. 2003). The current tendency to a decreased immune competence of human populations, as a consequence, for example, of HIV infections, chemotherapies, an increase in the elderly population in developed countries, etc. could enhance the risk of vector-borne diseases. On the other hand, the application of vector control measures, chemotherapy, and vaccination may be significant for reducing risk. These factors should be taken in account together with meteorological (or climatic, in climate change scenarios) and landscape variables when estimating risk of vector-borne diseases.

37.3 Climate and Arthropod-Borne Diseases

Over the last century, an increase in global temperatures has been recorded, more pronounced in elevated latitudes of the northern hemisphere. Consistent with this pattern, a decrease of the snow and mountain glacier cover has been recorded in both hemispheres, together with an expansion of the surface undergoing seasonal freezing. Since arthropods are small ectothermic organisms, that is, they control their body temperature through means that are external to their bodies, they are particularly sensitive to fluctuations of temperature and humidity levels. Consequently, their distribution and habitat use are frequently linked to climate. Temperature and soil moisture may affect arthropods' distribution directly through their physiological tolerance, but also indirectly, through their impact on the availability of adequate habitat and hosts (in the case of parasites).

Considering that climate limits species distribution, it is expected that climate change will lead to changes in several species, especially expansions or shifts to higher latitudes and altitudes. Climate change could particularly affect pathogens, since they spend an important part of their cycles in invertebrate hosts, whose temperature is maintained close to environmental conditions. In fact, successive reports from Intergovernmental Panel on Climate Change (IPCC) predict an expansion of the distribution of vector-borne diseases from tropical to temperate regions, an increase in the incidence of pathologies, and the emergence (or reemergence) of new diseases.

Temperature changes in a region may impact disease transmission by vectors through varying effects on the host, vector, and/or pathogen communities. Numerical responses include amplification and emergence of vector populations associated with changes in their development rates and survival and alterations in the seasonal disease transmission. Functional responses include biotic expansions of the hosts (and vectors) with potential to colonize new geographic regions, changes in habitat use, changes of ecotones and contact zones, local extinctions, changes in feeding behavior, and host exposure, modifying the susceptibility of the vector to the pathogens. Microevolutive responses include ephemeral patterns of local adaptation, directed changes in the frequency of genes through mutation, and selection of vectors or pathogens associated with the emergence.

At a regional or ocean basin scale, other climatic changes have been observed besides warming. In parts of North and South America, north of Europe, and Central-North Asia, rainfall will increase, while on the Sahel and south of Africa, Mediterranean, and South Asia, they are expected to decrease. Also, it is likely that the frequency of heavy rainfall increases. In general, it is assumed that increases in rainfall could enhance vector-borne disease transmission by diverse mechanisms. Higher rainfall could boost vegetation and allow the expansion of vertebrate host populations. Flooding may force more close contact between vertebrate hosts and humans or alternatively can eliminate both vector and their hosts' habitats. In the case of mosquitoes, the amount of rainfall affects the availability, persistence and dimensions of larval habitats, as well as the formation of new larval sites. On the other hand, rainfall could have a detrimental effect on larval survival, by flushing them out of their larval sites and killing them, mainly in early developmental stages. It has being observed that pupae developing in containers tend to spend more time away from the water surface than those dwelling in open water bodies. This kind of behavior would contribute to avoiding flushing from an overflowing container and also help the pupae survive from potential mechanic shocks from raindrops. During a rain episode, water currents can exhaust the larvae if they actively dip away from the surface to avoid being hit by raindrops or being swept away by the currents.

Not only volume of rainfall but also their frequency, that is, how isolated are rain events in time, may affect vector production. For example, explosive population peaks of the floodwater mosquito *Oc. albifasciatus* (vector of west equine encephalitis) are associated with fluctuations of the flood level of the larval habitats, depending not only on rainfall amount but also on rain frequency (Gleiser et al. 1997, 2000) (very frequent rains do not allow for the ground pool to shrink and expose enough mud/soil surfaces for egg laying). Wetland mosquito outbreaks following drought years could also be explained by the elimination of mosquito predators and competitors when (semi)permanent wetlands dry (Chase and Knight 2003). On the other hand, droughts reduce waterways flow, creating small stagnant pools where a number mosquito species can develop and where vectors and hosts congregate enhancing the potential for arbovirus transmission (Shaman et al. 2002).

Several studies show the convenience of considering only (or mostly) the effects of diverse meteorological variables or their estimators to predict changes in the vector population. Malaria research in Eritrea validates a combination of degree-days1[1] with a relation between rainfall and evapotranspiration rate as a biologically relevant measure of the habitat moisture conditions and temperature allowing the development of pathogens and vectors (Malone et al. 2006). Dynamic hydrologic models have been used to represent the availability of *An. walker* larval habitats. However, one should not forget that, even though temperature and humidity are important factors for the transmission of pathogens by

[1]Degree-days are temperature units above a threshold, accumulated over 24-h periods, that are used to measure or represent the physiological time (or age) of a poikilotherm organism, i.e., the amount of heat the organism requires to develop from one stage to the next.

arthropods, the relations between disease incidence and biophysical environment are not simple, and thus only using meteorological variables is usually not sufficient to explain development times, behavior and geographic distribution of the vectors, and the pathogens they transmit. Moreover, even conditions experienced during larval development may impact interactions between the emerging adult hosts and parasites. In the following sections, some examples are presented on relations between anthropic and environmental changes and some vector-borne diseases. These diseases (and other related) were selected from the IPCC (McCarthy et al. 2006) as those most sensitive to climatic changes.

37.3.1 Malaria

Even though it is usually assumed that malaria is a tropical disease, up to the mid-nineteenth century, this ailment affected latitudes as high as Sweden. This is because, even though climate limits where transmission may occur, other environmental, economical, political, and social factors may positively or negatively affect the risk of transmission. The deterioration of health and vector control programs, the emergence of drug resistance, changes in land use, construction of dams, and irrigation networks (which provide larval breeding) may affect the transmission. At the same time, this may only take place in spatial and temporal adequate zones in climatic terms.

The basic reproductive number (R_0) is generally defined as the expected number of new hosts who would be infected after one generation of the parasite by a single infectious person (host) introduced into an otherwise nonimmune population. Thus, the number of people infected by the plasmodium increases when R_0 is larger than one, while it declines if R_0 is less than one. The principal component of the reproductive rate is the vectorial capacity, which is determined by complex interactions between the host, vector, parasite, and environmental factors (see Chap. 2). Climate has a big influence on vector capacity, since variables such as mosquito population density and survival, blood meal frequency, and the mosquito intrinsic incubation period depend on temperature. In laboratory studies, it was observed that temperatures under which optimal survival levels were detected for *An. gambiae* were lower than those where development was faster, suggesting a critical relation between temperature and the insect's life cycle. Longevity of larval stages affects the survival rate of adults and, thus, the intensity of disease transmission. Rainfall may also have an effect on vector populations, for example, from their relation with larval habitat formation. Both in South America and in Asia, malaria prevalence generally varies in relation to El Niño events.[2] This relation is partly explained by increases in temperatures and partly by precipitation, increasing water collections, and consequently the availability of larval habitats for the mosquitoes. By contrast, in Peru a significantly lower *An. darlingi* human bite rate was detected during the rainy season, which may be due to flushing effects of larvae from the larval habitats and low temperatures that may decrease larval survival (Vittor et al. 2006).

In general, models produced to predict the impacts of global climate change on malaria risk distribution vary regarding the potential number of persons in risk, but they estimate few changes in the potential spatial distribution of the disease. These models agree that malaria incidence will increase in endemic tropical African areas and also in current areas of low endemicity. On the other hand, in some of the regions that are climatically adequate for disease transmission, such as the north of Australia, malaria has been eradicated. Simulations considering different climate change scenarios that also include non-climate components (e.g., a worldwide economy/energy model, a land use change model, a carbon-cycle model, etc.) predict a risk of reintroduction of the disease in developed countries such as the United States, Australia, and the south of Europe. However, the reestablishment of endemism would be more difficult because of the economy of these regions that allow the implementation of

[2]More information on the influence of climate on diseases, on R_0 and on vector capacity may be found in Chap. 2 (Epidemiology) (CBM).

control measures and access to better health services.

Because 90 % of malaria cases occur in Africa, it is in this continent where most studies are developed relating effects of climate and climate change with the prevalence and outbreaks of malaria. Climate changes recorded by remote sensors have shown to be particularly useful when there is little ground cover (Ceccato et al. 2007). Among the most interesting results, from different climate change scenarios, it was estimated that the increase in the number of people infected each month in 2100 will be 16–28 %, related to increases in the length of the transmission season. Changes will be more evident in terms of altitude than latitude. On the other hand, several countries of the west and south of Africa expect a decrease of the number of persons exposed monthly due to drier conditions. Regions where rainfall are a main limiting factor will be especially vulnerable to epidemics, as occurred in Ethiopia in 1958, in association to rainfall on higher quantities than the average for a long period.

Although climate has the leading impact on malaria changes, population growth, urbanization, and land use change impacts, among other factors, can also be significant on the local scale to determine the presence and distribution of vectors and risk of exposure to malaria vectors. For example, the house construction characteristics, such as the height above ground and ventilation openings, may affect exposure to *An. gambiae*, and plant cover may favor survival and affect fecundity rate of adult vectors such as *An. funestus* and *An. gambiae* s.s. (Zhou et al. 2007). In high areas of Kenya, it was reported that adult *An. arabiensis* (a species adapted to arid environments) increased in deforested areas, while *An. gambiae* s.s. (better adapted to humid environments) showed an inverse pattern. These studies showed that changes in land use might lead to the establishment of new vectors and thus increase the risk of malaria in previously non-endemic regions.

In the Brazilian Amazonia, *An. darlingi* is found mainly on sites that have been modified by human activity, such as hydroelectric dams and roads, but not in unaltered sites, suggesting that certain types of human activities increase the risk for malaria transmission. A similar pattern was reported from Peruvian Amazonia, where higher human bite rates were recorded in deforested areas. This could be explained by *An. darlingi* preferring larval habitats in pastures or crops with secondary growth and shrubs, leading to an increase of the adult mosquito population in these zones. Since the entomological inoculation rate (EIR) is proportional to *An. darlingi* biting rate, it is very likely that a higher bite rate by *An. darlingi* results in a higher malaria transmission when compared to forested areas. Time elapsed from deforestation to the occurrence of malaria epidemics in a region is variable, depending on the geographic region. For example, in Rondonia (Brazil), new towns had a higher case incidence than older communities, while in Amazonian regions of Peru, there was a malaria outbreak over a decade after intense deforestation periods.

Between 1930 and 1940, a serious malaria epidemic occurred in the northeast of Brazil, after the introduction and expansion of *An. gambiae* s.l. from Africa; the vector was removed by an eradication campaign in 1940. Models based on climate data from their native range in Africa indicated that *An. arabiensis* (one of the species of the *gambiae* complex) had an extremely limited potential distribution in Brazil, while the environment would be adequate for *An. gambiae* s.s., not only in its introduction site but also in a wider area of continental America and the Caribbean. Recent studies of DNA extracted from museum specimens collected in various localities during the 1930 decade confirmed that the invasive species was *An. arabiensis* (Parmakelis et al. 2008). Since this species is adapted to arid environments, it is probable that its expansion at the time was restricted by the rainforests surrounding the invaded area. The increasing destruction of tropical rainforest in South America, together with a transformation from forest to savanna of the oriental Amazonia as a consequence of global climate and land cover change, would provide the ideal conditions for the establishment and expansion of this vector and its sister species *An. gambiae* s.s. In the north of Argentina, the anti-malaria campaign ("antipalúdica") began during the 1930s and continued for 70 years, resulting in a pronounced reduction of the affected area. Particularly in tropical areas, the campaigns have had partial results due to impaired access, caused by climate and terrain, favoring discontinuity of control actions. During the 1990s, the endemic area suffered a small expansion. Malaria was considered endemic in the northwest of the country, where it persisted in sub-Andean valleys, and epidemic in the northeast. However, it should be noted that no autochthonous cases have been reported since 2010. Besides, there are sites with predominance of imported cases near the international frontier, associated with migrations. The overlay of maps of historical and recent positive localities with maps of the vector distribution may explain this endemic pattern in the northwest and epidemic pattern in the northeast. On the northwest, the vector associated with the disease is *An. pseudopunctipennis*, which in South America is distributed along the Andes mountain range (40 °N to 32 °S) and whose larval habitats are freshwater running mountain rivers with green algae year round. The stable presence of the species makes the natural continuity of transmission and the endemic condition of the disease possible in the northwest. On the northeast and center of the country, the epidemic character of the disease is linked to the sporadic presence of *An. darlingi*, the primary vector. This species is anthropophilic and domiciliary; however, the female, once blood fed, returns outdoors. In areas where the presence of *An. darlingi* has not been recorded, epidemics may be attributed to *An. albitarsis*, even though its vectorial capacity has not been well studied. In the center and south of the country, besides, transmission is interrupted by winter temperatures.

Environmental factors can be fundamental to predict malaria prevalence at large spatial scales (regional, continental), but at local scale individual level factors may be more relevant. A study of the spatial pattern and risk factors of *Plasmodium falciparum* prevalence at low spatial scale, in schoolchildren in Ivory Coast, detected relevant risk factors such as age, socioeconomic level, sleeping without mosquito nets, distance to primary health centers, and environmental variables (e.g., plant cover based on NVDI, rainfall, and distance to rivers representing larval habitats). These observations in principle are consistent with descriptions for other geographic regions. Besides these, children from schools located close to rivers (<500 m) have a lower risk than those from school at intermediate distances (500–1000 m), probably because the use of mosquito nets was more frequent in the former group to avoid mosquito nuisance. However, when spatial correlation between variables was considered, only age continued to be significant among the different associations found and the vegetation index only marginally so (Silue et al. 2008). A more detailed understanding of the spatial distribution of malaria to a local scale is paramount to focus control efforts.

In some countries, a reduction of malaria incidence was recorded, attributed in part to intense prevention and control programs at a national level, besides influence of climatic fluctuations, such as rainfall occurrence anomalies (Ceccato et al. 2007). Indoor residual spraying and impregnated mosquito nets have significantly contributed to reducing cases; larval control efficiency is usually only significant in non-endemic areas. A higher urbanization and human population density may influence on the vectors' life cycle and vector capacity, contributing to the observed transmission reduction. In Peru, since the eradication campaigns of the 1960s, prevalence of *P. falciparum* was very low. However, since the beginning of the 1990s, the number of cases increased in the Amazon region parallel to population growth, mainly in Andean regions, and with the economic and land use changes in the region. In 1993, Brazil, in collaboration with PAHO, established a new strategy to control malaria in the country, based on the concentration of control efforts in high-risk municipalities, which required a strengthening of surveillance and evaluation of activity, in order to direct more precisely control efforts. The program is based on the selective application of indoor residual spraying, environmental management, and anti-malaria treatment of suspect cases.

37.3.2 Arboviruses

Arboviruses are zoonosis whose natural maintenance and amplification involve susceptible vertebrate hosts and hematophagous arthropod vectors. Mosquitoes are the main vectors of arbovirus. Even though arboviruses have a worldwide distribution, climate has a strong influence on their transmission and on risk of disease outbreaks. Temperature and relative humidity influence the development rate, survival, and activity of the vectors. In turn, temperature markedly affects replication, maturation, and infectivity period of arbovirus. Laboratory studies indicate that the extrinsic incubation periods (or viral development rates) are linearly reduced with increasing temperatures, increasing the proportion of arthropods that become infective during a given period. The role of temperature on arbovirus transmission is corroborated by studies showing correlations between the detection of flavivirus or their transmission rates and cumulative temperatures above a degree-day threshold that depend on specific vector/pathogen complexes. Rainfall may affect larval habitat availability, and for mosquitoes, an increase in surface moisture associated with rainfall stimulates their flight activity and search for blood meal hosts.

In the Americas, the number of virus isolated from mosquitoes is increasing, while most of their transmission cycles and epidemiology remain little known. In the Amazonian region of Brazil, a fever caused by Oropouche virus manifests itself in cycles associated to the beginning of the rainy season (Githeko et al. 2000). Rocio virus circulates in birds and involves *Oc. scapularis* and *Psorophora ferox* as vectors. *Ochlerotatus scapularis* was also found infected with other arboviruses, such as Cache Valley, Ilhéus, Mayaro, SLE, VEE, and Minaçu. Since this species develops in transient aquatic environments, it is likely that rainfall has an effect on the epidemiology of these arboviruses, although these relations have not been explored yet.

In general, mosquito-borne viral encephalitides are very sensitive to climate. However, IPCC experts agree that predicting the influence of climate change on their distribution and incidence is difficult, due to the complexity of their transmission cycles. In the meantime, an increase of the global prevalence of diseases such as those from Zika and Chikungunya virus, previously restricted to tropical regions, justifies paying more attention to studies of the ecology and epidemiology of these pathogens and their vectors. Next are provided details on some of the best-known arboviruses.

37.3.3 Dengue and Yellow Fever

Dengue is the arthropod-borne disease of major world impact in terms of morbidity and mortality. An estimated burden of dengue in 2010 was 96 (67.1–135.6) million apparent infections globally and 13.3 (9.5–18.5) million in the Americas. This virus, original from tropical Asia, is transmitted between humans by *Aedes aegypti* in urban environments. Transmission is currently mostly restricted to the tropical and subtropical regions, because freezing temperatures kill the eggs and the mosquito larvae; however, outbreaks may occur at higher latitudes during the summer. The dispersal of *Aedes albopictus*, another potential vector, may contribute to changing this situation, because this species presents diapause. Considering the favorable or restrictive temperature limits for the survival of *Ae. aegypti* and a model that calculates the capacity of the virus to complete an extrinsic incubation period in a mosquito, as a function of air temperature, dengue risk maps were produced for Argentina (Carbajo et al. 2001) that could be applied to other countries. These maps showed a maximum transmission risk year round in the north and northeast of the country and on the summer time in the central region.

Using general circulation models (GCM) of climate change, it was found that the epidemic potential of dengue increased with relatively small temperature increases, indicating that fewer mosquitoes would be needed to maintain or transmit the disease in a vulnerable population (Patz et al. 1998). In turn, it has been observed that climate change could lead to longer periods of intensive transmission. This increase in transmission strength would result in greater seroprevalence and therefore a higher number of individuals at risk for an additional exposure to dengue virus, augmenting the risk of hemorrhagic dengue. The reduction of the age at which people are exposed to the virus for the first time increases the chance of a second exposure. This is particularly relevant for the preteen age group, which is the largest group with higher risk of having the hemorrhagic form of the disease. It is projected that the warming of 2 °C expected for 2100 will increase in latitude and altitude the potential risk areas and increase the length of the transmission season in the temperate regions. The rainfall regime may also influence the seasonal variations in the abundance of the vector, resulting in some regions in a "dengue season," limited to certain periods of the year. For example, in Trinidad a positive correlation was found between the incidence of dengue and rainfall, concentrated between June and November.

Another simulation study done for an area of annual dengue transmission in Far North Queensland, Australia, utilizing a field-validated mechanistic mosquito population model and climate projections, showed inconsistent patterns in dengue risk between climate change scenarios (Williams et al. 2014). Under one scenario (B1, lower carbon dioxide emissions, global temperature increases from 1.1–2.9 °C), *Ae. aegypti* abundance is predicted to increase, while it is predicted to decrease under a higher carbon emission scenario (A2 scenario, global temperature increases from 2.0 to 5.4 °C by the end of the twenty-first century), probably in relation to changes in rainfall patterns. A reduction in body size was predicted for both climate change scenarios compared to present climate, with the B1 scenario causing the largest decrease. However, these body size reductions were not linked to a reduction in oviposition rates, which were predicted to increase for both B1 and A2 scenarios. Shorter extrinsic incubation periods were projected, potentially increasing transmission.

Although climatic factors are relevant to the distribution and transmission of dengue, as with other vector-borne diseases, studies to date indicate that they are not sufficient by themselves to predict the occurrence of cases or the risk of disease in a region. For example, an analysis of the association between the incidence of hemorrhagic dengue and satellite estimates of land surface temperature (estimated by NOAA thermal infrared sensor, which combine land surface temperature and atmospheric vapor concentration) found positive correlations in only 21 % of the analyzed provinces, and correlations, in turn, were significant during the non-epidemic months, not on epidemic months (Nitatpattana et al. 2007). These results could be explained by the importance of non-climatic factors to the risk of incidence. In highly endemic tropical areas, the transmission of dengue virus could saturate the population, and thus the migration of human populations likely would have greater weight than climatic factors. The public health infrastructure contribution may reduce the risk of transmission, especially in developed countries. Considering that *Ae. aegypti* finds suitable larval habitat in water vessels of various kinds, both domestic and outdoors, water storage practices of a community may be more important in its impact on the mosquito population density than the rainfall regime itself (Patz et al. 1998).

Yellow fever originated in Africa and was initially dispersed together with its vector (*Ae. aegypti*), with the slave trade in the sixteenth and seventeenth centuries. This arbovirus is maintained in South America in wild cycles involving primates and forest species of mosquitoes. In urban environments, it can cause outbreaks, with *Ae. aegypti* as a vector, and, as dengue, is also likely susceptible to climate change. The invasion of urban centers by the African mosquito and the advancement of urbanization on the rain forests could increase the risk of epidemic outbreaks of this disease in tropical America.

Rogers et al. (2006) carried out an extensive review of point data records of yellow fever and dengue outbreaks worldwide, gleaned from archived reports and literature surveys. They used high-spatial-resolution environmental information derived from satellites (from the Advanced Very

High-Resolution Radiometer (AVHRR) on board the National Oceanographic and Atmospheric Administration (NOAA) satellites) to create risk maps of environmental suitability for both diseases. A digital elevation model (DEM) was also considered besides the satellite-derived environmental data (including land surface temperatures (LST) and NDVI). The maps resulting from the most suitable models predicted high-risk areas localized within the broad boundaries of the 2003 World Health Organization maps for these diseases and predicted areas of high risk outside these boundaries. These discrepancies could be due mostly to under-reporting of outbreaks but also to limitations of the models that did not include human population variables. Nevertheless, the results highlighted some interesting contrasts in environmental suitability for each pathology: yellow fever distribution seems particularly sensitive to the greenness or humidity of the environment and especially to variability of these factors. On the other hand, in the case of dengue, the variability of environmental temperature rather than of moisture appears to be very important to its distribution.

37.3.4 Arboviral Encephalitis

The Saint Louis encephalitis virus (SLE) has been detected from Canada to Argentina. In South America, this virus is widely distributed, with antibody prevalence in humans ranging from 3 to 50 %, and the occurrence of outbreaks is sporadic. Little is known about its ecology, and a marked association with weather conditions has not been determined yet. In contrast, in the northern hemisphere, this disease has been better studied. It is usually restricted to regions south of the isotherm of 20 °C in June, although it can extend further north in particularly warm years. The outbreaks of encephalitis by SLE tend to occur in the second half of the summer and early fall, after warm periods, when the temperature exceeds 30 °C for 7 consecutive days (Hunter 2003).

Both abundant rains and droughts have been associated with SLE epidemics, observing a significant association between years with unusually dry autumns and the activity of this virus. As an explanation, it was proposed that during drought periods, pregnant female vectors, particularly *Culex nigripalpus*, may not find suitable sites for oviposition or simply rest waiting for the right conditions. This resting period allows the virus incubation period to be completed in the infected females. When rains restart, soon after oviposition there would be a concentration of infected mosquitoes in search of new hosts. Moreover, precipitation stimulate the ingestion of blood in *Cx. nigripalpus* females, and therefore the abundant autumn rains could stimulate females to perform long flights seeking blood meals and reach hosts in locations that were harsh during the dry periods of the year. This could even be a cause of the change from bird to mammals' blood intake observed in the fall. A negative association was also detected between the abundance of adult *Cx. pipiens* (vector of SLE and West Nile virus) and surface moisture in the previous 10 days, modeled from a dynamic hydrological model based on topographic maps. This negative association is consistent with the preference of *Cx. pipiens* for contaminated water collections, as in times of drought, water collections reduce their dimensions and become eutrophic.

In 1999, West Nile virus (WNV) was first detected in North America and quickly dispersed, being found as far south as Argentina in 2005. This virus normally circulates in the bird population, mainly transmitted by mosquitoes, and can occasionally infect incidental hosts such as horses and humans. In subtropical and moderately temperate zones, significant correlations have been observed between the number of human cases of WNV and monthly temperature, while the correlations with rainfall are low. Mosquito catches increase with increases in monthly temperatures. In turn, human cases are correlated with positive *Cx. quinquefasciatus* and *Ae. albopictus* mosquitoes and infected birds (reservoirs of the virus).

There is an interesting example of interaction between vector species and climate in *Culex* populations from the north hemisphere. The populations of *Cx. restuans* increase from spring until they become the predominant population in the middle of summer, replaced by fall by *Cx. pipiens*. The seasonal change in *Cx. restuans* dominance for *Cx. pipiens* suggests that the first species begins the enzootic cycle of WNV, while *Cx. pipiens* amplifies the number of infected bird host. The time when both species are represented in similar proportions is called "crossover." Correlations between historical records, dates of crossovers, and parameters derived from temperature records (such as maximum and minimum temperatures, degree-days, etc.) suggest that high temperatures can accelerate the crossover and therefore lead to longer periods of transmission, encouraging the involvement of accidental hosts.

Associations have been found between urban characteristics and the prevalence of WNV, noting lower prevalence in suburbs with higher vegetation cover and diversity and higher prevalence in suburban areas dominated by houses over 40 years with moderate vegetation cover. This relationship could be explained by the conditions of the rainwater drainage system, which allows the accumulation of water (most relevant in times of drought) combined with organic matter, creating an enabling environment for the *Culex* species (Ruiz et al. 2007). On the other hand, as discussed in earlier sections of this chapter, low biodiversity areas could favor the set of birds and mosquito species necessary to keep transmission, since the mosquitoes can use containers and bodies of water (drainage, ditches) generated by humans.

37.3.5 Filariasis

The filarial worms are parasitic nematodes transmitted by several species of mosquitoes, which vary geographically. In each region, there is a close association between the frequency of filaremia and blood intake by mosquitoes. *Wuchereria bancrofti* predominates in urban and suburban areas of South and Central America, where *Cx. quinquefasciatus* is their main vector, though infective larvae have been isolated from *An. darlingi* and other species of mosquitoes as well. Human filariases are in general distributed in tropical and subtropical regions of the world. This could be due to climatic conditions and soil characteristics, influencing the availability and activity of vectors, as above, but also to its effect on the infectivity of the parasites. At least in laboratory tests, the amount of *Brugia pahangi* filarial larvae that penetrate a mammal host, bitten by the vector *Ae. aegypti*, is greater in high relative humidity conditions. But both the number of larvae leaving the female mosquitoes during feeding and the female's predisposition to feed are independent of relative humidity. When the filarial larva emerges from the proboscis of the mosquito, it is covered by a fluid, composed of the mosquito hemolymph. It has been proposed that evaporation of this protective hemolymph cover is the main factor regulating the penetration of filarial larvae in the host. The areas where this filariasis is endemic are limited to the tropics, where the evaporation rate is high. However, in these regions the mosquitoes feed predominantly at night and up to the sunrise, when temperatures are lower, the humidity increases and evaporation decreases. Local factors such as plant cover, the presence of large water surfaces, or other characteristics such as architecture, among others, that create or maintain the higher humidity conditions could increase the potential for transmission.

The larval habitats of *Cx. pipiens*, one of the principal filariasis vectors, can be identified by the analysis of the daily changes of land surface temperature, recorded by remote sensors, such as AVHRR NOAA (Malone et al. 2006), and soil moisture indicators derived from higher-spatial-resolution Landsat images. This is due to the buffering effect of water, which causes wetlands and other flooded land to have lower daily variations relative to neighboring non-flooded fields.

Dirofilaria immitis is widely distributed in tropical regions. In the Americas, potential vectors are species of *Culex*, *Anopheles*, *Mansonia*, and *Ochlerotatus*. Landscape features may affect the transmission of the disease, such as a negative relation reported between the degree of urbanization and the prevalence in dogs. Heartworm risk maps for Argentina were generated that considered the lower temperature thresholds for the development of the *D. immitis* in the mosquito (about 14-°C), the number of potential mosquito vector species, and a heat measurement (about 130-°C-days), necessary for the microfilariae to reach the infectious stage. These models indicated that the risk increases from the southwest to northeast, which is consistent with the highest values of prevalence of the disease in the north. The maps also stressed that the transmission in the Southern Cone of the subcontinent is markedly seasonal, even in the warmest regions (Vezzani and Carbajo 2006). This type of approach, complemented with more specific information on population parameters and habitat requirements of the vectors involved in various geographical regions, could serve as a starting point for evaluating the different climate change scenarios on the distribution of this disease.

37.3.6 Leishmaniasis

Leishmaniases are among the main reemerging diseases of the Americas in two main types of clinical manifestations: cutaneous and visceral (see Chap. 16). These diseases are caused by intracellular parasites of the genus *Leishmania*, transmitted between mammals by several species of *Lutzomyia* (Diptera: Psychodidae) with reservoirs varying according to the species of the parasite and the region. The dynamics of the disease are strongly associated with the presence of reservoirs, which act as sources of infection, and humans are generally regarded as accidental hosts. In Brazil, the main vector of zoonotic visceral leishmaniasis is *Lutzomyia longipalpis*, a species of endophilic and exophilic habitats, which bites humans during dusk and night rest (Ximenes et al. 2006). Mass migration of people from rural areas to the outskirts of cities, as a result of drought, malnutrition, and lack of land for agriculture, favors the formation of slums. In these settlements of predominantly young families, health and infrastructure conditions are poor. The combination of nonimmune hosts in malnutrition conditions with the habit of keeping domestic animals (dogs, chickens, horses) in or near households allows the proliferation of sand flies and increases the risk of infection by *L. chagasi* and eventually developing visceral leishmaniasis.

Weather conditions, besides affecting the migrations of hosts, have a more direct effect on development and proliferation of the vectors. The increases in visceral leishmaniasis are preceded by increases in population densities of *Lu. longipalpis*, which in turn are negatively correlated with relative humidity and wind speed and positively with average temperature in the previous months (Ximenes et al. 2006). In Rio Grande do Norte (Brazil), the insects have a peak incidence in May, when rain predominates and the temperatures are lower, and a second peak in November, when the rains are light and temperatures are higher, while the females only show a peak in May. The population peaks depend on the combination of appropriate temperature and humidity conditions for the development of the immature forms; high temperatures can accelerate the development of the larvae, but reducing the relative humidity can stop its development.

Time series analysis of the incidence of cutaneous leishmaniasis in Costa Rica showed that this disease has cycles of about 3 years, consistent with the temperature indexes and the El Niño-Southern Oscillation. Linear models were developed based on temperature estimates and El Niño-Southern Oscillation rates that could predict with accuracy greater than 70 % the incidence of cutaneous leishmaniasis, up to 12 months in advance (Chaves and Pascual 2006). In Brazil, visceral leishmaniasis has peaks approximately every 10 years, which probably reflect not only the vector responses to environmental change but also their interaction with human hosts and susceptible dogs.

As the vector species (sand flies) generally prefer forest and moist environments, data derived from remote sensors included in geographic information systems are suitable tools to generate vector distribution and leishmaniasis risk maps. For example, in East Africa distribution maps of the main vectors of visceral leishmaniasis, *Phlebotomus martini* and *Ph. orientalis*, were created using NVDI derived from AVHRR NOAA, maximum temperature analysis, agroclimatic data from FAO/CPSZ (Food and Agriculture Organization of the United Nations/ Crop Production System Zone), and FAO 1998 soil maps (Gebre-Michael et al. 2004). In the Brazilian state of São Paulo, the distribution of the five dominant species (*Pintomyia fischeri*, *Nyssomyia intermedia*, *Migonemyia migonei*, *Pi. pessoai*, and *Ny. whitmani*) was estimated from municipal data of occurrence of the species and ecological niche modeling approach using genetic algorithm to adjust the decision rule (GARP). This model considered aspects of topography (elevation, slope, tendency to accumulate surface water), climate (temperature parameters, precipitation, and relative air humidity), land use, and ground cover (estimated from AVHRR NOAA satellites). The resulting models allowed the production of general distribution maps of the species, covering areas not sampled or sparsely sampled (Peterson et al. 2004). On the other hand, in the state of Mato Grosso, *Ny. whitmani* s.l., originally associated with forest ecosystems, occurs in anthropogenic habitats and therefore may occur in different environmental profiles, sylvan or periurban. A supervised classification of NVDI MODIS images in four classes of vegetation/land cover showed that the spatial distribution of "species" is positively correlated with the rate of deforestation and negatively correlated with socioeconomic development indicators. Their adaptation to anthropogenic environments could explain why deforestation does not result in widespread loss of sand fly habitat in this state (Zeilhofer et al. 2008).

Whether driven by natural or human forcing, climate change together with changes in the ecosystems driven by human activities is expected to be reflected in the spatial distribution and dynamics of transmission of leishmaniasis, among other diseases. Potential geographical distribution of leishmaniasis vectors in South America based on ecological niche models indicates, for different climate change scenarios, the improvement of conditions for *Ny. whitmani* in the southeastern Brazil, influencing the transmission of cutaneous leishmaniasis (Peterson and Shaw 2003). The southern range of cutaneous and visceral leishmaniasis could be extended from northern Argentina to more austral latitudes. In addition, the improvement in conditions for *Ny. whitmani* on the eastern slopes of the Peruvian Andes could allow the expansion of cutaneous leishmaniasis in this region (Peterson and Shaw 2003). The reduction of natural habitats could favor the urbanization of the vectors and diseases, since, although the sand flies are more abundant with taller, dense vegetation and with high moisture content, some species may invade the home and surrounding dwellings, developing in the vegetation near the houses, and even in outdoor environments (latrines, pens). In the northeastern Brazil, a region subject to periodic droughts, a resurgence of visceral leishmaniasis has been noted in urban areas, and major epidemics coincided with more pronounced droughts caused by El Niño.

Salomón et al. (2012) argued that an increase in temperature or humidity due to climate change would change the transmission risk of leishmaniasis in Argentina (its southern range), by increasing duration of night activity of the females and human exposure from changes of the habits of humans. The increased vulnerability would be discriminated by socioeconomic variables, since, for example, families spending more time in the courtyard during very hot and humid nights would be more exposed than families spending this period indoors with air conditioning (a similar situation to dengue in Texas, see above). At a regional scale, an increase in precipitation and humidity by global trends and/or water reservoir-irrigation development projects could change the vector community and increase the area of dispersion of species such as *Lu. longipalpis*.

37.3.7 Chagas Disease

Chagas disease is widely distributed in Latin America, from Argentina to Mexico, and recently human cases were reported in the Southern United States. It is estimated to affect 15 million people and the population at risk of becoming infected is 28 million. Due to its slow development, it is considered the parasitic disease with the greatest economic importance in Latin America. As the parasite can also be transmitted by other means, such as blood transfusion and organ transplantation, even via contaminated food ingestion, including in countries where there is no vectorial transmission (e.g., Spain), Chagas disease shows up as a major public health problem due to the migration of infected people.

Considering only the regions where transmission is vectorial, this parasitosis is a complex zoonosis, involving several vertebrate reservoirs and various species of triatomines, which makes eradication complex or even impossible in most of its range. The number of triatomine species increases significantly from the poles to the equator and from the west to east, with a maximum of species between 5 and 10°S. The Amazon region has the greatest diversity of wild species, adapted to very humid microclimates of rainforest, and a variety of patterns and complexity of transmission cycles has been found. In this region, the home infestations are restricted to drier subregions of the Amazon, that is, those heavily modified by deforestation.

Outside the Amazon region, most of the transmission of Chagas disease depends on the household and outdoor premises' populations of vectors. The poor condition of houses, associated with poverty in both rural and "poverty belts" of big cities, facilitates the establishment of the vectors adapted to indoors and peridomestic environments. Domestic animals closely associated with the inhabitants, even living inside human homes, are a source of abundant blood and easy to access, allowing the insects to reach high densities.

Anthropogenic changes in land use can lead gradually to synanthropy of the triatomines. The pattern of invasion by *Rhodnius* in Amazon rainforests illustrates this. Selective removal of the forest, which only keeps palm trees in the peridomestic environment, is a common factor that reduces the populations of wild vertebrates. Although during the first stage of land use change the populations of wild triatomines (especially *Rhodnius*) suffer great reductions, individuals who persist adopt the palm trees as suitable refuges, where they coexist with marsupials and opportunistic rodents, which proliferate in environments modified by humans. Once the triatomine population saturates the palm trees' habitats, in search of new dwellings, they can invade nearby human homes and thus increase the risk of Chagas transmission. In the Alto Beni region, Bolivia, intensive deforestation and creation of new urban settlements have gone on for the last 20 years. The materials used to build houses, mainly palm leaves, may have facilitated the introduction of *Rhodnius stahli*, a normally wild species, to the home, where adults, nymphs, and eggs have been collected. *Triatoma dimidiata* is a widely distributed species in Mexico, Central America, Colombia, and Ecuador, both in wild habitats as inside and outside the house. They may colonize artificial environments in the Amazon, since their natural populations are present in very diverse ecoregions, including wet or dry forest environments.

Triatoma infestans is the main Chagas vector in the Southern Cone of South America, where it is closely associated with household and outdoor structures. The microenvironmental conditions of human habitation are better buffers of the largest outdoor temperature variations, compared to peridomestic sites such as pens of goats and chicken. The latter are open structures, exposed to sunlight, rain, and wind; thus, the differences in relation to external conditions are small. Therefore, temperature variations will affect more outdoor than indoor insect populations. Both empirical field data and ecogeographic models (Ceccarelli et al. 2015) suggest that *T. infestans* has a limited capacity to thrive in hot and humid climatic conditions of most forest regions of the Amazon. In fact, on a regional scale, the analysis suggests that the distribution of *T. infestans* is affected by environmental characteristics, mainly linked to annual cycles of air temperature and soil moisture, not necessarily linked to the availability of human habitation. In turn, temperature contributes to *T. infestans* dispersal, since it influences the proportion of specimens flying and also the distance they travel.

In Brazil, it is assumed that the characteristics of the indigenous settlements, added to their nomadic habits, prevent the establishment of *T. infestans* in the homes and thus reduce the likelihood of infection by *Trypanosoma cruzi* (at least by this vector). In fact, it was proposed that the beginning of protozoan infections was associated with domiciliation of *T. infestans*, due to the building of mud shacks from the European colonization. However, the detection of *T. cruzi* infection in 4500–7000 years human mummies shows that human infections, at least in Brazil, occurred in populations of hunter-gatherers, before *T. infestans* adopted indoor environments.

Not only the land use changes may affect the *T. cruzi* transmission cycles but also climate change could modify the risk of infection (McCarthy et al. 2001; Medone et al. 2015). Temperature affects the major components of the vectorial capacity: at higher temperatures, insects can accelerate development and achieve higher densities, and the development of *T. cruzi* is also accelerated. However, the relationship between the temperature and the infection rate is not necessarily linear, as when the densities of these insects are very high, competition density-dependent processes affect the amount of ingested blood, which reacts on the elapsed time between blood intake and the first defecation (lower ingestion → higher delay to excrete) and therefore on the infection risk.

It has been observed that large-scale climate disturbances, such as hurricanes, can increase the risk of transmission of Chagas disease, although the causes are still unclear. Among the possible explanations, it is speculated that the reduction of sources of wild hosts, due to the death of animals, could encourage the spread of vectors searching for a blood meal, favoring the invasion of houses and human contact. Alternatively, populations of some reservoirs, such as armadillos in Louisiana, can grow markedly after the hurricanes and encourage population growth of insects. Upon returning these reservoirs to their levels before the hurricane, the triatomine bugs seek alternative blood sources.

37.3.8 Tick-Borne Diseases

Tick-borne diseases, such as Lyme disease, ehrlichioses, encephalitis, and spotted fever, are common in temperate regions, mainly in the northern hemisphere. In general, species that transmit these diseases have complex cycles involving different hosts depending on the stage of development of the ticks. The distribution of these diseases is affected both by climate, which limits the distribution of vectors, their development, and their survival rates, as well as the habitat characteristics, which may condition the contact between the ticks and the reservoir hosts of different pathogens. For example, *Amblyomma americanum* (vector of *Ehrlichia chaffeensis*, the causative agent of human monocytotropic ehrlichiosis) and *Ixodes scapularis* (vector of *Borrelia burgdorferi*, which causes Lyme disease, and *Anaplasma phagocytophilum*, causing human granulocytic anaplasmosis) prefer higher temperatures and humidity. However, these relations are variable in space. Detailed analysis of plant cover (30-m resolution) pointed out that in the drier regions of the ticks' distribution, coverage by forests and fragmentation are important to predict the presence of *E. chaffeensis* and *A. phagocytophilum*. This would be because the tree cover would maintain adequate moisture conditions for the survival of the vector ticks. In turn, the presence of forest and agricultural land or urban areas fragments can affect the density of deer and small mammals (pathogen reservoirs) which are hosts of the various development stages of ticks (Wimberly et al. 2008).

Over the last two decades, increases in the abundance of ticks and incidence of diseases caused by them have been registered, which has been attributed to climate, land use change, and also greater awareness of these diseases. Models of the effects of weather on ticks in the temperate regions suggest that as the temperature rises, the larvae feed and are active earlier in the year, and a higher proportion of the nymphs feeds during autumn of the same year they fed as larvae. This growing overlap in the activity of larvae and nymphs could increase the pathogen transmission cycles that are transmitted efficiently between reservoirs and ticks only for short periods of the year, such as *A. phagocytophilum* (Ogden et al. 2006). An analysis of the distribution of ticks in different climate change scenarios (GCM) was consistent in presenting an expansion to lower latitudes of favorable climatic conditions for the establishment of *I. scapularis* populations and eventually the pathogen they transmit (Ogden et al. 2006). However, a study in the United Kingdom showed no correlation between cases of Lyme disease and average summer temperature (Hunter 2003). On the other hand, in Sweden, between 1980 and

1994, an expansion of the northern distribution limit and an increase in *I. ricinus* tick density were recorded, concomitant with milder winters. Another approach to elucidate the impact of climate change on the distribution and phenology of *I. ricinus* considered a stage structured Leslie matrix of a Western Palearctic region tick population and temperature, saturation deficit, and hours of light of the previous 2 weeks as independent variables. An advantage of this model is that the sensitivity of each stage to changes in climate as well as their impact on the whole life cycle of the tick can be assessed. Geographic variations were detected in the relative importance (elasticity) of the development processes, such as development and mortality rates, to the final population performance. For example, in most of the northern distribution range, the mortality rates associated with the molting processes had higher elasticity than development rates, while at southern latitudes, the duration and mortality of questing periods for adults were the greatest contributors to the final performance of the life cycle. Consistently, different sensitivity of the complete life cycle of the tick to proportional changes in temperature and saturation deficit was evidenced according to the performance of the tick in the geographic range (Estrada-Peña and Estrada-Sánchez 2014).

High-spatial-resolution remote sensing, such as Landsat™, can distinguish fine-scale features of the landscape composition that can be used to assess the risk of exposure to Lyme disease. The abundance of *I. scapularis* ticks can be associated with ground cover, distance to forests borders, elevation and soil type, environmental characteristics that can be estimated from satellite images, and/or modeled in geographic information systems (Bunnell et al. 2003). In urban and suburban environments, if reservoirs are present, the urbanization characteristics may influence the risk for diseases such as Lyme. Estimates from satellite images of structure, humidity, and "greenness" of vegetation showed a higher relative risk of exposure to the disease in the greener and most humid premises, containing a higher proportion of trees with broad leaves, while premises with lower risk had higher proportion of areas without vegetation or open grass cover (Dister et al. 1997).

The biological interactions involved in disease incidence are complex and include environmental components that are extrinsic to transmission cycles. Thus, for example, Lyme disease outbreaks can be predicted based on the productivity of oak seeds. The *Borrelia burgdorferi* spirochete complex causing this disease normally circulates in populations of a rodent (*Peromyscus leucopus*) and white-tailed deer, which also disperse the ticks. These acari can accidentally bite humans and transmit them the pathogen. Every 1 or 5 year's oaks produce large amounts of acorns (phenomena known as masting). The plenty availability of acorns leads to rodent population growth and attracts deer, with the consequent proliferation of ticks and facilitating contact with the infected rodents (Randolph 1998).

AVHRR satellite images, estimating vegetation and surface temperature, in combination with standard geostatistical methods can be used to analyze dynamically in time the suitability of habitat for *Boophilus microplus*. This type of study in central and south of North America indicated that the most suitable areas for the tick have relatively stable temperatures year round (20–25 °C) and have intermediate plant coverage. Gradual heating of the region would decrease the quality of habitats in areas where the tick is established and favor their establishment in other regions currently too cold to support their populations (Estrada-Peña 2001).

Considering the strong association between the ecology of the vectors, access to reservoirs, climate, and land cover, it is expected that climate and land cover changes have important repercussions on vector-borne diseases worldwide. Even though vaccines and other means of chemoprophylaxis are being developed for the more common/widespread diseases, the emergence or reemergence of novel vector-borne diseases is a widespread phenomenon. Thus, it is of paramount importance that more detailed studies on the biology and ecology of the vectors and the transmission cycles of these pathogens be encouraged. Deeper understanding of the processes and mechanisms involved should allow for more efficient application of vector control and disease prevention.

References

Achee NL, Grieco JP, Masuoka P, Andre RG, Roberts DR, Thomas J, Briceno I, King R, Rejmankova E (2006) Use of remote sensing and geographic information systems to predict locations of *Anopheles darlingi*-positive breeding sites within the Sibun River in Belize, Central America. J Med Entomol 43:382–392

Baylis M, Bouayoune H, Touti J et al (1998) Use of climatic data and satellite imagery to model the abundance of *Culicoides imicola*, the vector of African horse sickness virus, in Morocco. Med Vet Entomol 12:255–266

Bunnell JE, Price SD, Das A et al (2003) Geographic information systems and spatial analysis of adult *Ixodes scapularis* (Acari: Ixodidae) in the Middle Atlantic region of the U.S.A. J Med Entomol 40:570–576

Carbajo AE, Schweigmann N, Curto SI et al (2001) Dengue transmission risk maps of Argentina. Trop Med Int Health 6:170–183

Ceccarelli S, Balsalobre A, Susevich ML et al (2015) Modelling the potential geographic distribution of triatomines infected by *Triatoma* virus in the southern cone of South America. Parasit Vectors 8:153. doi:10.1186/s13071-015-0761-1

Ceccato P, Ghebremeskel T, Jaiteh M et al (2007) Malaria stratification, climate, and epidemic early warning in Eritrea. Am J Trop Med Hyg 77:61–68

Chase JM, Knight TM (2003) Drought-induced mosquito outbreaks in wetlands. Ecol Lett 6:1017–1024

Chaves LF, Pascual M (2006) Climate cycles and forecasts of cutaneous leishmaniasis, a nonstationary vector-borne disease. PLoS Med 3, e295

Conley AK, Watling JI, Orrock JL (2011) Invasive plant alters ability to predict disease vector distribution. Ecol Appl 21:329–334

Curtis A, Blackburn JK, Widmer JM et al (2013) A ubiquitous method for street scale spatial data collection and analysis in challenging urban environments: mapping health risks using spatial video in Haiti. Int J Health Geogr 12:21

Dister SW, Fish D, Bros SM et al (1997) Landscape characterization of peridomestic risk for Lyme disease using satellite imagery. Am J Trop Med Hyg 57:687–692

Diuk-Wasser MA, Brown HE, Andreadis TG et al (2006) Modeling the spatial distribution of mosquito vectors for West Nile virus in Connecticut, USA. Vector Borne Zoonotic Dis 6:283–295

Elias SP, Lubelczyk CB, Rand PW et al (2006) Deer browse resistant exotic-invasive understory: an indicator of elevated human risk of exposure to *Ixodes scapularis* (Acari: Ixodidae) in Southern coastal Maine woodlands. J Med Entomol 43:1142–1152

Estrada-Peña A (2001) Forecasting habitat suitability for ticks and prevention of tick-borne diseases. Vet Parasitol 98:111–132

Estrada Peña A, Estrada Sanchez D (2014) Deconstructing *Ixodes ricinus*: a partial matrix model allowing mapping of tick development, mortality and activity rates. Med Vet Entomol 28:35–49

Estrada-Peña A, Estrada-Sánchez A, Fuente J (2014) A global set of Fourier-transformed remotely sensed covariates for the description of abiotic niche in epidemiological studies of tick vector species. Parasit Vectors 7:302

Gebre-Michael T, Malone JB, Balkew M et al (2004) Mapping the potential distribution of *Phlebotomus martini* and *P. orientalis* (Diptera: Psychodidae), vectors of kala-azar in East Africa by use of geographic information systems. Acta Trop 90:73–86

Githeko AK, Lindsay SW, Confalonieri WE, Patz JA (2000) Climate change and vector-borne diseases: a regional analysis. Bull World Health Org 78:1136–1147

Gleiser RM, Zalazar LP (2009) Distribution of mosquitoes in relation to urban landscape characteristics. Bull Entomol Res 5:1–6

Gleiser RM, Gorla DE, Ludueña Almeida FF (1997) Monitoring the abundance of *Aedes (Ochlerotatus) albifasciatus* (Macquart 1838) (Diptera: Culicidae) to the south of Mar Chiquita Lake, central Argentina, with the aid of remote sensing. Ann Trop Med Parasitol 91:917–926

Gleiser RM, Gorla DE, Schelotto G (2000) Population dynamics of *Aedes albifasciatus* (Diptera: Culicidae) south of Mar Chiquita Lake, central Argentina. J Med Entomol 37:21–26

Hendrickx G, Napala A, Rogers D et al (1999) Can remotely sensed meteorological data significantly contribute to reduce costs of tsetse surveys? Mem Inst Oswaldo Cruz 94:273–276

Hunter PR (2003) Climate change and waterborne and vector-borne disease. J Appl Microbiol 94:S37–S46

Jacob BG, Shililu J, Muturi EJ et al (2006) Spatially targeting *Culex quinquefasciatus* aquatic habitats on modified land cover for implementing an Integrated Vector Management (IVM) program in three villages within the Mwea Rice Scheme, Kenya. Int J Health Geogr 5:18. doi:10.1186/1476-072X-5-18

Keesing F, Holt RD, Ostfeld RS (2006) Effects of species diversity on disease risk. Ecol Lett 9:485–498

Lee EC, Asher JM, Goldlust S, Kraemer JD, Lawson AB, Bansal S (2016) Mind the scales: Harnessing spatial big data for infectious disease surveillance and inference. arXiv:1605.08740v2 [q-bio.PE]

Linthicum KJ, Bailey CL, Tucker CJ et al (1991) Towards real-time predictions of Rift Valley fever epidemics in Africa. Prev Vet Med 11:325–334

Malone JB, Nieto P, Tadesse A (2006) A biology-based mapping of vector-borne parasites by geographic information systems and remote sensing. Parassitologia 48:77–79

Mangudo C, Aparicio JP, Gleiser RM (2015) Tree holes as larval habitats for *Aedes aegypti* in urban, suburban and forest habitats in a dengue affected area. Bull Entomol Res. doi:10.1017/S0007485315000590

Martinson HM, Raupp MJ (2013) A meta-analysis of the effects of urbanization on ground beetle communities. Ecosphere 4:art60

Matthews SD, Meehan LJ, Onyabe DY et al (2007) Evidence for late Pleistocene population expansion of the malarial mosquitoes, *Anopheles arabiensis* and *Anopheles gambiae* in Nigeria. Med Vet Entomol 21:358–369

McCann RS, Messina JP, MacFarlane DW et al (2014) Modeling larval malaria vector habitat locations using landscape features and cumulative precipitation measures. Int J Health Geogr 13:17. doi:10.1186/1476-072X-13-17

McCarthy JJ, Canziani OF, Leary NA et al (2001) Climate change 2001: impacts, adaptation, and vulnerability. Cambridge University Press, Published for the Intergovernmental Panel on Climate Change, Cambridge

Medone P, Ceccarelli S, Parham PE, Figuera A, Rabinovich JE (2015) The impact of climate change on the geographical distribution of two vectors of Chagas disease: implications for the force of infection. Phil Trans R Soc B 370:20130560

Mutuku FM, Bayoh MN, Hightower AW et al (2009) A supervised land cover classification of a western Kenya lowland endemic for human malaria: associations of land cover with larval *Anopheles* habitats. Int J Health Geogr 8:19. doi:10.1186/1476-072X-8-19

Nitatpattana N, Singhasivanon P, Kiyoshi H et al (2007) Potential association of dengue hemorrhagic fever incidence and remote senses land surface temperature, Thailand, 1998. Southeast Asian J Trop Med Public Health 38:427–433

Ogden NH, Maarouf A, Barker IK et al (2006) Climate change and the potential for range expansion of the Lyme disease vector *Ixodes scapularis* in Canada. Int J Parasitol 36:63–70

Parmakelis A, Russello MA, Caccone A et al (2008) Short report: historical analysis of a near disaster: *Anopheles gambiae* in Brazil. Am J Trop Med Hyg 78:176–178

Patitucci L, Mulieri P, Mariluis J et al (2010) The population ecology of *Muscina stabulans* (Fallén) (Diptera: Muscidae), along an urban-rural gradient of Buenos Aires, Argentina. Neotrop Entomol 39:441–446

Patz JA, Martens WJM, Focks DA et al (1998) Dengue fever epidemic potential as projected by general circulation models of global climate change. Environ Health Perspect 106:147–153

Peterson AT, Shaw J (2003) *Lutzomyia* vectors for cutaneous leishmaniasis in southern Brazil: ecological niche models, predicted geographic distributions, and climate change effects. Int J Parasitol 33:919–931

Peterson AT, Pereira RS, Camargo Neves VF (2004) Using epidemiological survey data to infer geographic distributions of leishmaniasis vector species. Rev Soc Bras Med Trop 37:10–14

Phillips PL, Welch JB, Kramer M (2014) Development of a spatially targeted field sampling technique for the southern cattle tick, *Rhipicephalus microplus*, by mapping white-tailed deer, *Odocoileus virginianus*, habitat in South Texas. J Insect Sci 14:88. doi:10.1093/jis/14.1.88, Available online http://www.insectscience.org/14.88

Rakotomanana F, Randremanana RV, Rabarijaona LP et al (2007) Determining areas that require indoor insecticide spraying using Multi Criteria Evaluation, a decision-support tool for malaria vector control programmes in the Central Highlands of Madagascar. Int J Health Geogr 6:2. doi:10.1186/1476-072X-6-2

Randolph SA (1998) Mighty theories from little acorns grow: is Lyme disease risk predictable from mast-seeding by oak trees? Trends Ecol Evol 13:301–303

Reiskind MH, Zarrabi AA (2011) The importance of an invasive tree fruit as a resource for mosquito larvae. J Vector Ecol 36:197–203

Reiter P, Lathrop S, Bunning M et al (2003) Texas lifestyle limits transmission of dengue virus. Emerg Infect Dis 9:86–89

Rogers DJ, Hay SI, Packer MJ (1996) Predicting the distribution of tsetse flies in West Africa using temporal Fourier processed meteorological satellite data. Ann Trop Med Parasitol 90:225–241

Rogers DJ, Wilson AJ, Hay SI et al (2006) The global distribution of yellow fever and dengue. Adv Parasitol 62:181–220

Ruiz MO, Walker ED, Foster ES et al (2007) Association of West Nile virus illness and urban landscapes in Chicago and Detroit. Int J Health Geogr 6:10

Salomón OD, Quintana MG, Mastrángelo AV et al (2012) Leishmaniasis and climate change—case study: Argentina. J Trop Med 11. doi:10.1155/2012/601242

Shaman J, Day JF, Stieglitz M (2002) Drought-induced amplification of Saint Louis encephalitis virus, Florida. Emerg Infect Dis 8:575–580

Silue KD, Raso G, Yapi A et al (2008) Spatially-explicit risk profiling of *Plasmodium falciparum* infections at a small scale: a geostatistical modelling approach. Malar J 7:111

Vazquez-Prokopec GM, Cecere MC et al (2005) Spatiotemporal patterns of reinfestation by *Triatoma guasayana* (Hemiptera: Reduviidae) in a rural community of northwestern Argentina. J Med Entomol 42:571–581

Vezzani D, Carbajo A (2006) Spatial and temporal transmission risk of *Dirofilaria immitis* in Argentina. Int J Parasitol 36:1463–1472

Vezzani D, Carbajo A (2008) *Aedes aegypti, Aedes albopictus*, and dengue in Argentina: current knowledge and future directions. Mem Inst Oswaldo Cruz 103:66–74

Vittor AY, Gilman RH, Tielsch J et al (2006) The effect of deforestation on the human-biting rate of *Anopheles darlingi*, the primary vector of *Falciparum* malaria in the Peruvian Amazon. Am J Trop Med Hyg 74:3–11

Williams CR, Mincham G, Ritchie SA et al (2014) Bionomic response of *Aedes aegypti* to two future climate change scenarios in far north Queensland,

Australia: implications for dengue outbreaks. Parasit Vectors 7:447

Wimberly MC, Yabsley MJ, Baer A et al (2008) Spatial heterogeneity of climate and land-cover constraints on distributions of tick-borne pathogens. Glob Ecol Biogeogr 17:189–202

Ximenes MFFM, Castellon EG, De Souza MF et al (2006) Effect of abiotic factors on seasonal population dynamics of *Lutzomyia longipalpis* (Diptera: Psychodidae) in Northeastern Brazil. J Med Entomol 43:990–995

Zeilhofer P, Kummer OP, Santos ES et al (2008) Spatial modelling of *Lutzomyia (Nyssomyia) whitmani* s.l. (Antunes & Coutinho, 1939) (Diptera: Psychodidae: Phlebotominae) habitat suitability in the state of Mato Grosso, Brazil. Mem Inst Oswaldo Cruz 103:653–660

Zhou G, Munga S, Minakawa N et al (2007) Spatial relationship between adult malaria vector abundance and environmental factors in Western Kenya Highlands. Am J Trop Med Hyg 77:29–35

Wolbachia: Influence on Pathogeny, Treatment, and Control of Arthropod-Borne Diseases

38

Eric P. Caragata and Luciano A. Moreira

Abstract

Wolbachia pipientis is a naturally occurring bacterial endosymbiont of insects and other invertebrates that has recently gained prominence as a means for the biological control of important disease vector species. In this chapter we describe the biology of *Wolbachia*, its varied relationship with the invertebrates it infects, and how this relationship is a useful tool that is currently being exploited by novel and inventive means to reduce the capacity of insects to serve as vectors for a variety of diseases including dengue fever, lymphatic filariasis, and potentially even malaria.

Keywords

Mosquito-transmitted disease • Mosquito control • Dengue • Malaria • Filariasis

38.1 The *Wolbachia*-Infected Insect

Biological methods of controlling insect vectors have long been used to supplement insecticide-based programs of control. Historical examples include the sterile insect technique—the release of sterilized male insects to mate with wild females and consequently eliminate the wild population, or animal predation, which involves the release of naturally predatory species such as fish and crustaceans into larval sites. While effective in controlling vector populations, these strategies possess a critical flaw, that being they are expensive and laborious to implement, requiring regular monitoring to identify major breeding sites in the target area, and regular release of insects to sustain population suppression.

Insecticides remain the most popular and effective form of vector control in the short term. However, the biological landscape into which they are deployed has changed drastically over the past 20 years. Repeated exposure to the most popular insecticidal compounds has led to the evolution of genetic resistance among target populations. There has also been an expansion in range and prevalence of key vector populations,

E.P. Caragata (✉) • L.A. Moreira (✉)
René Rachou Centre of Research,
Av. Augusto de Lima, 1715 Barro Preto, 30190-002
Belo Horizonte, MG, Brazil
e-mail: e.caragata@gmail.com;
luciano@cpqrr.fiocruz.br

C.B. Marcondes (ed.), *Arthropod Borne Diseases*, DOI 10.1007/978-3-319-13884-8_38

and this in turn has led to more frequent and more severe outbreaks of tropical diseases, which have likewise emerged in new areas. These changes have been driven by changing global climate and increased urbanization, providing increased access to favorable breeding sites and hosts, and also by an increase in tourism to areas of endemic disease transmission. Given the continued problem posed by these diseases, there is a necessity for novel forms of vector control to supplement insecticide usage. One of the most interesting and promising methods involves the bacterial endosymbiont *Wolbachia pipientis* (McGraw and O'Neill 2013).

38.1.1 The Prevalence of *Wolbachia*

Wolbachia are Gram-negative, maternally inherited, intracellular bacteria from the Alphaproteobacteria class. *Wolbachia* naturally infect a broad range of different arthropod species, including mosquitoes, flies, butterflies, ants, wasps, spiders, scorpions, terrestrial isopods, and filarial nematodes. As such, *Wolbachia* is possibly the most ubiquitous endosymbiotic bacterium on the planet (Zug and Hammerstein 2014). *Wolbachia* were first described by Hertig and Wolbach in 1924, as circular *Rickettsia*-like bacteria within the ovaries of *Culex pipiens* mosquitoes. Historical estimates have suggested that between 25 and 70% of all insect species are naturally infected with *Wolbachia*; however, more recent estimates based on statistical modeling of symbiont prevalence suggest that this number is actually closer to 40%. There is uncertainty over the exact number of extant insect species, but it is estimated that there could be upward of five million, which implies that there are likely more than two million *Wolbachia*-infected species. Among these species, *Wolbachia* has diversified genetically and in biological function into what are known as "strains," i.e., *w*Mel is the *Wolbachia* strain that infects *Drosophila melanogaster*. Critically, some key disease vectors are not naturally infected with *Wolbachia*. This includes the vast majority of mosquitoes from the genus *Anopheles* (Baldini et al. 2014), which transmit the malaria-causing *Plasmodium* parasites, and also the dengue and Zika mosquito, *Aedes aegypti*. However, other key vectors such as *Culex pipiens* and *Aedes albopictus* do possess natural infections.

There is a great deal of variability in effect of *Wolbachia* infection on the biology of its host. Some strains affect their host in a way that can be exploited for vector control purposes: to kill vector species before they can transmit pathogens, to suppress vector populations in the field, or to replace wild vector populations with pathogen-refractory insects. In order to understand how *Wolbachia* can be used in these ambitious strategies, it is first necessary to understand how the bacterium interacts with its host's biology.

38.1.2 The *Wolbachia*-Host Relationship

The relationship between *Wolbachia* and host is intricate and complex. The nature of which varies depending on the identity of the infecting *Wolbachia* strain, the invertebrate that they infect, and the time since the infection was first established. There are *Wolbachia*-host relationships that are mutualistic in nature, parasitic in nature, and others that involve aspects of both known as "Jekyll and Hyde" infections (Jiggins and Hurst 2011). The parasitism of *Wolbachia* is of great interest to evolutionary biologists as it contradicts one of the central theories of symbiosis, which suggests that vertically transmitted symbionts that behave parasitically must eventually develop a mutualistic relationship with their host or face elimination.

Wolbachia's parasitism comes about through manipulation of host reproductive biology to facilitate their own reproduction, usually to the detriment of the host. Only female insects transmit *Wolbachia*, so infection in males is largely irrelevant to their reproductive success, and consequently reproductive manipulations caused by the bacterium tend to result in a skew in the sex ratio to favor female progeny. Some examples of sex ratio-distorting phenotypes include feminization, male killing, and parthenogenesis. Feminization is the process whereby genetic male progeny are transformed into functional female progeny. This phenotype occurs through an unknown mechanism and is most prominent in isopod crustaceans and butterfly species such as *Eurema hecabe*. Male killing occurs when genetic male progeny die during embryogenesis, probably due to defects in chromatin remodeling and mitotic spindle formation. Male killing is present in multiple insect taxa and has evolved independently in numerous bacteria that infect insects. Parthenogenesis occurs when *Wolbachia* cause unfertilized eggs that would otherwise develop into males, to develop as females, as seen in *Trichogramma* wasps. In some parthenogenetic species, such as *Franklinothrips vespiformis*, *Wolbachia* is required in order to produce female progeny. All of these processes result in a proportionate decrease of male progeny, providing an advantage to the symbiont, as the *Wolbachia*-transmitting female progeny do not have to compete with male siblings.

The most prominent reproductive manipulation caused by *Wolbachia* is cytoplasmic incompatibility (CI). In the 1980s and early 1990s, biologists observed that mating of insects from the same species occasionally produced few to no viable eggs. This phenotype could be eliminated after treatment of the insects with antibiotics, with the causal factor determined to be *Wolbachia* infection in the gonads. CI occurs only when *Wolbachia*-infected males mate with uninfected females. In contrast, *Wolbachia*-infected females are capable of laying viable eggs after mating with either type of partner. This phenotype has been observed in a diverse range of taxa, including mosquitoes, flies, mites, and wasps. The incompatibility can be extended to matings between insects infected by two *Wolbachia* strains of different incompatibility types (known as bidirectional incompatibility). Here, normal fertility is only observed between same-strain matings. Bidirectional incompatibility has also been observed in nature between genetic variants of the same *Wolbachia* strain, as is the case with the *w*Pip strain that infects *Cx. pipiens*.

From a vector biocontrol perspective, CI is one of the more important phenotypes caused by *Wolbachia*. In a population of insects, with a small proportion of *Wolbachia*-infected individuals, CI will lead to an increased proportion of *Wolbachia*-infected individuals over successive generations. Consequently, this allows *Wolbachia* to invade naïve insect populations and quickly reach fixation. CI can also cause the spread of a *Wolbachia* strain across large geographic areas in a relatively short period of time. This was the case for the strain *w*Ri in *Drosophila simulans*, which was observed to invade populations across California between the 1980s and 2000s. CI provides utility for vector control programs by acting as a mechanism of "drive" that can be used to rapidly spread desirable *Wolbachia*-induced traits through an uninfected population. This gives *Wolbachia*-based control a comparative advantage over transgenic control, which inherently lacks a similar spreading mechanism. As *Wolbachia* have not yet been successfully genetically transformed, current control efforts are restricted to the use of naturally occurring *Wolbachia*-induced phenotypes.

In addition to parasitic manipulation of host reproductive biology, *Wolbachia* are also capable of mutualistic interactions with their hosts. These are likely to occur in the more established host-symbiont relationships and, in the absence of a symbiont-favoring reproductive parasitism, must provide a fitness benefit to the host in order for stable infections to persist. Examples of insects with *Wolbachia*-induced mutualisms include the parasitoid wasp *Trichogramma bourarachae* where *Wolbachia* increases fecundity. Other *Wolbachia* strains protect and benefit their host by altering iron metabolism in situations of high toxicity where uninfected insects struggle. The *Wolbachia*-host relationship can be mutually dependent, as for the filarial nematode *Brugia malayi* and its symbiont *w*Bm. Here the partners have evolved to contribute key metabolic resources required for each other's survival, which they cannot produce for themselves (Foster et al. 2005). Additionally, *Wolbachia* in some bed bugs live in a specialized organelle called the bacteriome and provide their hosts with B vitamins. Given that these phenotypes are often less pronounced and are typically less important to controlling pest insects, they are likely more abundant than currently described.

38.1.3 Integration with Host Biological Process

For a *Wolbachia* infection to be both sustainable and spreadable within and between populations, it must offer a fitness benefit to the host or at least have only minimal fitness costs. Given the varied range of phenotypes observed in *Wolbachia*-infected insects, it is clear that the symbiont must possess the capacity to affect many aspects of host biology, so it is not surprising that the bacteria infect a variety of tissues throughout their host's body. The ovaries are the most heavily infected tissue, which is logical given that *Wolbachia* biology is geared toward maternal transmission. Some *Wolbachia* strains are also found in numerous somatic tissues, and the extent and severity of the infection depends on the host-symbiont combination. Infected tissues include the digestive tract—including the midgut, hindgut, and malpighian tubules, the muscle, the ganglia and other nervous tissues, cells within the hemolymph, the salivary glands, and the fat body. Strains that infect a wide range of host tissues are more likely to cause numerous and diverse effects on host biology. Strains that cause reproductive parasitism can support these interactions having a mild deleterious effect without affecting the ability to maintain infection. However, without this reproductive drive, there would be high pressure to ensure that there was little cost to the host. Given *Wolbachia*'s strong maternal transmission, the role and purpose of infection in somatic tissues are unclear, but it may indirectly facilitate infection by altering host biological process to benefit *Wolbachia*. Certainly, *Wolbachia* infections in digestive tissues could alter the digestive process and enhance infections of reproductive tissues.

Further information on host-symbiont interactions can be derived through comparative genomics. Presently, a handful of draft *Wolbachia* genomes have been sequenced and annotated. These genomes are all small, on the order of 1000–1400 kb in size, and they contain a relatively small number of coding genes (Wu et al. 2004). Within these, there is a core set of *Wolbachia* genes shared by many of the strains. The typical *Wolbachia* genome is streamlined and contains large amounts of repeat and mobile elements. Variation in the pattern and number of these elements accounts for a large part of the variation between strains. Other key differences and areas that are possibly more likely to undergo rapid evolution are the ankyrin-coding genes and prophage regions. To date there is no clear link between a specific genetic profile and a *Wolbachia*-induced phenotype. Strains that induce vastly different responses in their hosts can have practically identical genotypes, while strains that induce the same phenotype in different hosts can have distinct genome structures.

The evolutionary history of *Wolbachia* is complex and unclear. The bacterium has currently been grouped phylogenetically into 13 distinct clades or supergroups. This figure is almost certainly an underestimate, given the diversity of host-symbiont combinations in existence. Critically, these supergroups do not conform to the phylogeny of their arthropod hosts, which suggests that *Wolbachia* did not arise simultaneously with the common insect ancestor. There is a strong body of evidence describing the horizontal transfer of *Wolbachia* strains between different hosts and genetic recombination between different strains and with their host. *Wolbachia* strains infecting diverse hosts can have surprisingly similar genomes, but one key difference is in the genetic architecture that they use to interact with their host. Some aspects of interaction, such as the presence of type IV secretion system genes, are strongly conserved. One interesting aspect uncovered by genomic analysis is the metabolic dependency of *Wolbachia* on its host. Many of the bacterium's genetic pathways for metabolite biosynthesis have either been lost or are incomplete. This suggests not only a dependency on the host to obtain key resources but also the presence of a system of biological architecture to promote these interactions. Together, these features suggest a complicated network of interactions between host and symbiont.

38.2 Transinfection of Disease Vectors

38.2.1 The Transinfection Process

Transinfection is the process of creating a *Wolbachia* infection in a new species by transferring the *Wolbachia* strain from another species. In terms of insect biocontrol, this can be a very useful technique if your target species is not naturally infected with *Wolbachia*, and it is desirable to have it infected. It may also be desirable to transfer an interesting or beneficial *Wolbachia*-induced phenotype from one species to another. Transinfection is also useful in other circumstances, such as when the target species already possesses a natural *Wolbachia* infection, and it might be desirable to introduce a second, for example, in order to generate bidirectional incompatibility. Transinfection can also be used to study how *Wolbachia* affect their hosts; if two host species are infected by the same *Wolbachia* strain, the physiological effects of infection can be compared between the two. From the perspective of biocontrol, the most important aspect to transinfection is that it is possible to transfer *Wolbachia* between distantly related hosts. This discovery has resulted in a major emphasis on the transinfection of vectors of major medical importance and the transinfection of pest species.

The process of transinfection involves microinjection of *Wolbachia* into early-stage embryos of the target species. This description is deceptively simple, as in reality the procedure is complicated, requires rigorous, delicate work, and has a high failure rate. The source of the donor material, the microinjection protocol, and the purification process can vary slightly. The material can come directly from the ovaries of the donor species, where *Wolbachia* is typically present at high density, or from cell culture where the preparation is typically cleaner. In both cases it is advantageous to have a closely related donor and target species, as there is evidence that this promotes a successful transinfection. In more distantly related taxa, it may be necessary to first generate a transinfected cell line from the target, as this is easier to create than a live insect line. *Wolbachia* can undergo adaptation during prolonged passage in cell culture and then can make subsequent in vivo transinfection a little easier (Walker et al. 2011). The purified *Wolbachia* is then promptly injected into the posterior pole of recently laid embryos using a glass capillary. Injection must be performed prior to embryo melanization in species where that process occurs. There is an alternative protocol, cytoplasm transfer, where embryonic cytoplasm from the donor species is withdrawn and then injected into the embryos of the target species. Both techniques remain in popular use.

38.2.2 Limiting Factors

To date, a large number of stable transinfected insects have been created, encompassing many different species (Hughes and Rasgon 2014). These successes conceal the inherent difficulties of transinfection, as there is only a very low success rate in generating stably infected lines. The vast majority of injected embryos fail to hatch due to mechanical damage caused by the capillary, while a high percentage of those embryos that actually develop to adulthood are sterile or, in the case of mosquitoes, will not take a blood meal. For *D. simulans* transinfected with the *Wolbachia* from *Ae. albopictus*, five lines were created from 695 injected eggs for a success rate of 0.72%. Oddly enough, this success rate is high in comparison to other attempts, likely because both donor and host had native *Wolbachia* infections. Transferring *Wolbachia* between closely related species is also more likely to have a high success rate, even if the target species is not naturally infected. But transinfection between more distantly related species is typically more problematic. This point is well illustrated by considering the case of *Ae. aegypti*. Even with a prolonged period of cell culture adaptation prior to transinfection, the success rates for transinfection with the *w*Mel and *w*MelPop strains were 0.12% and 0.10%, respectively, with more than 10,000 embryos injected. One further problem is that *Wolbachia* infection is commonly lost in the first few generations post-transinfection; six lines of *w*MelPop-infected *Ae. aegypti* were lost between G_0 and G_3 (McMeniman et al. 2009). This probably occurs because, in spite of their prevalence among insects, *Wolbachia* are actually not particularly proficient at establishing infections in new hosts (Zug and Hammerstein 2014).

38.2.3 Transinfected Host Biology

When a stable transinfection is eventually established, the resulting effect on host biology may be different to what was predicted. *Wolbachia* density can change with different hosts, and this can affect the strength of phenotypes such as CI. Alternatively, the strain can produce a completely different phenotype in its new host, i.e., male killing rather than CI, or simply cause no detectable effects. To that end it may be beneficial to create transient, somatic infections in the target insect in order to assess the effects of your desired strain (Hughes et al. 2011). However, stable infections of the germ line are unlikely to occur using this technique, so it cannot be used to study reproductive manipulations. The changes to host biology that result from transinfection need to be better understood, as they could potentially affect the relative competitiveness of infected insects in the field. Currently, the most important *Wolbachia*-transinfected insects are mosquitoes, given their roles as vectors of the most prominent tropical diseases.

There is an extensive project ongoing using the dengue and Zika vector *Aedes aegypti* (see *Mechanism* below), and consequently the effects of *Wolbachia* infection in this species are well characterized. There are currently four stable transinfected lines. Of these, *w*MelPop (a virulent, life-shortening *Wolbachia* strain from *D. melanogaster*) has the most extreme effect on host biology. This strain offers a unique opportunity to examine the extreme effects of symbiosis, as the pathogenicity and novelty of the infection have created many interesting manipulations of host biology, and CI coupled with a high rate of maternal transmission allows these phenotypes to persist rather than being purged.

The *w*MelPop infection causes strong life-shortening effects, very high maternal transmission, a small decrease in fecundity, and CI (McMeniman et al. 2009). It is present in numerous tissues, although the greatest density is in the ovaries (Moreira et al. 2009). Infection causes many unusual phenotypes including severe decreases in fecundity when the host feeds on nonhuman blood, decreased long-term viability of embryos, altered locomotor activity and metabolism rate, altered feeding and probing behaviors, altered transcriptional profile including immune gene activation (Kambris et al. 2009, Moreira et al. 2009), altered gene methylation profile, altered metabolic interactions with the host, and a strong anti-pathogenic response (see *Anti-pathogenic effects* below) (Moreira et al. 2009).

The other transinfections in *Ae. aegypti* are with more benign strains, *w*AlbB from *Ae. albopictus* and *w*Mel from *D. melanogaster*, and consequently the resulting effects of host biology are less extreme. The *w*AlbB strain causes CI and a decrease in fecundity and stimulates host immunity, increasing reactive oxygen species (ROS) production (Pan et al. 2012). The *w*Mel stain also causes CI, a modest life-shortening effect and no effect on fecundity (Walker et al. 2011), interacts with its host's metabolism, and likely has other as yet uncharacterized effects (Hoffmann et al. 2011). A further transinfection with both of these strains has recently been generated, and this double infection has fewer fitness costs than *w*MelPop but higher bacterial density than *w*Mel (Joubert et al. 2016). The fact that all these *Wolbachia* infections interfere with dengue infection in their host and cause CI makes them ideal candidates for biocontrol (see *Mechanism* below).

There have been three stable transinfections of the dengue and *chikungunya* vector *Ae. albopictus*. The effect of the *w*Mel strain is similar to what is seen with *Ae. aegypti*, as it causes CI and immune gene activation and has few fitness effects. It also interferes with both dengue and chikungunya virus infection. The *w*MelPop strain has virulent effects due to very high *Wolbachia* density, causing life shortening, reduced fecundity, and high embryonic mortality. Infections are easily lost due to the extreme pathogenicity and low maternal inheritance. The *w*Pip strain (from

Culex pipiens mosquitoes) causes CI, a decrease in fecundity and fertility, but has no effect on survival or male mating competitiveness. Both the *w*Mel and *w*Pip strains cause bidirectional incompatibility with the native *Ae. albopictus Wolbachia*, making these transinfected lines ideal candidates for the incompatible insect technique (Sect. 38.5.1).

Aedes polynesiensis is the vector of *Wuchereria bancrofti*, a lymphatic filariasis-causing nematode in the South Pacific. Like *Ae. albopictus*, it has a native *Wolbachia* infection, and populations have been infected with additional strains including *w*AlbB, in the hope that these will cause bidirectional incompatibility with the native strain (Sect. 38.4). A recent development is the first stable transinfection in an anopheline taxon: the infection of the primary vector of malaria in Asia—*Anopheles stephensi* with *w*AlbB (Bian et al. 2013). This strain causes CI, has high maternal transmission, and induces ROS. It also interferes with infection by the malaria-causing parasite *Plasmodium falciparum* and consequently represents a vital first step in the use of *Wolbachia* as a method of control against malaria. A further important step was the discovery of a natural *Wolbachia* infection in a field population of *An. gambiae* in Burkina Faso. Although the strain has a low maternal transmission rate, it may still play a role in limiting *Plasmodium* infection in its host. Additionally, its existence shows that *Wolbachia* can persist in human malaria vectors in the long term.

38.3 Anti-pathogenic Effects

38.3.1 Pathogen Interference

In addition to CI, the anti-pathogenic properties shown by some *Wolbachia* strains have the greatest utility to the bacterium's potential as a biocontrol agent. This phenotype is known by various names including pathogen interference, *Wolbachia*-mediated antimicrobial protection, virus protection, or pathogen blocking and refers to the ability of some strains to interfere with infection by some pathogens (for a comprehensive review of pathogen interference against arboviruses in mosquitoes, refer to Rainey et al. (2014)).

The phenotype was first described in *D. melanogaster* in 2008. It was observed that when flies were challenged with Drosophila C virus (DCV) infection by intraperitoneal injection, *Wolbachia*-infected flies experienced delays in virus-induced mortality in comparison to flies that had been cured of their *Wolbachia* infection. The implication was that infection by *Wolbachia* protected the flies against the pathogenic effects of DCV. It was shown that the strength of the protective effect of *Wolbachia* varied depending on the infecting virus. In challenge with two other insect viruses, Flock House virus and Nora virus, *Wolbachia*-infected flies were resistant and did not succumb to infection, while uninfected flies continued to die at a rapid rate. Examination of viral titers revealed that *Wolbachia*-infected flies had significantly lower levels than uninfected flies, indicating that the protective effect due to *Wolbachia* was because of inhibited or delayed virus accumulation. This link between *Wolbachia* infection, protection against virus-induced mortality, and decreased viral titers is not ubiquitous, as the *w*Ri strain in *D. simulans* prevents death in its host without affecting the viral titer (Rainey et al. 2014). This was the first indication of the complicated nature of the mechanism underlying the anti-pathogen phenotype. The interference effect also extends beyond traditional viruses of *D. melanogaster*, with similar effects observed against the medically important arboviruses *West Nile virus* and chikungunya virus. For both, *Wolbachia* infection in *D. melanogaster* led to significant reduction of viral titer.

The strength of the pathogen interference phenotype or the extent of protection observed within an individual fly is dependent on the identities of the infecting pathogen and the *Wolbachia* strain. In *D. simulans*, resistance or delays to virus-induced mortality are provided by both native *Wolbachia* infections and transinfections. *D. simulans* is an interesting model species for examining *Wolbachia* infection as there are multiple native *Wolbachia* strains. In a comparison between four of these strains, it was observed that the protective effect was very strongly correlated to *Wolbachia* density, with the two higher density strains providing protection to the host, while flies infected with either of the two lower density strains were affected by virus infection in the same manner as uninfected flies. Further analysis indicated that high *Wolbachia* density in the head, gut, and Malpighian tubes was important to a stronger protective effect. The effect was also directly correlated to the overall level of *Wolbachia* in the entire insect, as the protection decreases incrementally when flies were fed increasing doses of the *Wolbachia*-killing antibiotic tetracycline.

Protection by *Wolbachia* is common against viruses belonging to the single-stranded positive-sense RNA group. Examination of the phenotype using other types of pathogens for *Drosophila* has revealed that the effect is not ubiquitous. There is evidence that suggests that *Wolbachia* has no effect against a wide range of pathogenic bacterial species—both intracellular and extracellular—as both infected and uninfected flies die at a rapid rate when challenged. The lack of effect was not due to excessively high bacterial titer, as a reduction in the total number of colony-forming units used to infect the flies caused an equal reduction in death rate for both infected and uninfected flies. Preliminary testing has indicated that *Wolbachia* likely does not provide similar protection for the *Drosophila* host against DNA viruses. During challenge with insect iridescent virus 6 (IIV-6), *Wolbachia* infection actually enhanced the infection, causing flies to die at a faster rate than their uninfected counterparts. Further experimental infections with DNA viruses are necessary to determine whether this response is commonplace.

38.3.2 Effect Against Arboviruses

There has been extensive characterization of the pathogen interference effect in various *Wolbachia*-infected mosquito species, against viruses that they typically serve as vectors and to those that they do not. The first of these experiments occurred in *Ae. aegypti* (Moreira et al. 2009). Here, both the *w*MelPop and *w*Mel strains grow to high density and show a strong pattern of interference against a wide range of viruses.

Interference against DENV infection has been well characterized for *Wolbachia*-infected *Ae. aegypti* and *Ae. albopictus*. For infection with the *w*Mel and *w*MelPop strains in *Ae. aegypti*, there is practically complete blocking of DENV infection after oral feeding; the vast majority of mosquitoes are refractory to infection, while for those that do become infected, there is greatly decreased virus dissemination and no virus present in the saliva (Moreira et al. 2009, Walker et al. 2011). Similar results occur with the *w*AlbB-transinfected line, although the effect is not so strong. After oral feeding of DENV, *Wolbachia*-infected mosquitoes had greatly decreased susceptibility, decreased viral loads, and dissemination and transmission through saliva. *Ae. albopictus* transinfected with *w*Mel also show complete inhibition of DENV. This suggests that it is the *Wolbachia* strain rather than the host species, which is responsible for the strength of the interference effect.

These assays also showed that in *Wolbachia*-infected mosquitoes, DENV infections do become established within the first few days post-infection; however, they are typically eliminated within 7 days. In cases where DENV was able to establish infections in the tissues of *Wolbachia*-infected mosquitoes, it could only do so in tissues or cells where *Wolbachia* was not present (Moreira et al. 2009). Further experiments to test the strength of the interference involve bypassing the midgut barrier (which hinders DENV infection) by using intrathoracic injection into the hemolymph rather than oral feeding as a virus delivery method. Without the midgut to inhibit viral invasion, a greater proportion of *Wolbachia*-infected mosquitoes become infected, but the overall viral load was greatly reduced compared to *Wolbachia*-uninfected mosquitoes, and there was no detectable virus in the saliva (Moreira et al. 2009). This highlights the critical role of the midgut in the pathogen interference phenotype but also shows that other aspects of the *Wolbachia* infection can also contribute to the effect.

Interference by *Wolbachia* against chikungunya virus (CHIKV) has also been characterized for both *Ae. aegypti* and *Ae. albopictus*, two natural vectors of the virus. As for DENV, the strongest interference effect was observed with the *w*MelPop infection in *Ae. aegypti*, with that strain causing drastically reduced susceptibility and almost complete inhibition of viral dissemination (Moreira et al. 2009). For *w*Mel-infected *Ae. aegypti*, there is a strong but less pronounced effect, with both susceptibility and dissemination to the salivary glands decreased. That strain causes a similar effect against CHIKV in *Ae. albopictus*. In contrast to DENV, the interference effect is completely eliminated for *w*Mel by changing the mode of viral delivery to injection. These cases highlight the strong interference conferred by *Wolbachia* transinfections.

An interesting extension is to consider the potential effect of natural *Wolbachia* infections on pathogen transmission in their host. A prominent example is *Ae. albopictus*, which is a good vector for both DENV and CHIKV and yet possesses a natural *Wolbachia* infection. To that end, *Wolbachia* was cleared from *Ae. albopictus* using antibiotics, and then both infected and uninfected mosquitoes were challenged separately with DENV and CHIKV. Interestingly, it was shown that the native *Wolbachia* inhibited transmission of DENV from the salivary glands but did not significantly affect CHIKV infection, which suggests that native infections may provide some degree of interference to some pathogens.

The interference phenotype has also been examined for two additional arboviruses, *West Nile virus* (WNV) and yellow fever virus (YFV). The role of native *Wolbachia* strain (*w*Pip) in response to infection with WNV has been examined in *Cx. quinquefasciatus*. After oral feeding with a range of viral titers, there was similar susceptibility among *Wolbachia*-infected and *Wolbachia*-uninfected mosquitoes, but virus levels were lower in the presence of *Wolbachia* and greatly reduced in the saliva. The inhibitory effect was determined to be rare for *Cx. quinquefasciatus*, as the cohort where it was detected had an abnormally high *Wolbachia* density. Consequently, there is likely to be minimal effect of *Wolbachia* on WNV infection in most *Culex* clades. Interference against YFV has been examined in both the *w*MelPop- and *w*Mel-transinfected *Ae. aegypti* lines. Of the two, the strongest interference effect was observed for *w*MelPop, which significantly reduced susceptibility, dissemination, and viral load, as well as maintaining effective blocking at higher viral titers. By comparison, the *w*Mel strain only caused a reduction in viral titer (Rainey et al. 2014).

38.3.3 Effect Against *Plasmodium*

There is a growing body of work investigating the role of *Wolbachia* during infections of mosquitoes by malaria-causing *Plasmodium* parasites. Like *Ae. aegypti*, the typical vectors of malaria are not naturally infected with *Wolbachia*. Currently, there is no information on how the natural *Wolbachia* infection in *An. gambiae* affects *Plasmodium* infection (Baldini et al. 2014). In spite of numerous attempts, there is currently only one vector of human malaria that has been transinfected with *Wolbachia*—*An. stephensi*, a vector of *Plasmodium falciparum* from the Middle East and Southern Asia. These mosquitoes were transinfected with *w*AlbB from *Ae. albopictus*. It was shown that *Wolbachia* infection significantly reduced the prevalence of *P. falciparum* infection at the oocyst and sporozoite stages, resulting in a decreased pathogen load in the salivary glands. The creation of this transinfected line represented a key step forward for the use of *Wolbachia* to control malaria. Not only did it prove that it was possible to achieve a stable infection in an anopheline mosquito, but the pathogen interference response to *Plasmodium* was similar to what was seen for arboviruses (Bian et al. 2013). This suggests that *Wolbachia* could potentially be used to target the primary malaria vector *An. gambiae* at some point in the future. Currently, the effects of *Wolbachia* on *An. gambiae* have only been examined through the use of somatic *Wolbachia* infection. These infections are transient as they do not typically reach the ovaries and are thus not transmitted to the mosquito's progeny. In this technique, a solution of purified *Wolbachia* was injected directly into the hemolymph of adult mosquitoes. Infections with both *w*AlbB and *w*MelPop have been created, and although *Wolbachia* infections did not become established in the midgut, both strains caused a reduced pathogen load after feeding on a *Plasmodium*-infected blood meal. One additional consequence of these transient infections is an upregulation of immune gene expression, similar to what is seen with *Wolbachia*-transinfected *Ae. aegypti*. Such immune effects appear to be particularly important to the pathogen interference response in mosquitoes, particularly when the midgut is not heavily infected by *Wolbachia*. Recent evidence has shown that the natural gut microbiota of *Anopheles* mosquitoes may be impeding transinfection, suggesting that the concurrent use of antibiotics with somatic infection could actually lead to a stable transinfection.

Other significant work on interactions between *Wolbachia* and malaria has out of necessity occurred in species that do not normally serve as vectors for malaria or in those that are vectors for nonhuman malaria, as these species are more likely to be infected with *Wolbachia*. Transient infection of *An. gambiae* with *w*MelPop had variable effects on the rodent malaria parasite *P. berghei*. In some experiments, oocyst load was significantly increased; however, for others there were only small, nonsignificant effects. The effectiveness of such transient infection experiments is strongly dependent on both the *Wolbachia* strain and the type of pathogen. Transient infection of *An. gambiae* with *w*AlbB led to a comparative increase in *P. berghei* oocyst load compared to uninfected control mosquitoes. While such a result might give rise to concerns that *Wolbachia* could potentially have adverse effects upon exposure to some pathogens, it is important to note that a transient infection is not a stable transinfection; they affect their host and respond to pathogens in different ways. As such, while results obtained using this technique can be interesting and informative, they are not necessarily representative of how *Wolbachia* will affect *Plasmodium* infection in anophelines.

The interference effect has also been examined for two avian malarial parasites. The first observation of pathogen interference against malaria was with *w*MelPop-infected *Ae. aegypti* and *P. gallinaceum*. Oral feeding of the parasite resulted in significantly decreased susceptibility to infection and reduced oocyst loads. Similar experiments have been performed for *Aedes fluviatilis*, a South American mosquito with a natural *Wolbachia* infection that is thought not to be a natural vector of any human diseases. Here it was shown that *Wolbachia* did not have an inhibitory effect on *Plasmodium* infection. In fact in two of four experiments, the oocyst load was higher in the presence of *Wolbachia*. A further example of the unpredictability of natural systems is seen with *Cx. pipiens* and *P. relictum*, where *Wolbachia* infection increased host susceptibility to the parasite and also caused a small increase to oocyst load. The stark contrast in the pathogen interference effect between these two systems highlights the importance of the distinction between natural *Wolbachia* infections and transinfections, with the latter being more likely to cause strong blocking and the former likely to have a less predictable effect.

38.3.4 Other Pathogens

Interference by *Wolbachia* infections in mosquitoes has also been observed for other types of pathogens. Such is the case with the lymphatic filariasis-causing nematode *Brugia pahangi*, where *Wolbachia*-infected mosquitoes are more resistant to infection, and have a reduction in worm larvae, compared to uninfected mosquitoes (Andrews et al. 2012, Kambris et al. 2009). *Wolbachia* also protects its host against the pathogenic effects of some species of bacteria, but not others. In mosquitoes, *w*MelPop offers very strong protection against a wide variety or bacterial pathogens (after either stabbing or injection), both preventing death and decreasing bacterial load (Kambris et al. 2009, Ye et al. 2013), while the *w*Mel strain offers less effective protection. Yet in *D. melanogaster* infected by the same *Wolbachia* strains, and challenged with the same bacterial species, there was no protective effect (Ye et al. 2013). In fact protection against bacteria appears to be specific to mosquitoes, although it has not yet been determined if this is a consequence of *Wolbachia* transinfection or species-specific differences in pathogen response.

38.4 Mechanism

38.4.1 *Wolbachia* Density

There is still a great deal of uncertainty surrounding the mechanism that causes pathogen interference in *Wolbachia*-infected insects and the extent to which it is conserved between different host species and different *Wolbachia* strains and different pathogens. Amidst the persistent confusion, perhaps the most consistent feature affecting both the presence and strength of the pathogen blocking response is *Wolbachia* density: the amount of *Wolbachia* infecting a given insect. The general rule is that high density will produce a stronger blocking response. The rule is exemplified by the difference in blocking response between *w*MelPop- and *w*Mel-infected insects. In both mosquitoes and *Drosophila*, the higher density *w*MelPop infection provides stronger protection or interference. *Wolbachia* density is highly variable between strains, and also within insect populations, a fact that might explain some of the variability that is present in the pathogen interference effect. While whole body *Wolbachia* density is important, pathogen interference is also dependent on having high density in key tissues—the midgut and the salivary glands, both of which act as barriers to pathogen dissemination. Furthermore, the density-dependence effect extends down to the cellular level with more heavily infected cells offering stronger interference. Interestingly, as an indicator of the level of complexity in interaction between host, pathogen, and *Wolbachia*, pathogens are also capable of modulating *Wolbachia* density in the host, potentially through resource competition with *Wolbachia* or through stimulation of host immunity.

38.4.2 Immune Activation

Wolbachia density is an important component to the interference mechanism, but it should be considered more of a facilitator for the phenotype, rather than an actual biological cause of interference. This means that there are other means by which *Wolbachia* affects pathogen infection and replication.

After discovery of the pathogen interference phenotype, it was theorized that potentially one mechanism was responsible for the response against the entire range of pathogens. *Wolbachia* transinfection in *Ae. aegypti* causes a systemic upregulation of immune gene expression, affecting many aspects of immune response such as antimicrobial peptides, toll pathway genes, and the melanization response. This widespread upregulation plays a vital role in the pathogen interference response. Two of the more studied genes that are stimulated by *Wolbachia* are the cecropin D and defensin C antimicrobial peptide coding genes. When the activity of these two genes was suppressed using RNAi, the pathogen interference in *Wolbachia*-infected mosquitoes was weakened dramatically (Pan et al. 2012). Another feature of *Wolbachia* infection in mosquitoes is increased levels of reactive oxygen species. There is a direct link between increased ROS levels and the immune activation that is observed in *Wolbachia*-transinfected *Ae. aegypti* (Pan et al. 2012). Interestingly, among transinfected *Ae. aegypti*, the strength and breadth of the immune response appears to be dependent on the infecting *Wolbachia* strain, with the *w*MelPop strain having the highest density and causing the most widespread immune manipulation.

The case for the involvement of immune activation in the pathogen interference mechanism is strong; however, it appears that this phenotype is a response to transinfection rather than something typical of *Wolbachia* infections as a whole. Case in point, the same strains that stimulate increased immune activation in *Ae. aegypti*, *w*Mel and *w*MelPop, have no effect on the immune system of *D. melanogaster*. Yet, both strains cause pathogen interference against DENV in both species

(Rancès et al. 2012). This suggests that the notion of immune activation, while important to the mechanism, is insufficient to explain all occurrences of pathogen interference. Furthermore, rather than there being a ubiquitous, simplistic mechanism, it is likely that the mechanism is complex and multifaceted.

38.4.3 Nutrition and Other Factors

It follows that there must be additional components to the mechanism. Some of these may be unique to certain host species or *Wolbachia* strains, and additionally within a host-symbiont relationship there may be different mechanisms of response to different pathogens. One potential explanation relies on the nature of biological integration between *Wolbachia* and host. It has been well established that *Wolbachia* have parasitic qualities and require a wide range of nutritional resources from their hosts (Wu et al. 2004). As such, it is beneficial to think of *Wolbachia* as altering the "default state" of its host biology, with the extent of this alteration being dependent on the *Wolbachia* strain and the nature of its relationship with its host. Like *Wolbachia*, invading pathogens such as viruses also depend on metabolic resources derived from the host. When all three parties require the same scarce resource, there could plausibly be competition. When these resources are rate limiting for the pathogen and preferentially utilized by *Wolbachia*, it could restrict the pathogen's ability to invade cells and replicate.

In *Wolbachia*-infected *Drosophila*, there is such competition over host cholesterol. Increasing the amount of biologically available cholesterol leads to a decrease in the strength of the pathogen interference response, causing decreased host resistance to viral pathogenicity and increased viral titers in comparison to *Wolbachia*-infected flies with normal cholesterol levels. It is currently unclear if this is the only such competitive process affecting pathogen interference in flies or if similar processes occur in mosquitoes. There is a further question of the importance of such nutritional interactions when immune activation occurs. Furthermore, there is still a large volume of work remaining to understand the extent to which the blocking mechanism is conserved between species, and to what extent it varies in the presence of different pathogens (Rainey et al. 2014).

38.4.4 Natural Versus Artificial Infections

The difference between natural and artificial *Wolbachia*-host relationships, the later resulting from transinfections, is a key determinant of what kind of response might be expected after infection with a pathogen. The rough guide here is that a natural relationship, where the host and *Wolbachia* strain have coevolved together over a significant period of time, will produce little to no effect on a pathogen. The reason for this is based on the nature of symbiosis, with initial associations involving vertical transmitted symbionts, expected to evolve toward mutualism over time (Rainey et al. 2014). Some *Wolbachia* strains can subvert this rule to some extent, through reproductive manipulation of its host, which allows detrimental fitness effects to persist in populations. However, the trend is for infections to decrease in parasitism over time, leading to a reduction in density and the resultant ancillary effects on host biology. As these are the traits that influence the pathogen interference response, we would eventually expect to see a decrease in interference strength, although it is not likely that this evolution would occur rapidly.

Another factor influencing the expected pathogen interference response is the frequency with which a host-*Wolbachia* system encounters a particular pathogen. For example, with a typical pathogen such as DENV for the *Ae. albopictus*-*w*AlbB system, there is minimal inhibition by *Wolbachia*, as given the length of the relationship, the virus has likely overcome the blocking effects of the *Wolbachia* strain through repeated exposure. Simultaneously, the *w*AlbB strain likely persists in the host after a selection process for lower pathogenicity, which would select against high density and strong pathogen interference. A stronger interference response is expected for less frequently encountered pathogens. In this scenario some level of interference by native *Wolbachia* strains could be expected.

So consequently, the pathogen interference response in a *Wolbachia*-infected insect is dependent on three factors that potentially have some degree of interaction: (1) *Wolbachia* density, (2) a molecular mechanism which may or may not be held in common across different host-*Wolbachia* systems, and (3) the novelty of the host-*Wolbachia*-pathogen relationship, with stronger blocking occurring in situations of decreased familiarity between the parties.

38.5 Application in the Field

Wolbachia has been used or hypothesized for use in a number of different vector control projects in the field. Each of these projects takes advantage of the physiological manipulations that *Wolbachia* causes in its host. A further benefit, which assists both with facilitating the regulation of the projects and with public opinion, is the fact that control using *Wolbachia* does not involve transgenic organisms. Although there are differences in format, all of these projects seek to exploit the natural biological relationship between *Wolbachia* and host, with the aim being to target undesirable vector or pest populations.

38.5.1 Control Through Incompatibility

One of the earliest forms of vector control involving *Wolbachia* utilizes the reproductive incompatibilities of CI to crash or suppress populations of disease vectors. This is a variation on the early sterile insect technique (SIT) form of mosquito control but is instead called the incompatible insect technique (IIT). To date IIT has been mainly targeted against mosquito disease vectors. Like the SIT it only involves the release of male mosquitoes—not disease-transmitting females. The key to IIT is that mosquitoes containing *Wolbachia* are released into the same area as the target population. The released mosquitoes need to be infected with *Wolbachia*. If the target population is not infected with *Wolbachia*, the released males will mate with wild females that they encounter, and CI will then reduce mosquito numbers in the target area. If the wild population is also infected with *Wolbachia*, the released mosquitoes must contain a different strain; one that causes bidirectional incompatibility with the strain in the target population. This too will cause the wild population to crash.

One example of an IIT target is *Ae. polynesiensis*, the primary vector of the lymphatic filariasis-causing nematode *W. bancrofti* in the South Pacific. These mosquitoes have a natural *Wolbachia* infection and hence need to be controlled with another *Wolbachia* strain that causes bidirectional incompatibility. Currently, *Ae. polynesiensis* has been artificially infected with two strains that cause such an effect. Both show strong potential for field application, as there is little competitive difference in comparison to wild-type mosquitoes. In addition, one of these mosquito lines shows a high degree of refractoriness to filarial nematode infection (Andrews et al. 2012). There are also promising preliminary results for IIT against *Ae. albopictus*. IIT will likely prove to be particularly effective at suppressing mosquito populations in target areas; however, as with all such techniques, it will leave behind an ecological niche, suitable for future recolonization by other mosquitoes. Consequently, the long-term effectiveness of the IIT will likely be higher in more isolated areas where there is little evidence of mosquito immigration. Alternatively, frequent release of *Wolbachia*-infected mosquitoes will be required to prevent recolonization.

38.5.2 Control Through Life Shortening

Another form of *Wolbachia*-based control exploits the life-shortening effect of the *w*MelPop strain, which reduces the life span of infected adult mosquitoes by approximately half. The original target in this strategy was *Ae. aegypti*. From the perspective of vector competence, it is only the older disease-transmitting mosquitoes that are important. This is because there is a period of delay between the time when a mosquito first becomes infected with a virus and the time when it is capable of transmitting that virus (called the extrinsic incubation period). In this scenario, the introduction of a life-shortening *Wolbachia* could theoretically kill many mosquitoes before they became capable of transmitting pathogens that they became infected with near the start of their adult life. As with most forms of *Wolbachia* control, CI was a critical component of this strategy, being used to spread *w*MelPop through wild populations. One major problem with this strategy arose because there were a large number of unforeseen fitness costs to the host that occurred as a side effect of the *w*MelPop infection (see *Transinfected host biology* above). As a consequence, *w*MelPop-infected mosquitoes were less competitive than wild, uninfected mosquitoes, and mathematical modeling of the strain's invasive potential suggested that it would be extremely difficult for the strain to persist in wild populations. Fortunately, the discovery of the pathogen interference phenotype presented a more realistic alternative method of control, and subsequently efforts to influence disease transmission through life-shortening *Wolbachia* have been largely abandoned.

38.5.3 Control Through Obligate Mutualism

Another disease control strategy seeks to exploit the mutualistic relationship between the nematodes that cause lymphatic filariasis and onchocerciasis and the *Wolbachia* that infect them. This relationship is quite divergent from what is typically seen between *Wolbachia* and their hosts. In these systems, *Wolbachia* has evolved to have an obligate mutualistic relationship with its host—that is, the worm requires the presence of *Wolbachia* in order to survive. The two parties are metabolically mutually dependent, each requiring vital nutrients that are only produced by the other party. Furthermore, the presence of *Wolbachia* lipoproteins found in these strains can exacerbate the human inflammation response and produce more severe disease symptoms. The key to this control strategy is simple, eliminate *Wolbachia* from the worm, and the worm cannot survive and thus cannot cause disease. To that end, the nature of nematode-*Wolbachia* interaction has been closely studied to assist with the development of anti-*Wolbachia* drugs, including tetracycline and doxycycline (Taylor et al. 2010). Critically, a minimum treatment period of several weeks is required to clear the body of *Wolbachia*-infected worms. For other therapies the treatment period may be several months. Currently, efforts are underway to identify a drug candidate with long-term efficacy, which could be easily distributed en masse.

38.5.4 Population Replacement with Refractory Insects

The most advanced disease control strategy involving *Wolbachia* utilizes strains that cause both CI and pathogen interference in their host. Again the primary target of this strategy is mosquito vector species, with the current focus being on *Ae. aegypti* (McGraw and O'Neill 2013). Under this strategy, mosquitoes are infected with a *Wolbachia* strain that causes strong pathogen interference against dengue virus, leading to hosts with significantly reduced vector competence in comparison with uninfected mosquitoes. As these strains also typically cause CI, the desirable pathogen interference phenotype can be driven into a target mosquito population by releasing the mosquitoes in large numbers. This seems simple in principle, but there are complicated mathematics mechanics underlying the invasive potential of *Wolbachia* strains, which had to be characterized.

Semi-field cage experiments were conducted in order to provide an initial means of observing the ability of *Wolbachia* to invade and become fixed in a large mosquito population. Of particular importance is the initial frequency of *Wolbachia* infection required in order for the bacterium to spread to fixation. This depends in part on the relative fitness of infected mosquitoes, with strains that cause higher fitness costs requiring a higher initial volume of mosquitoes to be released in order to surpass the minimum frequency. Here, CI acts as the counterbalance; with the reproductive advantage, it provides overcoming fairly substantial effects on host biology. An example is seen with the *w*Mel strain, which has an estimated 20% fitness cost but is still capable of rapidly spreading through wild mosquito populations (Hoffmann et al. 2011). Here again the biological effects of high *Wolbachia* density act as something of a double-edged sword; a high density means a stronger effect against the target pathogen, but also a higher risk of unwanted negative effects on host biology, which could potentially interfere with the ability of the strain to invade. As such, the ideal *Wolbachia* strain has the right balance between strong CI to facilitate spreading and high *Wolbachia* density to yield strong blocking, yet not so high as to detrimentally affect host competitiveness. Currently, the strain with the best balance of these features is *w*Mel.

Subsequent to the semi-field trials, full-field trials involving *w*Mel-infected mosquitoes have been conducted in order to determine if the strain could invade and reach fixation in a wild population. The first releases were conducted around Cairns, in the north of Queensland, Australia, as here, there were relatively isolated locations, well-characterized *Ae. aegypti* populations, and minimal incidence of dengue transmission. Within the release area was a set of defined release points, from which small groups of adult mosquitoes were released weekly. Additionally, there was a network of mosquito traps that were used to monitor changes to the *Wolbachia* infection frequency. After a period of several months, *Wolbachia* had achieved fixation in the local mosquito populations (Hoffmann et al. 2011), and several years later similar infections levels have been recorded, showing that the infection persists in the field. Further releases are currently underway in additional locations in Australia, Indonesia, and Vietnam, and soon in Brazil and Colombia, as part of the Eliminate Dengue Program (EDP), and could also be used to target emerging arboviruses including CHIKV and Zika virus.

38.5.5 Community Engagement and Regulation

The successful projects involving the release of insects into the environment for biocontrol purposes (including both transgenic and non-transgenic insects) depend not only on the science involved but also on the acceptance of the methods and objectives of the project by the community within the test area. Therefore, all such projects must also adequately involve local community stakeholders and be able to address potential concerns on a local level.

The approach of the EDP is to prioritize transparency and the dissemination of information to the community, regardless of the country of operation. The view is that this will facilitate the development of strong support within the community toward the acceptance of the technologies being applied. The need for free and open dialogue between researchers and community members is particularly important in countries where dengue is endemic. Existing educational programs in these areas are typically focused on the destruction of mosquito breeding sites, and on adult extermination, so consequently utilizing a mosquito control technique that involves the release of adult mosquitoes could potentially seem confusing or counterintuitive. Generally, the idea of replacing a dengue susceptible mosquito population with one that potentially has lower vector competence has been well accepted by the residents, the stakeholders, and the wider population within release areas. As part of its communication strategy, the EDP works closely with residents, school children, community associations, local stakeholders, and leaders to engage them in each aspect of the project. Communication materials such as banners, newsletters, and activity books for children are widely used during the process.

Another important factor prior to the release of *Wolbachia*-infected insects in the environment is to obtain regulatory permission. One complicating issue was to identify the government regulatory body that would actually evaluate and approve these releases. As *Wolbachia* does not involve transgenic organisms, it should not be evaluated or regulated by agencies that deal with GMOs. As this approach involves deployment in an urban or public place, it does not necessarily fall under the purview of agriculture-related institutions.

In Brazil, the regulatory approval was eventually given jointly through a number of bodies. This included the National Review Board, due to the human ethics concerns as the project involved the release of mosquitoes into a populated area. The project was treated as a new insecticide or biological control product and was scrutinized by Government Bodies concerned with the environment, agriculture, and health, with the eventual approval being given through the health body due to the fact that dengue is a public health issue.

38.5.6 Measuring Success

As for other forms of vector control, the ultimate measure of success is a reduction in disease transmission. However, the potential success of control strategies involving *Wolbachia* can also be measured by other means, depending on the type of strategy. For population suppression strategies such as IIT, success is dependent on a rapid decrease in the size of the target insect population of the target area. For population replacement strategies, like those ongoing in Australia and Brazil, the first measure of success is the establishment of a stable *Wolbachia*-infected population. While these endpoints are different, both types of strategies require persistent monitoring of the target area, either to quantify insect numbers or to determine the *Wolbachia* infection frequency in the local population. In the case of mosquitoes, such monitoring is complicated by the fact that populations are typically in flux. Population sizes are typically dynamic and greatly affected by both short-term and seasonal changes to weather conditions. Populations are also subject to immigration, and typically this would be with *Wolbachia*-uninfected individuals. Both of these factors could negatively impact the successful application of the control strategy and as such need to be considered prior to mosquito release.

The second metric for success is a decrease in disease transmission in the target area, which is considerably more complicated to measure and to achieve. For example, in places like Northern Australia, dengue transmission is typically low with outbreaks involving only a few dozen cases, one circulating dengue virus strain, and only one mosquito vector species inhabiting the area. This makes the location ideal for testing the invasive potential of *Wolbachia*, but the potential to measure the impact on dengue transmission is limited. In other countries arbovirus transmission is notably more complicated with outbreaks that can involve thousands of infections, multiple viruses, and vector species. Under these conditions the key to measuring successful decrease in pathogen transmission due to the presence of *Wolbachia* is critically dependent on the effectiveness of local health monitoring systems. If these are inadequate, i.e., if it is not possible to track where infections arise and how they spread geographically, and link this to where *Wolbachia* is being deployed, it will be difficult to determine the overall efficacy of the control strategy. This could be addressed by deploying a network of adult mosquito collection traps (something which is typically done for dengue monitoring), which could be used to determine if local mosquitoes are infected with dengue, whether any are also infected with *Wolbachia*.

In the long term, the stable and persistent control of pathogen transmission using *Wolbachia* will depend on several factors. Perhaps the most important of these is the stability of the relationship between *Wolbachia*, host, and virus with respect to the strength of the interference against virus infection. It has theorized that these viruses will eventually be able to overcome the pathogen interference effects of *Wolbachia*. Whether this process is slow or rapid will determine the window of operability for *Wolbachia* in the field. One recent study has delivered some good news on this front. *Wolbachia*-infected mosquitoes were collected 1 year post-release in the field. These mosquitoes were challenged with three different DENV serotypes and showed a very strong level of interference against all of them (Frentiu et al. 2014). This suggests that any evolutionary changes to pathogen interference strength are not likely to occur in the short term. With strong anti-pathogen properties, a self-sustained spreading mechanisms, and advanced field trials currently ongoing, *Wolbachia* represents a viable and promising form of control that will likely be applied to combat a range of mosquito-borne diseases across the planet in the next decade.

References

Andrews ES, Crain PR, Fu Y et al (2012) Reactive oxygen species production and *Brugia pahangi* survivorship in *Aedes polynesiensis* with artificial *Wolbachia* infection types. PLoS Pathog 8, e1003075

Baldini F, Segata N, Pompon J et al (2014) Evidence of natural *Wolbachia* infections in field populations of *Anopheles gambiae*. Nat Commun 5:3985

Bian G, Joshi D, Dong Y et al (2013) *Wolbachia* invades *Anopheles stephensi* populations and induces refractoriness to *Plasmodium* infection. Science 340:748–751

Foster J, Ganatra M, Kamal I et al (2005) The *Wolbachia* genome of *Brugia malayi*: endosymbiont evolution within a human pathogenic nematode. PLoS Biol 3, e121

Frentiu FD, Zakir T, Walker T et al (2014) Limited dengue virus replication in field-collected *Aedes aegypti* mosquitoes infected with *Wolbachia*. PLoS Negl Trop Dis 8, e2688

Hoffmann AA, Montgomery BL, Popovici J et al (2011) Successful establishment of *Wolbachia* in *Aedes* populations to suppress dengue transmission. Nature 476:454–459

Hughes GL, Koga R, Xue P et al (2011) *Wolbachia* infections are virulent and inhibit the human malaria parasite *Plasmodium falciparum* in *Anopheles gambiae*. PLoS Pathog 7, e1002043

Hughes GL, Rasgon JL (2014) Transinfection: a method to investigate *Wolbachia*-host interactions and control arthropod-borne disease. Insect Mol Biol 23:141–151

Jiggins FM, Hurst GD (2011) Rapid insect evolution by symbiont transfer. Science 332:185–186

Joubert DA, Walker T, Carrington LB et al (2016) Establishment of a *Wolbachia* superinfection in *Aedes aegypti* mosquitoes as a potential approach for future resistance management. PLoS Pathog 12, e1005434

Kambris Z, Cook PE, Phuc HK et al (2009) Immune activation by life-shortening *Wolbachia* and reduced filarial competence in mosquitoes. Science 326:134–136

McGraw EA, O'Neill SL (2013) Beyond insecticides: new thinking on an ancient problem. Nat Rev Microbiol 11:181–193

McMeniman CJ, Lane RV, Cass BN et al (2009) Stable introduction of a life-shortening *Wolbachia* infection into the mosquito *Aedes aegypti*. Science 323:141–144

Moreira LA, Iturbe-Ormaetxe I, Jeffery JA et al (2009) A *Wolbachia* symbiont in *Aedes aegypti* limits infection with dengue, Chikungunya, and *Plasmodium*. Cell 139:1268–1278

Pan X, Zhou G, Wu J (2012) *Wolbachia* induces reactive oxygen species (ROS)-dependent activation of the Toll pathway to control dengue virus in the mosquito *Aedes aegypti*. Proc Natl Acad Sci U S A 109:E23–E31

Rainey SM, Shah P, Kohl A et al (2014) Understanding the *Wolbachia*-mediated inhibition of arboviruses in mosquitoes: progress and challenges. J Gen Virol 95:517–530

Rancès E, Ye YH, Woolfit M et al (2012) The relative importance of innate immune priming in *Wolbachia*-mediated dengue interference. PLoS Pathog 8, e1002548

Taylor MJ, Hoerauf A, Bockarie M (2010) Lymphatic filariasis and onchocerciasis. Lancet 376:1175–1185

Walker T, Johnson PH, Moreira LA et al (2011) A non-virulent *Wolbachia* infection blocks dengue transmission and rapidly invades *Aedes aegypti* populations. Nature 476:450–455

Wu M, Sun LV, Vamathevan J et al (2004) Phylogenomics of the reproductive parasite *Wolbachia pipientis w*Mel: a streamlined genome overrun by mobile genetic elements. PLoS Biol 2:327–341

Ye YH, Woolfit M, Rances E et al (2013) *Wolbachia*-associated bacterial protection in the mosquito *Aedes aegypti*. PLoS Negl Trop Dis 7, e2362

Zug R, Hammerstein P (2014) Bad guys turned nice? A critical assessment of Wolbachia mutualisms in arthropod hosts. Biol Rev Camb Philos Soc. doi:10.1111/brv.12098

Index

C.B. Marcondes (ed.), *Arthropod Borne Diseases*, DOI 10.1007/978-3-319-13884-8

W

MIX
Papier aus verantwortungsvollen Quellen
Paper from responsible sources
FSC® C105338

If you have any concerns about our products,
you can contact us on
ProductSafety@springernature.com

In case Publisher is established outside the EU,
the EU authorized representative is:
Springer Nature Customer Service Center GmbH
Europaplatz 3, 69115 Heidelberg, Germany

Printed by Libri Plureos GmbH
in Hamburg, Germany